physica status solidi (a)

ISSN 0031-8965 * VOL. 54 · NO. 1 · JULY 1979

AKADEMIE-VERLAG · BERLIN

THEORIA CUM PRAXI

Classification Scheme

1. Structure of Crystalline Solids
    1.1 Perfectly Periodic Structure
    1.2 Solid-State Phase Transformations
    1.3 Alloys. Metallurgy
    1.4 Microstructure (Magnetic Domains See 18; Ferroelectric Domains See 14.4.1)
    1.5 Films
    1.6 Surfaces
2. Non-Crystalline State
3. Crystal Growth
4. Bonding Properties
5. Mössbauer Spectroscopy
6. Lattice Dynamics. Phonons
7. Acoustic Properties
8. Thermal Properties
9. Diffusion
10. Defect Properties (Irradiation Defects See 11)
    10.1 Metals
    10.2 Non-Metals
11. Irradiation Effects (X-Ray Diffraction Investigations See 1 and 10)
12. Mechanical Properties (Plastic Deformations See 10)
    12.1 Metals
    12.2 Non-Metals
13. Electron States
    13.1 Band Structure
    13.2 Fermi Surfaces
    13.3 Surface and Interface States
    13.4 Impurity and Defect States
    13.5 Elementary Excitations (Phonons See 6)
        13.5.1 Excitons
        13.5.2 Plasmons
        13.5.3 Polarons
        13.5.4 Magnons
14. Electrical Properties. Transport Phenomena
    14.1 Metals. Semi-Metals
    14.2 Superconductivity. Superconducting Materials and Devices
    14.3 Semiconductors
        14.3.1 Films
        14.3.2 Surfaces and Interfaces
        14.3.3 Devices. Junctions (Contact Problems See 14.3.4)
        14.3.4 High-Field Phenomena, Space-Charge Effects, Inhomogeneities, Injected Carriers (Electroluminescence See 20.3; Junctions See 14.3.3)
    14.4 Dielectrics
        14.4.1 Ferroelectrics
15. Thermoelectric and Thermomagnetic Properties
16. Photoconductivity. Photovoltaic Effects
17. Emission of Electrons and Ions
    17.1 Field Emission Microscope Investigations
18. Magnetic Properties
    18.1 Paramagnetic Properties
    18.2 Ferromagnetic Properties
        18.2.1 Ferromagnetic Films
    18.3 Ferrimagnetic Properties
    18.4 Antiferromagnetic Properties

(Continued on cover three)

physica status solidi (a)

applied research

Volume 54 · Number 1 · Pages 1 to 436, K1 to K84, and A1 to A8

July 16, 1979

PSSA 54(1) 1—436, K1—K84, A1—A8 (1979)

ISSN 0031-8965

AKADEMIE-VERLAG · BERLIN

Subscriptions and orders for single copies should be directed

in the DDR:

to the Postzeitungsvertrieb,
to a book-shop, or
to the Akademie-Verlag, DDR-108 Berlin, Leipziger Straße 3 – 4;

in the other socialist countries:

to a book-shop for foreign language literature or
to the competent news-distributing agency;

in the BRD and BERLIN (WEST):

to a book-shop or
to the wholesale distributing agency Kunst und Wissen, Erich Bieber, D-7000 Stuttgart 1, Wilhelmstr. 4 – 6;

in ÖSTERREICH:

to the Globus-Buchvertrieb, A-1201 Wien, Höchstädtplatz 3;

in the other Western European countries:

to Kunst und Wissen, Erich Bieber GmbH, CH-8008 Zürich, Dufourstr. 51;

in USA and CANADA:

to Verlag Chemie International, Inc., 175 Fifth Avenue, New York, N. Y. 10010, USA;

in other countries:

to the international book- and journal-selling trade,
to Buchexport, Volkseigener Außenhandelsbetrieb der Deutschen Demokratischen Republik, DDR-701 Leipzig, Postfach 160, or
to the Akademie-Verlag, DDR-108 Berlin, Leipziger Straße 3 – 4.

Editorial Note:

"physica status solidi (a)" undertakes that an original paper *accepted for publication* before the 23rd of any month will be published within 50 days of this date unless the author requests a postponement. In special cases there may be some delay between *receipt* and *acceptance* of a paper due to the review and, if necessary, revision of the paper.

Schriftleiter und verantwortlich für den Inhalt:
Professor Dr. Dr. h.c. P. Görlich, DDR-102 Berlin, Neue Schönhauser Str. 20 bzw. DDR-69 Jena, Schillbachstr. 24.

Verlag:
Akademie-Verlag, DDR-108 Berlin, Leipziger Str. 3–4; Fernruf 2236221 und 2236229; Telex-Nr.: 114420; Bank: Staatsbank der DDR, Berlin, Kto.-Nr.: 6836-26-20712.

Chefredakteur:
Dr. H.-J. Hänsch.

Redaktionskollegium:
Prof. Dr. E. Gutsche, Dr. H.-J. Hänsch, Dr. H. Lange, Dr. S. Oberländer.

Anschrift der Redaktion:
DDR-102 Berlin, Neue Schönhauser Str. 20; Fernruf: 2 82 33 80.

Veröffentlicht unter der Lizenznummer 1620 des Presseamtes beim Vorsitzenden des Ministerrates der Deutschen Demokratischen Republik.

Gesamtherstellung:
VEB Druckerei „Thomas Müntzer", DDR-582 Bad Langensalza.

Erscheinungsweise:
Die Zeitschrift „physica status solidi (a)" erscheint jeweils am 16. eines jeden Monats. Jährlich erscheinen 6 Bände zu je 2 Heften.

Bezugspreis:
je Band 165,– M zuzüglich Versandspesen (Preis für die DDR: 130,– M).

Bestellnummer dieses Bandes: 1085/54.

Printed in the German Democratic Republic.

AN (EDV) 20735

Contents

Review Article

Original Papers

Short Notes

Systematic List

Contents of Volume 54 Continued on Page 439

Review Article

phys. stat. sol. (a) **54**, 11 (1979)

Subject classification: 19; 13; 18.1; 20; 22.2.1; 22.2.3; 22.2.4; 22.4.3; 22.4.4

Institute of Semiconductor Physics,
Academy of Sciences of the Lithuanian SSR, Vilnius,[1] *(a)*
and Department of Physics, Purdue University, W. Lafayette Indiana[2] *(b)*

Microwave Effects in Narrow-Gap Semiconductors (II)[3][4]

By

R. S. Brazis (a), J. K. Furdyna (b), and J. K. Požela (a)

Contents

5. Small Spheres, Ellipsoids, and Powders

The mathematical analysis used so far applies to plane waves normally incident either on an infinite half-space or on an infinite slab of thickness d. The application of the results for these unbounded geometries to finite samples, although frequently well justified, is strictly speaking an approximation. There are no exact analytic solutions for the interaction of an electromagnetic wave with a gyrotropic sample which is finite in all three dimensions. Nevertheless, some very good approximations are available for the case of a gyrotropic sphere and, in certain more restrictive limits, also for the gyrotropic ellipsoid. The major difficulties of this classical problem were in fact only recently overcome [83, 28]. These results have been applied to study NGS in the form of isolated spheres and disks in a microwave waveguide, and also can serve as a point of departure in analyzing the response of a granular medium (powder) to microwave excitation. There are two parameter ranges of interest: 1. The Rayleigh limit, in which the field of the wave is uniform both outside and within the bounded specimen. In this case, solutions for wave interaction exist for ellipsoidal samples, of which the sphere is a special case [28, 86, 87]. 2. The dipole limit in which the wave field is assumed uniform outside the sample, but can be arbitrary within the sample, i.e. the sample can effectively contain many wavelengths of propagating normal modes. This case has been only recently solved for the spherical geometry [83].

[1]) Vilnius 232600, USSR.

[2]) W. Lafayette, Indiana 47907, USA.

[3]) Work supported by the program of scientific exchange between the Academy of Sciences of the USSR and the US National Academy of Sciences.

[4]) Part I see phys. stat. sol. (a) **53**, 11 (1979).

5.1 Spheres, ellipsoids, and disks

5.1.1 Magnetoplasma resonance in the Rayleigh limit

Consider first the interaction of an isotropic ellipsoidal sample (described by scalar $\varkappa$) with the electric field of an electromagnetic wave. A good discussion of the response of isotropic spheres and a definition of basic concepts are given in monographs [84, 85]. If the electric field $\boldsymbol{E}_1$ of the incident wave (outside the sample) is assumed parallel to one of the principal axes of the ellipsoid, the internal field $\boldsymbol{E}_{\text{in}}$ is uniform in the Rayleigh limit, and is related to $\boldsymbol{E}_1$ by [88]

$$\boldsymbol{E}_{\text{in}} = \boldsymbol{E}_1 - L\frac{\boldsymbol{P}}{\varepsilon_0} = \boldsymbol{E}_1 - L(\varkappa - 1)\,\boldsymbol{E}_{\text{in}} = = \frac{\boldsymbol{E}_1}{1 + L(\varkappa - 1)} = \frac{1}{L}\frac{\boldsymbol{E}_1}{\varkappa + \gamma_L}, \tag{105}$$

where $\boldsymbol{P} = \varepsilon_0(\varkappa - 1)\,\boldsymbol{E}_{\text{in}}$ is the polarization of the ellipsoid medium, L is the depolarizing factor of the ellipsoid for the direction of $\boldsymbol{E}_1$, and $\gamma_L = (1/L) - 1$. For an oblate ellipsoid of revolution with the rotation axis perpendicular to the microwave electric field $\boldsymbol{E}$ the depolarizing factor (denoted for this field orientation as $L_\perp$) can be expressed in terms of the major axis a and the minor axis b [88]:

$$L_\perp = \frac{1}{2}\left[1 - \frac{1 + e^2}{e^3}\,(e - \arctan e)\right], \tag{106}$$

where $e = \sqrt{1 - (b/a)^2}$. When the rotation axis is parallel to the microwave electric field $\boldsymbol{E}_1$ then the depolarizing factor is $L_{\|}$:

$$L_{\|} = 1 - 2L_\perp\,. \tag{107}$$

For a sphere $L_{\|} = L_\perp = L = 1/3$. The dipole moment $\boldsymbol{p}$ excited by $\boldsymbol{E}_1$ is then

$$\boldsymbol{p} = \int\limits_V \boldsymbol{P}\,\mathrm{d}V = \varepsilon_0\frac{V}{L}\frac{\varkappa - 1}{\varkappa + \gamma_L}\boldsymbol{E}_1\,, \tag{108}$$

where V is the volume of the ellipsoid.

This result can be extended to gyrotropic ellipsoids with the dc magnetic field parallel to the minor axis by means of a polarizability tensor $\alpha^{(e)}$. The electric dipole moment can then be written

$$\begin{pmatrix} p_{+1} \\ p_{-1} \\ p_z \end{pmatrix} = \begin{pmatrix} \alpha_+^{(e)} & 0 & 0 \\ 0 & \alpha_-^{(e)} & 0 \\ 0 & 0 & \alpha_{zz}^{(e)} \end{pmatrix} \cdot \begin{pmatrix} E_{+1} \\ E_{-1} \\ E_z \end{pmatrix}, \tag{109}$$

where

$$\alpha_s^{(e)} = \varepsilon_0\frac{V}{L}\frac{\varkappa_s - 1}{\varkappa_s + \gamma_\perp}\,, \quad \alpha_{zz}^{(e)} = \varepsilon_0\frac{V}{L}\frac{\varkappa_{\|} - 1}{\varkappa_{\|} + \gamma_{\|}}, \tag{110}$$

and the subscript s can be $+1$ or -1 corresponding to the principal modes of excitation given by equations (52).

The power absorption associated with the s-th mode is given by

$$P_s = \frac{1}{2}\,\omega\,\mathrm{Im}\,(p_sE_s^*) = \frac{1}{2}\,\varepsilon_0\omega\,\frac{V}{L^2}\,\frac{\varkappa_s''}{(\varkappa_s' + \gamma_\perp)^2 + (\varkappa_s'')^2}. \tag{111}$$

For a single-component plasma and the constant-τ approximation, the above expression for the power absorbed can be reformulated in microscopic terms by substituting expressions (52) into (111). We then obtain [86, 87]

$$P_s = |E_s|^2 \frac{V}{2} \frac{1}{L^2} \frac{\sigma_0}{(\varkappa_1 + \gamma_L)^2} \frac{1}{1 + \left(\omega - \frac{\omega_0^2}{\omega} - s\omega_c\right)^2 \tau^2}, \tag{112}$$

where $\omega_0 = \sqrt{ne^2/[m^*\varepsilon_0(\varkappa_1 + \gamma_L)]}$ is the effective plasma frequency for the ellipsoid.

The interaction exhibits a resonance in P_s when

$$\omega - \frac{\omega_0^2}{\omega} - s\omega_c = 0\,. \tag{113}$$

Note that this condition can also be obtained by including the restoring depolarization fields directly into the microscopic equation of motion of the electrons, as was done by Dresselhaus et al. [89] in connection with cyclotron resonance in Ge. It is instructive to consider the evolution of this resonance as ω_0 increases. For low carrier concentration ($\omega_0 < \omega$) the resonance occurs in the CRA, or $s = +1$ polarization. At $\omega_0 = \omega$, the resonance occurs at $B = 0$ and corresponds to the plasma oscillation associated with the sample geometry. For high carrier concentration ($\omega_0 > \omega$) the resonance appears in the CRI ($s = -1$) polarization, with the resonance field increasing to very high fields as ω_0 increases. The resonance may be viewed alternatively as either plasma-shifted cyclotron resonance [89], or as cyclotron-shifted plasma resonance. We shall refer to it simply as "magnetoplasma resonance".

For spheres ($\gamma_L = 2$) and for parameters typical for narrow-gap semiconductors, the condition $\omega_0 \gg \omega$ is satisfied at microwave (and lower) frequencies. It can be seen from (113) that, for $\omega_0 \gg \omega$, magnetoplasma resonance in a sphere occurs in the CRI polarization at field $B_{\text{sphere}} = ne/[\omega\varepsilon_0(\varkappa_1 + 2)]$. This occurs near the "magnetoplasma cut-off" field given by equation (72) corresponding to $\varkappa_- = 0$. With the exception of very pure InSb, satisfying these conditions ($B = B_{\text{sphere}}$ or $B = B_{\text{MPC}}$) in NGS requires extremely high magnetic fields because of the high carrier concentration typical of these materials.

As can be seen from (113), the field at which resonance occurs,

$$B = B_{\text{CR}} - \frac{ne}{\omega\varepsilon_0(\varkappa_1 + \gamma_L)} \tag{114}$$

can be significantly lowered by increasing γ_L, i.e. by going to oblate ellipsoids with very high axial ratio, for which the depolarizing factor $L_\perp$ is very small [90]. Fig. 17 illustrates the magnetoplasma resonances in a sphere and in a thin disk. Fig. 18 represents the comparison of the experimental results obtained on disks with the theoretical predictions for an ellipsoid of revolution. The extent of agreement between the theory and the experiment is rather surprising. These results show that an effective depolarizing factor can be ascribed to the disk geometry, and the factor is given by the depolarizing factor of an ellipsoid of revolution having the same axial ratio. It is worth noting that for large γ_L, the resonances occur at large negative values of $\varkappa_-$, for which the waves are evanescent in the infinite medium. Such resonances then provide a probe of this otherwise inaccessible region.

The resonance condition is size-independent in the Rayleigh limit. It was shown by solving the dynamic boundary value problem for a small gyrotropic sphere [28] that equation (111) is correct to terms of the order of $\frac{1}{5}(2\pi a/\lambda_0)$.

Magnetoplasma resonance in n-type InSb spheres was studied experimentally by several authors [28, 89 to 92]. Typical microwave absorption data are shown in Fig. 19, illustrating the relative independence of this effect of size [28, 91]. Here the sample

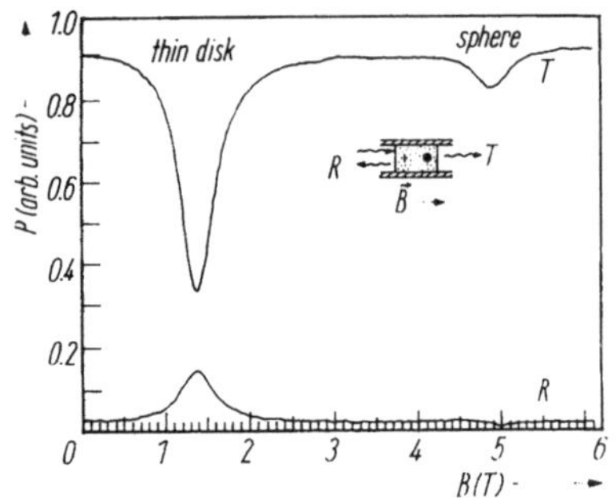

Fig. 17

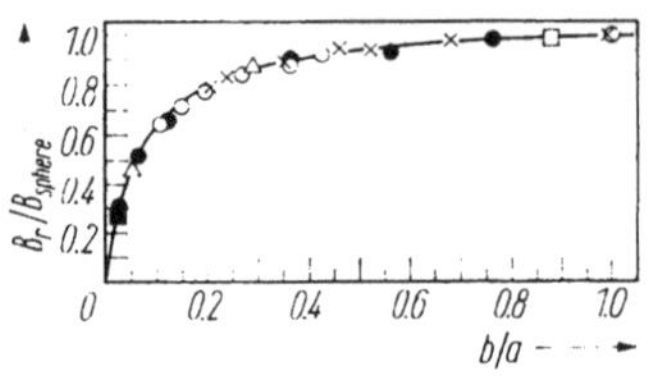

Fig. 18

Fig. 17. Magnetoplasma resonances observed in transmission (T) and reflection (R) of the CRI wave for a small disk and a sphere prepared from the same crystal of n-InSb ($n = 1.18 \times 10^{15}$ cm^{-3}, $T = 65$ K, $f = 35$ GHz). Both samples are placed symmetrically on the axis of a circular waveguide and separated from each other by approximately 1 cm. Sphere diameter $d = 0.793$ mm, disk diameter $a = 2.0$ mm ($b/a = 0.024$) (after [90])

Fig. 18. Experimental dependence of the magnetoplasma resonance field B_r on the thickness-to-diameter ratio for various cylindrical samples of n-InSb (points). The theoretical dependence of B_r on axial ratio b/a calculated for ellipsoids of revolution is given by the solid curve. B_{sphere} is the resonance field for the sphere geometry measured for each sample series of a given diameter. A rectangular sample is also included for comparison (□) (after [90])

	a (mm)	n (10^{14} cm^{-3})
△	2.74	1.7
×	1.92	2.3
○	1.38	2.5
●	1.29	2.8
■	2.00	11.8
□	2.40	2.5

was placed in a microwave cavity and excited by a linearly polarized electric field $\boldsymbol{E}$ transverse to $\boldsymbol{B}$, and a linearly polarized magnetic field $\boldsymbol{B}(\omega)$ parallel to $\boldsymbol{B}$. As pointed out later, the low-field non-resonant hump is associated with the ac magnetic field $\boldsymbol{B}(\omega)$ excitation.

As in the case of spheres, the above microwave interactions in disks can be used for the purpose of a quantitative characterization of materials. There are several advantages to the disk geometry. First, the disk is one of the easiest *bounded* geometries to prepare. This geometry is natural for epitaxial layers, and materials crystallizing in the form of thin platelets. Second, in samples involving high carrier concentrations, (e.g. $n > 10^{16}$ cm^{-3}, typical of NGS), where microwave magnetoplasma resonance in spheres occurs at exceedingly high fields, the resonance field can be significantly lowered by using a thin disk geometry, as seen in Fig. 17. Thus the response function of a material in the otherwise inaccessible evanescent range can be "mapped out" by observing the magnetoplasma resonances in disks of successively decreasing axial ratio. Magnetoplasma resonance, which depends on $\varkappa_l$, is of considerable importance as a possible tool for direct determination of the static dielectric constant in zero-gap materials (HgTe, HgSe, α-Sn). The ability to position the resonance at an arbitrary magnetic field by controlling the sample geometry offers the possibility of investigating the magnetic-field dependence of the static dielectric constant, predicted for zero-gap materials [30].

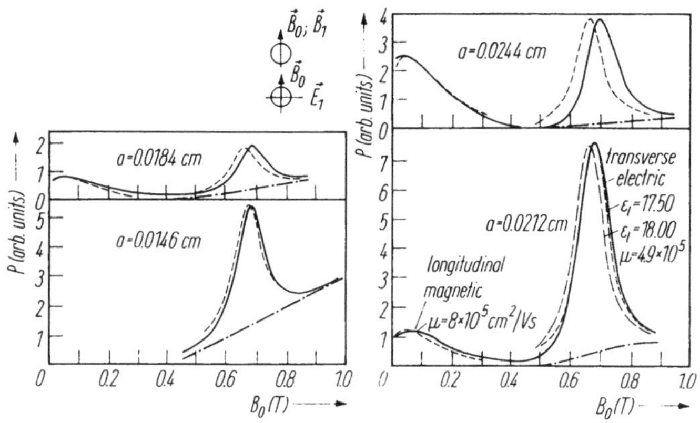

Fig. 19. Microwave absorption data for $\boldsymbol{E} \perp \boldsymbol{B}$, $\boldsymbol{B}(\omega) \parallel \boldsymbol{B}$ observed in a series of n-type InSb spheres of various sizes, obtained by successive etching of a single sample ($n = 1.7 \times 10^{14}$ cm^{-3}, $T = 77$ K, $f = 35$ GHz). The solid curve represents experimental data, the large peak corresponds to magnetoplasma resonance. For details of the fitting procedure (dashed curve) see [28]

We have concerned ourselves primarily with the one-electron Drude model with scalar effective mass. When the Drude model is inadequate (as for intrinsic or many-valley NGS), equations (111) and (112) are no longer equivalent. Equation (111) should then be used directly, with the resonance condition given by

$$\varkappa_s' = -\gamma_L \, . \tag{115}$$

This condition is applicable also to NGS with anisotropic carrier effective mass provided that the rotation axis of the ellipsoidal sample and the dc magnetic field are parallel to a high-symmetry crystallographic axis. Equations (111) and (115) were applied to the study of Rayleigh electric dipole resonances in many-valley semiconductors, and in semiconductors with more than one type of carriers [93].

5.1.2 Dimensional resonances in the dipole limit and their relation to the Fabry-Perot resonances in an infinite slab

When the sample size becomes comparable to the internal wavelength of a given mode, so that $|qa| \gtrapprox 1$, dimensional resonances can take place. The condition $|qa| \gtrapprox 1$ implies that the Rayleigh limit is violated, although the dipole limit may very well hold. In practice this can be easily satisfied for the CRA mode, particularly in the helicon region defined by inequalities (51). The problem is now inherently dynamic (non-static) and the corresponding boundary value problem of a gyrotropic sphere for arbitrary qa escaped rigorous solution until 1974. Even now the available formulation is only tractable by numerical methods [83].

Dimensional resonances occur in both magnetic and electric multipole excitations of the sample, with the magnetic dipole resonances becoming dominant as the sample size decreases. According to numerical studies, in the parameter range $|\varkappa_{\parallel}| \gg |\varkappa_{\pm 1}|$ the principal dipole resonances for a sphere occur approximately when

$$\left(\frac{\omega}{c}\, d\right)^2 \varkappa_s = 4N\left(N + \frac{1}{2}\right)\pi^2 \, , \tag{116}$$

where d is the sphere diameter,

$$N = \begin{cases} 1, 2, 3, \ldots & \text{for electric resonances,} \\ -1, -2, -3, \ldots & \text{for magnetic resonances.} \end{cases}$$

This behaviour can be viewed qualitatively as a series of Fabry-Perot-like resonances, associated with "fitting" of CRA ($s = +1$) or CRI ($s = -1$) wavelengths inside the sphere diameter.

Note that for an infinite plane-parallel slab of thickness d in the Faraday configuration, i.e. for $\boldsymbol{B} \parallel \boldsymbol{q}_s \parallel \boldsymbol{n}$, where $\boldsymbol{q}_s$ is the wave vector for the s-th mode and $\boldsymbol{n}$ is the normal to the surface of the slab, the conditions for Fabry-Perot resonance are

$$q_s d = \frac{\omega}{c} d \sqrt{\varkappa_s} = N\pi; \quad N = 1, 2, 3, \ldots. \tag{117}$$

Substituting (17) or (52) in (117) and taking into account (9) and (72) we obtain [90]

$$\frac{B_{\text{MPC}} + B_{\text{CR}}}{sB - B_{\text{CR}}} = \frac{\pi^2 c^2 N^2}{\varkappa_1 \omega^2} \frac{1}{d^2} - 1\,. \tag{118}$$

Equation (118) can be represented by a straight line in coordinates $(B_{\text{MPC}} + B_{\text{CR}})/(B - B_{\text{CR}})$ versus d^{-2}. This equation describes in a unified manner the resonance conditions in slabs for arbitrary magnetic field, including the helicon resonances in the CRA polarization, as well as resonances occurring at fields exceeding B_{MPC} which appear in both CRA and CRI polarizations. The advantage of this formulation is that it applies for arbitrary electron effective mass and concentration (which enter the expression implicitly through the characteristic fields B_{CR} and B_{MPC}). For a sphere we obtain from (116)

$$\frac{B_{\text{MPC}} + B_{\text{CR}}}{sB - B_{\text{CR}}} \approx \frac{4\pi^2 c^2}{\varkappa_1 \omega^2} \frac{N(N + \frac{1}{2})}{d^2} - 1\,. \tag{119}$$

The above resonances manifest themselves as absorption peaks when a plane electromagnetic wave is incident. Note that the lowest-order resonance which dominates the interaction is a magnetic one.

The Fabry-Perot-like dipole resonances observed for spheres mounted in a waveguide in a styrofoam holder and excited with CRA polarization are illustrated in Fig. 20. Note the strong dependence of dimensional resonances on the sphere dimension, while the magnetoplasma resonance is insensitive to the radius of the sphere. Similar transmission spectra were observed also for nearly ellipsoidal samples [90]. The resonances can be classified as electric or magnetic dipole resonances, marked E(N, 0) and M(N, 0), respectively, where N is the order of the resonance. (The labeling scheme is the same as that used in [29] and [83].) Selection of the resonances as electric or magnetic is obtained by observing the reflected signal when the sample section is terminated with a short circuit: When the small sample is located at an

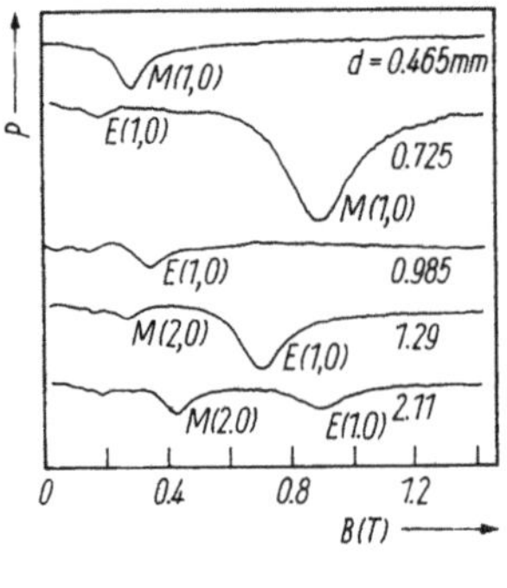

Fig. 20. CRA wave transmission curves for small disk samples of n-InSb ($n = 2.8 \times 10^{14}$ cm^{-3}, $f = 35$ GHz, CRA) of several thicknesses but the same diameter $a = 1.29$ mm, placed on the axis of circular waveguide. Each curve is shifted arbitrarily along the vertical axis for clarity. The minima are found to be electric and magnetic dipole resonances marked E(N, 0) and M(N, 0), respectively (after [90]; for the resonances in spheres see [29])

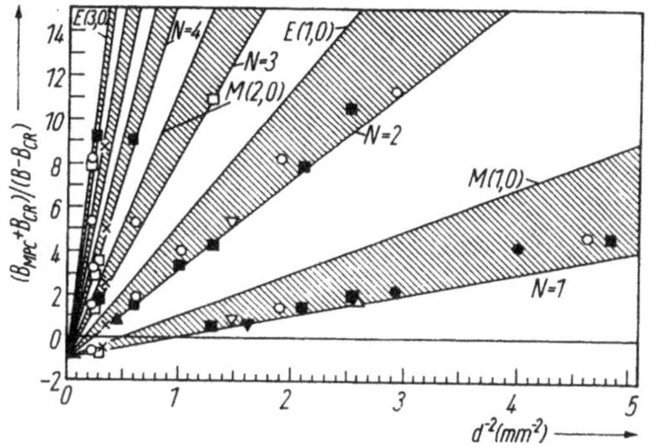

Fig. 21. Normalized magnetic field for Fabry-Perot-type resonances occurring at $B > B_{CR}$ in the CRA polarization and at $B > B_{MPC}$ in the CRI polarization as a function of sample thickness. Experimental points obtained on various disk samples 1 to 7 (●, ▽, ▼, △, ■, ▲, ○) and rectangular samples 8, 9 (□, ×) of n-InSb fall in the shaded regions between two straight lines, one of which (labeled N) represents the Fabry-Perot resonance conditions for an infinite slab and the other [labeled M(N, 0) or E (N,0)] represents the conditions of the magnetic or electric dipole resonances for the sphere of diameter d; frequency $f = 35.0$ GHz, $B_{CR} = 0.0175$ T. For details see [90]

electric node, the absorption peaks marked E (electric dipole resonances) disappear in the reflected signal; and, similarly, the magnetic dipole resonances (M) disappear when the sample is at a magnetic node. The same behaviour was observed for a rectangular bar [94].

The experimental and theoretical study of the standing-wave resonance conditions is summarized in Fig. 21. Theoretical conditions for successive Fabry-Perot resonances in a semi-infinite plate (marked $N = 1, 2, \ldots$) and successive magnetic and electric resonances in spheres, obtained using (118) and (119) are given by the straight lines.

Note that all experimental points for disk samples fall in a region between a Fabry-Perot resonance condition for an infinite plane-parallel slab and a sphere resonance. More specifically, *magnetic* resonances of the disks occur between *odd-N* Fabry-Perot resonances in slabs and *magnetic* resonances in spheres; *electric* resonances in disks lie between *even-N* slab resonances and *electric* sphere resonances.

The relationship of the disk resonances to the two limiting cases is worth emphasizing: While the resonance condition for a disk is very close to that for an infinite slab of the same thickness, the nature of excitation (electric *or* magnetic) is qualitatively identical to that of the sphere. On the other hand, a close analogy exists between the dipole resonances of the small disks and the slab resonances described by equations (91), (92) or (97), (98).

It is clear from the behaviour of $\varkappa$ that at fields corresponding to the helicon approximation, (51), the resonance position for a sphere measures essentially n. At higher fields, close to and above the magnetoplasma edge, the resonance conditions depend on both n and $\varkappa_1$. The resonance widths are primarily sensitive to the mobility and the free-carrier statistics. These effects can be taken into consideration via computer fitting. In addition we note that the theory for a sphere predicts a number of weak satellites associated with both the electric and magnetic dipole resonances [83].

5.1.3 *Magnetic dipole resonance in the Rayleigh limit*

As the sphere radius decreases toward the Rayleigh limit, the dominant magnetic dimensional resonance gradually transforms into a size-independent resonance. It was recently shown [28] that the magnetic dipole moment $\boldsymbol{m}$ of a small gyrotropic sphere can be written in terms of the magnetic polarizability tensor $\mathbf{a}^{(m)}$ as

$$\begin{pmatrix} m_{+1} \\ m_{-1} \\ m_z \end{pmatrix} = \begin{pmatrix} a_{+1}^{(m)} & 0 & 0 \\ 0 & a_{-1}^{(m)} & 0 \\ 0 & 0 & a_{zz}^{(m)} \end{pmatrix} \cdot \begin{pmatrix} B_{+1} \\ B_{-1} \\ B_z \end{pmatrix}, \tag{120}$$

where

$$a_s^{(m)} = \frac{a^5}{30} \left(\frac{\omega}{c}\right)^2 \frac{\tilde{\varkappa}_s}{1 - \frac{2}{21}\left(\frac{\omega}{c} a\right)^2 \tilde{\varkappa}_s}; \tag{121}$$

$$\tilde{\varkappa}_s = \frac{2\varkappa_s \varkappa_{zz}}{\varkappa_s + \varkappa_{zz}}, \quad \tilde{\varkappa}_{zz} = \frac{2\varkappa_{+1}\varkappa_{-1}}{\varkappa_{+1} + \varkappa_{-1}} \tag{122}$$

and the indices $s = +1, -1$, and (zz) correspond to the principal modes of excitation described previously.

When the dimensional corrections in (121) can be neglected, the response of the sample varies simply as $\tilde{\varkappa}$. Thus, in the small-sphere limit, the resonance condition for the transverse circularly polarized magnetic dipole excitation is

$$\varkappa_s' + \varkappa_{zz}' = 0\,. \tag{123}$$

It is easily shown that for $\omega_p^2/\omega^2 \gg 1$ and $\omega\tau \gg 1$ this predicts a resonance when [95 to 97]

$$|\omega_c| \approx 2\omega \tag{124}$$

independent of n and a. The existence of a resonance at $\omega_c = 2\omega$ was first suggested on the basis of a quantum model by Furdyna and Mycielski [98]. This feature is especially important for NGS because it opens the possibility of determining m^* in materials with high electron concentration, which was not possible in conventional

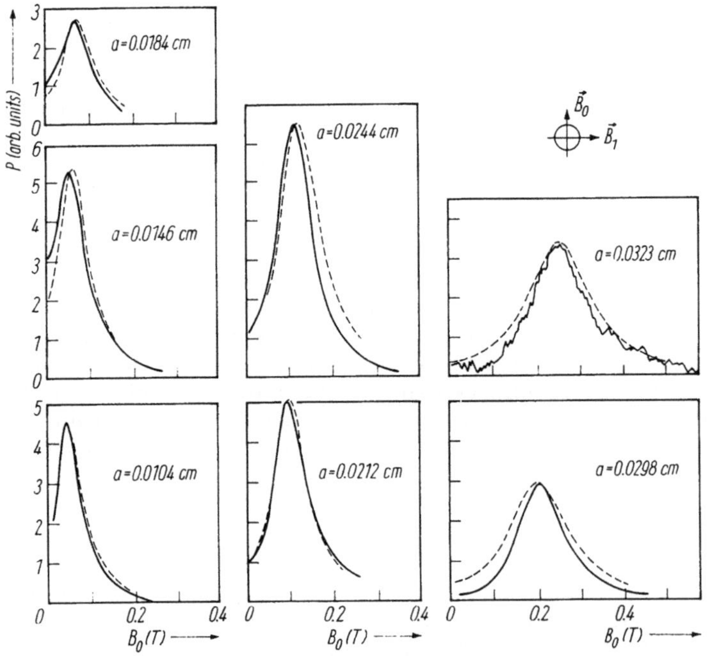

Fig. 22. Microwave absorption data for $\boldsymbol{B}(\omega) \perp \boldsymbol{B}$, $\boldsymbol{E} \| \boldsymbol{B}$ for the same InSb material, frequency, and temperature as that in Fig. 19. Experimental data (solid curve) illustrate the Rayleigh magnetic-dipole resonance. The dashed lines show the theoretical fit obtained with a single set of parameters. For details see [28]

microwave cyclotron resonance experiments (electric dipole interaction) because of the magnetoplasma shift discussed above.

The Rayleigh magnetic dipole resonance was observed experimentally by Evans and coworkers [91, 99, 100], and discussed using a semiquantitative, empirical model. A rigorous classical macroscopic analysis was carried out in [28], resulting basically in equations (120) to (122). An illustration of the fit of theory to experiment is shown in Fig. 22, showing magnetic dipole absorption for a linearly polarized microwave magnetic field transverse to the dc field $\boldsymbol{B}$. The electric field was polarized parallel to $\boldsymbol{B}$, and did not elicit magnetic-field-dependent response. The data and that in Fig. 19 were obtained on the same set of spheres, and the dashed curves represent theoretical fits using a single set of parameters in both figures. The low-field non-resonant hump in Fig. 22 corresponds to magnetic dipole absorption associated with $a_{zz}^{(m)}$, i.e. excited by $B_z(\omega)$. The theory also verified the experimental observations that the magnetic dipole resonance varies as the fifth power of the radius, is excited by CRA polarization, and in the small-size limit is independent of n and r [91]. Note also the low-field hump excited by B_z. In the small-size limit this response is related to the Voigt dielectric constants $\varkappa_\perp$, (cf. equation (122)), but again evolves into a dimensional resonance as the sample size is increased.

5.2 Powders

Microwave techniques can also be used to study electrical properties of semiconductors in powder form. An exact solution is, needless to say, lacking because of the complexity of the problem, which consists of two difficult steps: (i) the response of a single particle and (ii) summing up over the contributions of all grains. One must therefore resort to rather drastic approximations. The simplest approach, which nevertheless offers some rather useful features, is to assume that the grains are spherical and non-interacting. The powder response can then be discussed in terms of the foregoing discussion, and will manifest absorption related to the magnetoplasma and dimensional resonances in spheres.

Characteristic transmission spectra of circularly polarized microwaves through unsieved powders of InSb, InAs, and HgTe are shown in Fig. 23 [101 to 103]. All three materials satisfy helicon conditions (51) except near $B = 0$. The striking feature of the data is that the powder is opaque to the CRA polarization and relatively transparent to the CRI wave, in contrast to helicon transmission through slabs. The mechanism for CRA absorption is the dimensional magnetic dipole resonance in powder grains, corresponding to $N = 1$ in (116). Because of the wide distribution of grain sizes in unsieved powders, there are CRA resonances occurring throughout the magnetic

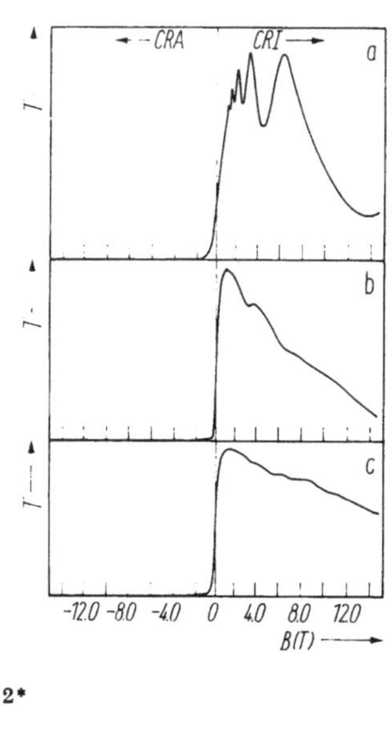

Fig. 23. Transmission of circularly polarized microwaves through unsieved powders of n-type a) InAs ($n = 9.6 \times 10^{16}$ cm^{-3}), b) InSb (5.1×10^{16} cm^{-3}), and c) HgTe (1×10^{17} cm^{-3}) at 35.0 GHz and $T = 4.2$ K in Faraday geometry. Here positive fields correspond to CRI, negative ones to CRA. Note quantum oscillations of the transmitted signal, seen especially well in the InAs data (after [101 to 103])

field range. In size-fractioned powders satisfying helicon conditions, CRA transmission shows a narrow magnetic dipole resonance line [103], as shown in Fig. 24. This powder resonance is in reasonable agreement with (116), and can be used to estimate the electron concentration in powder grains.

Electron concentrations corresponding to Fig. 23 are too high for magnetoplasma cut-off (equation (72)) to be observed in the available field range. Thus the powder is relatively transparent to the CRI polarization because neither dimensional nor magnetoplasma resonances are present. It was shown above that the power absorbed by a CRI-excited sphere satisfying conditions (51) is essentially proportional to the transverse dc magneto-conductivity σ_{xx} [103]. In the region of orbital quantization, σ_{xx} displays Shubnikov-de Haas oscillations, and this is seen particularly clearly in the powder transmission data for InAs in Fig. 23. The observed quantum oscillations in InSb and InAs powders give excellent agreement with single crystal data and with theory [103].

Fig. 25 illustrates the behaviour of magnetoplasma resonance in InSb powder of lower carrier concentration, investigated as a function of temperature in both the extrinsic and intrinsic ranges (see [104] and literature cited therein). The resonant field B_r is relatively constant while the sample remains extrinsic, and in the intrinsic temperature range, as n increases, B_r shifts to higher fields. This is qualitatively in agreement with the expected behaviour of the resonance field for an isolated sphere, cf. (114). Quantitatively, resonances in the powder data of Fig. 25 occur at fields approximately 40% below the values predicted by (114). This is a result of particle-particle interaction which is not fully understood.

A quantitative model of wave propagation in powders, including grain-grain interaction, is yet to be developed. The systematic, reproducible behaviour observed in NGS powders and its semiquantitative interpretation in terms of the independent sphere model appear encouraging in this context. When understood more fully, experiments with powders, in addition to their interest as such, will also offer a new approach to surface problems.

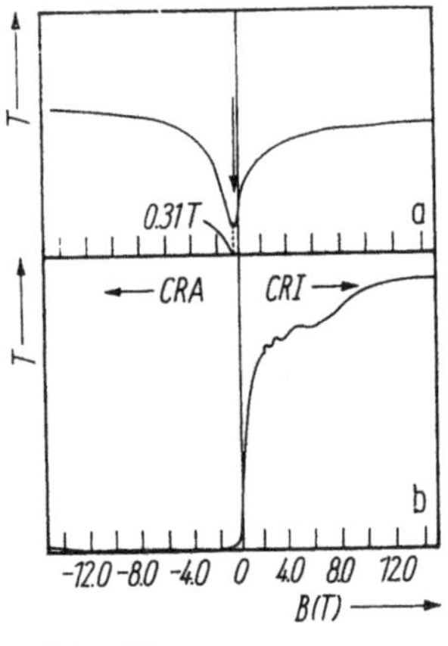

Fig. 24

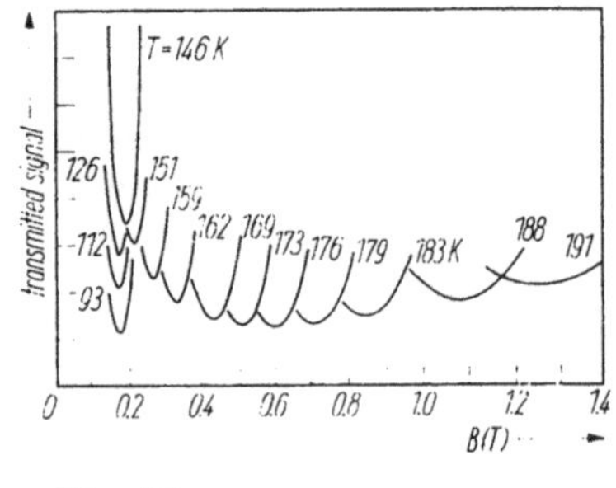

Fig. 25

Fig. 24. Comparison of transmission of circularly polarized microwaves through fine and coarse powders obtained from the same crystal of InSb ($n = 1.5 \times 10^{17}$ cm^{-3}, $T = 4.2$ K, $f = 35.0$ GHz). a) $d < 40$ μm, b) 40 μm $< d <$ 250 μm (after [101 to 103])

Fig. 25. Absorption of CRI-polarized microwaves in InSb powder in the immediate vicinity of magnetoplasma resonance at various temperatures ($n_D = 7.3 \times 10^{13}$ cm^{-3}, $f = 35.0$ GHz, $t \approx$ ≈ 2 mm, 250 μm $< d <$ 295 μm (after [104])

6. Electron Spin Resonance in Narrow-Gap Semiconductors

6.1 Helicon-excited spin resonance of paramagnetic impurities

The study of electron paramagnetic resonance (EPR) of localized magnetic impurities in conducting media is greatly complicated by lack of sufficient penetration of the microwave signal into the sample interior. This difficulty can be circumvented, however, when the electrical parameters in the vicinity of EPR satisfy the conditions for helicon wave propagation, equation (51). Electromagnetic waves can then penetrate the medium in the form of circularly polarized helicons and can resonantly interact with the paramagnetic ions contained within the bulk. These conditions will be satisfied when the free charge carriers have a relatively small effective mass, so that their cyclotron frequency $|\omega_c|$ exceeds the Larmor frequency $|\omega_L|$ of the localized magnetic moments. Furthermore, the free carrier collision time τ must be long enough so that $|\omega_c\tau| \gg 1$ in the EPR vicinity. Finally, the sense of helicon polarization must be compatible with the sense of precession of the localized magnetic moments. In the case of paramagnetic ions, the sense of precession is identical with the sense of polarization of helicons propagating in a gas of conduction electrons.

These conditions are easily satisfied in the case of NGS containing paramagnetic impurities. It is readily shown that in the Faraday geometry, propagation in such a medium is described by [105, 106]

$$q_\pm = \frac{\omega}{c}\, \varkappa_\pm^{1/2}(1 + \chi_\pm)^{1/2}\,, \tag{125}$$

where $\chi_\pm$ are the transverse components of the dynamic paramagnetic susceptibility of the spin system in the coordinate frame (7). Magnetic susceptibilities $\chi_\pm$ are conveniently expressed by the Bloch model (see, e.g. [107])

$$\chi_\pm = \mp i \frac{\chi_0 \omega_L T_2}{1 - i(\omega \pm \omega_L)\, T_2}\,, \tag{126}$$

where χ_0 is the static magnetic susceptibility of the system, T_2 the spin–spin relaxation time, and $\omega_L \equiv geB/2m_0$ the Larmor precession frequency with g the Landé g-factor, m_0 the free electron mass, B the magnitude of $\boldsymbol{B}$ (taken in the positive z-direction), and $e < 0$.

We are interested in the circularly polarized mode which elicits EPR. For $e < 0$, this corresponds to the q_+ (i.e. CRA) solution (note resonance denominator in χ_+). We are further concerned with the EPR region ($\omega \approx |\omega_L|$), which for microwave frequencies and $g = 2$ corresponds to fields of the order of several kG. In this region narrow-gap semiconductors typically satisfy the helicon conditions (51) owing to their relatively high carrier concentration ($n > 10^{15}$ cm^{-3}), small effective mass, and high electron mobility.

Assuming weak paramagnetism ($|\chi_\pm| \ll 1$), we can readily show that

$$\alpha_+ = (n\,|e|\,\omega\mu_0/B)^{1/2} \cdot (1 + \chi'_+)^{1/2} \approx \alpha_{0+}\,(1 + \tfrac{1}{2}\,\chi'_\pm)\,, \tag{127}$$

$$\beta_+ = \beta_{0+} + \tfrac{1}{2}\,\alpha_{0+}\chi''_\pm\,, \tag{128}$$

where single and double primes designate real and imaginary parts, and α_{0+} and β_{0+} correspond to values of α_+ and β_+ evaluated in the absence of paramagnetic ions (i.e. for $\chi_+ = 0$).

Thus the phase of the transmitted wave provides a maesure of χ'_+, and the transmitted amplitude measures χ''_+. We note in the expression for β_+ that the magnetic loss term χ''_+ is multiplied by α_{0+}, a rather large quantity, proportional to $n^{1/2}$. Thus the electron plasma appears to *enhance* the EPR interaction. The reason for this is that for large α_{0+} a helicon is a (practically) pure traveling magnetic field as can be understood from the characteristic impedance of the (+) mode

$$Z_+ = \frac{E_+}{H_+} = \left(\frac{\mu_+}{\varepsilon_+}\right)^{1/2} \approx \frac{\mu_0 \omega}{\alpha_{0+}}. \tag{129}$$

Helicon-excited EPR was investigated for Mn^{2+} ions in the ternary NGS $Hg_{1-x}Mn_xTe$ [105 to 108]. Experiments were carried out as a function of Mn^{2+} concentration, electron concentration, and temperature. The samples were single crystals, grown by the Bridgman technique, and annealed in Hg-vapour, with x ranging from 0.003 to 0.09. Some of the samples were doped with indium which acts as a donor for this material. The measurements were carried out in Faraday geometry at 35 GHz using a "Rayleigh"-type interferometer bridge which determines both amplitude and phase of the signal transmitted by the samples. Details of the experimental apparatus and procedure are similar to those described in [7]. Typical transmission data obtained for a pure (intrinsic) crystal, are shown in Fig. 26. The figure shows four superimposed interferograms obtained by beating the transmitted helicon signal with a reference signal at four phase settings indicated in the figure. The dashed line shows the envelope, and provides a measure of β_+. Cross-over points provide a precise measure of α_{0+}, required for the analysis of EPR. The resonance itself appears clearly at 12.46 kG, corresponding to a g-value of 2.0.

Real and imaginary parts of dynamic magnetic susceptibility corresponding to the data in Fig. 26, are shown in Fig. 27. The points represent experimental values; the curves show Lorentzian lines calculated for the observed resonance line intensity and half-width of χ''_+. The good Lorentzian fit is somewhat surprising: one would have expected more distortion of the Bloch-model lineshape due to hyperfine splitting.

The measured value of the dynamic susceptibility increases rapidly with decreasing temperature. This is illustrated in Fig. 28 which shows the helicon amplitude transmitted through an indium-doped (n-type, $n = 5.5 \times 10^{16}$ cm^{-3}) $Hg_{1-x}Mn_xTe$ crystal at three temperatures. The oscillatory nature of the curves outside of EPR arises from dimensional (Fabry-Perot) helicon resonances in the sample. Since in this crystal n does not vary significantly in the temperature range covered, the strong variation in the observed EPR intensity directly reflects the temperature dependence of χ_+. In undoped (intrinsic) crystals the observed resonance strength (relative amplitude change at EPR) actually *decreases* with decreasing temperature in spite of increasing χ_+, because of the simultaneous rapid drop of electron concentration. This is in accord with equations (128) and (129), which predict an enhancement of EPR absorption in dense electronic plasmas.

Our intention in this section was to focus on the *mechanism* of helicon-excited EPR, which is unique to NGS. We have not concerned ourselves with the details of EPR spectra, i.e. the precise values of the g-factor and the fine and hyperfine splitting

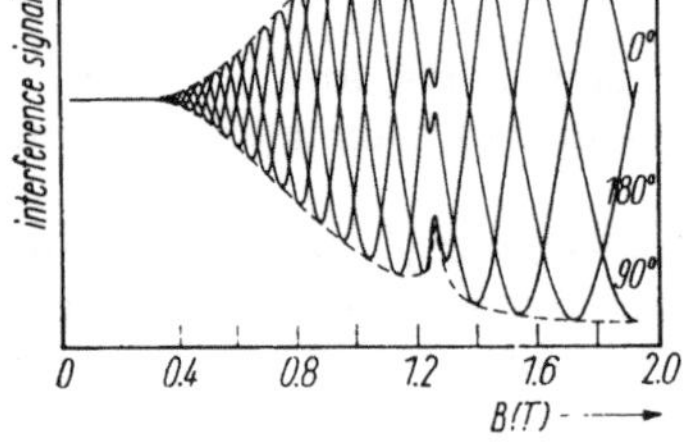

Fig. 26. Typical helicon interference data observed at 35 GHz in $Hg_{1-x}Mn_xTe$ at four phase settings of the interferometer bridge. Phase settings are shown in the figure, $d = 0.855$ mm, $x = 0.03$, $T = 80$ K, CRA. The dashed curve gives the envelope obtained by rapid variation of the phase shifter as the field is swept (see [7]). EPR is clearly visible at 1.246 T (after [105])

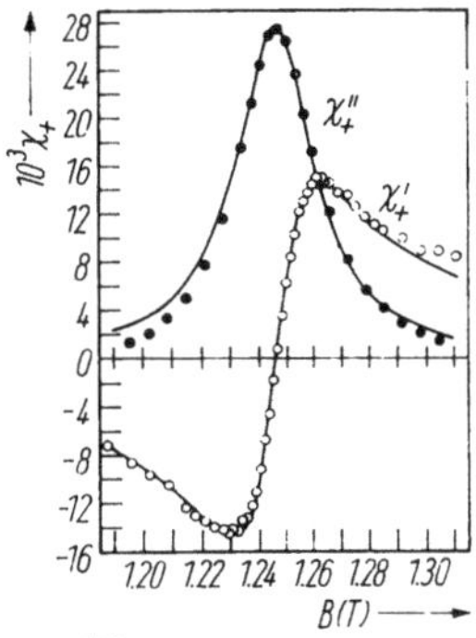

Fig. 27

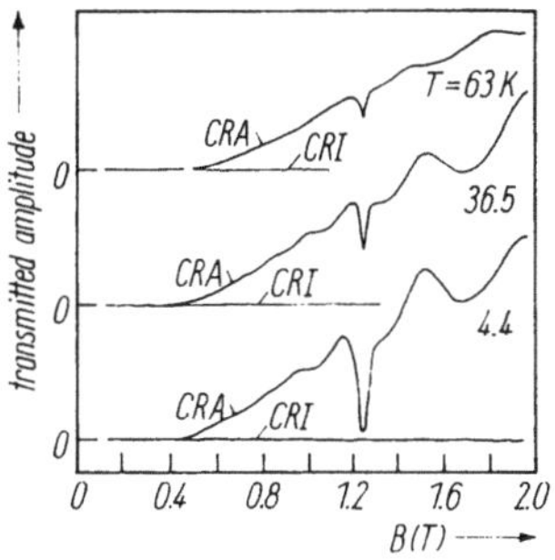

Fig. 28

Fig. 27. Real and imaginary parts of the dynamic susceptibility χ_+ as a function of B, corresponding to the data of Fig. 26 in the immediate vicinity of EPR (after [105])

Fig. 28. Transmitted helicon (CRA) amplitude in In-doped $Hg_{1-x}Mn_xTe$ crystal ($n = 5.5 \times 10^{16}$ cm^{-3}, $x = 0.03$, $d = 0.066$ cm) as a function of B at 4.4, 36.5, and 63 K. No transmission is observed for the CRI polarization (after [105])

constants usually determined by EPR. Extensive studies of Mn^{2+} in mercury II–VI compounds (mostly NGS) using conventional EPR spectrometry were carried out by Leibler et al. [109 to 111]. With the exception of one thin monocrystal of HgTe, the NGS were investigated in powder form because of the large conductivities of these materials. The details of the Mn^{2+} spectrum were found to depend on composition and band structure of the samples. Of particular interest is the correlation between the values of the g-factor of Mn^{2+} and of the energy gap, observed in $Hg_{1-x}Cd_xTe$ as a function of x [109 to 111]. We refer the reader to the original sources for details.

6.2 Conduction-electron spin resonance in NGS

Narrow-gap semiconductors are typically characterized by a relatively large number of conduction electrons. These can themselves lead to spin resonance absorption at microwave frequencies. Spin resonance arises because the Landau levels of the electrons are spin-split (i.e. Zeeman-split), as has been already remarked in connection with Shubnikov-de Haas data. As can be seen from equation (67), the energy difference between the spin-up and spin-down levels are given by $\Delta\varepsilon = g\mu_B B = \hbar\,|\omega_L|$ where $\omega_L \equiv geB/2m_0$ is the Larmor frequency of precession of conduction electrons, and $\mu_B = |e|\,\hbar/2m_0$ is the Bohr magneton. When the microwave frequency satisfies the condition

$$\hbar\omega = g\mu_B B \equiv \hbar\,|\omega_L|\,,$$

one observes resonant absorption of microwave power which is consumed in effecting spin-flip transitions *within* a given Landau level.

Unlike the case of bound electrons such as those in the Mn^{2+} resonance described above, the value of the g-factor of conduction electrons is far removed from the free-electron value of $g = 2$. The factor depends very strongly on the crystal lattice, as well as on the band structure, non-parabolicity, electron energy, and temperature [112 to 116]. Detailed theoretical studies of the g-factor in NGS were carried out by Zawadzki and co-workers [117 to 119].

Typically small values of the effective mass go hand-in-hand with high values of the g-factor. Thus, e.g. for InSb we have $m^* = 0.0139m_0$ and $g = 51$. This indicates that (as in the case of the free electron) spin resonance of conduction electron (CESR) occurs in the general vicinity of the cyclotron resonance condition. For InSb CESR will occur at fields where ω_c is of the order of ω, i.e. in a region of considerable reflection and electric losses. The standard ESR technique is therefore difficult because the microwave signal is limited to a very narrow skin-depth region. The skin depth is longer when the carrier concentration is small, but then the resonant signal intensity is low. As a result, extreme sensitivity is required in the conventional EPR procedure. Partly because of these limitations, experimental literature concerning CESR in NGS is not extensive and is limited to measurements on InSb [120 to 122] and InAs [123, 124]. Even here it is restricted almost exclusively to the determination of the g-factor. Measurements of the absolute magnetic susceptibility and a meaningful interpretation of the linewidth are essentially ruled out by the sensitivity required and by lineshape distortions associated with the non-uniformity of the microwave field throughout the sample.

Even with these restrictions, CESR has proved to be a useful and highly promising experimental tool in NGS. The g-factor of conduction electrons, which is measured by CESR *directly*, is a quantity of fundamental interest. By determining the variation of g as a function of experimental conditions, CESR also sheds light on a series of other physical parameters. Some of these are: electron concentration, lattice temperature in non-degenerate samples, energy relaxation time, non-parabolicity, and band deformation due to the presence of impurities [123 to 125].

The dependence of the g-factor on energy and the relative ease of heating electrons in NGS also gives the opportunity of distinguishing resonances associated with different Landau levels.

The observation of multiple CESR resonances associated with different Landau levels requires that the experiments are performed at magnetic fields which correspond to the "quantum limit" ($|\omega_c| \tau > 1$, $|\omega_c| > kT$). This typically requires fields in excess of 0.1 T and thus millimeter and submillimeter waves. Unfortunately experimental techniques in this frequency range usually do not meet the exacting sensitivity requirements of the experiments. A convenient and very sensitive technique consists of illuminating the sample with microwaves, but monitoring the sample resistivity [125]. In this case the changes in conductivity are observed by placing the sample in the vicinity of an inductor in a tank circuit, so that conductivity changes caused by absorption of microwaves manifest themselves as changes of Q of the tank circuit. The sample is, in effect, a microwave detector with response peaks at the absorption maximum, i.e. at $\omega = |\omega_L|$.

The method exploits specific properties of InSb, i.e. the fact that a relatively small absorption of power heats electrons sufficiently so that their mobilities (and thus the conductivity) show a detectable change. In the case of ESR, the power absorbed transforms into electron kinetic energy by spin–orbit coupling. The features necessary to observe this in InSb are characteristic of other NGS, and the technique should prove a powerful tool in the study of these materials in general.

7. Concluding Remarks

The above discussion enables us to conclude that the microwave magnetoreflection and magnetotransmission, as well as the small-particle absorption, provide wide opportunities for an electrical characterization of NGS. Such parameters as the carrier concentration n, mobility μ, effective mass m^*, lattice dielectric constant $\varkappa_l$, etc. can be determined both in "semi-infinite" ingots and in small disk-shaped or spherical samples. Note that the typical microwave spectrometer shown in Fig. 3, operating at a single frequency of 35 GHz covers a wide range of parameters (e.g. $n \approx 10^{11}$ to

10^{18} cm^{-3}) and sample shapes. Moreover, microwave helicons can be used for determining the inhomogeneity in the distribution of carrier concentration in NGS [126, 127].

Furthermore, MHz magnetoplasma wave methods combined both with hydrostatic and uniaxial pressure techniques are successfully used for the study of band-structure effects in many-valley semiconductors [128]. Effects arising from the non-ellipsoidal band model were recently observed at microwave frequencies [49].

Another interesting trend in the microwave studies of NGS is the wave propagation in periodic multiplayer structures. Stop-bands and pass-bands were observed in InSb structures [129] in accordance with theory [10]. These effects are of potential importance for the study of superlattices and growth inhomogeneities in NGS.

Other excitation types are studied at microwaves, e.g. the magnetoplasma surface polariton excitation using attenuated total reflection [130, 131] or the cyclotron waves [132] which are of potential interest for the NGS diagnostics. Gelmont et al. [133] predicted a new transparency window in zero-gap semiconductors at frequencies below the plasma frequency in the ordinary Voigt configuration using the quantum approach.

Recent results on the guided microwave interactions with bounded semiconductors are given in [134 to 137]. Microwave resonances in gyroelectric (semiconductor) spheroids were recently studied [138] employing the magnetostatic theory developed for ferromagnetics. On the other hand, the ideas of microwave interaction with a semiconductor sphere were applied to the study of resonances in a ferromagnetic sphere [139]. The general problem of electromagnetic wave interaction with ferromagnetic semiconductor spheroids in a dc magnetic field still awaits solution.

Finally, the submillimeter studies of NGS should be mentioned. Resonance submillimeter spectroscopy of $Hg_{1-x}Cd_xTe$ was used by Guldner et al. [140]. Such important parameters as the Landé factor, the effective mass including polaron effects, etc. were measured in this region, and reviewed by Lax [112], Button [141], Landwehr [142], Gershenzon [143], and others.

References (Part II)

References [1] to [82] see phys. stat. sol. (a) **53**, 11 (1979).
[83] G. W. Ford and S. A. Werner, Phys. Rev. B **8**, 3702 (1973).
[84] H. C. van de Hulst, Light Scattering by Small Particles, Wiley, New York 1957.
[85] M. Kerker, The Scattering of Light, Academic Press, New York 1969.
[86] F. L. Galeener, Phys. Rev. Letters **22**, 1292 (1969).
[87] B. Lax and J. G. Mavroides, Solid State Phys. **11**, 261 (1960).
[88] L. D. Landau and E. M. Lifshitz, Electrodynamics of Continuous Media, Addison-Wesley, Reading (Mass.) 1960 (p. 299), cf. also [46].
[89] G. Dresselhaus, A. F. Kip, and C. Kittel, Phys. Rev. **100**, 618 (1955).
[90] R. S. Brazis and J. K. Furdyna, J. appl. Phys. **48**, 2927 (1977).
[91] T. A. Evans and J. K. Furdyna, Phys. Rev. B **8**, 1461 (1973).
[92] J. R. Dixon, unpublished.
[93] J. K. Furdyna, Postepy Fiz. **25**, 67 (1974).
[94] E. A. Neifeld and A. B. Davydov, Fiz. Tekh. Poluprov. **5**, 1254 (1971) (Soviet Phys. — Semicond. **5**, 1107 (1971)).
[95] J. K. Furdyna and J. Mycielski, Bull. Amer. Phys. Soc. Ser. II, **19**, 246 (1974).
[96] R. S. Markiewicz, J. P. Wolfe, and C. D. Jeffries, Phys. Rev. Letters **32**, 1357 (1974).
[97] R. S. Markiewicz, Phys. Rev. B **10**, 1766 (1974).
[98] J. K. Furdyna and J. Mycielski, Proc. Topical Conf. II—VI Semiconductor Compounds, Jaszowiec 1973, IF-PAN, Warsaw 1974 (p. 178).
[99] F. L. Galeener, T. A. Evans, and J. K. Furdyna, Phys. Rev. Letters **29**, 728 (1972).
[100] T. A. Evans, F. L. Galeener, and J. K. Furdyna, Proc. Internat. Conf. Physics of Semiconductors, Warsaw 1972, PWN, Warsaw, 1972 (p. 357).

[101] J. K. FURDYNA and F. L. GALEENER, Proc. Internat. Conf. Semiconductor Physics, Moscow 1968, Vol. 2, Nauka, Leningrad 1968 (p. 870).
[102] F. L. GALEENER and J. K. FURDYNA, Appl. Phys. Letters **14**, 163 (1969).
[103] F. L. GALEENER and J. K. FURDYNA, Phys. Rev. B **4**, 1953 (1971).
[104] K. K. CHEN and J. K. FURDYNA, J. appl. Phys. **49**, 3363 (1978).
[105] R. T. HOLM and J, K. FURDYNA, Solid State Commun. **15**, 1459 (1974).
[106] R. T. HOLM and J. K. FURDYNA, Phys. Rev. B **15**, 844 (1977).
[107] C. P. SLICHTER, Principles of Magnetic Resonance, Harper and Row, New York 1963.
[108] D. P. MULLIN, J. R. DIXON, JR., and J. K. FURDYNA, Proc. III. Internat. Conf. Phys. Narrow-Gap Semicond., Warsaw 1977, PWN, Warsaw 1977.
[109] K. LEIBLER, W. GIRIAT, K. CHĘCIŃSKI, and Z. WILAMOWSKI, Coll. Ampère **16**, 987; 993 (1970).
[110] K. LEIBLER, W. GIRIAT, Z. WILAMOWSKI, and R. IWANOWSKI, phys. stat. sol. (b) **47**, 405 (1971).
[111] K. LEIBLER, W. GIRIAT, K. CHĘCIŃSKI, Z. WILAMOWSKI, and R. IWANOWSKI, phys. stat. sol. (b) **55**, 447 (1973); Praca IF-PAN 11 and 12, Warsaw (1971).
[112] B. LAX, Proc. Internat. Symp. Microwave Diagnostics of Semiconductors, Porvoo (Finland) 1977, R. Paananen, Swedish Acad. Engineering Sciences in Finland, Rep. No. 31, Helsinki, 1977 (p. 1).
[113] L. M. ROTH, B. LAX, and S. ZWERDLING, Phys. Rev. **114**, 90 (1959).
[114] R. BOWERS and Y. YAFET, Phys. Rev. **115**, 1165 (1959).
[115] Y. YAFET, Phys. Rev. **115**, 1172 (1959).
[116] B. LAX, J. G. MAVROIDES, H. J. ZEIGER, and R. J. KEYES, Phys. Rev. **122**, 31 (1961).
[117] W. ZAWADZKI, Phys. Letters **4**, 190 (1963).
[118] W. ZAWADZKI, phys. stat. sol. **3**, 1421 (1963).
[119] P. KACMAN and W. ZAWADZKI, phys. stat. sol. (b) **47**, 629 (1971).
[120] G. BEMSKI, Phys. Rev. Letters **4**, 62 (1960).
[121] M. GUERON, Proc. Internat. Conf. Semiconductor Physics, Paris 1964, Dunod, Paris 1964 (p. 433).
[122] R. A. ISAACSON, Phys. Rev. **169**, 312 (1968).
[123] J. KONOPKA, Phys. Letters A **26**, 29 (1967).
[124] D. KAPLAN and J. KONOPKA, Proc. Internat. Conf. Semiconductor Physics, Moscow 1968, Nauka, Leningrad 1968 (p. 1146).
[125] J. KONOPKA, Phys. Rev. Letters **24**, 666 (1970).
[126] A. LAURINAVICHIUS and YU. POZHELA, Prib. i Tekh. Eksper. **4**, 241 (1976).
[127] A. LAURINAVICHIUS, YU. POZHELA, and A. VITKUS, Proc. Internat. Symp. Microwave Diagnostics of Semiconductors, Porvoo (Finland) 1977, R. Paananen, Swedish Acad. Engineering Sciences in Finland, Rep. No. 31, Helsinki 1977 (p. 250).
[128] YU. K. POZHELA and R. B. TOLUTIS, in: Elektrony v Poluprovodnikakh, Vol. I, Ed. YU. K. POZHELA, Mintis, Vilnius 1978.
[129] R. S. BRAZIS, A. S. MIRONAS, and YU. K. POZHELA, Litovskii Fiz. Sbornik, **14**, 95 (1974) (Soviet Phys. — Collection **14**, 54 (1974)).
[130] V. AMBRAZEVIČIENĖ and R. BRAZIS, Solid State Commun. **18**, 415 (1975).
[131] V. AMBRAZEVIČIENĖ and R. BRAZIS, Fiz. Tekh. Polpurov. **12**, 1114 (1978) (Soviet Phys. — Semicond. **12**, 661 (1978)).
[132] R. HERRMANN, G. OELGART, and R. STEGMANN, Wiss. Z. Humboldt-Univ. Berlin, Math.-naturwiss. R., **25**, 3 (1976).
[133] B. L. GELMONT, V. I. IVANOV-OMSKII, and E. N. UKRAINTSEV, phys. stat. sol. (b) **87**, 381 (1978).
[134] YU. G. ARAPOV and A. B. DAVYDOV, Defektoskopiya **11**, 63 (1978).
[135] A. B. DAVYDOV, V. A. ZAKHAROV, and I. E. PODCHINYONOV, Fiz. tverd. Tela **19**, 1676 (1977) (Soviet Phys. — Solid State **19**, 978 (1977)).
[136] D. A. USANOV, V. B. FEKLISTOV, and A. YU. DERYAGIN, II. Vsesoyuz. Symp. Millimetrovym i Submillimetrovym Volnam, Kharkov 1978, Vol. I, Inst. Radiophys. Electronics, Acad. Sci. USSR (p. 139).
[137] L. I. KATS, V. V. POPOV, and R. M. REVZIN, Radiotekhnika i Elektronika **22**, 1107 (1977).

[138] A. B. DAVYDOV, B. B. PONIKAROV, and G. L. SHTRAPENIN, Fiz. Tekh. Poluprov. **12**, 866 (1978) (Soviet Phys. — Semicond. **12**, 511 (1978)).
[139] R. S. BRAZIS and J. K. FURDYNA, Phys. Rev. B **16**, 3273 (1977).
[140] Y. GULDNER, C. RIGAUX, A. MYCIELSKI, and Y. COUDER, phys. stat. sol. (b) **81**, 615 (1977); **82**, 149 (1977).
[141] K. J. BUTTON, Lectures Notes Internat. Conf. Application of High Magnetic Fields in Semiconductor Physics, Würzburg 1976, Phys. Inst. Univ. Würzburg (p. 326).
[142] G. LANDWEHR, II. Internat. Conf. and Winter School on Submillimeter Waves and Their Applications, 1976, Conf. Digest, Ed. S. PERKOWITZ, Inst. Electrical and Electronics Engineers, Inc. 1976 (p. 146); cf. also J. Opt. Soc. Amer. **67**, 922 (1977).
[143] E. M. GERSHENZON, Proc. Internat. Symp. Microwave Diagnostics of Semiconductors, Porvoo (Finland) 1977, R. Paananen, Swedish Acad. Engineering Sciences in Finland, Rep. No. 31, Helsinki 1977 (p. 242).

(Received October 19, 1978)

Note added in proof

A new theory of scattering of electromagnetic waves by a gyrotropic sphere [1] has appeared during the preparation of this paper. The theory is a generalization of the classic Mie formulation for dielectric spheres, and has direct relevance to the interaction of electromagnetic waves with spherical NGS samples. A series of microwave applications of the theory, such as helicon-like dimensional resonances in semiconductors, cyclotron resonances, and Alfvén resonances in electron–hole droplets, are presented [2].

Furthermore, new experimental results on the propagation of cyclotron waves in the electron–hole system of bismuth at frequencies up to 300 GHz have been published [3].

It is also necessary to mention earlier work on the theoretical [4] and experimental [5] evidence for the existence of a new type of wave propagation in metals, semimetals and semiconductors: the slow electromagnetic surface waves. These waves are of potential interest as a diagnostic tool for NGS.

Finally, we would like to mention a recent book [6] devoted to growth, structure and transport properties of NGS.

We hope that the reader will bear with us in our dual endeavor to give as complete a presentation as possible while confining the volume of the paper to reasonable limits, in consequence of which we had to leave unmentioned many other sources in this rapidly developing field.

[1] S. W. FORD and S. A. WERNER, Phys. Rev. B **18**, 6752 (1978).
[2] J. R. DIXON, JR. and J. K. FURDYNA, Phys. Rev. B **18**, 6770 (1978).
[3] W. BRAUNE, J. LEBECH and K. SAERMARK, phys. stat. sol. (b) **87**, 527 (1978).
[4] S. I. KHANKINA and V. M. YAKOVENKO, Fiz. tverd. Tela **9**, 2943 (1967) (Soviet Phys. — Solid State **9**, 2313 (1968)).
[5] V. I. BAIBAKOV and V. N. DATSKO, Fiz. tverd. Tela **15**, 1616 (1973) (Soviet Phys. — Solid State **15**, 1084 (1973)).
[6] D. R. LOVETT, Semimetals and Narrow-Bandgap Semiconductors, Pion Limited, London 1977.

Original Papers

phys. stat. sol. (a) 54, 29 (1979)
Subject classification: 10.2; 22.1.2

Institute of Solid State Physics, Chernogolovka[1]) *(a) and*
Institute of Crystallography, Moscow[2]) *(b) Academy of Sciences of the USSR*

Shape of Extinction Fringes and Determination of the Burgers Vector in Single-Crystal Interferometric Techniques

By
E. V. Suvorov (a), O. S. Gorelik (a), V. M. Kaganer (b),
and V. L. Indenbom (b)

An analysis of the shape of extinction contours on section topographs of dislocations in singular position, for high and low diffractive power of a dislocation is given. A comparison of experimental data with the conclusions of the theory enables to verify quantitatively the earlier developed concepts of formation of the dynamic image of dislocations. Methods are suggested to determine the Burgers vectors to an accuracy of $\approx 6\%$, based on analyzing the shape of interference fringes.

Проводится анализ формы экстинкционных контуров на секционных топограммах дислокаций в особом положении при малых и больших значениях дифракционной мощности дислокаций. Сравнение экспериментальных данных с выводами теории дает возможность количественной проверки развитых ранее представлений о формировании динамического изображения дислокаций. Предлагаются основанные на анализе формы полос методы определения величины вектора Бюргерса, обладающие точностью $\approx 6\%$.

1. Introduction

The progress in the theory of X-ray image formation resulted in growing interest in analyzing the possibilities of a quantitative interpretation of section topographs which in fact correspond to fundamental solutions (Riemann functions) of the equations of X-ray optics for real crystals. A number of authors suggested general semi-qualitative arguments concerning the use of section topographs [1, 2].

Using the techniques of geometrical optics the diffraction image arising at the edge of an oxide film which was produced on the surface of a silicon single crystal was analyzed in detail in [3]. This investigation confirmed the basic notions about the character of strains produced by the interface [4].

The behaviour of trajectories and phases of Bloch waves in the neighbourhood of dislocations in a crystal lattice was investigated by the present authors and reported in a series of papers [5 to 8]; it was demonstrated that a map of extinction fringes can be used to measure the characteristics of the elastic stress field of a dislocation. By analogy to standard optical interferometry this method was referred to as the single-crystal interferometer method, since the extinction fringes are formed on the section topographs as a result of interference of two types of Bloch waves arriving at each point on the crystal surface along two different paths. The basic qualitative features of dislocation images obtained for dislocations in special positions were investigated both theoretically and experimentally; the number of additional extinction contours and their position were determined, and the possibilities of experimentally determining the Burgers vectors using the topographs were discussed.

---

[1]) Chernogolovka 142432, Moscow district, USSR.
[2]) Leninsky Prospect 59, Moscow 117333, USSR.

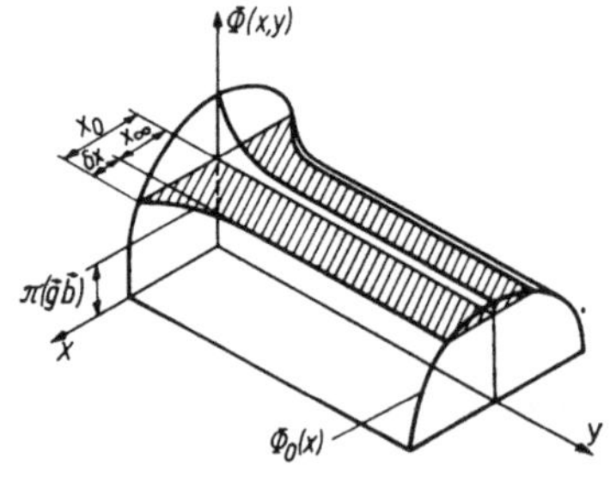

Fig. 1. The surface of total phase difference of Bloch waves of two types on the exit surface of a crystal. The additional phase difference generated in the tensile stress field of the dislocations results in shifting the interference fringe by $\delta x = x_0 - x_\infty$

The aim of the present paper is to analyze the shape of the extinction contours which enables one to check quantitatively the characteristics of the extinction fringes discussed earlier, and to develop techniques for determining the Burgers vectors with high accuracy.

2. Phases of Bloch Waves and Shapes of Extinction Fringes

Fig. 1. illustrates the general shape of the phase difference surface, $\Phi(x, y)$, of Bloch waves in the elastic stress field of a dislocation [8] (x denotes the distance along the reflection vector, and y the distance between the dislocation and the plane of scattering). Extinction contours correspond to horizontal cross-sections of this surface, and the contour shape, $y(x)$, is given implicitly by the equation $\Phi(x, y) = \text{const.}$

The phase difference between Bloch waves of two types in an ideal crystal is given by

$$\Phi_0(x) = \frac{2\pi}{\Lambda} \sqrt{d^2 - x^2}\,, \tag{1}$$

where d is the thickness of the crystal and Λ the extinction length; the scale is chosen so that the Bragg angle equals 45° (normalized coordinates). The elastic stress field of a dislocation produces an additional phase difference which at $y = 0$ reaches the maximum value of $\pi(\boldsymbol{gb})$, constant over the whole cross-section $y = 0$. Therefore, the displacement of an extinction fringe in the dislocation stress field is given by the expression

$$\Phi_0(x_\infty) = \Phi_0(x_0) + \pi(\boldsymbol{gb})\,, \tag{2}$$

where x_∞ and x_0 denote the fringe positions far from the dislocation and for $y = 0$, respectively (Fig. 1). Equation (2) gives the diffractive power of the dislocation, $(\boldsymbol{gb})$, as a function of the contour position in an ideal crystal and its total displacement in the elastic field of the dislocation, $\delta x = x_0 - x_\infty$, which enables one to find the Burgers vector experimentally from the topograph characteristics.

However, it should be noted that the range of small values of y is often inaccessible for experimental measurements, because a complex image due to focussing of the new wave field formed in the neighbourhood of the dislocation lies just in this range [5]. At the same time, the visual extrapolation to $y = 0$ cannot be sufficiently accurate, since the extinction fringes approach the $y = 0$ axis with zero slope. In order to use (2), one must also know the fringe position for $y \to \infty$; that demands the crystal thickness to be constant with high accuracy. Therefore, it is important to find methods of determining the Burgers vectors; it is sufficient to know the contour shape only in the central part of the topograph; an apparent generalization of (2) to the additional extinction contours close to the $y = 0$ plane and forming closed loops, makes it possible to reduce the requirement of constant thickness; however, the necessity to extrapolate the fringe to $y = 0$ is not eliminated.

In order to analyze the shape of extinction fringes in the range of small y, we differentiate the relationship $\Phi(x, y) = \text{const}$ and find the slope of the extinction fringe

$$\frac{\mathrm{d}x}{\mathrm{d}y} = -\frac{\partial\Phi/\partial y}{\partial\Phi/\partial x}. \tag{3}$$

It was established that, if a dislocation is in a singular position, i.e. parallel to the reflection vector, the local misorientation of the crystal is independent of x, and the generalized momentum p along the trajectories of Bloch waves is conserved [6, 8]. By virtue of the relationship $p = \partial S/\partial x$ (where S is the action [8]) we obtain

$$\frac{\partial\Phi}{\partial x} = -\frac{\pi}{\Lambda}[p^+(x, y) - p^-(x, y)], \tag{4}$$

where $p^\pm(x, y)$ denote the momenta corresponding to the two types of Bloch waves arriving at a point (x, y) of the exit surface. At small $(\boldsymbol{gb})$ the displacement of trajectories is small, and the momentum of the beam arriving at the point (x, y) can be found from the relationship between coordinate and momentum in an ideal crystal: $p^\pm = \pm x/\sqrt{d^2 - x^2}$. The additional phase difference between the beams of the two types for the central region of a topograph can be found, due to the smallness of trajectory displacement in the elastic stress field of a dislocation, as the additional phase difference of the two beams with $p = 0$ [6]

$$\Gamma_p(y) = 2\pi\frac{y}{\Lambda}\int\limits_{-d/2y}^{d/2y}\mathrm{d}\xi\left[\sqrt{1 + \left(\frac{A}{1 + \xi^2}\right)^2} - 1\right], \tag{5}$$

where $A = \Lambda(\boldsymbol{gb})/2\pi y$. Moreover, at small y the additional phase difference is constant over almost the whole base of the Borrmann delta [8]. The quantity $\partial\Phi/\partial y$ coincides with $\mathrm{d}\Gamma_p/\mathrm{d}y$, since the phase difference in an ideal crystal, Φ_0, is independent of y. By differentiating (5) for small y, we obtain

$$\frac{\mathrm{d}\Gamma_P}{\mathrm{d}y} = -\frac{4\pi}{\Lambda a}\{a^2[2E(\tau) - K(\tau)] - K(\tau)\}, \tag{6}$$

where $K(\tau)$ and $E(\tau)$ are complete elliptic integrals of first and second type, respectively, $a^4 = 1 + A^2$, $\tau^2 = (a^2 - 1)/2a^2$. In particular, for small y and small deviations of x from x_0, we integrate (3) and obtain for the contour shape the expression

$$\frac{x - x_0}{\Lambda} = -\eta\sqrt{(\boldsymbol{gb})\left[\left(\frac{d}{x_0}\right)^2 - 1\right]\frac{y}{\Lambda}}, \tag{7}$$

where

$$\eta = 2\sqrt{\frac{2}{\pi}}\left[2E\left(\frac{1}{\sqrt{2}}\right) - K\left(\frac{1}{\sqrt{2}}\right)\right] \approx 1.35.$$

It is clear now that for small y, $y(x)$ is a quadratic function of x, so that a plot of the extinction fringe in linearizing coordinates $(x, \sqrt{y})$ substantially increases the reliability of the extrapolation of the curve to the abscissa and improves the accuracy of finding x_0.

At high values of $(\boldsymbol{gb})$ a direct image of the dislocation is superimposed on the range of small y, so that most of the information on the value of $(\boldsymbol{gb})$ is extracted from the additional extinction contours. Most interesting is the shape of the additional contours in the neighbourhood of the apices. At $x = 0$, $p^+ = p^-$; according to [8], the difference $p^+ - p^-$ is propotional to x in a wide range of x-values. By virtue of (3), (4), the extinction contour is represented by a quadratic parabola in a considerable range around its apex.

At large $(\boldsymbol{gb})$ the displacements of beam trajectories cannot be neglected, since the beams of the two types with $p = 0$ arrive at essentially different points. Consequently, the additional phase difference is less than Γ_p, so that by taking into account the phase difference between the beam with $p = 0$ and the beam arriving at the point $x = 0$ we obtain the following expression [8] for the additional phase difference in the $x = 0$ section of the topograph:

$$\Gamma(y) = \Gamma_p - \frac{(\Gamma_p - R)^2}{2(2\pi(d/\Lambda) + \Gamma_p - R)}, \tag{8}$$

where $R = 2(\boldsymbol{gb}) \arctan(d/2y)$.

In order to find $\partial\Phi/\partial x$ we expand $p^{\pm}(x, y)$ into a series for small p

$$p^{\pm} = \frac{(\pm x - x_{p^+=0})}{\tilde{d}(y)}, \tag{9}$$

where

$$\tilde{d}(y) \equiv \left.\frac{\partial x}{\partial p^+}\right|_{p^+=0} = d - \frac{y}{a}\left\{\frac{3a^4 - 1}{a^2}[2E(\tau) - K(\tau)] - 2K(\tau)\right\}. \tag{10}$$

By substituting (9) into (4) and integrating the obtained expression, we derive the equation for the shape of the cross-sections of the phase difference surface for $y =$ $=$ const:

$$\Phi(x, y) = \Phi(0, y) - \frac{\pi x^2}{\Lambda\tilde{d}(y)}. \tag{11}$$

The curvature of these cross-sections is $2\pi/\tilde{d}$, while in an ideal crystal $\tilde{d}$ should have been replaced simply by d.

In particular, it follows from (11) that for y corresponding to the apex of the n-th contour, the $(n - 1)$-th contour is at a distance

$$\frac{x_{n-1}}{\Lambda} = \sqrt{\frac{2\tilde{d}(y)}{\Lambda}} \tag{12}$$

from the topograph axis. Equation (12) makes it possible to compare the curvature of the $y =$ const cross-sections of the surface of the phase difference with the experimental data.

For the shape of the additional contours in the central region of a topograph, we obtain from (3), (9)

$$\left(\frac{x}{\Lambda}\right)^2 = -\frac{1}{\pi}\tilde{d}(y)\,\Gamma'(y)\frac{y - y_0}{\Lambda}, \tag{13}$$

where $\Gamma'(y) = \mathrm{d}\Gamma/\mathrm{d}y$ is found from (8), (6). Hence, the additional contours in the central region of a topograph are parabolic, with curvature determined, according to (13), by the position of the contour apex and the value of $(\boldsymbol{gb})$.

3. Experiment: Shape of Contours and Determination of Burgers Vectors

Topographs were obtained from silicon single crystals with dislocation half-loops introduced by plastic bending as described in [9]. The samples were cut into four-facet prisms with (111), $(11\bar{2})$, and $(1\bar{1}0)$ faces. The dislocation type was found in a standard manner, by using the contrast extinction orientations. Fig. 2 illustrates typical section topographs of dislocations in singular position (dislocation axis parallel to the dif-

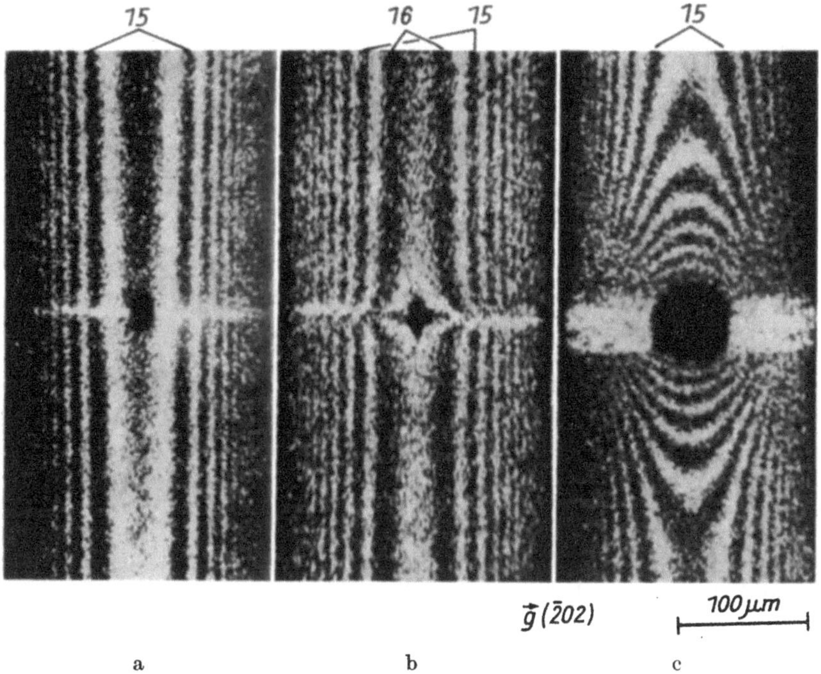

Fig. 2. Section topographs of screw and 60° dislocations. MoK_α a) 60° dislocation; b) screw dislocation; c) a bundle of 14 closely spaced screw dislocations

fraction vector). The positions of the interference fringes were found on the measuring microscope.

The simplest type of measurement on a topograph is that of the contour position far from the dislocation, x_∞, and its displacement in the dislocation field, δx; after that the value of the Burgers vector can be found by means of (2). The processing of more than ten topographs of screw and 60° dislocations gives the Burgers vector value

$$b = (3.79 \pm 0.39)\ \text{Å} \qquad (14)$$

which corresponds [10] to dislocations with $\frac{1}{2}\langle 110\rangle$-type Burgers vectors.

Fig. 2 a, b demonstrate that the image due to focussing of the new wave field obstructs an accurate determination of the coordinate x_0 of the contour in the vicinity of the dislocation trace. The accuracy of measuring x_0 can be improved by plotting the contour in linearizing coordinates $(x, \sqrt{y})$. Fig. 3a shows as an example the plot for the 15th and 16th contours of a topograph given in Fig. 2b. In accordance with (7), in the range $y \leqq 0.3(\boldsymbol{gb})\,\Lambda$ the function $x(\sqrt{y})$ is adequately approximated by a linear function whose extrapolation to the abscissa enables one to determine x_0 with an accuracy not worse than 3%. Reprocessing of the above-mentioned topographs by this method yields

$$b = (3.77 \pm 0.24)\ \text{Å}\,. \qquad (15)$$

The relative error is now 6%, i.e. less by a factor of nearly two than the value obtained by determining the Burgers vector by means of directly measuring the contour displacement on the topographs. It should be noted for comparison that the accuracy of measuring the Burgers vectors by means of electron microscopy and photoelastic techniques is normally not better than 10% [11 to 13].

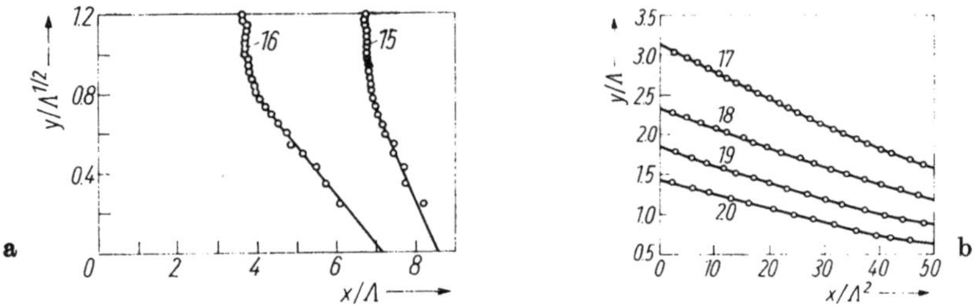

Fig. 3. a) Extinction fringe of order 15 and 16 plotted from the data of Fig. 2b in linearizing normalized coordinates $(x, \sqrt{y})$. b) Additional extinction contours plotted from the data of Fig. 2c in linearizing normalized coordinates (x^2, y)

The slope of the straight line obtained in $(x, \sqrt{y})$ coordinates is a function of x_0 and $(\boldsymbol{gb})$ (equation (7)), and hence, can also be used for the experimental determination of Burgers vector values. The slopes of the straight lines fitted to experimental data agree within 20% with the corresponding theoretical values from (7). Owing to its poor accuracy, this method of determining the Burgers vectors can be useful only when the competing techniques are inapplicable (for example, when the interference pattern far from the dislocation is distorted by other defects).

The maximum accuracy of measuring the Burgers vectors can be expected when higher-order interference fringes are used, since the displacement is maximum in this case. At the same time, the extinction fringes closest to topograph axis have largest half-width and, therefore, their positions are measured with maximum error. Our experience in processing the experimental topographs shows that an optimum is achieved when we use the fringes of the order $n_{\max} - 1$, since normally their half-widths are smaller and their displacements are sufficient for reliable measurements. In the case of mixed dislocations, when the screw component is not large and the fringe displacement is difficult to determine, $(\boldsymbol{gb})$ can be increased by increasing the order of reflection. However, this requires polarized radiation with increasing order of reflection the extinction fringes become more spread out due to the difference in extinction lengths corresponding to σ- and π-polarizations. While for low reflection orders this difference results in a rather small broadening of extinction fringes on topographs recorded from non-polarized X-ray sources, with high orders of reflection one has to take into account the consequences of overlapping of two non-identical systems of interference fringes; this complicates the analysis of the resulting interference pattern.

The range of small values of y on the topographs of dislocations with high values of $(\boldsymbol{gb})$ (cf. a bundle of dislocations in Fig. 2c) is overlapped by the central image of the dislocation. The region around the apices of the additional contours is more informative and accessible to measurements than other regions. According to (13) the additional contours are parabolic in a rather large range around the apices. As an example, Fig. 3b shows the contours of Fig. 2c in linearizing coordinates (x^2, y). The quantitative correspondence of the shape of the additional contours to the predictions of the theory can be checked directly via (13). However, as it follows from (13), the contour curvature is determined by the product of the slope of the phase difference surface, $\Gamma'(y)$, in the $x = 0$ cross-section, and the curvature radius $\tilde{d}(y)/2\pi$ of its $y = \text{const}$ cross-sections. The correspondence of these two quantities to the respective theoretical estimates can be checked independently.

In order to find the function $\Gamma(y)$, one has to identify the orders of contours; this is a complicated problem (the contours located too far from the dislocation may be missed [5]). But to find the derivative $\Gamma'(y)$, $\Gamma(y)$ must be known with the accuracy of an additive constant and hence no identification of the contour orders is needed. The processing of experimental topographs demonstrates that the value of $\Gamma'(y)$ found in this manner differs on the average, from that calculated by means of (6) and (8) by not more than 10%. This accuracy should be considered as rather high, in view of the approximate nature of the calculations based on replacing a bundle of closely spaced dislocations by a single dislocation with the Burgers vector being the sum of the individual Burgers vectors.

Whether the radius of curvature of the y = const cross-sections is adequately described by the theory can be verified, for instance, by determining the positions of the $(n - 1)$-th contour in the y = const cross-section corresponding to the apex of the n-th contour (see also [5], Fig. 4). The experimental data demonstrate that on the average the radius of curvature found by means of (12), differs from the theoretical estimate (10) by 15%. At the same time, the curvature of the additional contours found as a product of $\Gamma'(y)$ and $\tilde{d}(y)$, agrees with the theoretical predictions within 25%.

4. Conclusions

A quantitative analysis of the shape of extinction contours supports the earlier developed concepts on the formation of an dynamical image of dislocations, and permits a precise determination of the value of the Burgers vector. The methods suggested in the paper are useful in studying the images of screw and mixed dislocations; edge dislocations produce no displacement along the reflection vector (parallel to the dislocation axis) and thus do not produce distortions which would result in an additional phase difference between Bloch waves of two types and in shifts in the position of interference fringes. One can hope that the high accuracy of measuring Burgers vectors via (15) makes it possible to distinguish between true dislocations and pseudo-dislocations formed at the edges of monatomic impurity layers.

References

[1] A. AUTHIER, Adv. X-Ray Analysis **10**, 9 (1967).
A. AUTHIER, Collected Lect. Internat. Summer School on X-Ray Dynamical Theory and Topography, Limoges (France) 1975 (A12).
[2] V. L. INDENBOM and F. N. CHUKHOVSKII, Uspekhi fiz. Nauk **107**, 229 (1972).
[3] N. KATO, J. R. PATEL, and Y. ANDO, J. appl. Phys. **44**, 965 (1973); **44**, 971, 4405 (1973).
[4] I. A. BLECH and E. S. MEIERAN, J. appl. Phys. **38**, 2913 (1967).
[5] E. V. SUVOROV, V. I. POLOVINKINA, V. I. NIKITENKO, and V. L. INDENBOM, phys. stat. sol. (a) **26**, 385 (1974).
[6] V. L. INDENBOM V. I. NIKITENKO, V. I. POLOVINKINA, E. V. SUVOROV, F. N. CHUKHOVSKII, and A. A. SHTOLBERG, Collected Abstracts 2nd European Crystallography Meeting, Keszthely (Hungary) 1974 (p. 66).
[7] V. L. INDENBOM, E. V. SUVOROV, and V. I. NIKITENKO, Proc. IV. USSR Conf. Dynamic Effects of X-Ray and Electron Scattering, Leningrad 1977 (p. 36) (in Russian).
[8] V. L. INDENBOM, V. I. NIKITENKO, E. V. SUVOROV, and V. M. KAGANER, phys. stat. sol. (a) **46**, 379 (1978).
[9] V. I. NIKITENKO, V. N. EROFEEV, and N. M. NADGORNAYA, in: Dynamics of Dislocations, FTINT Akad. Nauk. UkrSSR, Kharkov 1968 (in Russian).
[10] J. HORNSTRA, J. Phys. Chem. Solids **5**, 129 (1958).
[11] V. L. INDENBOM, V. I. NIKITENKO, and L. S. MILEVSKII, in: Stress and Dislocations in Semiconductors, Moscow 1962 (p. 55) (in Russian).
[12] M. F. ASHBY and L. M. BROWN, Phil. Mag. **8**, 1083 (1963).
[13] V. I. NIKITENKO and L. M. DEDUKH, phys. stat. sol. (a) **3**, 383 (1970).

(Received February 5, 1979)

phys. stat. sol. (a) 54, 37 (1979)

Subject classification: 14.3.4; 22.6

Institute of Physics, Slovak Academy of Sciences, Bratislava[1])

Contact-Limited Conduction in V_2O_5 Single Crystals

By

Š. LÁNYI

The electrical properties of V_2O_5 single crystal–electrode interfaces are investigated. It is shown, that Au, Ag, Ti, Ga, and Al form blocking contacts with an approximately reverse dependence of the barrier height on the work function of metals as expected for Schottky barriers on n-type semiconductors. The phenomenon can be explained by downward bending of the polaron band, caused by injection of electrons from the less noble metals. This gives rise to a surface layer, in which either space charge limited conduction in presence of traps or more probably a Poole-Frenkel-type conduction occurs.

Die elektrischen Eigenschaften der Phasengrenze V_2O_5-Einkristall–Elektrode werden untersucht. Es wird gezeigt, daß Au, Ag, Ti, Ga and Al blockierende Kontakte bilden mit etwa umgekehrter Abhängigkeit der Barrierenhöhe von der Austrittsarbeit der Metalle, wie für eine Schottky-Barriere an n-Typ-Halbleitern erwartet wird. Diese Erscheinung läßt sich als durch Injektion von Elektronen aus den weniger edlen Metallen verursachte Herabbiegung des Polaronbandes erklären, die zur Entstehung einer Oberflächenschicht führt, in der entweder die Leitung durch Raumladung in Anwesenheit von Haftstellen begrenzt ist oder wahrscheinlicher Poole-Frenkel Leitung entsteht.

1. Introduction

Results on the electrical conductivity of vanadium pentoxide single crystals have been published in numerous papers [1 to 10]. Most authors suggest the material to be an n-type semiconductor with a low mobility of charge carriers, and the great difference between the optical gap width (2.2 to 2.3 eV [11]) and the small activation energy of electrical conductivity is interpreted in terms of transport in a narrow band arising from 3 d electrons [12], by hopping of small polarons [4, 5, 13], or by band conduction and excitation of electrons from V^{4+} donors [5]. On the contrary, Kislovskii and Abdullaev [6] came to the conclusion, that the material consists of microblocks with a mobility of electrons of 5.3×10^3 cm²/Vs, separated by barriers of low conductivity. They recommended conductivity measurements to be made at frequencies of at least 50 to 100 kHz.

The published conductivity measurements cover both dc and ac methods, two-, three-, and four-electrode arrangements, all giving essentially the same results. Such conclusions are at least partly contradicted by [9, 14], in which a steep increase of permittivity was observed above 150 K at 1 kHz, and above 250 K at 700 kHz, since a dispersion of the capacitance must be accompanied by an increase in conductance at high frequencies. In [10] the low-temperature conductivity was found to be frequency-dependent.

Various metals were used as electrodes, frequently gold, gallium, or a gallium–indium alloy. The absence of electrode-related effects is possible only if the electrodes are ohmic and regarding the work function range of the employed metals it is unexpected. The aim of the present investigation was to clarify this question.

[1]) CS-899 30 Bratislava, Czechoslovakia.

2. Experimental

The electrical conductivity of pure V_2O_5 single crystals was measured in the frequency range 10^{-3} to 10^6 Hz along the three crystal axes, preferentially in directions *c* (highest conductivity) and *b* (lowest conductivity).

Crystals obtained from the Institute of Crystallography of the Academy of Sciences, Moscow were employed. Some of their properties and their preparation have already been published [15]. Samples 0.025 to 0.4 cm, 0.02 to 0.05 cm, and 0.25 to 0.4 cm thick were cut in directions *a*, *b*, and *c* respectively, from close parts of the ingots by means of a wire saw, then polished using metallographic papers. Electrodes of Au, Ag, Ti, and Al were vacuum-deposited to opposite faces of the samples. Ga electrodes were prepared by pressing gallium droplets to the samples by means of tinned copper foils. The cross-section of the samples was 0.05 to 0.4 cm². The measurements were carried out in air at 25 °C.

In the frequency range from 10 to 10^6 Hz the in-phase component of the current flowing through the sample was measured by means of a lock-in amplifier. In the range 10^{-3} to 10^3 Hz an apparatus similar to that described in [16] was employed. The possible error was below 5% with the first apparatus and below 1% with the second one. The amplitude of the ac signal was 30 mV. Further the current–voltage characteristics of some samples was measured at 12, 102, 10^3, 10^4, and 10^5 Hz. In this case the measured current contained a contribution of the third harmonic $I_m = I_1 + I_3/3$, produced by the nonlinearity of the cell electrode–crystal–electrode.

3. Experimental Results

The frequency dependence of the apparent conductivity of the samples with different electrodes is seen in Fig. 1. In the range 10^{-3} to 10 Hz the conductivity was constant. Since the measurements revealed an effect of electrode material, the conclusion is

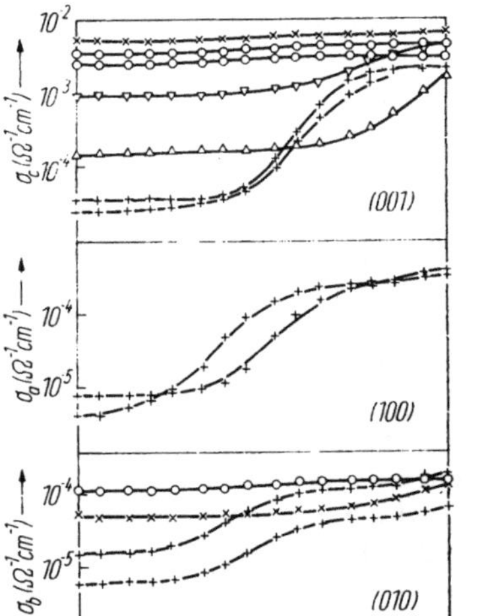

Fig. 1

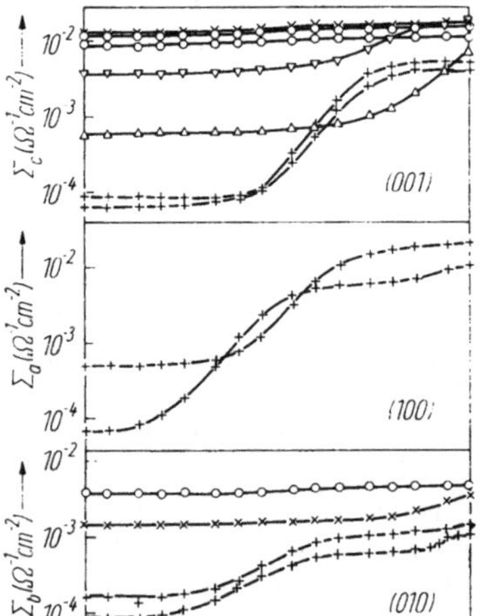

Fig. 2

Fig. 1. The apparent conductivity of V_2O_5 crystals with various electrodes: × Au, ○ Ag, ▽ Ti, △ Ga, and + Al

Fig. 2. The conductance per unit electrode area versus frequency of V_2O_5 crystals. Electrodes: × Au, ○ Ag, ▽ Ti, △ Ga, and + Al

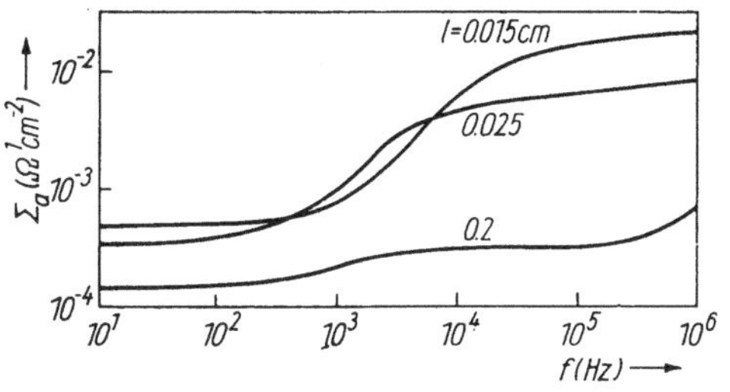

Fig. 3. Thickness dependence of conductance per unit area. (100) V_2O_5 with Al electrodes

justified, that at low frequencies the current is contact limited, first of all with Al electrodes. Therefore, the low-frequency conductance with a particular electrode material is expected to be independent of the sample thickness and a conductance per unit area, Σ, versus frequency plot in Fig. 2 is a more suitable representation of the low-frequency results. The thickness dependence of Σ is seen in Fig. 3.

The I–U characteristics of a sample with Al electrodes in the (001) plane are seen in Fig. 4. With other electrode metals the nonlinearity was less severe, indicating less blocking electrodes.

4. Discussion

The high-frequency conductivity is in reasonable accord with published data [5, 7]. The frequency dependence of the conductance may be formally interpreted as interfacial polarization. The nonlinearity of the conductance observed at relatively low voltages indicates a small total thickness of the insulating layers, causing polarization. The influence of electrode material allows the conclusion, that they are barriers at the crystal–electrode interfaces.

The interface conductance is expected to depend on the concentration of charge carriers, but not on their mobility in the bulk of the semiconductor, i.e. not on the crystallographic orientation. The results with the blocking Al electrodes in Fig. 2 and 3 show rather a scatter than a systematic dependence on the orientation of the samples.

The well expressed dispersion of the conductance with Al electrodes renders an approximate evaluation of interface capacitance possible. Using the simplified equivalent circuit (Fig. 5b) of the cell electrode–crystal–electrode the conductance measured at the angular frequency ω is

$$G(\omega) = (R_b + R_c + \omega^2 C^2 R_c^2 R_b)/[(R_b + R_c)^2 + \omega^2 C^2 R_c^2 R_b] .$$

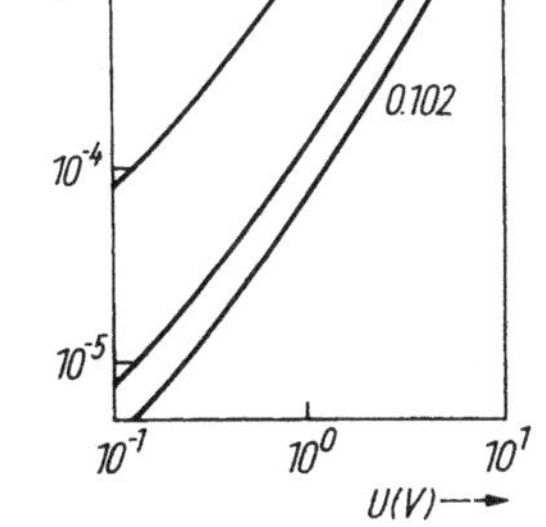

Fig. 4. I–U characteristic of (001) V_2O_5 with Al electrodes

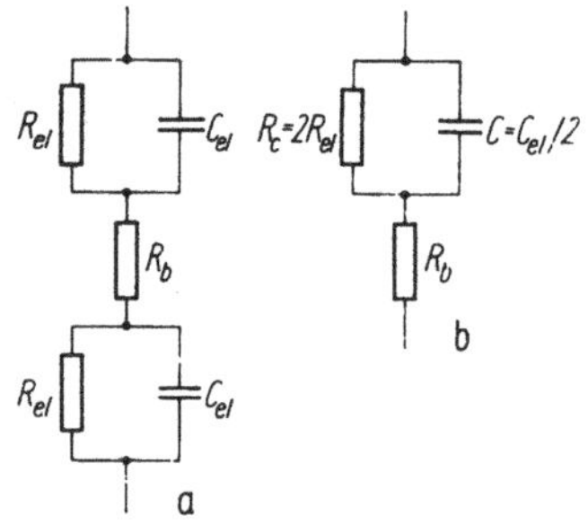

Fig. 5. Simple equivalent circuits of the cell electrode–V_2O_5–electrode. The effect of geometrical capacitance is negligible at the employed frequencies

Providing the bulk resistance R_b and the interface resistance R_c are known, the interface capacitance

$$C = \{(R_b + R_c)\,[G(R_b + R_c) - 1]/[R_c^2 R_b(1 - GR_b)]\}^{1/2}/\omega$$

can be calculated. The bulk resistance equal to the high-frequency resistance is constant at least above 5×10^3 Hz as it is seen from the measurements with gold and silver electrodes. A similar investigation of amorphous V_2O_5–P_2O_5 has shown that the interface resistance is frequency-dependent and the capacitance obtained under the assumption of constant R_c is generally too large, except at the highest frequencies of the conductance dispersion [17]. The following values result for 10^4 to 10^5 Hz:

$$C_a \approx 1 \times 10^{-7}\ \text{F/cm}^2\,, \quad C_b \approx 7.4 \times 10^{-8}\ \text{F/cm}^2\,, \quad C_c \approx 3.8 \times 10^{-8}\ \text{F/cm}^2\,.$$

The capacitance associated with the space charge in the vicinity of electrodes can be taken approximately as

$$C_{\text{el}} = \left(\frac{\varepsilon e^2 n}{4kT}\right)^{1/2}, \tag{1}$$

where ε is the permittivity, e the charge of a proton, n the concentration of charges, k the Boltzmann constant, and T the temperature. Expression (1) results with some modifications from most models of space charge. In V_2O_5 due to the high degree of localization of charge carriers in the polaron band, which is close to the Fermi level, n/kT may be interpreted as a density of states per kT near the Fermi level [18]. n obtained from (1) will be affected by band bending, therefore, it may not represent exactly the bulk values. Using the permittivity of V_2O_5 from [14] $\varepsilon_{ra} = 25$, $\varepsilon_{rb} = 15$, $\varepsilon_{rc} = 5$ the concentrations n are $n_a = 1.16 \times 10^{16}$ cm^{-3}, $n_b = 1.06 \times 10^{16}$ cm^{-3}, and $n_c = 8.4 \times 10^{15}$ cm^{-3}. With the less blocking Ti electrodes $n_c = 2.0 \times 10^{16}$ cm^{-3}. The capacitance of Ga electrodes is approximately one order of magnitude smaller than expected; this may be caused by insufficient wetting of V_2O_5 by gallium.

Assuming the bulk concentration of charge carriers is comparable to or higher than the calculated surface concentrations with Al electrodes ($\approx 1 \times 10^{16}$ cm^{-3}), the upper limits of their mobility may be estimated as $\mu_a = 0.3$ cm^2/Vs, $\mu_b = 0.07$ cm^2/Vs, $\mu_c = 1.8$ cm^2/Vs. The published Hall mobilities measured in direction a are around 0.1 cm^2/Vs at 300 K [11].

In some samples a further dispersion of the conductance was found above 10^5 Hz. It is associated with capacitances $< 10^{-9}$ F/cm^2 and it may result from accidental cracks in the (010) plane, which give rise to interfacial polarization regardless of the orientation of the electric field. Besides this there is no evidence of a significant increase of the conductivity above 50 to 100 kHz. If it should appear above 1 MHz, the corresponding capacitance would be below some 50 pF/cm^2 at a sample thickness of 1 mm. This is much less than the estimate of Kislovskii and Abdullaev ($\approx 5 \times$

$\times 10^3$ pF/cm^2) [6], and it would require a significant part of the sample to consist of the low-conductivity component. Thus we cannot confirm their conclusion that V_2O_5 is a high mobility semiconductor.

The work function of the employed electrode materials increases from Ti towards Au and since V_2O_5 is an n-type semiconductor [3, 5, 7, 12, 13] the expected interface barrier height should increase in this sequence. For many semiconductors such a dependence is not fulfilled and the barrier height is independent of the electronegativity of the metal [19], but these are usually nonpolar compounds. A recent investigation on InP has shown that reactive metals tend to form ohmic contacts, whereas the more inert ones form Schottky barriers [20]. Also from this point of view Al and Ti electrodes should not be blocking on V_2O_5. The present experiments have revealed an approximately reverse dependence. It is therefore interesting to find out the nature of the interfacial barriers. The attempts to treat the current–voltage characteristic as of two oppositely oriented Schottky barriers connected in series indicated either rather poor barriers or the unsuitability of such a model. Tunneling is out of question due to the high effective mass of polarons. The low-frequency I–U characteristics (Fig. 4) show a transition from linear to approximately quadratic dependence, which is known to occur in case of injection and space charge limited conduction [21]. Using the relation

$$I = 9\mu\varepsilon\theta U^2/(8s^3) ,$$

where θ is the ratio of free to trapped charge and s the thickness of the insulating layer and the relation [22]

$$U_x = en_0 s^2/(\theta\varepsilon) ,$$

where U_x denotes the voltage at which transition from ohmic to SCL conduction occurs and n_0 is the concentration of thermally generated free charges, two of the quantities s, n_0, and θ may be calculated assuming the third one is known. U_x is approximately 1 V and $I = 50\ \mu A/cm^2$ at 102 Hz (Fig. 4). The dependences obtained, taking for the mobility μ its bulk value, are shown in Fig. 6.

An alternative interpretation of experimental data assumes that electrons are trapped in donor-like centres present in the surface barriers and their emission is enhanced by the Poole-Frenkel effect. In this case the conductivity is

$$\sigma = \sigma_0 \exp\left(\frac{\beta_{PF} E^{1/2}}{kT}\right), \tag{2}$$

where

$$\beta_{PF} = 2\left(\frac{e^3}{4\pi\varepsilon}\right)^{1/2} \tag{3}$$

and E is the electric field strength. An anomalous PF effect requires $\beta'_{PF} = \beta_{PF}/2$

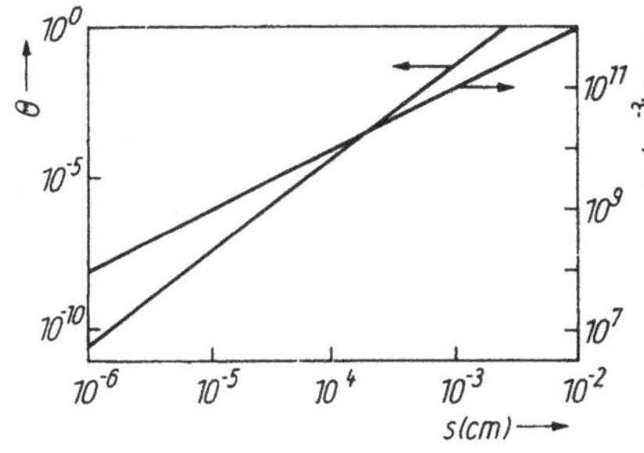

Fig. 6. The dependence of the ratio of free to trapped charge θ and of the concentration of thermally generated charges n_0 on the thickness of insulating layers calculated assuming SCL conduction through the surface barriers in (001) V_2O_5 with Al electrodes (cf. Fig. 4)

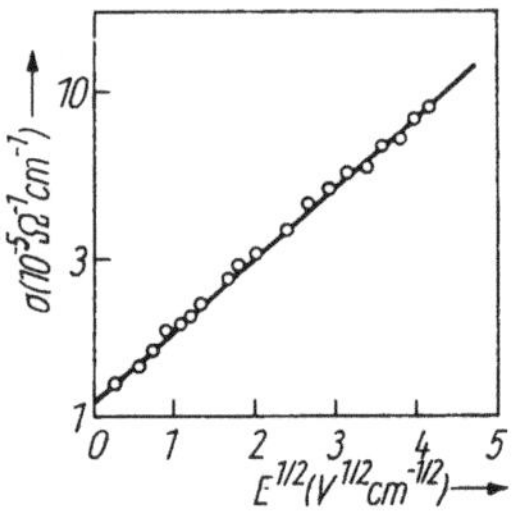

Fig. 7. lg σ vs. $E^{1/2}$ dependence for (001) V_2O_5 with Al electrodes (cf. Fig. 4)

[23]. A plot of the logarithm of the conductance versus $E^{1/2}$ is seen in Fig. 7 for a *c*-oriented sample 0.4 cm thick with Al electrodes. A least-squares fit yields the coefficient β by a factor of ≈ 76 greater than expression (3), which would mean that the effective thickness of the high-resistivity surface layer is 1.3 μm for the PF effect or 0.34 μm for the anomalous PF effect. The capacitance of a capacitor with such a plate spacing and $\varepsilon_r = 5$ is approximately 3.3×10^{-9} and 1.3×10^{-8} F/cm^2, respectively. However, the space charge should reside inside the surface barriers and the plate spacing of an equivalent capacitor should be equal to the distance from electrode to the centre of the space charge, if it is sufficiently mobile, or to some point, up to which it is able to respond to the ac signal of a given frequency. This can give a few times greater capacitance in reasonable agreement with $2C_c$ estimated from the frequency dependence of the conductance.

The concept of field-enhanced emission in the surface barriers is consistent with the idea of the bulk conductance by small polarons, as the interaction of the polaron with the ionic lattice is Coulombic. The validity of expression (2) requires a small overlap of neighbouring potential wells [24], i.e. large separation of empty sites. This can arise by such a bending of the polaron band near the electrodes that relatively few states in the band tail or in the forbidden gap remain available to thermal excitation.

Thus we can draw the following picture of the nature of the surface barriers: Electrons injected into the crystal from a low work function metal cause a downward band bending. Since the conduction in the crystal takes place in a narrow probably polaron-like band [5, 7, 12, 13], insulating layers would have to arise near the electrodes. The transport through these layers may be explained either as space charge limited conduction in presence of traps or by field enhanced hopping between Coulombic sites. The thickness of the barriers deduced from the interface capacitance should reach tenths of μm. This size supports the latter possibility since the conduction should occur outside the conductivity band, hence SCL conduction is not probable to a long distance as far as the effective mass of the electrons is as large as in the bulk of the crystal. However, a coexistence of both effects might be possible.

Acknowledgements

The author thanks Dr. G. F. Dobrzhanskii for supplying the V_2O_5 single crystals and Dr. G. Vlasák for preparation of electrodes.

References

[1] S. Kachi, T. Takada, and K. Kosuge, J. Phys. Soc. Japan **18**, 1839 (1963).
[2] T. S. Zolian and A. R. Regel, Fiz. tverd. Tela **6**, 1520 (1964).
[3] I. B. Patrina and V. A. Ioffe, Fiz. tverd. Tela **6**, 3227 (1964).

[4] T. ALLERSMA, R. HAKIM, T. N. KENNEDY, and J. D. MACKENZIE, J. chem. Phys. **46**, 154 (1967).
[5] D. S. VOLZHENSKII and M. V. PASHKOVSKII, Fiz. tverd. Tela **11**, 1168 (1969).
[6] L. D. KISLOVSKII and A. A. ABDULLAEV, Optika i Spectroskopiya **29**, 737 (1970).
[7] P. NAGELS and M. DENAYER, Proc. 10th Internat. Conf. Phys. Semiconductors, Cambridge 1970 (p. 321).
[8] J. H. PERLSTEIN, J. Solid State Chem. **3**, 217 (1971).
[9] A. B. SCOTT, J. C. MCCULLOCH, and K. M. MAR, Proc. 2nd Internat. Conf. Conduction in Low Mobility Materials, Eilat 1971 (p. 107).
[10] J. HAEMERS, E. BAETENS, and J. VENNIK, phys. stat. sol. (a) **20**, 381 (1973).
[11] N. KENNY, C. R. KANNEWURF, and D. H. WHITMORE, J. Phys. Chem. Solids **27**, 1237 (1966).
[12] A. A. VINOGRADOV and A. I. SHELIKH, Fiz. tverd. Tela **13**, 3310 (1971).
[13] V. A. IOFFE and I. B. PATRINA, phys. stat. sol. **40**, 389 (1970).
[14] I. M. CHERNENKO and A. I. IVON, Fiz. tverd. Tela **16**, 2130 (1974).
[15] A. A. ABDULLAEV, L. M. BELIAEV, I. V. VINAROV, G. F. DOBRZHANSKII, and R. G. YANKELEVICH, Kristallografiya **14**, 1095 (1969).
[16] Š. LÁNYI, J. Phys. Chem. Solids **36**, 775 (1975).
[17] Š. LÁNYI, prepared for publication.
[18] R. MEAUDRE and M. MEAUDRE, phys. stat. sol. (a) **37**, 633 (1976).
[19] C. A. MEAD, Solid State Electronics **9**, 1023 (1966).
[20] R. H. WILLIAMS, V. MONTGOMERY, and R. R. VARMA, J. Phys. C**11**, L735 (1978).
[21] A. ROSE, Phys. Rev. **97**, 1538 (1955).
[22] M. A. LAMPERT, Rep. Progr. Phys. **27**, 329 (1964),
[23] J. G. SIMMONS, Phys. Rev. **155**, 657 (1967).
[24] A. E. OWEN and J. M. ROBERTSON, IEEE Trans. Electron Devices **20**, 105 (1973).

(Received February 23, 1979)

phys. stat. sol. (a) **54**, 45 (1979)

Subject classification: 1.4 and 10

Institute of Crystallography, Academy of Sciences of the USSR, Moscow (a) and Erevan State University (b)

A Dynamical Analytical Theory of the Dislocation Image Contrast in Electron Transmission Microscopy

By

F. N. Chukhovskii (a) and A. M. Arustamyan (b)

A dynamical theory of dislocation contrast in electron transmission microscopy (DCETM) is discussed. A diagram method for the solution of the Howie-Whelan equations in the strongly deformed region of a crystal close to the dislocation core is developed. The solution is based on the general scattering matrix (S-matrix) theory. The complete analytical description of the dislocation image picture as a whole is proposed. It is shown that in the regions of slight and strong deformations the quasi-classical (eikonal) and the anti-classical approximations, respectively, are in good accordance with the well-known numerical results. The ranges of the applicability of each analytical approach are established and the general properties of the dislocation image profiles are explained.

Eine dynamische Theorie des Versetzungskontrasts im Elektronen-Transmissionsmikroskop (DCETM) wird diskutiert. Dazu wird eine Diagrammethode für die Lösung der Howie-Whelan Gleichungen in dem stark deformierten Bereich eines Kristalls in der Nähe des Versetzungskerns entwickelt. Die Lösung basiert auf der allgemeinen Streumatrix (S-Matrix)-Theorie. Eine vollständige analytische Beschreibung des Versetzungsabbildes als ganzes wird vorgeschlagen. Es wird gezeigt, daß in den Gebieten leichter und starker Deformationen die quasiklassische (Eikonal) bzw. die antiklassische Näherung sich in guter Übereinstimmung mit den gutbekannten numerischen Ergebnissen befinden. Die Anwendungsbereiche jeder analytischen Näherung werden festgestellt und die allgemeinen Eigenschaften der Versetzungsbildprofile erklärt.

1. Introduction

A dynamical theory of diffraction provides a quantitative guide to the interpretation of transmission electron micrographs of distorted crystals. It is widely used for the discussion of diffraction contrast effects due to strains around defects of a crystal lattice. A number of theoretical papers [1 to 5] are devoted to computer simulation of single defect images to solve the diffraction contrast problem by numerical methods. However, so far the straightforward transition from the dynamical equations to numerical results does not permit a comparable physical insight to describe analytically the features of the defect image contrast, in particular the contrast close to the dislocation core.

The analytical approach to the problem in question was firstly proposed by Wilkens [6] (see also [7]) using the geometric optics (eikonal) approximation (so-called "modified Bloch wave" theory). The most important conclusion obtained in [7] is that the asymmetry of the long-range contrast on two sides of the dislocation images is explained by the phase contrast due to the quasi-classical electron scattering in slightly deformed regions of a crystal outside the dislocation core.

In the present paper the consequent dynamical theory of diffraction contrast in electron transmission microscopy (DCETM) is proposed on the basis of the general formalism of the scattering matrix (S-matrix) [8 to 10]. It opens the possibility to investigate the physically different approaches, the quasi-classical and anti-classical ones, for the description of the electron propagation through slightly and strongly deformed regions of a crystal, respectively, and to determine the applicability ranges of each theoretical approximation. The principal conclusion is that the anti-classical approach describes DCETM taking into account the interband scattering involved, whereas the quasi-classical one quantitatively explains the "background contrast" of the image profiles outside the dislocation core, as it must be from physical speculations.

2. S-Matrix Formalism

The dynamical scattering of an electron in the imperfect absorbing crystal is described by the Howie-Whelan equations [2] (assuming a centre of symmetry which does not matter for the problem under consideration)

$$\left.\begin{aligned} \frac{d\Phi_0}{dz} &= \frac{i\pi}{\xi_0}\Phi_0 + \frac{i\pi}{\xi_g}\Phi_g \exp\left(2\pi isz + 2\pi i\boldsymbol{gR}\right), \\ \frac{d\Phi_g}{dz} &= \frac{i\pi}{\xi_0}\Phi_g + \frac{i\pi}{\xi_g}\Phi_0 \exp\left(-2\pi isz - 2\pi i\boldsymbol{gR}\right), \end{aligned}\right\} \tag{2.1}$$

where the atomic displacements from the perfect crystal positions at depth z in the crystal are given by $\boldsymbol{R}(z)$. Hereafter all notations are taken from [3].

Substitution of $\Phi_0 = \Phi_0'' \exp(i\pi sz + i\pi z/\xi_0)$, $\Phi_g = \Phi_g'' \exp(-i\pi sz + i\pi z/\xi_0)$ and elimination of Φ_g'' or Φ_0'' from (2.1) lead to second-order differential equations for the wave amplitudes Φ_0'' and Φ_g'', respectively (after dropping the primes for simplicity)

$$\frac{d^2\Phi_0}{dz^2} - 2\pi i\beta_g' \frac{d\Phi_0}{dz} + \pi^2\left[(\Delta k)^2 + 2\beta_g' \frac{w}{\xi_g}\right]\Phi_0 = 0\,, \tag{2.2}$$

$$\frac{d^2\Phi_g}{dz^2} + 2\pi i\beta_g' \frac{d\Phi_g}{dz} + \pi^2\left[(\Delta k)^2 + 2\beta_g' \frac{w}{\xi_g}\right]\Phi_g = 0; \tag{2.3}$$

$$w = s\xi_g\,, \qquad \Delta k = \xi_g^{-1}(1 + w^2)^{1/2}\,,$$

where $\beta_g'(z)$ is a local lattice deviation from the exact Bragg condition

$$\beta_g'(z) = \frac{1}{2\pi}\frac{d\alpha}{dz} \equiv \boldsymbol{g}\,\frac{d\boldsymbol{R}}{dz}(z)\,. \tag{2.4}$$

According to the general principles of the scattering theory [8 to 10] the formally exact solutions of (2.2), (2.3) may be written in a matrix form as

$$\Phi_0(z) = \sum_{j=1,2} C_0^{(j)}(z) \exp\left[(-1)^{j+1}\, i\pi\, \Delta kz\right]; \tag{2.5}$$

$$\begin{pmatrix} C_0^{(1)}(z) \\ C_0^{(2)}(z) \end{pmatrix} = S_0(z, 0) \begin{pmatrix} C_0^{(1)}(0) \\ C_0^{(2)}(0) \end{pmatrix},$$

$$\Phi_g(z) = \sum_{j=1,2} C_g^{(j)}(z) \exp\left[(-1)^{j+1}\, i\pi\, \Delta kz\right]; \tag{2.6}$$

$$\begin{pmatrix} C_g^{(1)}(z) \\ C_g^{(2)}(z) \end{pmatrix} = S_g(z, 0) \begin{pmatrix} C_g^{(1)}(0) \\ C_g^{(2)}(0) \end{pmatrix}.$$

The $S_{0(g)}$ matrices are 2×2 matrices relating the incident amplitudes $C_{0(g)}^{(j)}(0)$ to the exit ones $C_{0(g)}^{(j)}(z)$, and are expressed through the Pauli matrices σ_1, σ_2, σ_3 as follows:

$$S_{0(g)}(z, 0) = \exp\left[\pm \frac{i}{2} f(z)\right] Z \exp \Big\{\mp i\pi \int_0^z dz' \beta_g'(z') [\mp \cos \beta \sigma_3 + + \left(\sin (2\pi \Delta k z') \mp i \cos \beta \cos (2\pi \Delta k z')\right) \sigma_2 + + \left(\cos (2\pi \Delta k z') \pm i \cos \beta \sin (2\pi \Delta k z')\right) \sigma_1]\Big\}; \quad (2.7)$$

$$f(z) \equiv \alpha(z) - \alpha(0)\,, \qquad \cos \beta = w(1 + w^2)^{-1/2}\,.$$

The general expressions (2.5) to (2.7) are appropriate to the analytical investigation of DCETM for arbitrary distances from the dislocation core corresponding to the strongly deformed region as well as to the slightly deformed one. As an example, calculations will be carried out in cases of screw and edge dislocations lying parallel to the crystal (foil) surface.

Before considering the dislocation image in detail we shall describe the diagram method for the S-matrix calculation. Following [9], the exact S-matrix itself is graphically represented in the form (thick lines)

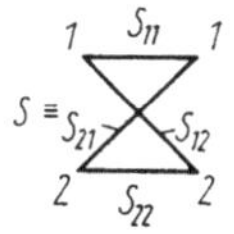

(2.8)

Then, its expansion series in powers of some parameter may be drawn as (thin lines)

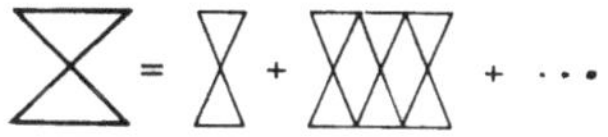

(2.9)

and any matrix element of the n-th order, ${}_nS_{0(g),ik}$, is the sum of all lines joining the points i and k ($i, k = 1, 2$).

From (2.7) it can be seen that for slight deformations, i.e. when $2\pi \Delta k |a| \gg 1$ ($|a|$ is the characteristic alternation length of the function $\beta_g'(z)$), the quasi-classical representation of the S-matrix is

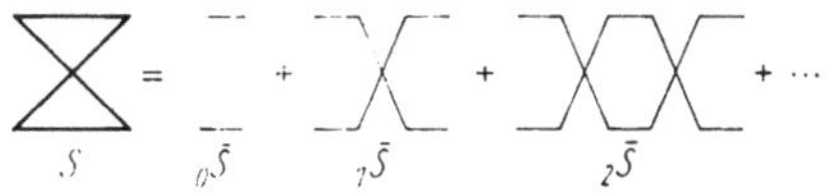

(2.10)

adequate to

$${}_0\bar{S}_{0(g)} = \exp\left[\pm \frac{i}{2} f(z) + \frac{i}{2} \cos \beta\, f(z)\, \sigma_3\right] \quad (2.11)$$

and one has to summarize the whole series of the pole diagrams from (2.10) [9].

As it was shown in [9], this leads to the quasi-classical (eikonal) representation of wave amplitudes, which in turn can be obtained from the dynamical equations (2.1) using the asymptotic solution method [11] (cf. [6]). In this case they consist of a linear combination of independent proper solutions of the kind

$$\Phi_{0(g)}(z) = \exp\left[\pm \frac{if(z)}{2}\right] \sum_{j=1,2} C^{(j)}_{0(g)} \left(\frac{dP}{dz}\right)^{-1/2} \exp\left[(-1)^{j+1}\, iP(z)\right]. \tag{2.12}$$

In (2.12) the eikonal integral is given by

$$P(z) = \frac{\pi}{\xi_g} \int_0^z dz' \left(1 + \overline{w}^2(z')\right)^{1/2}; \qquad \overline{w}(z) = w + \xi_g \beta_g'(z)\,, \tag{2.13}$$

and the constants $C^{(j)}_{0(g)}$ may be found from the boundary conditions of the DCETM problem.

In the opposite case of strong deformations, the anti-classical expansion series of the S-matrix is thought to be adequate, namely

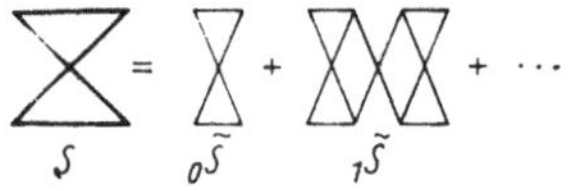

(2.14)

with the zero-order approximation term

$${}_0\tilde{S}_{0(g)} = \exp\left[\pm \frac{i}{2} f(z) \mp \frac{if(z)}{2} M_{0(g)}\right]. \tag{2.15}$$

Here the following notations have been introduced:

$$M_{0(g)} = \mp \cos \beta\sigma_3 + h(z)\,(\sigma_1 \mp i \cos \beta\sigma_2)\,; \tag{2.16}$$

$$h(z) = \left[\int_0^z dz'\, \beta_g'(z')\right]^{-1} \int_0^z dz'\, \beta_g'(z') \cos(2\pi\, \Delta k z')\,.$$

For small values $2\pi\, \Delta k\, |a| \ll 1$ one can neglect terms higher than ${}_1\tilde{S}$ (in special cases higher than ${}_2S$) on the right-hand side of (2.14). The next term in (2.14) is of the order of $(2\pi\, \Delta k a)^2$ at least, and therefore we expect that the anti-classical approximation should be qualitatively true up to $2\pi\, \Delta k\, |a| \lessapprox 1$. Indeed, as concerns the dislocation image contrast we shall show that the anti-classical approximation provides, in general, the correct dislocation profiles as a whole, except for the prolonged asymmetric "background contrast". The latter is consistent with the quas-classical electron scattering anywhere, except at the dislocation core.

3. Slight Deformations (Quasi-Classical Approximation)

In the slightly deformed region quasi-classical electron scattering occurs and interband scattering is exponentially negligible, of the order $\exp(-2\pi\, \Delta k\, |a|) \ll 1$ [9]. Taking into account the usual boundary conditions for the total wave amplitudes in electron transmission microscopy, one gets (cf. (2.12))

$$\left.\begin{aligned} \Phi_0(z) &= \exp\left[\frac{i}{2} f(z)\right] \left\{\left[\frac{dP(0)}{dz}\right]^{1/2} \frac{\cos(P(z))}{[dP(z)/dz]^{1/2}} - \right. \\ &\quad \left. - \frac{i\pi\overline{w}(0)}{\xi_g} \frac{\sin(P(z))}{[dP(z)/dz\; dP(0)/dz]^{1/2}}\right\}, \\ \Phi_g(z) &= \frac{i\pi}{\xi_g} \exp\left[-i\alpha(0) - \frac{if(z)}{2}\right] \frac{\sin(P(z))}{[dP(z)/dz\; dP(0)/dz]^{1/2}}, \end{aligned}\right\} \tag{3.1}$$

and the following investigation is similar as in [7]. The eikonal integral $P(z)$ in (3.1) is given by

$$P(z) \equiv \text{Re}\, P(z) + i \,\text{Im}\, P(z) \tag{3.2}$$

with the first-order approximation ($\xi_g \approx 0.1\xi_g'$)

$$P(z) = \frac{\pi}{\xi_g} \int_0^z \mathrm{d}z' \left(1 + \bar{w}^2(z')\right)^{1/2} + \frac{i\pi}{\xi_g'} \int_0^z \mathrm{d}z' \left(1 + \bar{w}^2(z')\right)^{-1/2} . \tag{3.3}$$

Bearing in mind to calculate the intensities of the transmitted and of the diffracted electron beams in the case of a thick foil, $(\pi t/\xi_g')\,(1 + w^2)^{-1/2} \gg 1$, which in turn means that the image profiles are determined by the imaginary part of the eikonal function only, one finds

$$\left.\begin{aligned} I_0(a) &= \frac{1}{4} \exp\left[-\frac{2\pi t}{\xi_0'} + 2\,\text{Im}\, P(t)\right] \frac{(\mathrm{d}P(0)/\mathrm{d}z + (\pi/\xi_g)\,\bar{w}(0))^2}{|\mathrm{d}P(t)/\mathrm{d}z\; \mathrm{d}P(0)/\mathrm{d}z|}, \\ I_g(a) &= \frac{1}{4} \exp\left[-\frac{2\pi t}{\xi_0'} + 2\,\text{Im}\, P(t)\right] \frac{\pi^2}{\xi_g^2} \left|\frac{\mathrm{d}P(t)}{\mathrm{d}z} \frac{\mathrm{d}P(0)}{\mathrm{d}z}\right|^{-1}. \end{aligned}\right\} \tag{3.4}$$

Further we shall restrict ourselves for simplicity to dislocations lying parallel to the foil plate. The diffraction vector $\boldsymbol{g}$ is parallel to the plate, too. Elastic isotropy and displacement fields as in an infinite continuum are assumed throughout. The dislocation is situated at the coordinate origin 0 at the depth z_0 below the top surface. The depth z_0 is supposed to be of the order of the foil thickness and is larger than the characteristic alternation length $|a|$ of the β_g' function and, hence, one can neglect the stress relaxation effect at the foil surfaces.

We shall consider two cases [3].

(i) screw dislocation:

$$\alpha(z) = n \,\text{arctg}\, \frac{z}{a}, \tag{3.5}$$

(ii) pure edge dislocation (Poisson's ratio $\nu = \frac{1}{3}$):

$$\alpha(z) = n \left(\text{arctg}\, \frac{z}{a} + \frac{3}{4} \frac{za}{z^2 + a^2}\right) \approx n \,\text{arctg}\left(2\, \frac{z}{a}\right). \tag{3.6}$$

The calculation of the imaginary part of the eikonal function (3.3) with (3.5) or (3.6) is straightforward. After direct integration one gets

$$\begin{aligned} \text{Im}\, P(t) = {} & \frac{\pi t}{\xi_g'} (1 + w^2)^{-1/2} - \frac{4\pi}{\xi_g'} (1 + w^2)^{-1/2}\, |A| \left\{\frac{v}{2\lambda} F\left(\frac{\pi}{2}, r\right) + \right. \\ & + (2u^2 - v - v^2) \left[\frac{\lambda}{2(\lambda^4 - \mu^4)} E\left(\frac{\pi}{2}, r\right) - \frac{1}{4\lambda(\lambda^2 - \mu^2)} F\left(\frac{\pi}{2}, r\right)\right] + \\ & \left. + 2(v + v^2 + u^2 v) \frac{\partial}{\partial(u^2)} \left[\frac{1}{2\lambda} F\left(\frac{\pi}{2}, r\right)\right]\right\}, \end{aligned} \tag{3.7}$$

where $F(\pi/2, r)$, $E(\pi/2, r)$ are the complete elliptic integrals [12]. Here the following notations have been introduced:

$$\left.\begin{aligned} v &= \frac{nw}{\pi(\Delta k)^2 \xi_g A}, \qquad u = \frac{|n|}{2\pi \Delta k |A|}, \\ \lambda &= (1 + v + u^2)^{1/4}, \qquad \mu = \left(1 + \frac{v}{2}\right)^{1/2}, \qquad r = \frac{1}{2}\left(1 - \frac{\mu^2}{\lambda^2}\right), \end{aligned}\right\} \tag{3.8}$$

where $A = a$ and $a/2$ for screw and edge dislocations, respectively.

From the above results the asymmetry of the background contrast on the two sides of the dislocation involved for both the bright- and dark-field images follows immediately as was first pointed out in [7]. Particularly, for $u < 1$ the intensity ratio

$$\eta_{0(g)} \equiv \frac{I_{0(g)}(a) - I_{0(g)}(-a)}{I_{0(g)}(a) + I_{0(g)}(-a)} = -\operatorname{tgh}\left[\frac{\pi\xi_g nw}{\xi_g'(1 + w^2)^{3/2}}\right] \tag{3.9}$$

characterizes the background contrast "jump" independent of the dislocation character, the foil thickness t, and the depth position z_0 of the dislocation inside the foil (compare with (4.5) from [7]). In a strict sense, the analytical formulae (3.4), (3.7), and (3.8) describe the prolonged "tails" of the dislocation images outside the dislocation core. It should be mentioned that they could, in general, be rewritten in order to take into account the stress relaxation at the foil surfaces.

4. Strong Deformations (Anti-Classical Approximation)

Of special interest is to investigate the DCETM problem in the region close to the dislocation core when the quasi-classical treatment (the eikonal theory) of the electron dynamical scattering fails and it is necessary to utilize the above diagram technique (see Section 2) to derive the S-matrix expansion in powers of the small anti-classical parameter $2\pi \Delta k |a| \ll 1$.

Confining ourselves to the first two terms of the S-matrix series,

$$S \approx {}_0\tilde{S} + {}_1\tilde{S}, \tag{4.1}$$

the first-order approximation terms are given by [10]

$$\begin{aligned} {}_1\tilde{S}_{0(g)}(z, 0) = {} & i\pi N_{0(g)} \exp\left[\pm \frac{i}{2} f(z)\right] \left\{\int_0^z dz' \beta_g'(z') \sin(2\pi \Delta k z') \times \right. \\ & \times \exp\left[\mp i M_{0(g)}\left(f(z') - \frac{f(z)}{2}\right)\right] \mp i \cos\beta M_{0(g)}^{-1} \times \\ & \left. \times \int_0^z dz' \beta_g'(z') [h(z') - \cos(2\pi \Delta k z')] \exp\left[\mp i M_{0(g)} \left(f(z') - \frac{f(z)}{2}\right)\right]\right\}, \end{aligned} \tag{4.2}$$

where the matrices $N_{0(g)}$ have the form

$$N_{0(g)} = \sigma_2 \pm i \cos\beta\sigma_1 \tag{4.3}$$

(the ${}_0\tilde{S}_{0(g)}$ terms are determined by (2.15) and (2.16)).

Now substituting the functions of the dislocation distortion fields (3.5) and (3.6) into (4.2) and taking into account the boundary conditions, we get the expressions for the wave amplitudes suitable for the physical analysis.

Before proceeding to this analysis we discuss the general properties of the solution obtained. As is easy to show, in the limit $a \to 0$ the first-order ${}_1\tilde{S}$-matrix tends to zero.

On the other hand, the zero-order ${}_0\tilde{S}$-matrix describes the phenomenon of the interband scattering due to the dislocation core term of the kind $\beta'_g(z) = \frac{1}{2}\,\delta(z)$. As a result the wave amplitudes are equal to the corresponding ones for a faulted crystal of thickness t with a relative displacement $\alpha = \pi n$ of the lower crystal of the thickness $t - z_0$ with respect to the upper one. In the opposite case of $|a| \to \infty$ the $\tilde{S}$-matrix is exponentially negligible: $\exp(-2\pi\,\Delta k\,|a|) \ll 1$, and the off-diagonal elements of the ${}_0\tilde{S}$-matrix also, whereas the diagonal ones are important and yield intensities of the transmitted and diffracted electron beams equal to those of a perfect crystal.

Let us derive the expressions for the dislocation profile for the region $2\pi\,\Delta k\,|a| \ll 1$, when an appreciable interband scattering occurs. For the practical analysis it is convenient to redetermine the zero-order ${}_0\tilde{S}$-matrix as the term corresponding to a crystal with a stacking fault exactly, in other words, ${}_0\tilde{S}$ for $a = 0$. This leads certainly to some reconstruction of the series (4.1) provided that the expressions (4.2) and (2.16) undergo obvious changes, too. It is clear from (3.5), (3.6) that changing the sign of n is equivalent to changing the sign of a, and, hence, one can consider only positive values of a.

As a result of the direct calculations we find the bright- and dark-field image profiles on the bottom surface of a thick crystal

$$\left.\begin{aligned} I_0(a) &= \frac{1}{4}\exp\left(-\frac{2\pi t}{\xi'_0}\right)\Big|(1+\cos\beta)\,\tilde{S}_{0,22}\exp(-i\pi\,\Delta k t) + \\ &\quad + (1+\cos\beta)\,\tilde{S}_{0,12}\exp[i\pi\,\Delta k(t-2z_0)] + (1-\cos\beta)\times \\ &\quad \times \tilde{S}_{0,21}\exp[-i\pi\,\Delta k(t-2z_0)]\Big|^2, \\ I_g(a) &= \frac{1}{4}\exp\left(-\frac{2\pi t}{\xi'_0}\right)\sin^2\beta\,\Big|\tilde{S}_{g,22}\exp(-i\pi\,\Delta k t) + \\ &\quad + \tilde{S}_{g,12}\exp[i\pi\,\Delta k(t-2z_0)] - \tilde{S}_{g,21}\exp[-i\pi\,\Delta k(t-2z_0)]\Big|^2. \end{aligned}\right\} \tag{4.4}$$

Here the $\tilde{S}$-matrix elements calculated accurately within the order of magnitude $(2\pi\,\Delta k a)^2 \ll 1$, $\tilde{S} \approx {}_0\tilde{S} + {}_1\tilde{S}$, take the form

$${}_0\tilde{S}_{0(g)} = \begin{pmatrix} \cos\frac{\pi n}{2} + i\cos\beta\sin\frac{\pi n}{2} & \mp i(1\mp\cos\beta)\sin\frac{\pi n}{2} \\ \mp i(1\pm\cos\beta)\sin\frac{\pi n}{2} & \cos\frac{\pi n}{2} - i\cos\beta\sin\frac{\pi n}{2} \end{pmatrix}, \tag{4.5}$$

$$\begin{aligned} {}_1\tilde{S}_{0(g)} &= \frac{in}{2}\,s(a)\begin{pmatrix} -\sin^2\beta & \mp\cos\beta(1\mp\cos\beta) \\ \mp\cos\beta(1\pm\cos\beta) & \sin^2\beta \end{pmatrix} + \\ &+ \frac{in}{2}\,q(a)\begin{pmatrix} -\cos\beta\cos\frac{\pi n}{2} - i\sin\frac{\pi n}{2} & \pm\cos\frac{\pi n}{2}(1\mp\cos\beta) \\ \pm\cos\frac{\pi n}{2}(1\pm\cos\beta) & \cos\beta\cos\frac{\pi n}{2} + i\sin\frac{\pi n}{2} \end{pmatrix}. \end{aligned} \tag{4.6}$$

The functions $s^{(s)}(a)$, $s^{(e)}(a)$ acquire different forms depending on whether the integral number n is odd or even.

(i) screw dislocation:

$$s^{(s)}(a) = \begin{cases} 2\pi^2\,\Delta ka(-1)^{n/2}\,(1 - \cos\beta)\,; & n \text{ even}, \\ 4\pi\,\Delta ka(-1)^{(n-1)/2}\,\ln(4\pi\,\Delta ka)\,; & n \text{ odd} \end{cases} \tag{4.7a}$$

and

$$q^{(s)}(a) = 2\pi^2\,\Delta ka, \tag{4.7b}$$

(ii) pure edge dislocation ($\nu = \frac{1}{3}$):

$$s^{(e)}(a) = \begin{cases} \dfrac{\pi^2\,\Delta ka}{2}\,(-1)^{n/2}\Big[1 - \cos\beta - J_n\Big(\dfrac{3}{8}\,n\Big) + \dfrac{3}{2}\Big(J_{n/2-1}\Big(\dfrac{3}{8}\,n\Big) - \\ \quad - J_{n/2+1}\Big(\dfrac{3}{8}\,n\Big)\Big) - 2\sum\limits_{m=1}^{\infty} J_{m+n/2}\Big(\dfrac{3}{8}\,n\Big)\Big]\,; & n \text{ even}, \\ \pi\,\Delta ka(-1)^{(n-1)/2}\,\ln(4\pi\,\Delta ka)\,; & n \text{ odd} \end{cases} \tag{4.8a}$$

and

$$q^{(e)}(a) = \frac{\pi^2\,\Delta ka}{2} \tag{4.8b}$$

($J_n(z)$ is the Bessel function).

It should be noticed that in some cases the second-order term of the anti-classical S-matrix expansion appears to be necessary for a correct description of the behaviour of the image profiles in the vicinity of the singularity (discontinuity) point $a = 0$. In particular, for even n and $w = 0$ one needs to know the diagonal elements $\tilde{S}_{0(g),22}$ with accuracy up to $(2\pi\,\Delta ka)^2$ [10]. By use of (4.4) to (4.8) one can analyse the general properties of screw or pure edge dislocations for arbitrary values of the scattering parameters n, t, z_0, w, ξ_g, ξ'_g. As an example, we consider the image profile of the screw dislocation versus either its depth position z_0 inside the crystal for $w = 0$ or the foil inclination w for $z_0 = 0.5t$.

In the case of $w = 0$ the image profiles (4.4) become

$$I^{(s)}_{0(g)}(a) = \frac{1}{4}\exp\Big[-2\pi t\Big(\frac{1}{\xi'_0} - \frac{1}{\xi'_g}\Big)\Big]\,|\Phi^{(s)}_{0(g)}(a)|^2\,; \tag{4.9}$$

$$\Phi^{(s)}_{0(g)} = \tilde{S}^{(s)}_{0(g),22} + S^{(s)}_{0(g),12}\times$$
$$\times\Big\{\exp\Big[2\pi i\Big(\frac{1}{\xi_g} + \frac{i}{\xi'_g}\Big)(t - z_0)\Big] \pm \exp\Big[2\pi i\Big(\frac{1}{\xi_g} + \frac{i}{\xi'_g}\Big)z_0\Big]\Big\},$$

where the matrix elements $\tilde{S}^{(s)}_{0(g),22}$ and $\tilde{S}^{(s)}_{0(g),12}$ are equal to

$$\left.\begin{aligned} \tilde{S}^{(s)}_{0(g),22} &= 2\pi i n\,\frac{a}{\xi_g}\,(-1)^{(n-1)/2}\,\ln\Big(4\pi\,\frac{a}{\xi_g}\Big)\,; \\ \tilde{S}^{(s)}_{0(g),12} &= \mp\, i(-1)^{(n-1)/2}\,; \end{aligned}\right\}\quad n \text{ odd}, \tag{4.10}$$

$$\left.\begin{aligned} \tilde{S}^{(s)}_{0(g),22} &= \Big(1 + i\pi^2 n\,\frac{a}{\xi_g} - \pi^4 n^2\,\frac{a^2}{\xi_g^2}\Big)(-1)^{n/2}\,; \\ \tilde{S}^{(s)}_{0(g),12} &= \pm\, i\pi^2 n\,\frac{a}{\xi_g}\,(-1)^{n/2}\,; \end{aligned}\right\}\quad n \text{ even}, \tag{4.11}$$

respectively.

From here all the conclusions with regard to the dependence of the image profiles on the depth position — in particular, whether they will be symmetrical or asymmetrical, on which side of the dislocation core the minimum will be — follow directly, and they are in agreement with the known numerical results [3]. However there exist some differences, which have a mathematical sense rather than a physical one. So, for the screw dislocation with $n = 1$ ($z_0 = 0.5t = 4\xi_g$), the curve $I_g^{(s)}(a)$ calculated numerically crosses the ordinate axis at a point lying higher than the curve $I_0^{(s)}(a)$ does, while the opposite must hold, since in this case $I_g^{(s)}(0) = 0$ and $I_0^{(s)}(0) = \exp(-2\pi t/\xi_0') > 0$.

In the case that the dislocation is in the centre of a thick foil (and $w \neq 0$), the corresponding expressions for the intensity profiles are given by (see (4.4))

$$I_0(a) = \frac{1}{4} \exp\left[-2\pi t\left(\frac{1}{\xi_0'} - \frac{1}{\xi_g'(1+w^2)^{1/2}}\right)\right] \Bigg| (1 + \cos\beta)\, \tilde{S}_{0,22}^{(s)} + \\ + \exp\left[-\frac{\pi t}{\xi_g'(1+w^2)^{1/2}} + \frac{i\pi t}{\xi_g}(1+w^2)^{1/2}\right] [(1+\cos\beta)\, \tilde{S}_{0,12}^{(s)} + \\ + (1 - \cos\beta)\, \tilde{S}_{0,21}^{(s)}] \Bigg|^2, \tag{4.12}$$

$$I_g(a) = \frac{\sin^2\beta}{4} \exp\left[-2\pi t\left(\frac{1}{\xi_0'} - \frac{1}{\xi_g'(1+w^2)^{1/2}}\right)\right] |\tilde{S}_{g,22}^{(s)}|^2, \tag{4.13}$$

and, for example, expressions for the matrix elements $\tilde{S}_{0(g),22}^{(s)}$, $\tilde{S}_{0(g),12}^{(s)}$ with odd n take the form

$$\left.\begin{aligned} \tilde{S}_{0(g),22}^{(s)} &= -i[\cos\beta - 2\pi n\, \Delta k a \sin^2\beta \ln(4\pi\, \Delta k a)]\, (-1)^{(n-1)/2}, \\ \tilde{S}_{0,12}^{(s)} &= -i(1 - \cos\beta)\, (-1)^{(n-1)/2}, \\ \tilde{S}_{0,21}^{(s)} &= -i(1 + \cos\beta)\, (-1)^{(n-1)/2}. \end{aligned}\right\} \tag{4.14}$$

It is interesting that in the case of a dislocation in the centre of a thick inclined foil the dark-field image is determined by the diagonal element $\tilde{S}_{g,22}^{(s)}$ only and therefore it is not sensitive to interband scattering.

Consider the case of a screw dislocation with $n = 1$. From (4.14) it follows that the dark-field profile is asymmetrical and the minimum position a_* of $I_g^{(s)}(a)$ is on the left (right) of the dislocation core depending on the sign of (nw). The situation is more complicated for the bright-field profile, since its minimum is located at a point a^*

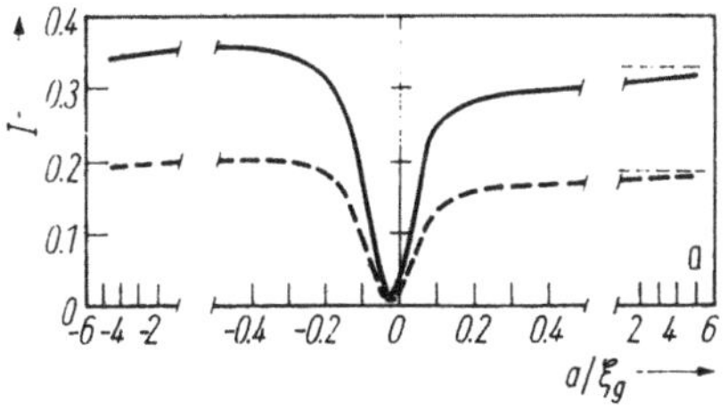

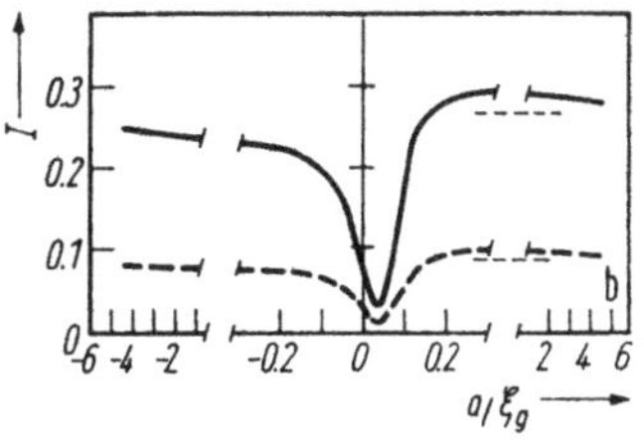

Fig. 1. Calculated images of a screw dislocation for the case $t/\xi_g = 8$, $z_0/\xi_g = 4$, $\xi_g/\xi_g' = 0.1$ $\xi_g' = \xi_0'$. a) $n = 1$, $w = 0.3$; b) $n = -1$, $w = 0.6$. Bright field — solid line, dark field — broken line. The quasi-classical and anti-classical approximations are utilized for calculations in the regions $|a|/\xi_g \gtrsim 1/2\pi$ and $|a|/\xi_g \lesssim 1/4\pi$, respectively

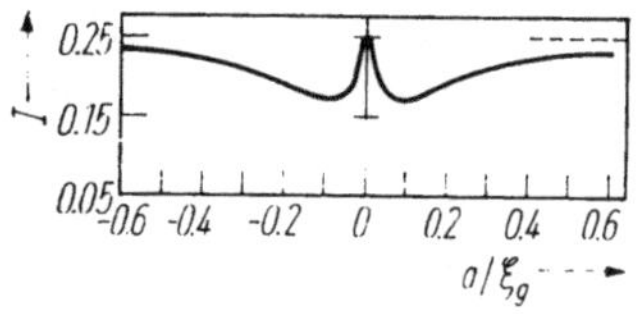

Fig. 2. Calculated images of a screw dislocation for the case $n = 2$, $t/\xi_g = 8$, $z_0/\xi_g = 4.25$, $\xi_g/\xi'_g = 0.1$, $\xi'_0 = \xi'_g$ $w = 0$. The solid and broken lines corresponding to the bright and dark fields, respectively, coincide. The quasi-classical and anti-classical approximations are utilized for calculations in the regions $|a|/\xi_g \gtrless 1/2\pi$ and $|a|/\xi_g \lessgtr 1/8\pi$, respectively

depending on the sign of

$$n\left\{\cos\beta + 2(1 - \cos\beta)\cos\left[\frac{\pi t}{\xi_g}(1 + w^2)^{1/2}\right]\exp\left[-\frac{\pi t}{\xi'_g}(1 + w^2)^{-1/2}\right]\right\}.$$

The total image profiles of the screw dislocations with $n = 1, -1$ and $w = 0.3$, 0.6 calculated according to (3.4) and (3.7) as well as (4.12) to (4.14) are shown in Fig. 1 a, b. The corresponding image profiles for $n = 2$ and $w = 0$ are given in Fig. 2. It is seen that the formulae of the anti-classical approximation valid for small distances from the dislocation core complement the quasi-classical description very well.

5. Conclusion

The most important conclusions to be made from the present investigation are the following ones:

1. The analytical treatment of the electron dynamical scattering by a strongly deformed region of a crystal is feasible. As a consequence the solution of the dislocation DCETM problem for the region close to the dislocation core is discussed. The solution is based on the general scattering matrix (S-matrix) theory.

2. In the region of slight deformations the quasi-classical (eikonal) approximation yields analytical results which agree with those already published by other authors.

Generally, the above theory permits to establish the range of applicability of each approach, the quasi- and anti-classical ones, and to give a quantitative description of the dislocation image picture as a whole.

Acknowledgements

The authors express their sincere thanks to Prof P. A. Besirganyan and Prof. Z. G. Pinsker for the encouraging support of the present study.

References

[1] H. Hashimoto, A. Howie, and M. J. Whelan, Phil. Mag. 5, 976 (1960).
[2] A. Howie and M. J. Whelan, Proc. Roy. Soc. A267, 206 (1962).
[3] P. B. Hirsch, A. Howie, R. B. Nicholson, D. W. Pashley, and M. J. Whelan, Electron Microscopy of Thin Crystals, Butterworths, London 1965.
[4] R. Gevers, Phil. Mag. 7, 59 (1962).
[5] R. Gevers, H. Blank, and S. Amelinckx, phys. stat. sol. 13, 449 (1966).
[6] M. Wilkens, phys. stat. sol. 13, 529 (1966).
[7] M. Wilkens, M. Rühle, and F. Häussermann, phys. stat. sol. 22, 689 (1967).
[8] V. V. Babikov, Phase Function Method in Quantum Mechanics, Izd. Nauka, Moscow 1976 (in Russian).
[9] A. M. Arustamyan, V. L. Vergasov, and F. N. Chukhovskii, Teoretich. i Matematich. Fizika 38, 380 (1979).
[10] F. N. Chukhovskii, A. M. Arustamyan, and Z. G. Pinsker, Kristallografiya 24, No. 4 (1979).
[11] F. N. Chukhovskii and A. A. Shtolberg, phys. stat. sol. 41, 815 (1970).
[12] Jahnke-Emde-Lösch, Tafeln Höherer Funktionen, B. G. Teubner Verlagsgesellschaft, Stuttgart 1960 (Chap. IX).

(Received February 28, 1979)

phys. stat. sol. (a) **54**, 55 (1979)

Subject classification: 2; 9; 11; 22.1.2

Istituto di Struttura della Materia, Istituto di Fisica, Università di Catania[1])

Laser Induced As Profile Broadening in Amorphous Silicon Layers[2])

By

M. G. Grimaldi, P. Baeri, S. U. Campisano, G. Foti, and E. Rimini

The profile broadening of 40 keV As atoms implanted in Si is measured after Q-switched ruby laser irradiation by MeV He scattering at grazing emergence. Two thicknesses of the amorphous layers on the Si single crystal substrates were used: 500 and 4500 Å. For the same incident energy density the measured impurity profile broadening is larger for the thicker amorphous layer sample. These results are interpreted in terms of matter transport in a molten region, whose spatial and time extent depends on the initial amorphous thickness. Heat and diffusion equations are solved by a numerical approach to calculate the impurity profiles. A good agreement is found with the experimental data.

Die Profilverbreiterung von Si implantierten 40 KeV-As-Atomen, wird nach Güte-geschalteter Rubinlaserbestrahlung mittels MeV-He-Streuung bei streifendem Einfall gemessen. Zwei Dicken der amorphen Schicht auf den Si-Einkristallsubstraten werden benutzt, 500 und 4500 Å. Bei der gleichen einfallenden Energiedichte ist die gemessene Störstellenprofilverbreiterung für die dickere amorphe Schichtprobe größer. Die Ergebnisse werden mit Materietransport in einem geschmolzenen Bereich erklärt, dessen räumliche und zeitliche Ausdehnung von der ursprünglichen Dicke der amorphen Schicht abhängt. Wärme- und Diffusionsgleichungen werden durch numerische Verfahren gelöst, um die Störstellenprofile zu berechnen. Gute Übereinstimmung mit den experimentellen Werten wird gefunden.

1. Introduction

Phase transitions have been induced by high-power laser irradiation in amorphous semiconductor layers produced by either ion implantation or vacuum deposition (see for example [1]). The transition to a single crystal involves the whole layer and the required energy density of the laser pulse must overcome a threshold value [2].

The formation of a liquid during laser irradiation accounts for the experimental results [3, 4]. The single-crystal transition occurs when at least the whole damaged layer is melted and wets the single-crystal substrate. Recently time resolved reflectivity measurements [5] clearly indicated the occurrence of a liquid surface layer during laser irradiation. Evidence of a molten layer arises also from the broadening of dopant profiles [6 to 9] and from the surface accumulation of impurities with a low segregation coefficient [10 to 11].

The thermal effects associated with the illumination depend mainly on the efficiency of the light to heat conversion and on the heat transport. A parameter of relevance is the absorption coefficient. Amorphous or disordered semiconductors absorb light more effectively than crystalline material, especially for photon energies near the band gap value [12]. Ion implanted material with different disordered layer thickness should then be characterized by a different thermal history. The extent in depth and time of the molten layer can be probed by the broadening of the impurity profile.

[1]) Corso Italia 57, 95129 Catania, Italia.

[2]) Work supported in part by GNSM-CNR and by CSFNSM, respectively.

In detail the evolution in depth and time of the liquid layer is studied by an experimental investigation of the redistribution of implanted arsenic atoms in silicon samples with different amorphous thickness. The results will be compared with calculations accounting for the liquid formation and impurity redistribution governed by diffusion and segregation coefficients.

2. Experimental

Silicon single crystal (111)-oriented substrates were implanted at RT with 40 keV As to a fluence of $5 \times 10^{15}/cm^2$. Some samples were further implanted at LN temperature with 300 keV P^+ to a fluence of $1 \times 10^{15}/cm^2$. The second implantation creates an amorphous layer 4500 Å thick.

The channeling effect in combination with Rutherford backscattering of MeV He ions was adopted to measure the thickness of the damaged layers and the residual disorder after laser irradiation. The As concentration profiles were measured by backscattering at grazing emergence. The overall electronic resolution was about 18 keV, corresponding to a depth resolution of ≈ 120 Å at 104° scattering angle [13].

The irradiations were performed in air with a Q-switched ruby laser ($\lambda = 0.69$ μm) single pulses of 15 ns duration. The energy density per pulse ranged between 0.5 and 2.5 J/cm². The beam spot was 1 cm² in area.

3. Results

The structure of the irradiated Si samples differs in dependence on the thickness of the overlaying amorphous layer. 40 keV As implantation creates an amorphous layer 500 Å thick, while the implantation with P^+ produces a 4500 Å thick amorphous region. The behaviour of the two samples depends on the ratio between the damaged layer thickness and the absorption length.

The absorption coefficient α of amorphous and crystalline silicon is 5×10^4 and 4×10^3 cm^{-1}, respectively, at $\lambda = 0.69$ μm [12]. If the thickness of the amorphous layer is larger than α^{-1}, all the incident power is converted into heat within this thickness in a small volume and the temperature rise can be very large. For amorphous layers thinner than α^{-1} only a fraction of the total power is converted into heat within the amorphous layer. The remaining energy fraction is converted into heat in a large volume due to the low absorbance of the underlaying single crystal. As soon as melting starts the absorption processes occur within a distance of a few hundredth Angstroms from the surface due to the metallic nature of the molten semiconductor. Thereafter the structure of the substrate does not play any role.

Laser effects can be described as a first approximation by the solution of the one-dimensional heat equation with a depth-dependent source term, to include absorption [14, 15]. The solution requires a numerical procedure to account for the changes in thermal and optical parameters with temperature and for phase transitions. By this method one can obtain quantitative information on the influence of the overlaying amorphous layer and the time and space evolution of the molten layer [15].

The results are shown in Fig. 1a and b for the 500 and 4500 Å thick amorphous layer, respectively. The following parameters were adopted for calculations: $R = 0.35$ and 0.7 for amorphous and liquid silicon, respectively [16], $\alpha = 5 \times 10^4$, 5×10^3, and 4×10^5 cm^{-1} for amorphous, crystalline, and liquid silicon [12]. The temperature dependence of thermal conductivity and reflectivity was also included [17, 18].

The curves report the time evolution of the melted thickness for several energy densities as a function of time for a 15 ns FWHM Gaussian profile of the Q-switched ruby laser pulse. The thickness of the melted region and the permanence of the surface in the liquid phase increase with the energy density of the laser pulse in both

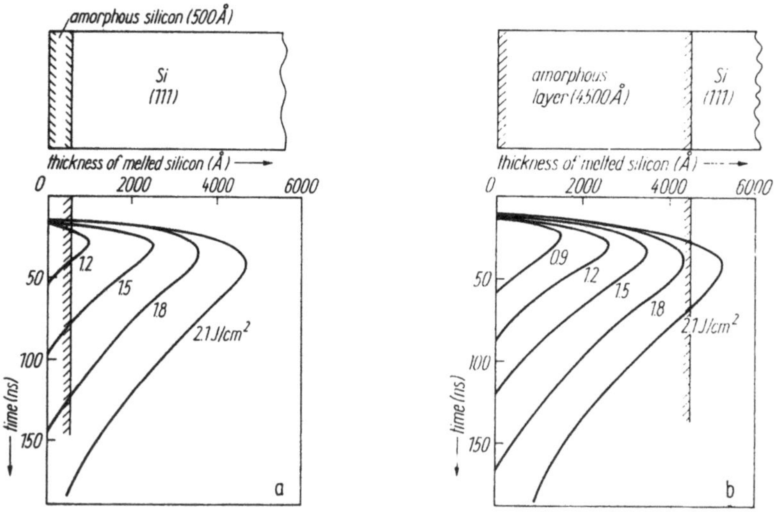

Fig. 1. Extent of the liquid layers as a function of time for several Q-switched ruby laser energy densities. a) 500 and b) 4500 Å thick amorphous layer on a Si(111)-oriented substrate, respectively

Fig. 1a and b. The main difference between the two irradiated samples is related to the threshold energy density for surface melting. The sample with a 4500 Å thick amorphous layer requires a threshold energy of 0.7 J/cm², while that with 500 Å 1.1 J/cm².

As a general feature the thickness of the molten layer and the permanence in the liquid phase is larger in the sample with a thicker amorphous region for the same incident energy density. These depth effects can be evidence in principle by a different broadening of the impurity profile. We should expect for instance an impurity redistribution occurring over greater distances for the 4500 than for the 500 Å thick amorphous layer.

The measured profiles of the 40 keV implanted arsenic atoms are shown in Fig. 2a after irradiation with different energy densities of the ruby laser pulse for the two considered structures. The as-implanted profile is shown by a dashed line. The measured projected range R_p, and the standard deviation, ΔR_p, are 260 and 275 Å, respectively, in good agreement with the calculated values 270 and 280 Å [19]. Curves labelled a and b represent the experimental profiles in the samples with 4500 and 500 Å thick amorphous layer, respectively.

The impurity redistribution is not symmetric around R_p. The near sample surface acts as a non-permeable boundary and the time duration of the liquid depends on the distance from the surface. The experimental profiles indicate that in the investigated range of densities the broadening of the dopant is larger in the samples overlaid with the thicker amorphous layer. With increasing energy density the difference between the two profiles becomes smaller and smaller.

These results reflect qualitatively the behaviour of the calculated spatial and time evolution of the liquid layer (compare Fig. 1a with Fig. 1b). To compute the impurity concentration profiles, $c(x)$, the diffusion equation, $\partial c/\partial t = (\partial/\partial x)\, D\, (\partial c/\partial x)$, was solved numerically by the standard finite-difference method [20], with typical values of $\Delta x = 50$ Å and $\Delta t = 0.025$ ns.

The diffusion, D, of As in solid Si near the melting temperature amounts to about 10^{-14} cm²/s [21]. In liquid Si impurities diffuse with a coefficient of about 10^{-4} cm²/s [22]. The value adopted in the calculation was then zero in the solid phase and

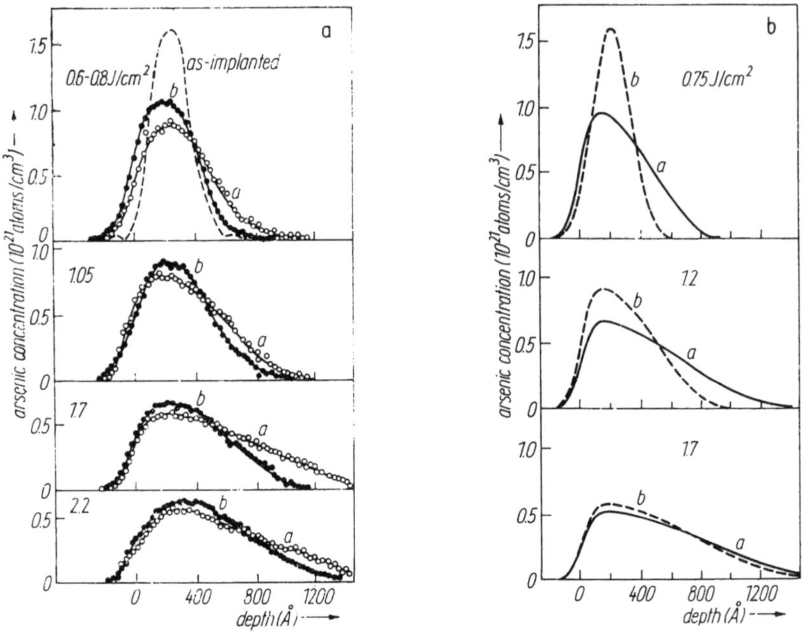

Fig. 2. a) Experimental and b) calculated arsenic profiles after laser irradiation at several energy densities. The curves labelled a and b refer to the 4500 and 500 Å thick amorphous layer, respectively

10^{-4} cm^2/s in the liquid layer. The low value of D in the solid phase causes broadening lower than 1 Å for a time interval of 100 ns.

The following boundary conditions were adopted:

a) $c(x, 0)$ Gaussian profile with parameters determined by backscattering measurements;

b) $J = 0$ at $x = 0$, any t, and at $x = s(t)$, i.e. at the sample surface and at the solid–liquid interface.

J is the arsenic atom flux and $s(t)$ is the time-dependent molten layer thickness.

The solid–liquid interface kinetics was derived from data reported in Fig. 1a and b, respectively. Segregation effects were neglected due to the value of the coefficient for As in Si (> 0.3) and to the high velocity of the solid–liquid interface (≈ 2 m/s) as evaluated from the negative slope of the curves of Fig. 1a and b.

Typical calculated results are shown in Fig. 2b for the same experimental conditions of the measured data reported in Fig. 2a. The difference between the calculated profiles a and b follows the same trend of the experiments. The main difference between the two profiles occurs for energy densities near the threshold value. For instance at 0.7 J/cm^2 below the calculated threshold for the 500 Å and above that for the 4500 Å layer, no broadening of the profile was computed of course in curve a, while it will occur in curve b.

4. Summary and Conclusion

Matter transport of impurities over a distance of a few hundredth Angstroms occurs after Q-switched laser irradiation. The movement can be explained in terms of diffusion in a liquid phase produced by the laser light. Experiments reported in this work indicate that the profile of implanted As atoms broadens more for a Si sample with a 4500 Å thick disordered surface layer than for a Si sample with a 500 Å thick damaged layer. These results were accounted for by solving the impurity equation in the liquid material. The broadening can then be correlated with the time and space evolution of the molten layer. The liquid evolution differs in the two structures because the efficiency of the laser light to heat conversion depends on the absorption coefficient which is an order of magnitude higher for amorphous than for crystalline silicon at 0.69 μm.

The agreement of the experimental and the calculated trends for the impurity profiles supports the description of the laser-solid interaction in terms of mainly thermal effects. The results point to the strong influence of the coupling on the thickness of the amorphous layer. This might be of relevance in view of the future possible applications of laser annealing in the semiconductor technology.

References

[1] E. Rimini (Ed.), Proc. Laser Effects in Ion Implanted Semiconductors, Istituto di Struttura della Materia, Catania 1978.
[2] G. Foti, W. S. Tseng, and J. W. Mayer, Appl. Phys. **15**, 365 (1978).
[3] J. Krymicki, J. Suski, S. Ugmiewski, R. Grötzschel, R. K. Klabes, U. Kreissig, and J. Rudiger, Phys. Letters A **54**, 157 (1977).
[4] I. B. Khaibullin, E. I. Shtyrkov, M. M. Zaripov, R. M. Bayazitov, and M. F. Galyatudinov, Radiat. Eff. **36**, 225 (1978).
[5] D. H. Auston, C. M. Surko, T. N. C. Venkatesan, R. E. Shusher, and J. A. Golovchenko, Appl. Phys. Letters **33**, 437 (1978).
[6] H.-D. Geiler, G. Götz, K.-D. Klinge, and N. Triem, phys. stat. sol. (a) **41**, K171 (1977).
[7] P. Baeri, S. U. Campisano, G. Foti, and E. Rimini, Appl. Phys. Letters **32**, 137 (1978).
[8] A. V. Dvurechenskii, G. A. Kachurin, and A. Kh. Antonenko, Radiat. Eff. **37**, 179 (1978).
[9] G. K. Celler, J. M. Poate, and L. C. Kimerlingh, Appl. Phys. Letters **32**, 464 (1978).
[10] G. Foti, E. Rimini, and S. U. Campisano, phys. stat. sol. (a) **47**, 533 (1978).
[11] P. Baeri, S. U. Campisano, G. Foti, and E. Rimini. Phys. Rev. Letters **41**, 1246 (1978).
[12] M. H. Brodsky, R. S. Title, K. Weiser, and G. D. Pettit, Phys. Rev. B **1**, 2632 (1970).
[13] J. S. Williams, Nuclear Instrum. and Methods **126**, 205 (1975).
[14] M. M. Zaripov, I. B. Khaibullin, and E. I. Shtyrkov, Soviet Phys. – Uspekhi **19**, 1032 (1976).
[15] P. Baeri, S. U. Campisano, G. Foti, and E. Rimini, J. appl. Phys. **50**, 788 (1979).
[16] T. E. Seidel, G. A. Pasteur, and J. C. C. Tsai, Appl. Phys. Letters **29**, 648 (1976).
[17] A. Goldsmith, Handbook of Thermophysical Properties of Solid Materials, Vol. I, Mac Millan Co., New York 1961.
[18] C. Y. Ho, R. W. Powell, and P E. Siley, Reference Data on the Thermal Conductivity of the Elements, National Standard Reference Data Series, N. B. S., 1969.
[19] B. Smith, Ion Implantation Range Data, Learned Information Ltd., Oxford 1977.
[20] B. W. Arden and K. N. Astill, Numerical Algorithms – Origins and Applications, Addison-Wesley, London 1970 (p. 280).
[21] C. S. Fuller and J. A. Ditzeenberger, J. appl. Phys. **27**, 544 (1956).
[22] J. O. McCaldin, J. Vacuum Sci. Technol. **11**, 910 (1974).

(Received February 26, 1979)

phys. stat. sol. (a) **54**, 61 (1979)

Subject classification: 19; 22.8.1

Institut de Physique des Solides, Université d'Etat de Lettonie, Riga

RPE de Mn^{2+} dans les polycristaux de basse symétrie

Le métaphosphate de magnésium

Par

J. G. Kliava, Z. A. Konstants, J. J. Purāns et A. I. Dimante

On a analysé les spectres de RPE de Mn^{2+} dans les polycristaux pour le hamiltonien de spin $\mathcal{H}_s = g\beta \boldsymbol{HS} + D\left(S_z^2 - \frac{S}{3}(S+1)\right) + E(S_x^2 - S_y^2) + \boldsymbol{ASI}$ pour $|D| \ll g\beta H, |E| \ll g\beta H, |A| \ll g\beta H$. Le calcul des spectres tient compte des transitions hyperfines permises $\Delta m = 0$ et interdites $\Delta m = \pm 1$ entre les sous-niveaux Zeeman $\pm 1/2$. Les probabilités de transition sont calculées suivant la méthode de Bir. Les spectres simulés à l'ordinateur sont présentés en fonction du rapport E/D. En comparant les spectres expérimentaux et théoriques pour le métaphosphate de magnésium $Mg(PO_3)_2$ dopé de manganèse on a déterminé les paramètres du hamiltonien de spin: $g = 2{,}000 \pm 0{,}002$; $D/g\beta = (220 \pm 10)$ G; $|E|/g\beta = (20 \pm 5)$ G; $A/g\beta = (-95{,}0 \pm 0{,}5)$ G. Il en résulte que la symétrie d'environnement de Mn^{2+} dans $Mg(PO_3)_2$ est inférieure à la symétrie axiale.

EPR spectra of Mn^{2+} in polycrystals are analysed for the spin-Hamiltonian $\mathcal{H}_s = g\beta \boldsymbol{HS} + D\left(S_z^2 - \frac{S}{3}(S+1)\right) + E(S_x^2 - S_y^2) + \boldsymbol{ASI}$. Spectra are calculated for $|D| \ll g\beta H$, $|E| \ll g\beta H$, $|A| \ll g\beta H$ taking into account the allowed $\Delta m = 0$ as well as the forbidden $\Delta m = \pm 1$ hyperfine transitions between the Zeeman sublevels $\pm 1/2$. The transition probabilities are accounted for using Bir's method. Computer simulated spectra are displayed for different values of the parameter ratio E/D. By comparing the experimental and theoretical spectra of polycrystalline magnesium metaphosphate $Mg(PO_3)_2$ doped with manganese the following values of the Mn^{2+} spin-Hamiltonian parameters are determined: $g = 2.000 \pm 0.002$; $D/g\beta = (220 \pm 10)$ G; $|E|/g\beta = (20 \pm 5)$ G; $A/g\beta = (-95.0 \pm 0.5)$ G. It follows from these values that the symmetry of the Mn^{2+} ion environment in $Mg(PO_3)_2$ is lower than axial.

1. Introduction

La méthode de RPE est particulièrement bien adaptée à l'analyse de la structure des solides grâce à la forte sensibilité des paramètres du hamiltonien de spin à l'environnement d'un centre paramagnétique. Cependant, pour les polycristaux ou pour les verres, une interprétation des spectres de RPE est compliquée à cause du désordre orientationnel.

Récemment, l'application des ordinateurs sur une grande échelle a rendu possible la simulation des spectres de RPE dans les systèmes désordonnés. Les paramètres du hamiltonien de spin peuvent être déterminés en comparant les spectres théoriques et expérimentaux, avec une haute précision conduisant à des renseignements sur les caractéristiques de structure. Ce procédé peut être très utile dans les cas où il est impossible ou difficile de fabriquer un monocristal et où l'on est donc contraint de se contenter d'un polycristal. Ceci peut s'avérer d'autant plus précieux pour étudier la structure des verres.

Comme sonde paramagnétique dans ce type d'étude, il est désirable de choisir un centre dont le spectre de RPE possède un grand nombre de raies. Quoique la simulation de tels spectres soit compliquée, elle permet de déterminer les paramètres du hamiltonien de spin sans ambiguité, ce qui n'est pas toujours possible pour un spectre plus simple, surtout s'il y a une distribution des paramètres.

Nous avons utilisé dans ce but l'ion manganèse Mn^{2+}. De nombreux auteurs ont effectué la simulation à l'ordinateur des spectres pour cet ion [1 à 4]. Cependant dans tous ces cas l'accord entre les spectres théoriques et expérimentaux était loin d'être excellent. Notre étude précédente [5] a été consacrée à l'analyse détaillée des spectres de RPE de Mn^{2+} de symétrie axiale dans les polycristaux. En utilisant la méthode de Bir [6] pour calculer les dépendances angulaires de la probabilité de transition, nous avons obtenu un bon accord avec le spectre expérimental de Mn^{2+} dans $CaWO_4$ polycristallin. Dans cet article, nous exposerons les résultats de l'extension de cette étude au cas d'une symétrie inférieure à la symétrie axiale. Les spectres théoriques seront comparés aux résultats expérimentaux obtenus sur le métaphosphate de magnésium $Mg(PO_3)_2$ polycristallin dopé de manganèse. A notre connaissance, c'est la première application de la RPE de Mn^{2+} à ce système. Les résultats des études de la RPE des phosphates de manganèse ont été publiés ailleurs [7].

2. Calcul des spectres

Le programme du calcul est rédigé en FORTRAN pour le hamiltonien de spin

$$\mathcal{H}_s = g\beta \boldsymbol{HS} + D\left[S_z^2 - \frac{S}{3}(S+1)\right] + E(S_x^2 - S_y^2) + A\boldsymbol{SI} \tag{1}$$

où $S = 5/2$, $I = 5/2$. Nous nous bornons au cas où les séparations fine et hyperfine sont faibles devant la séparation Zeeman ($|D| \ll g\beta H$, $|E| \ll g\beta H$, $|A| \ll g\beta H$) et ne considérons que la transition centrale $1/2 \leftrightarrow -1/2$ de la structure fine, les contributions relatives des autres transitions étant peu importantes dans les systèmes désordonnés. Le calcul est analogue à celui que nous avons développé pour la symétrie axiale [5] à l'exception de ce que, actuellement, les valeurs du champ magnétique de résonance H et les probabilités de transition dépendent non seulement de l'angle polaire ϑ mais aussi de l'angle azimutal φ.

Dans la littérature nous n'avons pas trouvé l'expression de H contenant tous les termes du troisième ordre des perturbations. Un calcul fastidieux conduit à l'expression suivante pour la transition $(1/2, m) \leftrightarrow (-1/2, m+i)$ (où $i = 0$ correspond aux transitions hyperfines permises et $i = \pm 1, \pm 2$ etc. correspond aux transitions interdites, $|2m + i| \leqq 5$):

$$H = H_0(m, i) + \frac{16}{H}(|\lambda|^2 - 2|\varrho|^2) - (2m+i)\frac{8A}{H^2}(9|\lambda|^2 - 2|\varrho|^2) - \\ - (2m+i)(1+i)\frac{4A^2}{H^2}\sigma \tag{2}$$

où

$$H_0(m, i) = \frac{h\nu}{g\beta} - (2m+i)\frac{A}{2} + [(2m+i)^2 + (17-i)^2 - 324]\frac{A^2}{8H} - \\ - [4(m+i)^3 + 4m^3 - (2m+i)(65+134i)]\frac{A^3}{4H^2},$$

$$|\lambda|^2 = \sin^2\vartheta[(D + E\cos 2\varphi)^2\cos^2\vartheta + E^2\sin^2 2\varphi],$$

$$|\varrho|^2 = \{[D\sin^2\vartheta - E(1+\cos^2\vartheta)\cos 2\varphi]^2 + 4E^2\cos^2\vartheta\sin^2 2\varphi\}/16,$$

$$\sigma = [D(3\cos^2\vartheta - 1) - 3E\sin^2\vartheta\cos 2\varphi]/2.$$

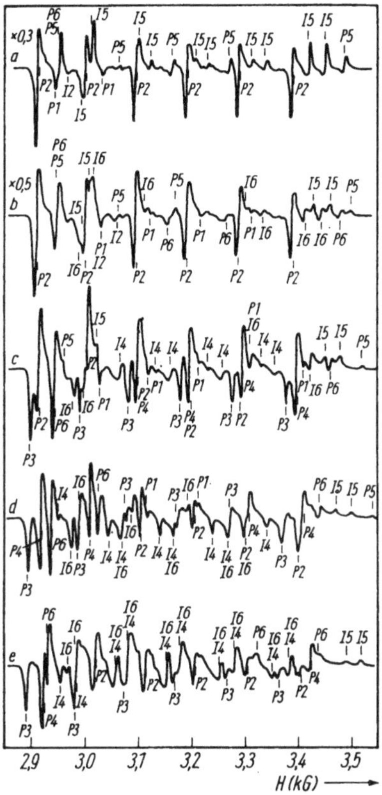

Fig. 1. Spectres de RPE de Mn^{2+} dans les polycristaux simulés à l'ordinateur (transition centrale $1/2 \leftrightarrow -1/2$ de la structure fine). Les valeurs des paramètres sont $g = 2{,}0$, $A/g\beta = -95$ G, $D/g\beta = 210$ G, $\nu = 8900$ MHz. Les valeurs de $E/g\beta$ pour les courbes différentes sont (en G): (a) 0, (b) 10, (c) 30, (d) 50, (e) 70. Le profil de raie est Lorentzien avec $\Delta H_{pp} = 3{,}5$ G. Les transitions hyperfines permises $\Delta m = 0$ et interdites $\Delta m = \pm 1$ sont marquées P ou I, respectivement. Les chiffres 1 à 6 indiquent les six types de singularités, en conformité avec le texte

La probabilité de transition est décrite par la même expression que pour le cas axial [5] où le cosinus de l'angle entre les axes de quantification du spin nucléaire peut être écrit au deuxième ordre sous la forme [6]

$$\mu = 1 - \frac{512\,|\lambda|^2}{(h\nu)^2}.$$

La simulation des spectres a été effectuée à l'aide d'un ordinateur M 6000. Les résultats sont illustrés sur la Fig. 1. On remarquera la forte variation de l'allure des spectres en fonction du paramètre E. Les "singularités" dans les spectres, c'est-à-dire les "raies" ou les "pics", apparaissent aux valeurs de H, pour lesquelles

$$\frac{\partial H(\vartheta, \varphi)}{\partial \vartheta} = 0 \quad \text{et} \quad \frac{1}{\sin\vartheta}\frac{\partial H(\vartheta, \varphi)}{\partial \varphi} = 0\,. \tag{3}$$

En portant l'expression (2) dans le système (3) nous obtenons pour chaque transition hyperfine six valeurs du champ de résonance correspondant à six paires d'angles ϑ et φ:

1. $\vartheta = 0$:

$$H = H_0(m, i) - 8\,\frac{E^2}{H} + 4(2m + i)\,\frac{AE^2}{H^2} - 4(2m + i)\,(1 + i)\,\frac{A^2D}{H^2};$$

2. et 3. $\vartheta = \frac{\pi}{2}$, $\varphi = \frac{\pi}{4} \mp \frac{\pi}{4}$ ou $\pi + \frac{\pi}{4} \pm \frac{\pi}{4}$:

$$H = H_0(m\,, i) - 2\,\frac{(D \mp E)^2}{H} + (2m + i)\,\frac{A(D \mp E)^2}{H^2} + $$

$$+ 2(2m + i)\,(1 + i)\,\frac{A^2(D \pm 3E)}{H^2}\,;$$

4. $\vartheta = \frac{\pi}{2}$, $\cos 2\varphi = \frac{\alpha D}{E}$

où

$$\alpha = \left[2 - (2m + i)\frac{A}{H} + 3(2m + i)(1 + i)\frac{A^2}{DH}\right] : \left[18 - 73(2m + i)\frac{A}{H}\right]$$

$$H = H_0(m, i) - 2(1 - 2\alpha + 9\alpha^2)\frac{D^2}{H} + 16\frac{E^2}{H} + (2m + i)(1 - 2\alpha +$$

$$+ 73\alpha^2)\frac{AD^2}{H^2} - 72(2m + i)\frac{AE^2}{H^2} + 2(2m + i)(1 + i)(1 + 3\alpha)\frac{A^2D}{H^2};$$

5. et 6. $\cos^2\vartheta = x$, $\varphi = \frac{\pi}{4} \mp \frac{\pi}{4}$ ou $\pi + \frac{\pi}{4} \pm \frac{\pi}{4}$

où

$$x = \left[10D \pm 6E - (2m + i)(37D \pm 35E)\frac{A}{H} - 3(2m + i)(1 + i)\frac{A^2}{H}\right] : \left[18 - 73(2m + i)\frac{A}{H}\right](D \pm E)$$

$$H = H_0(m, i) - 2(3x - 1)^2\frac{(D \pm E)^2}{H} + \frac{8D}{H}[x(D \pm E) \pm E] +$$

$$+ (2m + i)\{(73x^2 - 70x + 1)(D \pm E)^2 - 4D[x(D \pm E) \pm E]\}\frac{A}{H^2} -$$

$$- 2(2m + i)(1 + i)[3x(D \pm E) - (D \pm 3E)]\frac{A^2}{H^2}.$$

Il est évident que dans les cas 4 à 6 les singularités n'existent qu'à condition que $|\alpha D/E| \leqq 1$ ou $0 \leqq x \leqq 1$, respectivement.

Sur la Fig. 2 les champs de résonance calculés à l'aide de l'ordinateur sont présentés en fonction des paramètres D et E. En comparant les Fig. 1 et 2 on a attribué les singularités observées dans les spectres aux transitions hyperfines différentes ayant lieu aux valeurs des angles ϑ et φ qui correspondent aux cas 1 à 6. Cependant il faut noter que les intensités de ces singularités dépendent de la probabilité de transition aux mêmes valeurs de ϑ et φ. A cause de cela certaines singularités peuvent être considérablement réduites en hauteur ou même totalement disparues (voir Fig. 1).

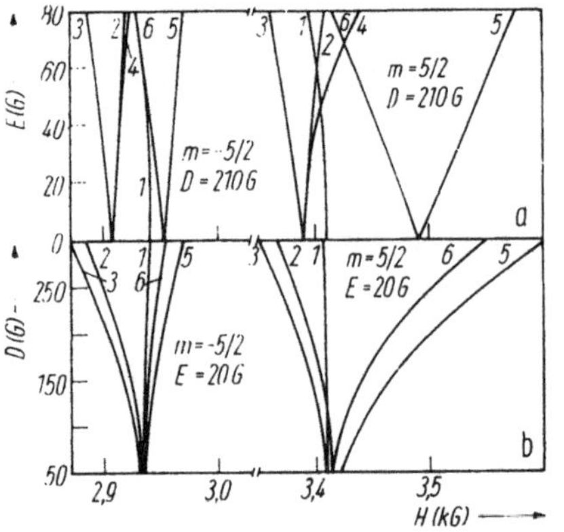

Fig. 2. Variation de la position des singularités en fonction a) des paramètres E ou b) D. Les chiffres indiquent le type de singularité. Les courbes sont calculées pour $g = 2{,}0$, $A/g\beta = -95$ G, $\nu = 8900$ MHz

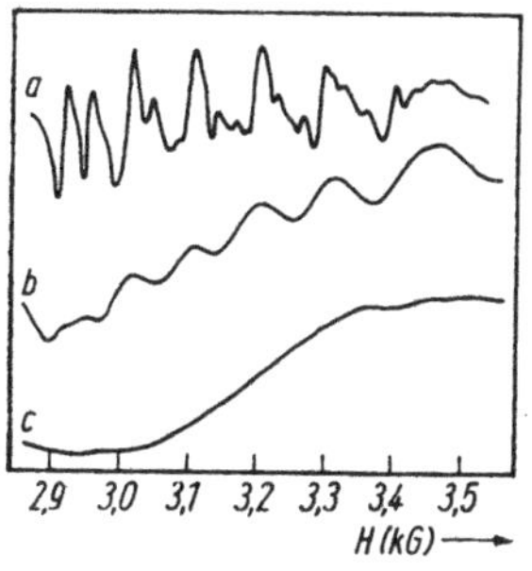

Fig. 3

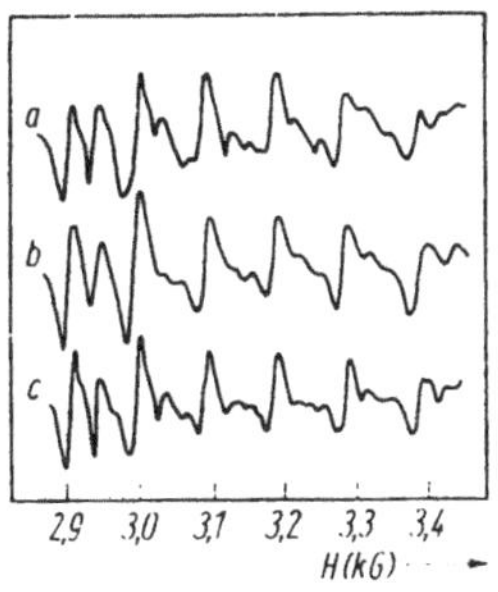

Fig. 4

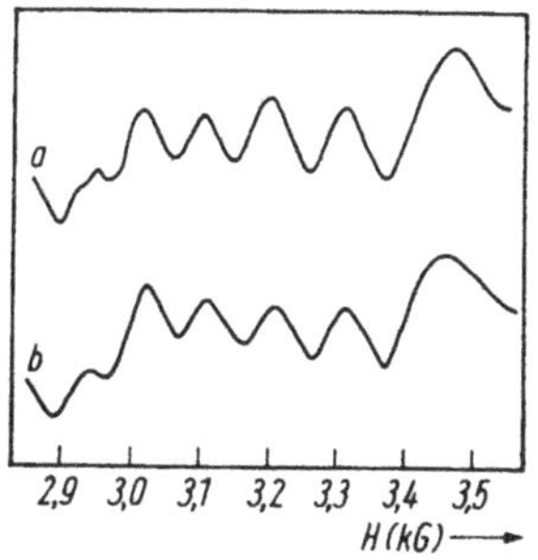

Fig. 5

Fig. 3. Spectres expérimentaux pour les échantillons 1 (a), 2 (b) et 3 (c) à la température ambiante; $\nu = 8900$ MHz

Fig. 4. Comparaison des spectres enregistrés et calculés pour l'échantillon 1. (a) Spectre expérimental après la soustraction du spectre de type II. (b) Spectre théorique; $g = 2,0$, $A/g\beta = -95$ G, $D/g\beta = 220$ G, $E/g\beta = -20$ G. Le profil de raie est Lorentzien avec $\Delta H_{pp} = 10$ G; $\nu = 8900$ MHz. (c) Spectre théorique calculé en tenant compte de la distribution des paramètres D et E; $g = 2,0$, $A/g\beta = -95$ G, $D/g\beta = 220$ G, $E/g\beta = 20$ G, $\Delta D/g\beta = 20$ G, $\Delta E/g\beta = 10$ G. Le profil de raie est Lorentzien avec $\Delta H_{pp} = 3$ G; $\nu = 8900$ MHz

Fig. 5. Comparaison des spectres enregistrés et calculés pour l'échantillon 2. (a) Spectre expérimental après la soustraction du spectre de type II. (b) Spectre théorique; $g = 2,0$, $A/g\beta = -95$ G, $D/g\beta = 220$ G, $E/g\beta = 20$ G. Le profil de raie est Lorentzien avec $\Delta H_{pp} = 60$ G; $\nu = 8900$ MHz

Remarquons que l'identification des singularités se révèle la plus simple pour la transition $(1/2, -5/2) \leftrightarrow (-1/2, -5/2)$. Dans beaucoup de cas on peut déduire de l'analyse de la partie correspondante du spectre les valeurs approchées des paramètres D et E, ce qui facilite la simulation des spectres expérimentaux.

3. Résultats expérimentaux et discussion

Les échantillons du métaphosphate de magnésium ont été fabriqués par frittage à 1200 °C des mélanges du phosphate d'ammonium monosubstitué et de l'oxyde de magnésium en rapport molaire 1:1. Pour doper les échantillons de manganèse, on a ajouté aux mélanges 0,05, 1,00 et 5,00% en masse de Mn_2O_3 (échantillons 1, 2 et 3, respectivement). Il est connu que, pour les températures de synthèse entre 1000 et 1300 °C, seuls les phosphates de Mn^{2+} sont formés, indépendamment du nombre d'oxydation initial du manganèse [8].

Les spectres de RPE étaient enregistrés en utilisant un spectromètre RE 1301 en bande X. Le champ magnétique était étalonné à l'aide d'un gaussmètre E 11-2.

La Fig. 3 représente les spectres expérimentaux. L'analyse montre que, pour tous les échantillons, on peut les décomposer en deux types de spectres. Le spectre de type I comprend un grand nombre de raies. Celui de type II représente la raie unique avec $g = 1,985 \pm 0,005$ et la largeur $\Delta H_{pp} = (510 \pm 20)$ G. A mesure que la teneur en Mn_2O_3 augmente, l'intensité du spectre de type II s'accroît par rapport à celle de type I.

Sur les Fig. 4 et 5 les spectres obtenus par soustraction des spectres de type II sont reproduits et la comparaison avec spectres simulés à l'ordinateur est illustrée pour les échantillons 1 et 2, respectivement. Il est intéressant de noter que l'accord entre les spectres théoriques et expérimentaux est amélioré si on prend en considération une certaine distribution des paramètres D et E. La courbe c sur la Fig. 4 représente le spectre théorique calculé par sommation des spectres avec les facteurs de pondération suivant la loi de répartition Gaussienne: $P(D) \sim \exp[-(D - D_0)^2/(\Delta D)^2]$, $P(E) \sim \exp[-(E - E_0)^2/(\Delta E)^2]$ où $\Delta D/g\beta = 20$ G et $\Delta E/g\beta = 10$ G. Une telle distribution sans doute est due aux perturbations de la structure cristalline provoquées par des défauts du réseau.

Les paramètres du hamiltonien de spin (1) fournissant le meilleur accord entre les spectres théoriques et expérimentaux de type I sont:

$$g = 2{,}000 \pm 0{,}002\,, \qquad \frac{D}{g\beta} = (220 \pm 10)\ \text{G}\,, \qquad \frac{|E|}{g\beta} = (20 \pm 5)\ \text{G}\,,$$

$$\frac{A}{g\beta} = (-95{,}0 \pm 0{,}5)\text{G}\,.$$

Entre les limites d'erreur, ces valeurs sont les mêmes pour les échantillons 1 et 2. Les changements du spectre de type I en fonction de la concentration en manganèse sont entièrement dus à la variation de ΔH_{pp} qui s'explique par l'interaction dipôle–dipôle. Donc on peut conclure que le spectre de type I est dû aux ions Mn^{2+} se substituant au magnésium dans $Mg(PO_3)_2$ et distribués au hazard.

Il ressort de la valeur du rapport $|E|/D \approx 0{,}1$ que la symétrie du champ cristallin est inférieure à la symétrie axiale. Une telle structure de l'environnement de Mn^{2+} dans $Mg(PO_3)_2$ est compatible avec les données de la structure des phosphates condensés y compris les métaphosphates de métaux bivalents [9].

En ce qui concerne le spectre de type II, on a observé des spectres semblables dans les phosphates de manganèse [7]. L'étude radiocristallographique montre que la seule phase cristalline qui existe dans nos échantillons est $Mg(PO_3)_2$ [10]. Par ailleurs, il est connu que le métaphosphate de manganèse forme une phase amorphe à la température de la synthèse de 1200 °C [8]. Donc on peut supposer que le spectre de type II est dû à la présence de cette phase dans les échantillons étudiés.

Il faut remarquer que des spectres de Mn^{2+} analogues au spectre de type II ont été observés précédemment dans les polycristaux et dans les verres [1, 3] et attribués à des sites caractérisés par un haut degré de désordre et donc par une très large distribution des paramètres du hamiltonien de spin. Cependant, il nous semble que cette explication soit peu probable, car l'existence d'un grand nombre de tels sites est assez problématique.

Bibliographie

[1] D. L. GRISCOM et R. E. GRISCOM, J. chem. Phys. **47**, 2711 (1967).
[2] N. N. TIKHOMIROVA, S. N. DOBRYAKOV et I. V. NIKOLAEVA, phys. stat. sol. (a) **10**, 593 (1972).
[3] N. J. STEINSCHNEIDER et G. M. ZHIDOMIROV, Zh. eksper. teor. Khim. **9**, 414 (1973).
[4] P. C. TAYLOR et P. J. BRAY, J. Phys. Chem. Solids **33**, 43 (1972).
[5] J. G. KLIAVA et J. J. PURĀNS, phys. stat. sol. (a) **49**, K43 (1978).
[6] G. L. BIR, Fiz. tverd. Tela **5**, 2236 (1963).
[7] A. I. DIMANTE, J. G. KLIAVA, Z. A. KONSTANTS et J. J. PURĀNS, Zh. neorg. Khim. **21**, 627 (1976).
[8] Z. A. KONSTANTS et A. I. DIMANTE, Izv. Akad. Nauk SSSR, Ser. neorg. Mater. **13**, 99 (1977).
[9] D. E. C. COLBRIDGE, The Structural Chemistry of Phosphorus, Elsevier Publ. Co, Amsterdam/London/New York 1974 (pp. 127 à 167).
[10] Powder Diffraction File, Joint Committee of Powder Diffraction Standards, Easton (USA) 1973 (pp. 11 à 41).

(Received February 23, 1979)

phys. stat. sol. (a) **54**, 67 (1979)

Subject classification: 2 and 14.3; 22

Moscow Engineering Physics Institute[1]) *(a)*
and Institute of Applied Physics, Academy of Sciences of the Moldavian SSR, Kishinev[2]) *(b)*

An Analysis of the Dispersive Charge Transport in Vitreous 0.55 As_2S_3:0.45 Sb_2S_3

By

V. I. Arkhipov (a), M. S. Iovu (b), A. I. Rudenko (a), and S. D. Shutov (b)

The measured charge transport characteristics of vitreous 0.55 As_2S_3:0.45 Sb_2S_3 are analysed on the basis of a theory developed for the anomalous carrier drift, controlled by capture in traps distributed in energy. For the case of an exponential trap distribution the dependence of the activation energy ε_a on the field E and the thickness L is calculated as well as the temperature dependence of the dispersion parameter $\alpha(T)$. A comparison of the theoretical dependence and experimental ε_a versus E/L and $1/\alpha$ versus $1/T$ curves obtained for 0.55 As_2S_3:0.45 Sb_2S_3 allows to draw conclusions about the magnitudes of some parameters of this material.

Результаты измерения характеристик транспорта заряда в стеклообразном 0,55 As_2S_3:0,45 Sb_2S_3 анализируются на основе развитой в работе теории аномального дрейфа носителей, контролируемого захватом на распределенные по энергии ловушки. Для случая экспоненциального распределения ловушек по энергии рассчитаны зависимость активационной энергии ε_a от поля E и толщины L, а также зависимость дисперсионного параметра α от температуры T. Теоретические зависимости сравниваются с экспериментальными кривыми ε_a от E/L и $1/\alpha$ от $1/T$, полученными на 0.55 As_2S_3:0,45 Sb_2S_3, что позволяет сделать выводы о величинах некоторых параметров данного материала.

1. Introduction

The time-of-flight method of drift mobility measurements [1] is up to now one of the few fruitful tools for charge transport investigations in semi-insulating amorphous materials. In this method a sample of thickness L is placed between two contacts to build up in it an electric field E. At the moment $t = 0$ at one of the contacts (say, anode) a sheet of carriers (holes) is generated with the surface density σ. The subsequent movement of the carriers produces a transient current $j(t)$ in the circuit.

Until recently, it was thought that in application to the amorphous materials the time-of-flight method could give useful information only in that case, when in the transient current signal a quasi-stationary plateau region was observed. This plateau corresponds to the charge packet propagation under the condition of thermal equilibrium, established between the localized and conductive states. The mean of the packet moves with constant velocity and the packet width is small compared to the travelled distance. The exit of the packet out of the sample leads to a sharp current drop, which allows to determine the transit time t_T and to find the drift mobility $\mu = L/Et_T$. Under equilibrium conditions the drift mobility is independent of the thickness L and the possible field dependence of $\mu(E)$ is thoroughly determined by the field dependence of the mobility μ_0 in the conductive state. It also appears that the temperature dependence of the drift mobility is thermally activated with the activation energy ε_a independent of the field E and the thickness L.

[1]) 115409 Moscow, USSR.
[2]) 277028 Kishinev, USSR.

Recently a new method has been developed to obtain information about transport characteristics in the case, when featureless current signals are observed [2, 3] (so-called anomalous or dispersive transport). It has been shown that if the monotonously decaying time dependence of the current usual for this case is represented in a double logarithmic scale $\log j$ versus $\log t$, then two linear portions can be defined: the first portion $\log j \sim -(1 - \alpha) \log t$ and the second one $\log j \sim -(1 + \alpha) \log t$ with a parameter α lying in the interval $0 < \alpha < 1$. The point of intersection of the linear portions is taken as the transit time t_T. The mobility determined with this transit time shows an algebraic dependence on the field and the thickness

$$\mu \sim \left(\frac{E}{L}\right)^{1/\alpha - 1} \tag{1}$$

at fixed temperature, and the value of α in the dependence μ versus E/L is the same as the earlier introduced parameter α characterizing the shape of the j versus t curve.

Great attention has been paid recently to the study of the temperature dependence of the drift mobility [4 to 7]. It was found that in a wide temperature interval the drift mobility is thermally activated

$$\mu \sim \exp\left(-\frac{\varepsilon_a}{kT}\right) \tag{2}$$

(k is the Boltzmann constant) with a field- and thickness-dependent activation energy. Some interest has been expressed to the possible change of α with temperature. As it is seen from (1) and (2), the effect of E and L on the activation energy ε_a may be explained, if we assume that the dependence $1/\alpha$ versus $1/T$ is linear. For materials as lexan [5] and TNF-PVK [8] there actually is an experimental evidence showing this dependence to exist. At the same time in As_2Se_3 at temperatures from 20 to 110 °C no definite change in α with temperature has been observed [4].

The first theoretical treatment of the dispersive transport was based on the so-called stochastic model [2], in which the charge transport is described as a sequence of carrier hops between neighbouring localized sites. The disorder in site distribution results in a broad hopping time dispersion, which in this model is the origin of the anomalous transport. The main advantage of the stochastic model is the fact that in its frames the shape of the current signal was first theoretically obtained together with the dependence of the transit time upon the field and the thickness, corresponding to the anomalous transport. But as the stochastic model does not operate with the band structure concepts, the comparison of the results of this model with experimental data does not give direct information about the energy distribution of localized states in the gap. This model cannot offer a complete explanation of the temperature dependence of transport characteristics. Thus the question of the activation energy and its field and thickness dependence lies outside this model.

In [9 to 11] a model of charge transport controlled by capture in traps distributed in a wide (compared with kT) energy interval in the gap was advanced for the explanation of the anomalous behaviour. In [9] the exact analytical solution of the dynamic equations for the current and free carrier density was derived for the case that transport was controlled by traps uniformly distributed in energy down to a depth $\varepsilon_0 \gg kT$. It was shown that under certain conditions the current has a form typical for the anomalous transport. In [10, 11] it was illustrated by numerical integration that the dispersive current shape could be obtained by approximating the trap energy distribution by several discrete levels lying in the energy interval $\Delta\varepsilon \gg kT$.

The analytical solution derived in [9] allowed to obtain in a subsequent paper [12] an analytical expression for the mobility as a function of field, thickness, and temperature in the case of a uniform energy distribution of traps. The temperature dependence found in [12] may be described in terms of an activation energy ε_a (see (2)) which is determined by the formula

$$\varepsilon_a = \varepsilon_0 \left(1 - \frac{\mu_0 \tau E}{L}\right), \tag{3}$$

where τ is the lifetime of a carrier in the conductive state. Note that in the experimental work [5] the linear dependence of ε_a versus E has been observed for the polymer lexan.

In the present work a method is developed for solving the problem of charge transport controlled by traps, the energy distribution of which is specified by an arbitrary function $n(\varepsilon)$. This method is applicable to the "deep" trap distribution $\varepsilon_0 \gg kT$, where ε_0 is the characteristic energy of the distribution $n(\varepsilon)$. With the aid of this method the case of an exponential energy distribution of traps is considered. For this distribution we have obtained the dynamics of the current $j(t)$ and the field and thickness dependences of the transit time and the mobility typical for the anomalous behaviour. The question about the field and thickness dependences of the activation energy ε_a and temperature dependence of the dispersion parameter α will be also especially considered. The dependences of ε_a versus E/L and $1/\alpha$ versus $1/T$ obtained for the exponential energy trap distribution are compared with the experimental results for 0.55 As_2S_3:0.45 Sb_2S_3 to get information about the characteristics of this material. It is also discussed to which extent the exponential distribution of traps describes the real trap distribution in the gap of the material.

2. Theory

The charge transport controlled by capture in traps distributed in energy is described by the equations

$$\frac{\partial}{\partial t} p(x, t) + v_0 \frac{\partial}{\partial x} p_0(x, t) = 0 , \tag{4}$$

$$\frac{\partial}{\partial t} \varrho(x, t, \varepsilon) = c(\varepsilon) \left[n(\varepsilon) p_0(x, t) - N_0 e^{-\varepsilon/kT} \varrho(x, t, \varepsilon)\right] , \tag{5}$$

$$p(x, t) = p_0(x, t) + \int_0^\infty d\varepsilon \, \varrho(x, t, \varepsilon) , \tag{6}$$

where x is the coordinate ($0 \leqq x \leqq L$, the anode is at $x = 0$, the cathode is at $x = L$), p the total carrier density (say, holes), p_0 the free carrier density, the velocity v_0 is determined by the free carrier mobility μ_0 as $v_0 = \mu_0 E$, ε is the trap energy, $n(\varepsilon)\, d\varepsilon$ the density of traps in the energy interval from ε to $\varepsilon + d\varepsilon$, $\varrho(x, t, \varepsilon)\, d\varepsilon$ the density of carriers in these traps, the function $c(\varepsilon)$ characterizes the probability of carrier transitions between the band of free states and traps having the energy ε, N_0 is the density of free states, k the Boltzmann constant, and T the temperature.

Boundary and initial conditions corresponding to the time-of-flight experiment have the form

$$p_0(0, t) = \frac{\sigma}{\mu_0 E} \delta(t), \qquad p_0(x, 0) = 0, \qquad \varrho(x, 0, \varepsilon) = 0 . \tag{7}$$

The main experimentally measured quantity is the total current

$$j(t) = \frac{e\mu_0 E}{L} \int_0^L \mathrm{d}x \, p_0(x, t) \,, \tag{8}$$

where e is the charge of a carrier. Using (4) and (7), one may write the current determined by (8) in terms of the total charge density

$$j(t) = -\frac{e}{L} \frac{\mathrm{d}}{\mathrm{d}t} \int_0^L \mathrm{d}x (L - x) \, p(x, t) \,. \tag{9}$$

An important characteristic of the energy trap distribution $n(\varepsilon)$ is the effective depth of this distribution

$$\varepsilon_0 = \frac{\int_0^\infty \mathrm{d}\varepsilon \, \varepsilon n(\varepsilon)}{\int_0^\infty \mathrm{d}\varepsilon \, n(\varepsilon)} \,. \tag{10}$$

In the present work we consider the case of a deep distribution $\varepsilon_0 \gg kT$, which is of practical interest. Under this condition the energy interval, in which the traps are distributed, may be divided into two characteristic regions: (i) The region of "shallow" traps is placed between the free state band edge and some boundary energy $\varepsilon_*(t)$. In this region the thermal equilibrium between the traps and the free state band has been established at a moment t. Thus in this region one can neglect the derivative $\partial\varrho/\partial t$ on the left side of (5). Then the solution of this equation takes the form

$$\varrho(x, t, \varepsilon) = \frac{1}{N_0} n(\varepsilon) \exp\left(\frac{\varepsilon}{kT}\right) p_0(x, t) ; \qquad \varepsilon < \varepsilon_*(t) \,. \tag{11}$$

(ii) The region of "deep" traps is located below the boundary energy $\varepsilon_*(t)$. At the moment t the carrier release from these traps has not yet begun. The carrier density in these traps is changed only due to capture. Thus in this region one can neglect the second term on the right side of (5) describing the release of carriers. The resulting solution of (5) has the form

$$\varrho(x, t, \varepsilon) = c(\varepsilon) \, n(\varepsilon) \int_0^t \mathrm{d}t' \, p_0(x, t') ; \qquad \varepsilon > \varepsilon_*(t) \,. \tag{12}$$

The boundary energy $\varepsilon_*(t)$ dividing the traps into two groups is determined by the condition

$$\left[N_0 c(\varepsilon_*) \exp\left(-\frac{\varepsilon_*}{kT}\right)\right]^{-1} = t \,. \tag{13}$$

Note that the quantity $[N_0 c(\varepsilon) \exp(-\varepsilon/kT)]^{-1}$ determines the release time for the traps with energy ε. Then condition (13) means that the time t exceeds the release time for traps lying above $\varepsilon_*(\varepsilon < \varepsilon_*)$. Thus a large number of capture and release events, in which these traps are involved, has occurred up to the moment t and caused the thermal equilibrium to establish (see (11)). In its turn, as it follows from (13), the release time for the deep traps, lying below the energy $\varepsilon_*(\varepsilon > \varepsilon_*)$, exceeds the time t. Thus it is unlikely for the carriers trapped in these levels to escape until the moment t (see (12)).

The boundary between "shallow" and "deep" traps is not abrupt. It is a layer of width $\Delta\varepsilon \approx kT$ centred at the energy $\varepsilon_*(t)$. This broadening of the boundary is not important for the considered deep trap distribution $\varepsilon_0 \gg kT$ (as it will be shown further, this approximation gives good results even for $\varepsilon_0 \approx 2kT$). Note that the procedure of dividing the traps into two groups is valid for times greater than the release time for the uppermost levels $t > 1/N_0c(0)$. But this limitation is quite unimportant as the time $1/N_0c(0)$ is many orders of magnitude less than all the characteristic charge transport times.

The solutions (11) and (12) allow to write the total charge density (6) as follows:

$$p(x, t) = \mu_0 E\left[\frac{1}{v(t)} p_0(x, t) + \frac{\lambda(t)}{v(t)} P_0(x, t)\right], \tag{14}$$

where we introduced the designations

$$P_0(x, t) = \int_0^t dt' \, p_0(x, t') , \tag{15}$$

$$v(t) = \mu_0 E\left[1 + \frac{1}{N_0}\int_0^{\varepsilon_*(t)} d\varepsilon n(\varepsilon) \exp\left(\frac{\varepsilon}{kT}\right)\right]^{-1}, \tag{16}$$

$$\lambda(t) = \frac{v(t)}{\mu_0 E}\int_{\varepsilon_*(t)}^{\infty} d\varepsilon \, n(\varepsilon) \, c(\varepsilon) . \tag{17}$$

Using the definition (15), we can write (4) in the following way:

$$\frac{\partial}{\partial t} p(x, t) + \mu_0 E \frac{\partial^2}{\partial x \, \partial t} P_0(x, t) = 0 \tag{18}$$

which with the initial condition (7) allows to find the connection between $p(x, t)$ and $P_0(x, t)$

$$p(x, t) = -\mu_0 E \frac{\partial}{\partial x} P_0(x, t) . \tag{19}$$

Substituting the function $P_0(x, t)$ found from (14) into (19) and using (4), we obtain an equation for the carrier density $p(x, t)$

$$\frac{\partial}{\partial t} p(x, t) + v(t) \frac{\partial}{\partial x} p(x, t) = -\lambda(t) \, p(x, t) \tag{20}$$

with the initial condition (7) and the boundary condition

$$p(0, t) = \frac{\sigma\lambda(t)}{v(t)} + \sigma \, \delta(t) \tag{21}$$

which results from (7), (14), and (15).

Equation (20) with the conditions (7) and (21) describes a spreading charge packet propagating with variable speed. Note that this equation conserves the full number of carriers in the sample until the exit of the packet: using (20) and (21) we obtain

$$\int_0^\infty dx \, p(x, t) = \sigma .$$

Equation (20) is the basic equation of the dispersive transport.

3. Results and Discussion

In this section we apply the above method to the analysis of the results obtained by the experimental study of the hole transport in vitreous 0.55 As_2S_3 : 0.45 Sb_2S_3. The time-of-flight experiment was performed with thin (from 0.6 to 6.0 μm) film samples prepared by vacuum evaporation and provided with aluminium electrodes. The transient current was excited by a short (1.5 μs) highly absorbed light pulse. The data were obtained over a range of temperatures (20 to 130 °C) and applied electric fields (8.5×10^4 to 8.5×10^5 V/cm).

The experiments have shown that the hole transport in this material manifests all the features of the anomalous behaviour [7]. According to the above described p rocedure of data representation in the case of dispersive transport, the transient time t_T was determined as the moment of intersection of the two transient current asymptotes: $j \sim t^{-(1-\alpha)}$ for $t < t_T$ and $j \sim t^{-(1+\alpha)}$ for $t > t_T$. The results of the transient time t_T measurements at various temperatures T, fields E, and thicknesses L, presented in Fig. 1, allowed to obtain the experimental dependence of ε_a versus E/L, which is shown in Fig. 2. The experimental variation of the dispersion parameter α with temperature was also determined (Fig. 3). The value of α was independently found both from the slopes of the linear portions of $\log j \sim -(1 \pm \alpha) \log t$ and from the dependence of $\log t_T \sim (1/\alpha) \log (L/E)$. The results of both methods are in good agreement.

To analyse the experimental results in the frame of the dispersive transport theory formulated in Section 2, we use the model of an exponential trap energy distribution

$$n(\varepsilon) = n_0 \exp\left(-\frac{\varepsilon}{\varepsilon_0}\right). \tag{22}$$

To analyse the carrier kinetics we have, besides the function $n(\varepsilon)$, also to specify the function $c(\varepsilon)$. For the first step of consideration we assume that this function is weakly energy-dependent

$$c(\varepsilon) = c_0 \,. \tag{23}$$

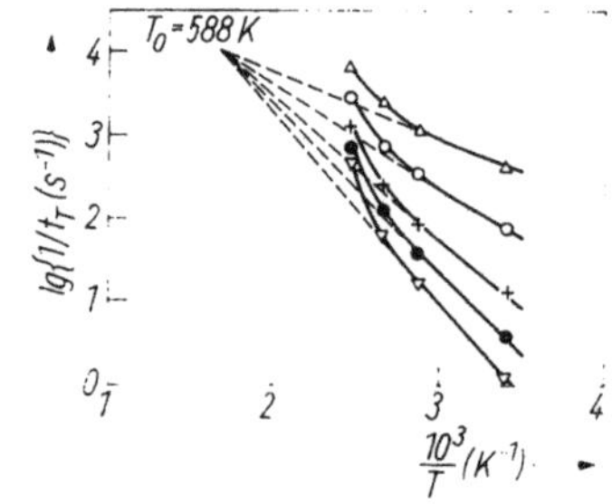

Fig. 1

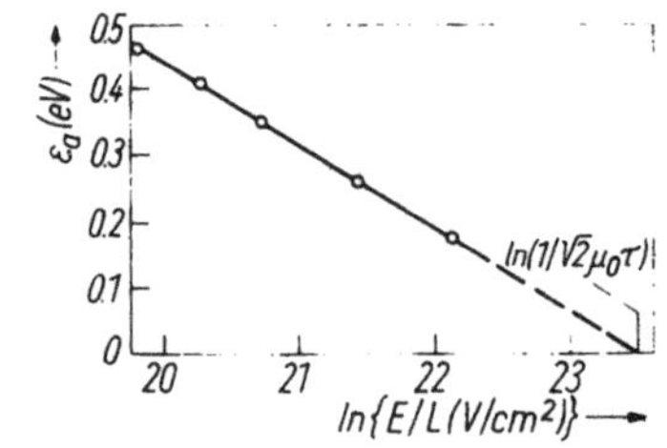

Fig. 2

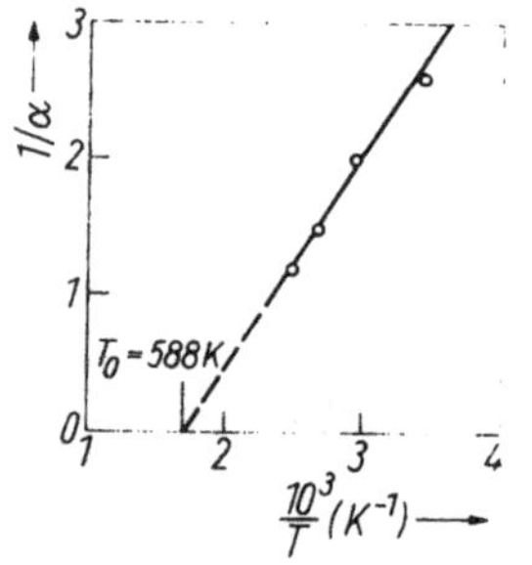

Fig. 3

Fig. 1. Semilogarithmic plot of the reverse transit time $1/t_T$ vs. $1/T$ for various values of log [(E/L) (V/cm²)]. △ 9.6, ○ 9.3, + 9.0, ● 8.8, ▽ 8.6

Fig. 2. Semilogarithmic plot of the drift mobility activation energy ε_a vs. E/L. The theoretical dependence (47) is shown by a straight line

Fig. 3. Temperature dependence of the dispersion parameter α. The theoretical dependence (44) is shown by a straight line

Substituting (22) and (23) into (16) and (17) we have

$$v(t) = \mu_0 E \frac{N_0}{n_0 kT} \left(\frac{N_0}{n_0 \varepsilon_0} \frac{t}{\tau}\right)^{-(1-kT/\varepsilon_0)}, \tag{24}$$

$$\lambda(t) = \frac{1}{\tau} \frac{N_0}{n_0 kT} \left(\frac{N_0}{n_0 \varepsilon_0} \frac{t}{\tau}\right)^{-1}, \tag{25}$$

where τ is the free carrier lifetime

$$\frac{1}{\tau} = \int_0^\infty \mathrm{d}\varepsilon\, c(\varepsilon)\, n(\varepsilon) = c_0 n_0 \varepsilon_0 \,. \tag{26}$$

To find the time-dependent space distribution of carriers $p(x, t)$ we solve (20) under the conditions (7) and (21), in which the functions $v(t)$ and $\lambda(t)$ are determined by (24) and (25)

$$p(x, t) = \frac{\sigma}{\mu_0 \tau E} \left(\frac{N_0}{n_0 \varepsilon_0} \frac{t}{\tau}\right)^{-kT/\varepsilon_0} \left[1 - \frac{x}{\mu_0 \tau E} \left(\frac{kT}{\varepsilon_0}\right)^2 \left(\frac{N_0}{n_0 \varepsilon_0} \frac{t}{\tau}\right)^{-kT/\varepsilon_0}\right]^{(\varepsilon_0/kT)^2 - 1};$$

$$0 \leqq x \leqq \mu_0 \tau E \left(\frac{kT}{\varepsilon_0}\right)^2 \left(\frac{N_0}{n_0 \varepsilon_0} \frac{t}{\tau}\right)^{kT/\varepsilon_0}, \tag{27}$$

$$p(x, t) = 0\,; \qquad \mu_0 \tau E \left(\frac{kT}{\varepsilon_0}\right)^2 \left(\frac{N_0}{n_0 \varepsilon_0} \frac{t}{\tau}\right)^{kT/\varepsilon_0} < x < \infty\,. \tag{28}$$

Equations (27), (28) describe a non-Gaussian carrier packet, propagating in the sample. Using the results (27), (28) we find the law of motion of the mean $\bar{x}(t)$ of the packet and its dispersion $\Delta x(t)$

$$\bar{x}(t) = \int_0^\infty \mathrm{d}x\, x p(x, t) = \mu_0 \tau E \left(\frac{N_0}{n_0 \varepsilon_0} \frac{t}{\tau}\right)^{kT/\varepsilon_0}, \tag{29}$$

$$\Delta x(t) = \left\{\int_0^\infty \mathrm{d}x\, [x - \bar{x}(t)]^2\, p(x, t)\right\}^{1/2} = \mu_0 \tau E \left(\frac{N_0}{n_0 \varepsilon_0} \frac{t}{\tau}\right)^{kT/\varepsilon_0}. \tag{30}$$

Note the following characteristic peculiarities of the packet (27), (28), which distinguish it from the "normal" Gaussian packet: (i) The maximum of the packet $p(x, t)$ is immobile and is located close to the anode $x = 0$. (ii) The mean of the packet $\bar{x}(t)$ moves non-uniformly with decreasing speed. (iii) The ratio of the dispersion to the mean position is time-independent and close to unity, whilst in the case of a Gaussian packet this ratio drops rapidly with time, proportional to the value $(\tau/t)^{1/2}$ and in real cases is much less than unity. Thus the non-equilibrium carrier packet (27), (28) is much more spread than the Gaussian packet. This leads to a wide dispersion of transit time and mobility and enables us to identify the transport described here as dispersive transport.

Substituting the obtained carrier density $p(x, t)$ in (9), we find the transient current for the model of the exponential energy trap distribution

$$j(t) = \frac{e\sigma\mu_0 E}{L} \frac{N_0 kT}{n_0 \varepsilon_0^2} \left(\frac{N_0}{n_0 \varepsilon_0} \frac{t}{\tau}\right)^{-(1-kT/\varepsilon_0)}; \qquad t \ll t_\mathrm{T}\,, \tag{31}$$

$$j(t) = \frac{e\sigma L}{\mu_0 E \tau^2} \frac{N_0 kT}{2 n_0 \varepsilon_0^2} \left(\frac{N_0}{n_0 \varepsilon_0} \frac{t}{\tau}\right)^{-(1+kT/\varepsilon_0)}; \qquad t \gg t_\mathrm{T}\,. \tag{32}$$

It is seen from (31) and (32) that the transient current $j(t)$ has two characteristic portions of decay: $j \sim t^{-(1-\alpha)}$ and $j \sim t^{-(1+\alpha)}$. Thus the shape of the current, calculated theoretically for the model of the exponential trap distribution, is consistent with the experimentally observed dependence of j versus t in the regime of the anomalous behaviour [3 to 8]. The dispersion parameter α turns out to be connected to the characteristic energy ε_0 of the distribution by

$$\alpha = \frac{kT}{\varepsilon_0}. \tag{33}$$

Following the experimental procedure of transit time t_T determination, we find it as the moment of intersection of the two transient current asymptotes

$$t_T = \tau \frac{n_0 \varepsilon_0}{N_0} \left(\frac{L}{\sqrt{2}\, \mu_0 \tau E} \right)^{\varepsilon_0/kT}. \tag{34}$$

The mobility corresponding to this transit time has to be written as

$$\mu = \sqrt{2}\, \mu_0 \frac{N_0}{n_0 \varepsilon_0} \left(\frac{\sqrt{2}\, \mu_0 \tau E}{L} \right)^{\varepsilon_0/kT - 1}. \tag{35}$$

Note that the transit time t_T corresponds to the exit of the mean of the packet out of the sample $\bar{x}(t_T) \approx L$.

We wish to note the following characteristic peculiarities of the transit time (34) and the drift mobility (35):

(i) The transit time is non-linearly dependent on the ratio E/L: $t_T \sim (L/E)^{\varepsilon_0/kT}$ and hence the drift mobility depends on the field and the thickness $\mu \sim (E/L)^{\varepsilon_0/kT - 1}$. This kind of field and thickness dependence of the mobility was observed at not too high fields for a-Se [3], As_2Se_3 [4], lexan [5], and $As_2S_3{:}Sb_2S_3$ [7].

(ii) The activation energy of the drift mobility ε_a is non-linearly dependent on the field and the thickness

$$\varepsilon_a = \varepsilon_0 \ln \left(\frac{L}{\sqrt{2}\, \mu_0 \tau E} \right). \tag{36}$$

In [4] the experimental results of the study of the ε_a versus E/L dependence were interpreted by the formula

$$\varepsilon_a = \varepsilon_a^{(0)} \ln \left(\frac{L}{L_0} \frac{E}{E_0} \right). \tag{37}$$

Formulae (36) and (37) are completely consistent if one defines the intercalating parameters $\varepsilon_a^{(0)}$, E_0, and L_0 in terms of the exponential trap distribution model in the following way: $\varepsilon_a^{(0)} = \varepsilon_0$, $E_0/L_0 = 1/\sqrt{2}\, \mu_0 \tau$.

As it follows from (36), the experimentally observed activation energy ε_a may considerably exceed the characteristic depth ε_0 of the trap distribution. For example, if we take for μ_0 and τ the values $\mu_0 = 1\ cm^2/Vs$, $\tau = 10^{-10}$ s, then for the field $E = 10^4$ V/cm and the thickness $L = 50\ \mu m$ we obtain $\varepsilon_a = 8.2\varepsilon_0$.

The above results show that there is qualitative agreement between the theoretical dependences obtained in the frame of the exponential trap distribution model and the experimental data for several amorphous materials. However, a quantitative comparison of the theoretical and experimental results requires some modification of the considered model. The dependence $1/\alpha$ versus $1/T$ obtained for $As_2S_3{:}Sb_2S_3$ (see Fig. 3) can be described by the formula

$$\frac{1}{\alpha} = \varepsilon_0' \left(\frac{1}{kT} - \frac{1}{kT_0} \right) \tag{38}$$

which disagrees with (33). Furthermore, according to (34), the dependence log $(1/t_T)$ versus $1/T$ is a straight line, the slope of which depends on the value E/L. The extrapolations of these lines for different E/L intersect at a single point at $1/T = 0$. As it is seen from Fig. 1, the experimental results at not too high temperatures are described by linear dependences of log $1/t_T$ versus $(1/T)$. But the extrapolations of these lines intersect at $1/T = 1/T_0$ with the value of T_0 being the same as in (38).

These discrepancies may be removed, if in the theoretical model of the exponential trap distribution we take into account the possible dependence of the carrier capture probability on the trap energy. We assume that the capture probability $c(\varepsilon)$ increases with energy at least in the interval $0 \leqq \varepsilon \lessapprox \varepsilon_a$. This increase is approximated by

$$c(\varepsilon) = c_0 \exp\left(\frac{\varepsilon}{kT_0}\right), \tag{39}$$

where kT_0 is the characteristic energy of the $c(\varepsilon)$ variation and the proper numerical value of the parameter T_0 is chosen using the experimental dependence (38). The substitution of the functions (22) and (39) in (5) gives

$$\frac{\partial}{\partial t}\varrho(x, t, \varepsilon) = c_0\left[n_0 \exp\left(-\frac{\varepsilon}{\varepsilon_{\text{eff}}}\right) p_0(x, t) - N_0 \exp\left(-\frac{\varepsilon}{kT_{\text{eff}}}\right) \varrho(x, t, \varepsilon)\right], \tag{40}$$

where

$$\frac{1}{\varepsilon_{\text{eff}}} = \frac{1}{\varepsilon_0} - \frac{1}{kT_0}, \tag{41}$$

$$\frac{1}{kT_{\text{eff}}} = \frac{1}{kT} - \frac{1}{kT_0}. \tag{42}$$

Note that in the case considered above, when $n(\varepsilon)$ and $c(\varepsilon)$ are determined by (22) and (23), equation (5) has the form

$$\frac{\partial}{\partial t}\varrho(x, t, \varepsilon) = c_0\left[n_0 \exp\left(-\frac{\varepsilon}{\varepsilon_0}\right) p_0(x, t) - N_0 \exp\left(-\frac{\varepsilon}{kT}\right) \varrho(x, t, \varepsilon)\right]. \tag{43}$$

Equations (4), (6) to (8) are independent of the form of the functions $c(\varepsilon)$ and $n(\varepsilon)$. Thus accounting for the energy dependence of the capture probability reduces to the following replacement of parameters: ε_0 by ε_{eff} and T by T_{eff} in all the above results for the model defined by (22) and (23). Particularly, the formulae for the dispersion parameter α, the transit time t_T, and the mobility μ will be written as

$$\frac{1}{\alpha} = \varepsilon_{\text{eff}}\left(\frac{1}{kT} - \frac{1}{kT_0}\right), \tag{44}$$

$$t_T = \tau \frac{n_0 \varepsilon_{\text{eff}}}{N_0}\left(\frac{L}{\sqrt{2}\,\mu_0 \tau E}\right)^{\varepsilon_{\text{eff}}(1/kT - 1/kT_0)}, \tag{45}$$

$$\mu = \sqrt{2}\,\mu_0 \frac{N_0}{n_0 \varepsilon_{\text{eff}}}\left(\frac{\sqrt{2}\,\mu_0 \tau E}{L}\right)^{\varepsilon_{\text{eff}}(1/kT - 1/kT_0) - 1}. \tag{46}$$

From (45) and (46) one can derive the formula for the activation energy

$$\varepsilon_a = \varepsilon_{\text{eff}} \ln\left(\frac{L}{\sqrt{2}\,\mu_0 \tau E}\right). \tag{47}$$

The obtained theoretical formula (44) completely agrees with the experimental dependence (38), and the experimentally determined parameter ε_0' gives the value of ε_{eff}. This agreement is illustrated in Fig. 3: the experimental points obtained for

$As_2S_3:Sb_2S_3$ fit well the theoretical dependence with the parameters $T_0 = 588$ K and $\varepsilon_{eff} = 0.136$ eV. This fact allows to find the depth of the trap distribution $n(\varepsilon)$ and the characteristic energy of the capture probability variation $c(\varepsilon)$ for this material: $\varepsilon_0 = 0.037$ eV, $kT_0 = 0.051$ eV. Note, that the same value of T_0 may be found independently from the data presented in Fig. 1. The extrapolations of the straight lines $\log (1/t_T)$ versus $1/T$ intersect at the point $1/T = 1/T_0$ — note the theoretical formula (45), which describes well the experimental data of Fig. 1 at not too high temperatures. For a more detailed comparison of theory and experiment the dependence of the activation energy ε_a upon E/L was obtained from the data presented in Fig. 1. The experimental points found in this way fall on the theoretical curve described by (47) (see Fig. 2). In this figure (47) is presented together with the experimental points in the scale ε_a versus $\ln (E/L)$. In this scale the slope of the straight line defines ε_{eff} and the point of intersection of this line and the abscissa gives the value of $\ln \left(1/\sqrt{2}\, \mu_0\tau\right)$. The obtained value of ε_{eff} is coincident with the ε_{eff}-value found independently from the slope of the line $1/\alpha$ versus $1/T$. For the value of $\mu_0\tau$ we obtain $\mu_0\tau = 4.4 \times 10^{-11}$ cm²/V.

Note that a temperature dependence of the dispersion parameter α analogous to the dependence (38) has been observed, for example, in such a material as lexan [5]. At the same time for vitreous As_2Se_3 one observes the following temperature dependence of the dispersion parameter α [4]:

$$\frac{1}{\alpha} = \varepsilon_0' \left(\frac{1}{kT} + \frac{1}{kT_0'}\right). \tag{48}$$

Formula (48) differs from (38) by the sign of the second term in the right-hand side. The dependence (48) can be explained under the assumption that the capture probability $c(\varepsilon)$ decreases in energy according to the formula

$$c(\varepsilon) = c_0' \exp\left(-\frac{\varepsilon}{kT_0'}\right). \tag{49}$$

In this case all the results for the exponential trap distribution model are valid after replacing ε_0 by ε_{eff}' and T by T_{eff}', where the values of ε_{eff}' and T_{eff}' are determined by

$$\frac{1}{\varepsilon_{eff}'} = \frac{1}{\varepsilon_0} + \frac{1}{kT_0'}, \tag{50}$$

$$\frac{1}{T_{eff}'} = \frac{1}{T} + \frac{1}{T_0'}. \tag{51}$$

It appears that the general behaviour of the function $c(\varepsilon)$ is the following: In the initial energy interval $0 < \varepsilon < \varepsilon_1$ the function $c(\varepsilon)$ increases. Then an interval $\varepsilon_1 < \varepsilon < \varepsilon_2$ is possible, where $c(\varepsilon)$ is weakly energy dependent. Finally, beginning from a characteristic energy ε_2, the function $c(\varepsilon)$ decreases with increasing energy ε. An example of a function exhibiting this kind of behaviour is the following model function:

$$\left.\begin{aligned} c(\varepsilon) &= c_0 \exp\left(\frac{\varepsilon}{kT_0}\right); && 0 \leqq \varepsilon \leqq \varepsilon_1 ,\\ c(\varepsilon) &= c_0' \exp\left(-\frac{\varepsilon}{kT_0'}\right); && \varepsilon_2 \leqq \varepsilon < \infty ,\\ c(\varepsilon) &= c_0''; && \varepsilon_1 < \varepsilon < \varepsilon_2 . \end{aligned}\right\} \tag{52}$$

In the case $\varepsilon_1 > \varepsilon_a$ only the first section of the function $c(\varepsilon)$ affects the charge transport. Under this condition formula (39) is a good approximation for the function $c(\varepsilon)$.

If the conditions $\varepsilon_1 \ll \varepsilon_a$, $\varepsilon_2 > \varepsilon_a$ are valid, then the function $c(\varepsilon)$ may be considered as a constant (see (23)). Finally, in the case $\varepsilon_2 \ll \varepsilon_a$ the third interval dominates and thus the function $c(\varepsilon)$ may be approximated by (49).

All the presented considereations allow to conclude that the model of the exponential trap distribution with account of the energy dependence of the capture probability gives a satisfactory description of the experimental data for vitreous As_2S_3:Sb_2S_3 in the considered interval of field, thickness, and temperature variations. At the same time there is some experimental evidence that several amorphous materials are outside the frames of the model at hand. In [8], a discrepancy was detected between the dispersion parameter α found from the initial portion of the current decay $\log j \sim -(1 - \alpha_i) \log t$ and the final stage $\log j \sim -(1 + \alpha_f) \log t$: $\alpha_i \neq \alpha_f$. It was also found that the values α_i and α_f show different experimental temperature dependences. Besides, for some of the materials (see, for example, [5] for lexan and [6] for TNF-PVK) it was detected that at high fields the dependence of μ versus E/L becomes exponential rather than algebraic. This suggests that it is advantageous to consider in terms of the general method developed in Section 2 some other kinds of trap distribution $n(\varepsilon)$. The authors are planning to do this in subsequent papers.

Acknowledgements

The authors are indebted to Prof. B. T. Kolomiets, Prof. A. M. Andriesh, and Dr. E. A. Lebedev for stimulating discussions.

References

[1] W. E. Spear, J. non-crystall. Solids **1**, 197 (1969).
[2] H. Scher and E. W. Montroll, Phys. Rev. B **12**, 2455 (1975).
[3] G. Pfister, Phys. Rev. Letters **36**, 271 (1976).
[4] G. Pfister and H. Scher, Phys. Rev. B **15**, 2062 (1977).
[5] G. Pfister, Phys. Rev. B **16**, 3676 (1977).
[6] W. D. Gill, in: Photoconductivity and Related Phenomena, Ed. J. Mort and D. M. Pai, Elsevier, Publ. Co., New York 1976 (p. 303).
[7] S. D. Shutov and M. S. Iovu, phys. stat. sol. (a) **50**, K59 (1978).
[8] G. Pfister and C. H. Griffiths, Phys. Rev. Letters **40**, 659 (1978).
[9] V. I. Arkhipov and A. I. Rudenko, Phys. Letters A **61**, 55 (1977).
[10] F. W. Schmidlin, Phys. Rev. B **16**, 2362 (1977).
[11] J. Noolandi, Phys. Rev. B **16**, 4474 (1977).
[12] A. I. Rudenko and V. I. Arkhipov, J. non-crystall. Solids **30**, 163 (1978).

(Received March 2, 1979)

phys. stat. sol. (a) 54, 79 (1979)

Subject classification: 20; 21; 21.4

Institute of Physics, Polish Academy of Sciences, Warsaw

Soft X-Ray Emission Spectra of Aluminium and Manganese from $Gd(Al_{1-x}Mn_x)_2$ Intermetallic Compounds

By

A. ŚLEBARSKI[1]), K. ŁAWNICZAK, and J. AULEYTNER

X-ray emission spectra of manganese, aluminium, and gadolinium in $Gd(Al_{1-x}Mn_x)_2$ intermetallic compounds are investigated. Changes in the position and intensity of the Mn and Al L spectra and Gd M_ζ spectra are observed. The discussion of the obtained results leads to the following conclusions: 1. manganese is a donor of the 3d electrons in the investigated compounds in agreement with the results of magnetic measurements, 2. 3d electrons of manganese are mainly concentrated in the vicinity of gadolinium atoms, 3. the gadolinium–manganese bond has an ionic character, 4. the presence of the manganese atoms does not change the character of the aluminium–gadolinium bond.

Röntgenemissionsspektren von Mangan, Aluminium und Gadolinium in den intermetallischen Verbindungen $Gd(Al_{1-x}M_x)_2$ werden untersucht. Es werden Änderungen der Lage und der Intensität der L-Spektren von Mangan und Aluminium und M_ζ-Spektren von Gadolinium festgestellt. Die Diskussion der Resultate führt zu den folgenden Schlüssen: 1. Mangan verhält sich in den untersuchten Verbindungen wie ein Donator von 3d-Elektronen, 2. 3d-Elektronen von Mangan sind hauptsächlich in der Nähe der Gd-Atome lokalisiert, 3. die Gadolinium–Mangan-Bindungen haben Ionen-Charakter, 4. Mangan verursacht keine Änderung in den Aluminium–Gadolinium Bindungen.

1. Introduction

The Laves phases comprise a large group of intermetallic compounds. The main factors which influence the character of these phases are the ratio of atomic radii and the concentration of free electrons [1]. These phases may have one of the three following structure types: $MgCu_2$ (C15), $MgZn_2$ (C14), and $MgNi_2$ (C36). The RAl_2 and RT_2 compounds (R and T denote the rare-earth element and the transition metal, respectively) with a cubic structure of the $MgCu_2$ type have very interesting magnetic and structural properties. The ternary $R(T, Al)_2$ intermetallic alloys also represent one type of the Laves phases. In general, the structure of the $R(T, Al)_2$ system changes with the increase of the free electron number from the C15 type to again the C15 type through the C14 type. Only the RMn_2–RAl_2 systems with R = Ho, Tb, Dy, Gd [2] have the C15 structure in the whole concentration range. Detailed information about magnetic and structural properties of the $GdMn_2$–$GdAl_2$ series is reported in the literature [3 to 6]. The following conclusions result from those studies on $Gd(Al_{1-x}Mn_x)_2$ compounds:

1. In the case of large concentration of $GdAl_2$ the magnetic moment per unit cell is practically equal to zero.

2. The effective magnetic moment increases with the increase of the $GdMn_2$ concentration, which indicates the decrease of the 3d electron concentration in manganese.

The change of the Mn valency and thus the magnetic properties in the investigated alloys should be connected with the electronic band structure of these compounds.

---

[1]) Present address: Institute of Physics, Silesian University, Katowice, Poland.

Very important information on the electronic band structure can be deduced from the X-ray emission spectra. The object of this work was to check if the conclusions formulated on the basis of the magnetic and structural investigations will be confirmed by the X-ray emission spectra. The features of L_l, $L_{\alpha_{1,2}}(L_{III})$ emission lines of manganese, the $L_{2,3}$ emission line of aluminium, and the M_ξ emission lines of gadolinium were investigated.

2. Experiment

The polycrystalline metallic $Gd(Al_{1-x}Mn_x)_2$ alloys were prepared by arc melting in argon atmosphere. The purities of the Gd, Al, and Mn were 99.5, 99.999, and 99.99%, respectively. The X-ray analysis of the compounds obtained did not reveal the presence of any foreign phases.

The emission bands were investigated with the grating spectrometer RSM-500 [7]. Two different gratings were used, each of them with 600 lines/mm. A 6 m grating and slits of 10 μm were employed to measure the manganese L and gadolinium M_ξ emission lines. The resolution in this region was about 1 eV.A 2 m grating was used for investigation of the aluminium $L_{2,3}$ lines. The slits were 100 μm wide and

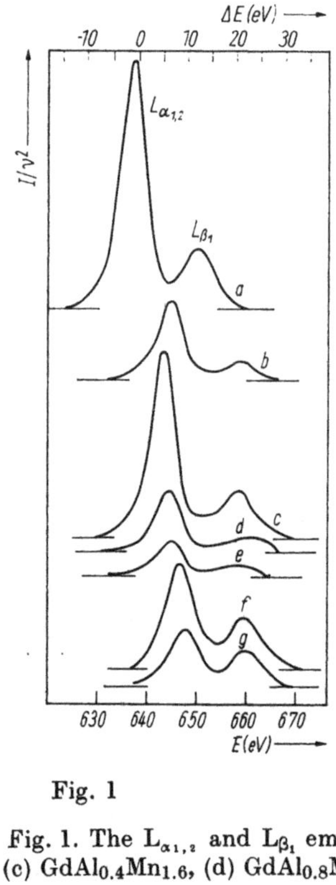

Fig. 1

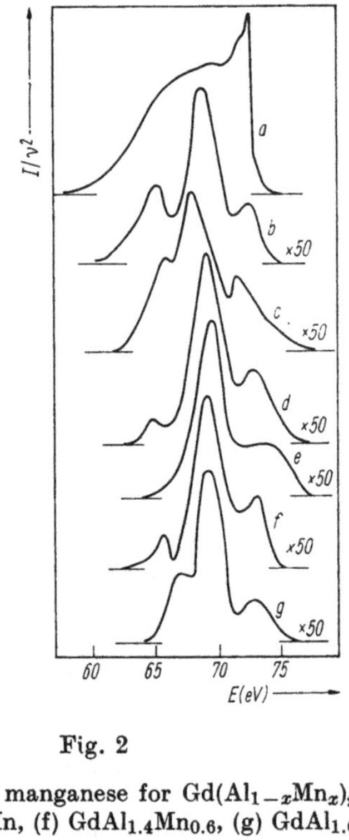

Fig. 2

Fig. 1. The $L_{\alpha_{1,2}}$ and L_{β_1} emission bands of manganese for $Gd(Al_{1-x}Mn_x)_2$. (a) Mn, (b) $GdMn_2$ (c) $GdAl_{0.4}Mn_{1.6}$, (d) $GdAl_{0.8}Mn_{1.2}$, (e) GdAlMn, (f) $GdAl_{1.4}Mn_{0.6}$, (g) $GdAl_{1.6}Mn_{0.4}$

Fig. 2. The $L_{2,3}$ emission bands of aluminium for $Gd(Al_{1-x}Mn_x)_2$. Intensities of the $L_{2,3}$ lines in the investigated samples were 50 times smaller than the intensity of the $L_{2,3}$ line in pure aluminium. (a) Al, (b) $GdAl_2$, (c) $GdAl_{1.6}Mn_{0.4}$, (d) $GdAl_{1.4}Mn_{0.6}$, (e) GdAlMn, (f) $GdAl_{0.8}Mn_{1.2}$, (g) $GdAl_{0.4}Mn_{1.6}$

the resolution was about 0.6 eV. The X-ray photons were registered using an electron photomultiplier. The compounds under investigation were placed on a water-cooled copper anode. The soft X-rays were excited by electron bombardment of the target material. The acceleration voltage was 5 kV and the current 12 mA. The normal operating pressure was 10^{-6} Torr. Measurements of the emission bands for each compound were repeated several times. A very good reproducibility of energy positions, full widths at half maximum, and intensities was observed in the case of manganese and gadolinium lines. The measurements of these quantities in the case of Al $L_{2,3}$ emission bands were less reproducible due to small signal-to-noise ratio.

3. Results

The investigated spectra are shown in Fig. 1 and 2. Distinct changes of intensity and energetic positions of manganese $L_{\alpha_{1,2}}$ and L_{β_1} bands were observed, depending on the concentration of this element. Considerable positive shifts in energetic scale of about 7 eV with respect to pure manganese, were confirmed in the case of the $L_{\alpha_{1,2}}$ and L_{β_1} bands. This shift increased with the decrease of the Mn concentration in the investigated $Gd(Al_{1-x}Mn_x)_2$ compounds. Also, the width at half maximum intensity of $L_{\alpha_{1,2}}$ and L_{β_1} bands increased as compared with pure manganese. The Al $L_{2,3}$ spectra presented in Fig. 2 showed considerable change of the shape in comparison with pure aluminium. The $L_{2,3}$ band shows three characteristic peaks. The intensity of the high-energy part of the spectrum decreased rapidly. On the other hand, two peaks in the low-energy range were observed. The main change of the shape of the Al $L_{2,3}$ band was observed already in the $GdAl_2$ compound. In the case of $Gd(Al_{1-x}Mn_x)_2$ compounds the general character of these spectra does not change with increasing manganese concentration (see Fig. 2). Only small changes in the intensity of the three maxima are observed. These spectra were less reproducible which makes them difficult to interpret.

4. Discussion

We shall try to interpret the experimental results in terms of the electronic structure of the investigated alloys. At first, we calculated the effect of alloying on the number of 3d electrons at the manganese atoms.

The integrated intensity of two X-ray lines involving the same inner level was measured. In the considered case the Mn $L_{\alpha_{1,2}}$ and Mn L_ι lines were chosen. These lines correspond to the transitions 3d(s) → 2p and 3s → 2p, respectively. In fact, the $L_{\alpha_{1,2}}$ band characterizes mainly the number of 3d electrons because the 3d → 2p transition probability is about one order of magnitude greater than that of 4s → 2p and furthermore the number of 3d electrons is several times larger than that of 4s electrons [8]. On the other hand, the L_ι line appears as a measure of the number of Mn atoms ionized in the 2p level and thus, it can be used to normalize the intensity of the L_α band. In the case of this transition it is assumed that the wave functions of 3s and 2p levels (which are the core levels) are only weakly changed by alloying and that the population of these levels is unaffected [9]. With this assumption the ratio of intensities becomes

$$\frac{\{I(L_\alpha)/I(L_\iota)\}_{\text{alloy}}}{\{I(L_\alpha)/I(L_\iota)\}_{\text{Mn}}} = \frac{n_d(\text{alloy})}{n_d(\text{Mn})},$$

where n_d is the number of d-symmetry electrons (at the Mn atom) [10]. Thus, we measure the ratio of the number of electrons with d-symmetry inside the ion. The calculations were performed under the assumption that pure manganese has six 3d electrons. The results are summarized in Table 1.

Table 1

compound	change in the number of 3d electrons (Δn_d)	the number of Mn 3d electrons calculated from	
		soft X-ray spectroscopy	magnetic measurements [6]
$GdMn_2$	-4.14 ± 0.2	1.86	2
$GdAl_{0.4}Mn_{1.6}$	-1.56 ± 0.2	4.44	4
$GdAl_{0.8}Mn_{1.2}$	-4.56 ± 0.3	1.44	2
GdAlMn	-5.16 ± 0.2	0.84	3
$GdAl_{1.4}Mn_{0.6}$	-3.42 ± 0.2	2.58	0
$GdAl_{1.6}Mn_{0.4}$	-4.50 ± 0.2	1.50	0

In this table the numbers of 3d electrons per manganese atom calculated from magnetic measurements are also presented [6]. In the case of $GdAl_2$ with a low concentration of $GdMn_2$ a discrepancy of results occurs which is difficult to explain at the moment. The number of 3d manganese electrons was obtained from magnetic data assuming that the effective magnetic moment of gadolinium in the investigated compounds is equal to the effective magnetic moment of gadolinium observed in $GdAl_2$. This assumption is probably not correct because of the great change in the number of valence electrons observed in manganese and gadolinium atoms. On the other hand, the decrease of the manganese concentration in these alloys also decreases the accuracy of the X-ray spectral analysis.

The considerable energy shift of the gadolinium M_ξ line shows that the number of valence electrons of gadolinium is changing. The shift in the direction of lower

Table 2

compound	Mn $L_{\alpha_{1,2}}$			Gd M_ξ	
	δE_{max} (eV)	ΔE (eV)	I	δE_{max} (eV)	ΔE (eV)
Gd				0	8.9
Mn	0	5.6	10		
$GdMn_2$	7.0	5.0	3.1	-6.0	9.3
$GdAl_{0.4}Mn_{1.6}$	5.8	4.9	7.4	-4.0	9.0
$GdAl_{0.8}Mn_{1.2}$	6.1	5.4	2.4	-6.0	9.4
GdAlMn	7.4	5.5	1.4		
$GdAl_{1.4}Mn_{0.6}$	8.7	7.6	4.3		
$GdAl_{1.6}Mn_{0.4}$	9.7	8.2	2.5		

energies (see Table 2) shows that the gadolinium atom is an acceptor of 3d electrons from manganese.

All characteristic features of Mn $L_{\alpha_{1,2}}$ and Gd M_ξ lines are presented in Table 2 and 3. δE_{max} denotes the shift of the line in the case of alloys with respect to the same line in the case of pure manganese, ΔE is the width at half maximum intensity, and I the intensity of the $L_{\alpha_{1,2}}$ band normalized to the L_ι line.

The positive shift in energy scale of the $L_{\alpha_{1,2}}$ line indicates that manganese is a donor of electrons in the investigated compounds. For example, the same shift towards higher energy was observed in the case of MnO_2 [11] in contrast to Al–Mn

Table 3

compound	$L_{\alpha_{1,2}}$ (eV)	L_{β_1} (eV)	L_ι (eV)	L_η (eV)	$L_{\beta_1} - L_{\alpha_{1,2}}$ (eV)
Mn	637.4	648.8	556.3	567.5	11.4
$GdMn_2$	644.4	658.1	520.5		13.7
$GdAl_{0.4}Mn_{1.6}$	643.2	658.3	521.1		15.1
$GdAl_{0.8}Mn_{1.2}$	643.5	659.5	520.0		16.0
GdAlMn	644.8	659.1	520.0		14.3
$GdAl_{1.4}Mn_{0.6}$	646.1	658.8	520.5		12.7
$GdAl_{1.6}Mn_{0.4}$	647.1	659.2	520.3		12.1
errors	±0.2	±0.2	±0.2	±0.2	±0.4

alloys [10]. The ionic character of these bonds is also confirmed by the increase of the distance in energy scale between L_{III} and L_{II} levels (Table 3, $L_{\beta_1} - L_{\alpha_{1,2}}$).

Fig. 2 shows the $L_{2,3}$ spectra of aluminium in $Gd(Al_{1-x}Mn_x)_2$ alloys. These spectra correspond to the transition 3s(d) → 2p. The $L_{2,3}$ emission band of pure aluminium differs from the band obtained for alloys. The main change of the character of the spectra occurs already in the $GdAl_2$ compound. The spectra for the remaining compounds of the series under investigation have a similar shape. Thus, it seems that the presence of manganese slightly effects the character of bonding of aluminium in those compounds. Only the intensity of the three observed maxima changes. This is difficult to interpret, because the signal-to-noise ratio of the aluminium line was too low. It is probably caused by the presence of the heavy Gd atoms in the alloys, which causes a strong absorption. The Al $L_{2,3}$ spectra in $GdAl_2$ were also investigated by Wiech and Zöpf [12]. They obtained similar results.

Acknowledgement

The authors would like to express their sincere thanks to Prof. A. Chełkowski for helpful discussion.

References

[1] F. Laves and H. Witte, Metallwirtsch., Metallwiss., Metalltech. **14**, 645 (1935); **15**, 840 (1936).
[2] H. Oesterreicher, J. Phys. Chem. Solids **33**, 1031 (1972).
[3] A. Ślebarski and A. Chełkowski, phys. stat. sol. (a) **36**, K171 (1976).
[4] A. Ślebarski, phys. stat. sol. (a) **46**, K29 (1978).
[5] A. Ślebarski and A. Chełkowski, J. less-common Metals **57**, 125 (1978).
[6] A. Ślebarski, J. less-common Metals **59**, 1 (1978).
[7] T. M. Zimkina and V. A. Fomichev, Ultramyagkaya rentgenovskaya spektroskopiya, Izd. Leningradskogo Universiteta, 1971.
[8] S. Steinemann and A. Wenger, Helv. phys. Acta **41**, 1299 (1968).
[9] A. T. Shubaev, Izv. Akad Nauk SSSR **24**, 424 (1960).
[10] A. Wenger, G. Bürri, and S. Steinemann, Solid State Commun. **9**, 1125 (1971).
[11] D. W. Fischer, J. appl. Phys. **36**, 2048 (1965).
[12] G. Wiech and E. Zöpf, Band Structure Spectroscopy of Metals and Alloys, Academic Press, London 1973 (p. 173).

(Received March 2, 1979)

phys. stat. sol. (a) 54, 85 (1979)

Subject classification: 17; 10.2; 22.5.2

Department of Physics and Astronomy, University of Oklahoma, Norman[1])

Optically Stimulated Exoelectron Emission from Potassium Chloride Single Crystals

By

G. M. RENFRO[2]) and H. J. FISCHBECK

Potassium chloride single crystals are irradiated with a 3 keV electron beam, and the optically stimulated exoelectron spectra (OSEE) are studied as a function of dose, temperature, and energy in the range from 300 to 650 nm. Optical extinction spectra are measured simultaneously with the OSEE spectra. The exoemission spectra seem to show exoemission peaks corresponding to the F and K optical absorption bands. In addition an emission peak appears at about 400 nm within the high-energy tail of the K-band. Emission lineshapes are narrower than extinction lines and peak positions do not exactly correspond. With increasing surface damage the F-band emission diminishes. Increasing temperature produces a shift in the F-band to longer wavelengths and an increase in intensity.

Kaliumchlorid-Einkristalle werden mit einem 3 keV-Elektronenstrahl bestrahlt, und die optisch angeregten Exoelektronenspektren (OSEE) werden in Abhängigkeit von Dosis, Temperatur und Energie im Bereich von 300 bis 650 nm untersucht. Die optischen Extinktionsspektren werden gleichzeitig mit den OSEE-Spektren gemessen. Die Exoemissionsspektren scheinen Exoemissionsmaxima entsprechend den optischen F- und K-Absorptionsbanden zu zeigen. Zusätzlich erscheint ein Emissionsmaximum bei etwa 400 nm innerhalb des hochenergetischen Ausläufers der K-Bande. Die Emissionslinien sind schmaler als die Extinktionslinien, und die Lagen der Maxima entsprechen sich nicht exakt. Mit zunehmendem Oberflächendamage verschwindet die Emission der F-Bande. Das Anwachsen der Temperatur ruft eine Verschiebung der F-Bande zu längeren Wellenlängen und einen Anstieg der Intensität hervor.

1. Introduction

Stimulated exoelectron emission refers to the emission of low-energy electrons from the surface of materials when they are stimulated by heat or light. Thermally stimulated exoelectron emission (TSEE) occurs at temperatures well below the threshold for thermionic emission, and optically stimulated emission (OSEE) occurs at photon energies much lower than those required for the normal photoelectric effect. Exoemission has been observed from many materials including minerals, ceramics, glasses, metal oxides, and semiconductors.

Optically stimulated exoelectron emission was first noticed as a shifting of the photoelectron emission edge to longer wavelengths. This phenomenon was referred to as the "external photoeffect" and was reported by Lukirsky et al. [1] in 1926 and by Fleischmann [2a] in 1933. After Apker and Taft [2b] reported a correlation between OSEE and the α absorption band, in 1950 investigations were started by Bohun [3, 4] and by Sujak [5] culminating in a series of papers by Petrescu [6 to 9] in which the OSEE peaks were correlated with the optical absorption of F, M, and R color centers in many of the alkali halides, thus relating these electrons to lattice defects of the crystal.

---

[1]) Norman, Oklahoma 73019, USA.

[2]) Present address: Physics Department, Oklahoma State University, Stillwater, Oklahoma 74074, USA.

In these early investigations samples were irradiated with low-energy (less than 10 keV) electron beams and then placed into a gas flow counter where they could be stimulated by light from a monochromator and their exoemission could be recorded. Petrescu reported that the observed emission rate was very small, on the order of 50 counts/s, and that there was often difficulty in reproducing the emission peaks [7]. Because of the low emission intensity he did not correct his emission spectra for the spectral shape of his light source. Polycrystalline layers of salt precipitated or melted onto a metal substrate were used as samples rather than single crystals. Petrescu reported that sometimes emission peaks were observed that had no relationship to the published color center levels, and sometimes only an emission edge was observed — similar to the "external photoeffect" [7].

The question arises to what extent the peaks observed by Bohun [4], Sujak [5], and Petrescu [6] are influenced by surface effects. Recently, Brunsmann and Scharmann [10] have monitored the surface composition of NaF using Auger electron spectroscopy during a TSEE experiment. They found no signals except those of the host lattice. However, Sujak [11] pointed out that when OSEE experiments were performed in a vacuum, rather than in the gas atmosphere of the counter, no emission peaks were observed. The connection between the OSEE peaks and the optical absorption maxima has only been made via the close agreement of the wavelength at which the maximum occurs. No information is available as to how they agree in shape, or in behavior under thermal annealing. Since the optical absorption of irradiated crystals is known to arise from lattice defects only, a simultaneous observation of the OSEE and the optical absorption on the same crystal should, in principle, provide information as to the extent of the relationship between the two processes. It was our aim to explore the feasibility of such an experiment.

If single crystals are bombarded with low-energy electrons, then lattice defects are produced in a very thin layer on the side exposed to the electron beam. The stimulating light can be passed from the unexposed side through the crystal, and electron emission and optical absorption will take place in the same thin layer, provided the undamaged crystal is transparent over the range of wavelengths where color center absorption occurs. We chose KCl for this experiment since its F-center absorption falls conveniently in the middle of the visible region. Furthermore, KCl is an example where the OSEE peak observed by Petrescu [7] at 576 nm does not agree too well with the maximum of the optical absorption spectrum for the F-band which at 300 K occurs at 557 nm [12]. Petrescu detected F and R emission bands in KCl, but he did not report investigating the region on the high-energy side of the F-band.

2. Experimental Method

The experimental arrangement is shown in Fig. 1. Thin KCl crystals (X), cleaved in air, were mounted an a stainless steel sample holder (S). Light from a monochromator

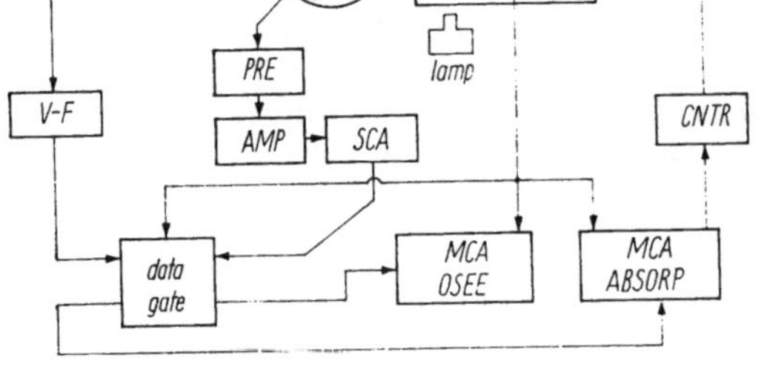

Fig. 1. Schematic representation of experimental apparatus. The crystal (X) is mounted on a conical sample holder (S) within the vacuum chamber. Light from the monochromator illuminates the rear of the crystal. Transmitted light intensity is detected by the phototube (PMT) and stored in the multichannel analyzer (MCA ABSORP). Simultaneously, exoelectrons are detected by the channeltron (C) and stored (MCA OSEE). Advancing of channels is controlled by pulses from the monochromator (λ)

traverses the crystal and is observed with a photomultiplier tube (PMT). Electrons stimulated by the light and leaving the crystal are detected with a channeltron (C). The signals from the two detectors are then stored in two multichannel analyzers (MCA) operated in the multiscale mode. The data recorded in this experiment are thus the intensity of the light transmitted through the sample and, simultaneously, the intensity of electron emission from the sample surface. Both these intensities are recorded as a function of the wavelength of the light selected by the monochromator. The motor-driven monochromator generates a channel advance pulse every 2 nm (λ). The power to the monochromator motor is controlled by a relay which is energized by a gate signal from a presetable up/down counter (CNTR), which counts the channel advance pulses and herefore controls the length of the scan.

The OSEE spectra are assembled by counting pulses from the channeltron. After amplification and noise discrimination (SCA) these pulses go directly into the MCA via the data gate. Optical absorption data are collected by a two-step process. First, a reference spectrum is taken of intensity versus wavelength for light passing unobstructed through a hole in the sample holder. Then the intensity of light transmitted through the crystal is recorded. The output of the phototube is measured by a microammeter. In order to record this signal in the MCA the output of the meter is then converted to a pulse train whose frequency varies linearly with the current from the phototube (V-F).

A spectral scan of a crystal consists, therefore, of three data sets. Two of these record the intensity of the light with and without the crystal in the light beam, and one is the emission spectrum of the sample. After these three spectra are accumulated in the MCA's, the values are transferred digitally to a minicomputer where they are recorded on magnetic tape and processed for plotting. The data processing algorithms consist of a routine for calculating the extinction coefficient for the sample and a routine to adjust the electron spectrum for relative light intensity.

The extinction coefficient is defined as $D = \lg (I/I')$, where I is the incident light intensity and I' is the intensity transmitted through the sample. Since the sample intercepts the light beam at 45°, an appreciable fraction of the incident light is reflected from the crystal surfaces. This reflected light will produce a non-zero value for D even when the crystal is uncolored. This base line shift is compensated by subtracting from D a value D', which is the extinction of the freshly cleaved sample before exposure to any radiation.

Petrescu reported that he did not correct his OSEE data to account for the spectral shape of the stimulating light [6]. He noted that the count rate of detected electrons was so low that the lineshapes were probably limited by the number of emission centers rather than the intensity of the light. In the present experiment, however, normal exoemission rates were 100 to 1000 electrons/s, and the characteristic emission lines of the xenon lamp were clearly reflected in the raw exoemission data. It was therefore necessary to normalize the spectra with respect to light intensity. Since both light intensity and exoemission are recorded simultaneously, this would merely require a point-by-point division of the two data sets, except for the fact that the recorded lamp spectrum has been convoluted by the response characteristics of the photomultiplier tube. Therefore, the lamp spectrum must be corrected for the phototube response before it is used to normalize the exoemission spectrum. The photomultiplier response curve was obtained by dividing a lamp spectrum taken with the phototube by a second spectrum taken with a thermopile.

Irradiations of the samples were performed by rotating the sample holder so it faced an electron gun. A picoammeter was connected to a copper grid in the electron beam in order to measure the irradiation current. The output of the picoammeter was connected to the V-F. Accurate dose control was obtained by counting the output pulses of the V-F to determine the irradiation time. Since the frequency of the V-F depends linearly upon the beam current, the number of pulses counted during irradiation is proportional to the product of the beam current and the irradiation time, which is itself proportional to energy deposited. Typical irradiations were 20 μA for 10 s, which amounts to 0.3 J deposited on an area of 0.78 cm^2. The surface temperature of the crystal, measured with a thermocouple, was observed to rise about 6 °C on the average during this irradiation.

3. Experimental Results

Typical OSEE spectra together with the optical absorption of an electron-bombarded KCl crystal are shown in Fig. 2. Curve b is the OSEE spectrum of a crystal cleaved in air after irradiation with 3 keV electrons of 20 μA beam current for 10 s. The surface temperature during the scan was 28 °C. A peak observed at 525 nm is close to the maximum of the optical absorption at 554 nm, which was measured simultaneously (curve a). The optical absorption maximum is in good agreement with the measurements of Konitzer and Markham [12] who found the maximum for the F-band at 557 nm when the temperature was 27 °C. It is clear from the figure that the emission peak is much narrower than the absorption peak. Repeated electron bombardment of the crystal produced a slight shift and reduction in intensity of the OSEE F-peak and had a drastic effect on the structure below 500 nm. This is shown in curve c which gives the OSEE spectrum of the same crystal after three additional irradiations with the same dose. The crystal temperature was again 28 °C during the scan. No shift in the peak position of the extinction was observed.

The crystal temperature was then raised to 75 °C. The extinction maximum (curve a) shifted by about 6 nm to the red in agreement with the temperature dependence of the F-band observed by Konitzer and Markham [12]. Since, except for a reduction due to thermal bleaching, there was no other change in the extinction spectrum, it is not shown explicitly in the figure. At a surface temperature of 75 °C electron emission below 500 nm shows a marked increase in intensity (curve d). The peak in the region of the F-band has shifted to 558 nm with a slight increase in intensity, and now coincides closely with the maximum of the F absorption band. This agrees with Petrescu's observation [7]. We prepared a polycrystalline KCl sample by vacuum evaporation on a thin glass substrate similar to Petrescu's samples. In Fig. 3 (curve a) a scan of this sample is shown to be very similar to the single crystal spectrum. The

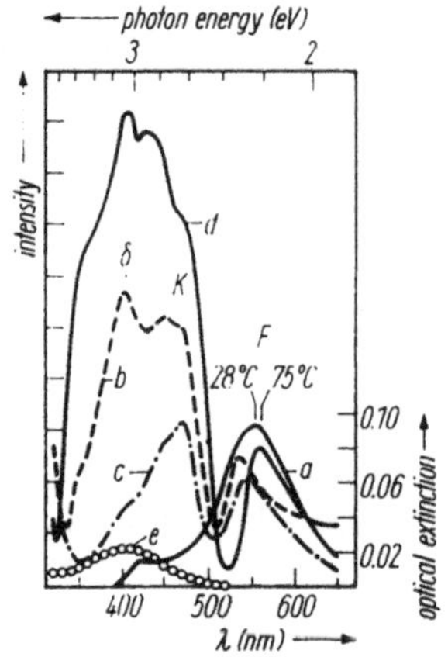

Fig. 2. Curve a is the optical extinction curve of the irradiated crystal. The F-band shifts to longer wavelength as the temperature is increased from 28 to 75 °C. Curve b is the OSEE spectrum at 28 °C taken simultaneously with curve a. Curve c is the OSEE spectrum at 28 °C after six successive irradiations. Curve d is the OSEE spectrum at 75 °C. Curve e is the OSEE spectrum with an external 5 G magnetic field applied

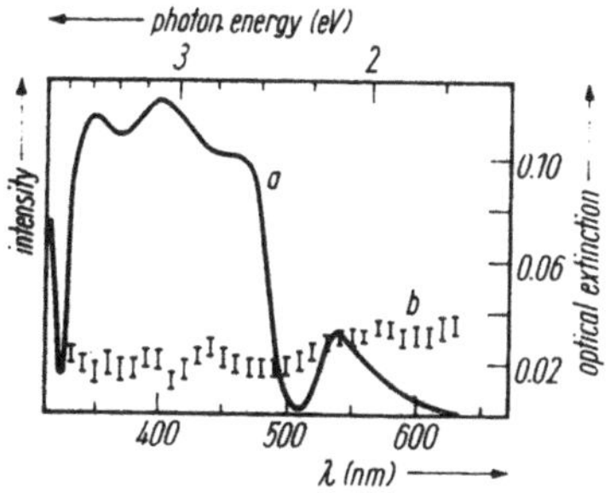

Fig. 3. Curve a is the OSEE spectrum of a polycrystalline sample of KCl. Curve b is the optical extinction spectrum of the polycrystalline sample

optical absorption shown in curve b is difficult to measure because of the large amount of scattered light from the polycrystalline surface.

The effect of bleaching on the absorption spectrum and the OSEE spectrum was investigated by scanning the crystal from red to blue six times in succession. The area under the emission and extinction peak was obtained by summing the counts in the channels of the MCA corresponding to the wavelength interval from 490 to 610 nm. The areas were then normalized to their respective values at the first scan. The results are listed in Table 1. Both peaks bleach at about the same rate.

Table 1

scan number	relative peak OSEE	area extinction
1	1.0	1.0
2	0.85 ± 0.08	0.82 ± 0.08
3	0.84 ± 0.08	0.76 ± 0.08
4	0.79 ± 0.08	0.77 ± 0.08
5	0.71 ± 0.07	0.57 ± 0.06
6	0.62 ± 0.06	0.62 ± 0.06

In interpreting any exoemission data it is essential to keep in mind that the surface of the sample is severely damaged by the irradiation. Sputtering of halogen ions is known to be a very efficient process for electron beams with energies as low as a few hundred electron volts. In fact Palmberg and Rhodin [13] have shown that irradiating KCl with a 1 keV electron beam at room temperature for several minutes produces a surface layer made up almost entirely of potassium. This fact would seem to account for the upper limit of F-center production observed by Soul and Tubbs [14], and is probably responsible for the decrease in exoemission intensity observed in Fig. 2.

Since Soul and Tubbs [14] have shown the F-center concentration to remain essentially constant under repeated irradiation, we can conclude from Palmberg and Rhodin's data [13] that repeated irradiations must alter the work function of the crystal. We may expect the escape of low-energy exoelectrons to become less likely as the surface damage increases. This may imply that electrons emitted in the F-band have less energy than those emitted in the region of the K-band. This view would be consistent with the model of Mott and Gurney [15] that the K-band is due to electronic transitions to higher excited states of the F-center. It would also account for the fact that the F absorption band remains nearly constant for repeated irradiation while the exoemission decreases. Moreover, Petrescu noted that in several alkali halides including KCl the intensity of the F emission peak was enhanced by increasing the temperature. This also suggests that these electrons are more likely to escape if given additional energy. The spectra at 28 and 75 °C (curve c and d in Fig. 2) confirm this observation. At the same time that the exoemission intensity is increasing, the F-band optical absorption is being thermally bleached away. The exoemission process must therefore account for only a small fraction of the optical absorption processes occurring at the F-center.

In addition to the increase of emission intensity in the region of the F-band, the peak also shifts to lower energy, suggesting that with more thermal energy available less optical energy is needed to release the electrons. Emission in the region of the K-band also grows with increasing temperature, but the growth in the region of 400 nm is most pronounced. It is significant that there is no detectable optical absorption in the region of 400 nm.

A crude energy discrimination scheme was tried. An external magnetic field of 5 G was applied to the region between the crystal and the electron detector, such that all electrons with energies of less than 1 eV were deflected more than the width of the detector aperture and were not counted. The result is the spectrum shown as curve e in Fig. 2. A single broad emission peak with a maximum at 400 nm persists. These counts are either electrons of more than 1 eV or photo-desorbed ions similar to those observed by De Muer and Maenhout-van der Vorst [16]. To establish that this observation was not caused by some effect of the magnetic field on the channeltron, the detector was moved to a distance of 4 cm from the crystal (the maximum distance possible), so that the 0.5 G component of the earth's field would prevent all electrons of < 0.5 eV from entering the detector aperture. Again only the same broad feature at 400 nm was observed. Increasing the light intensity in order to compensate for the reduced solid angle did not produce any detectable electron emission in the region of the F- or K-bands. We conclude that the electron energies are less than 0.5 eV, which agrees with TSEE measurements made by Gordan and Scharmann [17].

4. Discussion

If we assume that at least part of the exoemission is in some way connected to bulk defects in the crystal and that the peak in the region of 540 nm is related to the F-center optical absorption band, then the OSEE should be explainable in terms of an F-center model. Mott and Gurney [15] have treated the F-center as a solid state analogue to the hydrogen atom. They proposed that the F absorption band arises from the 1s–2p transition of the trapped electron and that the K-band is formed from a superposition of 1s–np transitions where $n \geqq 3$. Based upon this model predictions were made for the energy levels and transitions associated with the F-center [18]. Wood and Öpik [19]have published Hartree-Fock calculations on KCl, which predicts not only the energy levels of the F-center, but also their relationship to the conduction band edge and the effects of lattice relaxation upon the energies of the excited states. This calculation has been very successful in predicting the observed positions of the F and K absorption bands, and also predicts that the ground state of the unexcited F-center should lie about 2.86 eV below the conduction band.

In Fig. 4 we have taken the position of the energy levels from the Wood and Öpik paper [19]. In addition the position of the vacuum level for direct photoemission in KCl was obtained from Taft and Philipp [20]. It is clear that at least two different mechanisms for OSEE must operate in this model. If the stimulating light has an energy greater than about 3.06 eV, then direct photoemission from the F-center ground state should occur. This process should begin to occur in the high-energy tail of the K-band about where the peak marked δ is seen in Fig. 2. It is interesting to note that this sort of direct escape from the F-center ground state has also been seen in the photoconductivity spectra of Wild and Brown [21]. In that case, however, the

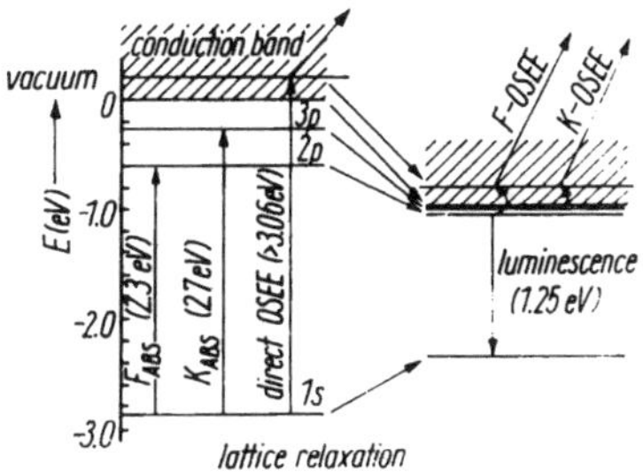

Fig. 4. Energy level scheme of the F-center in KCl taken from the calculations of Wood and Öpik [19], showing possible OSEE mechanisms

emission edge corresponds to a promotion of F-center electrons to the bottom of the conduction band, whereas for exoemission to occur the transition must end at or above the vacuum level.

For exoemission to occur in the F- and K-bands a two-step process is required. Stimulating light will promote the trapped electron to the 2p or 3p state of the F-center, and lattice relaxation will occur, lowering the energy of these states. In order for the excited electron to escape, sufficient thermal energy must be available to raise it to the vacuum level. The question immediately arises as to the position of the vacuum level with respect to the relaxed excited F-center. Since OSEE is observed to occur at room temperature, it is clear that the vacuum level must also relax. Otherwise a thermal activation of more than 1.16 eV would be required and exoemission from these bands would be highly improbable at room temperature.

From this model we would expect that the kinetic energy of exoelectrons emitted directly from the F-center ground state should be larger than the energy of electrons emitted by the two-step process. Moreover, we would expect the probability of escape, and hence the OSEE intensity, to increase for the F- and K-bands as the temperature increases. This behavior was observed in Fig. 2, curves c and d. We might expect that as the temperature increases, the average displacement of the ions surrounding the F-center will shift to larger values and the F-center will appear to be partially relaxed resulting in a shift to longer wavelengths of the OSEE peaks. This is also observed in Fig. 2; however, the shift amounts to about 33 nm, which is much larger than the 6 nm shift of the absorption peak. It is difficult to understand why the emission peaks should be narrower than the absorption band unless an energy discrimination occurs as the electrons leave the crystal surface.

This model does not predict that the intensity of the direct exoemission in the region above 3.06 eV should increase with temperature. Nor is it easy to see why the direct photoemission process above 3.06 eV should produce a peak rather than an emission edge as in the normal photoeffect. Indeed the photoconductivity data of Wild and Brown [21] does show an edge at about 2.7 eV with no significant structure toward higher energies. The OSEE spectrum of Fig. 2 seems to show a peak at about 3.41 eV. One could speculate that this peak represents an auto-ionizing resonance state degenerate with the conduction band. On the other hand, the possibility that these electrons have nothing to do with the F-center is just as likely. The present experiment cannot provide any definite clue as to their origin.

Based upon the F-center model for exoemission, we would predict that as the temperature of the sample is lowered, first the F emission band and then the K emission band should disappear as the necessary thermal activation becomes unavailable. However, direct exoemission above 3.06 eV should be independent of temperature since no thermal activation is required. The photoconductivity spectra of Wild and Brown [21] were taken at 10 K and display no F- or K-band conductivity. Furthermore, since exoemission in the F- and K-bands requires a two-step process in this model, we expect exoemission from these states to have long lifetimes, while exoemission above 3.06 eV should be direct and therefore short-lived. From this exploratory experiment it seems feasible and of interest to study exoemission and optical absorption on alkali halide crystals at low temperatures, and to attempt OSEE lifetime measurements.

References

[1] P. Lukirsky, N. Gudris, and L. Kulikowa, Z. Phys. **37**, 308 (1926).
[2a] R. Fleischmann, Z. Phys. **84**, 717 (1933).
[2b] L. Apker and E. Taft, Phys. Rev. **79**, 964 (1950).
[3] A. Bohun, Czech. J. Phys. **4**, 139 (1954).
[4] A. Bohun, Czech. J. Phys. **5**, 64 (1955).
[5] B. Sujak, Acta phys. Polon. **12**, 241 (1953).
[6] P. Petruscu, phys. stat. sol. **3**, 950 (1963).
[7] P. Petruscu, phys. stat. sol. **9**, 539 (1965).
[8] P. Petruscu, phys. stat. sol. **29**, 333 (1968).
[9] P. Petruscu, phys. stat. sol. **29**, K1 (1968).
[10] U. Brunsmann and A. Scharmann, phys. stat. sol. (a) **43**, 519 (1977).
[11] B. Sujak, Acta phys. Austriaca **10**, 460 (1957).
[12] J. D. Konitzer and J. J. Markham, J. chem. Phys. **32**, 843 (1960).
[13] P. Palmberg and T. Rhodin, J. Phys. Chem. Solids **29**, 1917 (1968).
[14] P. Soul and M. Tubbs, Solid State Commun. **7**, 397 (1969).
[15] N. Mott and R. Gurney, Electronic Processes in Ionic Crystals, Oxford University Press, 1940.
[16] D. De Muer and W. Maenhout-van der Vorst, Physica (Utrecht) **49**, 157 (1970).
[17] P. Gordan and A. Scharmann, Z. Phys. **217**, 309 (1968).
[18] W. Fowler, Physics of Color Centers, Academic Press, New York/London 1968.
[19] R. F. Wood and U. Öpik, Phys. Rev. **197**, 783 (1969).
[20] E. Taft and M. Philipp, J. Phys. Chem. Solids **3**, 1 (1957).
[21] R. L. Wild and F. C. Brown, Phys. Rev. **121**, 1296 (1961).

(Received March 14, 1979)

phys. stat. sol. (a) **54**, 93 (1979)

Subject classification: 2 and 14.3.1; 22.9

Institute of Physics, Nicholas Copernicus University, Toruń

Preparation and DC Conductivity of the Low-Temperature Glassy Carbon[1])

By

K. ROZWADOWSKA

The electrical properties of the low-temperature glassy carbon are determined. The activation energy decreases from about 0.6 eV for heat treatment temperature (HTT) 400 °C to 0.08 eV for HTT 700 °C. For the temperature range 80 to 200 K Mott's law is fulfilled, and the density of states which is equal to $N(E) = 3 \times 10^{19}$ eV^{-1} cm^{-3} is calculated. The implications of the conduction mechanism are discussed.

Es werden die elektrischen Eigenschaften des bei niedrigen Temperaturen behandelten glasartigen Kohlenstoff untersucht. Die Aktivierungsenergie nimmt von etwa 0,4 eV für eine thermische Behandlung bei 400 °C bis auf 0,08 eV bei 700 °C ab. Im Temperaturbereich von 80 bis 200 K ist das Mottsche Gesetz erfüllt. Die errechnete Niveaudichte beträgt 3×10^{19} eV^{-1} cm^{-3}. Der elektrische Leitungsmechanismus im untersuchten Material wird diskutiert.

1. Introduction

Low-temperature glassy carbon has a number of interesting features. It shows switching effect [1], superconductivity at room temperature [2 to 4], and is a suitable object for investigations of the transition high polymer–carbon. For this reason it is interesting to find the mechanism of conduction in such a material.

The electrical properties of amorphous films of semiconductors have been widely investigated. The main attention has been given to the elemental semiconductors a-Ge, a-Si, a-C, and to chalcogenide glasses [5 to 9]. The known theoretical models were proposed mainly for these materials [10, 11]. Considering a model in which charge is transported via phonon-induced tunneling of electrons between localized states with energy near the Fermi level, Mott has obtained the electrical conductivity of the form [12]

$$\sigma = \sigma_0 \exp\left(\frac{-E}{kT}\right), \tag{1}$$

where σ is the conductivity at temperature T, σ_0 a constant, k the Boltzmann constant' E the activation energy. At lower temperatures variable-range hopping of charged carriers between localized states is more probable. The observed temperature dependence has been fitted to the Mott law:

$$\sigma(T) = \sigma_0 \exp\left(-BT^{-1/4}\right), \tag{2}$$

where $\sigma(T)$ is the dc conductivity at the temperature T and B is given by

$$B = c\alpha^3 k^{-1} N_F^{-1}, \tag{3}$$

where c is a numerical constant of order 2 to 10 [13, 14], N_F the density of localized

[1]) Sponsored by the Institute of Low Temperatures and Structural Research of the Polish Academy of Sciences.

states near the Fermi level, α^{-1} the characteristic fall-off length of the wave functions of the localized charge carriers, σ_0 a pre-exponential factor which depends on α, N_F, and the electron–phonon coupling.

Results of such an investigation can help us to understand the mechanism of electrical conduction in organic semiconductors and should be able to demonstrate how strongly the conductivity of disordered (or glassy) systems is affected by chemical composition, and the physical structure.

Similar works have been done for other materials: phenolic resin [15], phenol formaldehyde resin [16, 17], and polyvinilidence chloride [18].

2. Samples

Samples were prepared by gradual thermal degradation of the polymer starting material. The most interesting materials for these investigations are highly cross-linked polymers or polymers with the ability to develop a three-dimensional aromatic network.

Furfuryl alcohol was chosen as a starting material, because previously examined samples had shown many interesting features [1 to 4]. Samples were produced in the following way: a gold electrode was evaporated on to carefully cleaned mica sheets. The electrode was covered by prepolymerized polyfurfuryl alcohol and samples were heated up to temperatures 400, 500, 600, and 700 °C with the rate 8.5 K/min in a quartz tube. From 300 °C the tube was evacuated to 10^{-3} hPa. In this way samples with gradually changing chemical composition were prepared.

The thermal degradation is a complex process. Bonds are broken in the starting material and volatile decomposition products like H_2O, CO, CO_2, CH_4, and H_2 are evolved [19].

Following this heat treatment, the upper gold electrode was deposited to get a sandwich. The thickness of the glass-like carbon layers varied from 80 nm to 2 μm.

3. DC Conductivity

The dc conductivity was determined by simple current–voltage measurements. All measurements were made in vacuum and are reproducible.

3.1 Dependence of the conductivity on the heat treatment temperature

The room temperature electrical resistivity of glassy carbon derived from phenolic resin was measured by Kawamura for heat treatment temperatures (HTT) 500 to 2500 °C and for pyrolized furfuryl alcohol by Turło [20]. Fig. 1 shows the conductivity of the thermally degraded polyfurfuryl alcohol versus the HTT. This figure demonstrates that the dc conductivity for HTT 400 to 700 °C covers more than ten orders of magnitude, and for further investigations samples from this range of heat treatment were chosen.

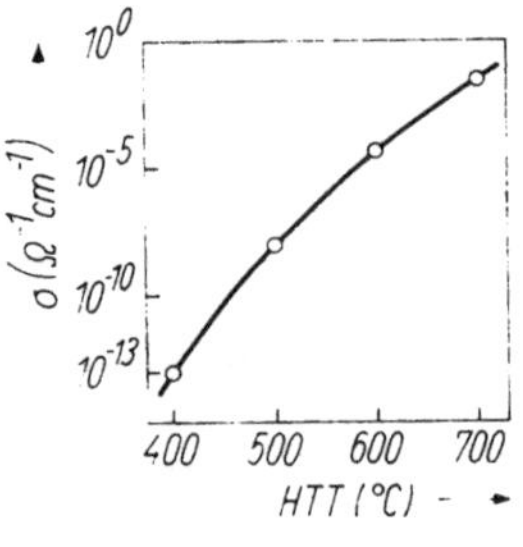

Fig. 1. Conductivity of the pyrolized furfuryl alcohol versus HTT

3.2 Time dependence of the conductivity

The early stages of pyrolysis are characterized by a marked time dependence. In low-temperature glassy carbon, as well as in other materials [21] conductivity varies in time. Typical characteristics for samples with HTT 500 and 600 °C are shown in Fig. 2. Similar characteristics were observed for samples with HTT 400 and 700 °C. From Fig. 2 we can see that current decreases in time, and after 10 to 15 min it reaches its equilibrium value. For higher electrical fields (10^4 V/cm for samples with HTT 600 °C) current starts to increase in time, and for fields higher than 5×10^5 V/cm the sample switches to a high-conductivity state.

Current–voltage characteristics were taken after reaching the equilibrium state. Various parts of each characteristic can be approximated by straight lines with different slopes (Fig. 3). From this figure we can find

$$n = \frac{d(\lg I)}{d(\lg U)} .$$

For voltages $U < 1.5$ V, $n = 1$ and Ohm's law is fulfilled ($I \sim U^1$). For $U > 1.5$ V we determined for n the values 2, 3, 5.

3.3 Temperature dependence of the conductivity

Fig. 4 shows examples of variation of the electrical conductivity on a logarithmic scale as a function of the reciprocal of ambient temperature in samples of pyrolized furfuryl resin from the boiling point of nitrogen to 300 °C in a vacuum 10^{-6} hPa.

From this figure we find that the plot of lg I against T^{-1} cannot be resolved into linear portions between 80 and 500 K, but is a continuous curve. A straight line occurs only between about 200 and 400 K. An activation energy can be calculated from the slope in Fig. 4 at test temperatures above 250 K.

For lower temperatures we can see (Fig. 6) that the plot of lg I against $T^{-1/4}$ yields a straight line between 80 and 200 K.

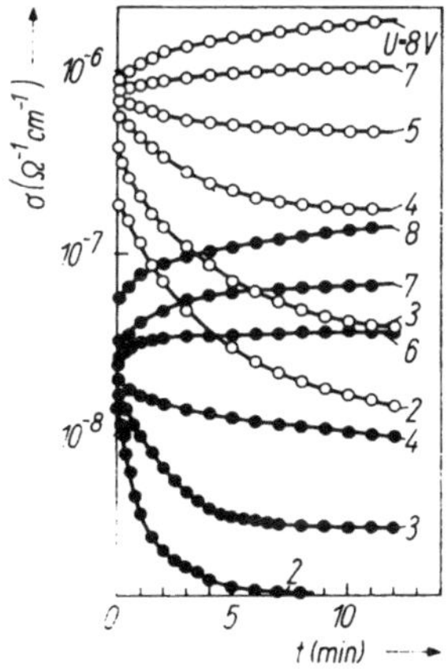

Fig. 2

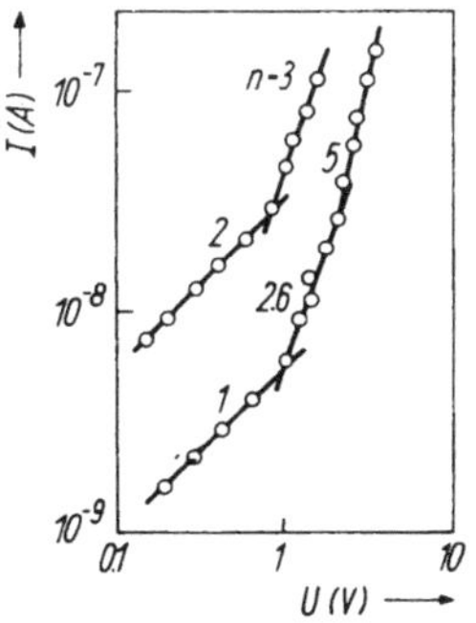

Fig. 3

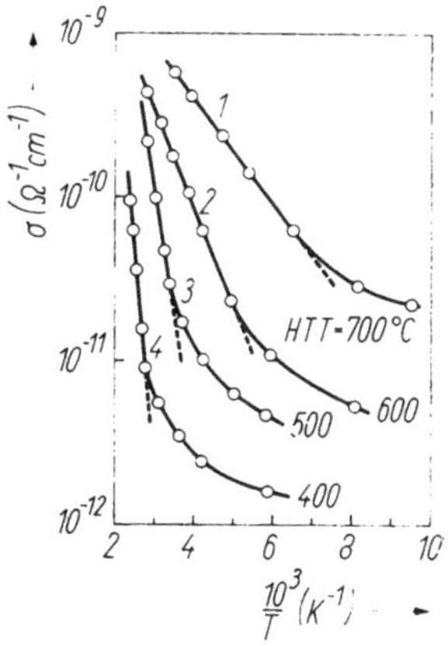

Fig. 4

Fig. 2. Time dependence of conductivity for different applied voltages for samples with HTT 500 (●) and 600 °C (○)

Fig. 3. Current–voltage characteristics

Fig. 4. The dependence lg σ versus $1/T$. (1) $E_a = 0.064$, (2) 0.116, (3) 0.282, (4) 0.565 eV

4. Results

The plot of lg I of thermally degraded polyfurfuryl alcohol with HTT 400 °C versus the reciprocal temperature is not continually curved. One can approximate the curve by tangent segments. Within these regions the conductivity can be described by (1)

$$\sigma = \sigma_0 \exp\left(-\frac{E}{kT}\right).$$

The slope of the tangents corresponds to the formally introduced activation energy

$$E_a = \frac{d\,(\lg \sigma)}{d\,(kT)^{-1}}.$$

E_a is strongly dependent on the heat treatment temperature (Fig. 5). It decreases steeply down to HTT of 700 °C.

An important reference to the mechanism of electrical conduction in solids is supplied by the knowledge of the relation between the conductivity and the temperature in the low-temperature region. From Fig. 6 we can see that the conductivity follows equation (2). Similar relations for evaporated layers of Ge [9] and carbon [22] were found.

From the linear part of Fig. 6 we can determine the coefficient B: (from equation (2)).

HTT (°C)	B ($K^{1/4}$)
400	54.34
500	40.21
600	36.14
700	37.73

A suitable model for the physical interpretation of the experimental results is based on the assumption of localized states near the Fermi level [12]. They may arise from dangling bonds arising during the decomposition process. The carrier transport occurs by thermally assisted hopping between states at the Fermi level. If one assumes

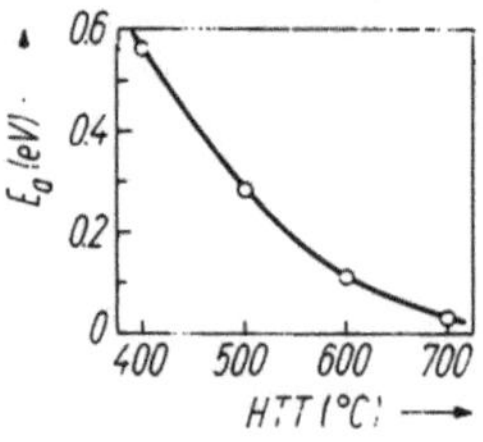

Fig. 5

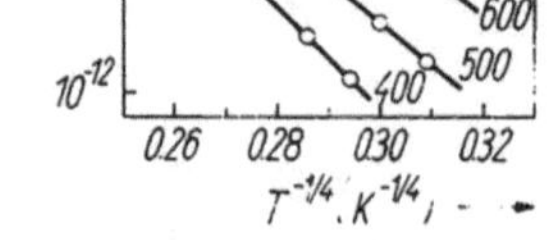

Fig. 6

Fig. 5. The decrease of the activation energy versus HTT

Fig. 6. The dependence lg σ versus $T^{-1/4}$

variable-range hopping processes, it is possible to calculate the state density $N(E)$. Assuming $c = 2$ [23, 24] and $\alpha = 7.5\ c \times 10^6\ \text{cm}^{-1}$ [25] we obtain:

HTT (°C)	$N(E)$ ($\text{eV}^{-1}\ \text{cm}^{-3}$)
400	9.0×10^{18}
500	3.0×10^{19}
600	4.6×10^{19}
700	3.8×10^{19}

This is in fair agreement with EPR measurements [26].

5. Conclusions

We have interpreted the obtained electrical data in terms of the model described by Mott [10] and Cohen et al. [11].

We can divide the conductivity curves into two regions. As indicated in Fig. 4, region 1 extends from about 200 K to higher temperatures. We suggest that conductivity in this region is associated with the transport of electrons by the excitation through the mobility gap to the extended states. In region 2 (between 80 and 200 K) the conduction can be due by variable-range hopping between localized states.

References

[1] K. Antonowicz, L. Cacha, and J. Turło, Carbon **11**, 1 (1973).
[2] K. Antonowicz, Nature (London) **247**, 358 (1974).
[3] K. Antonowicz, phys. stat. sol. (a) **28**, 497 (1975).
[4] K. Antonowicz and K. Rozwadowska, Fizyka Cieńkich Warstw, Państwowe Wydawnictwo Naukowe, 1977 (p. 467) (in Polish).
[5] S. Orzeszko, Preprint Inst. Phys., Nicholas Copernicus Univ., No. 194, Toruń 1972 (in Polish).
[6] A. H. Clark, Phys. Rev. **154**, 750 (1968).
[7] A. H. Clark, J. non-crystall. Solids **2**, 52 (1970).
[8] M. Morgan and P. A. Walley, Phil. Mag. **23**, 661 (1971).
[9] R. C. Chittick, J. non-crystall. Solids **3**, 255 (1970).
[10] N. F. Mott, Phil. Mag. **19**, 835 (1969).
[11] M. H. Cohen, H. Fritzsche, and S. Ovshinsky, Phys. Rev. Letters **22**, 1065 (1969).
[12] N. F. Mott and E. A. Davies, Electronic Processes in Non-Crystalline Materials, Clarendon Press, Oxford 1971.
[13] V. Ambegaokar, B. I. Halperin, and J. S. Langer Phys. Rev. B **4**, 2612 (1971).
[14] M. H. Brodsky, G. H. Döhler, and P. J. Steinhardt, phys. stat. sol. (b) **72**, 761 (1975).
[15] G. M. Jenkins and K. Kawamura, Polymeric Carbons – Carbon Fibre, Glass, and Char, Cambridge University Press, 1976.
[16] W. Bücker, J. non-crystall. Solids **12**, 115 (1973).
[17] W. Bücker, J. non-crystall. Solids **18**, 11 (1975).
[18] H. E. Blayden and D. T. Westcott, Proc. V. Conf. Carbon, Vol. 2, Pergamon Press, New York 1961.
[19] E. Fitzer and W. Schäfer, Carbon **8**, 597 (1970).
[20] J. Turlo, Preprint Inst. Phys., Nicholas Copernicus Univ., No. 277, Torun 1974 (in Polish).
[21] R. Andreichin, Amorphous Semiconductors 76, Akademiai Kiadó, Budapest 1977 (p. 389).
[22] C. J. Adkins, S. M. Freake, and E. M. Hamilton, Phil. Mag. **22**, 183 (1970).
[23] V. Ambegaokar, B. I. Halperin, and J. S. Langer, J. non-crystall. Solids **8/10**, 492 (1972).
[24] M. Pollak, J. non-crystall. Solids **11**, 1 (1972).
[25] R. M. Hill, Phil. Mag. **24**, 1307 (1971).
[26] J. Turło, to be published.

(Received March 7, 1979)

phys. stat. sol. (a) 54, 99 (1979)

Subject classification: 1.2; 14.4.1; 20.1; 22.8.1

Institute of Physics, Czechoslovak Academy of Sciences, Prague[1])

Changes of Optical Properties at an Incommensurate–Commensurate Phase Transition in $(NH_4)_2BeF_4$ Crystals

By

Č. KOŇÁK

The optical properties of ammonium fluoberyllate in the incommensurate phase, namely birefringent and electrooptical ones, are discussed within the framework of a thermodynamic theory. The analysis of experimental results show that the incommensurate–commensurate phase transition in AFB is nearly of second order. The soft-mode and electrooptic contributions to the spontaneous birefringence are found to be small in comparison with the elastooptic one.

Die optischen Eigenschaften von Ammoniumfluoberyllat in der inkommensurablen Phase, nämlich doppelbrechende und elektrooptische, werden im Rahmen einer thermodynamischen Theorie diskutiert. Die Analyse der experimentellen Ergebnisse zeigt, daß der Phasenübergang inkommensurabel–kommensurabel in AFB nahezu von zweiter Ordnung ist. Die Soft-Mode- und die elektrooptischen Beiträge zur spontanen Doppelbrechung sind klein im Vergleich zu den elastooptischen.

1. Introduction

Ammonium fluoberyllate (AFB) undergoes two phase transitions (PT): a second-order nonferroelectric PT at $T_\text{I} = 183$ K [1] and a first-order ferroelectric PT with spontaneous polarization $\boldsymbol{P}_\text{s}$ arising in b-direction at $T_\text{C} = 177$ K [2, 3]. The phases arising successively with decreasing temperature will be labelled by the numbers I, II, III in this work. The space group D_{2h}^{16} [4] was found in the phase I and C_{2v}^{9} with the a-axis doubled in the polar one (III) [5]. Thermodynamical study and some experimental results [6] showed the phase III to be an improper ferroelectric. Recently Levanyuk and Sannikov [7] have predicted the existence of an incommensurate structure in phase II on the basis of the Lifshits invariants [8], and neutron scattering [5] has clearly confirmed it; several thermodynamic theories have been devoted to the problem of the incommensurate (IC)–commensurate (C) PT in AFB [7, 9 to 11]. Besides AFB the IC phases have been observed and studied in other materials (see, e.g., [12]).

Two theoretical concepts were worked out to describe general features of an IC phase [9]. The starting point of the "modulation" concept [7] is the expansion of a thermodynamic potential Φ in terms of variables transforming according to the representation with a particular wave vector $\boldsymbol{k}_\text{c}$ of the Brillouin zone and the Lifshits invariants [8]. The IC phase can be understood as a macroscopic space modulation of a periodic structure of the C phase. The same problem can be described in $\boldsymbol{k}$-space by means of the "mode" concept [12]. This means that the thermodynamic potential of the high-symmetry phase is expanded in terms of normal modes of lattice vibrations. Normal modes with $\boldsymbol{k} = \boldsymbol{k}_\text{c}$ are used for the description of the low-symmetry C phase and normal modes with $\boldsymbol{k}_0 = \boldsymbol{k}_\text{c} + \Delta\boldsymbol{k}$ for the IC phase. Mutual relationships between

[1]) Na Slovance 2, 18040 Prague 8, Czechoslovakia.

7*

these concepts have been discussed by Dvořák and Ishibashi [9]. In our opinion the former concept is more suitable to interpret the optical data in the low-symmetry phases; therefore it will be used in this work.

The optical properties of AFB, namely birefringent and electrooptic properties, have already been published by several authors [13 to 16]. We have shown [15] that the optical properties in the ferroelectric phase can be described satisfactorily in terms of the phenomenological theory for improper ferroelectrics. Since optical results in the IC phase have not yet been discussed within the framework of a thermodynamic theory, we present such a discussion here.

The basic results of the thermodynamic theory for the C and IC phases of AFB are summarized in Section 2. A phenomenological description of spontaneous and induced optical effects in the low-temperature phases of AFB is given in Section 3, which is followed by a survey of experimental results (Section 4). The temperature dependence of spontaneous birefringence and electrooptic effect is discussed in the last section of this paper. New data on the behaviour of spontaneous quantities in the low-temperature phases have been obtained from the experimental results, especially in the vicinity of the IC–C phase transition of AFB.

Throughout this paper, all tensor quantities are related to the coordinate system (x, y, z) of the phase I, chosen so that $\boldsymbol{x} \parallel \boldsymbol{a}$, $\boldsymbol{y} \parallel \boldsymbol{b}$, and $\boldsymbol{z} \parallel \boldsymbol{c}$, where a, b, and c are the crystallographic axes [4].

2. Thermodynamic Theory for the C and IC Phases of AFB

The thermodynamic potential Φ adequately describing the PT's from the high-symmetry phase (I) into the IC [7, 10] or C [15] phase, respectively, has the form

$$\Phi = \Phi_0 + \int \mathrm{d}x \left\{ \frac{\alpha_0(T - T_0)}{2} r^2 + \frac{\beta_1}{4} r^4 + \frac{\beta_2}{4} r^4 \cos 4\varphi + \frac{\gamma}{6} r^6 - \sigma r^2 \frac{\mathrm{d}\varphi}{\mathrm{d}x} + \right.$$
$$+ \delta \left[\left(\frac{\mathrm{d}r}{\mathrm{d}x}\right)^2 + r^2 \left(\frac{\mathrm{d}\varphi}{\mathrm{d}x}\right)^2 \right] + \mu r^2 P \sin 2\varphi + \tau r^2 u_{13} \cos 2\varphi + \eta r^2 P^2 +$$
$$+ \sum_{i=1}^{3} \varrho_i r^2 u_{ii} - \sum_{i=1}^{3} \varkappa_i u_{ii} r^2 \frac{\mathrm{d}\varphi}{\mathrm{d}x} + \frac{k}{2} P^2 + \frac{C_{55}}{2} u_{13}^2 + \sum_{i=1}^{3} \frac{C_{ii}}{2} u_{ii}^2 +$$
$$\left. + \sum_{i=1}^{3} Q_{i2} u_{ii} P^2 \right\}; \qquad \alpha_0, \delta > 0\,, \tag{1}$$

where P is the y-component of the polarization vector, u_{ij} are the strain tensor components, r and φ are polar coordinates of the TP's p, q ($p = r \sin \varphi$, $q = r \cos \varphi$).

In order to describe the IC phase, so-called gradient invariants of p, q permitted by symmetry were introduced in (1). If the σ-term (so-called Lifshits invariant), the $\varkappa_i$-, and the δ-terms are missing, then the minimum of Φ corresponds to homogeneous (x-independent) p, q.

Let us first look for homogeneous solutions of (1), i.e. put all derivatives equal to zero. The ferroelectric phase ($u_{13} = 0$, $P \neq 0$) stable at $T < T_0$ corresponds to values $\varphi = (2n + 1)\,\pi/4$ ($n = 0, 1, \ldots$) for both zero and non-zero electric field [17]. These solutions correspond to a pair of antiparallel ferroelectric domains and two antiphase domains. Neglecting sixth-order terms in (1), the stability conditions of phase III are: $\beta_2 + (\mu^2/k) - (\tau^2/C_{55}) > 0$, $\beta_1' = \beta_1 - \beta_2 - 2\mu^2/k > 0$ [17]. For small values of $T - T_0$ from $\partial\Phi/\partial r = 0$ we obtain [17]

$$(r_s^h)^2 = -\frac{\alpha_0(T - T_0)}{\beta_1'}\,. \tag{2}$$

(In this paper any quantity labelled h, m, or p relates to the ferroelectric C phase III, the IC phase II, or the high-symmetry phase I, respectively; the spontaneous values will be denoted by a subscript or superscript s.)

In case the interactions between P and u_{ii} are neglected ($Q_{i2} = 0$), the following relations between r_s^h and P_s^h, u_{ii}^{hs} can be found from the equilibrium conditions:

$$P_s^h = \pm \frac{\mu (r_s^h)^2}{k + 2\eta (r_s^h)^2}\,, \qquad u_{13}^{hs} = 0\,, \qquad u_{ii} = -\frac{\varrho_i (r_s^h)^2}{C_{ii}}. \tag{3}$$

If the Q_{i2} terms are taken into consideration, then for the spontaneous u_{ii}^{hs} we get

$$u_{ii}^{hs} = -\frac{\varrho_i}{C_{ii}} (r_s^h)^2 - \frac{Q_{i2}}{C_{ii}} (P_s^h)^2\,. \tag{4}$$

Taking into account the gradient invariants ($\sigma \neq 0$, $\delta \neq 0$) in the thermodynamic potential (1), the phase I already loses stability at $T_I = T_0 + \sigma^2/\alpha_0\delta$ in respect of the non-homogeneous parameters p and q, with wave vector magnitude $k = k_0$ [7]:

$$k_0 = \frac{|\sigma|}{\delta}. \tag{5}$$

From the equilibrium conditions the relations for P_s^m, u_{13}^{ms}, and u_{ii}^{ms} can be found easily. Neglecting Q-, $\varkappa$-, and η-terms, one obtains

$$P_s^m = -\frac{\mu}{k} (r_s^m)^2 \sin 2\varphi\,, \quad u_{13}^{ms} = -\frac{\tau}{C_{55}} (r_s^m)^2 \cos 2\varphi\,, \quad u_{ii}^{ms} = -\frac{\varrho}{C_{ii}} (r_s^m)^2\,. \tag{6}$$

If $Q_{i2} \neq 0$ and $\varkappa_i \neq 0$, then we get

$$u_{ii}^{ms} = -\frac{\varrho_i}{C_{ii}} (r_s^m)^2 - \frac{Q_{i2}}{C_{ii}} (P_s^m)^2 + \frac{\varkappa_i}{C_{ii}} (r_s^m)^2 \frac{d\varphi_s^m}{dx}. \tag{7}$$

Putting $\tilde{\beta}_2 = \beta_2 + (\mu^2/k) + (\tau^2/C_{55}) = 0$, i.e. neglecting anisotropic and $\varkappa_i$-terms in (1) (and also anisotropic invariants of higher orders of r), then we get simple solutions for the IC phase [7]:

$$(r_s^m)^2 = \frac{\alpha_0(T_I - T)}{\tilde{\beta}_1}\,, \qquad \varphi_s^m = k_0 x\,, \tag{8}$$

where k_0 is given by relation (5) and $\tilde{\beta}_1 = \beta_1 - (\mu^2/k) - (\tau^2/C_{55})$. If $\tilde{\beta}_2 \neq 0$, then the solution (8) is exact in the limiting case $T \to T_I$, and therefore T_I can be identified as the phase transition temperature into the IC phase; the transition is a second-order one. The solution (8) simply describes a rotation of the vector $\boldsymbol{r}_s^m$ with angular velocity k_0, with x in the p, q space, and it is called "harmonic approximation".

If the anisotropic energy is not negligible, then with decreasing temperature below T_I the anisotropy energy will prefer large regions of nearly homogeneous phase (different domains), which will be separated by transient regions (domain walls); thus a domain-like solution will be obtained (we believe the solution is valid near T_C). The representation points in p, q space will be more dense near the points corresponding to homogeneous stable solutions. Finally, the homogeneous C phase will be reached at $T = T_C$ (for details see, e.g., [7, 9 to 11]). The transition should be a first-order one [12].

In the IC phase the spontaneous polarization P_s^m and deformation u_{13}^{ms} are modulated and on an average they are equal to zero. The components u_{ii}^{ms} are according to (7) partly independent of x, and therefore they can give a macroscopic effect.

An influence of an external electric field $\boldsymbol{E} = (0, y, 0)$ on the IC structure can be studied easily near T_I only. Starting from the corresponding Gibbs function, changes δp, δq of TP's p, q due to an external electric field $\boldsymbol{E}$ can be derived. From the Euler equations we have

$$\delta p = p_1 \cos k_0 x\,, \qquad \delta q = q_1 \sin k_0 x\,, \tag{9}$$

where

$$p_1 = q_1 = \frac{\Delta\chi\, E}{2\chi_p r_s^m \mu}\,;$$

χ_m and χ_p are the free dielectric susceptibilities and $\Delta\chi = \chi_m - \chi_p$.

3. Phenomenology of Optical Effects in C and IC Phases of AFB

The optical properties of AFB will be related to the optical indicatrix $B_{ij}x_i x_j = 0$, where $B_{ij} = \delta E_i/\delta D_j$ at optical frequencies. From symmetry considerations the functional dependence of the changes ΔB_{ij} of the optical permittivity below T_C on the polarization P, the strain u_{ij}, and the polar coordinates r, φ of the TP's may be found [15, 18]. To the first approximation we have

$$\left.\begin{aligned}
\Delta B_1 &= \omega_1^* r^2 + R_{12}^* P^2 + \sum_{i=1}^{3} p_{1i} u_i = \omega_1 r^2 + R_{12} P^2\,,\\
\Delta B_2 &= \omega_2^* r^2 + R_{22}^* P^2 + \sum_{i=1}^{3} p_{2i} u_i = \omega_2 r^2 + R_{22} P^2\,,\\
\Delta B_3 &= \omega_3^* r^2 + R_{32}^* P^2 + \sum_{i=1}^{3} p_{3i} u_i = \omega_3 r^2 + R_{32} P^2\,,\\
\Delta B_5 &= \omega_5^* r^2 \cos 2\varphi + p_{55} u_5 = \omega_5 r^2 \cos 2\varphi = 0\,,
\end{aligned}\right\} \tag{10}$$

where R_{ij}^* and R_{ij} are the clamped and free quadratic electrooptic coefficients in phase I, respectively, the terms $\omega_i^* r^2$ represent the contribution of TP's p and q, and p_{ij} are elastooptic coefficients in the phase I; we use the Voigt notation. The free coefficients ω_i include the elastooptic contribution to ΔB_i due to the part of the strain produced by direct coupling of strain to the p, q.

The theoretical expression for the spontaneous birefringence Δn_{31}^{hs} in phase III can be derived from the preceding formulas (10):

$$\Delta n_{31}^{hs} = \tfrac{1}{2}\, R(P_s^h)^2 + \tfrac{1}{2}\, (r_s^h)^2\,, \tag{11}$$

where $R = n_3^3 R_{32} - n_1^3 R_{12}$ and $\omega = n_3^3\omega_3 - n_1^3\omega_1$.

In order to describe properly the optical changes in the IC phase, we must add gradient terms $v_i^* r^2 (\mathrm{d}\varphi/\mathrm{d}x)$ to (10). Since the wavelength λ_I of the modulated structure in AFB is still much shorter than λ of the visible light [5], the proper changes of the optical permittivity, ΔB_i, can be obtained by space averaging of all the contributions in (10).

In order to carry out this averaging and to determine $\mathrm{d}\varphi/\mathrm{d}x$, we have to know r_s^m, φ_s^m, and P_s^m as functions of x. According to the preceding section this is possible only for the temperature range close to T_I (harmonic approximation) or close to T_C (domain-like structure). In the former case we get

$$\Delta n_{31}^{ms} = \tfrac{1}{4}\, R(P_{0s}^m)^2 + \tfrac{1}{2}\, \omega'(r_s^m)^2\,, \tag{12}$$

where $P_{0s}^m = -\mu(r_s^m)^2/k$ and $\omega' = \omega + (n_3^3 v_3^* - n_1^3 v_1^*)\, k_0$, and in the latter one

$$\Delta n_{31}^{ms} = \tfrac{1}{2}\, R(P_{ds}^m) + \tfrac{1}{2}\, \omega(r_{ds}^m)^2\,, \tag{13}$$

where r_{ds}^m and P_{ds}^m are the spontaneous values of these variables in the domain-like structure (the domain walls are assumed to be very thin: $\ll \lambda_I$).

The linear electrooptic effect is allowed by symmetry in the phase III only (for details see [15]). In the phases I and II only the quadratic electrooptic effect can be measured. Using relation (9) we get from (10) in the temperature range close to T_I

$$Z = \frac{d^2\Delta n_{31}}{dE^2} = R(\chi_m)^2 + \frac{1}{4}\,\omega'\left(\frac{\Delta\chi}{\chi_p}\right)^2 \frac{1}{\mu^2(r_s^m)^2}, \tag{14}$$

and in phase I

$$Z = R(\chi_p)^2\,. \tag{15}$$

4. Survey of Experimental Results

Our published experimental data on the birefringence Δn_{31} [15] and Anistratov's ones on the quadratic electrooptic effect Z [14] will be used for a discussion.

As the spontaneous birefringence Δn_{31}^s was not explicitly published in [15], it is plotted in Fig. 1. These data have been obtained by subtracting the linearly extrapolated temperature dependence of Δn_{31} in phase I from data for phases II and III. The mark at T_C indicates the temperature of the ferroelectric PT obtained from our electric and electooptic measurements. T_I designates the temperature of the IC phase transition according to the specific heat measurements [11]. Our temperature scale is shifted by ≈ 2 K upwards in comparison with other works [1 to 3], which is within the accuracy of thermometer calibration. The measurements were carried out on cooling.

It can be seen on Fig. 1 that Δn_{31}^s is (within experimental error) a continuous function of temperature in the vicinity of T_C; it agrees well with results of other authors [13, 14, 16]. On the other hand, the first derivative $d\Delta n_{31}^s/dT$ is discontinuous at T_C. The tail at temperatures $T > T_I$ is attributable partly to IC phase fluctuations and partly to a blurring due to mechanical local strains.

As for the quadratic electrooptic effect, Anistratov found that the coefficient Z has an anomaly near T_C. R was found to be practically independent of temperature in the phases I and II.

5. Discussion and Conclusions

Equations (11), (12), and (13) form the basis for the interpretation of the spontaneous values $\Delta n_{31}^s(T)$ in the phases II and III; they contain the electrooptic contribution, the elastooptic one, and the contribution of TP's.

Let us first deal with the phase III for which sufficient data on spontaneous quantities are available. The electrooptic and elastooptic contributions can be estimated from coefficients R_{ij} and p_{ij}, which are measurable in phase I, and spontaneous values P_s^h and u_{ii}^{hs}; the contribution of TP's can be found by subtracting these from the measured Δn_{31}^{hs}. It was shown in [14, 15] that the electrooptic contribution to Δn_{31}^{hs} is very small, only 0.5%, and the elastooptic one could account for the whole spontaneous optical effect. Unfortunately, there are no data on the sign of p_{ij}; therefore two situations can occur: (i) The elastooptic contribution has the same sign as Δn_{31}^{hs}; then the PT contribution is small. (ii) The elastooptic contribution has the opposite sign to Δn_{31}^{hs}; then TP's play a prevailing role in the origin of the spontaneous birefringence. This problem can be solved by means of electrooptic measurements and will be discussed further.

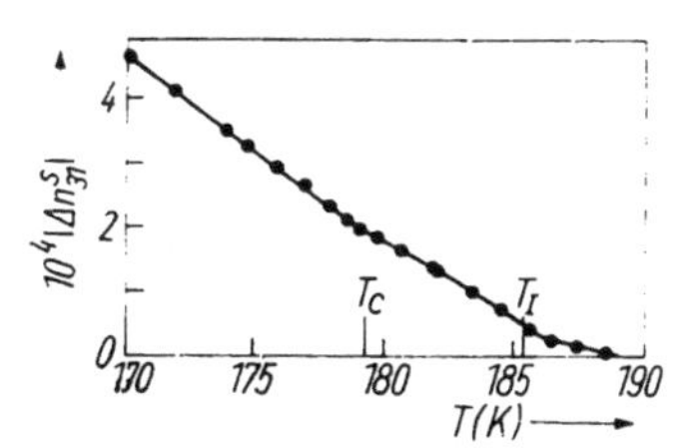

Fig. 1. Temperature dependence of the spontaneous birefringence Δn_{31}^s in the vicinity of both phase transitions

Let us discuss the IC phase. Starting from (10) and taking into account the results (3), (4) and (6), (7) one arrives at the conclusion that the relative contributions to Δn_{31}^{ms} close to T_C are the same as in phase III. The experimental fact that Δn_{31}^{s} is a continuous function of temperature at T_C allows us to draw the conclusion that $|r_s^h| \doteq |r_{ds}^m|$ at T_C. In other words, the IC–C phase transition in AFB should be nearly of second order.

Equations (14) and (15) can be used to discuss the temperature dependence of Z in phase I and II, respectively. The experimental result $Z = R\chi^2$ in both phases leads to the conclusion that $\omega' \doteq 0$ in (14). Thus the case (i) mentioned above could better explain all optical data known up to now.

As for the anomaly of $d\Delta n_{31}^{s}/dT$ at T_C, it can be qualitatively explained by generalized Pippard relations [19], considering the specific heat data showing an anomaly at T_C [1]. Since the coefficients of thermal expansion and thus the first derivatives of refractive indices should behave at PT like the specific heat, the anomaly mentioned above presents no mystery.

In the author's opinion the most interesting result is the equality $|r_s^h| \doteq |r_{ds}^m|$ at T_C. Unfortunately we have no complete set of optical experimental data for other IC–C phase transitions to look whether this result is accidental or not. Nevertheless, measurements of the spontaneous deformation can be discussed in the same way as the spontaneous optical data using equations (4) and (7). Very interesting experimental data on spontaneous deformation were published for K_2SeO_4 crystals [20]. Similar as in our Δn_{31}^{s} measurements, u_{ii}^{s} was found to be a continuous function of T at T_C, which gives the same equality for K_2SeO_4.

Acknowledgements

The author wishes to express his thanks to Dr. V. Dvořák for stimulated discussions, Dr. B. Březina for providing him with crystals of AFB, and Dr. J. Petzelt for critically reading the manuscript.

References

[1] B. A. Strukov, T. L. Skomorokhova, V. A. Koptsik, A. A. Boiko, and A. N. Izrailenko, Kristallografiya 18, 143 (1973).
[2] R. Pepinsky and J. Jona, Phys. Rev. **105**, 344 (1957).
[3] S. Hishino, K. Vedam, J. Okaya, and R. Pepinsky, Phys. Rev. **112**, 405 (1958).
[4] Y. Makita and Y. Yamauchi, J. Phys. Soc. Japan **37**, 1470 (1974).
[5] M. Iizumi and K. Gesi, Solid State Commun. **22**, 37 (1977).
[6] A. P. Levanyuk and D. G. Sannikov, Zh. eksper. teor. Fiz. **55**, 256 (1968).
[7] A. P. Levanyuk and D. G. Sannikov, Fiz. tverd. Tela **18**, 423 (1976).
[8] L. D. Landau and E. M. Lifshits, Statistical Physics, Pergamon Press, 1959 (p. 444).
[9] V. Dvořák and Y. Ishibashi, J. Phys. Soc. Japan **45**, 775 (1978).
[10] Y. Ishibashi and V. Dvořák, J. Phys. Soc. Japan **44**, 32 (1978).
[11] H. Shiba and Y. Ishibashi, J. Phys. Soc. Japan **44**, 1592 (1978).
[12] J. D. Axe, Proc. Conf. Neutron Scattering, Gatlinburg (Tenn.) 1976.
[13] B. A. Strukov, Kristallografiya **6**, 635 (1961).
[14] A. T. Anistratov, Rev. Phys. appl. **7**, 77 (1972).
[15] Č. Košák and J. Matras, Czech. J. Phys. **B26**, 577 (1976).
[16] A. T. Anistratov and S. V. Melnikova, Kristallografiya 18, 1289 (1973).
[17] Č. Košák, Czech. J. Phys. **B25**, 1073 (1975).
[18] J. Fousek and M. Glogarová, Solid State Commun. **17**, 97 (1975).
[19] V. Janovec, IV. Internat. Meeting Ferroelectricity, Leningrad, September 1977 (to be published in Ferroelectrics (1979)).
[20] Sh. Siozaki, A. Sawada, Y. Ishibashi, and Y. Takagi, J. Phys. Soc. Japan **42**, 353 (1977).

(Received March 22, 1979)

phys. stat. sol. (a) 54, 105 (1979)

Subject classification: 18.3; 22.8.2

Institute of Physics, Czechoslovak Academy of Sciences, Prague

Magnetocrystalline Anisotropy of Neodymium-Substituted Yttrium–Iron Garnet

By

E. Kratochvílová, V. Roskovec[1]), and V. Nekvasil

An experimental study of the low-temperature magnetocrystalline anisotropy in $Nd_{0.343}Y_{2.657}Fe_5O_{12}$ is presented. The method chosen at 4.2 K is the determination of anisotropy constants from magnetization curves, supplemented by torque measurements at temperatures up to 77 K. The resulting values of the anisotropy constants are compatible with the literature data obtained by different methods and on samples with widely differing neodymium concentration, confirming thus the essentially additive character of the rare-earths contribution. Special attention is paid to the assumptions used when evaluating the experimental data.

Была экспериментально исследована магнитокристаллическая анизотропия граната $Nd_{0,343}Y_{2,657}Fe_5O_{12}$. При 4,2 K были константы анизотропии определены из кривых намагничивания, при температурах до 77 K вращательным методом. Результаты согласаются с данными в литературе, полученными другими методами на образцах в широком диапазоне содержания неодима. Таким образом подтверждается адитивный характер вклада редкоземельных ионов к анизотропии. В дискуссии разобраны некоторые предложения лежащие в основе вычисления констант анизотропии из экспериментальных данных.

1. Introduction

The magnetocrystalline anisotropy of the garnet system $Nd_xY_{3-x}Fe_5O_{12}$ was studied by several authors [1 to 4].

Volkova and Rajcis [1] explored this system for $0 < x < 0.09$ in the temperature range 4.2 to 77 K, Krishnan and Rivoire [2] for $x = 0.002$, 0.006, and 0.01 at 77 K, both using the resonance method. The torque measurements in magnetic field up to 7 T were performed by Egami and co-workers [3, 4] for $x = 1.0$, 1.3, 1.6, and 1.8 at 77 K.

An advantage of the measurements carried out on the samples with low concentration of rare-earth ions is the possibility to achieve magnetic saturation even at liquid helium temperatures by readily accessible magnetic fields. A disadvantage appears to be the difficulty in the determination of low x, making uncertain the conclusions concerning the contribution of the rare-earth ions to the magnetocrystalline anisotropy.

The aim of this paper is to determine the constants of magnetocrystalline anisotropy at low temperatures for a sample with higher content of neodymium ions, for which a standard chemical analysis is suitable. The method chosen at 4.2 K was the determination of anisotropy constants from magnetization curves ($B_{max} = 5$ T), supplemented by torque measurements ($B_{max} = 1.5$ T) in the temperature range up to 77 K. In order to achieve saturation by our magnet we have chosen the Nd concentration in the sample $x \doteq 0.3$.

[1]) Now at Research Institute for Mathematical Machines, Loretánské nám. 3, 11855 Prague 1, Czechoslovakia.

As it is shown in Section 2 of this paper, we were able to determine:

1. the constants K_1 and K_2 at 4.2 K from the magnetization curves measured in directions [001], [110], and [111];

2. the values of $(K_1 + \frac{1}{3} K_2)$ in the temperature range 4.2 to 50 K from the straight parts of the "unsaturated" torque curves in the vicinity of the easy direction [111];

3. the constants K_1 and K_2 by Fourier analysis of the "saturated" torque curves observed at 77 K.

In Section 3 we discuss our results with emphasis on the assumptions used in the evaluation of the experimental data.

2. Experimental Methods and Results

Single crystals of the ferrimagnetic garnet of the composition $Nd_{0.343}Y_{2.657}Fe_5O_{12}$ were prepared by the standard flux method. Two spherical samples were ground, then polished and oriented by the X-ray method in the (110) plane. One of the samples (diameter $d = 2.87$ mm) was used in the torque measurements, the other one ($d = 2.29$ mm) in magnetization studies.

The magnetization curves at 4.2 K were determined by the induction method on an apparatus with a ballistic galvanometer, in directions [001], [110], and [111]. Magnetic fields up to 5 T were produced by a superconducting solenoid. The anisotropy constants were obtained from the magnetization curves shown in Fig. 1 by the following procedure.

It is well known [5] that as long as the magnetization process in the cubic crystal consists of the rotation of the magnetization vector only, the anisotropy constants K_1 and K_2 are given by the relations

$$K_1 = 4(W_{110} - W_{100}) , \qquad K_2 = 27(W_{111} - W_{100}) - 36(W_{110} - W_{100}) , \tag{1}$$

where W_{hkl} is the work required to turn the magnetization from the easy direction to the direction $[hkl]$.

In real conditions the Bloch wall displacements take place at the beginning of the magnetization process. Assuming that their contribution does not depend strongly on the applied field direction we can approximate the differences $W_{110} - W_{100}$ and $W_{111} - W_{100}$ by the differences of the total magnetization work in the corresponding directions. These differences are equal to the area between the magnetization curves and were obtained by numerical integration. This way we got the anisotropy constants $K_1 = -2.47 \times 10^5$ J m^{-3} and $K_2 = -4.22 \times 10^5$ J m^{-3}. We note that the results of the magnetization measurements do not contradict the above assumption about the nature of the magnetization process. If the remanent magnetizations M_{100} and M_{110} in the directions [100] and [110], respectively, result from the rotation of the magnetization vector from the easy direction, we have

$$\left. \begin{aligned} M_{100} &= M_{111} \cos 54.7^\circ = M_{111}\, 0.5773 , \\ M_{110} &= M_{111} \cos 35.3^\circ = M_{111}\, 0.8165 . \end{aligned} \right\} \tag{2}$$

For the experimental value of the saturation magnetization $M_{111} = 193$ kA m^{-1} and

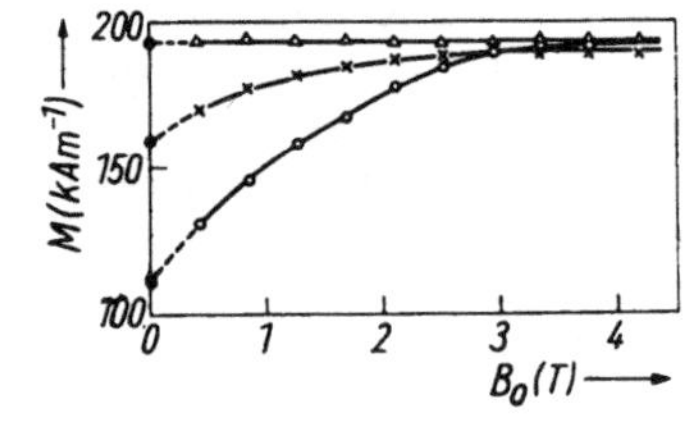

Fig. 1. Magnetization curves of $Nd_{0.343}Y_{2.657}Fe_5O_{12}$ at 4.2 K. ○ [100], × [110], △ [111], ● extrapolated value of M for $B_0 = 0$

the extrapolated values of the remanent magnetization $M_{100} = 111$ kA m^{-1} and $M_{110} = 159$ kA m^{-1} (see Fig. 1), we obtain the ratios

$$\frac{M_{100}}{M_{111}} = 0.575 \quad \text{and} \quad \frac{M_{110}}{M_{111}} = 0.824 \tag{3}$$

which satisfactorily agree with (2).

The torque measurements were performed on the compensated magnetometer in the magnetic field 1.5 T at temperatures 4.2 to 77 K. The anisotropy constants K_1 and K_2 of a cubic crystal are related to the torque in (110) plane by the equation

$$L_{110}(\varphi) = -\frac{K_1}{8}(2\sin 2\varphi + 3\sin 4\varphi) - \frac{K_2}{64}(\sin 2\varphi + 4\sin 4\varphi - 3\sin 6\varphi)\,, \tag{4}$$

where L_{110} represents the torque exerted by a unit volume of the specimen and φ is the angle between the magnetization vector and [001] direction.

At liquid nitrogen temperature the "saturated" torque curves in the (110) plane were obtained (the magnetization vector pointed nearly in field direction). Their Fourier analysis based on (4) yields at 77 K the anisotropy constants $K_1 = -3.7 \times 10^4$ J m^{-3} and $K_2 = -0.6 \times 10^4$ J m^{-3}.

At lower temperatures (4.2 to 50 K) the anisotropy field is so high that only the straight parts of the "unsaturated" torque curves in the vicinity of the easy direction [111] were accessible. If φ_0 denotes the angle between the easy direction [111] and the direction [001] ($\varphi_0 = 54.7°$), then for a small deviation $\Delta\varphi$ of the direction of the magnetization from φ_0 we get from (4)

$$L_{110}(\varphi_0 + \Delta\varphi) \doteq L_{110}(\varphi_0) + \left(\frac{\mathrm{d}L_{110}}{\mathrm{d}\varphi}\right)_{\varphi=\varphi_0} \Delta\varphi = \frac{4}{3}\left(K_1 + \frac{1}{3}K_2\right)\Delta\varphi\,. \tag{5}$$

During the experiment we do not know the angle φ but the angle Φ between the applied magnetic field $\boldsymbol{B}_0$ and the direction [001]. The angles $\Delta\Phi \equiv \Phi - \varphi_0$ and $\Delta\varphi \equiv \varphi - \varphi_0$ are connected by the relation

$$\Delta\varphi = \Delta\Phi - \frac{\bar{L}_{110}(\Phi)}{MB_0}\,, \tag{6}$$

where $\bar{L}_{110}(\Phi)$ is the measured torque and M the magnitude of the magnetization vector in the applied magnetic field B_0.

Using (6) we were able to construct the required functions $L(\varphi) = \bar{L}(\Phi)$ (see Fig. 2). From the straight parts of the $L_{110}(\varphi)$ curves in the vicinity of the easy direction [111] we determined, using (5), the values of $K_1 + \frac{1}{3}K_2$ in the temperature range 4.2 to 50 K. The results are shown in Fig. 3. Also shown in this figure are the values $K_1 + \frac{1}{3}K_2 = -3.88 \times 10^5$ J m^{-3} obtained at 4.2 K from the magnetization measurements and -3.9×10^4 J m^{-3} obtained at 77 K by complete Fourier analysis.

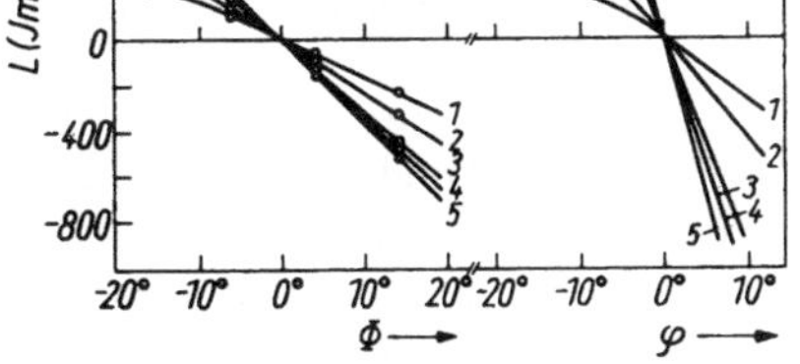

Fig. 2. The measured torque curves $\bar{L}_{110}(\Phi)$ in the vicinity of easy direction [111] and the corresponding functions $L(\varphi)$ for $Nd_{0.343}Y_{2.657}Fe_5O_{12}$. (1) $T = 50$, (2) 40, (3) 27.5, (4) 23, (5) 4.2 K

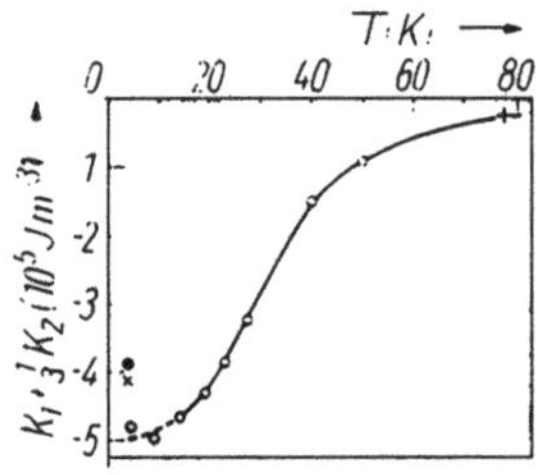

Fig. 3 Temperature dependence of $K_1 + \frac{1}{3} K_2$ for $Nd_{0.343}Y_{2.657}Fe_5O_{12}$; ○ torque method, + Fourier analysis of the torque curve at 77 K, ● from magnetization curves, × resonance method by Volkova and Rajcis [1]

We estimate that the experimental error associated with our values of the anisotropy constants is $\leqq 5\%$ in the case of the magnetization study and somewhat smaller in other cases.

To compare our results with the data obtained by other authors for different Nd concentrations we have calculated the anisotropy constants per Nd^{3+} ion. We have neglected the contribution of the iron sublattice as the anisotropy constants for $Y_3Fe_5O_{12}$ are at 4.2 K: $K_1 = -2.48 \times 10^3$ J m^{-3}, $K_2 = -1.0 \times 10^3$ J m^{-3}; at 77 K: $K_1 = -2.20 \times 10^3$ J m^{-3} [6].

As some authors describe the magnetocrystalline anisotropy by means of the constant $\tilde{K}_4 = K_1 + K_2/11$ only, we have rewritten all available data to $\tilde{K}_4$ and summarized them in Table 1. (The unit used is K/Nd^{3+} ion.)

Table 1
Constant $\tilde{K}_4$ per Nd^{3+} ion in $Nd_xY_{3-x}Fe_5O_{12}$ garnets

reference	x	$\tilde{K}_4$ at 77 K (K/ion)	$\tilde{K}_4$ at 4.2 K (K/ion)
this work	0.343	− 1.9	−14.5
[4]	independent of x	− 2.09 ± 0.01	
[3]	1.8	− 2.5*)	—
[1]	0.03	− 2.3**)	
	0.015	− 3.2**)	−14.1**)
[2]	0.002	−31.6***)	
	0.01	−25.3***)	—

*) $\tilde{K}_4$ calculated from K_1 and K_2 given in [3].

**) $\tilde{K}_4$ calculated from K_1/M_s and K_2/M_s given in [1] taking $M_s = 185$ kA m^{-1}.

***) $\tilde{K}_4$ calculated from K_1/M_s ($M_s = 185$ kA m^{-1}) only, as in [2] the value of K_2 is not given.

3. Discussion

From Fig. 3 a good consistency between our results obtained from the magnetization curves at 4.2 K and by the torque method can be seen. The difference can be ascribed to the experimental errors given in Section 2.

As seen from the data summarized in Table 1, there is a reasonable agreement of the results obtained by other authors [1, 3, 4] using various methods and samples of different provenance widely differing in the neodymium content. The only exception are the data of Krishnan and Rivoire [2] at 77 K. The values of $\tilde{K}_4$ calculated from K_1/M_s taken from Fig. 1 in [2] are an order of magnitude greater than the other ones.

We have no explanation for this discrepancy. The results given in Table 1 also indicate that the neodymium contribution to the magnetocrystalline anisotropy in ferrimagnetic garnets is essentially independent of its concentration.

In the further discussion we shall focus our attention to some tacit assumptions made during our evaluation of the anisotropy constants from experimental data, namely that K_1, K_2 suffice the description of anisotropy and that they are field-independent. As the present status of the theory does not allow to account for the experimental data on Nd^{3+}-substituted YIG quantitatively, we base our discussion mainly on the comparison of our results with those obtained in ytterbium iron garnet (YbIG).

First, we wish to test the effect of the higher-order anisotropy constants appearing in a general expansion for the free energy, neglected in our considerations so far. The need to include these terms appears whenever the exchange splitting of the ground doublet is strongly anisotropic [7]. It was shown by Henderson and White [8], however, that in the case of Yb^{3+} in the dodecahedral sites of YbIG the first two constants K_1, K_2 suffice to reproduce the anisotropy of the free energy fairly well.

Here, we wish to point out that the ground state splittings of Nd^{3+} and Yb^{3+} do not differ appreciably. The exchange splitting of ≈ 26 and $\approx 35\ cm^{-1}$ in two inequivalent sites of Nd^{3+} in YIG for magnetization along the easy direction [111] reported by Faulhaber et al. [9] is comparable with the splitting of 22 and 25 cm^{-1} observed by Wickersheim in [12]; the paramagnetic g-tensors in different Nd^{3+} and Yb^{3+} garnets [10] are of similar anisotropy. We may reasonably expect that K_1, K_2 suffice in the case of Nd^{3+} in YIG as well.

Before adopting the above arguments we would mention the anomaly in the resonance spectra observed by Volkova and Rajcis [1]. It is known that the appearance of such an anomaly is tied to the near-coincidence of the two lowest levels for the applied field along certain directions, as it is e.g. the case for Yb^{3+} in the octahedral sites of garnet [11]. In view of the fact that no anomaly phenomena have been observed in the resonance experiments of Clarke [13], however, we take the presence of the other rare-earth ions in the samples as the most probable explanation of the anomaly reported by Volkova and Rajcis [1].

The values of K_1, K_2 derived in Section 2 from the magnetization curves are generally effected by a high applied field. In order to get a rough estimate of its effect we have used the effective Hamiltonian $\mathcal{H}_{\text{eff}}$ with the spin $s = 1/2$ to describe the ground state splitting of Nd^{3+} in YIG

$$\mathcal{H}_{\text{eff}} = \mu_B \sum_i (\boldsymbol{H}_0 \mathbf{g}_i \boldsymbol{s}_i + \boldsymbol{H}_e \mathbf{G}_i \boldsymbol{s}_i)\,, \tag{7}$$

where $\boldsymbol{H}_0$ is the applied field, $\boldsymbol{H}_e$ a unit vector parallel to the iron sublattice magnetization, $\mathbf{G}$ the anisotropic exchange tensor, and the sum in i runs over inequivalent sites. The free energy associated with one rare-earth ion is then given by

$$F = -kT \sum_i \ln \left[\sum_n \exp\left(-\frac{E_n^i}{kT}\right)\right], \tag{8}$$

where E_n^i, $n = 0, 1$, are the eigenvalues of (7). We wish to emphasize here that the Hamiltonian (7) represents in our case a rather crude approximation [12] and all the results obtained using this Hamiltonian are only of a qualitative validity. In the calculation we have taken the g-values reported in yttrium gallium garnet by Wolf et al. [10] and the G-values reported by Clarke [13]. Considering the experimental field 0 to 5 T, substituting the eigenvalues of the Hamiltonian (7) into the expression (8) for the free energy F, and putting the differences $(F_{hkl} - F_{h'k'l'})$ instead of $(W_{hkl} - W_{h'k'l'})$ into (1) we find that the magnetic field contribution to our experimental values of K_1, K_2 is $\leqq 10\%$.

Acknowledgements

The authors are indebted to Dr. S. Krupička for a critical reading of the manuscript, Mr. P. Pechatý for growing the crystals, Mr. J. Vaněk for X-ray orientation, and Dr. F. Zounová for the assistance at the magnetization measurement.

References

[1] N. V. VOLKOVA and V. I. RAJCIS, Zh. eksper. teor. Fiz. **65**, 688 (1973).
[2] R. KRISHNAN and M. RIVOIRE, AIP Conf. Proc. **18**, 1227 (1973).
[3] P. J. FLANDERS, T. EGAMI, E. M. GYORGY, and L. G. VAN UITERT, AIP Conf. Proc. **24**, 666 (1974).
[4] T. EGAMI, P. J. FLANDERS, E. M. GYORGY, and L. G. VAN UITERT, Physica (Utrecht) **86-88 B+C**, 1463 (1977).
[5] B. D. CULLITY, Indroduction to Magnetic Materials, Addison-Wesley Publ. Co., Reading (Mass.) 1972 (p. 232).
[6] G. P. RODRIGUE, H. MEYER, and R. V. JONES, J. appl. Phys. **31**, 376 S (1960).
[7] B. A. CALHOUN, M. J. FREISER, and R. F. PENOYER, J. appl. Phys. **33**, 1259S (1962).
[8] J. W. HENDERSON and R. L. WHITE, Phys. Rev. **123**, 1627 (1961).
[9] R. FAULHABER, S. HÜFNER, E. ORLICH, H. SCHMIDT, and H. SCHUCHERT, in: Optical Properties of Ions in Crystals, Ed. H. M. CROSSWHITE and H. W. MOOS, Interscience Publishers, Inc., New York 1967 (p. 329).
[10] W. P. WOLF, M. BALL, M. T. HUTCHINGS, M. J. M. LEASK, and A. F. G. WYATT, J. Phys. Soc. Japan, Suppl. **17**, **B-1**, 443 (1962).
[11] R. S. ALBEN, Phys. Rev. **167**, 249 (1968).
[12] K. A. WICKERSHEIM, in: Magnetism, Ed. G. T. RADO and H. SUHL, Academic Press, Inc., New York 1963 (p. 269).
[13] B. H. CLARKE, Phys. Rev. **139**, A1944 (1965).

(Received March 29, 1979)

phys. stat. sol. (a) 54, 111 (1979)

Subject classification: 8; 21.1

Institute of Physics, Polish Academy of Sciences, Warsaw[1]) *(a)*
and Institute of Physics, Warsaw Technical University[2]) *(b)*

Rounding of Specific Heat of Nickel near the Critical Temperature

By

M. MASZKIEWICZ (a), B. MRYGOŃ (a), and K. WENTOWSKA (b)

Measurements of the specific heat of nickel crystals of different crystalline and chemical perfection are made. A shape dependence of $C(T)$ peaks on imperfections is observed. However, the effect of rounding is still present even for a very high purity single crystal, suggesting that it may be an intrinsic property of phase transition. The specific heat calculations which are based on a theory taking into account the fluctuations of magnetic moments are satisfactorily fitted to the experimental data.

Es werden Messungen der spezifischen Wärme von Nickelkristallen verschiedener kristalliner und chemischer Perfektion durchgeführt. Eine Formabhängigkeit der $C(T)$-Maxima von den Defekten wird beobachtet. Jedoch bleibt der Einfluß der Verrundung noch für Einkristalle sehr hoher Reinheit erhalten, was auf eine Grundeigenschaft des Phasenübergangs hinweist. Die Berechnungen der spezifischen Wärme, die auf einer Theorie beruhen, die die Fluktuationen der magnetischen Momente berücksichtigt, lassen sich befriedigend an experimentelle Werte anpassen.

1. Introduction

It has been pointed out many times that the experimental data of specific heat measurements near the temperature of a magnetic second-order phase transition demonstrate the effect of rounding. It means that the experimental data deviate close to the critical temperature T_c from the customary power-law relation

$$C = \frac{A^\pm}{\alpha^\pm}\,\varepsilon^{-\alpha^\pm} + B^\pm\,, \tag{1}$$

where $\varepsilon = |T - T_c|/T_c$, $\alpha^\pm$ are critical exponents, $A^\pm$ and $B^\pm$ are constants. Signs "+" and "−" correspond to $T > T_c$ and to $T < T_c$, respectively. According to [1] for $\alpha^\pm > 0$ the specific heat diverges to infinity, for $\alpha^\pm < 0$ has a sharp but finite cusp, and for $\alpha^\pm = 0$ has a logarithmic dependence.

In most cases for magnetics the value of ε at which the effect of rounding becomes evident is about 10^{-3}, when for comparison the experimental possibilities allow us to measure ε up to values of 10^{-5} to 10^{-6}.

Typically one argues that the main source of rounding are the chemical or physical imperfections in an ideal pure sample. Generally the compositional disorder is divided into two categories: annealed and quenched impurities [1]. In the first case, which seems to be satisfactorily described by Fisher and Scesney [2], one expects the ordinary critical behaviour with renormalized values of the critical exponents only. If the values α for the pure system are positive, like for Ising models, one obtains for the non-ideal system modified values α_R which are negative, and thus finite values of the

---

[1]) Al. Lotników 32/46, 02 668 Warsaw, Poland.
[2]) Koszykowa 75, Warsaw, Poland.

specific heat at T_c. However, the procedure of renormalization is not applicable for Heisenberg models which exhibit negative values of α just for ideal systems [3]. In the second case of quenched impurities the situation is less clear.

McCoy and Wu [4] have solved exactly a disordered two-dimensional Ising model in which the impurities exhibit long-range correlations. As a result they obtained finite and rounded specific heat peaks instead of the logarithmic divergence.

On the other hand, the Syozi and Miyazima model [5] for a dilute ferromagnet displays finite values of specific heat but with a cusp with vertical tangent at T_c.

For the case of short-range disorder, Harris [6] has considered Ising models with local fluctuations in interaction strength. One of his conclusions is that if α is positive, the specific heat divergence is really rounded, but if α is negative, then the addition of disorder to a system will not affect the sharpness of the second-order phase transition.

Recently Lubensky [7] and Grinstein and Luther [8] have studied the critical behaviour of spin systems with quenched short-range disorder by renormalization-group methods. Their calculations confirm fully the rather heuristic argument of Harris.

An experimental support of the above-cited theoretical predictions was given by Schowalter et al. [9]. They showed that the specific heat in an amorphous ferromagnet has critical exponents and the value ε at which rounding becomes evident is almost the same like for pure Heisenberg systems.

Similarly the measurements [10 to 12] performed on DAG, $RbMnF_3$, and $MnBr_2 \cdot 4\,H_2O$ do not indicate any significant effect of impurities or defects on the specific heat near T_c. In turn, the recent experiment by one of us [13] on nickel showed that rounding is still present even for high-purity monocrystals. So, the additional conclusion which may be drawn is that as local and quenched defects and impurities are more characteristic of real physical systems than non-local ones, we do not obtain from the renormalization-group approach an answer about the nature of this rounding which is observed in most experiments on critical phenomena.

The possible answer could be a presumable subdivision of the sample into an array of macroscopic clusters of impurities each with slightly different critical temperature. However, the situation still remains unclear in the case of high-purity monocrystals.

The consequence of lattice compressibility on the critical behaviour of magnetics has been considered by several authors (see, e.g., [14]). However, Guttman [15] has estimated for the Baker-Essam model that this effect can have only a negligibly small influence on the specific heat near T_c for a wide range of real crystals. This conclusion, however, cannot be very general as the model is notalthogether realistic.

In this paper we present the results of our measurements of the specific heat of nickel samples having different degree of crystalline or chemical perfection. It is shown that the effect of rounding may be described qualitatively by the fluctuation theory of phase transitions which is based on the ideas of Smoluchowski [16] and Einstein [17] connected with the theory of critical opalescence, and then developed by Kociński and co-workers, (see e.g., [18]). Calculations of the specific heat have been done by Mrygoń and Wentowska [19] for a ferromagnet and by Maszkiewicz [20] for an antiferromagnet.

2. Experiment and Samples

A standard ac calorimetry was used. The apparatus and the sample preparation method used in this experiment are the same as those described previously [13]. The description of the samples is given in Table 1.

Sample 1 was prepared from a different nickel rod than samples 2 and 3 and has a not very uniform distribution of the cobalt impurity. The spectral X-ray analysis done in four points on its surface showed 0.0, 0.1, 0.2 and 0.2 at% Co. Sample 3 was the same as in [13]. Sample 2 was identical with sample 3, but before its mounting in the calorimeter it was rolled up and then flattened again.

Table 1

sample		main impurities
1	single crystal	0.2 at% Co
2	single crystal, deformed	0.001 at% Si, Fe 0.002 at% Co, Mn
3	single crystal	—

3. Experimental Results and Power-Law Analysis

A direct comparison of the experimental data of specific heat of all samples near the critical temperature is given on Fig. 1 and 2. One can observe a rather strong dependence of the shape of the specific heat curve on the kind of sample. The highest and sharpest peak corresponds to the "best" sample 3. The qualitative comparison of the other two samples is easier on Fig. 2 which shows that for $T > T_c$ the peak of $C(T)$ of the deformed sample 2 is sharper although earlier rounded than in the case of the "dirty" sample 1.

The quantitative analysis of the experimental data was done in the way described in [13], mainly in order to determine the range $t = \lg \varepsilon$ over which (1) is valid. The results are presented in Table 2.

Table 2

sample	$t = \lg \varepsilon$		T_c	$\alpha^+ = \alpha^-$
	$T < T_c$	$T > T_c$		
1	$-2.9 \leqq t \leqq -2.0$	$-3.2 \leqq t \leqq -2.0$	$T_m + 0.45$ K	-0.10
2	$-1.7 \leqq t \leqq -1.0$	$-2.6 \leqq t \leqq -1.2$	T_m	-0.10
3	$-2.8 \leqq t \leqq -1.5$	$-3.0 \leqq t \leqq -1.6$	T_m	0.0

T_m in Table 2 denotes the position of the maxima of the specific heat curves on the temperature scale. The values of T_c given in Table 2 are those which produced the best least-squares fit.

The measurements of specific heat performed on nickel by other authors [21, 22] gave $-3.2 \leqq t \leqq -1.5$ for 99.8% pure nickel polycrystalline foil and $-3.2 \leqq t \leqq$ $\leqq -1.6$ for 99.9% pure nickel single crystal. These results, together with that for sample 3, show that the value $t = -2.8$ to -3.2 may be a limiting one for the case of nickel, independently of the crystalline or chemical perfection of the sample, providing the disorder is homogeneous, that is no macroscopic clusters of impurities or defects exist. The biggest rounding which was displayd by sample 2 may be caused by macroscopic damages of the structure. Similarly, for sample 1 in which macroscopic clustering of impurities was observed, the low value of $t = -3.2$ for $T > T_c$ was caused by its low value $t = -2$ at the other end of the critical region and by the very high displacement of T_c versus T_m.

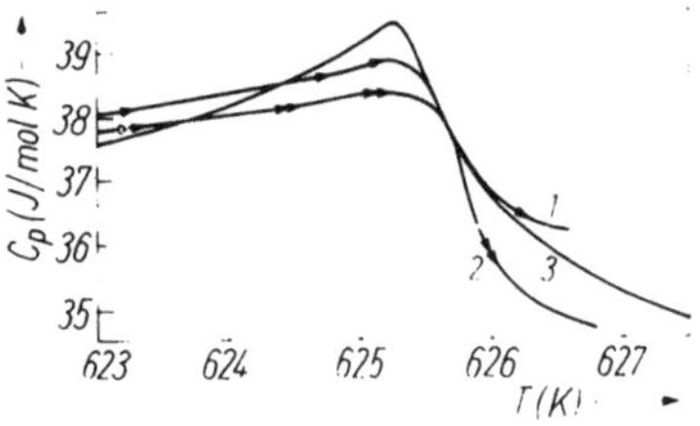

Fig. 1

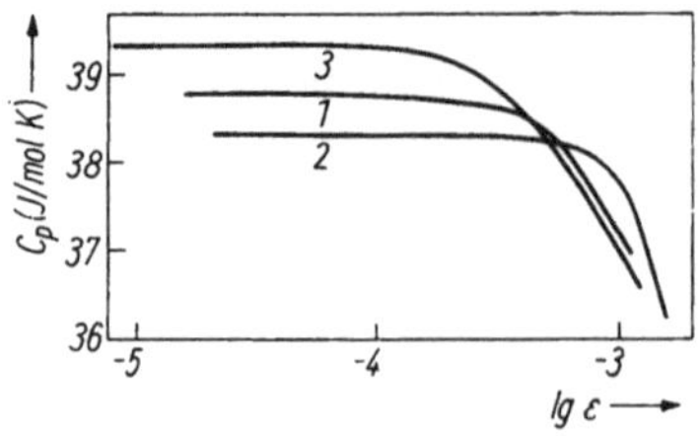

Fig. 2

Fig. 1. Comparison of $C(T)$ data near the critical temperature. (1) Sample 1, (2) 2, (3) 3

Fig. 2. The same as Fig. 1 but for C versus lg ε. Region of rounding $T > T_c$; $T_c = T_m$

4. Fluctuation Model of Phase Transition

As the details of calculations were described previously [19] we mention in the present paper only the main features of our model.

The starting point is the observation that near the critical temperature the system should show up a grain-like structure which grows rapidly when the critical point is approached. Each grain is identified as a single fluctuation of magnetic moment and is treated as a subsystem immersed in a reservoir. The fluctuations remain in a non-equilibrium state with respect to the whole system, but we assume local equilibrium.

Each fluctuation is sufficiently large to apply statistical mechanics to described its properties. The fluctuations do not interact, which is equivalent to the statement that they are not correlated. Correlation only inside the fluctuations is taken into account by the assumption that the magnetic moment distribution in space and the space dependence of the correlation function are equivalent. This correlation function is derived as a solution [23] of a non-linear differential equation instead of a linear one which corresponds to the Ornstein-Zernike approximation. As a result a correlation function of finite range at the critical point is obtained. As a simplification the whole system is assumed to be the sum of N_f equal fluctuations. So, the total free energy ΔF of the system due to fluctuations is equal to

$$\Delta F = N_f \langle \Delta F_1 \rangle \,, \tag{2}$$

where $\langle \Delta F_1 \rangle$ is the mean value of the free energy of a single fluctuation. The mean number of fluctuations N_f can be expressed by the mean number of spins in the fluctuation, which in turn can be determined from the condition for the maximum probability for the occurrence of a fluctuation.

5. Comparison of Theory with Experiment and Conclusions

Detailed calculations of specific heat due to fluctuations were performed according to

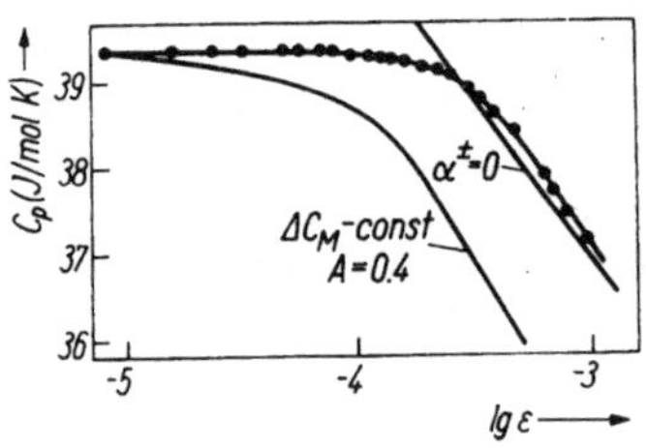

Fig. 3. The specific heat of sample 3 in the region of rounding (dotted line) together with a power-law fit (1) for $\alpha^{\pm} = 0$ (solid line) and with the result of our calculations ($T > T_c$, $T_c = T_m$). ΔC_M – const denotes that a constant was subtracted from the absolute values of ΔC_M in order to obtain the best qualitative agreement with experimental data. The experimental and theoretical values are equated at the point lg $\varepsilon = -5$. A is the single parameter of fitting

the relation

$$\Delta C_M = -T \frac{\partial^2}{\partial T^2} (\Delta F) , \tag{3}$$

using the constant coupling approximation, for $T > T_c$ only.

A comparison of the experimental data very near T_c with the theory is given on Fig. 3 for the case of the "best" sample 3. Our theoretical dependence of specific heat versus temperature near T_c contains only one fitting parameter A, which has the meaning of the mean value of the magnetic moment at the origin of the fluctuation.

It appears that the results obtained for the fluctuation model of a ferromagnet are in much better qualitative agreement with the experimental data than those based on the method of critical exponents. In particular, the effect of rounding was predicted as an intrinsic property of the phase transition. As we believe, this means that our fluctuation model of the phase transition may be generally adequate to the physical reality commonly observed experimentally.

References

[1] R. Brout, Phys. Rev. **115**, 824 (1959).
[2] M. E. Fisher, Phys. Rev. **176**, 257 (1968).
M. E. Fisher and P. E. Scesney, Phys. Rev. A **2**, 825 (1970).
[3] E. Brezin, J. C. Le Guilloui, and J. Zinn-Justin, Phys. Letters A **47**, 285 (1974).
[4] B. M. McCoy and T. T. Wu, Phys. Rev. **176**, 631 (1968).
[5] I. Syozi and S. Miyazima, Progr. theoret. Phys. (Kyoto) **36**, 1083 (1965).
[6] A. B. Harris, J. Phys. C **7**, 1671 (1974).
[7] T. C. Lubensky, Phys. Rev. B **11**, 3573 (1975).
[8] G. Grinstein and A. Luther, Phys. Rev. B **13**, 1329 (1976).
[9] L. J. Schowalter, M. B. Salamon, C. C. Tsuei, and R. A. Craven, Solid State Commun. **24**, 525 (1977).
[10] D. P. Landau, B. E. Keen, B. Schneider, and W. P. Wolf, Phys. Rev. B **3**, 2310 (1971).
[11] D. T. Teaney, V. L. Moruzzi, and B. E. Argyle, J. appl. Phys. **37**, 1122 (1966).
[12] R. D. Hempstead and J. M. Mochel, Phys. Rev. B **7**, 287 (1973).
[13] M. Maszkiewicz, phys. stat. sol. (a) **47**, K77 (1978).
[14] H. Wagner, Phys. Rev. Letters **25**, 31, 261E (1971);
G. A. Baker, Jr. and J. W. Essam, Phys. Rev. Letters **24**, 447 (1970); J. chem. Phys. **55**, 861 (1971).
[15] L. Guttman, Phys. Rev. B **2**, 1432 (1970).
[16] M. Smoluchowski, Ann. Phys. (Leipzig) **25**, 205 (1908).
[17] A. Einstein, Ann. Phys. (Leipzig) **33**, 1275 (1910).
[18] J. Kociński, L. Wojtczak, and B. Mrygoń, Acta Phys. Polon. **A43**, 425 (1973).
B. Mrygoń, J. Stat. Phys. **10**, 471 (1974).
J. Kociński, in: Summer School on the Theory of Magnetism of Metals, Zakopane 1970, Polish Scientific Publishers, Warszawa 1971 (p. 167).
[19] B. Mrygoń and K. Wentowska, Acta Phys. Polon. **A51**, 207 (1977).
[20] M. Maszkiewicz, Acta Phys. Polon. **A54**, 337 (1978).
[21] P. Handler, D. E. Mapother, and M. Rayl, Phys. Rev. Letters **19**, 356 (1967).
[22] D. L. Connelly, J. S. Loomis, and D. E. Mapother, Phys. Rev. B **3**, 924 (1971).
[23] B. Mrygoń and L. Wojtczak, phys. stat. sol. (b) **93**, 105 (1979).

(Received March 7, 1979)

phys. stat. sol. (a) 54, 117 (1979)

Subject classification: 1 and 18.3; 22.8.2

Research Institute of Electrical Communication, Tôhoku University, Sendai[1])

Phenomenological Theory of Magnetostriction and Growth-Induced Anisotropy in Garnet Films

By

H. SZYMCZAK[2]) and N. TSUYA

A phenomenological theory of magnetostriction in LPE garnet films is presented. In calculations the second- and fourth-order terms in the direction cosines of the normal to the substrate are included. The theory is model-independent, based only on symmetry considerations. Simultaneously, on the basis of the theory developed, the phenomenological model of growth-induced anisotropy in LPE garnet films is extended.

Es wird eine phänomenologische Theorie der Magnetostriktion in LPE-Granatschichten dargestellt. In die Berechnung werden die Terme zweiter und vierter Ordnung in den Richtungscosinus der Substratnormale einbezogen. Die Theorie ist modellunabhängig und beruht nur auf Symmetriebetrachtungen. Gleichzeitig wird auf Grundlage der entwickelten Theorie das phänomenologische Modell der wachstumsinduzierten Anisotropie in LPE-Granatschichten erweitert.

1. Introduction

The recent broad interest in the growth-induced magnetic anisotropy in LPE (liquid phase epitaxy) garnet films has produced a number of questions about the origin of that anisotropy. Two mechanisms have been proposed to explain this phenomenon: pair-ordering model [1] and site-preference model [2, 3], but it has not been clarified which mechanism is responsible for the induced noncubic anisotropy in garnet films. All the mechanisms are subject to the same symmetry considerations which, in turn, determine the symmetry properties of the induced anisotropy. Therefore, the growth-induced anisotropy can be treated in a phenomenological manner [4] by using the symmetry properties of the system. It was shown [4] that under certain plausible assumptions all models must reduce to only two adjustable parameters. The "two-parameter" model gives comparatively good quantitative agreement with experimental data. In order to explain the origin of the noncubic anisotropy in LPE garnet films one should analyze other phenomena, determined by the same mechanism as the anisotropy. It is well established that mechanisms determining magnetostriction and magnetocrystalline anisotropy are just strongly related. There is also a second reason to study more attentively magnetoelastic interactions in garnets. It was observed that magnetostriction can give a significant contribution to the anisotropy of LPE films. LPE films are normally in a state of mechanical stress due to mismatch between lattice constants and thermal expansions of film and substrate. If the film is magnetostrictive, this stress produces a uniaxial magnetic anisotropy which is superimposed on the unstrained anisotropy [5]. From general considerations one can expect that the magnetoelastic energy in LPE garnet films has noncubic character. In fact, Yang and Muller [6] have reported indirect evidence for the noncubic character of the magnetoelastic energy in the LPE (111) GdTmY garnet film. Further,

[1]) Sendai, Japan.

[2]) Permanent address: Institute of Physics, Polish Academy of Sciences, Warsaw, Poland

Muller [7 to 9] has theoretically studied the noncubic magnetostriction in LPE garnet films using the site-preference model. Since Muller's model is formulated rather for some particular growth facets, the main purpose of this paper is to develop a phenomenological theory of magnetostriction in LPE garnet films, which from its origin will be model independent. Simultaneously, on the basis of the developed theory, the two-parameter model of growth-induced anisotropy in LPE garnet films will be extended.

2. Theory of Magnetostriction in LPE Garnet Films

The magnetoelastic energy of LPE garnet films is assumed to be a function of strain ε_{ij}, direction cosines of magnetization α_k, and direction cosines of the normal to the substrate β_l:

$$E_{\mathrm{me}} = E_{\mathrm{me}}(\varepsilon_{ij}, \alpha_k, \beta_l) \,. \tag{1}$$

If we now expand E_{me} and assume that the lowest-order terms are of first order (linear) in ε_{ij} and of second and fourth order in α_k and β_l, the magnetoelastic energy E_{me} may be written as (odd orders of α_i and β_j are, of course, forbidden):

$$E_{\mathrm{me}} = \sum_{ijklmn} A_{ijklmn}\alpha_i\alpha_j\beta_k\beta_l\varepsilon_{mn} + \sum_{\substack{ijkl\\mnpr}} B_{ijklmnpr}\alpha_i\alpha_j\beta_k\beta_l\beta_m\beta_n\varepsilon_{pr} + \\ + \sum_{\substack{ijkl\\mnpr}} C_{ijklmnpr}\alpha_i\alpha_j\alpha_k\alpha_l\beta_m\beta_n\varepsilon_{pr} \,. \tag{2}$$

The form of the tensors **A**, **B**, and **C** is determined by crystal symmetry. The following considerations will be restricted to cubic symmetry (classes T_d, O and O_h) only. Furthermore, to reduce the number of independent constants the "hidden" symmetries, which imply relationships between some of the parameters not given by the crystal symmetry alone, are to be examined. For example, conditions for traceless tensors

$$\sum_{i=1}^{3} A_{iiklmn} = 0 \,, \qquad \sum_{k=l}^{3} A_{ijkkmn} = 0 \quad \text{(and similarly for tensors } \mathbf{B} \text{ and } \mathbf{C}) \,,$$

lead to some additional relations between the parameters of theory. Finally, particular terms in (2) may be written in the following form:

$$\sum_{ijklmn} A_{ijklmn}\alpha_i\alpha_j\beta_k\beta_l\varepsilon_{mn} = \varepsilon_{xx}\{A_1\beta_1^2(3\alpha_1^2 - 1) + A_2[\beta_2^2(3\alpha_3^2 - 1) + \\ + \beta_3^2(3\alpha_2^2 - 1)] + A_3\alpha_2\alpha_3\beta_2\beta_3 + A_4(\alpha_1\alpha_3\beta_1\beta_3 + \alpha_1\alpha_2\beta_1\beta_2)\} + \\ + \varepsilon_{yz}\{A_5(\alpha_1\alpha_3\beta_1\beta_3 + \alpha_1\alpha_2\beta_1\beta_2) + (A_3 - A_4)\,[\beta_2\beta_3(\alpha_1^2 - \tfrac{1}{3}) + \\ + \alpha_2\alpha_3(\beta_1^2 - \tfrac{1}{3})]\} + \text{c.p.} \,, \tag{3}$$

where c.p. means cyclic permutations. Equations (3) to (7) are given in a coordinate system the axes of which are parallelto the cube edges.

One can compare expression (3) with the results of the site-preference model [7] for growth directions [001] and [111]. From the comparison it follows:

a) Calculations in [7] appear to contain incidental errors.[3])

[3]) There are errors in the expressions for β_{12}, β_{13}, and β_{31} in equation (A2) (and consequently in (18)). The correct forms of these terms are

$$\beta_{12} = \tfrac{1}{2}\,(B_{11} + B_{12} - B_{13} + B_{21} + B_{22} - B_{23} - B_{31} - B_{32} - 4B_{66}) \,,$$
$$\beta_{13} = \tfrac{1}{4}\,(-B_{11} - B_{12} + 4B_{13} - B_{21} - B_{22} + 4B_{23} - 2B_{31} - 2B_{32} + 4B_{66}) \,,$$
$$\beta_{31} = \tfrac{1}{4}\,(-B_{11} - B_{12} - 2B_{13} - B_{21} - B_{22} - 2B_{23} + 4B_{31} + 4B_{32} + 4B_{66}) \,.$$

The site preference model [7] is based on results of measurements given in [10]. It is rather unexpected that traceless conditions are used in [10] but not in [7].

b) It is not sufficient to take into account in expansion (2) only the lowest terms in β_i.

The second term in (2) has the following form:

$$\sum_{\substack{ijkl\\mnpr}} B_{ijklmnpr}\alpha_i\alpha_j\beta_k\beta_l\beta_m\beta_n\varepsilon_{pr} = \varepsilon_{xx}\{B_1\beta_1^4(3\alpha_1^2-1) + B_2[2\beta_1^2\beta_2^2(\alpha_3^2-\alpha_1^2) +$$
$$+ 2\beta_1^2\beta_3^2(\alpha_2^2-\alpha_1^2) + (\beta_3^4-\beta_2^4)(\alpha_2^2-\alpha_3^2)] + B_3[\beta_3^4(\alpha_2^2-\alpha_1^2) + \beta_2^4(\alpha_3^2-\alpha_1^2) +$$
$$+ 2\beta_1^2(\beta_3^2-\beta_2^2)(\alpha_2^2-\alpha_3^2)] + B_4[\beta_2^2\beta_3^2(-3\alpha_1^2+1) + 2\beta_2\beta_3\alpha_2\alpha_3(-3\beta_1^2+1)] +$$
$$+ B_5[2\beta_1\beta_3(\beta_1^2-\beta_3^2)\alpha_1\alpha_3 + 2\beta_1\beta_2(\beta_1^2-\beta_2^2)\alpha_1\alpha_2 + \beta_1^2\beta_3^2(\alpha_1^2-\alpha_3^2) +$$
$$+ \beta_1^2\beta_2^2(\alpha_1^2-\alpha_2^2)] + B_6(\beta_3^2-\beta_2^2)[2(\beta_1\beta_2\alpha_1\alpha_2-\beta_1\beta_3\alpha_1\alpha_3) + \beta_1^2(\alpha_2^2-\alpha_3^2)]\} +$$
$$+ \varepsilon_{yz}\{2B_4\beta_1^2\beta_2\beta_3(-3\alpha_1^2+1) + 2B_5\beta_2\beta_3(\beta_3^2-\beta_2^2)(\alpha_3^2-\alpha_2^2) +$$
$$+ 2B_6\beta_2\beta_3[\beta_3^2(\alpha_1^2-\alpha_2^2) + \beta_2^2(\alpha_1^2-\alpha_3^2)] + \tfrac{1}{2}\alpha_2\alpha_3[(-5B_4-\tfrac{1}{3}B_5+\tfrac{4}{3}B_6)\times$$
$$\times(\beta_1^4-2\beta_2^2\beta_3^2) + (B_4+\tfrac{5}{3}B_5-\tfrac{2}{3}B_6)(\beta_2^4+\beta_3^4-4\beta_2^2\beta_3^2) +$$
$$+ 2(B_4-\tfrac{1}{3}B_5+\tfrac{4}{5}B_6)(\beta_1^2\beta_2^2+\beta_1^2\beta_3^2-2\beta_2^2\beta_3^2)] + B_7\beta_2^2\beta_3^2\alpha_2\alpha_3 +$$
$$+ B_8[\alpha_2\alpha_3\beta_1^2(\beta_2^2+\beta_3^2) + 2\beta_1\beta_2\beta_3(\alpha_1\alpha_3\beta_3+\alpha_1\alpha_2\beta_2)]\} + \text{c.p.} \qquad (4)$$

It can be seen from expressions (3) and (4) that magnetoelastic interactions in LPE garnet films have to be described in general case by 13 parameters (including cubic components). But for some particular growth directions this number will be considerably reduced. For example in the case of (111) films the pure noncubic magnetostriction is described only by four parameters and for (100) films by three (according to [7] in both cases the number of independent parameters is equal to five). The last term (probably important in some special cases, e.g. at low temperatures) has the following form:

$$\sum_{\substack{ijkl\\mnpr}} C_{ijklmnpr}\alpha_i\alpha_j\alpha_k\alpha_l\beta_m\beta_n\varepsilon_{pr} = \varepsilon_{xx}\{C_1\alpha_1^4(3\beta_1^2-1) + C_2[2\alpha_1^2\alpha_2^2(\beta_3^2-\beta_1^2) +$$
$$+ 2\alpha_1^2\alpha_3^2(\beta_2^2-\beta_1^2)] + (\alpha_3^4-\alpha_2^4)(\beta_2^2-\beta_3^2)] + C_3[\alpha_3^4(\beta_2^2-\beta_1^2) + \alpha_2^4(\beta_3^2-\beta_1^2) +$$
$$+ 2\alpha_1^2(\alpha_3^2-\alpha_2^2)(\beta_2^2-\beta_3^2)] + C_4[\alpha_2^2\alpha_3^2(3\beta_1^2-1) + 2\alpha_2\alpha_3\beta_2\beta_3(3\alpha_1^2-1)] +$$
$$+ C_5[2\alpha_1\alpha_3\beta_1\beta_3(\alpha_3^2-\alpha_1^2) + 2\alpha_1\alpha_2(\alpha_2^2-\alpha_1^2)\beta_1\beta_2 + \alpha_1^2\alpha_3^2(\beta_3^2-\beta_1^2) +$$
$$+ \alpha_1^2\alpha_2^2(\beta_2^2-\beta_1^2)] + C_6[2(\alpha_2^2-\alpha_3^2)(\alpha_1\alpha_2\beta_1\beta_2-\alpha_1\alpha_3\beta_1\beta_3) +$$
$$+ \alpha_1^2(\alpha_3^2-\alpha_2^2)(\beta_3^2-\beta_2^2)]\} + \varepsilon_{yz}\{C_4[2(3\alpha_1^2-1)(\alpha_1^2\beta_2\beta_3 +$$
$$+ \alpha_2\alpha_3\beta_1^2)] + 2C_5(\alpha_2^2-\alpha_3^2)[\alpha_2\alpha_3(\beta_3^2-\beta_2^2) + \beta_2\beta_3(\alpha_3^2-\alpha_2^2)] +$$
$$+ 2C_6[(\alpha_3^2-\alpha_1^2)(\alpha_2\alpha_3\beta_2^2+\alpha_2^2\beta_2\beta_3) + (\alpha_2^2-\alpha_1^2)(\alpha_2\alpha_3\beta_3^2 +$$
$$+ \beta_2\beta_3\alpha_3^2)] + \tfrac{1}{2}\beta_2\beta_3[(5C_4-\tfrac{1}{3}C_5+\tfrac{4}{3}C_6)(\alpha_1^4-2\alpha_2^2\alpha_3^2) +$$
$$+ (C_4+\tfrac{5}{3}C_5-\tfrac{2}{3}C_6)(\alpha_2^4+\alpha_3^4-4\alpha_2^2\alpha_3^2) + 2(C_4-\tfrac{1}{3}C_5+\tfrac{4}{3}C_6)(\alpha_1^2\alpha_2^2 +$$
$$+ \alpha_1^2\alpha_3^2-2\alpha_2^2\alpha_3^2)] + C_7\alpha_2^2\alpha_3^2\beta_2\beta_3 + C_8[\alpha_1^2(\alpha_2^2+\alpha_3^2)\beta_2\beta_3 +$$
$$+ 2\alpha_1\alpha_2\alpha_3(\beta_1\beta_3\alpha_3+\beta_1\beta_2\alpha_2)]\} + \text{c.p.} \qquad (5)$$

3. Phenomenological Theory of Noncubic Anisotropy in LPE Garnet Films

We have shown in the previous section that for a correct description of magnetoelastic interactions in LPE garnet films it is necessary to take into account fourth-order terms in direction cosines of the normal to the substrate. It seems reasonable to suppose that a similar treatment should be applied also in the case of growth-induced anisotropy in LPE garnet films. This problem can be solved in the same manner as the problem of magnetoelastic interactions. The following expression for magnetic anisotropy in LPE garnets can be easily deduced:

$$E_{\text{an}} = A[\alpha_1^2\beta_1^2] + B[\alpha_1\alpha_2\beta_1\beta_2] + C[\beta_1^4(3\alpha_1^2 - 1)] + \\ + D[(\beta_1^2\beta_2^2 3\alpha_3^2 - 1)] + E[\beta_1^2\beta_2\beta_3\alpha_2\alpha_3] + \text{c.p.} \qquad (6)$$

The first and second terms in (6) are identical to those given by the two-parameter model [4]. Although the two-parameter model gives rather good quantitative agreement with experimental data, in some cases considerable deviations from this simple theory are observed [11]. Dillon et al. [11] have shown, in the frame of the two-parameter model, the lack of a discernible correlation between experimental results for $(10\bar{1})$ and (112) planes. Such a situation is obvious in the five-parameter theory presented here. Unfortunately experimental data from which an accurate estimation of phenomenological parameters can be made using the proposed model are very limited. It is also possible that in somes case (slow temperatures) fourth-order noncubic anisotropy has to be taken into account. Such terms have the following form:

$$E_{\text{an}}^4 = F[\beta_1^2\alpha_1^4] + G[\alpha_1^2\alpha_3^2\beta_2^2] + H[\alpha_1^2\alpha_2\alpha_3\beta_2\beta_3] + \text{c.p.} \qquad (7)$$

4. Conclusions

We have presented the phenomenological theory of magnetostriction and growth-induced anisotropy in LPE garnet films. This theory, based on symmetry considerations, is model independent. It has to be noticed that the above developed theory gives simple relations between the parameters for different facets. Such relations cannot be evident in the site-preference or pair-preference models.

Acknowledgement

One of us (H.S.) wishes to thank the Japan Society for the Promotion of Science who sponsored the visit to Tôhoku University.

References

[1] A. Rosencwaig and W. J. Tabor, J. appl. Phys. **42**, 1643 (1971).
[2] H. Callen, Appl. Phys. Letters **18**, 311 (1971).
[3] A. Rosencwaig, J. W. Tabor, and R. D. Pierce, Phys. Rev. Letters **26**, 779 (1971).
[4] E. M. Gyorgy, A. Rosencwaig, E. J. Blount, W. J. Tabor, and M. E. Lines, Appl. Phys. Letters **18**, 479 (1971).
[5] W. T. Stacy, M. M. Janssen, J. M. Robertson, and M. J. G. van Hout, AIP Conf. Proc. **10**, 314 (1973).
[6] M. H. Yang and M. W. Muller, J. Magnetism magnetic Mater. **1**, 251 (1976).
[7] M. W. Muller, phys. stat. sol. (b) **83**, 177 (1977).
[8] M. W. Muller, phys. stat. sol. (b) **86**, 345 (1978).
[9] M. H. Muller, J. appl. Phys. **49**, 1992 (1978).
[10] T. G. Phillips and R. L. White, Phys. Rev. **160**, 316 (1967).
[11] J. F. Dillon, Jr., E. J. Blount, and E. M. Gyorgy, AIP Conf. Proc. **10**, 1953 (1973).

(Received February 23, 1979)

phys. stat. sol. (a) **54**, 121 (1979)

Subject classification: 13.4 and 20.3; 22.2.1

Institute of Semiconductors, Academy of Sciences of the Ukrainian SSR, Kiev

Study of Non-Linear Extrinsic Luminescence in GaAs

III. Sublinear Emission Due to Auger Transitions in Radiative Centres[1])

By

K. D. GLINCHUK, A. V. PROKHOROVICH, and V. I. VOVNENKO

The excitation intensity dependence of the 0.94 and 1.0 eV extrinsic emission in n-type GaAs crystals ($n_0 \approx 10^{17}$ cm^{-3}) is studied at $L \lessgtr 10^{18}$ quanta/cm^2 s and $T = 77$ K. Linear and sublinear 0.94 and 1.0 eV extrinsic emission is observed depending on the strength of the non-radiative electronic transitions in the radiative centres studied. The main features of the non-linear extrinsic emission (the sublinear emission increase with excitation intensity, a decrease in the internal quantum efficiency of the extrinsic emission and in the luminescence decay time as exciting power is increased) are satisfactorily explained by a model which takes into account an increase in the rate of non-radiative electronic transitions in 0.94 and 1.0 eV radiative centres as the exciting light considerably changes the hole occupancy of defects associated with the centres studied.

Изучена в кристаллах n-GaAs ($n_0 = 10^{17}$ cm^{-3}) при $L \lessgtr 10^{18}$ квантов/cm^2 s и $T = 77$ K зависимость интенсивности полос излучения с $h\nu_m = 0{,}94$ и 1,0 eV от интенсивности возбуждения. Наблюдались линейные и сублинейные зависимости интенсивности изучаемых полос люминесценции от интенсивности возбуждения при слабой и значительной роли безызлучательных электронных переходов в центрах люминесценции, соответственно. Основные черты нелинейного примесного излучения (сублинейное возрастание интенсивности излучения с интенсивностью возбуждения, уменьшения внутренней квантовой эффективности и времени затухания излучения при повышении уровня возбуждения) удовлетворительно объясняются моделью, учитывающей увеличение скорости безызлучательных электронных переходов в 0,94 и 1,0 eV излучающих центрах вследствия значительного изменения при освещении заполнения дырками дефектов, ассоциированных с изучаемыми центрами люминесценции.

List of Symbols

c_{pr}, c_{pd}, c_{ps} hole capture coefficients by radiative centres, defects, s-centres, respectively;

c_{n1} and c_{n2} radiative and non-radiative electron capture coefficients by radiative centres, respectively;

c_{nd} electron capture coefficient by defects;

E_c and E_v energies of the bottom of the conduction and valence bands, respectively;

ε_r and ε_d energy positions of radiative centres and defects, respectively;

f_d and $1 - f_d$ probability for the defect to be filled by an electron and by a hole, respectively;

f_r electron occupation of radiative centres;

g_r' fraction of charge carriers recombining via radiative centres;

k absorption coefficient for the exciting light;

L intensity of the exciting light;

---

[1]) Part I and II see phys. stat. sol. (a) **48**, 593 (1978) and **51**, 645 (1979), respectively.

n_0 and δn equilibrium and excess free electron concentration;
$n = n_0 + \delta n$ electron concentration;
δp excess free hole concentration;
$N_1^{-'}$ concentration of pairs which include the radiative centre filled by an electron and the defect filled by a hole;

δN^- concentration of pairs which include the radiative centre filled by a hole and the defect filled by an electron;

$N^{=\prime}$ and δN^0 concentration of pairs in which both constituents are filled by electrons and holes, respectively;

$N = N_1^{-'} + N^{=\prime} + \delta N^- + \delta N^0$ pair concentration;
N_s' concentration of deep s-centres;
I_{r1} and I_{r2} rates of radiative electron transitions via δN^- and δN^0 pairs per unit volume, respectively;

$I_r = I_{r1} + I_{r2}$ extrinsic emission intensity;
r separation between the radiative centre and the defect;
t_i duration of a light pulse;
u_{r1} and u_{r2} rates of electron transitions via δN^- and δN^0 pairs per unit volume, respectively;

$u_r = u_{r1} + u_{r2}$ rate of electron transitions via radiative centres;
u_{nd} and u_{pd} rates of electron and hole capture by defects, respectively;
u_s and u_d rates of excess carrier capture by s-centres and defects, respectively;
η internal quantum efficiency of the extrinsic emission;
τ emission decay time.

1. Introduction

It is known that in semiconductors at high excitation intensities one can observe a non-linear extrinsic luminescence (a non-linear increase in the extrinsic luminescence intensity I_r with exciting power L) [1, 2]. As the extrinsic emission intensity is $I_r = \eta g_r' L$, the luminescence non-linearity could be due to an excitation dependence of the fraction of charge carriers g_r', recombining via radiative centres [1], or of the quantum efficiency η of the extrinsic emission [2].[2]) The emission non-linearity caused by the excitation dependence of g_r' was investigated in detail previously and is well understood now [1, 2]. On the contrary, scattered data exist now about the emission non-linearity caused by the excitation dependence of η, so further attempts are needed to study it [2 to 4]. The origin of this non-linearity is the following: it is caused by a non-linear change with excitation in the rate of non-radiative electronic transitions in radiative centres [2]. This is due to the fact that not only radiative, but non-radiative electronic transitions in radiative centres could exist, too [2 to 4]. These radiationless transition are of Auger type in which the energy available is imparted by collisions to other localized carriers — electrons (GaAs [2, 3, 5, 6], CdSe [7]) or holes (CdP_2 [4], CaP [8 to 10]). Then excitation could non-linearly change the rate of Auger transitions involving traps because of a considerable change in the concentration of localized carriers which are absorbing the energy released by impact ionization [2]. Earlier [2] we showed that a strong superlinear extrinsic emission could be observed in n-type GaAs crystals due to a decrease with excitation in the concentration of localized electrons which could absorb the energy released by a recombining electron by impact

---

[2]) The η-value shows which fraction of charge carriers of those recombining via radiative r-centres emits photons, i.e. it is the ratio of the rate of radiative recombination I_r via r-centres to the sum rate of radiative and non-radiative recombination u_r via r-centres — $\eta = I_r/u_r$.

ionization. This leads to a decrease with excitation in the rate of non-radiative (Auger) transitions in deep radiative centres; as a result an excitation increase in the quantum efficiency η and the emission decay time τ arises [2]. In the present work we will show that a strong sublinear extrinsic emission caused by an increase with excitation in the rate of non-radiative (Auger) transitions in deep radiative centres could also be observed in n-type GaAs crystals. This is caused by an increase with excitation in the concentration of localized holes which could absorb the energy released by a recombining electron by impact ionization; as a result an excitation decrease in η and τ arises.

2. Sublinear Emission Due to Auger Recombination of Free Electrons (Theory)

2.1 Recombination model. Kinetic equations

Let us find the steady-state extrinsic emission intensity I_r versus exciting power L (it generates kL hole–electron pairs per unit volume) for an n-type GaAs crystal which contains the radiative r-centre associated with the defect (this is the real structure of deep luminescence centres studied below [5, 6]) (Fig. 1). If the defect is filled by a hole radiative (with the coefficient c_{n1}) and non-radiative (with coefficient c_{n2}) free electron transitions to radiative centres are possible (Fig. 1). In the first case photon emission of energy $h\nu$ occurs. In the second case the energy released by a recombining free electron is imparted by impact collision to a localized hole of the associated defect, so another free hole appears in the valence band (Fig. 1). If the defect is filled by an electron, then only the radiative electronic transition exists (the energy released by a recombining free electron is not sufficient for the impact ionization of a defect electron as $E_c - \varepsilon_r < E_c - \varepsilon_d$) (Fig. 1). So for the discussed case the emission intensity I_r is determined by the relative role of two types of pairs in the recombination processes shown schematically in Fig. 1 as δN° and δN^- pairs. These two types of pairs include the radiative acceptor centre filled by a hole and associated acceptor defect filled by a hole or by an electron (Fig. 1). The relative role in recombination of δN^- and δN^0 pairs is determined by the rates of free electron transitions via these pairs: $I_{r1} = c_{n1}n\,\delta N^- = u_{r1}$, $I_{r2} = c_{n1}n\,\delta N^0$, $u_{r2} = (c_{n1} + c_{n2})\,n\,\delta N^0$ [2, 5]. Then the extrinsic emission intensity I_r is

$$I_r = I_{r1} + I_{r2} = u_{r1} + \frac{c_{n1}}{c_{n1} + c_{n2}}\,u_{r2} = c_{n1}n(\delta N^- + \delta N^0)\ . \qquad (1)$$

If all radiative centres are associated with defects forming pairs with constant r, then the δN^0, δN^- values and, consequently, I_r could easily be found from the kinetic equations for δN^0, δN^-, Nf_d, and δp. The following must be taken into account [2]. The defects capture free electrons, holes with a rate $u_{nd} = c_{nd}\,N(1 - f_d)\,n$, $u_{pd} = c_{pd}Nf_d\,\delta p - c_{n2}n\,\delta N^0$. Radiative centres capture free electrons more effectively than the defects do ($c_{n1,2} > c_{nd}$). The δN^- and δN^0 pairs are mainly created when the free excess holes are captured by $N^{='}$ and $N_1^{-'}$ pairs in which both their constituents or one (namely the radiative centre) are filled by electrons. Other than pair recombina-

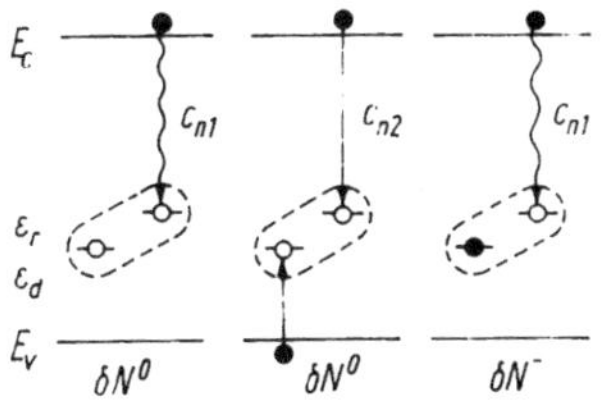

Fig. 1. Scheme of radiative (wavy lines) and non-radiative (straight lines) electronic transitions in radiative centres associated with a defect filled by a hole (δN^0 pairs) or by an electron (δN^- pairs). ● electrons, ○ holes

tion routes (s-centres) can exist in GaAs crystals. Radiative centres do not saturate before the defects do ($f_r = (N^{=\prime} + N_1^{-\prime})/N \approx 1$ exceeds considerably f_d and $1 - f_r = (\delta N^0 + \delta N^-)/N$), so $N_1^{-\prime} \approx N(1 - f_d)$, $N^{=\prime} \approx N f_d$ and $N_1^{-\prime} + N^{=\prime} \approx N$. For simplicity we limit ourselves to the real case of low excess electron concentration $\delta n (n \approx n_0)$, dominating role of non-saturable s-centres in recombination ($u_s = c_{ps} N_s' \delta p > u_r$, $u_{pd,nd}$ and $N_s' = N_s + \delta N_s \approx N_s$), deep radiative centres and defects (in thermal equilibrium radiative centres and defects are filled by electrons, so $N^{=} \approx N$ and $f_d = f_r = 1$), low temperatures (no thermal release of captured electrons and holes from the defects and radiative centres is possible) [1, 5, 6]. So for the case discussed the kinetic equations for δN^0, δN^-, $N f_d$ and δp are [2, 5]

$$\frac{d\delta N^-}{dt} \approx c_{pr} N^{=\prime} \delta p - c_{n1} n_0 \delta N^- = c_{pr} N^{=\prime} \delta p - u_{r1} \,, \tag{2}$$

$$\frac{d\delta N^0}{dt} \approx c_{pr} N_1^{-\prime} \delta p - (c_{n1} + c_{n2}) n_0 \delta N^0 = c_{pr} N_1^{-\prime} \delta p - u_{r2} \,, \tag{3}$$

$$\frac{d(N f_d)}{dt} \approx c_{nd} n_0 N (1 - f_d) + c_{n2} n_0 \delta N^0 - c_{pd} N f_d \delta p = u_{nd} - u_{pd} \,, \tag{4}$$

$$\frac{d\delta p}{dt} \approx kL - c_{ps} N_s \delta p = kL - u_s \,. \tag{5}$$

Below we discuss the most important consequences of non-linear extrinsic luminescence due to non-radiative electronic transitions in δN^0 pairs following from (2) to (5).

2.2 *Extrinsic emission intensity versus exciting power*

As follows from (1) to (5) at $d\delta N^-/dt$, $d\delta N^0/dt$, $d(N f_d)/dt$, $d\delta p/dt = 0$, $u_{r1} = c_{pr} N^{=\prime} \delta p$, $u_{r2} = c_{pr} N_1^{-\prime} \delta p$, $u_r = c_{pr} N \delta p$, $u_{nd} = u_{pd} = u_d$ and the steady-state intensity of the extrinsic emission I_r per unit volume is

$$I_r = \eta g_r' kL = \left(N^{=\prime} + \frac{c_{n1}}{c_{n1} + c_{n2}} N_1^{-\prime}\right) c_{pr} \delta p = \left[f_d + \frac{c_{n1}}{c_{n1} + c_{n2}} (1 - f_d)\right] c_{pr} N \delta p \,, \tag{6}$$

where $\delta p = kL/c_{ps} N_s \sim L$, $g_r' = u_r/(u_r + u_d + u_s) \approx c_{pr} N / c_{ps} N_s = \text{const}$ and

$$f_d = \left(1 + \frac{c_{pd} \delta p}{c_{nd} n_0 + \dfrac{c_{n2} c_{pr} \delta p}{c_{n1} + c_{n2}}}\right)^{-1} . \tag{7}$$

As one can see from (1), (6), (7) a non-linear extrinsic emission could be observed as the varying excitation intensity changes the ratio between the $N^{=\prime}(\delta N^-)$ and $N_1^{-\prime}(\delta N^0)$ pairs. This non-linear (sublinear) extrinsic emission is due to non-radiative transitions in radiative centres ($g_r' = \text{const} \neq f(L)$): $I_r \sim L$ at low excitation intensities ($u_{r2} < u_{r1} \approx u_r$, $I_{r2} \ll I_{r1} \approx I_r$, $\delta N^-(N^{=\prime}) \gg \delta N^0(N_1^{-\prime})$) and $I_r \sim [c_{n1}/(c_{n1} + c_{n2})] L$ at high excitation intensities ($u_{r1} \ll u_{r2} \approx u_r$, $I_{r1} \ll I_{r2} \approx I_r$, $\delta N^0(N_1^{-\prime}) \gg \delta N^-(N^{=\prime})$) and $I_r \sim L^\beta$ ($\beta < 1$) at intermediate excitation intensities. A considerable change in a ratio $\delta N^-/\delta N^0 \sim N^{=\prime}/N_1^{-\prime} \approx f_d/(1 - f_d)$ with the exciting power occurs due to a change in f_d with L (see equation (7)): $f_d \approx 1$ at low excitation intensities ($c_{nd} n_0 > c_{pd} \delta p$) and $f_d \ll 1$ at high excitation intensities ($c_{pd} \delta p > c_{nd} n_0$ and $c_{pr} < c_{pd} \times$

$\times c_{n2}/(c_{n1} + c_{n2})$ or $c_{pd}\,\delta p > c_{nd} n_0$ and $c_{pd} > c_{pr}$ if $c_{n2} \gtrapprox c_{n1}$).[3]) So a linear emission at low excitation intensities will be followed by a sublinear emission at intermediate excitation intensities and then one again observes a linear emission at high excitation intensities.

2.3 *Excitation dependence of the internal quantum efficiency of the extrinsic emission*

As follows from (2), (3), the internal quantum efficiency of the extrinsic emission studied $\eta = I_r/u_r$ is given by

$$\eta = \frac{c_{n1}\,\delta N^- + c_{n1}\,\delta N^0}{c_{n1}\,\delta N^- + (c_{n1} + c_{n2})\,\delta N_0} \approx \frac{N^{=\prime}}{N} + \frac{c_{n1}}{c_{n1} + c_{n2}} \frac{N_1^{-\prime}}{N} =$$

$$= f_d + \frac{c_{n1}}{c_{n1} + c_{n2}} (1 - f_d)\,. \tag{8}$$

As one can see if radiationless transitions in radiative centres could exist ($c_{n2} \gtrapprox c_{n1}$), then the excitation dependence of η could be observed (if $c_{n2} \ll c_{n1}$, then $\eta \approx 1$ at all excitation intensities). Really in this case a decrease in f_d with L leads to a decrease in η — the quantum efficiency $\eta \approx 1$ at $f_d \approx 1$ (low excitation intensities $\delta N^- > \delta N^0$) and $\eta = c_{n1}/(c_{n1} + c_{n2})$ at $f_d \ll 1$ (high excitation intensities $\delta N^0 > \delta N^-$). The detailed $\eta(L)$ dependence could easily be found using (7). By comparing (6), (8) one can see that $I_r = \eta c_{pr} N_r\,\delta p$ and so the I_r non-linearity ($I_r \sim L^\beta$) is completely caused by a non-linear change in η with L — $\beta < 1$ at $\eta = f(L)$ and $\beta = 1$ at $\eta \neq f(L)$.

2.4 *Extrinsic emission decay versus exciting power*

After light impulse of duration t_i is switched off an emission decay occurs: $I_r(t) = [\delta N^-(t) + \delta N^0(t)]\,c_{n1} n_0$. The emission decay is mainly determined by the rate of free electron transitions to acceptor radiative centres, as excess holes including those released by an impact ionization in δN^0 pairs recombine much more rapidly than excess electrons ($\delta p \to 0$ at $t > t_i$ as c_{ps}, $c_{pr} > c_{n1,2}$, and $u_s > u_r$, u_d [6]). Then as follows from (2), (3), $\delta N^-(t) = \delta N^-(t_i) \exp(-c_{n1} n_0 t)$ and $\delta N^0(t) = \delta N^0(t_i) \times \exp[-(c_{n1} + c_{n2})\,n_0 t)]$ and the emission decay is

$$I_r = \left[\delta N^-(t_i) \exp\left(-\frac{t}{\tau_1}\right) + \delta N^0(t_i) \exp\left(-\frac{t}{\tau_2}\right)\right] c_{n1} n_0\,, \tag{9}$$

where the emission decay times τ_1 and τ_2 are:

$$\tau_1 = (c_{n1} n_0)^{-1}\,, \qquad \tau_2 = [(c_{n1} + c_{n2})\,n_0]^{-1}\,. \tag{10}$$

If non-radiative (Auger) transitions in radiative centres exist ($c_{n2} \gtrapprox c_{n1}$), an increase in exciting power will lead to an decrease in τ from τ_1 at low excitation intensities ($\delta N^-(t_i) > \delta N^0(t_i)$) to τ_2 at high excitation intensities ($\delta N^0(t_i) > \delta N^-(t_i)$), i.e. with a variation of the $\delta N^-(t_i)/\delta N^0(t_i)$ ($\sim N^{=\prime}/N_1^{-\prime} \sim f_d/(1 - f_d)$) ratio.

2.5 *Concluding remarks*

As one can see from (6) to (10) the sublinear $I_r(L)$ dependence and the decrease of η and τ with L are the consequences of simultaneous radiative and non-radiative electronic transitions in radiative centres caused by a strong association between the radiative centre and the defect ($c_{n2} \gtrapprox c_{n1}$). If the association between the radiative centre and the defect is weak ($c_{n2} \ll c_{n1}$), i.e. there are practically no radiationless transitions in radiative centres (only radiative transitions exist), then, as follows from (6) to (10), only linear extrinsic luminescence and no excitation dependences of η and τ will be observed. So the characteristics of non-linear emission caused by a non-linear intensity dependence of radiationless transitions in radiative centres, depend

[3]) $c_{pd} > c_{pr}$ means that the defects could carry a recombination flow more efficient than the radiative centres do, especially at low excitation intensities ($u_d = \{c_{pd} f_d - [c_{n2}/(c_{n1} + c_{n2})] \times c_{pr}(1 - f_d)\}\,N\,\delta p > u_r = c_{pr} N\,\delta p$ and $g_r' < 1$ even if s-centres are absent, see (2), (3), (4)).

on the value $c_{n1}/(c_{n1} + c_{n2})$. This quantity is the internal quantum efficiency η of the extrinsic emission measured at high excitation intensities (at $f_d \ll 1$, see (8)). It is convenient to check the main conclusions which result from the above theory by measuring $I_r(L)$, $\eta(L)$, and $\tau(L)$ in crystals with different η.

3. Experimental

The main experimental details were the same as given in [1, 2, 5, 6], so only a few remarks will be made here. The 0.94 (1.0) eV emission bands excited by radiative electron recombination via $V_{Ga}V_{As}(Cu_{Ga}V_{As})$ centres [1, 5, 6] were studied. Measurements were performed at 77 K on a variety of n-type GaAs(Te) crystals of nearly the same doping ($n_0 \approx 10^{17}$ cm^{-3}) in which radiationless transitions in radiative centres were weak ($c_{n2} \ll c_{n1}$) or strong ($c_{n2} \gtrapprox c_{n1}$) enough. In crystals in which emission sublinearity was observed non-radiative s-centres dominated in recombination. The front face emission intensity per unit surface I_r (in quanta/cm²s) excited by a He–Ne laser was measured [2]. To exclude non-linear emission effects due to a substantial change with L in the occupancy of radiative (non-radiative) centres or due to high δn ($> n_0$) values excitation intensities were used which could not lead to saturation of radiative (non-radiative) centres or to high δn values ($N^{-\prime} + N_1^{-\prime} \approx N \gg \delta N^0 + \delta N^-$, $g_r' \approx \text{const} \neq f(L)$ and $\delta n < n_0, N$) [1, 11]. Linear excitation dependences of the intrinsic emission intensity I_{cv} confirm the fact that non-radiative centres do not saturate and that at the excitations used δn did not exceed n_0 (Fig. 2) [1]. The η-values were found from the analysis of I_r/L dependences and the τ-values from the analysis of the emission decay excited by a pulse light with $t_i \approx 3$ μs superimposed on a constant bias light [2].

4. Results

Fig. 2 and 3 illustrate the main typical characteristics of non-linear extrinsic luminescence for the 0.94, 1.0 eV emission bands in n-type GaAs with increasing excitation intensity. Data (extrinsic emission intensity I_r, internal quantum efficiency η, and emission decay time τ versus exciting power) are given for crystals in which the role of the non-radiative transitions in 0.94 and 1.0 eV radiative centres differs considerably — from strong ($c_{n1}/(c_{n1} + c_{n2}) \approx 0.3$) to weak ($c_{n1}/(c_{n1} + c_{n2}) \approx 1$). The following main features deserve special attention: If strong radiationless transitions in radiative centres exist ($c_{n2} \gtrapprox c_{n1}$), then a) the extrinsic emission intensity I_r could vary sublinearly with exciting power ($I_r \sim L^\beta$ where $\beta = 0.5$ and 0.7)[4]; b) the internal quantum efficiency η decreases with exciting power L until it reaches a plateau value less than unity; c) the emission decay time τ decreases with exciting power at a definite

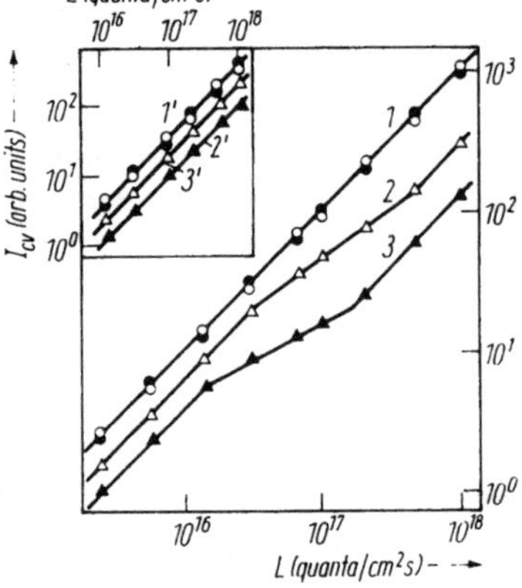

Fig. 2. 77 K dependence of the extrinsic (1 to 3) and intrinsic (1′ to 3′) emission intensities on the exciting light intensity for n-type GaAs ($n_0 \approx 10^{17}$ cm^{-3}) with weak and strong non-radiative electronic transition in radiative centres ○, △ $h\nu_m = 0.94$ eV and ●, ▲ 1.0 eV; (1), (1′) $c_{n1}/(c_{n1} + c_{n2}) = 1$, (2), (2′) 0.5, (3), (3′) 0.33

[4]) $\beta = 0.5$ to 0.9 could be observed in other samples.

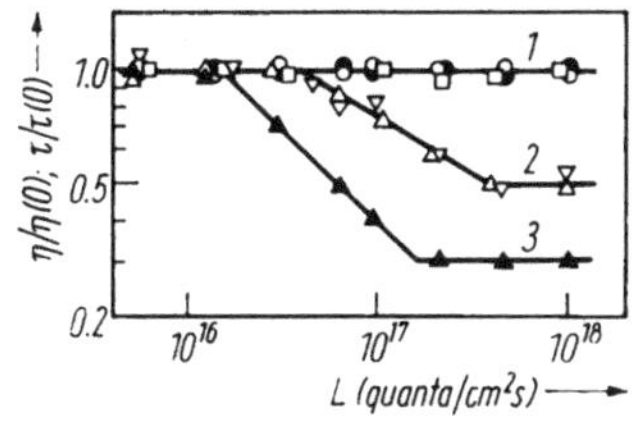

Fig. 3. 77 K dependence of the internal quantum efficiencies (○, ●, △, ▲) and emission decay times (□, ▽) on the exciting light intensity for n-type GaAs ($n_0 \approx 10^{17}$ cm^{-3}) with weak and strong non-radiative electronic transitions in radiative centres. ○, □, △, ▽ $h\nu_m$ 0.94 eV; ●, ▲ 1.0 eV; (1) $c_{n1}/(c_{n1} + c_{n2}) = 1$; (2) 0.5; (3) 0.33; $\eta(0)$ and $\tau(0)$ are measured at $L \approx 10^{16}$ quanta/cm² s

interval of L values. If radiationless transitions in radiative centres are weak enough ($c_{n2} \ll c_{n1}$), then (i) I_r changes linearly with $L(\beta = 1)$; (ii) η and τ do not depend on L.

5. Discussion

All the data observed agree well with theory (see Section 2) and are explained by the same arguments discussed above, i.e. the observed sublinear excitation dependence of the 0.94, 1.0 eV emission bands in n-type GaAs is due to the non-linear intensity dependence of the radiationless electron transitions in 0.94 and 1.0 eV radiative centres. The picture is the following. Excitation gradually increases the localized hole concentration on the defect levels associated with radiative centres. This results in an increase of the rate of non-radiative electronic transitions in radiative centres as the concentration of localized holes which could absorb the energy released by a recombining free electron by collision is greatly increased. All this leads to a sublinear extrinsic emission, to a decrease in η and τ with excitation (Fig. 2, 3). At sufficiently high excitation intensities all defect levels are mostly filled by holes. In this case I_r again linearly changes with L and η (<1) and τ does not depend on L (Fig. 2, 3). If non-radiative transitions in radiative centres are weak enough ($c_{n2} \ll c_{n1}$), then only linear luminescence is observed at all excitation intensities (Fig. 2, 3). Naturally, the micro-characteristics of radiative centres could differ in various crystals; as a result, various emission sublinearities could be observed ($\beta = 0.5$ to 0.9).[5]

The validity of the assumptions of the theory, see Section 2.1 (low c_{nd} values, $c_{pd} > > c_{pr} > c_{n1,2} > c_{nd}$, weak dependence of c_{n1}, c_{pr}, c_{nd}, c_{pd} on the pair charge, $1 \gg g_r' \neq \neq f(L)$, $f_r > f_d$) is confirmed in experiment. Really the observed coincidence of the $\eta(L)$ and $\tau(L)$ dependences and no excitation dependences of η and τ, if radiationless transitions are absent, testify, as one can easily see, the validity of the above assumptions.[6]

[5]) GaAs heat treatment generating many types of efficient local centres usually promotes the appearance of a variety of nonlinear emission effects (a change of β-values) due to radiationless transitions in radiative centres [13].

[6]) If coefficients c_{pr}, c_{n1} depend on a pair charge (hole capture coefficients $c_{pr}^{=}$, c_{pr}^{-} by $N_1^{=\prime}$, $N_1^{-\prime}$ pairs and electron capture coefficients c_{n1}, c_{n1}' by δN^-, δN^0 pairs differ), i.e.

$$I_{r1} = c_{n1} n_0 \, \delta N^- = c_{pr}^{=} N^{=\prime} \delta p = u_{r1} ,$$

$$I_{r2} = c_{n1}' n_0 \, \delta N^0 = [c_{n1}'/(c_{n1}' + c_{n2})] \, c_{pr}^{-} N_1^{-\prime} \, \delta p = [c_{n1}'/(c_{n1}' + c_{n2})] \, u_{r2} ,$$

then (6), (8), (10) are changed for the following:

$$I_r \approx \left[c_{pr}^{=} f_d + \frac{c_{n1}'}{c_{n1}' + c_{n2}} \, c_{pr}^{-}(1 - f_d) \right] N \, \delta p , \tag{11}$$

$$\eta = \frac{c_{pr}^{=} f_d + \left[\frac{c_{n1}'}{c_{n1}' + c_{n2}} \right] c_{pr}^{-}(1 - f_d)}{c_{pr}^{=} f_d + c_{pr}^{-}(1 - f_d)} , \tag{12}$$

$$\tau_1 = (c_{n1} n_0)^{-1} , \qquad \tau_2 = [(c_{n1}' + c_{n2}) \, n_0]^{-1} . \tag{13}$$

So in this case at $c_{pr}^{=} \gg c_{pr}^{-}$ the sublinear $I_r(L)$ emission could be observed even if non-radiative transitions in radiative centres are absent ($c_{n2} \ll c_{n1}'$, i.e. $\eta \approx 1$) (the $I_r(L)$-sub-linearity is due to excitation changes in $g_r' = [c_{pr}^{=} N f_d + c_{pr}^{-} N(1 - f_d)]/c_{ps} N_s$). But here no coincidence of $I_r/L = g_r'(L)$ and $\tau(L)$ dependences will be observed.

From the η- and τ-values found follows that $c_{n2} \lessapprox 10^{-13}$ cm³/s. Such c_{n2}-values are also predicted theoretically for Auger transitions involving traps for the pair radius $r = 10^{-6}$ cm [12].

Comparing the above results and those given in [2] one sees that a variety of $I_r(L)$ non-linearities due to non-linear Auger transitions in radiative centres could be observed in GaAs crystals ($1 \lessgtr \beta = 0.5$ to 1.3).[7]) This is due to the fact that deep radiative centres could be associated with different defects [5, 6].

In conclusion we note that the data obtained confirm the origin of non-radiative transitions as Auger transitions involving traps (radiationless multiphonon emission should not show any excitation dependence of η, τ) and are important for solving the complex problem of non-radiative electronic transitions in radiative centres [8, 14, 15].

References

[1] K. D. GLINCHUK, A. V. PROKHOROVICH, V. E. RODIONOV, and V. I. VOVNENKO, phys. stat. sol. (a) **48**, 593 (1978).
[2] K. D. GLINCHUK, A. V. PROKHOROVICH, and V. I. VOVNENKO, phys. stat. sol. (a) **51**, 645 (1979).
[3] YU. V. VOROBEV and M. K. SCHEINKMAN, phys. stat. sol. (a) **21**, 369 (1974); **39**, 717 (1977).
[4] W. MANDEMA and N. J. TRAPPENIERS, Physica B (Utrecht) **76**, 123 (1974).
[5] YU. GACHEGOR, I. V. ISMESTIER, and G. B. SOIFER, Radiospectroscopiya Nr. 10, Perm State Univ. Publ., Perm 1976 (p. 33).
[6] K. D. GLINCHUK, A. V. PROKHOROVICH, V. E. RODIONOV, and V. I. VOVNENKO, phys. stat. sol. (a) **41**, 659 (1977).
[7] M. K. SHEINKMAN, A. V. LUBCHENKO, and A. I. FEDOROV, Fiz. Tekh. Poluprov. **11**, 955 (1977).
[8] A. A. BERGH and P. J. DEAN, Proc. IEEE **60**, 156 (1972).
[9] J. C. TSANG, P. J. DEAN, and P. T. LANDSBERG, Phys. Rev. **173**, 814 (1968).
[10] J. M. DISHMAN, Phys. Rev. B **3**, 2588 (1971).
[11] D. F. BAISA, A. V. BNODAR, A. YA. GORDON, and S. V. MALTSEV, Ukr. fiz. Zh. **24**, 128 (1979).
[12] E. I. TOLPIGO, K. B. TOLPIGO, and M. K. SHEINKMAN, Fiz. Tekh. Poluprov. **8**, 509 (1974).
[13] K. D. GLINCHUK, A. V. PROKHOROVICH, and V. I. VOVNENKO, Fiz. Tekh. Poluprov. **10**, 2167 (1976).
[14] P. T. LANDSBERG, phys. stat. sol. **41**, 457 (1970).
P. T. LANDSBERG and M. J. ADAMS, J. Lum. **7**, 3 (1973); Proc. Royal. Soc. London, **A 334**, 523 (1973)
[15] C. H. HENRY, J. Lum. **12/13**, 47 (1976).

(Received April 12, 1979)

[7]) Note here that as for the discussed case the emission decay is:

$$I_r(t) = I_{r1}\left[1 - \exp\left(-\frac{t_i}{\tau_1}\right)\right]\exp\left(-\frac{t}{\tau_1}\right) + I_{r2}\left[1 - \exp\left(-\frac{t_i}{\tau_2}\right)\right]\exp\left(-\frac{t}{\tau_2}\right) \tag{14}$$

then the transient (measured at $t = t_i \ll \tau_1, \tau_2$) emission intensity $I_r(t_i)$ is not a function of c_{n2} (see also [6])

$$I_r(t_i) \approx \left(\frac{I_{r1}}{\tau_1} + \frac{I_{r2}}{\tau_2}\right) t_i \approx [c_{n1}c_{pr}^{=} f_d + c'_{n1}c_{pr}^{-}(1 - f_d)]\, n_0 N\, \delta p t_i\,. \tag{15}$$

So comparing (6), (11), and (15) one can see that radiationless transitions in radiative centres could lead to the $I_r(L)$-non-linearity and could not lead to the $I_r(t_i)(L)$ non-linearity (for example, $I_r(t_i) \sim L$ at $c_{n1}c_{pr}^{=} \approx c'_{n1}c_{pr}^{-}$ and any c_{n2}-values).

phys. stat. sol. (a) 54, 129 (1979)

Subject classification: 13.4 and 20.3; 20.1; 22.5

Department of Physics, Harcourt Butler Technological Institute, Kanpur[1] *(a) and Department of Physics, Indian Institute of Technology, Kanpur*[2] *(b)*

On the Origin of the 375 nm Emission Band in KCl:Tl

By

V. B. SHARAN (a), D. N. S. SRIVASTWA (a), and S. C. SEN[3] (b)

Fluorescence spectra of KCl:Tl, RbCl:Tl, and NH_4Cl:Tl for various thallium concentrations when excited at different wavelengths in the A band region are recorded at room temperature (300 K) and liquid nitrogen temperature (77 K) in order to ascertain the origin of the 375 nm emission band of KCl:Tl. The fluorescence spectra are presented and from a comparative analysis of excitation spectra obtained for the 375 nm emission of KCl:Tl and those obtained for the 375 nm emission of NH_4Cl:Tl at higher temperature, it is concluded that the centre responsible for the 375 nm emission band in KCl:Tl is a Tl^+ monomer having local CsCl like environment. The 375 nm emission band is attributed to the electronic transition $^3T_{1u} \rightarrow {}^1A_{1g}$ in such Tl^+ monomers.

Fluoreszenzspektren von KCl:Tl, RbCl:Tl und NH_4Cl:Tl werden für verschiedene Thallium-konzentrationen bei Zimmertemperatur (300 K) und der Temperatur des flüssigen Stickstoff (77 K) nach Anregung bei verschiedenen Wellenlängen im Bereich der A-Bande aufgenommen, um den Ursprung der 375 nm-Emissionsbande von KCl:Tl aufzuklären. Die Fluoreszenzspektren werden angegeben, und aus einer Vergleichsanalyse der Anregungsspektren für die 375 nm-Emission von KCl:Tl und der 375 nm-Emission für NH_4Cl:Tl bei höheren Temperaturen wird geschlossen, daß das Zentrum, das für die 375 nm-Bande in KCl:Tl verantwortlich ist, ein Tl^+-Monomer ist und eine lokale CsCl-ähnliche Umgebung besitzt. Die 375 nm-Emissionsbande wird dem elektronischen Übergang $^3T_{1u} \rightarrow {}^1A_{1g}$ in solchen Tl^+-Monomeren zugeordnet.

1. Introduction

Although a great amount of work has been done on the optical absorption and fluorescence spectra of KCl:Tl, the origin of various absorption and emission bands is still not completely understood. At very low concentration of thallium, the Tl^+ ion enters the KCl lattice substitutionally [1 to 3]. The substitution is possible even by simple mortar grinding because K^+ (1.33 Å) and Tl^+ (1.44 Å) have nearly the same ionic size [4]. Substitutional Tl^+ introduces the well-known C, B, and A absorption bands at 196, 209, and 248 nm, which are attributed to $^1S_0 \rightarrow {}^1P_1$ ($^1A_{1g} \rightarrow {}^1T_{1u}$), $^1S_0 \rightarrow {}^3P_2$ ($^1A_{1g} \rightarrow {}^3E_u$ or $^3T_{2u}$), and $^1S_0 \rightarrow {}^3P_1$ ($^1A_{1g} \rightarrow {}^3T_{1u}$) electronic transitions, respectively, within Tl^+. The emission spectrum of the KCl:Tl phosphor as recorded by von Meyeren [5] gave an emission band at 298 nm which was found to shift to 302 nm on warming to 150 °C. Buenger and Flechsig [6] observed an additional emission band at 470 nm. Later Pringsheim [7] reported that for samples with 2.68 mol% thallium content emission bands occurred in the regions 360 to 405 nm, 430 to 500 nm, and 500 to 600 nm in addition to the prominent 300 nm emission band. After a reexamination of the spectra Hutten and Pringsheim [4] reported that the emission bands peak at 300, 385, 457, 540, and 610 nm. Since then many reports have appeared on the emission spectra of KCl:Tl for various concentrations of thallium and under various excitations [8 to 17]. They gave contradicting data both about the number of emission bands and their peak positions.

---

[1]) Kanpur-208002, India.

[2]) Kanpur-208016, India.

[3]) With whom all correspondence is to be made.

Edgerton and Teegarden [18] observed an emission band around 250 nm at low concentration so that the assignment $^1T_{1u} \rightarrow {}^1A_{1g}$ accepted for many years [8] for the origin of the visible emission bands in isolated Tl^+ in cubic symmetry lost validity. Now, it is believed that absorption at 196 and emission at 250 nm are due to the transitions $^1A_{1g} \rightleftarrows {}^1T_{1u}$ whereas the absorption at 248 and emission at 305 nm are due to $^1A_{1g} \rightleftarrows {}^3T_{1u}$ transitions in substitutional Tl^+. These assignments call for an explanation for the visible emission bands. Various propositions which have been made are summarized in Table 1. Most of the workers believe that there is at least one more

Table 1
Explanations for the 475 nm emission band of KCl:Tl

sample No.	excited states responsible for the emission	authors (year)	references
1	$^1P_1$ or mixing of $^1P_1$ and $^3P_1$	Johnson and Williams (1952);	[8]
2	unspecified level underlying level responsible for 305 nm emission	Knox (1959)	[19]
3	X-minimum on $^3P_1$ state produced by Jahn-Teller effect	Kamimura and Sugano (1959)	[20]
4	305 and 475 nm bands to different thallium centres having slightly different environment	Patterson and Klick (1957)	[21]
5	pairs or aggregates of thallium ions	Patterson (1960)	[22]
6	perturbed excited state	Alderson (1964)	[23]
7	thallium dimers	Delbecq et al. (1966)	[16]
8	$^3P_0$ after excitation in $^3P_1$ state	Sturge (1967)	[24]
9	$^3A_{1u}$ (or $^3P_0$) directly excited by $^1A_{1g} \rightarrow {}^3A_{1u}$ transition	Ranfagni et al. (1972)	[25]
10	thallium dimers	Tsuboi (1978)	[17]
11	$^3E_u \rightarrow {}^1A_{1g}$ electronic transitions in the dimer centres having D_{4h} symmetry	this work	—

type of luminescence centre besides the substitutional Tl^+ which was proposed by Seitz. This other luminescence centre is considered to be the origin of the visible emission. However, Ranfagni et al. [25] have recently proposed that the visible emission in KCl:Tl at 475 nm may be due to the $^3A_{1u} \rightarrow {}^1A_{1g}$ forbidden transition in substitutional Tl^+ itself, becoming weakly allowed due to mixing of the $^3A_{1u}$ state with ^3E_u and $^3T_{2u}$ states through lattice vibrations.

It is interesting that relatively little attention has been paid to the emission band around 375 nm which is also reported to be present in the emission spectrum of KCl:Tl even at moderate thallium contents [8, 9]. In order to ascertain the origin of this emission band we have recorded photoluminescence spectra of KCl:Tl for various concentrations of thallium under excitation in the A absorption region. The excitation spectra have also been recorded for different emission bands. The photoluminescence spectra of RbCl:Tl and NH_4Cl:Tl have also been obtained for comparison, because RbCl and NH_4Cl (above 185 °C) have the same structure as KCl, and also the ionic sizes of K^+, Rb^+, and NH_4^+ are nearly the same. Thus one would expect similar environments of Tl^+ in these hosts. Our experimental results lead to a model consistent also with the data of most of the earlier workers. An attempt is made here to explain the origin of the 375 nm emission band using this model.

2. Experimental Procedure

2.1 Method of sample preparation

The fluorescence spectra are known to depend on the method of sample preparation. Keeping this in view all the powder phosphors studied in the present investigation were prepared by simple mortar grinding of AR grade host with AR grade thallous chloride. Samples of various concentrations were obtained by successive dilution [4]. For convenience the KCl:Tl samples were labelled K-10, K-11, K-12, K-13, K-14, and K-15 which contained 13.4, 2.7, 0.66, 0.16, 0.04, and 0.01 mol% thallium, respectively. Similarly the RbCl:Tl samples labelled Rb-10, Rb-11, Rb-12, Rb-13, Rb-14, and Rb-15 had thallium concentrations 11.8, 2.7, 0.66, 0.16, 0.04, and 0.01 mol%, and NH_4Cl:Tl samples labelled NH_4-10, NH_4-11, NH_4-12, NH_4-14, NH_4-16, and NH_4-18 had 11.8, 2.25, 0.53, 0.032, 0.002, and 10^{-4} mol% thallium, respectively.

2.2 Measurement of emission and excitation spectra

The spectra were recorded using a double monochromator arrangement. The exciting radiation obtained from an AH4 lamp was monochromatized by a Gaertner L-234-150 quartz monochromator. The emission was analysed by another Gaertner monochromator with RCA IP28 photomultiplier and dc electrometer. The fluorescence at room temperature (RT) and at liquid nitrogen temperature (LNT) was measured using the same LNT cell. For temperatures above RT the temperature in the sample holder could be controlled to ±5 K. The spectra have been reproduced after corrections for factors concerning the source, the dispersion device, and the detector response.

3. Experimental Results

In the present investigation when KCl:Tl with Tl^+ concentration below K-15 was excited in the A-region only a single bell-shaped fluorescence band at 305 nm was obtained which was most efficiently excited at 248 nm. Fig. 1 shows the emission spectra obtained under 265 and 248 nm excitation for samples K-15 to K-10. For K-15 (0.01 mol%) under 265 nm excitation a weak emission extending from 340 to 620 nm is observed in addition to the 305 nm band. The 305 nm band grows in intensity up to K-13 and then starts becoming weaker, almost vanishing for K-10. The emission beyond 340 nm steadily gains intensity with increase of thallium content,

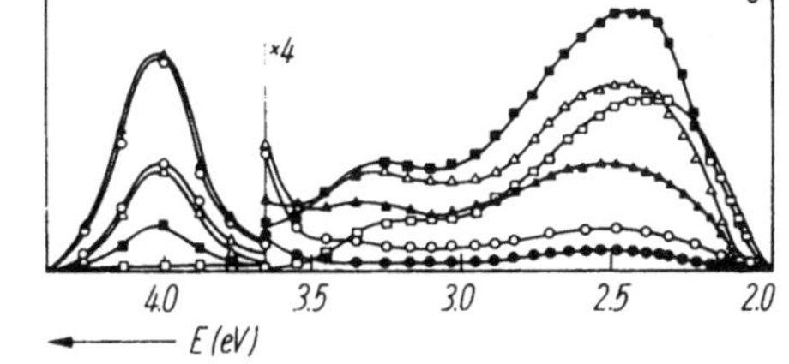

Fig. 1. Emission spectra of K-15 to K-10 under a) 248 and b) 265 nm excitation at 300 K. ● K-15, ○ K-14, ▲ K-13, △ K-12, ■ K-11, □ K-10

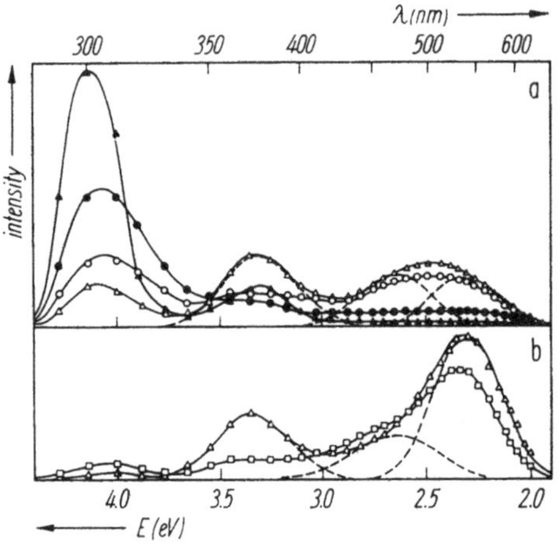

Fig. 2

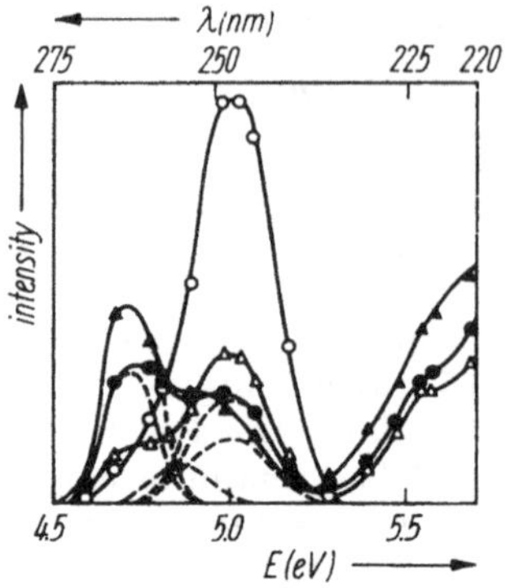

Fig. 3

Fig. 2. a) Emission spectra of K-12 (●, ▲) and K-11 (○, △) at 300 K (●, ○) and 77 K (▲, △) under 248 nm excitation; b) emission spectra of K-11 at 300 K (□) and 77 K(△) under 258 nm excitation

Fig. 3. Excitation spectra of K-11 for ○ 305, △ 375, ● 470, and ▲ 540 nm emission bands at 300 K

a distinct peak appearing at 375 nm as we go to K-11. This peak also develops for this sample under 248 nm excitation.

The emission spectra for K-12 and K-11 obtained at RT and LNT under 248 nm excitation are shown in Fig. 2a. On lowering the temperature to LNT the 305 nm emission band becomes stronger and shifts to 300 nm. The 375 nm emission band which appears only as shoulder at RT develops distinctly at LNT, the peak shifting to 372 nm. The asymmetric emission band in the region 400 to 650 nm which is not substantially affected by temperature lowering can be resolved into two Gaussian components with peaks at 470 and 540 nm. The spectra under 258 nm excitation, shown in Fig. 2b, exhibit similar changes in the 375 nm and visible emission bands.

The excitation spectra for the emission bands peaking at 305, 375, 470, and 540 nm at RT are presented in Fig. 3. The main excitation band for both 305 and 375 nm emission bands is around 248 nm. The excitation band for 470 and 540 nm emission bands are at 255 and 263 nm, respectively.

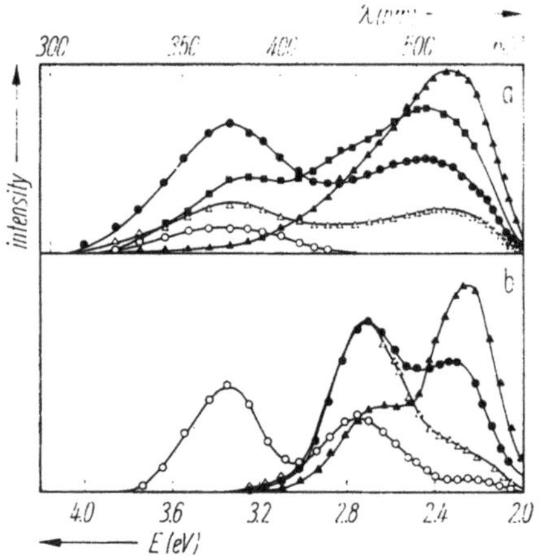

Fig. 4. a) Emission spectra of ○ NH_4-18, △ NH_4-16, ● NH_4-14, ■ NH_4-12, and ▲ NH_4-10 under 248 nm excitation at 300 K; b) emission spectra of NH_4-11 (2.25 mol% Tl) under ○ 240, △ 248, ● 258, and ▲ 265 nm excitation at 77 K

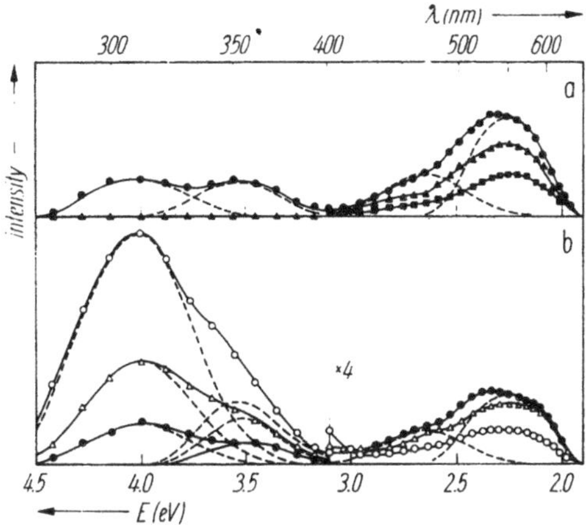

Fig. 5. a) Emission spectra of Rb-10 (11.8 mol% Tl) under ● 248, ▲ 258, and ■ 265 nm excitation at 300 K; b) emission spectra of ○ Rb-14, △ Rb-13, and ● Rb-12 under 248 nm excitation at 300 K

The emission spectra of NH_4Cl:Tl for five samples from NH_4-18 to NH_4-10 under 248 nm excitation are shown in Fig. 4a. It is reported that the emission spectrum is very sensitive to the excitation wavelength [26]. So we have also recorded the emission spectra of NH_4-11 under 240, 248, 258, and 265 nm excitation at LNT (Fig. 4b). The latter shows three emission bands peaking at 370, 460, and 550 nm though the last two appear only as composite band at RT.

Fig. 5a shows the emission of Rb-10 at RT under 248, 258, and 265 nm excitation. Two bands peaking at 310 and 350 nm are obtained under 248 nm excitation. These appear even for traces of Tl^+ and increase in intensity with rise in Tl^+ concentration. Fig. 5b shows that for concentrations above that of Rb-14 the intensity of these bands starts falling. In the visible region a weak asymmetric emission band, which extends from 400 to 650 nm, strengthens with rising thallium content and can be resolved into two Gaussian bands with peaks at 470 and 550 nm.

4. Discussion

The emission spectra of phosphors prepared with the same activator ion in different hosts having similar structure are expected to be similar. A change in the lattice constant of the host would only cause a shift of the peaks of the emission bands, a red shift being expected as the lattice constant decreases [27 to 29]. Among the hosts chosen in the present investigation the ionic size of K^+, Rb^+, and NH_4^+ match well with Tl^+ and the ions surrounding the activator Tl^+ also remain the same (Cl^-). Since KCl and RbCl have a common structure (f.c.c.), a similarity in the emission spectra of KCl:Tl and RbCl:Tl is expected. In contrast, NH_4Cl at RT has s.c. structure, so that substitutional Tl^+ in NH_4Cl:Tl is coordinated to eight chlorine ions, as distinct from six ions in KCl:Tl and RbCl:Tl. The Stokes shift, which occurs in fluorescence, is known to increase with higher coordination [30, 31]. The characteristic emission band for NH_4Cl:Tl at 370 nm is on the long-wavelength side of the corresponding emission band of KCl:Tl (305 nm), which is in line with expectation. It may be noted that the A-band in NH_4Cl:Tl falls at 240 nm, whereas that for KCl:Tl falls at 248 nm. Thus the Stokes shift in NH_4Cl:Tl is 73 nm larger than in KCl:Tl.

In all three phosphors studied additional emission bands appear with increasing Tl^+ content, and they have their excitation bands on the longer-wavelength side of the A-band. In KCl:Tl three additional emission bands occur, peaking at 375, 470, and 540 nm. In RbCl:Tl at RT a band at 350 nm (besides that at 310 nm) also appears even for traces of Tl^+; the additional bands appearing at higher Tl^+ content peak at 470 and 550 nm.

Our measurements (Fig. 1) show that the 375 nm emission band of KCl:Tl depends on the Tl^+ content and its temperature dependence (Fig. 2a) is not similar to that of the 470 and 540 nm emission bands in as much as the 375 nm emission enhances considerably at LNT. Its excitation band is around 248 nm (Fig. 3) near the A-absorption. In NH_4Cl:Tl the 450 to 600 nm band enhances in intensity with rising thallium content (Fig. 4a) and at LNT it splits into two bands peaking at 460 and 550 nm. Their excitation bands are reported at 248 and 260 nm, respectively [32]. In contrast, the ultraviolet emission is quenched as the Tl^+ content is above 0.032 mol% (NH_4-14).

Since the excitation spectra of the ultraviolet and visible emission bands differ in their dependence on the thallium content and temperature it will be appropriate to attribute the ultraviolet and visible emission bands to different luminescence centres, rather than to different excited states of the same centre as suggested by some earlier investigators [19, 20, 23 to 25]. If these different emission bands would come from a single centre, one should observe the visible emission even in the samples containing traces of thallium. The only emission band obtained at very low Tl^+ content is at 370 nm in NH_4Cl:Tl and at 305 nm in KCl:Tl. These, therefore, represent the emission in monomer centres due to the $^3T_{1u} \rightarrow {}^1A_{1g}$ electronic transition.

The visible fluorescence bands obtained for NH_4Cl:Tl disappear at temperatures above 185 °C, whereas the 370 nm characteristic emission band shifts to 345 nm [33, 34]. This fact also shows that the centres responsible for ultraviolet and visible emission bands must be different. In the mechanical grinding method Tl^+ goes largely on the surface of the host crystallites, and therefore the probability for the formation of dimers is large. Brauer [35] has reported that in NH_4Br:Tl prepared by grinding the green band attributed to dimers got considerably weakened on washing with water. Herb et al. [36] have found the symmetry of dimers for ⟨100⟩ orientation to be D_{4h}. In this symmetry the $^3T_{1u}$ level splits into nondegenerate $^3A_{2u}$ and a doubly degenerate ^3E_u and the electronic transitions from these states to the $^1A_{1g}$ state are allowed. We may therefore attribute the 460 and 550 nm additional emission bands of NH_4Cl:Tl to $^3E_u \rightarrow {}^1A_{1g}$ and $^3A_{2u} \rightarrow {}^1A_{1g}$, respectively, in these dimers. Correspondingly, it will be reasonable to attribute the additional emission bands around 470 and 540 nm for KCl:Tl as well as RbCl:Tl to similar transitions in dimers.

As regards the origin of the other emission bands in KCl:Tl and RbCl:Tl, we note that in RbCl:Tl even at the lowest thallium content we get two emission bands (310 and 350 nm). RbCl undergoes a phase transformation from f.c.c. to s.c. at LNT. Recently we have observed [37] that no new emission band is obtained at LNT and both emission bands show a blue shift, the peak positions becoming 304 and 340 nm. Even more significant was the fact that the intensity ratio of these two bands which was 3:1 at RT changed to 1:2 at LNT. Double emission is also observed in KBr:Tl and KI:Tl and is attributed to double minima in the adiabatic potential for the $^3T_{1u}$ state [38]. However, for RbCl:Tl Pathak [39] has proposed an off-centre position of Tl^+ and has attributed the splitting to a lowering of the site symmetry from O_h to D_{4h}. This origin is improbable because Rb^+ and Tl^+ present an excellent matching of ionic radii. On the other hand, the drastic change in intensity ratio is difficult to explain on the basis of double minima. Following Schleede [40] and Patterson [21, 22] we have proposed [37] that the two emission bands may be due to the simultaneous coexistence of two Tl^+ monomers with different local environments — one of NaCl-type and the other of CsCl-type. Such double centres may be easily formed in RbCl:Tl because the host is known to transform from f.c.c. to s.c. at relatively small pressure

[41], and also because in the grinding method of preparing the sample, Tl^+ is incorporated largely near the surface of RbCl crystallites. The 310 and 350 nm bands may be attributed to monomers with NaCl-type and CsCl-type environments, respectively. The phase transformation in RbCl at LNT shifts the balance considerably in favour of the higher coordination monomer (CsCl-type local environment) which explains the drastic change in intensity ratio on lowering the temperature from RT to LNT.

The 375 nm emission band of KCl:Tl cannot be assigned to any transition in Tl^+ [19 to 21, 25] or a transition of transfer type [23] or higher aggregates of Tl^+ [17], because then its counterpart must appear in the RbCl:Tl and NH_4Cl:Tl phosphor. It cannot be due to undissociated TlCl molecules because then also we would expect its appearance in the RbCl:Tl phosphor [42]. Our present investigation reveals that with rising thallium content no fluorescence band similar to the 375 nm emission band of KCl:Tl appears in RbCl:Tl. This indicates that the origin of the 375 nm band in KCl:Tl may be the same as that of the 350 nm band in RbCl:Tl, viz. a monomer in CsCl-like environment. The enhancement of the 375 nm band with decreasing temperature would indicate that lower temperature favours the formation of local centres with CsCl-type environment. With increasing thallium concentration at first the population of both monomers would increase, but later dimer formation would become more probable. This is in line with experimental data (Fig. 1a).

The fluorescent emission obtained at 375, 470, and 540 nm in KCl:Tl and the emission bands obtained at 370, 460, and 550 nm for NH_4Cl:Tl at RT under various excitations are compared in Fig. 6 after matching the emission around 375 nm for the phosphors NH_4-11 and K-11 under 248 nm excitation. The similarity of the emission spectra and their change with excitation wavelength provides a strong evidence for Patterson's model [21, 22].

However, it remains to explain why only one absorption band appears for the two monomers in KCl:Tl. Firstly, we recall that the A-bands in NH_4Cl:Tl and KCl:Tl differ by only 7 nm, whereas the emission bands differ by as much as 65 nm. Secondly, the excitation spectrum for 375 nm fluorescence in NH_4Cl:Tl at various temperatures

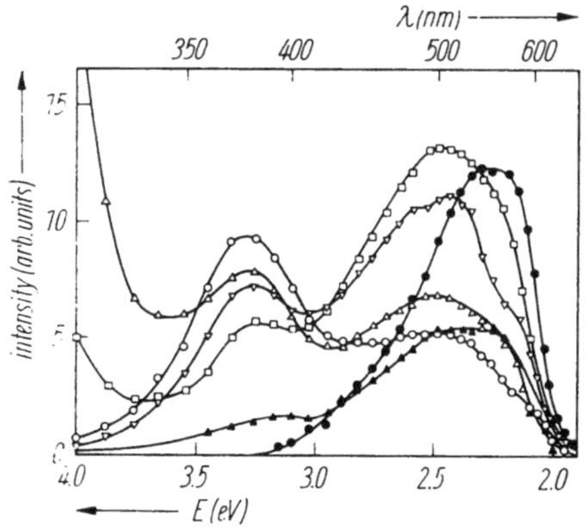

Fig. 6

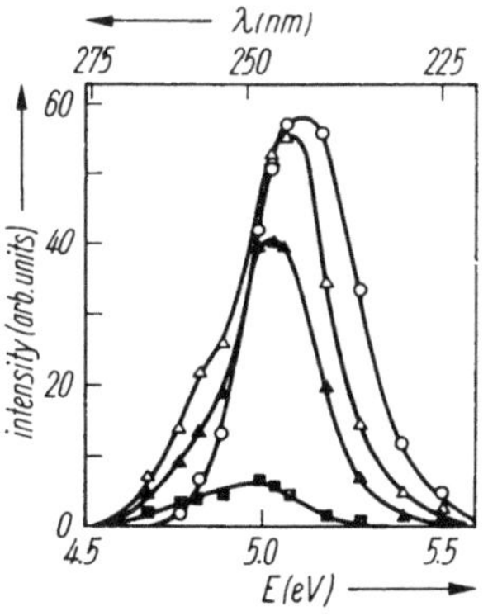

Fig. 7

Fig. 6. Emission spectra of K-11 (KCl:Tl) under △ 248, □ 260, and ▲ 265 nm excitation and of NH_4-11 (NH_4Cl:Tl) under ○ 240, ▽ 248, and ● 265 nm excitation at 300 K

Fig. 7. Excitation spectra for 375 nm emission of NH_4-11 at higher temperatures. ○ $T = 20$, △ 160, ▲ 185, ■ 200 °C above RT (Fig. 7) shows that the A-band does not show a sharp change with change in the host structure but only a gradual drift. Therefore, it would not be surprising if in KCl:Tl the A-bands of the two monomers were too close to be resolved at RT.

Our proposition agrees well with the findings of Joshi [43] who also suggests two types of species in KCl:Tl from his thermoluminescence data. Absorption bands of Tl^+ in a CsCl-like environment (NH_4Cl:Tl) are reported around 222, 240, 248, and 260 nm [26]. The A-absorption of KCl:Tl is at 248 nm. So any shift in excitation to lower or higher wavelengths should favour emission bands other than the characteristic emission of KCl:Tl at 305 nm. Pringsheim [7] did observe this in his measurements. Buenger and Flechsig's [6] stimulation data which have been confirmed by Aoyagi and Kuwabara [11] can also be tentatively explained on the basis of the present two-centre model. Johnson and Williams [44] have ascribed the 1150 and 1550 nm stimulation peaks to the $^3P_0 \rightarrow {}^1P_1$ and $^3P_0 \rightarrow {}^3P_1$ transitions in Tl^+ in NaCl-like environment. No explanation still exists for the stimulation peaks observed at 360, 460, and 620 nm. These may be attributed to $^3P_0 \rightarrow {}^1P_1$, $^3P_0 \rightarrow {}^3P_2$, and $^3P_0 \rightarrow {}^3P_1$ in a Tl^+ centre having a local CsCl-like environment.

Abdusadykov [10] has also proposed the coexistence of at least two types of centres, one of which is of Seitz-type. He observed that the conditions for the formation of the other centre differ considerably. The other centre which has consistent spectral characteristics is not only favoured by rising thallium content, but also by all processes making the crystal heterogeneous. If the other centre is a Tl^+ ion in local CsCl-like environment it would show the behaviour observed by Abdusadykov. Moreover, if with increase in thallium content local centres grow in number, then at higher thallium contents the host lattice cannot hold itself and should collapse. This has also been observed [45].

5. Conclusion

Thallous chloride with highly homopolar bonding is able to form mixed crystals with alkali chlorides of purely heteropolar bonding and of different lattice structure. However, under conditions of strain, which are favoured by rising thallium content and all processes which create heterogeneity in the phosphor, it prefers to have its own structure in the host. Accordingly two types of monomer centres manifest themselves simultaneously apart from dimers. One of the monomers is a Tl^+ ion in a NaCl-like environment and the other is a Tl^+ ion in a local CsCl-like environment. This proposition is amply supported by the fact that the local centres do not exist in CsCl-like hosts. On the basis of this two-centre model the 375 nm emission band of KCl:Tl is attributed to the $^3T_{1u} \rightarrow {}^1A_{1g}$ transition in Tl^+ with local CsCl-like environment.

Acknowledgements

We are grateful to the University Grants Commission, New Delhi, India, for providing financial assistance and for granting a Junior Research Fellowship to V. B. Sharan. We are also thankful to Dr. D. P. Khandelwal for valuable suggestions and helpful discussions.

References

[1] F. Seitz, J. chem. Phys. **6**, 150 (1938).
[2] W. A. Runciman and E. G. Steward, Proc. Phys. Soc. **A66**, 484 (1953).
[3] N. K. Banerjee and G. B. Mitra, Indian J. Phys. **7**, 13 (1968).
[4] E. H. Hutten and P. Pringsheim, J. chem. Phys. **16**, 241 (1948).
[5] W. von Meyeren, Z. Phys. **61**, 321 (1930).

[6] W. BUENGER and W. FLECHSIG, Z. Phys. **67**, 42 (1931); **69**, 637 (1932).
[7] P. PRINGSHEIM, Rev. mod. Phys. **14**, 132 (1942).
[8] P. D. JOHNSON and F. E. WILLIAMS, J. chem. Phys. **20**, 126 (1952).
[9] K. H. BUTLER, J. Electrochem. Soc. **103**, 508 (1956).
[10] T. ABDUSADYKOV, Soviet Phys. — Optics and Spectroscopy **7**, 158 (1959).
[11] K. AOYAGI and G. KUWABARA, J. Phys. Soc. Japan **15**, 2334 (1960).
[12] J. EWLES and R. V. JOSHI, Proc. Roy. Soc. **A254**, 358 (1963).
[13] N. I. IVANOVA, L. I. TARASOVA, and A. P. ZHUKOVSKII, Izv. Akad. Nauk SSSR, Ser. fiz. **25**, 331 (1961).
[14] N. I. IVANOVA and A. P. ZHUKOVSKII, Soviet Phys.—Optics and Spectroscopy **12**, 56 (1961).
[15] H. N. BOSE and S. C. SEN, Z. Phys. **167**, 20 (1962).
[16] C. J. DELBECQ, A. K. GHOSH, and P. H. YUSTER, Phys. Rev. **151**, 599 (1966).
[17] T. TSUBOI, Physica (Utrecht) **93B**, 379 (1978).
[18] R. EDGERTON and K. TEEGARDEN, Phys. Rev. **129**, 169 (1963).
[19] R. S. KNOX, Phys. Rev. **115**, 1095 (1959).
[20] H. KAMIMURA and S. SUGANO, J. Phys. Soc. Japan **14**, 1612 (1959).
[21] D. A. PATTERSON and C. C. KLICK, Phys. Rev. **105**, 401 (1957).
[22] D. A. PATTERSON, Phys. Rev. **119**, 962 (1960); **112**, 296 (1958).
[23] J. E. A. ALDERSON, phys. stat. sol. **7**, 21 (1964).
[24] M. D. STURGE, Solid State Phys. **20**, 91 (1967).
[25] A. RANFAGNI, G. P. PAZZI, P. FABENI, G. VILLIANI, and M. P. FONTANA, Phys. Rev. Letters **28**, 1035 (1972).
[26] D. N. S. SRIVASTWA, Ph. D. Thesis, Kanpur University, 1973.
[27] A. SCHLEEDE and A. GRUHL, Z. Electrochem. **29**, 411 (1923).
[28] S. H. LINWOOD and W. A. WEYL, J. Opt. Soc. Amer. **32**, 443 (1942).
[29] G. R. FONDA, J. Opt. Soc. Amer. **47**, 877 (1942).
[30] R. HILL and O. R. HOWELL, Phil. Mag. **48**, 833 (1924).
[31] W. R. BRODE, J. Amer. Chem. Soc. **55**, 939 (1933).
[32] N. L. PATHAK and S. C. SEN, phys. stat. sol. (a) **24**, 415 (1974).
[33] A. YA. PAE, Trudy Inst. Fiz. i Astron. Akad. Nauk. Est. SSR 8, 47 (1958); **11**, 45 (1960).
[34] F. D. KLEMENT, Izv. Akad. Nauk SSS, Ser. fiz. **26**, 481 (1962).
[35] P. BRAUER, Z. Naturf. **13a**, 1002 (1958); **15a**, 418 (1959).
[36] G. K. HERB, M. P. FONTANA, and W. J. VAN SCIVER, Phys. Rev. **168**, 1000 (1968).
[37] V. B. SHARAN, D. N. S. SRIVASTWA, and S. C. SEN, unpublished.
[38] A. FUKUDA, Phys. Rev. B **1**, 4161 (1970).
[39] N. L. PATHAK, Indian J. pure appl. Phys. **13**, 707 (1975).
[40] A. SCHLEEDE, Komplexbildung, Absorption und Lumineszenz, in: Leuchten und Struktur fester Stoffe, Ed. R. TOMASCHEK, R. Oldenburg, München 1943 (p. 50).
[41] R. W. BRIDGMAN, Phys. Rev. **57**, 237 (1940).
[42] R. V. JOSHI and L. H. H. PRASAD, J. Lum. **15**, 105 (1977).
[43] R. V. JOSHI, Physica (Utrecht) **27**, 1119 (1961).
[44] P. D. JOHNSON and F. E. WILLIAMS, J. chem. Phys. **21**, 125 (1953).
[45] L. M. SHAMOVSKII and I. N. ZHVANKO, Izv. Akad. Nauk SSSR, Ser. fiz. **21**, 551 (1957).

(Received January 31, 1979)

phys. stat. sol. (a) **54**, 139 (1979)

Subject classification: 1.2 and 5; 22.7

Institute of High Pressure Physics, Academy of Sciences of the USSR, Akademgorodok

Mössbauer Study of Phase Transitions under High Hydrostatic Pressures

II. The Phase Transition of SnTe[1])

By

E. V. KAPITANOV and E. N. YAKOVLEV

Experimental results reflect a linear increase of the recoilless fraction and a linear decrease of the electron density at the tin nuclei of SnTe with increasing pressure to 16 kbar. Beginning with 16 kbar the appearance of the high pressure phase is observed, the recoilless fraction of which is greater and the electron density at the tin nuclei is less than the low pressure phase ones. For powdered and single crystal specimens the phase transition is accompanied by a large two-phase region. The result of the recoilless fraction change are used for the calculation of the phase-transition heat and the slope of the equilibrium line.

Полученные экспериментальные результаты указывают на линейное увеличение вероятности эффекта Мёссбауэра f и линейное уменьшение плотности электронов на ядре олова в SnTe с ростом гидростатического давления P. Начиная с 16 kbar наблюдалось появление новой модификации SnTe, вероятность эффекта f в которой больше, а плотность электронов на ядре олова меньше, чем у фазы низкого давления. Как для порошкообразных, так и для монокристаллических образцов фазовое превращение сопровождается большой двухфазной областью. Результат изменения вероятности эффекта Мёссбауэра использован для вычисления теплоты фазового превращения и наклона линии равновесия.

1. Introduction

Tin telluride crystallizing in a cubic lattice of NaCl type can be classified as narrow-band p-type semiconductor with energy gap of 0.3 eV at 4.2 K and 0.18 eV at 300 K [1]. At 18 kbar SnTe undergoes a phase transition from cubic to rhombic structure of SnS type [2, 3]. The density of the high-pressure phase is higher than of the low-pressure phase by 3% [3].

A bend on the melting curve of tin telluride has been discovered at 12 kbar and 844 °C [4]. This result was interpreted by the authors as the appearance of the phase transition discovered in [2, 3].

By X-ray investigations it was found that the phase transition of SnTe is accompanied by a large two-phase region up to 100 kbar [5].

Theoretical calculations of the SnTe P–T phase diagram have led to a good coincidence with experimental observations of the melting curve [6]. According to the result of this work, the slope of the equilibrium line of this structural high-pressure phase transition of SnTe is positive, $\mathrm{d}P/\mathrm{d}T > 0$.

On the other hand, if one draws the equilibrium solid–solid phase boundary from the triple point at 12 kbar and 844 °C [4] towards the room temperature, 18 kbar position [2], then the slope appears to be negative, $\mathrm{d}P/\mathrm{d}T < 0$. There is a contradiction between theoretical and experimental results. The application of such a sensitive method as the Mössbauer effect gives a possibility to find the pressure ranges at

---

[1]) Part I see phys. stat. sol. (a) **53**, 473 (1979).

which the various modifications of SnTe are existing, and to receive data on the pressure dependence of the recoilless fraction of the high- and low-pressure phases. The slope of the equilibrium dP/dT can also be obtained from Mössbauer data.

Moreover, in difference from results of [2, 3, 5], in the present work SnTe was investigated under pure hydrostatic conditions up to 50 kbar. Thus, the influence of the uncontrollable shear stresses is eliminated and a more complete knowledge of the phase transition character is available.

The method of creating high hydrostatic pressures for the Mössbauer experiment is described elsewhere [7, 8]. All spectra were least-squares fitted by a computer. The quality of fitting was controlled by χ^2, the quality of the fit [9].

2. Results

The Mössbauer study of powdered specimens was carried out on pulverized single crystals of SnTe with particle dimensions of 50 to 100 μm. The particles were precipitated onto paper with effective thickness of 40 mg cm^{-2}. Single crystals of SnTe, 50 μm thick and of arbitrary orientation were used for the single crystal Mössbauer experiment.

Mössbauer spectra of powdered SnTe specimens are shown in Fig. 1 at some pressures. According to the experimental results, the structural phase transition led to a progressive nonsymmetrical broadening of absorption lines with pressure increasing above 16 kbar. After fitting the experimental Mössbauer spectra, corresponding to $P > 16$ kbar, with two Lorentzian lines a linear dependence of the relative high-pressure phase Mössbauer line position ΔE was found (Fig. 2, curve 2). The χ^2/n values indicate that such a fitting is more reliable than one-line fitting. The pressure dependence of the relative low-pressure phase Mössbauer line position was drawn using one-line fitting results up to 16 kbar and two-line ones for $P > 16$ kbar (Fig. 2, curve 1).

The influence of hydrostatic pressure on the recoilless fraction of SnTe was investigated with single crystal specimens. Fig. 3, curve 1 shows that the recoilless fraction is linearly increasing up to 16 kbar. At larger pressures the Mössbauer spectra of the single crystal specimen consisted of two lines, as the powdered specimen. The intensity of the high-pressure phase absorption line increased, but the low-pressure one decreased with increasing pressure.

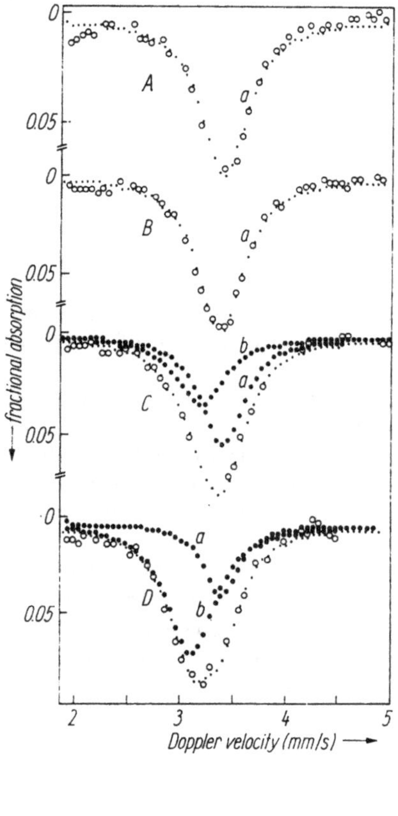

Fig. 1. Mössbauer spectra of powdered SnTe at various pressures. A) 1 atm, $\chi^2/n = 1.2$; B) 10 kbar, $\chi^2/n = 1.7$; C) 20.7 kbar, $\chi^2/n = 1.5$; D) 43 kbar, $\chi^2/n = 1.6$; (a) and (b) absorption lines of low- and high-pressure phases, respectively; source: $^{119m}CaSnO_3$

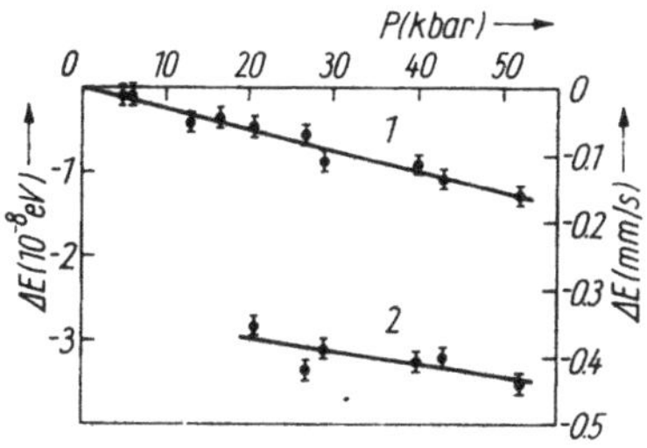

Fig. 2

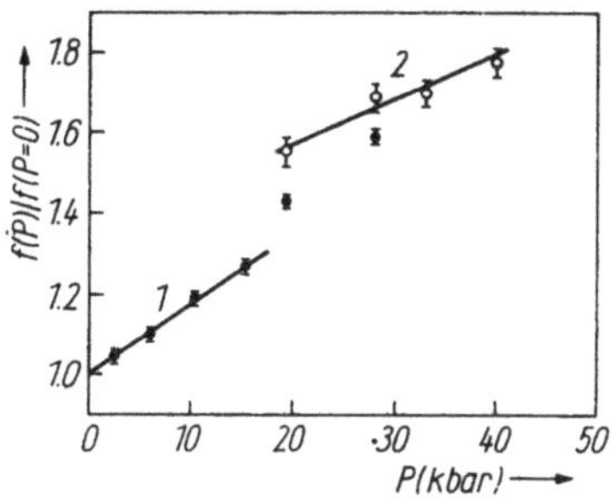

Fig. 3

Fig. 2. Relative shifts of Mössbauer lines of (1) low and (2) high-pressure phases ($\Delta E = E(P) - E_0$) for powdered SnTe with increasing pressure

Fig. 3. Pressure dependence of the relative recoilless fractions $f(P)/f_0$ of (1) low- and (2) high-pressure phases of single crystal SnTe Black points are the results of fitting with one Lorentzian line; circles represent results of calculations

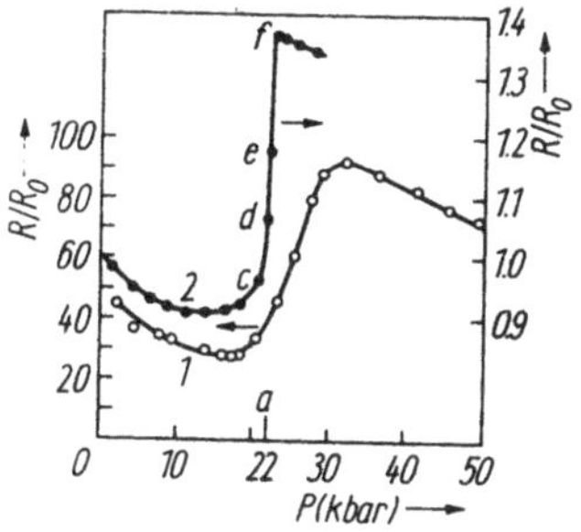

Fig. 4. Pressure dependence of the relative electrical resistance $R(P)/R_0$ of single crystal SnTe under different conditions. (1) Nonhydrostatic pressure [2]; (2) hydrostatic pressure (our results); a: pressure of the beginning of the phase transition with increasing pressure; point d corresponds to the value of R/R_0, measured after 1 h; e after 2 h; f 51 h after measuring point c

Curve 2 in Fig. 3 has been calculated using two-line fitting data at $P > 16$ kbar, depths and areas of lines of the coexisting phases, and extrapolated data on the linear pressure dependence of the recoilless fraction of the low-pressure phase up to 50 kbar. It is known that the recoilless fraction of SnTe at 1 atm is $f_0 = 0.070 \pm 0.006$ [10], $f_0 = 0.080 \pm 0.005$ [11].

The pressure dependence of the relative electrical resistance $R(P)/R_0$ of the SnTe single crystal is shown in Fig. 4 under different conditions. At first, the resistance decreased and then, from 11 to 18 kbar the resistance was practically unaltered. At a fixed pressure of 22 kbar the resistance slowly increased.

3. Discussion

The results of the Mössbauer investigation of the cubic phase of SnTe indicate a linear decrease of the electron density at the tin nuclei and a linear increase of the recoilless fraction with increasing pressure. Above 16 kbar a nonsymmetrical broadening of the experimental spectra was observed as the result of the appearance of the high-pressure phase in the specimen, the quantity of which increased with pressure. The recoilless fraction of the high-pressure phase is larger, but the electron density at tin nuclei is smaller than the cubic low-pressure phase ones.

The measurement of the recoilless fractions of low- and high-pressure phases in the point of the beginning of the phase transition gives a possibility to determine the ratio of Debye temperatures θ_1/θ_2 of the initial and the new phases of SnTe. According to the expression obtained in [12], this ratio is equal to exp $(Q/3RT)$, where Q is the phase transition heat, R the gas constant. For SnTe this ratio is $\theta_1/\theta_2 = 0.96$ at 16 kbar and 300 K. Using this value we obtained $Q = -70$ cal mol^{-1}. By using the Clapeyron-Clausius equation and the known volume change after the phase transition one can calculate the derivative $\mathrm{d}P/\mathrm{d}T$ of the equilibrium line at 16 kbar and 300 K. The calculations led to $(\mathrm{d}P/\mathrm{d}T) = +9 \times 10^{-3}$ kbar K^{-1}. This value is in agreement with the value of $+1.2 \times 10^{-3}$ kbar K^{-1}, determined from theoretical calculations of the P–T diagram of SnTe [6].

On the other hand, if one draws the equilibrium line using data of [2] and [4] as mentioned above, a negative slope is received. The difference can be explained by the fact that the pressure of 18 kbar, determined in [2, 3], is not an equilibrium pressure. Really, the intensity corresponding to the new phase nucleus reaches a considerable value, when $\Delta G = G_1 - G_2 \approx \Delta V(P_e - P)$ is increased to some value, where G_1 and G_2 are the thermodynamic potentials of the initial and the new phases, P_e is the equilibrium pressure, ΔV the volume change as a result of the phase transition. It follows that a phase transition is always observed at $P > P_e$, and the temperature is the lower, the larger the difference $P - P_e$.

By assuming that the difference $P - P_e$ can be neglected in the triple point of the P–T phase diagram of SnTe at 844 °C, the triple point pressure of 12 kbar determined in [4] can be considered to be equal to P_e. Taking into account $(\mathrm{d}P/\mathrm{d}T) = +9 \times 10^{-3}$ kbar K^{-1} the equilibrium pressure for the phase transition of SnTe at 300 K is obtained equal to approximately 6 kbar.

The present investigation was carried out with powdered and single crystal specimens. Differences have not been discovered. For powdered as well as single crystal specimens the phase transition of SnTe is accompanied by a large two-phase region. However, in accordance with results of the Mössbauer investigation around the melting point of tin [13], small particles of tin with a diameter of (5 ± 1) μm isolated from each other by the BN matrix showed a pronounced hysteresis, resulting from an undercooling of the specimen. The intensity of the resonance line shows a hysteresis loop which is completely reversible and extends over a wide temperature range of 80 K. Because the Sn particles are isolated from each other the existence of a solidification nucleus in one or a few Sn particles does not affect an instantaneous solidification of the whole specimen.

The analysis of Mössbauer spectra explained the presence of a part in the $R(P)$ dependence of SnTe (11 to 18 kbar) in which the electrical resistance is unaltered by the appearance of the high-pressure phase and showed the coexistence of new and old phases at $P > 16$ kbar, although the character of the $R(P)$ dependence at 22 kbar may be interpreted as the result of the complete transformation of SnTe to the new phase.

This discrepancy may be explained by assuming that at 22 kbar the intensity of the arising nucleus of the high-pressure phase of SnTe suddenly reaches a maximum and as a result the electrical resistance of SnTe is noticeably increased and in the specimen spring stresses were created which turned the initial single crystal into a fine-grained polycrystal. The subsequent growth of the new phase begins to be inhibited by crystal boundaries.

The decrease of the electron density at the tin nuclei with increasing pressure for both phases of SnTe confirms a common characteristic of all tin systems studied so far. This decrease may be explained by a stronger 5p electron shielding of 5s electrons and an overcompensation of the comparatively small direct contribution from the compressed 5s electrons. The decrease of the distance of 5p electrons from the nuclei, expressed by the increasing quadrupole splitting of the Mössbauer spectrum of SnO with increasing pressure [14], agrees with this interpretation of the above-mentioned common characteristic. The same behaviour has been observed in Mg_2Sn under pressure [15].

The increase of the recoilless fraction and the decrease of the electron density at the tin nuclei after the phase transition of SnTe indicate that the high-pressure phase of SnTe is more rigid than the initial one, because the increase of 5p electrons at the tin atoms strengthens the bond formation.

The values of the least-squares pressure coefficients for the linear dependence of the recoilless fraction $f(P)$ and the relative absorption line position $\Delta E(P)$ of low-(L.P.P.) and high-pressure phases (H.P.P.) of SnTe are listed in Table 1. The recoilless fraction f_0 and the absorption line position E_0 at 1 atm are taken to be equal to (0.080 ± ± 0.005) and (3.406 ± 0.005) mm s^{-1} (relative to the $^{119m}CaSnO_3$ source), respectively.

Table 1

compound	$(1/f_0)\ df/dP$ (10^{-2} kbar^{-1})	$(1/E_0)\ dE/dP$ (10^{-4} kbar^{-1})
SnTe L.P.P.	1.8	−9.3
SnTe H.P.P.	1.3	−5.6

References

[1] L. Esaki and P. J. Stiles, Phys. Rev. Letters **16**, 1108 (1966).
[2] J. A. Kafalas and A. N. Mariano, Science (Washington) **143**, 952 (1964).
[3] S. S. Kabalkina, N. R. Serebrynaya, and L. F. Vereshchagin, Fiz. tverd. Tela **10**, 733 (1968).
[4] W. Klement and L. H. Cohen, Science (Washington) **154**, 753 (1966).
[5] Z. V. Malushitskai, S. S. Kabalkina, and L. F. Vereshchagin, Fiz. tverd. Tela **14**, 1219 (1972).
[6] A. M. Altshuler, Yu. Kh. Vekilov, and G. R. Umarov, Fiz. tverd. Tela **18**, 2577 (1976).
[7] E. V. Kapitanov and V. N. Panyushkin, Prib. i Tekh. Eksper. **3**, 203 (1976).
[8] E. V. Kapitanov and E. N. Yakovlev, Prib. i Tekh. Eksper. **6**, 149 (1978).
[9] W. G. Cochran, Ann. Math. Stat. **23**, 315 (1952).
[10] V. A. Brukhanov, N. N. Delyagin, R. N. Kuzmin, and V. S. Shpinel, Zh. eksper. teor. Fiz. **46**, 14 (1964).
[11] V. Fano, G. Fedeli, and I. Ortalli, Solid State Commun. **22**, 467 (1977).
[12] C. Zener and S. Bilinsky, Phys. Rev. **50**, 101 (1936).
[13] W. Wildner, H. D. Pfannes, and U. Gonser, J. Physique **35**, C6-381 (1974).
[14] E. V. Kapitanov and E. N. Yakovlev, phys. stat. sol. (a) **51**, 641 (1979).
[15] E. V. Kapitanov and E. N. Yakovlev, phys. stat. sol. (a) **53**, 473 (1979).

(Received February 20, 1979)

phys. stat. sol. (a) 54, 145 (1979)

Subject classification: 10.1; 21.1

Physico-Technical Institute of Low Temperatures, Academy of Sciences of the Ukrainian SSR, Kharkov

Jump-Like Plastic Deformation and Dislocation Structure Instability in Zinc Crystals with Forest Dislocations

By

F. F. Lavrentev, O. P. Salita, and S. V. Sokolski

Jump-like plastic deformation and work softening of zinc crystals previously subjected to work hardening by dislocations in the slip plane $\{11\bar{2}2\}$ $\langle 11\bar{2}3\rangle$ are investigated, and the dislocation structures arising from deformation of such crystals by basal slip. Based on the results obtained, a mechanism is proposed to account for the work softening of such crystals, namely the instability of the pre-formed dislocation structure which causes the work hardening.

Es werden die sprungähnliche plastische Deformation und die Erweichung von Zinkkristallen untersucht, die vorher einer Härtung durch Versetzungen in der Gleitebene $\{11\bar{2}2\}$ $\langle 11\bar{2}3\rangle$ unterworfen wurden, sowie die Versetzungsstruktur untersucht, die von der Deformation der Kristalle durch Basalgleiten verursacht wird. Auf Grund der erhaltenen Ergebnisse wird ein Mechanismus vorgeschlagen, der für die Erweichung verantwortlich ist und auf der Instabilität der vorher gebildeten Versetzungsstruktur beruht, die die Erhärtung bewirkt.

1. Introduction

The production of the hardened state is closely associated with work softening revealed in many metals with different lattices. Work softening may appear in different ways: as a sharp yield drop in the work hardening curve, elongation of the easy glide region resulting from the change of the acting slip system during single crystal deformation, decrease of the critical shear stress in case of repeated deformations, etc. It has been disclosed that work softening is related to a rearrangement of the material dislocation structure [1]. However, until recently, no direct systematic investigations of the stability of the dislocation structure and no qualitative descriptions of its evolution were carried out. The reason is the difficulties involved in structure studies which require an estimation of the densities of various interacting dislocations and of their changes during the rearrangement. In f.c.c. and b.c.c. crystals containing several equivalent slip systems, it is very difficult to measure and separate the dislocation densities according to the slip systems. Therefore, we thought it suitable to use h.c.p. crystals for the investigation of the dislocation structure and the work-hardened state stability, because their anisotropic mechanical properties permit tracing the stability and evolution of each of various dislocation structure components and defining the plastic deformation mechanisms acting. In [2] the stability of the pre-formed work-hardening structure in Zn single crystals to repeated thermomechanical action was studied in detail. In particular, it was found that such treatments bring about essential work softening of the crystal: significant lowering of the yield stress and decrease of the forest dislocation density in the sample test region.

This work is an attempt to explore how plastic deformation is realized in a crystal previously hardened by forest dislocations. To this end, the authors examined peculiarities of the work hardening curves as well as the associated patterns of slip bands and dislocation structures in crystals work-hardened by forest dislocations.

2. Experimental

The experiments were carried out on zinc single crystalline samples of 99.997% purity. The method of growing single crystals the shape and crystallographic orientation of the samples prepared for shear deformation are presented in [2]. The initial grown-in density of basal dislocations was not in excess of 8×10^5, that of pyramidal dislocations 6×10^3 cm^{-2}, the yield stress τ_e proper to this structure was 30 to 40 p/mm^2. The hardened state was produced by a programmed compressive strain along $[2\bar{1}\bar{1}0]$, with a specified ($\approx 2 \times 10^6$ cm^{-2}) forest dislocation density introduced into the single crystal in the systems $(\bar{2}112)$ $[2\bar{1}\bar{1}3]$ and $(2\bar{1}\bar{1}2)$ $[2\bar{1}\bar{1}\bar{3}]$. Note that the introduction of forest dislocations resulted in an appreciable increase of the basal dislocation density (up to about 2×10^7 cm^{-2}) [2]. Subsequent deformation was simple shear in the system (0001) $[2\bar{1}\bar{1}0]$ at 293 K, the test machine rod velocity was 10^{-4} cm/s. The yield stress in samples with well-formed forest structure was ≈ 300 p/mm^2. Then the samples which had been deformed to different amounts of plastic deformation were subjected to a metallographic investigation of the formation and development of slip bands. The observation used the $(2\bar{1}\bar{1}0)$ plane which in this case was the plane of basal dislocation edge components shown on the surface. By using selective chemical etching of the (0001) and $(01\bar{1}0)$ planes, the pyramidal and basal dislocation densities, respectively, were determined, both in and outside the slip bands.

3. Experimental Results

Fig. 1 present work-hardening curves for the initial sample (Fig. 1a) and the sample with the pyramidal dislocation forest introduced (the forest sample) (Fig. 1b). They are essentially quantitatively and qualitatively different. In the initial sample, the start of macro-yield and the transition to the easy glide are smooth and monotonic, the stress continuously rising. No work softening is observed, that is, as the sample deformed to τ_c is again loaded, the new yield stress is never smaller than τ_c, as obtained during the first loading.

In the forest sample, there are sharp repeated load jumps which begin to occur in the region of a certain critical shear stress and cease to occur, as the sample is deformed further. The yield stress in this case is the stress of the first load jump in the deformation curve. The depth of each jump in our case was not above $0.1\,\tau_c$ (20 to 30 p/mm^2) In the jump-like deformation region, the upper level of the deformation stress is practically constant. Not before the jumps disappear this stress starts to display a sen-

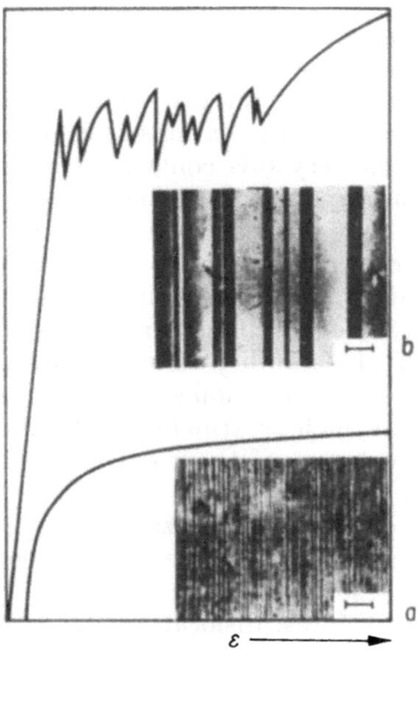

Fig. 1. Work hardening curves. a) Initial sample; b) sample with pyramidal dislocation forest introduced. The scale marks mean 0.1 mm

sible increase. Unlike the case of the initial sample, here essential work softening takes place which is disclosed during the second loading and manifests itself in a yield stress reduction by about 15% as compared with the initial sample.

The processes of plastic deformation are also different in initial and forest samples. In the initial sample, the macroscopic yield onset is attended by multiplication of basal dislocations and increase in their density almost by an order of magnitude, and the appearance of thin uniformly distributed slip bands. This state of the initial sample corresponds to the yield stress (Fig. 1a).

In the forest sample, slip is highly localized, involving the formation of large slip bands with clear-cut edges, rather large areas between them remaining unaffected by the plastic deformation at all. The basal dislocation density increases but slightly in the slip bands as compared with the high density of dislocations introduced during sample preparation ($\approx 2 \times 10^7$ cm^{-2}) [2]. At the same time, the sample deformation is so rapid that crystal parts move with respect to each other at a much higher speed than that of the test machine rod, thereby causing load jumps in the work-hardening curve.

During deformation of the initial sample, the rod and shear speeds were probably equal, therefore, the work-hardening curve is monotonic.

If during the forest sample experiment the test machine is stopped and the load relieved directly after the first jump in the curve, then a single slip band, several hundredth parts of a millimeter in width, is observed in the test region. Fig. 2a shows a 0.02 mm slip band and the corresponding load jump. It is reasonable to assume a correlation between the curve jump and the slip band. Repeated loading results in the development of plastic deformation in the single slip band which formed during the first loading (Fig. 2b, shown by arrows). In this case, before resumption of the jump-like deformation, the work-hardening curve deviates from the linear shape. As the

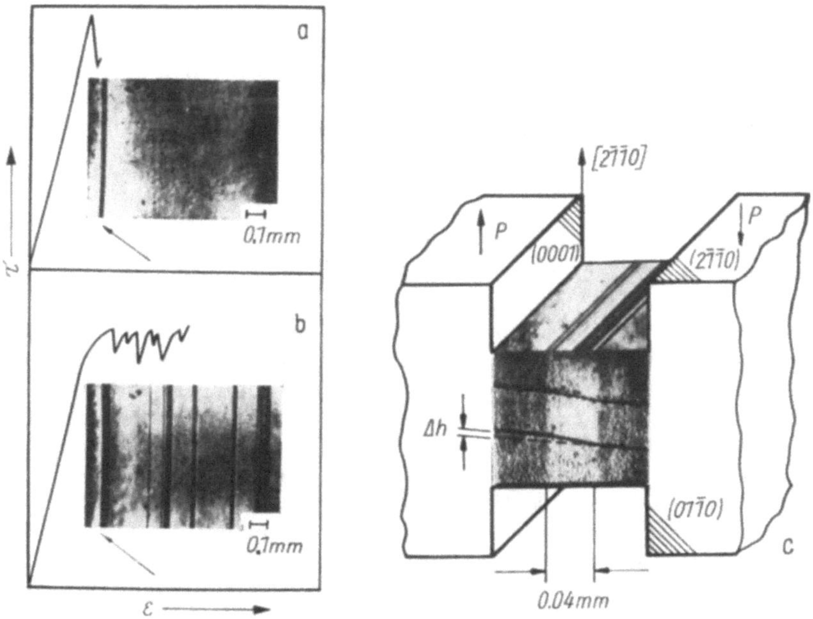

Fig. 2. Plastic deformation localization and jump-like character in the forest sample. The scale marks in a) and b) mean 0.1 mm

load is increased, the band becomes broader and thin slip lines become increasingly pronounced therein. Markers scratched perpendicular to the shear direction $[2\bar{1}\bar{1}0]$ in the sample test region of the plane $(01\bar{1}0)$ before deformation confirm the local character of the plastic deformation within the slip band (Fig. 2c). They also can be used to estimate the true magnitude of the plastic deformation in this slip band ($\approx 15\%$). Thus, the experiment shows that in case of repeated loading, the weakest place in the crystal is the slip band (suggested by work softening in the slip band). In the initial sample, slip bands are the most hardened regions after their origin.

Further loading of the forest sample brings about new bands which gradually fill all the test region; simultaneously, the work-hardening curve ceases to make jumps. Thus, there occurs homogenization of plastic yield with deformation in the forest sample.

The forest dislocation density measurement by selective chemical etching on the initial and forest samples led to the following results. In the initial sample, the pyramidal dislocation density is constant during deformation to the easy glide stage and equal to the initial grown-in density. In the forest sample, in the regions with abundant slip bands, the forest dislocation density reduces by a factor of 2 to 3 as compared with the introduced density (about 2×10^6 cm^{-2}).

4. Discussion

Thus, the analysis of the work-hardening curve and of the dislocation structure of the crystal with pre-formed hardening structure allows one to assert that the work softening observed (yield stress decrease during repeated loading) is associated with the destruction of the dislocation structure formed in the pyramidal slip system. The results obtained corroborate the conclusion of [2] that the dislocation structure rearrangement in intersecting slip systems is essentially dependent on the plastic deformation processes in the primary basal slip system.

The correlation between the deformation stress jumps, the slip band localization, and the pyramidal dislocation "sweep-out" of the slip bands indicates that in crystals with forest dislocations an avalanche of basal dislocations occurs which is responsible for the abrupt rise of the plastic deformation rate of the crystal.

Let us consider the candidate mechanisms to account for such an avalanche.

Evidently, in crystals with high forest dislocation density, the appearance of a large density of basal dislocations may be due to the following transformation reaction:

$$b^{4.46}_{1\,1/3[2\bar{1}\bar{1}3](\bar{2}112)} + b^{4.46}_{2\,1/3[2\bar{1}\bar{1}\bar{3}]\,(2\bar{1}\bar{1}2)} \rightarrow b^{4.00}_{3\,2/3[2\bar{1}\bar{1}0](0001)}\,, \tag{1}$$

where subscripts at the Burgers vectors $\boldsymbol{b}_n$ indicate the magnitude, direction, and plane of occurrence of the reacting dislocations, and the figures above them show the relative dislocation energies. This reaction is obviously energetically advantageous. The scheme in Fig. 3 shows the Burgers vector arrangement for reaction (1), whence it is seen that the pyramidal dislocation interaction results in a slipping basal dislocation.

This scheme shows not only one of the mechanisms responsible for the excessive basal dislocation density, but also the possibility for the decrease of the pyramidal dislocation density by virtue of the decrease of their total length.

Besides the above-considered case (reaction (1)) the mobile dislocation density could increase if it is true that crystal deformation is attended by liberation of a large number of basal dislocations as a result of annihilation of the stoppers, i.e. forest dislocations, which has been observed in the pre-formed slip bands. Motion of pyramidal dislocations within the crystal is in this case an indispensable condition of the annihilation and transformation. Pyramidal dislocations are hard to move, and therefore the large stresses comparable with those applied to introduce forest dislocations (3000 p/mm^2)

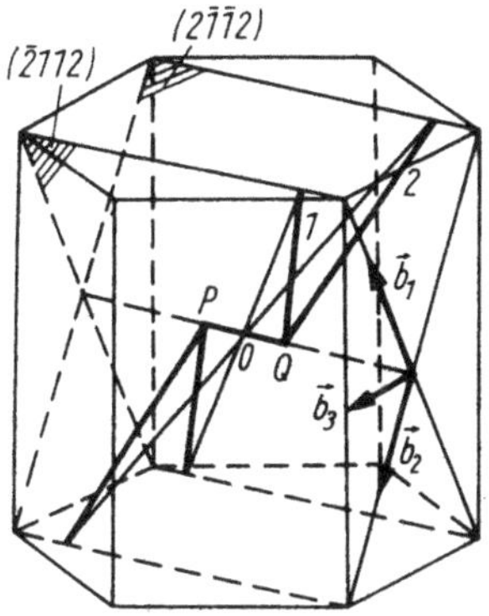

Fig. 3. Pyramidal-pyramidal interaction scheme, according to reaction (1). PQ is the basal dislocation

are needed. The stresses specified are obviously much higher than the stress component acting in the system $\{11\bar{2}2\}$ $\langle 11\bar{2}3\rangle$ during the jump-like deformation in the system (0001) $[11\bar{2}0]$. Therefore, even at this stress, stress concentrators should develop in the crystal which can cause the pyramidal dislocations to move. In our case, only basal dislocation pile-ups can act as stress concentrators. Conditions are favourable for such pile-ups, since the introduction of forest dislocations results in a large basal dislocation density ($\approx 2 \times 10^7$ cm^{-2}).

One can estimate the number of basal dislocations in a pile-up sufficient for the appropriate stress concentration to develop in the pyramidal plane with the use of the results of Svabova's calculation [3].

The stresses arising in the pyramidal plane, if a dislocation pile-up is present in the basal plane, are

$$\tau_p \approx \tau_b n \cos 2\alpha \;, \tag{2}$$

where $\tau_b \approx 300$ p/mm^2, n is the basal dislocation number in a pile-up, α the angle between the basal plane and the pyramidal one. If assume that the stress required for the pyramidal dislocations to start moving is roughly equal to the stress applied for the introduction of pyramidal forest dislocations into the crystal ($\tau_p \approx 3000$ p/mm^2), then $n = 18$.

The pile-ups which have formed may be discharged by instantaneously forcing their way up through the obstacle. In a crystal with high forest dislocation density, pairwise annihilation in the vicinity of a pile-up (in its stress field) is quite realistic; therefore the pile-up discharge occurs in this way. As this takes place, the basal dislocation due to pyramidal–pyramidal interaction experiences the action of the pile-up stress, and thus a high-speed motion of basal dislocations starts. With recourse to this assumption, we can understand the high slip localization in the previously hardened crystal, since an avalanche may be triggered by a single discharge which would affect the pile-ups in the plane where it occurs and in neighbouring planes. The stress decreases and slip occurs in the formed band till it hardens. Later, when the appropriate stress has been recovered as a result of the band hardening, a new avalanche starts elsewhere in the crystal where there is a critical basal dislocation pile-up.

The results of this work allow to estimate the density of basal dislocation breakthrough sites during the load jump attended by localization of about 10% of the plastic deformation within the slip band. If the slip band width is $h \approx 2 \times 10^{-3}$ cm, this deformation produces a shear of $\Delta h \approx 2 \times 10^{-4}$ cm. To realize this shear, it is necessary that $N = \Delta h/b \approx 8 \times 10^3$ basal dislocations ($\boldsymbol{b}$ is the Burgers vector of the basal dislocation) should move and leave the crystal. The estimated basal dislocation number in a pile-up is $n \approx 18$. Therefore, the break-through site density in the basal dislocation pile-up during the load jump should be $\varrho = N/nhL \approx 0.4 \times 10^6$ cm^{-2} ($L = 0.5$ cm is the sample test region size along [$2\bar{1}\bar{1}0$]). The estimated value of ϱ and the forest dislocation density have close orders of magnitude. This allows one to conclude that basal dislocation pile-ups develop mostly near forest dislocations and stimulate their motion and eventual annihilation or transformation into basal ones, that is to say their "sweeping".

If the number of dislocations performing shear in the slip band is known ($\approx 8 \times 10^3$), then the mobile basal dislocation density can be estimated

$$\varrho_b = \frac{N}{hL} \approx 0.8 \times 10^7 \text{ cm}^{-2} . \tag{3}$$

The value of ϱ_b is of the same order of magnitude as the initial basal dislocation density measured by counting etch pits, that is to say, practically all the basal dislocations present in the slip band are mobile during the load jump.

Let us estimate the plastic deformation rate in the slip band

$$\varepsilon \geqq \frac{\Delta h}{h} \frac{1}{t} . \tag{4}$$

Here $\Delta h \approx 2 \times 10^{-4}$ cm is the shear amount, $h \approx 2 \times 10^{-3}$ cm the slip band width, $t = 3 \times 10^{-2}$ s the load drop time during the jump. Then $\dot{\varepsilon} \geqq 3$ s^{-1}.

Then, using the following relation,

$$\dot{\varepsilon} = \varrho_b b \bar{v} \geqq 3 \text{ s}^{-1} , \tag{5}$$

where ϱ_b is the density and $\bar{v}$ the velocity of dislocations performing plastic deformation, we estimate the average dislocation velocity in the slip band to be $\bar{v} \geqq 14$ cm/s.

The estimated $\bar{v}$ is the minimum dislocation velocity, for the time t deliberately overrated. The estimated velocity $v \geqq 14$ cm/s corresponds to the quasi-viscous region of the dislocation motion densities [4].

The above analysis of the experimental data of this work permits to propose a certain physical model of work softening for the crystal with pre-formed hardening structure. It is obvious that, irrespective of whether the sharp rise of the plastic deformation rate is due to an abrupt increase of the density of the mobile dislocations or to an increase of their velocity, in both cases this process requires the presence of dislocation pile-ups. The pile-ups stimulate annihilation and dislocation reactions of transformation responsible for the large numbers of mobile basal dislocations that come into play in the crystal. Besides, the pile-ups, when discharged, give high velocities to dislocations.

Besides the above-proposed mechanism explaining the jump-like deformation and work softening in h.c.p. crystals, three more models have been developed for the destruction of the hardened state in f.c.c. [5, 6] and b.c.c. [7] crystals. The Washburn [5] model for f.c.c. metals based on experimental data seems the most realistic one. In this model, work softening is related to annihilation of metastable configurations of forest dislocations which takes place as they intersect with primary dislocations having the same Burgers vector. This process leads to a decrease of the forest dislocation density and, therefore, to work softening which should be accompanied by jump-like changing of the deformation stress and slip band localization. The principles of this model are in qualitative agreement with experimental results on latent hardening in f.c.c. metals [8]. Washburn's and our models have a common feature, i.e. the assumption of forest dislocation instability to plastic deformation in the primary

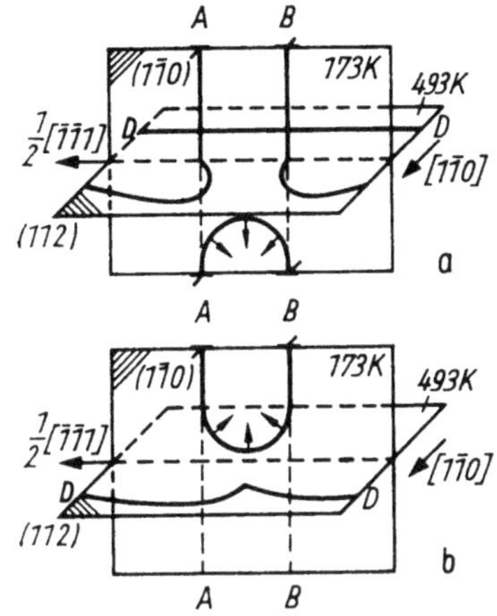

Fig. 4. Scheme of dislocation density decrease in the system (1$\bar{1}$0) [$\bar{1}\bar{1}$1] due to an interaction of dislocations of this system and (112) [$\bar{1}\bar{1}$1]

slip system. However, the mechanisms of this instability are different: in the h.c.p. lattice, it is the transformation of forest dislocations into basal ones which are mobile whereas in the f.c.c. lattice, it is the annihilation of forest dislocation parts intersected with by primary dislocations.

The concepts and mechanisms proposed in this work and in [5] are of very general nature. In terms of these concepts, the data due to Lüft et al. [1] on jump-like deformation, work softening, and formation of zones with low dislocation density in Mo b.c.c. single crystals can be explained. Fig. 4a, b shows the Washburn scheme modified for b.c.c. crystals and allowing to account for the data given by Lüft. In case of low temperature deformation (173 K), in Mo single crystals the slip system (1$\bar{1}$0) [$\bar{1}\bar{1}$1] acts which is represented in the scheme 4a by edge dislocations of different sign AA and BB in the plane (1$\bar{1}$0). As we go on to high temperature deformation (493 K), the slip system (112) [1$\bar{1}$1] comes into play in which the DD dislocation moves (scheme 4b). For this slip system, AA and BB are forest dislocations. The Burgers vector is equal for all the mentioned dislocations and is $\frac{1}{2}$ [$\bar{1}\bar{1}$1]. The DD dislocation, as it intersects with the AA and BB dislocations, causes the latter two ones to vanish, as is seen on Fig. 4a, b. This process should lead to a decreasing forest dislocation densiy and to work softening jumps and slip band localization.

Although the actual interaction processes taking place in the crystals are, of course, much more complicated than the models proposed, these models can adequately describe the processes responsible for work softening.

5. Conclusion

The jump-like character of plastic deformation during basal slip in zinc single crystals hardened by a dislocation forest in the system {11$\bar{2}$2} ⟨11$\bar{2}$3⟩ has been ascertained.

The association of deformation stress jumps and localized slip bands, disclosed in this study, and crystal work softening are shown to be due to the instability of the introduced forest dislocation system to stresses applied in the primary slip system and to the decrease of the density of these forest dislocations during subsequent deformation.

The number of basal dislocations in a pile-up has been estimated ($n \approx 18$). The estimated densities of pile-ups, i.e. of basal dislocation assembly break-through sites, suggest that these pile-ups form mainly in the vicinity of forest dislocations.

The plastic deformation rate increase observed in the slip band ($\dot{\varepsilon} \geqq 3\ s^{-1}$) and the average basal dislocation velocity in it ($v \geqq 14$ cm/s) evidence avalanche-type motion of mobile dislocations at the moment of the load jump.

A mechanism has been proposed to account for work softening and other effects observable, based on transformation and annihilation of forest dislocations stimulated by stresses produced by basal dislocation pile-ups.

References

[1] A. LÜFT, J. RICHTER, K. SCHLAUBITZ, CH. LOOSE, and CH. MÜHLHAUS, Mater. Sci. Engng. **20**, 113 (1975).
[2] F. F. LAVRENTEV, O. P. SALITA, and S. V. SOKOLSKY, Mater. Sci. Engng. **33**, 199 (1978).
[3] M. SAXLOVA-ŠVABOVA, Z. Metallk. **58**, 266 (1967).
[4] N. NAGATA and T. VREELAND, JR., Phil. Mag. **25**, 1137 (1972).
[5] J. WASHBURN and G. MURTY, J. appl. Phys. **37**, 2511 (1966).
[6] E. J. H. WESSELS and F. R. N. NABARRO, Acta metall. **19**, 915 (1971).
[7] V. I. VLADIMIROV and A. A. KUSOV, Fiz. Metallovi Metallovedenie **43**, 1127 (1977).
[8] Z. S. BASINSKI and P. J. JACKSON, phys. stat. sol. **9**, 805 (1965).

(Received January 18, 1979)

phys. stat. sol. (a) 54, 153 (1979)

Subject classification: 11; 14.2

Institut für Experimentelle Kernphysik der Universität Karlsruhe[1]) *(a) and Département de Physique de la Matière Condensée, Université de Genève*[2]) *(b)*

Summation and Saturation Behaviour of the Volume Pinning Force in Neutron Irradiated V_3Si[3])

By

H. KÜPFER (a) and A. A. MANUEL (b)

Neutron irradiated V_3Si is used as a model system for the investigation of the dependence of the volume pinning force on the defect structure in A15 superconductors. The dislocation loop concentration determined by TEM is varied within three orders of magnitude. Up to the highest concentration the system is still dilute, but there is absolutely no field region where the volume pinning force is proportional to the concentration of pinning centres as predicted by the summation models and proved in low $\varkappa$-material like Nb. This unexpected behaviour is accompanied by the peak effect and by a saturation of the volume pinning force in the whole field region at a fluence of 10^{18} cm^{-2}. The explanation is based on the threshold criterion which is not satisfied in these microstructures.

V_3Si wird durch Bestrahlung mit Neutronen als Modellsystem zur Untersuchung der Abhängigkeit der Volumenhaftkraft von der Defektstruktur in A15-Supraleitern präpariert. Die Konzentration der Versetzungsringe wird mittels TEM bestimmt und um drei Größenordnungen variiert. Obwohl die Haftzentren selbst bei der höchsten Konzentration hinreichend verdünnt sind, gibt es keinen Feldbereich, in dem die Volumenhaftkraft proportional zur Konzentration ist, wie es von den Summationsmodellen vorausgesagt wird und in Supraleitern mit kleinem $\varkappa$ wie Nb nachgewiesen wurde. Dieses unerwartete Verhalten ist begleitet von einem Peakeffekt und oberhalb einer Dosis von 10^{18} cm^{-2} von einer Sättigung der Volumenhaftkraft im gesamten Feldbereich. Die Erklärung für dieses Verhalten geht vom Thresholdkriterium aus, das für diese Mikrostruktur nicht erfüllt ist.

1. Introduction

Non-dissipative transport currents in the mixed state of a type II superconductor are caused by the pinning of the flux line lattice at crystal defects. The questions about the influence of the superconducting properties and the microstructure on this interaction are far from being solved. There is particular interest in the summation law of the elementary pinning forces, in the dependence of the volume pinning force on the defect concentration, and, in close connection with this, in the peak effect and the saturation of the volume pinning force with increasing strength of the defect structure. Investigations on model systems in which one kind of defects is predominant, seem to be the only way to get a better understanding of these problems. Up to now measurements on such microstructures are made on superconductors with small Ginzburg-Landau parameters and large coherence length like Nb or NbTa alloys. An extrapolation of these results to the A15 superconductors is not possible for two reasons: First, the ratios between the mean distance of the crystal defects and the superconducting coherence length or the penetration depth, respectively, are completely different. Further, the field region where the elementary pinning force does not depend on the field and the shear modulus of the flux line lattice increases with

[1]) Postfach 3640, 7500 Karlsruhe, BRD.

[2]) 32, Boulevard d'Yvoy, 1211 Genève 4, Suisse.

[3]) Work supported by the Institut für Technische Physik of the Kernforschungszentrum Karlsruhe.

increasing field, is very small and may be neglected for high-T_c A15 superconductors, whereas in Nb this area extends to a reduced field of about 0.5. Studying the pinning mechanism in A15 superconductors, one is therefore compelled to investigate a model system of these intermetallic compounds. V_3Si seems to be a very promising candidate for that purpose because it is obtainable with a low concentration of crystal defects and the corresponding reversible properties are well known. Starting from single crystals and polycrystalline material, the microstructure was changed by irradiation with high-energy neutrons. Dislocation loops are identified as the part of the radiation induced defects responsible for the enhanced pinning. The important condition — a reliable estimation of the elementary pinning force — is realized for this kind of defects because this interaction has been calculated by several authors [1, 2]. The concentration of the loops has been varied by different fluences. Comparisons of the dependence of the volume pinning force on temperature, field, and concentration with theoretical models and concepts give a more extended view of the above-mentioned problems.

2. Microstructure of the Samples

2.1 Sample preparation

A single crystal of ⟨100⟩ orientation was prepared by the zone melting technique as described by Manuel [3]. The rod was sliced by spark cutting techniques in cylinders of 0.2 cm in diameter and 1 cm in length. These samples could be divided in two groups with respect to the transition temperature T_c, the residual resistance ratio RRR $= R_{293\,K}/R_{20\,K}$, and the critical current density j_c. Samples with $T_c = 16.45$ K and RRR $= 43$ had an j_c increasing monotonically with decreasing field, whereas samples with higher $T_c = 16.70$ K and smaller RRR $= 18$ showed a small maximum in j_c close to the upper critical field H_{c2}. We attribute this different behaviour to the extent of the transformation from the cubic into the tetragonal phase [4], which should be much more incomplete in low-RRR samples. A structural transformation of only 10% was observed on samples with a similarly small RRR by Viswanathan

Table 1

	fluence $E > 1$ MeV (cm^{-2})	N (cm^{-3})	$\varrho_{20\,K}$ (μΩcm)	RRR	T_c (K)	$-dH_{c2}/dT\|_{T\to T_c}$ (TK^{-1})	j_c^{max} $(t = 0.84)$ (A cm^{-2})
S0	0	—	1.5	47.9	16.49	1.70	50 ($b = 0.48$)
S1	1.68×10^{16}	4×10^{11}	1.5	43.9	16.71	1.77	310 ($b = 0.92$)
S2	1.71×10^{17}	4×10^{12}	2.5	31.7	16.53	1.78	11×10^3 ($b = 0.69$)
S3	1.63×10^{18}	4×10^{13}	8	10.7	16.13	2.28	100×10^3 ($b = 0.29$)
S4	4.37×10^{18}	9×10^{13}	11	7.65	15.90	2.40	72×10^3 ($b = 0.17$)
S5	1.17×10^{19}	2×10^{14}	30	3.37	14.08	2.73	60×10^3 ($b = 0.13$)
P0	0	—	—	25	16.82	1.80	150 ($b = 0.92$)
P2	1.71×10^{17}	—	—	21.1	16.70	1.89	3.9×10^3 ($b = 0.78$)
P4	4.37×10^{18}	—	—	6.97	16.05	2.59	70×10^3 ($b = 0.23$)

and Caton [5]. The larger amount of phase boundaries between the tetragonal and the cubic state in these samples decreases the electron mean free path and may be responsible for the different field dependence of the critical current. This explanation is in accordance with the smaller T_c of the tetragonal phase in comparison with the cubic phase [6].

All samples were annealed at 1400 °C and at a pressure of 2×10^{-9} Torr for 100 h. Then they were slowly cooled to room temperature. After this treatment the above-mentioned differences between the two groups vanished. Characteristic values of these samples, marked with S0, are summarized in Table 1.

The mean dislocation density of S0 estimated by transmission electron microscopy was about 10^6 cm^{-2}. Most of the dislocations were lumped together.

Some polycrystalline samples P0, additionally investigated, were not annealed. They have large V_5Si_3 precipitates of a concentration 7×10^7 cm^{-3} and an average size of 4 μm. Their dislocation density is about 10^7 cm^{-2}, with an inhomogeneous distribution as observed in the sample S0.

The neutron irradiation was performed at 240 °C in the reactor FR2 of the Kernforschungszentrum Karlsruhe. The samples were wrapped up in Cd foil and irradiated in the same position in the reactor core. The fluence was determined by the (n, p) reaction of Ni^{58}. In Table 1 the fluence is given for neutrons with energies higher than 1 MeV. This value has to be multiplied by 2.5 in order to get the fluence above 0.1 MeV.

2.2 *Transmission electron microscopy*

Transmission specimens were prepared by spark cutting small discs from the cylindrical samples. The discs were thinned electrolytically as described by Essmann et al. [7] and investigated at a voltage of 125 kV using a double tilting stage. Bright-field images at magnifications up to 200000 revealed black–white contrast figures of resolved loops and unresolved black dots. For the determination of the size distribution and the loop density the whole region of the thin edge of the hole had to be covered by several pictures, because of the low density and the small size of the loops. Fig. 1 shows a micrograph of the sample S5. The mean thickness of the observed area was determined by extinction patterns. The extinction length was taken from Goringe and Valdrè [8]. Fig. 2 shows the size distribution in the samples S3 and S5. The root mean square diameter is about (160 ± 30) Å for S3 and (180 ± 30) Å for S5. The total density in S3 and S5 is $(4 \pm 2) \times 10^{13}$ and $(2 \pm 1) \times 10^{14}$ cm^{-3}, respectively.

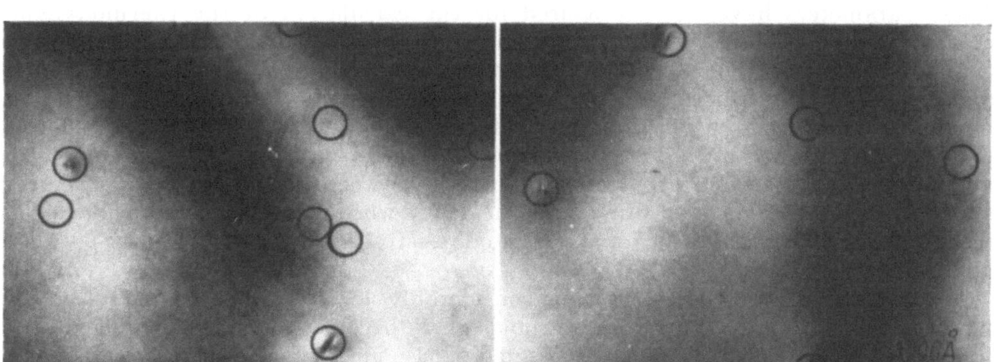

Fig. 1. Bright-field electron micrographs (62000×) showing dislocation loops of sizes between 400 and 50 Å in the irradiated sample S5. In the right picture the sample is slightly tilted against he position in the left one

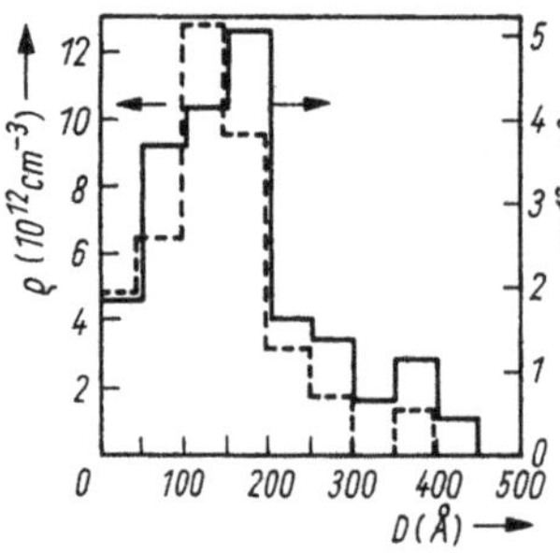

Fig. 2. Size distribution of loops for the samples S3 (dashed line) and S5 (solid line)

With these results we derived the dependence of the total loop density N on the fluence as follows: At low fluences up to 10^{18} cm^{-2} it is reasonable to assume no interaction between primary point defect cascades. This leads to a proportionality between fluence and loop density and to a constant root mean square diameter. With increasing fluence the mean loop diameter increases, and therefore the density increases less than linear with fluence as shown by S3 and S5. From this we get the relation $N = 4.65 \times \times 10^{-2}\Phi^{0.82}$ for fluences 10^{18} cm$^{-2} \leqq \Phi \leqq 1.2 \times 10^{19}$ cm^{-2}. The values are given in Table 1. The error results mainly from the uncertainties in recognizing loops below 100 Å in diameter. An additional annealing of sample S1 at the irradiation temperature of 240 °C for a time much longer than the irradiation time did not change the superconducting parameters. From this we deduce that agglomeration of the point defect occurred already during the irradiation. This supports the extrapolation of the relation between fluence and loop concentration to the lower fluences of samples S1 and S2. This holds for single crystals which have the same impurity content. The preparation of the polycrystalline samples turned out to be much more difficult due to the large precipitates. Therefore a reliable estimation of the loop concentration of P4 was not possible. Sample S3 was additionally investigated in a crystal direction perpendicular to ⟨100⟩. The total loop density was within the above results. Small disordered regions as detected by Pande [9, 10] in irradiated Nb_3Sn were not observed probably because the specimens were not thin enough.

3. Measurements and Results

3.1 Critical temperature, resistivity, and upper critical field

These parameters have been measured for the calculation of the thermodynamic critical field H_c and of the Ginzburg-Landau parameter $\varkappa$ in Section 4 and for the characterization of the microstructure of the samples. The H_{c2} was determined inductively. T_c was measured both inductively and resistively. The error is about 0.03 K for the temperature and 0.1% for the field. From the dependence of H_{c2} on the temperature we obtained dH_{c2}/dT at T_c. The values given in Table 1 correspond to the middle of the transition. The transition width between 10 and 90% is 0.1 K for S0 and 0.25 K for S5. The resistance and the RRR were measured by the four-probe technique. The absolute error for the residual resistivity $\varrho_{20\,K}$ is of the order 50% and approximately 5% for the RRR.

Fig. 3 shows that RRR decreases considerably for low fluences monotonically with increasing fluence. This does not hold for T_c because it is barely changed up to fluences of 4×10^{18} cm^{-2}. Such a 'threshold behaviour' is in accordance with observations of several authors [5, 11] and with a theory given by Pande [9, 10] based on very small disordered regions. The reduction of T_c at high fluences is lower than that reported by [5, 11] for V_3Si. This may be due to a less distorted starting material documented by RRR and $\varrho_{20\,K}$.

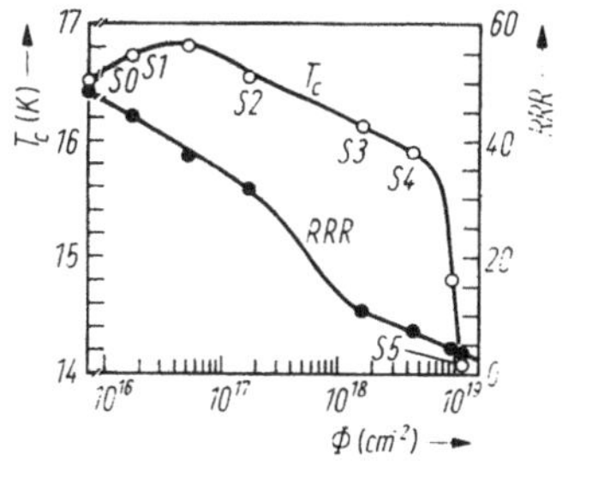

Fig. 3. Superconducting transition temperature T_c and resistivity ratio RRR as a function of the neutron fluence Φ ($E > 1$ MeV)

Up to fluences of 3×10^{17} cm^{-2} the expected, but small decrease of T_c with increasing fluence is not observed. On the contrary, T_c shows a maximum of 16.82 K at 5×10^{16} cm^{-2}. As described in Section 2.1 this behaviour may be correlated with the stabilization of the cubic phase by the irradiation induced defects.

Comparing the samples S0, S1, and S2 in Fig. 3 with the samples of the group showing the much lower RRR of 18, but a similar T_c of 16.70 K, one can conclude that a clear correlation between T_c and RRR — often used in literature — does not exist in spite of the single-phase state of these materials. This correlation seems to be valid only far below the optimal transition temperature where the specimen is in a highly disordered state.

3.2 Critical current measurements and volume pinning force

The radial distribution of the magnetic induction was measured by means of an ac method developed by Rollins et al. [12]. The measurement was performed as a function of the temperature and of the external field directed parallel to the cylinder axis of the sample. One important advantage of this technique as compared to integral measurements is the recognition of surface pinning, which otherwise is often correlated erroneously with the microstructure in the bulk. The annealed and the irradiated samples did not show an enhanced pinning in the surface region. The critical current density j_c was calculated from the slope of the local magnetic induction versus radial position. This was done always at the same distance about 50 μm from the surface of the samples. The relative error of the j_c determination is between 5% and 10% for a reduced field $b \leq 0.1$ and less than 5% for $b > 0.1$. The absolute error may be twice as large as the relative one.

Small differences in the values of j_c at the same field and the same temperature appeared for different paths taken to arrive at the state (B, T). This property was observed at low fields for the samples S1, S2, and P2 and is in full accordance with

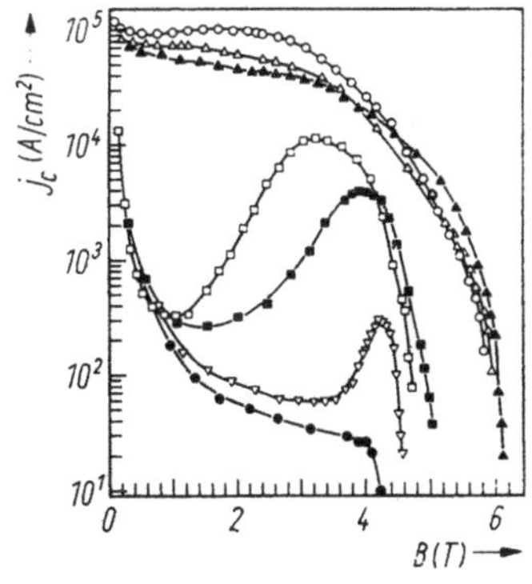

Fig. 4. Critical current density j_c versus magnetic field B at reduced temperature 0.84 for the samples S0 (●), S1 (▽), S2 (□), S3 (○), S4 (△), S5 (▲), and P2 (■)

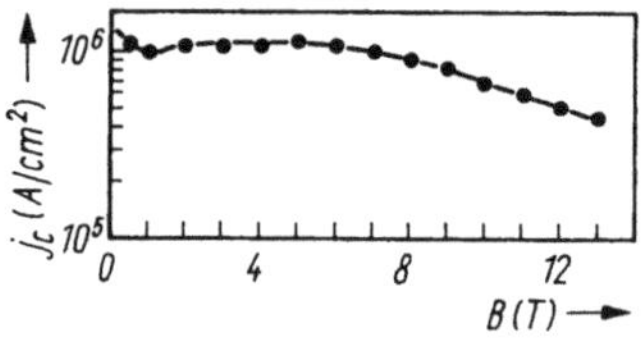

Fig. 5. Critical current density j_c versus magnetic field at 4.2 K for the sample S3

the explanation of the 'history effect' given by Küpfer and Gey [13]. All measurements discussed in this paper were performed in the following manner: First the temperature was adjusted, and then the field was swept from the normal state through H_{c2} to the desired value.

Fig. 4 shows the critical current density of the single crystals and of P2 in dependence on the magnetic field at a reduced temperature $t = T/T_c$ of 0.84. The specimens S1, S2, and P2 with a fluence less than 10^{18} cm^{-2} show a pronounced peak at high fields. This peak effect is discussed in Section 4. In Fig. 5 the critical current density is plotted up to 13 T at 4.2 K for the sample S3. Remarkable are the very high values of j_c exceeding 10^6 A cm^{-2} at 7 T. These values are obtained only by neutron irradiation, leaving the sample in a much better crystalline state than specimens produced by diffusion techniques, sputtering, or evaporation, which show lower critical currents. This means, if there is no intrinsic saturation of the pinning mechanism, it should be possible to get much higher currents and a smaller reduction of the other critical parameters by a more careful preparation.

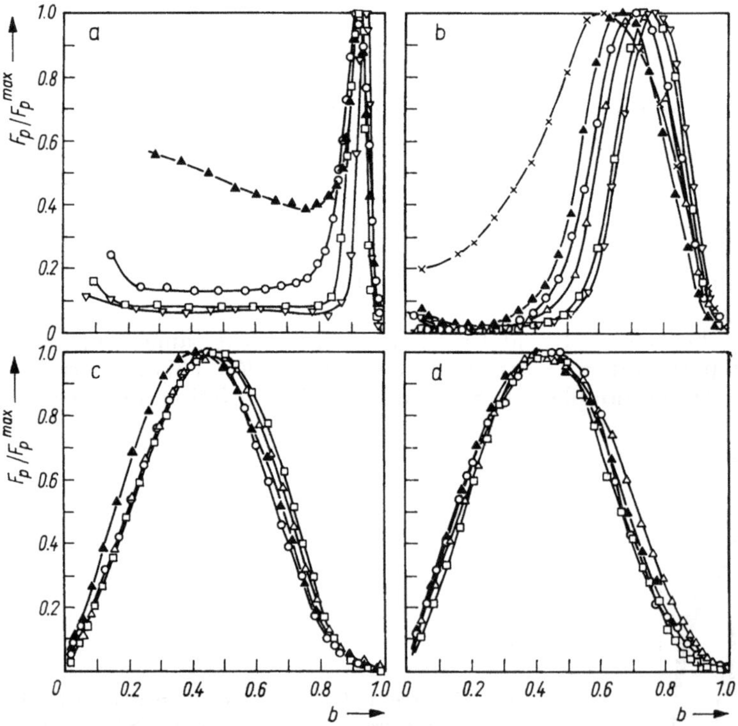

Fig. 6. Reduced volume pinning force F_p/F_p^{max} versus reduced field $b = B/H_{c2}$ at the reduced temperatures 0.97 (×), 0.92 (▲), 0.84 (○), 0.76 (△), 0.66 (□), and 0.53 (▽), a) S1, b) S2, c) S3, d) S4

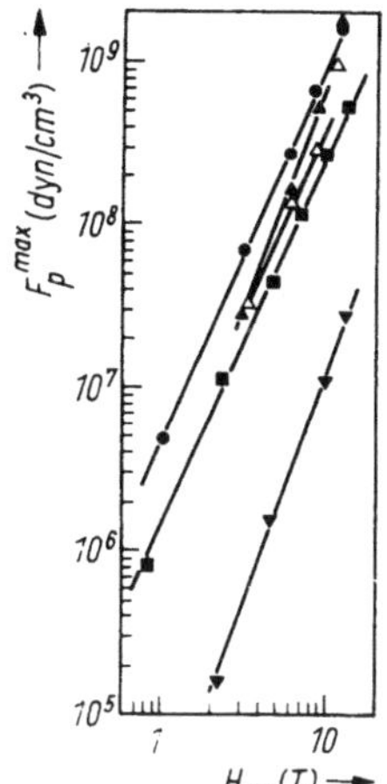

Fig. 7. Maximum volume pinning force F_p^{max} versus H_{c2} for the samples S1 (▼), S2 (■), S3 (●), S4 (▲), and S5 (△)

Fig. 6 shows the reduced volume pinning force $F_p(F_p^{max})^{-1}$ in dependence on the reduced field at different temperatures. In the following we list only some remarkable properties of these pictures; a comparison with theory is made in Section 4. All samples show a scaling behaviour $F_p \sim H_{c2}^n(T)\, f(b)$ with satisfying accuracy. With decreasing temperature the maximum of the volume pinning force is slightly shifted to higher fields and the width at half maximum of the $F_p(F_p^{max})^{-1}$ function decreases. This behaviour is emphasized by sample S2. The origin of this deviation from the temperature scalign is the paramagnetically limited H_{c2}, which is used for the calculation of the reduced field. The paramagnetic effect vanishes for temperatures close to T_c [14]. This is the reason why a fairly good scaling is observed in the measured temperature region.

An additional deviation from the scaling law is observed at reduced temperature $t \geqq 0.9$ and at low fields in the samples S1 and S2. This is due to the unaffected j_c at low fluences in the field region below the maximum of j_c, as can be recognized in Fig. 4. But the temperature dependence of F_p in S0 is different from that induced by irradiation. Therefore the ratio of these two mechanisms varies with temperature.

Fig. 4 and 6 show that the dependence of j_c, respectively of F_p, on the field is approximately unchanged for fluences above 10^{18} cm^{-2} in the samples S3, S4, and S5. The absolute value of F_p decreases for fluences larger than 4×10^{18} cm^{-2} compared at the same reduced temperature.

The exponent n of the scaling law is taken from a lg–lg plot of F_p^{max} versus H_{c2} (Fig. 7). The values are between 2.3 (S3) and 2.8 (S1). The results at the lowest temperature for the specimens S4 and S5 are not considered. Probably the influence of the paramagnetically limited H_{c2} is more pronounced in these samples. This is not in contradiction to the apparent better temperature scaling of these samples compared to S1 and S2, because their maximum shows a smaller curvature, i.e., they are less sensitive to a too small H_{c2}.

4. Discussion

In this section the elementary pinning force f_p is calculated (4.1), and then we enter into a discussion of the summation of these elementary pinning forces (4.2), the dependence of the volume pinning force on the concentration of the pinning centres (4.3), the peak effect and the saturation behaviour (4.4), and the threshold criterion (4.5).

4.1 Calculation of the elementary pinning force

For an introduction we briefly discuss the possible influence of the disordered regions (agglomeration of antisite defects) on j_c. The threshold behaviour of T_c at high fluences (Fig. 3) has been attributed to these defects by Pande [9, 10]. According to a fairly good agreement between this theory and the experiments, these disordered regions may influence T_c, but they are ineffective as pinning centres because their size is smaller than the coherence length. Further, the defects may cause only a very small elastic interaction with the flux line lattice because they have the same lattice parameter as the matrix [10]. This is confirmed by the different dependence of T_c

and j_c on the fluence, on the irradiation temperature, and on the post-irradiation annealing. This means that T_c and j_c are influenced by different kinds of defects: j_c is predominantly governed by loops and only indirectly by $\varkappa$ and T_c, which are taken into account in the following considerations.

For the calculation of the elementary pinning force f_p between a dislocation loop and a flux line, we have to know H_c and $\varkappa$. Neglecting the strong coupling of V_3Si, the relation

$$H_c(0) = 244 \sqrt{\gamma}\, T_c \tag{1}$$

is used. The electronic heat capacity coefficient γ and its dependence on the fluence was taken by interpolation from the data of Viswanathan and Caton [5]. These data measured on very similar specimens seem to be more reliable than those calculated from the Maki parameter and the normal-state resistivity. Using H_c and $\mathrm{d}H_{c2}/\mathrm{d}T|_{T=T_c}$ from Table 1, we calculate

$$\varkappa = \frac{-(\mathrm{d}H_{c2}/\mathrm{d}T|_{T=T_c})\, T_c}{2\sqrt{2}\, H_c(0)}. \tag{2}$$

This equation is valid in the temperature range of our experiment, where the paramagnetic limitation can be neglected. The data are given in Table 2.

With $H_c(0)$ and $\varkappa$ we estimate the elementary pinning interaction as follows: After Kramer [1], the first-order elastic interaction corresponds to the predominant elementary pinning force between a dislocation loop and a flux line. Using the equations in Section 2.1 and Fig. 4 in reference [1], we obtain the maximum of this elementary pinning force. Values of the elastic moduli have been taken from Testardi [15] at T_c for a non-transforming sample. After Guha et al. [16], fluences up to 3.5×10^{18} cm^{-2} do not affect these moduli very much. For this reason, the same values are used for all specimens. In order to get the maximum elementary pinning force, we did not use the soft acoustic shear mode $(C_{11} - C_{12})/2$, but C_{44} for the calculation. Using equation (1) the simplification was made that the relative volume dependence of H_{c2} was assumed to be primarily given by the volume dependence of H_c or T_c, respectively, and not by the volume dependence of $\varkappa$. This is justified since the resistance is constant below 25 K, and thus the phonon influence on the mean free path of the electrons, and there fore on $\varkappa$, can be neglected. With the pressure dependence of 3×10^{-5} Kbar^{-1} from Smith et al. [17], the elementary pinning force was estimated for a loop of a root mean square diameter of 170 Å. The values of f_p for $b = 0.5$ and $t = 0.84$ are given in Table 2.

4.2 The summation of the elementary pinning forces

Comparing the mean diameter of the loops and the interaction length ξ with the mean spacing of the flux line lattice, f_p can be regarded as a point force: In the temperature region $0.66 \leq t \leq 0.92$ the coherence length varied between 50 and 120 Å, whereas the minimum flux line distance is 140 Å at H_{c2} $(t = 0.66)$ and 300 Å at H_{c2} $(t = 0.92)$. The mean distance between the loops is about 1700 Å for sample S5 with the highest fluence. As this distance is much larger than the interaction length, it is obvious that all samples are in the dilute limit, where neighbouring centres do not interact. Therefore a proportionality between the volume pinning force and the concentration of pins is expected [18, 19]. In the following we shall compare our results with different summation models. We discuss the simple direct summation of the elementary pinning forces [20],

$$F_p^D = N f_p \sim (1 - b)\,(1 - t^4)\,, \tag{3}$$

Table 2

	γ $\left(\frac{mJ}{cm^3\,K^2}\right)$	$H_c(0)$ (T)	$\varkappa$	f_p ($b = 0.5$, $t = 0.84$) (dyn)	F_p^D ($b = 0.5$, $t = 0.84$) (dyn cm^{-3})	F_p^L ($b = 0.5$, $t = 0.84$) (dyn cm^{-3})	F_p^B ($b = 0.5$, $t = 0.84$) (dyn cm^{-3})	F_p^{exp} ($b = 0.5$, $t = 0.84$) (dyn cm^{-3})
S0	2.05	0.576	17.2	—	—	—	—	—
S1	2.05	0.584	17.9	8.7×10^{-7}	3.5×10^5	1.0×10^3	1.8×10^3	2.0×10^5
S2	2.06	0.579	18.0	8.7×10^{-7}	3.5×10^6	1.1×10^4	1.9×10^4	1.0×10^7
S3	1.86	0.537	24.2	7.7×10^{-7}	3.1×10^7	8.8×10^4	2.1×10^5	2.6×10^8
S4	1.63	0.495	27.3	6.7×10^{-7}	6.0×10^7	1.6×10^5	4.3×10^6	1.6×10^8
S5	1.24	0.383	35.5	4.5×10^{-7}	9.0×10^7	2.1×10^5	7.4×10^6	1.35×10^8

the statistical summation of Labusch [21], which takes into consideration the rigidity of the flux line lattice,

$$F_p^L = \frac{1}{8\sqrt{\pi}} N f_p^2 \left(\frac{B}{\Phi_0}\right)^{2/3} \xi \left(\frac{1}{\sqrt{C_{66}C_{44}}} + \frac{1}{\sqrt{C_{44}C_{11}}}\right) \sim \sqrt{b}\,(1-b)\,(1-t^4)^2\,, \tag{4}$$

and the same statistical summation, but using the recalculated modulus of the flux line lattice based on the non-local theory by Brandt [22],

$$F_p^B = F_p^L \frac{\varkappa}{\sqrt{2}\sqrt{1-b}} \sim \sqrt{b(1-b)}\,(1-t^4)^2\,. \tag{5}$$

Calculations are made for $t = 0.84$ and $b = 0.5$; the values are given in Table 2. The results of the direct summation (3) are below the experimental values. This is probably caused by too small a value of the calculated elementary pinning force, as has also been assumed in [23]. The absolute values of the statistical summations (4) and (5) are far below the experimental ones. This fact is correct in spite of the considerable simplifications made in the calculation of f_p. Furthermore, there does not exist any similarity in the field dependence or in the temperature dependence between the measured and the calculated volume pinning forces. This holds for the three summation models discussed here as well as for the model recently presented by Campbell [24].

4.3 The dependence of the volume pinning force on the concentration of obstacles

A comparison of the experimental results with the summation models seems not to be significant in that field region where the peak in the critical current density occurs, because it is commonly believed that the mechanism in the peak region is a different one (see, e.g., [25 to 27]). With this concept Kramer [27] has shown by plotting F_pN^{-1} versus N at a reduced field of 0.55 that the volume pinning force for irradiated niobium is indeed proportional to the concentration provided N is smaller than about ξ^{-3}. Fig. 8 shows F_pN^{-1} versus N for our V_3Si samples at reduced fields of 0.6, 0.4, and 0.2. Though the concentrations are far below ξ^{-3}, which is about 2.6×10^{18} cm^{-3}, the diagram clearly demonstrates that there is no field region which shows a proportionality between F_p and N. Therefore none of the models proposing a linear dependence on N is appropriate for the microstructure of these samples, although they are in the dilute limit.

4.4 Peak effect and saturation behaviour

A constant volume pinning force at high fields in spite of an increasing concentration is an important feature of the peak effect. This saturation behaviour is manifested in Fig. 9 which shows the volume pinning force versus reduced field for three reduced

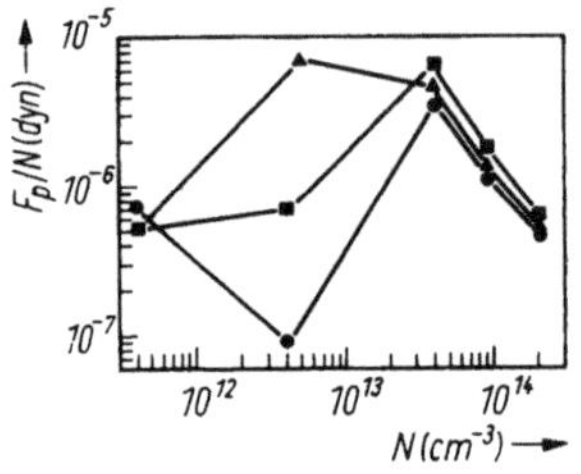

Fig. 8. Volume pinning force divided by the loop concentration versus N at the reduced fields 0.2 (●), 0.4 (■), and 0.6 (▲) at reduced temperature 0.84. The values are corrected for different H_{c2} and $\varkappa$

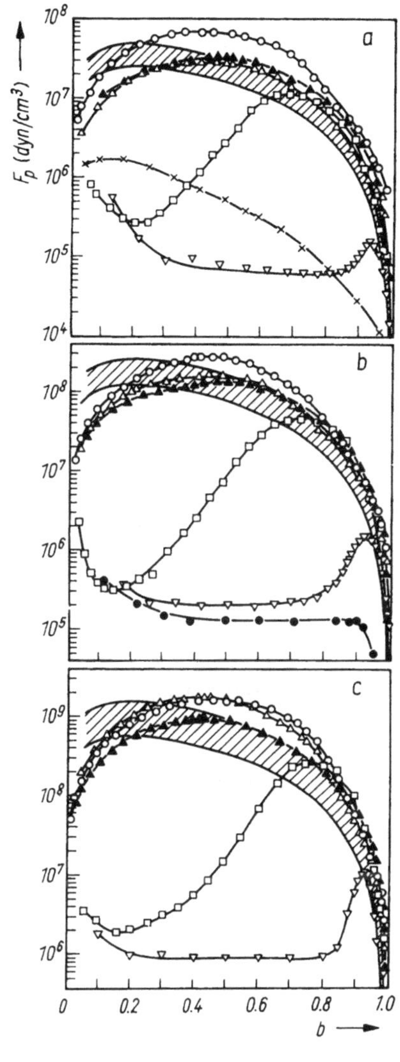

Fig. 9. Volume pinning force F_p versus reduced field b for the samples S0 (●), S1 (▽), S2 (□), S3 (○), S4 (△), S5 (▲) and for the data of Pulver [29] (×). a) $t = 0.92$, b) 0.84, and c) 0.66

temperatures. All models, except that of Kramer [28], which are proposed for an explanation of the peak effect, can easily be rejected by comparing with our results. In this model the volume pinning force is given by the equation

$$F_p = 0.56 \frac{H_{c2}^{5/2}}{\varkappa^2} \sqrt{b}\,(1 - b)^2 \tag{6}$$

which is valid above that field where the volume pinning force shows a maximum. The hatched regions in Fig. 9 cover the calculated values for all samples. The agreement with the experimental results in the field dependence is satisfying for reduced fields above 0.8. There is, however, a less satisfying agreement for samples S3, S4, and S5 which show saturation also at lower fields. The temperature dependence given in Section 3.2 is fairly well described by equation (6), but this similarity is much less conclusive.

Kramer's model predicts independence of the volume pinning force of the microstructure in the field region where equation (6) is valid. A measurement of Pulver [29] on a V_3Si wire produced by diffusion technique contradicts this prediction. The volume pinning force of this sample shown in Fig. 9 is about a factor of 50 smaller than the calculated values. In spite of this discrepancy the field dependence is in good agreement with equation (6). This comparison shows that only a qualitative similarity in $F_p(b)$ between experiment and theory is not conclusive. A complete saturation of F_p in the whole field range is approximately reached in sample S3. The effect that F_p passes through a maximum, cannot be understood neither in Kramer's model nor by a superposition of increasing concentration and of decreasing H_c or increasing $\varkappa$, respectively. Further, the increase in the concentration itself does not give an explanation by an overlapping of the strain fields of the loops because the system is still dilute. The question of the dependence of the saturation of F_p on the concentration and on the kind of the pinning centres is an aim of further investigations.

4.5 The threshold criterion

The low-field mechanism in Kramer's model [28] as well as the summation models discussed above are completely inadequate to explain our results in the low-field region. In this field range the interaction between the flux line lattice and the microstructure is in samples S1 and S2 almost the same as in the unirradiated state. This is demonstrat-

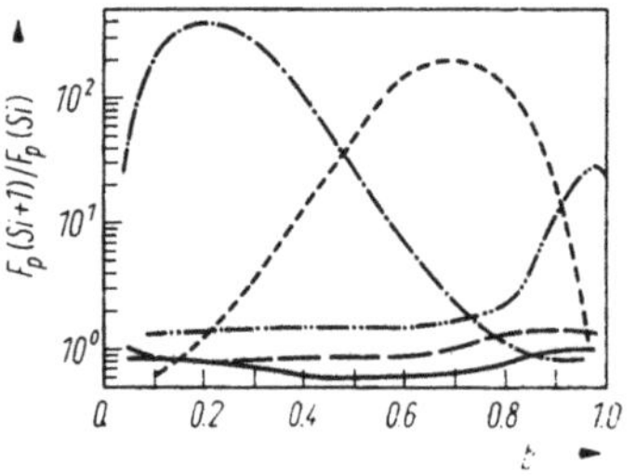

Fig. 10. Ratio of the volume pinning forces of samples with successive loop concentration in dependence on the reduced field at reduced temperature 0.84 for S1/S0 (– · · –), S2/S1 (- - - - -), S3/S2 (— · —), S4/S3 (———), and S5/S4 (— — —)

ed in Fig. 10 by the ratio of the volume pinning forces as a function of the reduced field for samples with successively increasing concentration. The volume pinning force in sample S1 is about the same as in the unirradiated sample S0 below a reduced field 0.8. With increasing concentration the field region where the volume pinning force is enhanced, is shifted to lower fields, whereas it is saturated above. The unaffected volume pinning force in the low-field region may be due to the threshold criterion [18, 21], which is not satisfied for this microstructure. This threshold criterion predicts a zero volume pinning force if the elementary pinning force is below a certain value which is about 1.5×10^{-4} dyn for $b = 0.5$ and $t = 0.84$ in these samples. The calculated values of f_p are about 200 times lower than this threshold value. From this, one expects no influence of the loops on F_p. This prediction is in accordance with the experimental results at low concentrations and low fields, but it fails more and more as B and N increase.

One possibility for overcoming this discrepany was proposed by Fietz and Webb [30], and recently reexamined by Labusch in the discussion part of [27]. According to this idea the threshold may be decreased by a suitable arrangement of pinning centres interacting as one pin with the flux line lattice. The probability for the formation of such groups increases with increasing density of pinning centres. This explains an overproportional dependence of F_p on N and is in accordance with the observed rapid enhancement of the volume pinning force with increasing N. Another explanation considers defects in the flux line lattice. After Evetts [31] it is possible to decrease the threshold by flux line lattice dislocations. The density of dislocations may increase with increasing field [27] and with increasing density of pinning centres [32]. Therefore an increasing concentration of pinning centres should lower the field range where the threshold criterion is satisfied. Above this field range the volume pinning force becomes saturated and is no longer dependent on the density of the flux line lattice defects. This interpretation as well as the first one is consistent with the observed concentration behaviour of F_p. But additionally, it may explain the drastic variation of the field dependence of F_p with increasing concentration as demonstrated in Fig. 9 and 10. It is further in agreement with the explanation of the history effect observed in microstructures like the ones discussed here [13] with elementary pinning forces not satisfaying the threshold criterion.

In reference [27] the validity of the relation $F_p \sim f_p^m$ is demonstrated for $1 < m < 2$, and it is shown that the exponent m depends on the field and even on f_p. A plausible explanation for this behaviour is based on a lowered threshold with increasing strength of the elementary pinning force. We have found that the threshold may also be influenced by the concentration. Combining these two results, the relation between the volume pinning force and the microstructure represented by N and f_p can be written in a simple generalized form $F_p = N f_p b(f_p, C_{ij}(B, T, N, f_p))$. The function b describes the influence of the rigidity of the flux line lattice and includes the threshold criterion as shown by Campbell [24]. Microsctructures with elementary pinning forces below the threshold value influence the function b by a decreasing threshold via a softening of the elastic moduli of the flux line lattice, whereas in a system with strong elementary pinning forces the function b may only be slightly increased by this softening. This is probably the reason why strong pins do not cause a peak effect, as expected by Kramer [27], and satisfy the linear dependence of F_p on N. Such a dependence is often observed for the pinning on grain boundaries in A15 superconductors, thus indicating a strong interaction between a grain boundary and a flux line.

The present stage of knowledge about the influence of the microstructure on the defect structure of the flux line lattice and on its elastic behaviour, respectively, allows only to give a qualitative summary. The volume pinning force of superconductors with high $\varkappa$ and weak pinning centres is determined by the threshold condition, which seems to be lowered with increasing concentration and increasing field, leading to a complex dependence of F_p on N. Above the field where the volume pinning force shows a maximum, F_p is no longer dependent on the concentration. This saturation behaviour is in rough agreement with the high-field mechanism in Kramer's model.

Acknowledgements

The authors whish to thank Prof. Dr. H. Wühl for stimulating discussions and Mrs. B. Runtsch for the resistivity measurements and for technical assistance.

References

[1] E. J. Kramer, Phil. Mag. **33**, 331 (1976).
[2] C. S. Pande, Appl. Phys. Letters **28**, 426 (1976).
[3] A. A. Manuel, Ph. D. Thesis, Geneva Univ., 1978.
[4] B. W. Battermann and C. S. Barrett, Phys. Rev. **145**, 296 (1966).
[5] R. Viswanathan and R. Caton, Phys. Rev. B **18**, 15 (1978).
[6] L. R. Testardi, Rev. mod. Phys. **47**, 637 (1975).
[7] U. Essmann, H. Haas, and G. Zerweck, Commun. Phys. **2**, 127 (1977).
[8] M. J. Goringe and U. Valdrè, Proc. Roy. Soc. **A295**, 192 (1966).
[9] C. S. Pande, Solid State Commun. **24**, 241 (1977).
[10] C. S. Pande, J. nuclear Mater. (Amsterdam) **72**, 83 (1978).
[11] A. R. Sweedler, D. E. Cox, and S. Moehlecke, J. nuclear Mater. (Amsterdam) **72**, 50 (1978).
[12] R. W. Rollins, H. Küpfer, and W. Gey, J. appl. Phys. **45**, 5392 (1974).
[13] H. Küpfer and W. Gey, Phil. Mag. **36**, 859 (1977).
[14] R. R. Hake, Phys. Rev. **158**, 356 (1967).
[15] L. R. Testardi, Phys. Rev. B **3**, 95 (1971).
[16] A. Guha, M. P. Sarachik, F. W. Smith, and L. R. Testardi, Phys. Rev. B **18**, 9 (1978).
[17] T. F. Smith, T. R. Finlayson, and R. N. Shelton, J. less-common Metals **43**, 21 (1975).
[18] A. M. Campbell and J. E. Evetts, Adv. Phys. **21**, 199 (1972).
[19] H. Ullmaier, Springer Tracts mod. Phys. **76**, 68 (1975).
[20] D. Dew-Hughes, Phil. Mag. **30**, 293 (1974).
[21] R. Labusch, Crystal Lattice Defects **1**, 1 (1969).
[22] E. H. Brandt, Frühjahrstagung der DPG, Münster 1977 (p. 406).
[23] D. C. Agrawal, E. J. Kramer, and B. A. Loomis, Phil. Mag. **33**, 343 (1976).
[24] A. M. Campbell, Phil. Mag. **37**, 149 (1978).
[25] H. C. Freyhardt, Phil. Mag. **23**, 345 (1971).
[26] L. Schultz, Phil. Mag. **36**, 1171 (1977).
[27] E. J. Kramer, J. nuclear Mater. (Amsterdam) **72**, 5 (1978).
[28] E. J. Kramer, J. appl. Phys. **44**, 1360 (1973).
[29] M. Pulver, Z. Phys. **257**, 22 (1972).
[30] W. A. Fietz and W. W. Webb, Phys. Rev. **178**, 657 (1969).
[31] J. E. Evetts, TMS-AIME Symposium on Superconducting Materials and Applications, Niagara Falls 1976.
[32] C. P. Herring, J. Phys. F **6**, 99 (1976).

(Received March 15, 1979)

phys. stat. sol. (a) 54, 167 (1979)

Subject classification: 18.2.1; 21.1; 21.1.1

Institut für Metallforschung der Universität Münster[1])

Analysis of the Extraordinary Viscous Magnetic Domain Wall Motion in NiFe Films

By

E. Nembach

Experimental data, published by Bostanjoglo and Gemünd, on the extraordinary viscous domain wall motion in thin oxidized films of 83 at % Ni–Fe are analyzed. It is assumed that the velocity v of the wall is governed by the interaction of the wall with defects present in the film and that this interaction can be overcome by thermal activation. The activation energy E is calculated from the experimental data $v(H, T)$ (H applied magnetic field, T temperature). E is a unique smooth function of H. From the obtained relation $E(H)$, information on the force–distance profile of the interaction between defect and wall is derived.

Von Bostanjoglo und Gemünd publizierte Meßdaten zur „außergewöhnlich viskosen" Domänenwandbewegung in dünnen, oxidierten 83 At % Ni–Fe-Filmen werden analysiert. Es wird angenommen, daß die Wandgeschwindigkeit v durch die Wechselwirkung der Wand mit Defekten im Film bestimmt wird und daß diese Wechselwirkung durch thermische Aktivierung überwunden werden kann. Die Aktivierungsenergie E wird aus den experimentellen Daten $v(H, T)$ (H anliegendes Magnetfeld, T Temperatur) berechnet. E ist eine eindeutige, stetige Funktion von H. Aus der erhaltenen Funktion $E(H)$ wird Information über das Kraft–Orts-Profil der Wechselwirkung zwischen Defekt und Wand gewonnen.

1. Introduction

Bostanjoglo and Gemünd [1] recently published extensive measurements of the extraordinary viscous domain wall motion in thin films of oxidized 83 at% Ni–Fe. These authors measured the wall velocity v as a function of the applied magnetic field H and the temperature T by Lorentz microscopy. In Fig. 1 their data for $v(H, T)$ of one film are reproduced. v covers the enormous range of more than nine orders of magnitude from 10^{-7} to $10^{2.3}$ m/s. Below 9 K, v is independent of T. In the following only velocities up to 1.5 m/s will be discussed because at higher ones — i.e. above the marks in Fig. 1 — the mechanism of the magnetization reversal changes: the wall becomes disrupted, incoherent rotation occurs, and phonon damping is no more negligible.

In the range between about 10^{-5} and 1 m/s, Bostanjoglo and Gemünd approximated v (H, T = const) by

$$\ln v = a(T)\,H + b(T)\,. \tag{1}$$

The parameters a and b depend only on temperature, but not on the applied field. If T is less than half the blocking temperature T_b, which will be defined in the next paragraph, a is independent of temperature and equals $0.43a(T_b)$. Two films of different thickness, but of the same T_b, yielded at $T = 0.35T_b$ identical slopes a, but different intercepts b in (1).

The authors gave the following interpretation of their experimental results. The moving domain wall is coupled to relaxing oxide inclusions in the film; these inclusions are assumed to be ferrimagnetic. T_b is their blocking temperature [1]. Their relaxation times τ cover the range from τ_1 to τ_2, they are distributed according to

$$P(\tau, T) = c(T)/\tau\,; \tag{2}$$

[1]) 4400 Münster, BRD.

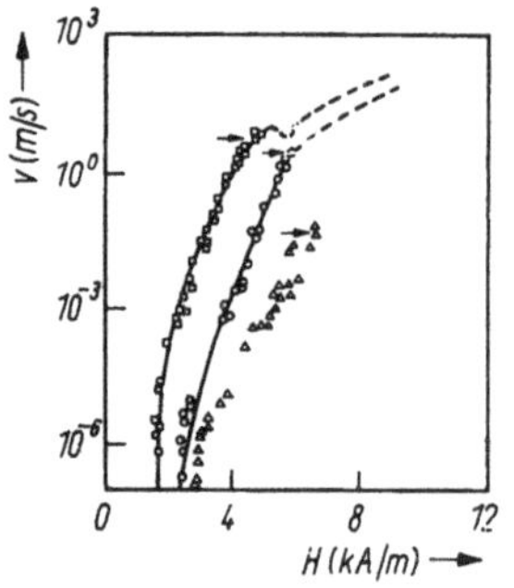

Fig. 1. Wall velocity v versus magnetic field H, semilogarithmically. □ $T = 25$, ○ 17, △ 9, and 4.2 K. These experimental data have been taken from a publication by Bostanjoglo and Gemünd [1]. The curves have been calculated as described in Section 3. The marks denote the velocities beyond which the mechanism of magnetization reversal changes. The blocking temperature T_b equals about 31 K

c depends only on T. Equation (2) implies that there is — at given H and T — a range of activation energies. Under these assumptions the authors are able to derive (1) and also to explain the sharp drop of $\lg v$ for $v < 10^{-6}$ m/s in Fig. 1. The fact that $a(T)$ does not vary with T for $T < T_b/2$ (Fig. 1), is attributed to "flattening of $P(\tau, T)$" of (2).

In this paper an alternative view is presented. The domain wall is assumed to interact with defects in the film, probably with the oxide inclusions. This interaction is overcome by thermal activation. Using a model, which had been developed for the analysis of thermally activated domain wall motion in compact picture frame single crystals [2, 3], the activation energy $E(H)$ is derived from the experimental data $v(H, T)$. This analysis can be applied to Bostanjoglo and Gemünd's data of Fig. 1, if $T > T_b/2$ and $v < 1.5$ m/s. At given H and T, only *one* activation energy is used, it is independent of T.

2. Model of Thermally Activated Domain Wall Motion

The following assumptions are made:

1. There are defects in the film which interact with the wall. These obstacles, which impede the wall motion, are represented by an average obstacle. The wall can overcome this interaction by thermal activation.
2. The interaction is independent of temperature.
3. A wall piece of area A can move independently. A will depend, e.g. on the wall flexibility.

The interaction force $F(x)$ between an obstacle and the wall of area A is sketched in Fig. 2. The applied field H drives the wall towards $+x$, it exerts the force $2AM_sH$ upon the wall. M_s is the film's saturation magnetization. If $2AM_sH$ is smaller than F_0 (Fig. 2), ballances of forces are established at x_1 and x_2:

$$F(x_1) = F(x_2) = -2AM_sH\,. \tag{3}$$

At x_1 the total energy of the wall has a minimum and at x_2 a maximum. Thus the position at x_2 is unstable. The wall is trapped at x_1. It can only escape from the obstacle if the energy difference $E(H)$ between the maximum and the minimum is supplied by

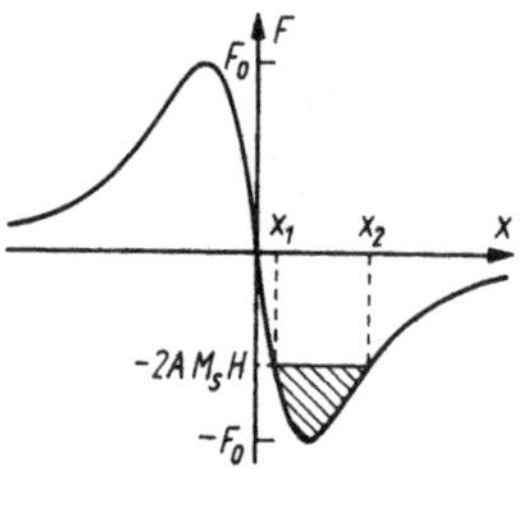

Fig. 2. Force–distance profile of an attractive obstacle–domain wall interaction, schematically. If the field H is applied, the total energy of the wall has a minimum at x_1 and a maximum at x_2

thermal fluctuations. One has not to specify whether the strength of the obstacle or the force $2AM_sH$, which the applied field provides, fluctuate. The activation energy $E(H)$ is given in Fig. 2 by the hatched area. In the following the difference $x_2(H) - x_1(H)$ will be abbreviated by $\Delta(H)$. From Fig. 2 the following equation is obvious:

$$\partial E(H)/\partial H = -2AM_s \Delta(H) . \quad (4)$$

In Fig. 2 the interaction force is assumed to be attractive. A repulsive one, however, yields the same results; in this case $F(x)$ of Fig. 2 has to be replaced by $-F(x) = F(-x)$. Using (3), H can be eliminated from (4). Introducing $F(x_1)/A$ instead, yields $F(x_1)/A$ as a function of $A\Delta$. This relation characterizes the interaction force profile.

The mean waiting time τ of the wall at the obstacle is given by

$$\tau = \tau_0 \exp\left(E(H)/k_BT\right); \quad (5)$$

k_B is Boltzmann's constant and τ_0 is related to the vibration frequency ν of the wall by $\tau_0 = 1/\nu$. After release from the obstacle, the wall will move freely during the time τ' and cover the distance d. τ' and d may be averages. If τ is much larger than τ', one obtains for the wall velocity v

$$v(H, T) = v_0 \exp\left(-E(H)/k_BT\right); \quad (6)$$

v_0 equals d/τ_0. In the case of strong phonon or eddy current damping, the relation $\tau \gg \tau'$ cannot be taken for granted.

3. Data Analysis

In the following, it will be shown how the function $E(H)$ and v_0 can be derived from experimental data $v_i(H_i, T_i)$, e.g. from those of Fig. 1. Inversion of (6) yields

$$E(H) = k_BT(\ln v_0 - \ln v) . \quad (7)$$

Treating v_0 as an unknown parameter, $E_i(H_i, v_0)$ can be calculated from $v_i(H_i, T_i)$. Only if v_0 is assigned its correct value, E is a unique smooth function of H. Wrong choices of v_0 make E temperature-dependent; this is obvious from (7). In this way, the data for $v(H, T = 17\text{ K})$ and $v(H, T = 25\text{ K})$ of Fig. 1 have been analyzed for $v < 1.5$ m/s. The two temperatures below 17 K could not be used, since for them v was independent of T. Setting $v_0 = 1.365 \times 10^5$ m/s yields the smooth function $E(H)$ plotted in Fig. 3. Data taken for 17 and 25 K fall upon the same curve, proving that v_0 has been chosen correctly. The scatter of the points $E(H)$ reflects the scatter of the data of Fig. 1. v_0 has been calculated in the following way. The data $E_i(H_i, v_0)$ have been fitted to a polynomial of the third degree: $H_i = \sum_{j=0}^{3} \alpha_j(v_0)\, E_i^j(v_0)$. For numerical reasons, this inverse function has been used. The correct choice of v_0 yields the best fit. The third degree of the polynomial proved to be the most suitable degree for this test. Differentiation of $E(H)$ of Fig. 3 yields according to (4) and (3), $F(x_1)/A$ versus $A\Delta$. This result is shown in Fig. 4. Thus information on the interaction profile between an obstacle and the domain wall of area A is obtained.

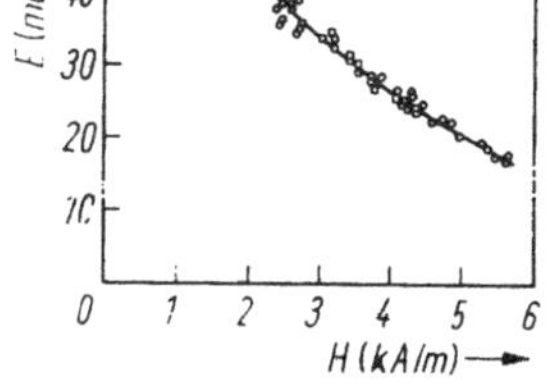

Fig. 3. Activation energy E versus applied magnetic field H. The data points have been derived from the experimental ones of Fig. 1 as described in Section 3. The symbols are the same as in Fig. 1, i.e. □ represents activation energies derived from measurements performed at 25 K and ○ refers to those of 17 K

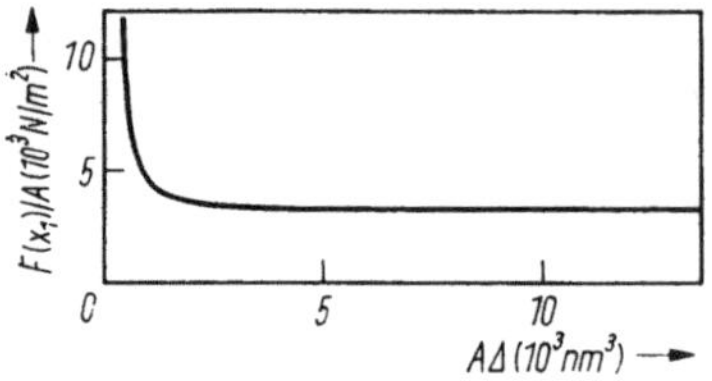

Fig. 4. $F(x_1)/A$ versus $A\Delta$. This curve represents the force-distance profile of the interaction between the domain wall of area A and the obstacle

Using (6), the curve of Fig. 4, $v_0 = 1.365 \times 10^5$ m/s and one point of Fig. 3, the relation $v(H, T)$ can be reconstructed. In this way, the two curves in Fig. 1 for 17 and 25 K have been calculated. The one point of Fig. 3 is needed as an integration constant, when $E(H)$ is derived from Fig. 4. The two curves of Fig. 1 represent the data points very well for v between 1.6×10^{-7} and 1.5 m/s; i.e. over nearly seven orders of magnitude. Equation 1 applies only to the smaller range above 10^{-5} m/s. For high velocities the mechanism of magnetization reversal changes (Section 1). The two curves of Fig. 1 are not independent of one another. If one approximates their middle part by (1), the constants a and b of one curve are related to a and b of the other curve. $v(H, T)$ of any field and temperature can be calculated as long as the activation energy falls into the range covered by the curve of Fig. 3. Bostanjoglo and Gemünd [1] did not specify the temperature dependence of b of (1).

The low temperature region, for which $T < T_b/2$ holds, remains to be discussed. This applies e.g. to the data for 9 and 4.2 K of Fig. 1; these temperatures equal $0.29T_b$ and $0.14T_b$, respectively. For $T < T_b/2$, v varies with H, but is independent of T. Within the framework of the formalism of thermal activation, this cannot be explained without assuming an explicit temperature dependence of the interaction between the domain wall and an obstacle. According to (6), v is independent of T only if $E(H)$ is proportional to T. A more reasonable assumption seems to be that there is another athermal mechanism by which the domain wall can overcome the obstacle, e.g. tunneling. Then τ of (5) has to be replaced by $1/(1/\tau_{th} + 1/\tau_x)$, where τ_{th} is governed by thermal activation, and τ_x by the unknown mechanism. τ_x is independent of temperature, but varies with the applied magnetic field. At high temperatures $\tau_x \gg \tau_{th}$ holds and (6) is reproduced. At low temperatures, where thermal activation is negligible, τ_x is much smaller than τ_{th} and v becomes independent of T. It should be mentioned that Bostanjoglo and Gemünd [1] did not believe in domain wall tunnelling because τ_x becomes dominant at rather high temperatures. It is, however, difficult to imagine another mechanism which makes v dependent on H, but independent of T. Simple friction would make v a linear function of H.

In conclusion, it can be stated that the model presented in Section 2 can be successfully applied to the analysis of the extraordinary viscous domain wall motion in thin Ni–Fe films. Information on the activation energy E as a function of H and on the interaction force–distance profile is obtained. In order to find out, which type of defects governs the domain wall velocity in Bostanjoglo and Gemünd's oxidized Ni–Fe films, theoretical force–distance profiles of various types of defects have to be compared with the one of Fig. 4. Moreover, the knowledge of the function $F(x_1)/A$ versus $A\Delta$ makes the calculation of $v(H, T)$ possible, if the activation energy falls between 16 and 60 meV.

References

[1] O. Bostanjoglo and H. P. Gemünd, phys. stat. sol. (a) 48, 481 (1978).
[2] D. Siemers and E. Nembach, Phys. Letters 65A, 345 (1978).
[3] D. Siemers and E. Nembach, J. appl. Phys., in print.

(Received March 8, 1979)

phys. stat. sol. (a) 54, 171 (1979)

Subject classification: 11; 10.1; 21

School of Engineering and Applied Science, University of California, Los Angeles[1])

The Simultaneous Clustering of Point Defects during Irradiation

By
N. M. GHONIEM and D. D. CHO

The simultaneous clustering of irradiation produced vacancies and interstitials in metals is investigated using the rate theory approach. Rate equations are used to describe the time-dependent concentrations of point defect clusters. Size-dependent bias factors and self-consistent reaction rate constants are used to evaluate the feed-back effects between the evolving vacancy cluster and interstitial loop populations. It is found that the concurrent presence of vacancy clusters depletes the free vacancy population and thereby reduces mutual point defect recombination. This has the effect of enhancing interstitial loop formation; increasing loop concentration, nucleation rate, average size, and growth speed of the average size. In 316 stainless steel, a diinterstitial binding energy of 1.19 eV results in a substantial decrease in loop concentrations at high temperatures, which is in qualitative agreement with experimental data.

Es wird die gleichzeitige Clusterbildung von durch Strahlung erzeugten Leerstellen und Zwischengitterstörstellen mit der Methode der Ratengleichungen untersucht. Bilanzgleichungen werden benutzt, um die zeitabhängigen Konzentrationen der Punktdefektcluster zu beschreiben. Größenabhängige Vorfaktoren und selbstkonsistente Reaktionskonstanten werden benutz, um die Rückkopplungseffekte zwischen den sich entwickelnden Leerstellencluster und den Besetzungen der Interstitialschleifen zu berechnen. Es wird gefunden, daß die konkurrierende Anwesenheit von Leerstellencluster die Besetzung freier Leerstellen verringert und dabei die gegenseitige Punktdefektrekombination verringert. Dies führt zu einer Erhöhung der Zwischengitterschleifenbildung und vergrößert die Schleifenkonzentration, Keimbildungsrate, mittlere Größe und die Wachstumsgeschwindigkeit der mittleren Ausdehnung. In rostfreiem 316-Stahl führt eine Bindungsenergie von 1,19 eV der Di-Interstitials zu einer beträchtlichen Verringerung der Schleifenkonzentration bei hohen Temperaturen, die in qualitati er Übereinstimmung mit experimentellen Ergebnissen ist.

1. Introduction

It has become important to investigate the detailed effects of material and irradiation variables on the nucleation and growth of interstitial loops. Recently, Johnson [1] showed that the diinterstitial binding energy has a significant influence on the formation of interstitial loops in metals irradiated at high temperatures. On the other hand, Hall and Potter [2] examined the kinetics of interstitial loop nucleation and growth during low dose irradiation, paying special attention to impurity and trapping effects.

The aim of the present paper is to give an understanding of the coupling between simultaneously forming vacancy and interstitial clusters during irradiation. The salient features of the investigation are summarized as: (1) size-dependent bias factors for point defect diffusion to interstitial loops and microvoids are used to calculate reaction rates, (2) mixed rate reactions (diffusion and surface limited) are assumed for the computation of point defect impingement and emission fluxes, (3) point defect conservation equations are utilized to determine the necessary number of rate equations at any irradiation time, (4) the relationship between vacancy and interstitial cluster formation is analyzed.

[1]) Los Angeles, California 90024, USA.

2. A Model for the Simultaneous Formation of Defect Clusters

2.1 Rate constants

In the *jump* method [3], the reaction between diffusing species proceeds by a surface control mechanism. On the other hand, the rate constants in the *diffusion* method are obtained by solving steady-state diffusion equations [3]. Defining the instability radius of a particular cluster, R, as the radius at which point defects will inevitably be attracted towards the cluster (interaction energy $\approx kT$), then

$$R = Z(r)\, r\,, \tag{1}$$

where r is the physical cluster radius and $Z(r)$ is a size-dependent bias factor (capture efficiency). Results of the detailed elasticity calculations of Wolfer and Ashkin [4] have been used in (1). The overall reaction rate constant is obtained in a manner similar to that used to determine the overall heat flux with series resistance [5].

Define x as the cluster size in atomic units, i for an interstitial, v for a vacancy, l for a loop, and c for a cavity. For a spherical cluster in f.c.c. lattices the following impingement frequencies, K (s^{-1}), are used:

$$K^{c}_{v,i}(x) = \frac{2.216[Z^{c}_{v,i}(x)]^2\, x^{2/3}}{1 + 0.1128 Z^{c}_{v,i}(x)\, x^{1/3}}\, \nu_{v,i} \exp\left(\frac{-E^{m}_{v,i}}{kT}\right), \tag{2}$$

where ν is the atomic vibrational frequency, $E^{m}_{v,i}$ the vacancy–interstitial migration energy, k the Boltzmann constant, and T the absolute irradiation temperature. For large size cavities the reaction rate is purely diffusion controlled and the impingement frequency is proportional to $x^{1/3}$. Interstitial clusters are treated as small spherical inclusions with the application of (2) up to a maximum size x^{max}_{i} of few interstitials. However, larger sizes behave as two-dimensional disks and the interaction with point defects is only surface controlled. The relevant impingement frequency is thend

$$K^{l}_{v,i}(x) = 1.555 Z^{l}_{v,i}(x)\, x^{1/2} \nu_{v,i} \exp\left(\frac{-E^{m}_{v,i}}{kT}\right). \tag{3}$$

The emission of vacancies from the surface of vacancy clusters is included in this model. However, point defect thermal emission from interstitial loops is not considered because of the small probability of this process for small size loops. The only thermal dissociation process treated is that for diinterstitial break-up. The emission *rate* of vacancies from a vacancy cluster size x is given by

$$\gamma^{c}_{v}(x) = K^{c}_{v}(x)\, C^{e}_{v} \exp\left[\left(\frac{1.28 g a^2}{kT}\right) x^{-1/3}\right], \tag{4}$$

where a (cm) is the lattice constant, g (eV/cm^2) the surface energy, and C^{e}_{v} is the thermal matrix fractional vacancy concentration. Diinterstitial and divacancy thermal dissociation rates are given as

$$\gamma^{c,l}_{v,i}(2) = K^{c,l}_{v,i}(2) \exp\left(-\frac{E^{B}_{2v,2i}}{kT}\right), \tag{5}$$

where E^{B}_{2v} and E^{B}_{2i} are the binding energies for divacancies and diinterstitials, respectively.

2.2 Rate Equations

In this section, a set of rate equations is used to describe the nucleation and growth of both voids and interstitial loops. Using the assumptions homogeneous rate theory, and the nomenclature of chemical kinetics, one can write the following set of equations for the fractional concentrations:

$$\left.\begin{aligned}
\frac{dC_v}{dt} &= P + K_i^c(2)\, C_i C_{2v} + (2\gamma_v^c(2) - K_v^c(2)\, C_v)\, C_{2v} + \\
&+ \sum_{x=3}^{x_{max}} (\gamma_v^c(x) - K_v^c(x)\, C_v)\, C_{xv} - \sum_{x=3}^{x_{max}} K_v^l(x)\, C_v C_{xi} + \\
&+ Z_v \varrho_d D_v (C_v^e - C_v) - \alpha C_v C_i - K_v^c(1)\, C_v^2 - K_v^l(2)\, C_v C_{2i}\,, \\
\frac{dC_{2v}}{dt} &= \frac{1}{2} K_v^c(1)\, C_v^2 + \gamma_v^c(3)\, C_{3v} + K_i^c(3)\, C_i C_{3v} + \varrho_d D_{2v} C_{2v}^e - \\
&- K_v^c(2)\, C_v C_{2v} - K_i^c(2)\, C_i C_{2v} - \gamma_v^c(2)\, C_{2v} - \varrho_d D_{2v} C_{2v}\,, \\
&\vdots \\
\frac{dC_c(x)}{dt} &= K_v^c(x-1)\, C_v C_c(x-1) - [K_i^c(x)\, C_i + K_v^c(x)\, C_v + \\
&+ \gamma_v^c(x)]\, C(x) + [K_i^c(x+1)\, C_i + \gamma_v^c(x+1)]\, C_c(x+1)\,, \\
\frac{dC_i}{dt} &= P + K_v^l(2)\, C_v C_{2i} - K_v^c(1)\, C_i^2 - \alpha C_v C_i - \\
&- K_i^l(2)\, C_i C_{2i} - K_i^c(2)\, C_i C_{2v} - \sum_{x=3}^{x_{max}} K_i^l(x)\, C_i C_{xi} - \\
&- \sum_{x=3}^{x_{max}} K_v^c(x)\, C_i C_{xv} - Z_i \varrho_d D_i C_i\,, \\
\frac{dC_{2i}}{dt} &= \frac{1}{2} K_i^l(2)\, C_i^2 + K_v^l(3)\, C_v C_{3i} - K_i^l(2)\, C_i C_{2i} - K_v^l(2)\, C_v C_{2i}\,, \\
&\vdots \\
\frac{dC_l(x)}{dt} &= K_i^l(x-1)\, C_i C_l(x-1) + \gamma_v^l(x-1)\, C_l(x-1) - \\
&- \{K_i^l(x)\, C_i + K_v^l(x)\, C_v + \gamma_v^l(x)\}\, C_l(x) + \\
&+ K_v^l(x+1)\, C_v C_l(x+1)
\end{aligned}\right\} \quad (6)$$

with the parameters defined as P (at/at/s) irradiation production rate of Frenkel pairs, α (s^{-1}) point defect recombination coefficient,

$$\alpha = 48\nu_i \exp\left(-\frac{E_i^m}{kT}\right), \quad (7)$$

$Z_{v,i}$ point defect–straight dislocation bias factors, ϱ_d (cm^{-2}) straight dislocation density, $D_{v,i}$ ($cm^{-2}\,s^{-1}$) point defect diffusion coefficients, D_{2v} ($cm^{-2}\,s^{-1}$) divacancy diffusion coefficient [6],

$$D_{2v} = \nu_v a^2 \exp\left(-\frac{E_m^{2v}}{kT}\right), \quad (8)$$

E_m^{2v} migration energy of a divacancy and C_{2v}^e (at/at/s) divacancy thermal fractional concentration [6],

$$C_{2v}^e = 6 \exp\left(-\frac{2E_F^v - E_B^{2v}}{kT}\right). \quad (9)$$

The set of rate equations (6) are derived by considering the balance between various production and destruction processes for a certain cluster size. A detailed description of this derivation is given in [7]. The analytical solution of this set of equations can be very complicated, and would have to involve numerous approximations, therefore a numerical approach is followed.

3. Method of Solution

Two distinct problems arise when one attempts to numerically solve a large number of coupled rate equations such as that described in the previous section. The first complication is due to the time scales of the problem while the second stems from handling a large set of equations. Some reaction rates are very fast while others are very slow giving rise to a stiff system of ordinary differential equations. For stability requirements, most of the conventional methods limit the integration time step to small values. This can considerably increase the computing time, and is impractical for solving our system. On the other hand, storage requirements can seriously discourage handling large systems of equations. Fortunately, the Gear [8] computer package was designed to handle both large and stiff systems of ordinary differential equations (ODEs).

The rate equations described in the previous section have been numerically integrated using the GEAR computer package. Up to 200 rate equations were solved with no storage problems on the IBM 360/90 computer. On the other hand, a total of 178 CPU seconds were required to integrate 100 rate equations simultating the irradiation behavior of 316 stainless steel at 450 °C, a dose rate of 10^{-6} dpa/s and a dislocation density of 10^9 cm/cm^3.

The total number of rate equations required to give information on the system has been arbitrarily chosen by various investigators [1, 2, 9]. In this work, we use a conservation principle for irradiation produced defects to determine the necessary number of equations. For interstitial clusters, and at any irradiation time t^*, the integrated rate of interstitial accumulation should be equal to the total number of interstitials in loops; or,

$$\left| \sum_{x=2}^{N} \left\{ \int_0^{t^*} [K_i^1(x-1,t')\, C_i(t') - K_v^1(x,t')\, C_v(t')]\, C_l(x,t)\, dt' - xC_l(x,t^*) \right\} \right| \leqq \varepsilon', \quad (10)$$

where ε' is an absolute error. Dividing by the total number of interstitials in loops $\left(f(t^*) \equiv \sum_{x=2}^{N} xC_l(x, t^*)\right)$, one gets

$$\left| \frac{1}{f(t^*)} \int_0^{t^*} \left\{ \sum_{x=2}^{N} [K_i^1(x-1,t')\, C_i(t') - K_v^1(x,t')\, C_v(t')]\, C_l(x,t') \right\} dt' - 1 \right| \leqq \varepsilon, \quad (11)$$

where ε is a relative error. Here, the number of necessary equations, N, is determined at any time t^* such that the relative error in (11) is typically small.

4. Result and Discussion

4.1 Analysis of a reference case

Many material and irradiation variables can affect the clustering kinetics of both interstitial loops and voids. For practical considerations, however, 316 stainless steel irradiated in fusion reactor first wall environment will be considered. In addition to the material parameters of Table 1, the following input parameters are adopted fot

Table 1
Model input parameters for 316 stainless steel

parameter	symbol	numerical value	units	ref.
interstitial-interstitial combinatorial number	$C_i(1)$	84		[9]
vacancy–vacancy combinatorial number	$C_v^c(1)$	84		[9]
vacancy migration energy	E_v^m	1.4	eV	[11]
interstitial migration energy	E_i^m	0.2	eV	[11]
vacancy formation energy	E_v^f	1.6	eV	[11]
interstitial formation energy	E_i^f	4.08	eV	[11]
lattice parameter	a	3.63	Å	[9]
interstitial frequency	ν_i	5×10^{12}	s^{-1}	[1]
vacancy frequency	ν_v	5×10^{13}	s^{-1}	[1]
divacancy migration energy	E_{2v}^M	0.9	eV	[6]
divacancy binding energy	E_{2v}^B	0.25	eV	[6]
trivacancy binding energy	E_{3v}^B	0.76	eV	[6]
vacancy–dislocation bias factor	Z_v	1.0		[11]
interstitial–dislocation bias factor	Z_i	1.08		[11]
surface energy	g	6.24×10^{14}	eV/cm^2	[11]

the analysis of this reference case: irradiation temperature 450 °C, initial dislocation density 10^9 cm/cm³, dose rate 10^{-6} dpa/s.

First, we consider the evolution of interstitial clusters under these conditions with only single vacancies and interstitial loops allowed to exist, and then the effects of simultaneously forming microvoids will be explored.

Fig. 1 depicts the general characteristics of the clustering process at 450 °C. During the early stages of irradiation (times $\leq 10\,\mu s$) the vacancy concentration remains around the thermodynamic equilibrium value ($\approx 7 \times 10^{-12}$ at/at) while the interstitial concentration builds up linearly with time. Point defects accumulate in the matrix, since no major loss mechanism is active yet. Between 10 μs and 1 ms, the interstitial concentration levels off and then decreases, due to the fast diffusion of interstitials to straight dislocations. However, single vacancies keep on accumulating in the matrix as a result of their low diffusion rate. Single interstitials are mainly removed by diffusion to straight dislocations and by mutual recombination with vacancies. Diffusion to straight dislocations is the major removal process at the start

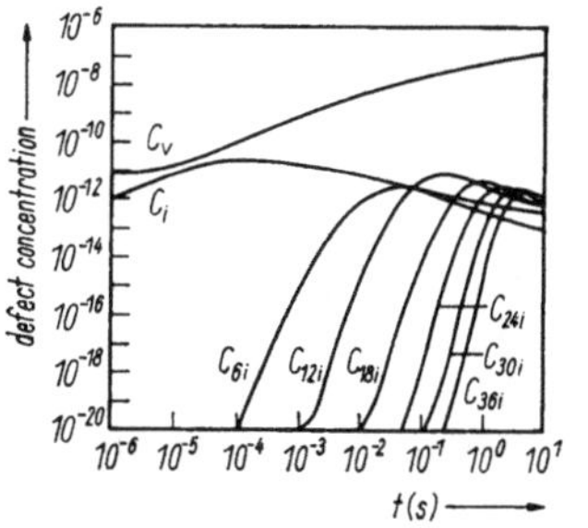

Fig. 1. Interstitial loop fractional concentrations as functions of irradiation time in the absence of microvoids. $T = 450$ °C, $\varrho_d = 10^9$ cm/cm³, dose rate 10^{-6} dpa/s

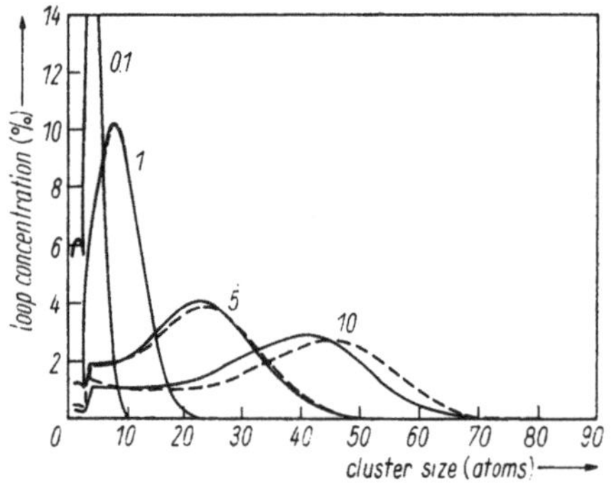

Fig. 2. Interstitial loop size distributions at various irradiation times. ——— without vacancy clusters. - - - - with vacancy clusters. $T = 450$ °C, $\varrho_d = 10^9$ cm/cm³, dose rate 10^{-6} dpa/s

of irradiation, and then recombination dominates at longer times. For example, at 5 ms, about 50% of interstitials recombine, 49.3% are absorbed at straight dislocations, and 0.8% contribute to the formation of interstitial clusters. It is observed from Fig. 1 that the drop in the single interstitial concentration is in keeping with a loss rate of ≈ 100.1%. The loss rate greater than 100% of the production rate indicates a continuous bleeding from matrix interstitial inventory, contributing to the evolution of larger size clusters. Smaller clusters start to linearly accumulate at short times, then larger sizes grow on the expense of the small size clusters. This process leads to concentration maxima of ever increasing larger sizes as irradiation proceeds.

When microvoids are included, the general trends remain the same as described above. However, the magnitude of characterizing quantities change. Ten equations representing vacancy clusters were numerically solved with up to 190 equations for interstitial loops. Only ten equations were necessary for vacancy clusters because the irriadiation times considered in this work are too short to allow larger vacancy cluster formation. The interstitial size distribution, at different irradiation times, is shown in Fig. 2 for the two cases investigated here. It can be seen that including vacancy clusters has considerable effects on the size distribution, especially for longer irradiation periods. Results of the calculations indicate that larger loops grow faster in the presence of microvoids. Consequently, a broader size distribution is created when vacancy clusters are included.

Due to a larger sink density in the presence of microvoids, one might intuitively expect slower growth kinetics of interstitial loops. On the contrary, when microvoids are included, the nucleation rate and average growth speed are significantly enhanced. The presence of microvoids reduces the free vacancy concentration, thus reducing matrix point defect recombination and increasing free interstitial concentration. This will tend to promote interstitial loop nucleation increasing average growth speed, average loop size, nucleation rate, and total loop concentration, as can be seen from Table 2.

4.2 *Effect of initial dislocation density*

The calculations indicate that a broader interstitial loop distribution is produced when the initial dislocation density is lowered to 10^8 cm/cm³. On the contrary, slower growth kinetics are associated with high dislocation densities resulting in a narrower loop distribution. An important result of the computations is that interstitial loop formation has distinct transient and steady state regimes. Linear loop accumulation starts 5 to 10 s after the beginning of irradiation, as can be observed from Fig. 3. The same figure shows that the total loop concentration is roughly inversely proportional to the initial dislocation density. On the other hand, the average loop size and its growth speed have been found to increase as the initial dislocation density decreases.

Table 2

Comparison of the reference case with and without microvoids after 10 s of irradiation for 316 stainless steel at 450 °C, $\varrho_d = 10^9$ cm/cm³ and 10^{-6} dpa/s

parameter	with microvoids	without microvoids	% change
vacancy concentration (at/at)	1.718×10^{-7}	2.1891×10^{-7}	−21.5
interstitial concentration at/at()	5.926×10^{-13}	4.665×10^{-13}	+27.0
average loop size (at)	37.13	34.63	+ 7.2
growth speed (at/s)	2.82	2.35	+20.0
nucleation rate (loops/cm³/s)	2.5×10^{11}	1.58×10^{11}	+37.2
total loop concentration (loops/cm³)	1.42×10^{13}	1.36×10^{13}	+4.6

4.3 Diinterstitial binding energy effect

Previous theoretical investigations [9] of interstitial loop formation and growth assumed that diinterstitial binding energy is so large that once they form, diinterstitials will never break up to single interstitials. Recently, Johnson [1] studied the effect of diinterstitial binding on *small* interstitial loops. In this section, we investigate diinterstitial binding energy effects on clusters of sizes up to 190 interstitials.

Total loop concentrations as a function of irradiation time are shown in Fig. 4 for $E_{2i}^B = 0.76$, 1.19, 1.5, and $\geqq 2.0$ eV. It is interesting to note that binding energies $\geqq 2$ eV will give identical results, indicating that $E_{2i}^B = 2$ eV is just as strong as any arger binding energy at 450 °C. It is also clear that E_{2i}^B has a dramatic effect on the

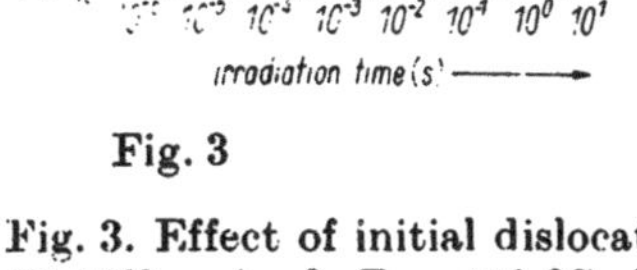

Fig. 3

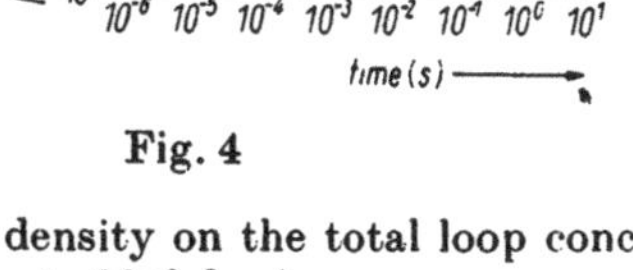

Fig. 4

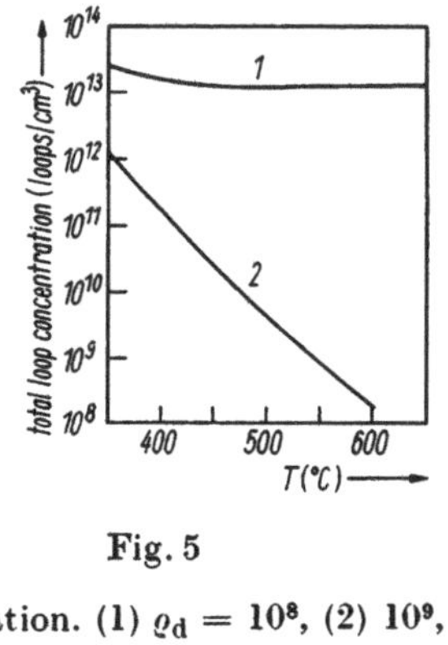

Fig. 5

Fig. 3. Effect of initial dislocation density on the total loop concentration. (1) $\varrho_d = 10^8$, (2) 10^9, (3) 10^{10} cm/cm³. $T = 450$ °C, dose rate 10^{-6} dpa/s

Fig. 4. Effect of diinterstitial binding energy on the total loop concentration. (1) $E_{2i}^B \geqq 2.0$ eV, (2) $E_{2i}^B = 1.5$, (3) 1.19, and (4) 0.76 eV

Fig. 5. Total loop concentration as a function of irradiation temperature. (1) $E_{2i}^B \geqq 2.0$ eV, (2) $E_{2i}^B = 1.19$ eV total interstitial loop population, which becomes large for high binding energies and small for low binding energies.

4.4 Temperature dependence

The temperature dependence of interstitial loop clustering has been studied using the model parameters and a dislocation density of 10^9 cm/cm^3 at 10^{-6} dpa/s. Total interstitial loop concentration is shown in Fig. 5 as a function of temperature for $E_{2i}^B = 1.19$ eV and $\geqq 2$ eV. For $E_{2i}^B \geqq 2$ eV., the interstitial loop concentration is large and insensitive to the irradiation temperature. However, a smaller diinterstitial binding energy ($E_{2i}^B = 1.19$ eV) has the effect of a sharp drop in loop concentrations at higher temperatures. This behavior has been experimentally observed [10].

5. Conclusions

The detailed rate theory model presented in this paper has indicated that the total interstitial loop concentration is reduced when vacancy clustering is simultaneously taken into consideration. This is found to be due to a higher free interstitial concentration which results from a small recombination rate with free vacancies. It is shown that the total interstitial loop concentration will increase almost linearly with irradiation dose after an initial transient. It is also concluded that the nucleation rate, growth speed, and average size of interstitial loops are all enhanced by the simultaneous microvoid formation.

High initial dislocation density has been found to considerably slow down the kinetics of interstitial loop formation. The large sink strength of straight dislocations reduces matrix point defect fluxes. This constricts the formation of interstitial loops reducing their total concentration, average size, and growth speed.

Studying the diinterstitial binding energy effects on the temperature dependence of interstitial loop concentrations revealed that a binding energy of 1.19 eV results in lowering the total interstitial loop concentration at high temperatures due to diinterstitial dissociation.

Acknowledgement

This work was supported by the National Science Foundation through Grant No. ENG78-05413.

References

[1] R. A. JOHNSON, J. nuclear Mater. (Amsterdam) **75**, 77 (1978).
[2] B. O. HALL and D. I. POTTER, Proc. IX. Internat. Symp. Eff. Rad. Structural Mat., Richland, WA 1978.
[3] U. GOSELE, J. nuclear Mater. (Amsterdam) **78**, 83 (1978).
[4] W. G. WOLFER and M. ASHKIN, J. appl. Phys. **46**, 547 (1975).
[5] D. R. OLANDER, Fundamental Aspects of Nuclear Fuel Elements, NTIS publication, Tennesse 1976 (p. 215).
[6] A. C. DAMASK and G. DIENES, Phys. Rev. **120**, 99 (1960).
[7] N. M. GHONIEM and D. D. CHO, Univ. Calif. Eng. Rep., UCLA-ENG-7845, Los Angeles 1978.
[8] A. C. HINDMARCH, Lawrence Livermore Lab. Rep., UCID-30001, Rev. 3, Livermore 1974.
[9] M. R. HAYNS, J. nuclear Mater. (Amsterdam) **56**, 267 (1975).
[10] H. R. BRAGER and J. L. STRAALSUND, J. nuclear Mater. (Amsterdam) **46**, 134 (1973).
[11] N. M. GHONIEM and G. L. KULCINSKI, Radiat. Eff., in the press.

(Received March 7, 1979)

phys. stat. sol. (a) 54, 179 (1979)
Subject classification: 1.4; 10.2; 22.1.2

Institut für Festkörperphysik und Elektronenmikroskopie der Akademie der Wissenschaften der DDR, Halle (Saale)

Identification of Small Defects in Silicon

By
M. PASEMANN and P. WERNER

In silicon monocrystals grown both by Czochralski and zone-floating method small defects (diameter about 100 Å) in surface-near layers, with typical black–white contrast in the dynamical image in the transmission electron microscope are observed. The identification is carried out by means of diffraction contrast analysis and high resolution lattice plane imaging. According to this analysis the defects are found to be spherical amorphous particles which compress the silicon matrix due to an internal strain $\varepsilon \approx 0.03$ to 0.04. Three possibilities of the structure of these interstitial type clusters are discussed.

In einkristallinem Silizium, das sowohl mittels Czochralskiverfahren als auch tiegelfrei gezüchtet worden war, werden kleine Defekte (Durchmesser ca. 100 Å) in oberflächennahen Schichten mit typischem Schwarz–Weiß-Kontrast in der dynamischen Abbildung im Transmissionselektronenmikroskop beobachtet. Die Identifizierung erfolgt mittels Beugungskontrastanalyse und hochauflösender Netzebenenabbildung. Danach sind diese Defekte sphärische amorphe Teilchen, die die Silizium-Matrix mit einer Spannung von $\varepsilon \approx 0{,}03$ bis 0,04 komprimieren. Drei Möglichkeiten für die Struktur dieser „interstitial"-Typ Cluster werden diskutiert.

1. Introduction

During the last few years an increasing interest has been directed to the investigation of microdefects in as-grown silicon. Especially the characterization of the so-called swirl defects in zone-floated silicon was the aim of a lot of papers (see e.g. [1, 2]). In the case of the A-defects the identification was successful by means of high-voltage electron microscopy in connection with a special sample preparation [3]. It was proved that A-swirls are extrinsic prismatic dislocation loops with $\boldsymbol{b} = \frac{1}{2}\langle 110\rangle$, which are decorated with impurities in most cases. Concerning the B-defects, up to now the attempts for identification failed and it is not clear whether they are vacancy or interstitial clusters or if there is some impurity precipitation, too.

A quite different behaviour was found in Czochralski crystals. Here the dominant microdefects are due to microsegregation (see e.g. [4]). Extensive studies have been carried out to characterize the behaviour of oxygen in Czochralski-grown silicon and it seems to be obvious that there is some precipitation of a silicon oxide [5, 6].

Whether there are defects due to thermal point defects of the silicon in Czochralski material, too, comparable to swirl defects has not been decided.

The investigations of microdefects in silicon are very interesting because they themselves can influence the electrical parameters of electronic devices or some more extended defects can be generated at the microdefects during the device manufacturing process.

In this work we made an attempt to identify very small defects (diameter about 100 Å) with a typical black–white contrast in TEM micrographs which we have found both in Czochralski and in zone-floated silicon crystals.

As will be shown later, it has been proved to be advantageous to define the surface density of the defects. In the as-grown state the surface density of the defects was very low (10^2 to $10^3/\mathrm{cm}^2$), but after the thermal processes of the device manufacturing the density increased and it reached a value of about 10^5 to $10^6/\mathrm{cm}^2$ in MOS devices.

But here we found a difference in the two starting materials. The above-mentioned value is only valid for Czochralski starting material, in zone-floated material the defect density is smaller.

The defects occur in (111), (100), and (110) oriented silicon wafers and they occur in specimens which were prepared from sections cut perpendicular to the plane of the wafers. We observed them after chemical thinning from one side, the other side being protected against the chemical reagent, and after chemical thinning from both sides of the specimens.

Rühle et al. [7] studied extensively black–white contrast defects in irradiated copper foils. They referred to the possibility of the formation of such defects due to the preparation process for transmission electron microscopy. For excluding such an explanation for the defects described here we prepared some samples by ion beam thinning and got the same defects with the same density as after chemical thinning. So we must conclude, they are independent of the preparation process.

For identifying the defects observed firstly we used the diffraction contrast analysis both in a high-voltage electron microscope (1000 kV accelerating voltage) and for better resolution in a JEM 100 C high-resolution electron microscope equipped with a top entry goniometer stage.

Secondly we used the lattice plane imaging method at the JEM 100 C microscope with the standard stage.

2. Diffraction Contrast Analysis

2.1 Dynamical and quasi-kinematical images

The experiments described here are based on the papers by Rühle et al. [7, 8] and Chik et al. [9]. Therefore, we will generally use their nomenclature. For explanation we want to refer to their works where the calculations and the necessary experimental conditions are described and proved in a very comprehensive manner:

1. In the dynamical case, i.e. $w = 0$ the defects have a clear black–white contrast (Fig. 1).

2. Using kinematical conditions, i.e. $w > 0.6$ the defects appear as black dots (Fig. 2a) and under weak-beam conditions as white dots (Fig. 2b).

3. In bright-field images the black–white vector $\boldsymbol{l}$ is always parallel or antiparallel to the diffraction vector $\boldsymbol{g}$ applied (Fig. 1). This means the defects have a spherical distortion field.

4. In bright-field images the amount of defects with $\boldsymbol{l}$ antiparallel to $\boldsymbol{g}$ was higher than that with $\boldsymbol{l}$ parallel to $\boldsymbol{g}$.

5. In dark-field images $\boldsymbol{l}$ was always antiparallel to $\boldsymbol{g}$ (Fig. 3). This means either the defects have a large misfit (in this case the Ashby-Brown rule [11] can be applied) or the defects have a small misfit and must show depth oscillations. In the case of depth oscillations the vector $\boldsymbol{l}$ changes its sign every half extinction distance. Therefore, in this case the defects would lie in two layers, each having the same distance from both foil surfaces (for explanation see [8]).

To decide between these two possibilities one can estimate the size of the particles and the amount of the internal strain. According to a diagram given by Chick et al. [9] concerning the contrast behaviour of spherical particles one can then find out which region is valid for the defects observed.

2.2 Size of the defect and estimation of the internal strain

According to the paper of Ashby and Brown [11] one can determine the size of the particles from their dynamical images. A more accurate method is to use the kinematical or the weak-beam image of the defects (Fig. 2). We found typical sizes between about 70 and 100 Å diameter.

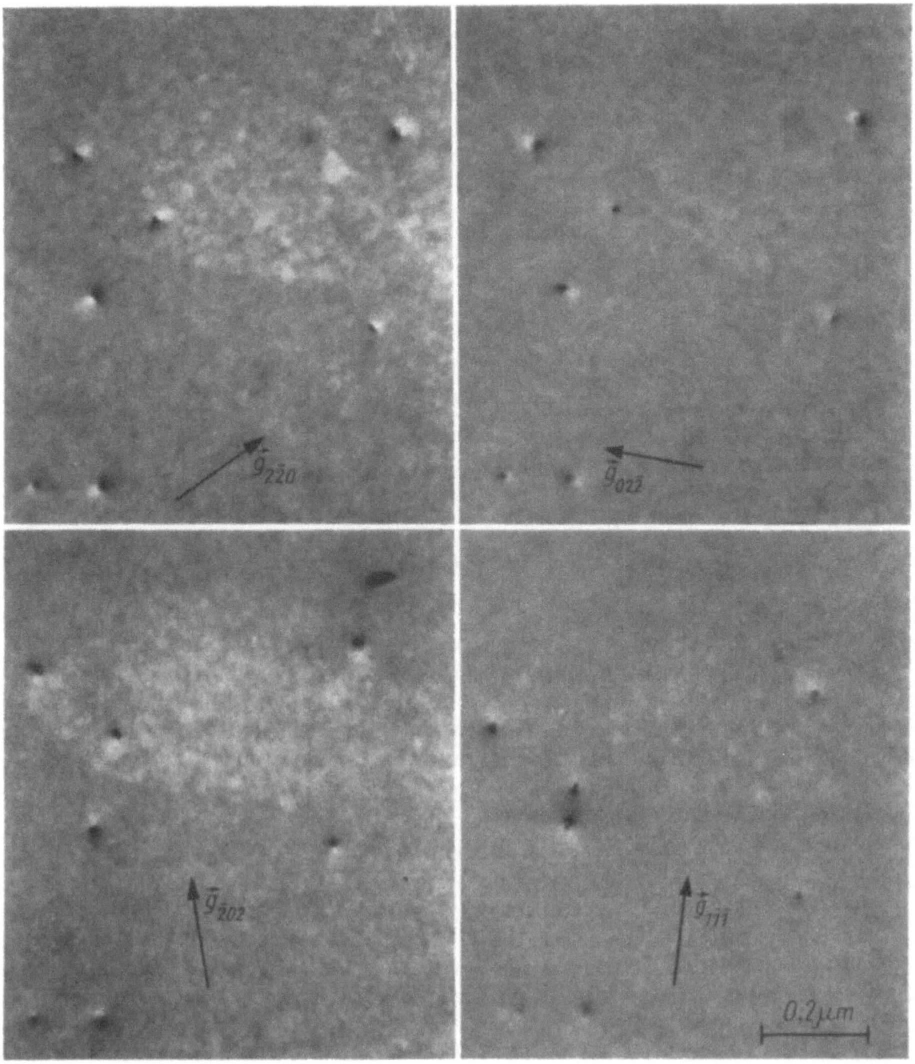

a b c d

Fig. 1. Black–white contrast defects in silicon. The same place is imaged under dynamical bright-field conditions with different **g**-vectors, showing that the black–white vector **l** is always parallel or antiparallel to **g**

One needs these sizes of the particles for an estimation of the internal strain in the matrix. This can be only very rough because of the influence of the surface which is very critical in our case. According to Ashby and Brown [11] the following equation for the evaluation of the internal strain ε can be used:

$$\frac{\text{image width}}{\xi_g} = f\left(\log \frac{\varepsilon g r_0^3}{\xi_g^2}\right),$$

where r_0 is the particle radius, ξ_g the extinction distance, **g** the diffraction vector applied. As to the problem how to measure the special image width we refer to the original work [11].

a  b

Fig. 2. a) Kinematical image of the defects ($w > 0.6$); b) weak-beam image of the defects, $(\bar{2}02)$ excited, imaged with $(20\bar{2})$

Using the following values, $\boldsymbol{g} = (220)$, $r_0 = 50$ Å, $\xi_g = 757$ Å, and an image width of 150 Å for 50% intensity, we got an internal stress parameter of $\varepsilon = 0.023$.

Using another vector, $\boldsymbol{g} = (111)$, $\xi_g = 602$ Å, and an image width of 185 Å for 50% intensity, we got a value of $\varepsilon = 0.04$. As these values can only be taken as a clue, another method of estimation was used as described in Section 3.

2.3 Depth and nature of the defects

Using these values for r_0 and ε one finds out that the defects lie in the region of depth oscillations (see [9]). But we only observed defects in the first layers of the depth oscillation, i.e. down to a depth of maximum $0.3\xi_g$ from both surfaces.

For an accurate depth measurement we tried to carry out stereo experiments and marked the surface of the silicon samples by gold evaporation, as described by Diepers and Diehl [10] (Fig. 4). Fig. 4a shows a specimen region in dynamical diffraction contrast, and in Fig. 4b one sees the same region under quasi-kinematical conditions which are necessary for an accurate measurement in the stereo experiment.

In fact, using these depth measurements we only found defects down to depths of

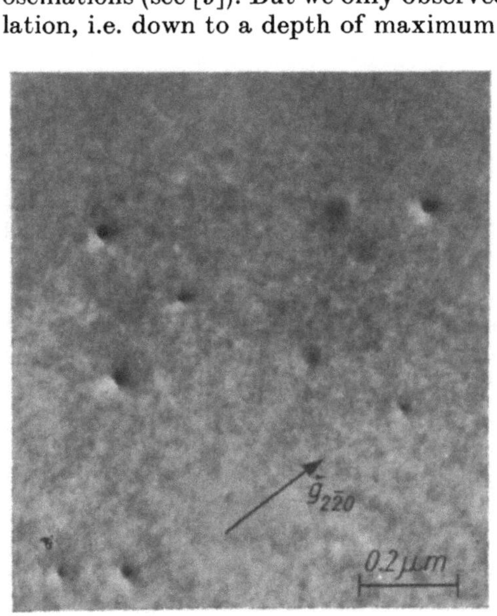

Fig. 3. Dynamical dark-field image of the defects showing that the black–white vector $\boldsymbol{l}$ is always antiparallel to $\boldsymbol{g}$

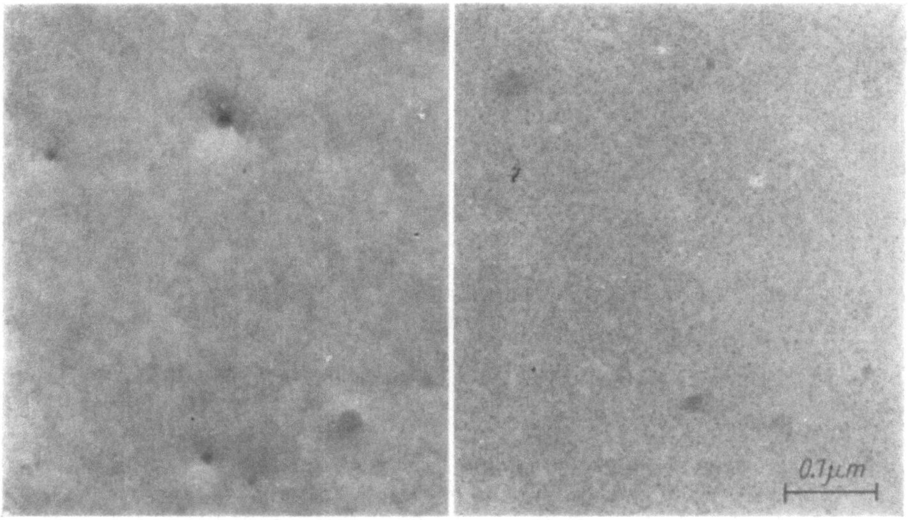

a b

Fig. 4. Gold evaporated silicon sample with black–white contrast defects imaged a) under dynamical, b) under kinematical bright-field conditions

$0.3\xi_g$ distance from both surfaces. Within these two layers the defects were lying at different distances from the surfaces.

So we can state that the defects observed are situated within surface-near regions down to a maximum depth of about 230 Å from both specimen surfaces ($0.3\xi_g$ for (220) at 100 kV $\approx$ 230 Å). Another effect is worth mentioning. We observed a higher density of defects at that side of the sample which is turned towards the electron source. If one takes a photograph of a special place of the specimen and then turns the specimen round and takes a micrograph at the same specimen position, one finds more defects than before. This means, some defects must have been generated due to the electron beam.

To summarize the observed black–white contrast effects one can state: As in dark-field images the defects always have an $\boldsymbol{l}$-vector antiparallel to $\boldsymbol{g}$ and the defects are found to lie in the first layers of the depth oscillation they must be spherical particles of the interstitial type.

2.4 Some remarks concerning the elastic anisotropy

Lepski [12] calculated the figures of black–white contrast of small coherent particles in a matrix and took into consideration the elastic anisotropy of the matrix. As different distortions exist in the different crystallographic directions there is also an influence of the black–white contrasts on the contrast figures. The contrast behaviour is determined by the parameter of the elastic anisotropy $(d - c)/(1 + a)$, where

$$d = \frac{1 + a - ca}{\sqrt{1 - \left(\frac{c}{4}\right)(1 + a)}}, \qquad c = \frac{c_{12} - c_{11} + 2c_{44}}{c_{44}}, \qquad a = \frac{c_{12}}{c_{11}};$$

c_{ij} being the elastic constants.

If for a given substance this parameter differs very much from unity, then the black–white contrasts exhibit a contrast connected with the elastic anisotropy of the matrix. Lepski distinguishes between substances where $c > 0$ and $c < 0$, respectively. For $c > 0$ (e.g. iron) the displacements have maxima along the $\langle 110 \rangle$ and $\langle 111 \rangle$ directions, and minima along the $\langle 100 \rangle$ directions. In the case of $c < 0$ (e.g. molybdenum) the

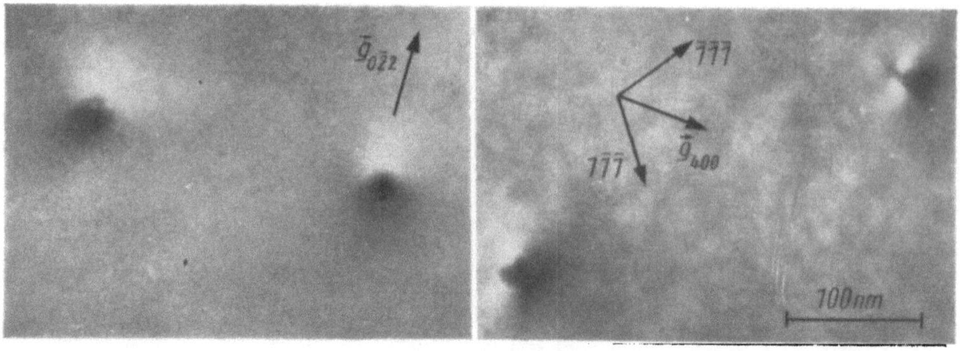

a b

Fig. 5. Black–white contrast figures in silicon in the bright-field image, a) $\boldsymbol{g}$ = (0$\bar{2}$2), b) $\boldsymbol{g}$ = (400). For explanation see text

behaviour is reversed. The contrast calculations confirmed these results. As to the authors' knowledge, up to now there was no experimental proof for these calculations.

As the parameter for elastic anisotropy of silicon yields to 0.4, which is quite different from unity, the black–white contrasts observed in this material seemed to be suitable to confirm the results of Lepski.

For silicon is valid $c > 0$, i.e. one has to expect extended contrasts along the ⟨110⟩ and ⟨111⟩ directions and a waist in the contrast figure along ⟨100⟩ directions.

Fig. 5 shows black–white contrasts at a high magnification, in a) a (0$\bar{2}$2) $\boldsymbol{g}$-vector was used and one sees an extended contrast in this direction, in b) a (400) vector was used and one can observe a strong waist in this direction, but an extended contrast along the indicated ⟨111⟩ directions.

So the elastic anisotropy of the silicon matrix determines the contrast behaviour of the black–white contrast of the defects to a high degree.

3. Analysis of the Defects by the Lattice Plane Imaging Technique

An additional possibility of identification of such small defects in crystalline silicon is given by the method of lattice plane imaging. By this imaging technique in the last few years different kinds of defects in silicon could be analysed by several authors [13 to 15]. However, for imaging the crystal structure with high contrast some important assumptions have to be fulfilled concerning the preparation of the specimens as well as the imaging conditions of the electron microscope. For this method (110)-oriented silicon specimens of the same material as employed for the examination by diffraction contrast (Section 2) were used. The specimens were thinned down to a thickness smaller than 400 Å by chemical polishing. The investigations were performed in a 100 kV electron microscope JEM 100 C (f_{obj} = 1.6 mm, C_s = 1.4 mm).

To get an optimum imaging process an objective aperture of 10^{-2} rad was applied allowing an image formation by seven beams. Fig. 6 shows the diffraction pattern of (110)-oriented silicon, the position of the objective aperture is indicated.

The small defects which had been identified before by diffraction contrast as interstitial-like precipitations were imaged by axial illumination and by an underfocus of the objective lens of about 100 nm. Fig. 7 shows a lattice plane image of such a defect, which has a spherical shape, its greatest diameter being about 80 Å. The (111) planes of the silicon matrix with a distance of 3.14 Å are distorted in a positive manner on the particle boundary, i.e. the planes are bended outwardly, which is well observable by holding the picture in a slanting position.

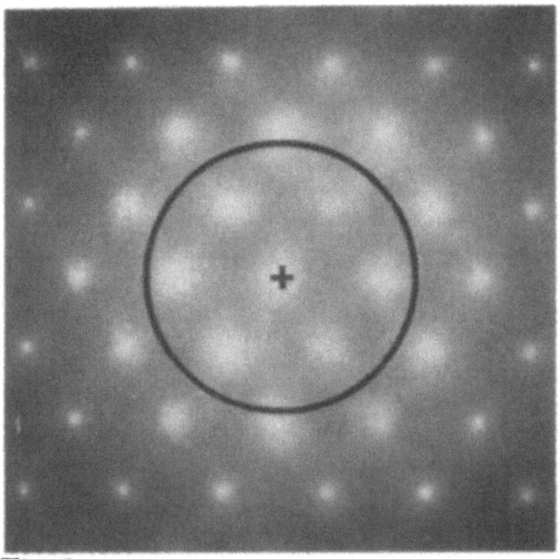

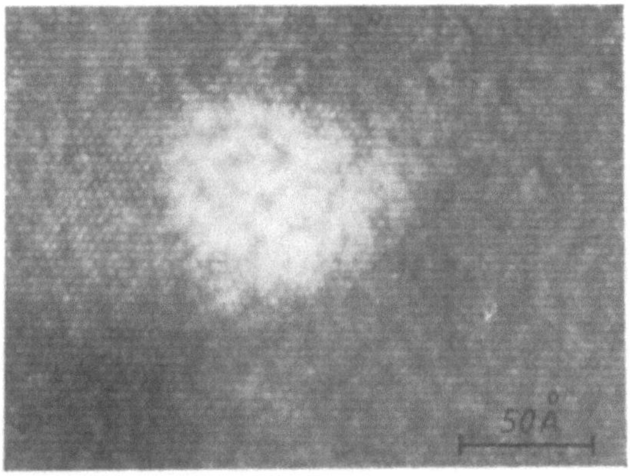

Fig. 6 Fig. 7

Fig. 6. Diffraction pattern of (110)-oriented silicon. The position of the objective aperture for high-resolution imaging is indicated

Fig. 7. High-resolution image of a small defect of spherical shape with a greatest diameter of about 80 Å in silicon. The (111) planes have a distance of 3.14 Å

While the lattice planes of the matrix are imaged with good contrast, a granular contrast appears inside the particle. Further lattice plane images of such particles identified by diffraction contrast as black–white contrast defects give similar results. Fig. 8 shows a spherical defects with a diameter of about 100 Å. In this case also a positive strain of the (111) planes appears, i.e. the particles behave like interstitial clusters.

For the (220) planes of the matrix with a distance of 1.92 Å a measurement of the lattice plane distances turns out to be too incorrect. From the investigations it follows that the diameter of the small defects usually varies between 70 and 100 Å. Particles of greater dimensions are exceptions. As an example Fig. 9 shows a particle with a diameter of about 200 Å.

For 15 spherical particles imaged by the high-resolution technique the so-called constrained strain of the matrix was determined. These experimental results were compared with values calculated by means of the theory of the elastic distortion field around such strain centres, first explained by Mott and Nabarro [16] and Eshelby [17] and which are also the basis for the calculations of Ashby and Brown [11] used in Section 2.

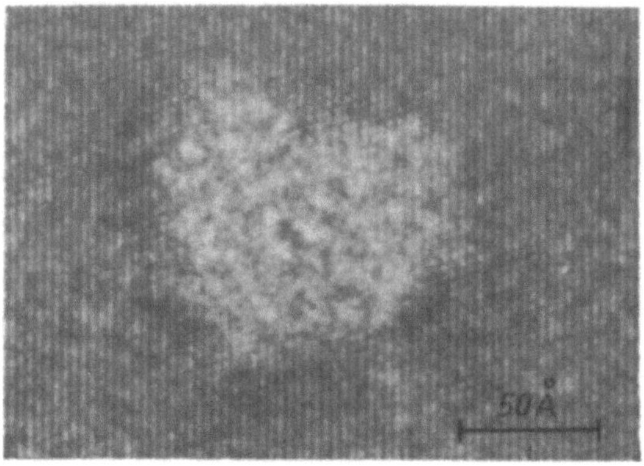

Fig. 8. Spherical defect in silicon with a diameter of about 100 Å in the lattice plane imaging mode

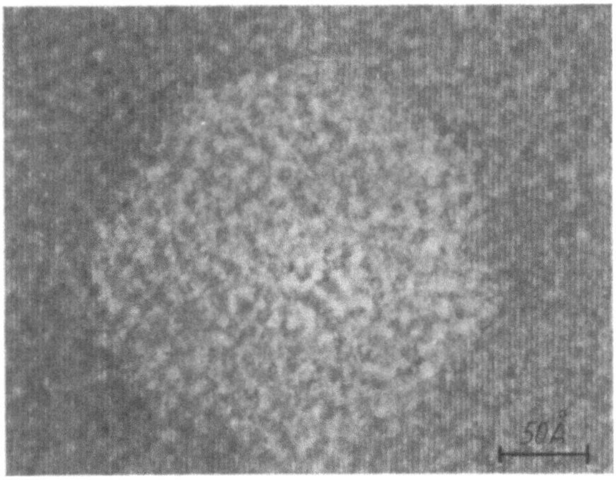

Fig. 9. Lattice plane image of a spherical particle with a diameter of about 200 Å

The investigations show that in most cases a positive distortion of the (111) planes exists around a particle. In two cases a strain of the matrix could not be detected.

The diagrams represented in Fig. 10 demonstrate the measured relative changes of the lattice plane distance of the (111) planes $\Delta d/d$ as a function of the distances x from the surface of the particle for different particle diameters (see circles). Additionally, the diagrams contain the theoretical curves of the variation of the lattice plane distance of the (111) planes for different values of the constrained strain.

We calculated the displacement of the matrix using the relation (see also Ashby and Brown [11])

$$\boldsymbol{U}(\boldsymbol{r}) = \varepsilon \left(\frac{r_0}{r}\right)^3 \boldsymbol{r} ,$$

where $\boldsymbol{U}(\boldsymbol{r})$ is the displacement vector, r_0 the particle radius, and ε the constrained strain.

From the diagrams in Fig. 10 it can clearly be seen that depending on the particle diameter the measured values of the matrix strain lie between $\varepsilon \approx 0.03$ and ≈ 0.04.

Inside the particles the lattice plane image shows a granular contrast similar to that which appears in thin amorphous films. The thickness of the specimen in the regions where high-resolution imaging was carried out was about 100 to 300 Å. From this it follows that in many cases the diameters of the particles were comparable with the thickness, so that the present thin layers of the silicon matrix above and below the particle are too small to produce a sufficient contrast. Assuming that the particles have a spherical shape the images show that on the boundary the lattice planes go a short way into the defect in the case of small particles.

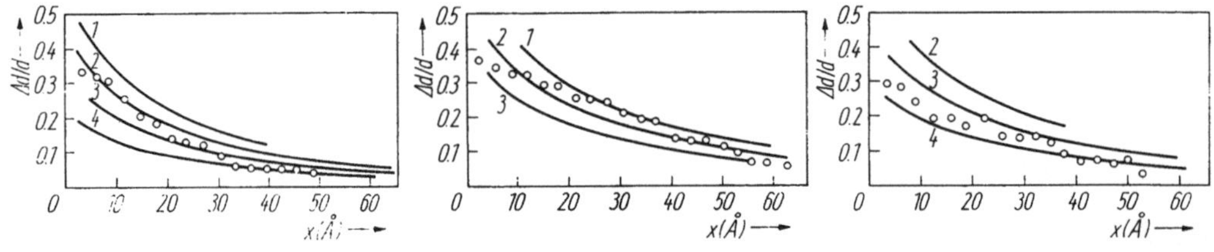

a b c

Fig. 10. Measured relative changes of the lattice plane distance $\Delta d/d$ as a function of the distance from the particle, particle radius being a) 35, b) 40, c) 45 Å. The diagrams also show the theoretical curves for different constrained strain parameters ε. (1) $\varepsilon = 0.05$, (2) 0.04, (3) 0.03, (4) 0.02

The imaging conditions, like the amount of defocus and the position of the objective aperture, were selected in such a way that the acting phase contrast transfer function enabled imaging of object structures with distances between 10 and 3 Å. The interpretation of the phase contrast of such small particles is very difficult. For this, additional simulation of the image contrast is necessary which should be based on models of the microstructure of such particles. From the present results of the experiments one can only conclude that if the investigated precipitates consist of small crystalline particles the dimensions of those particles should be smaller than 15 Å.

Therefore, we suppose that the particles are of an amorphous state.

4. Conclusions

In both Czochralski and zone-floated silicon small defects have been observed which were identified by two methods, diffraction contrast analysis and high-resolution imaging, to be spherical particles with an interstitial-type strain field. The internal strain of the matrix was estimated by the two methods independently to be about $\varepsilon \approx 0.03$ to 0.04. But it has to be emphasized that the observed defects lie near the surfaces within a depth of $0.3\xi_g$ from both surfaces. So in this case the influence of the surface will be strong. But to carry out simple contrast experiments this influence had to be neglected. The values given for ε are therefore just a rough estimation.

Concerning the imaging of the defects observed we want to consider two cases of the influence of the surface:

Firstly, defects which are invisible within the crystal volume due to their small strain ε (see Chik et al. [9]) become visible near the surface as a consequence of the changed stress conditions.

Secondly, impurity atoms diffusing from surface layers, like for instance SiO_2, into the surface-near regions lead to the generation of observable defects. Nuclei for such particles can be defects of the crystal volume which are not visible due to their small strain ε. These considerations have to be taken into account in the discussion of the structure of the defects analysed.

By means of high-resolution imaging an additional information was obtained. The inner structure of the particles seems to be amorphous because there was only a granular phase contrast observable, although due to the phase contrast transfer function of the objective selected a periodic structure with a periodicity between 3 and 10 Å could be imaged if it were present.

So we suggest the defects have an amorphous structure and in this connection discuss three possibilities:

1. The defects consist of amorphous silicon. In this case we propose that a special crystal volume suffers a transition into the amorphous state, the nearest neighbour distance is the same as in the crystalline state [18] and amounts to 2.35 Å. Furthermore, we use an estimation of Guttman [19] who got a Young modulus of the amorphous silicon of 90% of that of the crystalline silicon.

Under these assumptions the calculation results in a concentration of about 20 vol% interstitial atoms inside the particles to get a strain of 0.03 to 0.04. These interstitial atoms could be self-interstitials as well as impurities, like oxygen and/or carbon, or impurity clusters.

In this connection it is important to refer to the fact that the electron beam generates such defects at that specimen surface which looks towards the beam. This supports the first interpretation.

2. The defects consist of amorphous SiO_2. In this case according to Maher et al. [5] the internal strain amounts to about $\varepsilon \approx 0.07$. Although this estimation seems to be

as rough as ours a concentration of vacancies had to take place inside the particle or at its boundaries to lower the internal strain.

But as we observed nearly the same density of defects in Czochralski and zone-floated silicon starting material this interpretation does not seem to be probable, because the oxygen content is at least two orders of magnitude higher in Czochralski silicon than in zone-floated one.

3. The defects are connected with the droplet model proposed by Föll et al. [20] which should explain the B-swirls. In this model B-swirls are imaged as droplets of liquid silicon inside the crystalline matrix. The density of liquid silicon is about 10% higher than that of crystalline silicon, i.e. such a droplet could contain an excess of interstitials without generating a strong strain.

According to [20] in the consideration of the droplet model carbon plays an important role in a twofold manner, first, it influences the stability of the droplet and second, the droplets form heterogeneously at nuclei probably consisting of carbon clusters.

It should be pointed out that large droplets in silicon could experimentally be established by an in-situ X-ray topography by Chikawa and Shirai [21], although these droplets probably have another generation mechanism than B-swirls.

The droplet model would explain the defects observed if one admits an additional concentration of interstitials inside the droplets to obtain the estimated strain of 0.03 to 0.04. In this case the B-swirls would be present in Czochralski and zone-floated silicon starting material of about the same density.

Acknowledgements

The authors are very indebted to Prof. H. Bethge and Prof. J. Heydenreich for their interest in the work and to Mr. D. Baither and Mr. H. Bartsch for helpful discussions.

References

[1] H. FÖLL and B. O. KOLBESEN, Appl. Phys. 8, 319 (1975).
[2] A. J. R. DE KOCK, P. J. ROKSNOER, and P. G. T. BOONEN, J. Crystal Growth **28**, 125 (1975).
[3] L. J. BERNEWITZ, B. O. KOLBESEN, K. R. MAYER, and G. E. SCHUH, Appl. Phys. Letters **25**, 277 (1974).
[4] A. MURGAI, H. C. GATOS, and A. F. WITT, Proc. Third Internat. Symp. on Silicon Materials Science and Technology, Ed. R. HUFF and E. SIRTL, Electrochem. Soc., Princeton, N. J., 1977 (p. 72).
[5] D. M. MAHER, A. STAUDINGER, and J. R. PATEL, J. appl. Phys. **47**, 3813 (1976).
[6] K. H. YANG, R. ANDERSON, and H. F. KAPPERT, Appl. Phys. Letters **33**, 225 (1978).
[7] M. RÜHLE, M. WILKENS, and U. ESSMANN, phys. stat. sol. **11**, 819 (1965).
[8] M. RÜHLE, phys. stat. sol. **19**, 263 (1967).
[9] K. P. CHIK, M. WILKENS, and M. RÜHLE, phys. stat. sol. **23**, 113 (1967).
[10] H. DIEPERS and J. DIEHL, phys. stat. sol. **16**, K109 (1966).
[11] M. F. ASHBY and L. M. BROWN, Phil. Mag. 8, 1083 (1963).
[12] D. LEPSKI, phys. stat. sol. (a) **24**, 99 (1974).
[13] V. A. PHILLIPS and J. A. HUGO, Acta metall. **18**, 123 (1970).
[14] V. A. PHILLIPS and J. A. HUGO, Acta metall. **20**, 1143 (1972).
[15] O. L. KRIVANEK and D. M. MAHER, Appl. Phys. Letters **32**, 451 (1978).
[16] N. F. MOTT and F. R. N. NABARRO, Proc. Phys. Soc. **52**, 86 (1940).
[17] J. D. ESHELBY, Progress in Solid Mechanics, Vol. II, Chap. III, North-Holland Publ. Co., Amsterdam 1961 (p. 96).
[18] S. C. MOSS and J. F. GRACZYK, Proc. X. Internat. Conf. Phys. Semicond., Cambridge 1970 (p. 658).
[19] L. GUTTMAN, Solid State Commun. **24**, 211 (1977).
[20] H. FÖLL, U. GÖSELE, and B. O. KOLBESEN, J. Crystal Growth **40**, 90 (1977).
[21] J. CHIKAWA and S. SHIRAI, J. Crystal Growth **39**, 328 (1977).

(Received March 23, 1979)

phys. stat. sol. (a) 54, 189 (1979)

Subject classification: 17.1; 14.4

Laboratoires de Physico-Chimie des Surfaces, Département de Chimie, Université de Liège[1])

Field-Emission Microscopy from Glass-Coated Tips

Observation of Fowler-Nordheim Emission Regime and of Periodic Current Oscillations

By

N. Rihon

Injection of electrons into glass by tunneling from metallic tips — specially platinum tips — is investigated in a field-emission microscope. No current is detected at applied voltages up to 30 kV when the thickness of glass layers is larger than a few microns. On the contrary, in the case of very thin glass coatings ($\lessapprox 0.1$ μm), current–voltage characteristics can be measured; at lowest currents, the emission obeys the Fowler-Nordheim equation, while periodic current oscillations are observed at higher currents ($i \approx 10^{-8}$ A). Field-electron and field-ion micrographs from platinum tips whose glass coating is removed by various treatments (heating in vacuum, dissolution of glass in hydrofluoric acid) are also presented and discussed.

On a examiné, à l'aide de la microscopie à émission de champ, l'injection d'électrons dans le verre par effet tunnel à l'interface métal–verre. Dans le cas de pointes enrobées par des couches de verre dont l'épaisseur est de quelques microns, aucun courant n'est détecté, même à des voltages aussi élevés que 30 kV. Par contre, des caractéristiques courant–tension ont été mesurées dans le cas d'enrobages de verre minces ($\lessapprox 0{,}1$ μm); aux faibles courants, l'émission obéit à l'équation de Fowler-Nordheim, tandis que des oscillations périodiques de courant ont été observées dans le domaine des courants plus élevés ($i \approx 10^{-8}$ A). On présente également des micrographies électroniques et ioniques de champ obtenues avec des pointes de platine dont l'enrobage de verre a été enlevé par divers traitements (chauffage sous vide, dissolution du verre dans l'acide fluorhydrique).

1. Introduction

The study of field-electron emission at the metal–dielectric interface is of great interest in various branches of physics, chemistry, electronics, and electrical engineering [1 to 3]. Most of the experimental results published in the literature are obtained with sandwich structures where a thin layer of a dielectric is placed between two plane electrodes (metals or semiconductors). This measurement set-up gives many useful data about current injection and current flow in dielectrics, but it sometimes presents a major drawback: electrons may be field-emitted from small asperities located on the cathode, so that the interpretation of results becomes delicate and controversial [4]. Field-emission microscopy (F.E.M.; for details, see for instance the review articles [1, 5]) allows to circumvent this difficulty by using a cathode in form of a small metallic tip whose atomic structure and crystallographic orientation are well defined. In comparison with the sandwich technique, F.E.M. has three other advantages. First, it gives the possibility of a better control of pre-breakdown phenomena in dielectrics, because the field strength decreases very rapidly with distance from the metallic tip surface and, as a consequence, the pre-breakdown processes are not directly converted in a destructive breakdown. A second attractive feature of F.E.M. is that only electrons are injected into the dielectric. At last, F.E.M. is adequate for the use of high sensitivity counting techniques in the detection of emitted electrons; this point is

[1]) Sart Tilman, 4000 Liège I, Belgium.

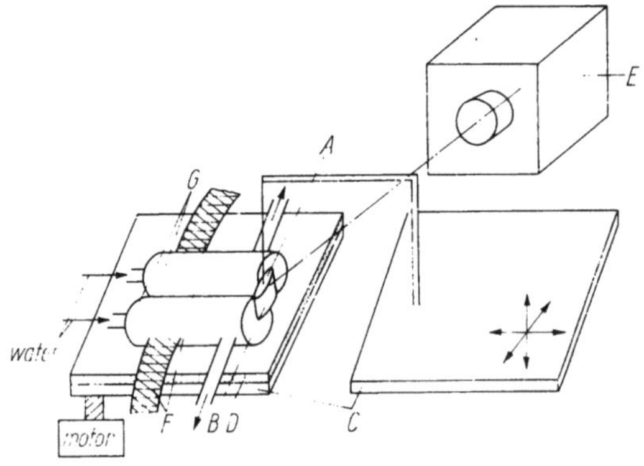

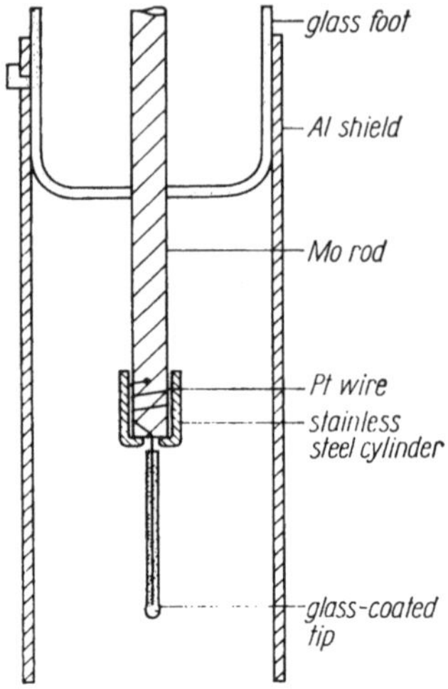

Fig. 1 Fig. 2

Fig. 1. Schematic diagram of experimental set-up used for coating of tips. A metallic tip; B platinum crucible containing molten glass; C micromanipulators; D plexiglass plate; E optical pyrometer; F copper leads; G copper cylinders cooled by flow of water

Fig. 2. Schematic diagram of tip support for work at high emission voltage (30 kV)

very interesting in order to avoid space charge effects within dielectrics and emission instabilities due to high current densities.

In the literature of field-emission microscopy, there are a few results obtained with tips surrounded by liquid dielectrics [1, 6, 7]. As regards the solid dielectrics, a study has been published by Roynet and Mignolet [8] about field emission from tungsten tips coated with tungsten glass. That study was affected by limitations and deficiencies which have been removed in the present work.

This paper reports results of field-electron emission from platinum tips coated with relatively thick (≈ 1 μm) and thin ($\lessapprox 0.1$ μm) glass layers. We have taken platinum as tip material hoping to reduce, during the coating operation, the corrosion of the tip and the formation of oxide layers at the metal–glass interface. A part of the paper deals with both field-electron and field-ion images obtained from tips whose glass layer has been removed by heating in vacuum or by dissolution in hydrofluoric acid.

2. Experimental

Platinum tips are prepared from 0.05 mm platinum wire (spectroscopically pure) by electrolytic etching in NaCl–$ZnCl_2$ molten bath; details are given elsewhere [8]. They are shaped by field evaporation in a field-ion microscope of usual design until a perfect crystalline structure is obtained. The emitters are then coated with glass as follows (Fig. 1). A small platinum crucible contains glass; it is heated to about 1200 °C by Joule's effect. The tip is slowly immersed in molten glass with the help of a micromanipulator whose vertical displacements are governed by a motor. The coating operations are visually controlled with an optical pyrometer (Leeds-Northrup) which indicates also the glass temperature. When the latter is lower than ≈ 1100 °C, glass is too viscous and the tips are destroyed, while at temperatures higher than ≈ 1250 °C, the apex of the tip is generally not covered because glass moves up by the capillary effect along the shank of the emitter. It is possible to realize glass coatings with variable thicknesses by acting on the extraction speed of the tip and on the heating temperature of glass. Before insertion into a conventional field-emission microscope, the coated tips are examined with an optical microscope which allows to characterize the quality, shape, and thickness of the coating. The field-electron microscope is equiped with an electron multiplier (Balzers SEV 117) well suited for the detection of electrons by high-sensitivity counting technique. A special care was given to the tip support in view to avoid discharges and spurious emissions at applied voltages as high as 30 kV. This support (Fig. 2) consists essentially of a molybdenum rod (diameter 2 mm) sealed into glass. The tip is spot-welded on a platinum wire (diameter 0.4 mm) covered with glass and rolled around the Mo rod. A stainless steel cylinder, whose edges have been rounded by electrochemical polishing, claps the Pt wire against the Mo rod. The temperature of the emitter can be varied by putting various refrigerants (ice, liquified gases) in the cold finger. An aluminium electrostatic shield insures that the electric field strength is well defined around the tip.

3. Results and Discussion

No field-electron emission was observed from platinum tips coated with glass layers whose thickness exceeds a few microns, at emission voltages as high as 30 kV. This result is quite contrary to what is expected on the basis of the data previously published by Roynet and Mignolet [8]. Indeed, these authors measured electron currents, of the order of 10^{-8} A, from glass-coated tungsten tips at emission voltages (≈ 10 kV) about five times greater than the emission voltages before coating, the glass thicknesses ranging from a few μm to more than 1000 μm. We have checked that the platinum emitters were not destroyed during the coating operations and, besides, that their mean radius was not increased to such a degree that the electric field strength at the metal–glass interface was too low to produce electron tunneling. Tips whose glass coating had been previously dissolved in hydrofluoric acid were studied in a field-ion microscope. The following results were obtained. After field evaporation (evaporation voltage ≈ 20 kV) of some disordered surface layers in a mixture of neon and hydrogen as imaging gas (pressure of each gas 10^{-4} Torr), the field-ion micrograph of Fig. 3 is observed. The main feature is the presence of three bright rings located around the {100} planes (tip orientation ⟨111⟩). The electron image (Fig. 4) exhibits also three strongly emitting areas corresponding to the bright regions in the ion image.

While the electron image represents the surface distribution of the work function and of the local field strength, the ion image reflects the distribution of the ionization

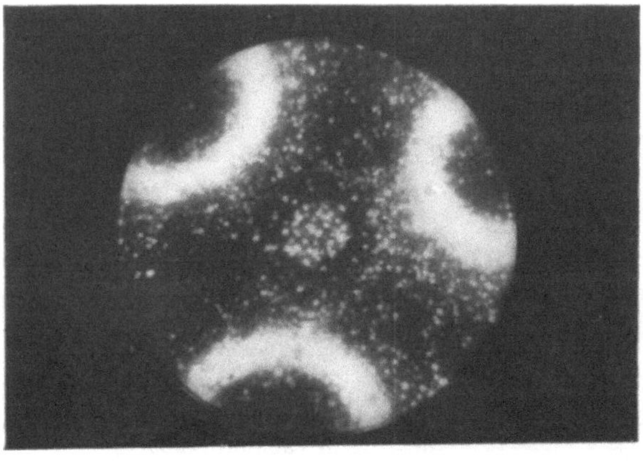

Fig. 3. Field-ion emission image of a platinum tip after dissolution of glass coating in hydrofluoric acid and field evaporation of disordered surface layers (imaging gas Ne–H_2; 77 K)

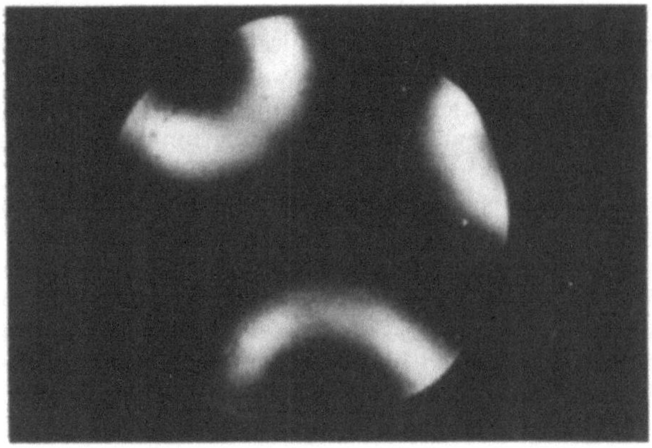

Fig. 4. Field-electron emission image corresponding to Fig. 3 (1900 V; 77 K)

probability of imaging gas atoms very near the surface (more precisely at a critical distance of the order of 5 Å [10]). This ionization probability depends mainly on the local field strength. Thus, the bright regions in the ion micrograph are probably due to the presence of protruding atoms, atom clusters, or steps which enhance locally the field strength. It is apparent from the above considerations that the platinum tips are not destroyed during coating and their mean radius is not increased to such a degree that field-electron injection from platinum into glass is impossible. These conclusions are supported by another significant observation. When the coated tips are heated (by Joule's effect) in vacuum, glass is partly or entirely removed and an electron image (Fig. 5) is observed at low emission voltages (≈ 2000 V): the ternary symmetry is again characteristic of field emission from a ⟨111⟩ oriented platinum emitter.

Some experiments were also carried out with tungsten tips coated with relatively thick layers of tungsten glass. No electron emission was measured at applied voltages up to 30 kV. In our view, the fact that no emission is observed from metallic tips covered by thick layers of glass can be explained as follows. Electrons are first field-injected into the dielectric but they are rapidly trapped by defects. High-field conduction mechanisms such as those occurring in sandwich structures [3], for example Poole-Frenkel process where conduction is controlled by field-enhanced thermal emission of electrons from discrete traps levels into the conduction band of the insulator, cannot occur because the field strength decreases rapidly with distance from

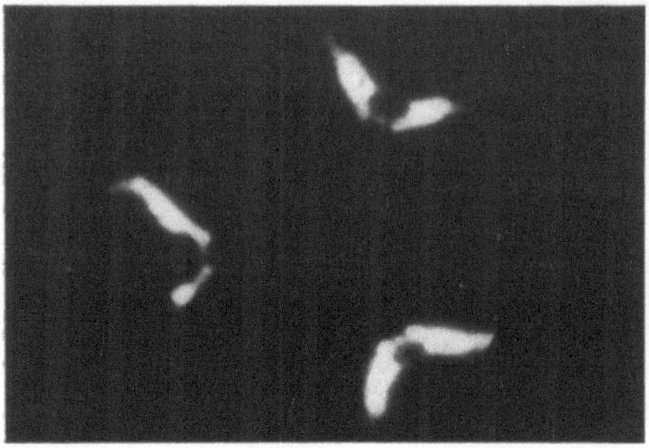

Fig. 5. Field-electron emission image of a glass-coated platinum tip after heating in vacuum (2200 V)

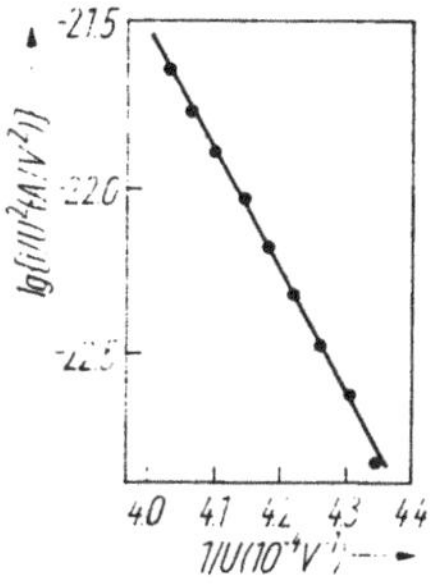

Fig. 6. Fowler-Nordheim plot for emission from a platinum tip coated with a thin layer of glass (77 K)

the cathode surface (this is a consequence of the geometrical configuration in field-emission microscopy [1, 6]). On the other hand, a direct tunneling from the metal to vacuum is not possible because the glass layers are too thick.

Platinum tips coated with thin layers (< 0.1 μm) of glass have also been studied. Here, an electron emission was measured at applied voltages of the order of 3 kV; the field-electron image consisted of bright spots located on the fluorescent screen without any symmetry. As shown in Fig. 6, the current–voltage characteristic is in rather good agreement with the Fowler-Nordheim equation which establishes the following relation between the field-emitted current (i) and the applied voltage (U) when some correction factors due to image forces are neglected [1, 5]:

$$i = A U^2 \exp\left(-\frac{B}{U}\right), \tag{1}$$

where A and B represent coefficients depending essentially on the height of the potential barrier at the metal–dielectric interface and on the proportionality factor k between the field strength F at the emitter surface and U ($F = kU$). As expected from (1), it was found that the slope of Fowler-Nordheim plots does not vary when the tip temperature is changed by putting various refrigerants (ice, liquid nitrogen, liquid helium) in the cold finger. From these results, it seems well that emission is controlled by tunneling at the platinum–glass interface, while bulk effects within glass such as space charge and electron trapping effects do not play an important role. At higher voltages, an oscillatory behaviour of current was sometimes observed. Fig. 7a gives an experimental plot of current versus time at constant applied voltage (2700 V). Each current oscillation consists of a rapid increase to a maximum current, followed by a decrease to a minimum current. As Fig. 7b reveals, the frequency of the oscillations increases with the applied voltage.

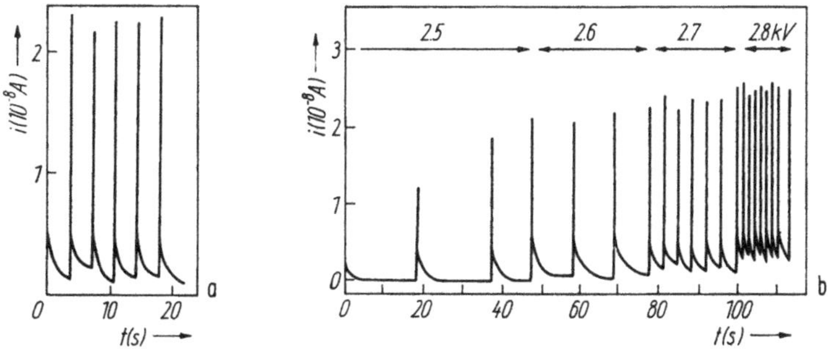

Fig. 7. Current oscillations from a glass-coated platinum tip ($T = 77$ K)

Similar current fluctuations have been measured in sandwich structures with several dielectrics, particularly SiO_2 [11] and organic polymers [12]. They are probably associated with transitions from a low conductivity state to a high conductivity state. Since the discovery of the Ovshinsky effect in chalcogenide glasses [13], the switching behaviour of dielectrics submitted to very high fields ($\approx 10^7$ V/cm) has received a great attention from physicists as well as from engineers [14]. Several interpretations based on electronic interface states, ionic displacements, thermal processes, polarization and space charge effects have been suggested. It might be that the current oscillations described here result from thermal devitrification of glass; the transition from insulating state to conducting state would be related to al ocal glass–crystal phase transition due to heating by Joule's effect in thin channels of glass (filamentary conduction [15]).

In conclusion, it was shown that, in convenient experimental conditions, field-emission microscopy from metallic tips coated with glass can give valuable information about the behaviour of glass submitted to very high electric fields. Various emission regimes were observed. While no emission occurs when the glass layer thickness is greater than about 1 μm, pure Fowler-Nordheim field emission and bulk effects in glass have been reported in the case of thin glass coatings.

Acknowledgements

The author wishes to gratefully acknowledge Prof. J. C. P. Mignolet for his stimulating interest in this work and for helpful discussions. He is very much indebted to "Le Fonds de la Recherche Fondamentale Collective" for financial support to the laboratory and to "Les Services de Programmation de la Politique Scientifique, Centre de Catalyse, Université de Liège", for a research grant. The skilful technical assistance of G. Quoilin and M. Fedyniak is also greatly appreciated.

References

[1] R. GOMER, Accounts chem. Res. 5, 41 (1971).
[2] M. A. LAMPERT and P. MARK, Current Injection in Solids, Academic Press, New York 1970.
[3] J. J. O'DWYER, The Theory of Electrical Conduction and Breakdown in Solid Dielectrics, Oxford University Press, 1973.
[4] R. M. ANDERSON and D. R. KERR, J. appl. Phys. **48**, 4834 (1977).
[5] L. W. SWANSON and A. E. BELL, Adv. Electron. Electron Phys. **32**, 193 (1973).
[6] R. COELHO and P. SIBILLOT, Rev. gén. Electr. 79, 29 (1970).
[7] B. HALPERN and R. GOMER, J. chem. Phys. **51**, 1031, 3043 (1969).
[8] H. ROYNET and J. C. P. MIGNOLET, Bull. Soc. Roy. Sci. Liège 5/6, 274 (1969).
[9] J. MARIEN, J. C. P. MIGNOLET, and N. RIHON, Bull. Soc. Roy. Sci. Liège **11/12**, 614 (1970).
[10] E. W. MÜLLER and T. T. TSONG, Field Ion Microscopy — Principles and Applications, Elsevier Publ. Co., Amsterdam 1969.
[11] M. SHATZKES, M. AV-RON, and R. M. ANDERSON, J. appl. Phys. **45**, 2065 (1974).
[12] N. SWAROOP, J. appl. Phys. **42**, 863 (1971).
[13] S. R. OVSHINSKY, Phys. Rev. Letters **21**, 1450 (1968).
[14] R. GOFFAUX, Conf. Ageing Phenomena in Solid Dielectrics and Methods of Their Testing, Wrocław, 16 to 19 September 1975; Bull. Sci. A.I.M. Université de Liège **4**, 299 (1975).
[15] H. J. STOCKER, Appl. Phys. Letters **15**, 55 (1969).

(Received March 19, 1979)

phys. stat. sol. (a) 54, 195 (1979)

Subject classification: 10.2; 12.2; 22.1.1; 22.1.2; 22.2.1; 22.2.3

Institute of Materials Science, Academy of Sciences of the Ukrainian SSR, Kiev

Analysis of Dislocation Mobility under Concentrated Loads at Indentations of Single Crystals

By

I. V. GRIDNEVA, YU. V. MILMAN, V. I. TREFILOV, and S. I. CHUGUNOVA

An approximate solution is given for the dependence of the dislocation track length, l, around a microhardness indentation in single crystals on the load applied to the indenter (P), the temperature (T), and the duration under load (t): $l \sim P^{m/(2m+1)} t^{1/(2m+1)} \exp[-U/kT(2m+1)]$ The assumption is made that dislocations move from the indenter in plane arrays and that the dislocation velocity is $v \sim \tau^m \exp(-U/kT)$, where τ is an effective stress, U is the activation energy. In a definite range of parameters the equation agrees satisfactorily with the literature data and those obtained in the present work, and may be used to estimate the dislocation mobility from the dislocation track length. The value of the ratio l/d (d is the indentation diagonal) is constant for ionic crystals but depends on the temperature and the load in the case of covalent crystals. l is connected with the yield strength τ_s as $l\tau_s \approx$ const for covalent crystals ($m \approx 1$), but $l^2\tau_s \approx$ const for ionic crystals ($m \gg 2$).

Es wird eine Näherungslösung für die Abhängigkeit der Versetzungslänge l um einen Mikrohärteeindruck in Einkristallen von der angewandten Last auf den Stempel (P), der Temperatur (T) und der Dauer der Last (t) angegeben: $l \sim P^{m/(2m+1)} t^{1/(2m+1)} \exp[-U/kT(2m+1)]$. Es wird die Annahme gemacht, daß Versetzungen sich vom Stempel in ebenen Anordnungen bewegen und daß die Versetzungsgeschwindigkeit $v \sim \tau^m \exp(-U/kT)$ ist, wobei τ eine effektive Spannung und U die Aktivierungsenergie ist. In einem definierten Bereich von Parametern stimmt die Gleichung befriedigend mit Literaturwerten überein und solchen, die in der vorliegenden Arbeit erhalten werden, und kann zur Berechnung der Versetzungsbeweglichkeit aus der Länge des Versetzungspfades benutzt werden. Der Wert des Verhältnisses l/d (d ist die Eindruckdiagonale) ist für Ionenkristalle konstant, hängt jedoch von der Temperatur und Last im Fall der kovalenten Kristalle ab. l ist mit der Flußstärke τ_s wie $l\tau_s =$ const für kovalente Kristalle ($m \approx 1$), jedoch wie $l^2\tau_s \approx$ const für Ionenkristalle ($m \gg 2$) verknüpft.

1. Introduction

The indentation and etching of single crystals with the aim to reveal dislocations leads usually to the appearance of so-called dislocation "rosettes", that is a sharp increase in the dislocation density around the impression with a symmetry centre in the very point.

The character of the dislocation distribution in the rosette and the length of dislocation tracks have been used to estimate the dislocation mobility in crystals and to evaluate strength properties of crystals. Investigations on semiconductors with a diamond lattice have revealed dislocation loops expanding from a surface in {111} planes [1]. If the hardness is measured on the (111) plane the glide planes intersect it in the ⟨110⟩ direction and the etching shows the existence of dislocations (typically 60°) on the surface as "rays" spreading from the impression (Fig. 1).

Dislocation rosettes have been studied on Ge [1 to 6], Si [3, 7], and $A^{III}B^{V}$-type crystals [8, 10]. Most of the above-cited work deals with the effect of the temperature on the indentation track, l.

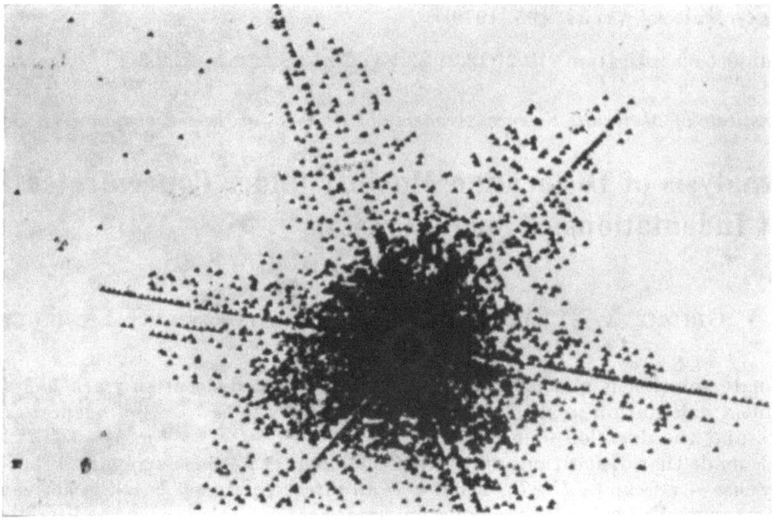

Fig. 1. A typical dislocation rosette around a microhardness impression on a Ge single crystal. Indenter load $P = 2.34$ N, $T = 750$ °C (50 ×)

It is shown in [6, 8, 9] that l is an exponential function of temperature which grows with the stress P applied to the indenter as $l \sim P^n$, where n differs only slightly from $\frac{1}{3}$ (see [6]).

In contrast to semiconductors the hardness and dislocation track length in ionic crystals were measured [11 to 17] at room temperature, the dependence $l(P)$ being approximated by a power function with $n \approx 0.5$ [16] (experimental data [14] give $n \approx 0.57$). For ionic crystals a relationship was clearly established between the length of dislocation tracks, l, and the yield strength, τ_s, in single crystals [11, 16] l decreases monotonously as τ_s rises.

No attempts have been made to find the relation between l and the dislocation mobility. The knowledge of such an even approximate relation would be a great help in the study of the dislocation dynamics because the method is very simple.

The aim of the present work is the development of theoretical arguments which would allow to estimate the dislocation mobility from the length of their tracks. The obtained relation is proved by literature data and some additional experiments on covalent single crystals.

A theoretical justification of such an estimate was tried in [4, 6, 18, 19]. The subject of [6, 18, 19] was the motion of a single dislocation in the field of elastic forces produced by the indenter, but without allowance for the interaction between the dislocations. In [4] the motion of a leading dislocation was considered in the stress field produced by other dislocations, but external forces were ignored.

It is easy to see that both these cases are only rough approximations since the motion of a leading dislocation, and hence the dislocation track length l, depend on the superposition of forces produced by the indenter and the interaction of the elastic fields of dislocations.

We shall consider the motion of a leading dislocation at the head of a dislocation pile-up originating from a single source on the indenter surface. An approximate solution for this case will be obtained with the use of the results of [18, 20 to 22].

2. Equation of the Motion of a Leading Dislocation

The stress acting on the leading dislocation is given as

$$\tau = \tau_i + \Delta\tau \,, \tag{1}$$

where τ_i is the stress created directly by the indenter under the load P; $\Delta\tau$ is the stress produced on the leading dislocations in the pile-up.

When the dislocation pile-up is aligned along a straight line through the centre of the hardness impression and when we deal with the motion of dislocations situated not very far from the surface, τ_i may be given as [6, 18, 21, 23]

$$\tau_i = \frac{AP}{r^2} \,, \tag{2}$$

where r is the spacing from the leading dislocation to the indenter, $A = (1 - 2\mu)/\pi\sqrt{6}$, μ Poisson's ratio, $\Delta\tau$ may be calculated from

$$\Delta\tau = \frac{Gb\,\{[\cos^2\alpha\,(1-\mu) + \sin^2\alpha]\}}{2\pi(1-\mu)} \sum_{i=2}^{n} \frac{1}{r - r_i} \,. \tag{3}$$

Here G is the shear modulus, b the magnitude of the Burgers vector, α the angle made by the dislocation line and its Burgers vector, and n the dislocation number in a given pile-up (it increases with time).

It follows from the results in [21, 22] that the motion of the dislocation at the head of a dislocation pile-up originating from one source may be treated as the motion of a single dislocation in the effective stress field τ whose magnitude is proportional to the stress τ_a applied in the slip plane. The velocity of a dislocation moving at the head of a pile-up may be much higher than that of a single dislocation. Numerical calculations in [22] show that in a very short time interval (≈ 0.01 s) after the constant stress τ_a was applied along the slip plane, the leading dislocation attains a constant velocity equivalent to the motion of a single dislocation under the constant effective stress defined by (1).

$\Delta\tau$ is independent of the sensitivity of the dislocation velocity with respect to the stress and may be found as [22]

$$\Delta\tau = \frac{3}{4}\left(\cos\alpha + \frac{\sin\alpha}{1-\mu}\right)(\tau_a - \tau_b) \,, \tag{4}$$

where τ_b is the source activation stress.

Our case differs from that considered in [22] by the fact that the stress along the slip plane is variable and is given by (2). Therefore, for τ_a in (4) one may, in a first approximation, substitute the average value of the stress along the slip plane τ_{av} which is estimated as

$$\tau_{av} = \frac{\int_{l_1}^{l} \tau_i \, dr}{l - l_1} \,.$$

Here l is the dislocation track length, l_1 some effective value of l beyond which dislocations affect the leading one.

Substitution of τ_i from (2) and integration yield

$$\tau_{av} = \frac{l}{l_1} \tau_i = \beta\tau_i \,, \quad \text{where} \quad \beta > 1 \,.$$

We shall also put $\tau_b = 0$ since dislocation generation near the indenter in the region of high stress concentration occurs without difficulty.

Making use of (1), (2), (4) we get

$$\tau = \frac{ANP}{r^2}, \tag{5}$$

where

$$N = 1 + \frac{3}{4}\beta\left(\cos\alpha + \frac{\sin\alpha}{1-\mu}\right).$$

If the dislocation motion is controlled by a single thermally activated process with symmetrical potential barrier, the dislocation velocity v in the presence of the stress τ is approximately given by the relation [24]

$$v = v_0 \exp\left(-\frac{U}{kT}\right) \sinh\frac{V\tau}{kT}. \tag{6}$$

Here v_0 is a frequency factor, U the activation energy, and V the activation volume.
In some cases the dislocation velocity may be conveniently described as

$$v = B\left(\frac{\tau}{\tau_0}\right)^m \exp\left(-\frac{U}{kT}\right), \tag{7}$$

where $\tau_0 = 10^7$ N/m^2 (kp/mm^2), B and m are material parameters. Comparison of (6) and (7) shows that m may depend on temperature; it grows with decreasing temperature.

For covalent crystals the m-value found in some experiments at high temperatures differs weakly from unity (see [25]), but at room temperature a very strong $v(\tau)$ dependence is characteristic of ionic crystals and the m-value may reach 20 to 30 [26, 27].

An intermediate position is taken by refractory b.c.c. transition metals (Cr, Mo, W, Nb) for which direct velocity measurements at 20 °C have given $m \approx 5$ to 6 [20] while according to the velocity dependence of the yield strength the m-value may be 7 to 20 [28].

Substituting τ from (5) into (7) we get

$$v = \frac{dr}{dt} = \frac{B(AN)^m P^m}{\tau_0^m r^{2m}} \exp\left(-\frac{U}{kT}\right). \tag{8}$$

On integrating we find

$$\int_0^l r^{2m}\,dr = \int_0^t B(AN)^m \left(\frac{P}{\tau_0}\right)^m \exp\left(-\frac{U}{kT}\right) dt\ . \tag{9}$$

During integration we ignore the time required to load and unload the indenter as compared to the time when the indenter is under load, t. The size of the impression diagonal, d, is also neglected compared to l.

It follows from (9) that

$$l = CP^{m/(2m+1)}\, t^{1/(2m+1)} \exp\left[-\frac{U}{kT\,(2m+1)}\right], \tag{10}$$

where

$$C = [(2m+1)\,B]^{1/(2m+1)} \left(\frac{AN}{\tau_0}\right)^{m/(2m+1)}.$$

3. Results and Discussion

On the basis of this equation we shall now consider the dependence of the dislocation track length, l, and the dislocation velocity on the parameters involved in (10), as well as the relation between l and the crystal strength.

We have carried out a series of additional experiments to estimate the correction of the obtained equation. The experiments were made on covalent crystals with characteristics given in the figure captions. The Vickers indenter was used, the loading was made in the apparatus for hot hardness [10]; the dislocations were revealed by chemical etching in accordance with [7, 10].

3.1 The effect of the load value on the indenter, P

It follows from (10) that $l \sim P^n$. Since $m \geqq 1$ the relation $n = m/(2m + 1)$ yields $\frac{1}{3} \leqq n < \frac{1}{2}$. For covalent crystals at sufficiently high temperatures and $m \approx 1$ the condition $n = \frac{1}{3}$ must therefore be fulfilled. Such a dependence has been observed in [8] where GaAs single crystals with alloying additions of various types were studied. The same result is deduced from the data obtained by Lange et al. [9] on InSb, and from the data of the present investigation on Ge and Si single crystals (Fig. 2). But for ionic crystals where m is large, some authors [16, 29] have shown that $n = 0.5$ (Fig. 3).

3.2 The effect of the temperature of indentation

According to (10), an exponential dependence l on T^{-1} is experimentally observed for Si, Ge, and InSb single crystals (Fig. 4). From (10) and the measured data in Fig. 4 with the assumptions $m = 1$, $\beta = 2$, $\alpha = 60°$ (that is for the case of the motion of 60° dislocations) the activation energy for dislocation motion, U, and the constant B in (7) for the dislocation velocity were found. Table 1 summarizes the results for single crystals with low densities of grown-in dislocations. As may be seen the agree-

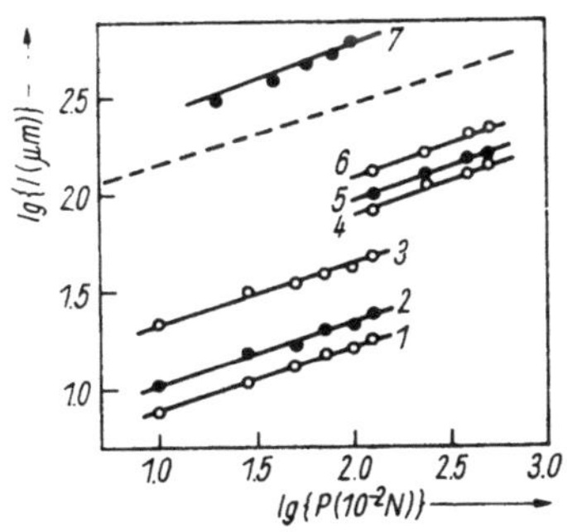

Fig. 2

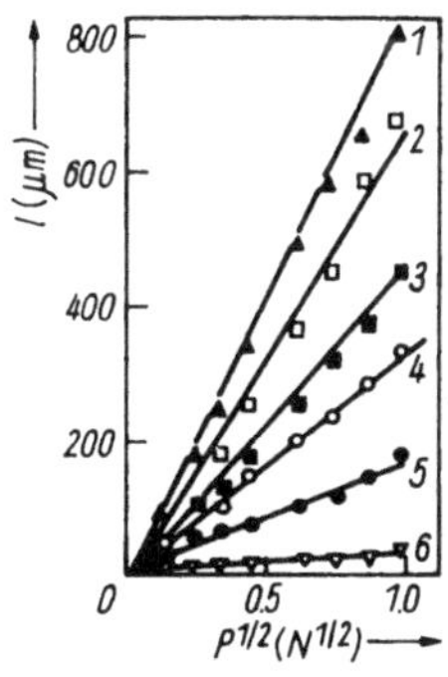

Fig. 3

Fig. 2. Dislocation track length, l, versus indenter load, P. (1) GaAs alloyed with Te, n-type, $n = 40 \times 10^{18}$ cm^{-3}, $T = 320$ °C [8]; (2) GaAs alloyed with Zn, p-type, $n = 1.5 \times 10^{19}$ cm^{-3}, $T = 320$ °C [8]; (3) GaAs non-alloyed, n-type, $n = 2 \times 10^{15}$ cm^{-3}, $T = 320$ °C [8]; (4) Ge alloyed with As, n-type, $n = 3.4 \times 10^{19}$ cm^{-3}, $T = 500$ °C; (5) Ge non-alloyed, n-type ,$\varrho = 40\,\Omega$ cm, $T = 400$ °C; (6) Si non-alloyed, n-type, $\varrho = 20\,\Omega$ cm, $T = 700$ °C; (7) InSb non-alloyed, n-type, $\varrho = 0.005\,\Omega$ cm, $T = 250$ °C [7]. Dashed line refers to l = const $P^{1/3}$

Fig. 3. Dislocation track length, l, versus indenter load, P, for (1) KBr, (2) KCl, (3) NaCl single crystals, and three LiF crystals with different yield strength: (4) $\tau_s = 10^6$, (5) 7×10^6, (6) $7.7 \times \times 10^7$ N/m^2. Data from [16]

Table 1

material	U (eV) from direct measurements of dislocation velocity	U (eV) from temperature dependence of l (Fig. 4) and equation (10) for $m = 1$	U (eV) from temperature dependence of l/d (Fig. 8) and equation (19)	B (mm/s) from the data in Fig. 4 and equation (10) with $\alpha = 60°$, $m = 1$, $\beta = 2$	B (mm/s) from literature data
Si	2.2 [34] 2.2 [35]	2.24	2.32	1.35×10^8	—
Ge	1.6 [34] 1.47 screw dislocation [36] 1.49 to 2.25 60° dislocation 1.62 [36] 1.47 60° dislocation [38] 1.55 [38] 1.45 screw dislocation 1.64 60° dislocation [39]	1.64	1.61 1.64 (data in [4])	0.8×10^8	(1 to 0.5) $\times 10^8$ [37]
InSb	0.7 [34] 0.72 to 0.93 [40] 1.6 In dislocation [41] 1.3 Sb dislocation	0.6	0.86	0.9×10^5	—

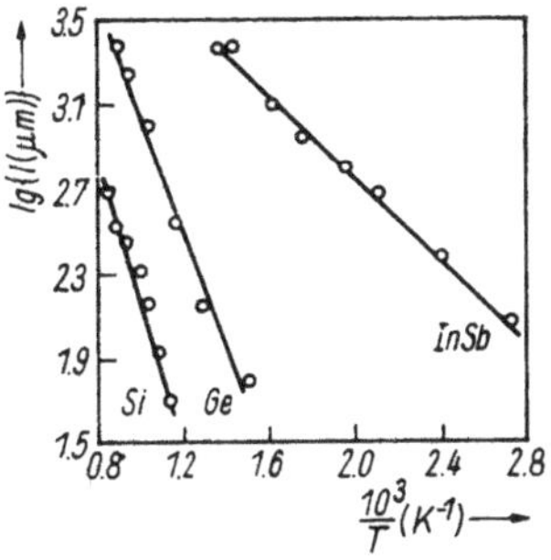

Fig. 4. Dislocation track length, l, around a hot hardness impression versus inverse temperature. Single crystal samples. Indentation on the (111) plane, $P = 2.34$ N, $t = 15$ s. Dislocation-free Si, n-type, $\varrho = 20\ \Omega$ cm; dislocation-free Ge, n-type, $\varrho = 40\ \Omega$ cm; InSb, initial dislocation density 10^2 cm^{-2}, n-type, $\varrho = 0.005\ \Omega$ cm

ment between these results and the values obtained by direct measurements of the dislocation velocity is satisfactory. It follows from (10) that for large m the dependence $l(T)$ must be weak. This is true for most b.c.c. transition metals and ionic crystals when microhardness indentations are made at temperatures close to room temperature

3.3 *The effect of the time under load*

For covalent crystals (10) gives $l \sim t^{1/3}$ for $m \approx 1$, but $l \sim t^{1/2m}$ for large m, that is the $l(t)$ dependence, e.g. for ionic crystals, must be relatively weak. Experimental data on the $l(t)$ dependence may in principle be used to determine the parameter m. However, it is difficult to obtain reliable experimental data on $l(t)$, because in deducing (10) for small t we ignored the time of loading and, besides this, for large t one must consider the possibility that dislocations may be pinned by impurities and that long-range stress fields may be created by other dislocation arrays in the crystal. Our experimental investigations of the $l(t)$ behaviour in dislocation-free silicon (Fig. 5) suggest that the relation $l \sim t^{1/3}$ is satisfactorily fulfilled only up to a certain time t_{cr}. For 700 °C, $t_{cr} = 200$ s and decreases with temperature loading. Beyond t_{cr} the dependence $l(t)$ becomes weaker than that expected from (10).

When dislocation tracks are long and the stress from the indenter is low, the presence of a back stress due to the long-range fields produced by dislocations in other rays of the rosette as well as the impurity pinning effect may be taken into account by the assumption that dislocations move under the action of some effective stress $\tau_{eff} = \tau - \tau_T$ (where $\tau_T \approx$ const refers to an athermal resistance to dislocation motion). Substitution of this expression into (7) makes the integration more difficult and the result more complex. In the special case $m = 1$, a simple analytical solution has been obtained:

$$t = -\frac{l}{Q\tau_T} + \frac{l_0}{2Q\tau_T} \ln \frac{1 + l/l_0}{1 - l/l_0}, \tag{11}$$

where

$$Q = \frac{B}{\tau_0} \exp\left(-\frac{U}{kT}\right); \quad l_0 = \sqrt{\frac{ANP}{\tau_T}}.$$

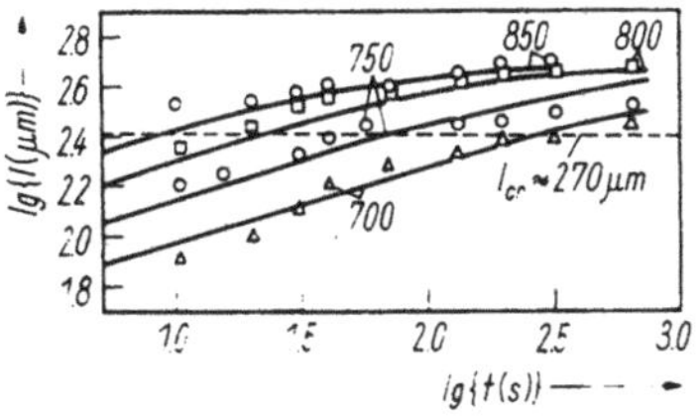

Fig. 5. Dislocation track length, l, versus loading time for Si dislocation-free single crystal. Numbers refer to temperature (in °C) at which the indentations were made. Solid lines are calculated from (11)

For the case where l/l_0 is small (that is τ_T is small or l is small) expression (11) leads (on series expansion $[\ln (1 + x)/(1 - x) \approx 2 (x + x^3/3)]$) to the dependence

$$t = \frac{1}{3} \frac{l^3}{QANP} \tag{12}$$

which coincides completely with (10) for $m = 1$. Thus for small l the obtained result does not depend on τ_T. However, for long periods under load (for long t) and large l, equation (11) leads to a weaker $l(t)$ dependence in accord with the experimental data in Fig. 5 where curves are also given as calculated from (11) for silicon. Here $U = 2.2$ eV, $B = 1.35 \times 10^8$ mm/s (see Table 1), $\tau_T = 2 \times 10^6$ N/m², $P = 2.34$ N according to the experiment; $AN = 0.2$ (at $\beta = 2$, $\mu = 0.3$, $\alpha = 60°$).

The value of τ_T is chosen in such a way that through the whole range of l-values the condition $\tau_{eff} > 0$ is satisfied and $l/l_0 < 1$.

Theoretical curves for various temperatures in Fig. 5 may be deduced from each other by simple shifting parallel to the abscissa axis. In fact, from (11) we find

$$\lg t_2 = \lg t_1 + \lg \frac{Q_1}{Q_2} = \lg t_1 + \frac{0.434\, U}{k} \left(\frac{1}{T_2} - \frac{1}{T_1}\right), \tag{13}$$

where t_2 and t_1 are the periods under load until the same value of l is reached at the temperatures T_2 and T_1, respectively. This means that the value of l_{cr} at which a deviation from the simple relationship given by (10) and (12) is observed must not depend on temperature. An analysis of the experimental data according to (10) using the above ideas should be carried out for $l < l_{cr}$.

3.4 The velocity of the leading dislocation

Differentiation of (10) yields an expression for the leading dislocation velocity

$$v_l = \frac{dl}{dt} = \frac{C}{2m + 1} P^{m/(2m+1)} t^{-2m/(2m+1)} \exp\left[-\frac{U}{kT(2m + 1)}\right]. \tag{14}$$

The dependence of v_l on load and temperature must be the same as for l, but the time dependence will be different. From (14) we get $v_l \sim t^{-n}$, where $\frac{2}{3} \leqq n < 1$. In particular, for ionic crystals $n \to 1$. The instantaneous velocity of the leading dislocation was determined in [15] for LiF single crystals. The analysis of these experimental data actually gives $v_l \sim t^{-1}$ for not very long periods of loading (Fig. 6). When these periods are large ($t_{cr} > 100$ s for the data in [16]) the velocity of the leading dislocation decreases more rapidly than is expected from 14).

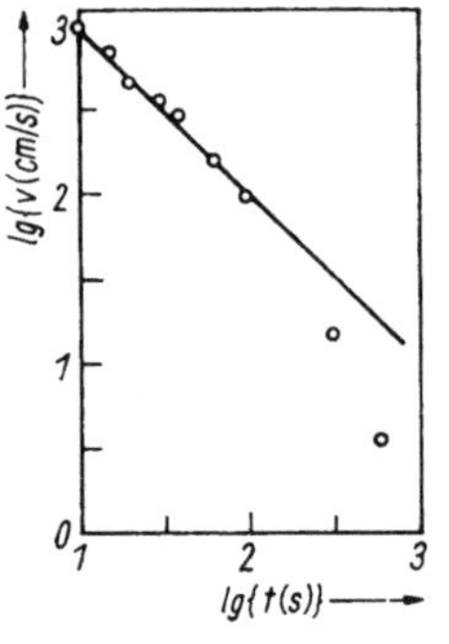

Fig. 6. The change in the instantaneous velocity of the leading dislocation versus indentation period for a LiF single crystal. The straight line corresponds to (11), the points refer to experimental data [16]

Hence in ionic crystals, as well as in covalent ones, at long loading periods dislocation motion is decelerated compared to the velocity calculated from (10). But for $l < l_{cr}$ and $t < t_{cr}$ this equation gives an appropriate description of the $l(t)$ dependence. It must be said that a significant change in the indentation temperature can modify the character of the dislocation rosettes and the slip planes operative at indentation [30]. The above relations describe naturally the change in dislocation track length with indentation temperature, indentation load, and loading period, provided the dislocation rosettes in a given range of parameters (temperature, load, time) are of the same character.

3.5 Correlation of dislocation track length, l, with strength characteristics

If we assume that the dislocation velocity is given by (7) in accordance with Haasen's theory [25, 31], the yield strength, τ_s, may be described as

$$\tau_s = \text{const}\ \dot{\varepsilon}^{1/(2+m)} \exp\left[\frac{U}{kT(m+2)}\right], \tag{15}$$

where $\dot{\varepsilon}$ is the strain rate.

Comparison of (10) and (15) for covalent crystals with $m \approx 1$ yields

$$[\tau_s]_{\dot{\varepsilon}}\ [l]_{P,t} = \text{const}\ . \tag{16}$$

Hence at constant ε, P, t, the product $\tau_s l$ should be independent of the measurement temperature and the activation energy U. This relation is to some extent supported by the fact that the activation energies found from the temperature dependence of the critical shear stress [25] and the temperature dependence of l practically coincide (see Table 1).

For a rather wide temperature range it may be assumed that $H_V \sim \tau_s$. In this case the relation

$$H_V l \approx \text{const} \tag{16a}$$

must be satisfied.

Table 2

Si	T (°C)	650	700	800	900
	$H_V l$ (N/m)	5×10^5	4.5×10^5	5×10^5	4.4×10^5
Ge	T (°C)	400	500	600	700
	$H_V l$ (N/m)	3.2×10^5	3.8×10^5	4.8×10^5	10×10^5
InSb	T (°C)	240	290	355	430
	$H_V l$ (N/m)	5.4×10^5	5.3×10^5	5.0×10^5	4.6×10^5

The results of the present investigation given in Table 2 show that (16a) holds well enough. At high temperatures a deviation from this dependence is observed (see the data for Ge) due perhaps to the presence of an athermal component of the flow stress and the hardness, this component becoming important at high temperatures. We believe that alloying and increasing defect density in the crystal cause a departure from (16) at high temperatures.

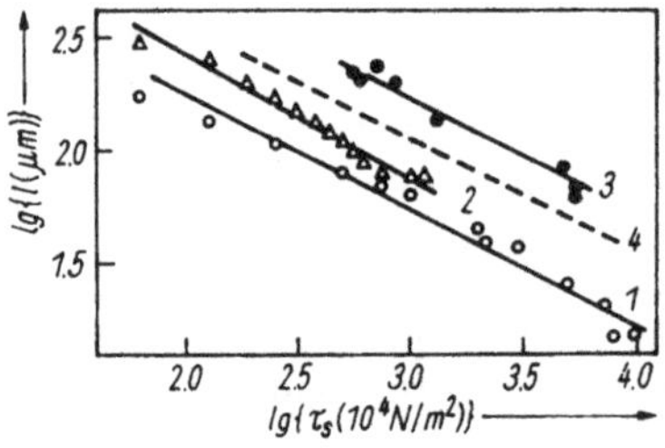

Fig. 7. l versus yield strength, τ_s, for (1) LiF, (2) NaCl, KCl, KBr single crystals [16], and for (3) LiF from [11]. The dashed line (4) corresponds to the relation $l^2\tau_s =$ const

When the temperature is low the approach to the athermal part in the curve $H(T)$ does not make the hardness reflect the change in the yield strength [32, 33], and therefore (16a) also fails.

Comparison of (10) and (12) for large m gives

$$[\tau_s]_{\dot\varepsilon}\, [l]^2_{P,t} = \text{const}\,. \tag{17}$$

To our knowledge there are no experimental data on $l(T)$ for ionic crystals with simultaneous evaluation of their yield strength. But in [11, 16] the relation between τ_s and l was studied for LiF, NaCl, KCl, KBr single crystals of different "rigidity", that is with different yield stress due to the presence of radiation defects and various impurity concentrations. All the experiments were done at 20 °C. $\dot\varepsilon$, P, and t were kept constant. It follows from the data obtained in [11, 16] that (17) is well fulfilled for all the crystals studied (Fig. 7).

3.6 Correlation between the dislocation track length, l, and the indentation diagonal, d

The obtained results permit to explain the well-known experimental fact that l/d is constant for ionic crystals [14, 15].

Actually from the definition of the hardness, H_V, it follows

$$d = \text{const}\ \frac{P^{1/2}}{H_V^{1/2}}\,. \tag{18}$$

If to put $H_V \sim \tau_s$ and use (12), we get

$$d = \text{const}\ \dot\varepsilon^{1/2(2+m)} P^{1/2} \exp\left[-\frac{U}{2(m+2)\,kT}\right]. \tag{18a}$$

The comparison of this expression with (10) for the case $m \gg 2$ yields $l/d =$ const. In other words, l/d is independent of indenter load, temperature and activation energy U for ionic crystals. But in [4] an experimental evidence is given for the dependence of l/d on load P and temperature T for Ge.

Keeping in mind that $m \approx 1$ for covalent crystals and comparing (10) with (18a) we obtain

$$\frac{l}{d} = \text{const}\ P^{-1/6} \exp\left(-\frac{U}{6kT}\right). \tag{19}$$

The experimental results in [3, 4] and the present study for covalent crystals are in satisfactory agreement with (16) (Fig. 8, 9). Really, lg (l/d) depends linearly on the inverse temperature for Si, Ge, InSb, and the values of the activation energy U, obtained from the analysis of the straight lines in Fig. 8 with the use of (16), differ little from those obtained from determinations of the dislocation velocity by other methods (Table 1). The l/d ratio decreases with the load on the indenter, P (Fig. 9), the exponent k in $l/d \sim P^k$ being close to the calculated value, $k = -\frac{1}{6}$.

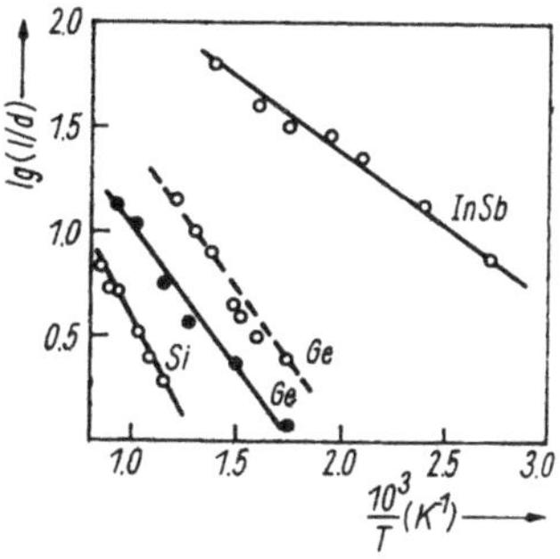

Fig. 8

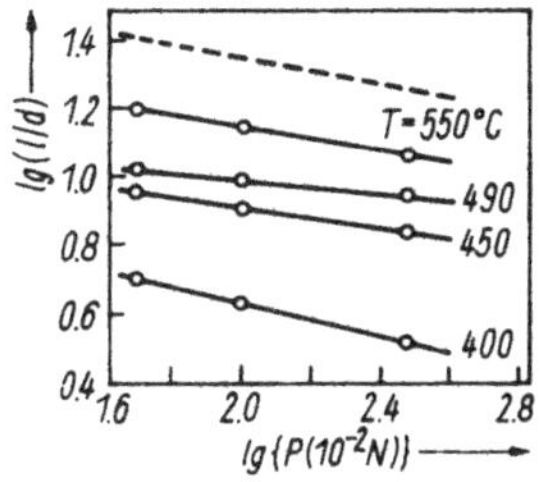

Fig. 9

Fig. 8. lg (l/d) versus inverse temperature for covalent crystals. Solid line corresponds to the results of the present study, dashed line refers to the data from [4]

Fig. 9. lg (l/d) versus lg P for Ge single crystal [4]. Numbers refer to the temperature at which the indentation is made. The dashed line corresponds to (16)

4. Conclusions

Thus the obtained approximate solution for the dependence of the dislocation track length, l, around a microhardness impression on the indenter load, P, indentation duration, t, and temperature, T, describes satisfactorily the available experimental data in a certain range of parameters and may be used to estimate the dislocation mobility in crystals and to explain some experimentally found regularities.

Acknowledgement

The authors would like to acknowledge the helpful criticism by A. I. Landau of the Physico-Technical Institute of Low Temperatures, Academy of Sciences of the Ukrainian SSR, Kharkov.

References

[1] V. I. Trefilov, L.I. Datsenko, Yu. V. Milman, M. Ya. Skorokhod, and I. V. Gridneva, Sb. Metallofizika, Vol. 24, Izd. Naukova dumka, Kiev 1968 (p. 18 7).
[2] V. I. Trefilov and Yu. V. Milman, Dokl. Akad. Nauk SSSR **153**, 824 (1963).
[3] V. I. Trefilov and Yu. V. Milman, Sb. Voprosy fiziki metallov i metallovedeniya, Vol. 17, Izd. Akad. Nauk Ukr. SSR, Kiev, 1963 (p. 45).
[4] K. Sumino and H. Hasegawa, Proc. Internat. Conf. Strength of Metals and Alloys, Tokyo 1967; Trans. Japan. Inst. Metals 9, 1968 (749).
[5] S. S. Gorelik, M. G. Lozinskii, Yu. S. Kleinfeld, and Yu. M. Litvinov, Izv. Akad. Nauk SSSR, Ser. neorg. Mater. **4**, 984 (1968).
[6] L. I. Datsenko, I. V. Gridneva, Yu. V. Milman, Yu. N. Petrov, V. I. Trefilov, and S. I. Chugunova, Ukr. fiz. Zh. **15**, 4 (1970).
[7] I. V. Gridneva, Yu. V. Milman, and V. I. Trefilov, Sb. Issledovaniya v oblasti izmereniya tverdosti, Trudy meteorologicheskikh institutov SSSR, Vol. 91 (151), Moskva/Leningrad 1967 (p. 151).
[8] M. G. Milvidskii, V. B. Osvenskii, and O. G. Stolyarov, Kristallografiya **12**, 1967 (1030).
[9] V. N. Lange, T. I. Lange, and S. I. Simashko, Sb.Issledovaniya po poluprovodnikam, Akad. Nauk Moldavskoi SSR, Izd. Kartya Moldovenyaske, Kishinev 1968.
[10] I. V. Gridneva, Yu. V. Milman, and V. I. Trefilov, Sb. Nesovershenstva kristallicheskogo stroeniya, Izd. Naukova Dumka, Kiev 1968.
[11] M. P. Shaskolskaya and G. F. Dobrzhanskii, Kristallografiya **7**, 103 (1962).
[12] J. J. Gilman and W. G. Johnston, Proc. Lake Placid Conf., Lake Placid (USA) 1957 (p. 116).
[13] S. A. Varcheniya, F. O. Muktepavel, and G. P. Upit, phys. stat. sol. (a) **1**, K165 (1970).

[14] S. A. VARCHENIYA, F. O. MUKTEPAVEL, and G. P. UPIT, Fiz. tverd. Tela **11**, 2841 (1969).
[15] A. A. SHPUNT and O. A. NABUTOVSKAYA, Raboty po fizike tverdogo tela, Vol. 1, Novosibirsk 1967 (p. 161).
[16] A. M. SOIFER, M. G. BURAVLEVA, and Z. A. SHCHEGOLEVA, Ukr. fiz. Zh. **15**, 1107 (1971).
[17] I. V. STRATAN, A. A. PREDVODITELEV, and V. M. STEPANOVA, Fiz. tverd. Tela **12**, 767 (1970).
[18] B. RÖSNER, Acta Phys. Acad. Sci. Hungar. **20**, 289 (1966).
[19] S. M. HU, J. appl. Phys. **46**, 1470 (1975).
[20] A. R. ROSENFIELD and M. F. KANNINEN, Phil. Mag. **22**, 175, 143 (1970).
[21] E. SMITH, Phil. Mag. **16**, 144, 1285 (1967).
[22] A. R. ROSENFIELD and G. T. HAHN, Dislocation Dynamics, Mac Graw-Hill Book Co., New York 1968 (p. 255).
[23] M. M. FILONENKO-BORODICH, Teoriya uprugosti, Izd. fiz.-mat. Lit., Moskva 1959.
[24] YU. V. MILMAN and V. I. TREFILOV, Sb. Metallofizika, Mekhanizm razrusheniya metallov, Izd. Naukova Dumka, Kiev 1964 (p. 59).
[25] H. ALEXANDER and P. HAASEN, Solid State Phys. **22**, 28 (1968).
[26] W. G. JOHNSTON and J. J. GILMAN, J. appl. Phys. **30**, 129 (1959).
[27] E. YU. GUTMANAS and E. M. NADGORNYI, Fiz. tverd. Tela **10**, 2285 (1968).
[28] DZH. G. BUT, R. I. DZHAFFI, and E. I. SALKOVITS, Sb. Tugoplavkie metallicheskie materialy dlya kosmicheskoi tekhniki, Izd. Mir, Moskva 1966 (p. 315).
[29] YA. E. GEGUZIN, Sb. Fizika shchelochno-galloidnykh kristallov, Izd. Latviiskogo Universiteta, Riga 1962 (p. 468).
[30] YU. S. BOYARSKAYA, S. S. SHUTOVA, and R. P. ZHITARU, in: Deformirovanie kristallov pri deistvii sosredotochennoi nagruzki, Izd. Shtintsa, Kishinev 1978.
[31] P. HAASEN, Z. Phys. **167**, 461 (1962).
[32] I. V. GRIDNEVA, YU. V. MILMAN, and V. I. TREFILOV, phys. stat. sol. (a) **14**, 177 (1972).
[33] V. I. TREFILOV, YU. V. MILMAN, and S. A. FIRSTOV, Fizicheskie osnovy prochnosti tugoplavkikh metallov, Izd. Naukova Dumka, Kiev 1975.
[34] A. R. CHAUDHURI, J. R. PATEL, and L. G. RUBIN, J. appl. Phys. **33**, 2736 (1962).
[35] V. I. NIKITENKO, V. N. EROFEEV, and N. M. NADGORNAYA, Sb. Dinamika dislokatsii, FTINT Akad. Nauk Ukr. SSR, Kharkov 1968 (p. 34).
[36] M. N. KABLER, Phys. Rev. **131**, 54 (1963).
[37] S. SCHÄFER, phys. stat. sol. **19**, 297 (1967).
[38] S. S. GORELIK and YU. M. LITVINOV, Fiz. Khim. obrabotki Materialov **2**, 131 (1968).
[39] H. SCHAUMBURG, phys. stat. sol. **40**, K1 (1970).
[40] YU. A. OSIPYAN and S. A. EROFEEVA, Fiz. tverd. Tela. **11**, 944 (1969).
[41] M. MIHARA and T. NINOMIJA, J. Phys. Soc. Japan **25**, 1198 (1968).

(Received March 16, 1979)

phys. stat. sol. (a) 54, 207 (1979)

Subject classification: 14.3.1; 13.4; 22.4.2

Laboratory of Electronics, Ghent State University[1])

Barrier-Limited Conductivity in Thin Semiconducting Films

By

A. VAN CALSTER

It is shown that for the calculation of polycrystalline field-effect thin film structures, such as thin film transistors, one may use, to a good approximation, the homogeneous field-effect calculations given in the literature. The polycrystalline nature of the film is described by introducing in the calculations a barrier-limited mobility, which is characterized by an activation energy.

Es wird gezeigt, daß man für die Berechnung von polykristallinen Feldeffekt-Dünnschichtstrukturen, wie Dünnschichttransistoren, in guter Näherung die in der Literatur angegebenen homogenen Feldeffektberechnungen benutzen kann. Die polykristalline Natur der Schicht wird durch die Einführung einer barrierenbegrenzten Beweglichkeit beschrieben, die durch eine Aktivierungsenergie charakterisiert ist.

1. Introduction

Several authors [1 to 3] have investigated the influence of intercrystalline boundaries on the transport parameters of thin polycrystalline semiconducting films. They concluded that the conductivity was limited by potential barriers at the intergrain boundaries. The potential barriers are due to the depletion layers caused by trapping of the charge carriers at the grain boundaries. On the other hand, an impurity concentration of 10^{17} cm^{-3} gives an average spacing between the impurity atoms of 216 Å, which is of the same order as the grain size of very thin films [4, 5]. Thus the potential will not only fluctuate along the film due to the trapping at the intergrain boundaries, but also due to the impurity distribution. Normally experimentally measured quantities involve large areas, so that often the fluctuations of the potential about the average, due to traps and impurities, may be neglected [6]. This gives rise to the question whether the potential barriers at the boundaries caused by traps may be neglected in thin films when large areas are involved.

Another problem which has to be examined in thin films is the fact that the band gap at the grain boundary is different from the band gap in the grain [7], due to elastic strain and/or due to incorporated inhomogeneities at the grain boundaries [8 to 10].

The purpose of this paper is to examine to which extent the potential barriers and band-gap changes at the grain boundaries influence the conduction properties of thin films. The results of our calculations are compared with the experiments.

2. The Conductivity of a Thin Semiconducting Film

Before calculating the conductivity, we make the following assumptions:

1. The polycrystalline film is made of more or less cubic grains, as shown in Fig 1. The average grain size is equal to the film thickness.

2. The polycrystalline film can be described by a band structure which resembles the band structure of a single crystal inside the grain, and which shows a band-gap widening at the grain boundaries caused by elastic strain and incorporated inhomogeneities [7].

3. It is accepted that the theory, developed by Marshak and Van Vliet[9, 10], can be applied to calculate the current through the film.

[1]) 9000 Ghent, Belgium.

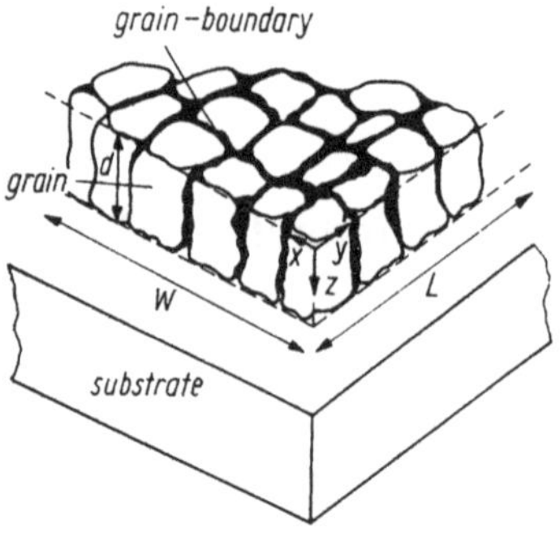

Fig. 1. Model of the polycrystalline film

4. The effect of the film thickness on the mobility is neglected and only one type of charge carriers (e.g. electrons) takes part in the conduction process. All minority effects, such as inversion layers, will be ignored.

Although all the assumptions made above cannot be justified properly, we believe that the results obtained have a quality validity.

Marshak and Van Vliet [9, 10] showed that, even for a position-dependent band structure, the electron current J_e is given by the relation

$$J_e = n\mu_e \nabla E_F , \tag{1}$$

where n the electron concentration, μ_e the electron mobility, and E_F the quasi-Fermi level of the electrons. The electron concentration n is given by the relation

$$n = \bar{N}_C \exp\left(\frac{E_F - E_C}{kT}\right), \tag{2}$$

where E_C is the bottom of the conduction band and $\bar{N}_C$ the density of states. $\bar{N}_C$ takes the value N_C inside the grain and the value N_C' at the grain boundary. E_C can be written as

$$E_C = E_0 + \Phi - \chi , \tag{3}$$

where E_0 is the vacuum level of a field-free crystal, χ the electron affinity, and Φ the potential energy. The electron affinity χ has different values inside the grain and at the grain boundary. Φ is composed of

$$\Phi = \Phi_0(\boldsymbol{r}) - qU(\boldsymbol{r}) . \tag{4}$$

Φ_0 is the potential energy of the elastic strain in the grain boundary and $U(\boldsymbol{r})$ the electrostatic potential. If the current J_e flows along the y-direction, E_F will not depend on x and z. From (1) and (2) one obtains

$$J_e = \mu_e \frac{dE_F}{dy} \bar{N}_C \exp\left(\frac{E_F - E_C}{kT}\right). \tag{5}$$

As the semiconducting film is composed of many grains along the y-axis, a voltage applied along the y-axis will give only a small change ΔE_F of the quasi-Fermi level over one grain. If we multiply (5) by $\exp(E_C/kT)/\bar{N}_C$ and integrate over the average grain size D along the y-axis, we find

$$J_e \int_0^D \frac{\exp(E_C/kT)}{\bar{N}_C} dy = \mu_e \exp\left(\frac{E_F}{kT}\right) \Delta E_F . \tag{6}$$

After integrating (6) along x one obtains

$$J_e \int_0^W dx \int_0^D \frac{\exp(E_C/kT)}{\bar{N}_C} dy = \mu_e \exp\left(\frac{E_F}{kT}\right) \Delta E_F W . \tag{7}$$

The dependence of E_C on the position is due to the dependence of the band structure (χ, Φ_0) and the dependence of the electrostatic potential U on the position. As mentioned in the introduction we might ignore the fluctuations of U and use an average electrostatic potential $\bar{U}$. $\bar{U}$ will normally not depend on x and according to the gradual channel approximation [11], the dependence of $\bar{U}$ on y is weak and might thus be neglected over one grain. The evaluation of the following part of (7):

$$\int_0^W dx \int_0^D \frac{\exp(E_C/kT)}{\bar{N}_C} dy \tag{8}$$

is carried out by keeping z or thus $\bar{U}$ constant. First we integrate over the grain size D and sum afterwards all these results over x. The integrand of (8) will thus only change when a grain boundary is crossed. But as we integrate over the average grain size D, only one grain boundary will be crossed on the average when all the results are summed over x. We therefore assume that the difficult evaluation of (8) can be done by replacing the actual film by the average grain of Fig. 2. According to Fig. 2 and relations (3) and (4), (8) becomes

$$W \exp\left(\frac{E_0 - q\bar{U}}{kT}\right) \int_0^D \frac{\exp[(-\chi + \Phi_0)/kT]}{\bar{N}_C} dy . \tag{9}$$

Making use of the mean-value theorem for integrals, (9) becomes

$$WD \exp\left(\frac{E_0 - \chi_0 - q\bar{U}}{kT}\right) \frac{\exp[(-\Delta\chi(\xi) + \Phi_0(\xi))/kT]}{\bar{N}_C(\xi)} , \tag{10}$$

where ξ is an appropriately chosen value of the interval (0, D), χ_0 the electron affinity inside the grain, and $\Delta\chi$ the change in electron affinity at the grain boundary. In the case of a band widening at the grain boundary, $\Delta\chi$ becomes negative. Taking for $\Phi_0 - \Delta\chi$ the simple rectangular energy barrier of Fig. 3, expression (10) becomes

$$W \frac{D}{\bar{N}_C} \exp\left(\frac{E_0 - \chi_0 - q\bar{U}}{kT}\right) \exp\left(\frac{\Phi_b}{kT}\right) , \tag{11}$$

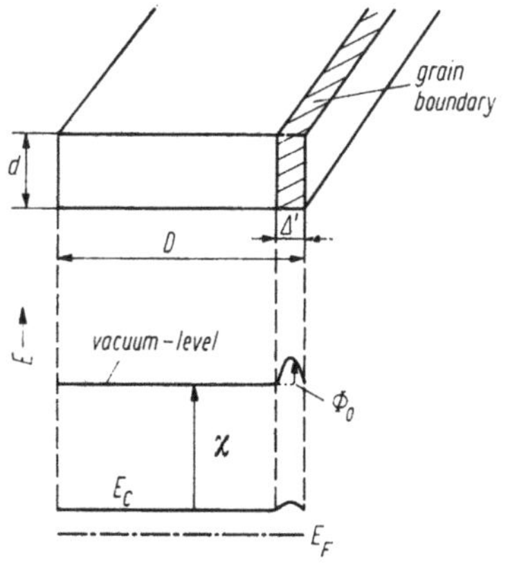

Fig. 2. Band structure of the average grain

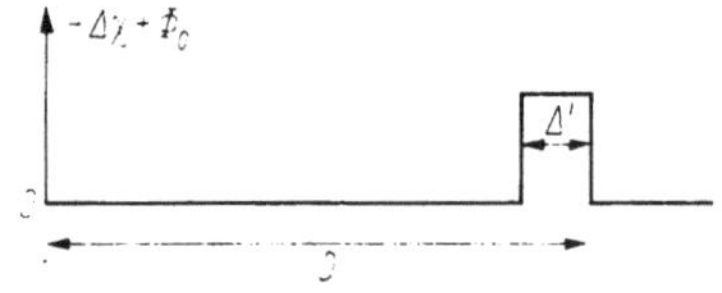

Fig. 3. Rectangular grain barrier

where

$$\exp\left(\frac{\Phi_b}{kT}\right) = 1 + \frac{\Delta'}{D}\left[\exp\left(\frac{\Phi_0 - \Delta\chi}{kT}\right) - 1\right] = \exp\left[\frac{-\Delta\chi(\xi) + \Phi_0(\xi)}{kT}\right]. \tag{12}$$

$\bar{N}_C(\xi)$ is taken for the sake of simplicity equal to N_C. As a numerical example we take a 500 Å thick CdSe with $D = 500$ Å [4, 5] and the grain boundary $\Delta' \approx 20$ Å. For these values Φ_b behaves like an activation energy if $\Phi_0 - \Delta\chi$ is larger than 0.1 eV in the temperature range 0 to 500 K. We may thus conclude that normally $-\Delta\chi(\xi) + \Phi_0(\xi)$ of expression (10) may be considered as the potential barrier Φ_b of the grain boundary. According to (11), expression (7) becomes

$$J_e = \mu_e \exp\left(-\frac{\Phi_b}{kT}\right) n \frac{\Delta E_F}{D}, \tag{13}$$

where

$$n = N_C \exp\left(\frac{E_F - E_0 + \chi_0 + q\bar{U}}{kT}\right). \tag{14}$$

n is the electron concentration in the average grain, calculated with the average potential $\bar{U}$. As the film is made of cubic grains shown in Fig. 1, Φ_b is not expected to depend on z and thus (13) becomes after integrating over z

$$I = \mu_e \exp\left(-\frac{\Phi_b}{kT}\right) Q_e W \frac{\Delta E_F}{D}, \tag{15}$$

where I is the total current and Q_e the total amount of electrons per unit area in the average grain,

$$Q_e = \int_0^d n \, dz \,. \tag{16}$$

Along the y-axis the film contains a large number of grains and thus $\Delta E_F/D$ may be approximated by $\Delta E_F/\Delta y$, with $\Delta y = D$. This means that along y the film is considered to be composed of a large number of sufficiently small sections Δy equal to the average grain size, with an increment ΔE_F of the Fermi level over the distance Δy. Expression (15) becomes thus

$$I = \mu_e \exp\left(-\frac{\Phi_b}{kT}\right) W Q_e \frac{\Delta E_F}{\Delta y}. \tag{17}$$

When only a small voltage U is applied along y, Q_e will not depend on y and thus we find for the conductivity of the film G

$$G = \frac{I}{U} = q\mu^* \frac{W}{L} Q_e, \tag{18}$$

where L the length of the film and $\mu^* = \mu_e \exp(-\Phi_b/kT)$. Volger [12] showed that the measured Hall coefficient R of inhomogeneous material corresponds to the low-resistivity regions (grains), and thus that $R = 1/nq$. With a negligible field effect perpendicular to the surface $Q_e = nd$ and thus $G = \sigma WL/d$, where $\sigma = q\mu^* n$. This means that μ^* has to be considered as a barrier-limited mobility of a polycrystalline film. This barrier-limited mobility is due to the potential barriers at the grain boundaries, caused by elastic strain and/or incorporated inhomogeneities.

If an important field effect is present, as is the case in thin film structures [1, 2, 11], the calculation of the amount of electrons in the grain Q_e becomes somewhat more complicated. According to (16) we have to know the dependence of n on z, or according to (14) the dependence of the average potential $\overline{U}$ on z. This average potential $\overline{U}$ is obtained by solving the Poisson equation

$$\frac{\mathrm{d}^2\overline{U}}{\mathrm{d}z^2} = \frac{q}{\varepsilon\varepsilon_0} (\overline{N}_\mathrm{D} - \overline{N}_\mathrm{A} - n) , \tag{19}$$

where we made use of the above-mentioned gradual channel approximation. $\overline{N}_\mathrm{D}$ and $\overline{N}_\mathrm{A}$ are the average donor and acceptor concentrations. $\overline{N}_\mathrm{D}$ and $\overline{N}_\mathrm{A}$ are related to the real concentrations by the relations

$$\overline{N}_\mathrm{D} = \frac{1}{W} \int\limits_0^W N_\mathrm{D}^+(\boldsymbol{r}) \,\mathrm{d}x , \qquad \overline{N}_\mathrm{A} = \frac{1}{W} \int\limits_0^W N_\mathrm{A}^-(\boldsymbol{r}) \,\mathrm{d}x . \tag{20}$$

The donor and acceptor impurities N_D and N_A are normally distributed in a random way in the film, which means that the average concentrations $\overline{N}_\mathrm{D}$ and $\overline{N}_\mathrm{A}$, calculated from (20) where W is a large quantity, are not expected to depend on y and z. This means that $\overline{N}_\mathrm{D}$ and $\overline{N}_\mathrm{A}$ may be considered uniformly distributed in the yz plane. However, in the case of the frequently used semiconductor CdSe the acceptor states, which act as traps, are assumed to be located at the intergrain boundaries [1, 2]. Nevertheless, the traps are distributed at random in the grain boundaries, which means that also in the case of CdSe $\overline{N}_\mathrm{A}$ may be considered independent of y and z. This means that (19) is the Poisson equation used for the calculation of the field effect in homogeneous MIS structures, and the solution of this equation can be found in the literature [6, 11]. Furthermore we can also refer to the literature [11] for the calculation of the current I when a large voltage U is applied. From (17) I becomes

$$I = \mu^* \frac{W}{L} \int\limits_0^{qU} Q_\mathrm{e} \,\mathrm{d}E_\mathrm{F} . \tag{21}$$

We may summarize that for the calculation of the current I of a thin semiconducting film, we may use the results given in the literature for a homogeneous conducting channel [6, 11], provided the mobility is replaced by the barrier-limited mobility μ^*.

We will now verify our assumption of neglecting the potential fluctuations due to the distribution of donors and acceptors. The Poisson equation one should normally solve is

$$\nabla^2 U = -\frac{\varrho(\boldsymbol{r}, U)}{\varepsilon\varepsilon_0}, \tag{22}$$

where

$$\varrho(\boldsymbol{r}, U) = q[N_\mathrm{D}^+(\boldsymbol{r}) - N_\mathrm{A}^-(\boldsymbol{r}) - n(U)] .$$

N_D^+ and N_A^- can also depend on the potential U, if these levels are incompletely ionized. The dependence of U on x is only due to the impurity and trap distribution, and thus if the fluctuations u of U are small, (22) can be replaced by (19) and (20), if also the gradual channel approximation is used

$$\frac{\partial^2 \bar{U}}{\partial z^2} \gg \frac{\partial^2 \bar{U}}{\partial y^2}. \tag{23}$$

In the Appendix an expression is derived for the calculation of the fluctuations u

$$u(x, y, z) = U - \bar{U} = -\int\limits_x^W \frac{(\varrho - \bar{\varrho})(x' - x)}{\varepsilon\varepsilon_0} \mathrm{d}x' + \frac{1}{2W}\int\limits_0^W x'^2 \frac{\varrho - \bar{\varrho}}{\varepsilon\varepsilon_0} \mathrm{d}x', \tag{24}$$

where

$$\bar{\varrho} = \frac{1}{W}\int\limits_0^W \varrho \, \mathrm{d}x,$$

and ϱ is calculated with U replaced by $\bar{U}$. According to (24) we may calculate the fluctuations due to N_D^+ and N_A^- separately. In order to calculate the fluctuations u due to N_D^+, we have to put $\bar{\varrho}$ equal to $\bar{N}_D$ in (24) and ϱ equal to N_D^+. $\bar{N}_D$ is assumed independent of y, z, as is explained above. To calculate u along the x-axis for a particular point in the yz plane, we have to know the distribution of N_D^+ along the x-axis for this particular point. After the calculations are carried out, we choose another point y', z' and calculate for this new point the fluctuations u, and after repeating this procedure many times we obtain finally a total picture of the fluctuations u. Instead of doing this tedious work we remark that the donor impurities are separated "on the average" from each other by the distance $l = (1/N_D)^{1/3}$, where N_D is the donor concentration. We assume now that we obtain "an average" picture of the fluctuations u along the x-axis by calculating the fluctuations for the average donor distribution of Fig. 4. In Fig. 4 the donors are distributed in a periodic way and are separated from each other by the distance l. The width of the donors Δ is taken equal to the Bohr radius of the donors. This periodic donor distribution leads to the periodic fluctuating potential u of Fig. 5. The maximum and minimum values of u are given by the relations

$$\left.\begin{aligned} \frac{qu}{kT}(\max) &= \frac{1}{24}\left(\frac{l}{L_D}\right)^2 \frac{(l - \Delta)(2l + \Delta)}{l^2}, \\ \frac{qu}{kT}(\min) &= -\frac{1}{24}\left(\frac{l}{L_D}\right)^2 \frac{(l - \Delta)(l + \Delta)}{l^2}, \end{aligned}\right\} \tag{25}$$

where L_D is the Debye length, defined by

$$L_D = \left(\frac{kT\varepsilon\varepsilon_0}{q^2 N_D^+}\right)^{1/2}. \tag{26}$$

The potential fluctuations represented in Fig. 5 are calculated for the semiconductor CdSe, with a donor concentration $N_D = 10^{17}$ cm^{-3}, $L_D = 118$ Å, $l = 216$ Å, and a Bohr radius of the donor impurities Δ equal to 40 Å. From Fig. 5 it follows that the amplitude of the fluctuations is very small ($\approx 0.2kT$). As a criterion to judge the importance of the fluctuations u, we take the error introduced in the calculation of

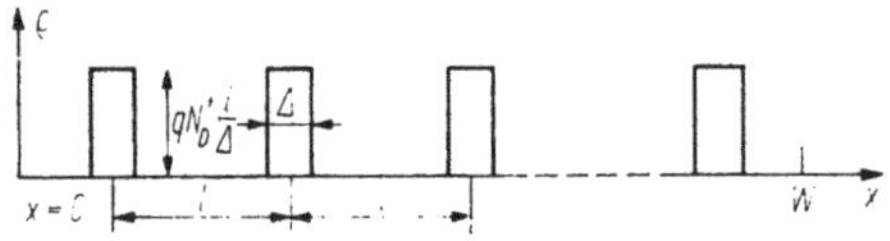

Fig. 4. Average donor distribution

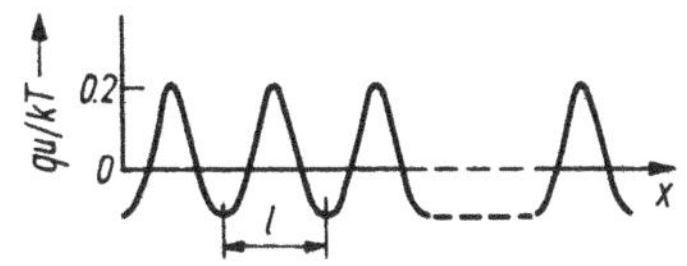

Fig. 5. Potential fluctuations due to the average donor distribution

the total number of electrons by neglecting the fluctuations. This error E is given by the expression

$$E = \left| 1 - \frac{1}{W} \int\limits_0^W \exp\left(\frac{qu}{kT}\right) dx \right| . \tag{27}$$

For our numerical example E becomes 0.7%. From relation (25) it is seen that the dependence of the fluctuations on the donor concentration is not straightforward. In the case of CdSe the fluctuations become maximum for $N_D \approx 2 \times 10^{18}$ cm^{-3}, which gives an error smaller than 2%. We conclude thus that the fluctuations due to the donor concentration may normally be neglected.

The fluctuations due to the acceptor impurities N_A^- can be calculated in an analogous way. However, the acceptor states in CdSe are most likely located at the grain boundaries [1, 2], which poses the problem of choosing an appropriate average acceptor spacing l. We chose l equal to the grain size D, which is equal to the film thickness [4, 5]. The Debye length of the acceptors is given by the expression

$$L_D = \left(\frac{kT\varepsilon\varepsilon_0}{q^2 N_A^-}\right)^{1/2} . \tag{28}$$

The Bohr radius Δ of an acceptor level in CdSe is about 11 Å, which means that, for sufficiently large grains and thus for sufficiently thick films (> 500 Å), the relations (25) can be simplified to

$$\frac{qu}{kT}(\max) \approx \frac{1}{12}\left(\frac{D}{L_D}\right)^2 , \qquad \frac{qu}{kT}(\min) \approx -\frac{1}{24}\left(\frac{D}{L_D}\right)^2 . \tag{29}$$

According to (29) limiting the fluctuations to $1kT$, causing a still allowed error E of 10%, means that the trap concentration N_A must be smaller than

$$N_A \leqq 1.7 \times 10^{22}/D^2 , \tag{30}$$

where D is expressed in Å and N_A in cm^{-3}. From (30) it follows that for films 500 Å thick, which are often used in thin film transistors [1, 2, 11], N_A must be smaller than 7×10^{16} cm^{-3}, if we want to neglect the fluctuations. We may conclude that for a given grain size D (film thickness), the maximum trap concentration, for which the fluctuations may be neglected, is given by (30). If the trap concentration is larger than the one obtained from (30), the fluctuations may no longer be ignored. Consequently one must take into account the potential barriers caused by the traps (acceptors) at the grain boundary. This means also that the conclusions made above for the derivation of the current I are no longer valid. For these situations we refer to the literature [1, 2].

In an analogous way we can obtain an idea of the fluctuations u due to the changes of the electron concentration in the grain and at the grain boundary, by putting in (25) l equal to the grain size D and Δ equal to $D - \Delta'$, where Δ' the width of the grain boundary. In the Debye length L_D we introduce the average electron concentration n

$$n = N_C \exp\left(\frac{E_F - E_0 + \chi_0 + q\bar{U}}{kT}\right) .$$

Taking $D = 500$ Å, $\Delta' = 20$ Å, and $n = 10^{17}$ cm^{-3} gives according to (25) and (27) a fluctuation of $0.1kT$ and an error E of 0.5%. Also other values of n gave negligible errors.

We may thus conclude that a thin semiconducting film (e.g. CdSe) with a not too high trap concentration may be considered as a homogeneously doped film for the calculation of the electrostatic potential. This potential is obtained by solving the Poisson equation (19).

3. Comparison with Experiments

Hall experiments carried out on thin semiconducting films [3, 4, 13] indicate that the Hall mobility is characterized by an activation energy. This activation energy is normally attributed to the grain barriers caused by traps [1, 2]. We consider foı instance the CdSe films used in thin film transistor structures. These films are a few hundred Å thick (e.g. 500 Å) and have normally a donor concentration equal to or smaller than 10^{17} cm^{-3} [11]. The average grain size D may be estimated around 500 Å. It seems thus reasonable to consider these films homogeneous from the point of view of donor and trap distribution, as discussed in the previous section. Consequently the activation energy of the Hall mobility should be attributed to the changes in the band gap at the grain boundary, due to strain and inhomogeneities, as shown in Section 2.

We consider now the thin film transistor structure of Fig. 6. According to the homogeneous field-effect theory [11], Q_e may be approximated for thin semiconductors with a not too large concentration of surface states by

$$Q_e = a_s C_i \cdot \frac{U_G - U_{FB}}{q} + n_b d\,, \tag{31}$$

where C_i is the insulator capacitance per unit area, U_G the gate voltage, U_{FB} the flatband voltage, n_b the bulk concentration of electrons (in the average grain), and a_s the surface state trapping factor. a_s gives the fraction of injected electrons which are not trapped by surface states. From expressions (18) and (31) it follows that a plot of the channel conductance G (or transconductance g_m which has a similar behaviour [11]) versus $1/T$ has to be interpreted carefully, because not only μ^* is characterized by an activation energy but also the surface states (a_s) introduce an activation energy. When bulk traps are present, another a has to be defined, which will probably also introduce an activation energy. This means that the activation energy detected in the literature [1, 13, 14] is hard to identify, and certainly not field-effect experiments alone will allow the identification of the possible different activation energies.

4. Conclusions

We may conclude that in the case of thin films with small grains, the film may be considered to have homogeneous donor and trap concentrations. Only in the case of large grains, as is the case in the layers studied by Petritz [15], one has to consider

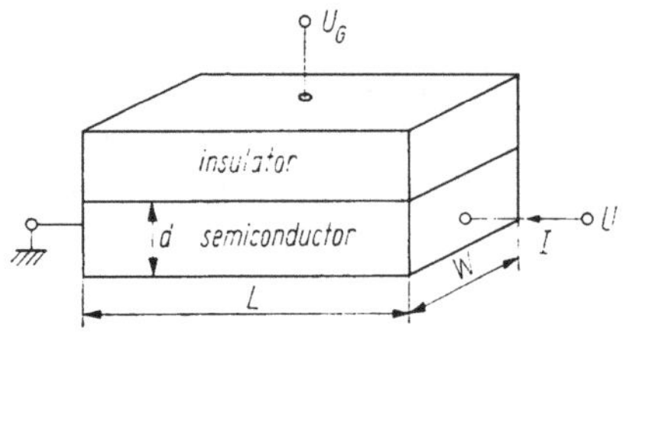

Fig. 6. Thin film transistor

the inhomogeneity caused by the potential barriers of the traps at the grain boundaries. A criterion for this inhomogeneity is given in Section 2.

On the other hand, some arguments are given to state a band widening at the grain boundaries of polycrystalline films. This change in band gap can explain the activation energy found in the mobility determined by Hall experiments. The activation energies found in field-effect measurements are more difficult to identify.

Finally it is shown that the theoretical behaviour of thin film transistor structures can be predicted by the homogeneous field-effect theory, provided a barrier-limited mobility μ^* is introduced in the calculations.

Appendix

We introduce the function $G(x - x')$

$$\left.\begin{aligned} G(x - x') &= x - x' \quad \text{for } x \geqq x' \,, \\ &= 0 \qquad\quad\; \text{for } x < x' \,, \end{aligned}\right\} \tag{A1}$$

where x' is a point chosen in the interval $(0, W)$. It follows from (A1) that $G(x - x')$ obeys the equation

$$\frac{\partial^2 G}{\partial x^2} = \delta(x - x') \,. \tag{A2}$$

The electrostatic potential $U(x', y, z)$ of the semiconductor can be written as

$$U(x', y, z) = \int_0^W U(x, y, z)\, \delta(x - x')\, dx = \left(U \frac{\partial G}{\partial x} - G \frac{\partial U}{\partial x}\right)_{x=0}^{x=W} + \\ + \int_0^W \frac{\partial^2 U}{\partial x^2} G(x - x')\, dx \,. \tag{A3}$$

According to (22) and the boundary conditions

$$\left.\frac{\partial U}{\partial x}\right|_{x=0} = \left.\frac{\partial U}{\partial x}\right|_{x=W} = 0 \,,$$

(A3) becomes

$$U(x', y, z) = U(W, y, z) - \int_{x'}^{W} \frac{\varrho}{\varepsilon\varepsilon_0} (x - x')\, dx - \int_{x'}^{W} \left(\frac{\partial^2 U}{\partial z^2} + \frac{\partial^2 U}{\partial y^2}\right) (x - x')\, dx \,. \tag{A4}$$

We write $U(x, y, z)$ as a sum of the average potential $\bar{U}(y, z)$ and the fluctuations $u(x, y, z)$, where

$$\bar{U}(y, z) = \frac{1}{W} \int_0^W U(x, y, z)\, dx \,. \tag{A5}$$

From (A5) and (A4) $\bar{U}(y, z)$ becomes

$$\bar{U}(y, z) = U(W, y, z) - \frac{1}{W} \int_0^W dx' \int_{x'}^W \frac{\varrho}{\varepsilon\varepsilon_0} (x - x')\, dx -$$

$$-\frac{1}{W}\int_0^W dx' \int_{x'}^W \left(\frac{\partial^2 U}{\partial z^2}+\frac{\partial^2 U}{\partial y^2}\right)(x-x')\,dx\,. \tag{A6}$$

By interchanging the order of integration we obtain

$$\bar{U}(y,z) = U(W,y,z) - \frac{1}{2W}\int_0^W \frac{\varrho}{\varepsilon\varepsilon_0}x^2\,dx - \frac{1}{2W}\int_0^W \left(\frac{\partial^2 U}{\partial z^2}+\frac{\partial^2 U}{\partial y^2}\right)x^2\,dx\,. \tag{A7}$$

If we assume the fluctuations u small, we can replace U in the right-hand terms of (A4) and (A7) by $\bar{U}$. If we also take in mind that

$$\frac{\partial^2 \bar{U}}{\partial y^2}+\frac{\partial^2 \bar{U}}{\partial z^2} = -\frac{\varrho}{\varepsilon\varepsilon_0}\,, \tag{A8}$$

we find

$$\bar{U}(y,z) = U(W,y,z) - \frac{1}{2W}\int_0^W x^2\,\frac{\varrho-\bar{\varrho}}{\varepsilon\varepsilon_0}\,dx\,, \tag{A9}$$

$$U(x',y,z) = U(W,y,z) - \int_{x'}^W \frac{\varrho-\bar{\varrho}}{\varepsilon\varepsilon_0}(x-x')\,dx\,. \tag{A10}$$

From (A9) and (A10) u becomes

$$u(x',y,z) = -\int_{x'}^W \frac{\varrho-\bar{\varrho}}{\varepsilon\varepsilon_0}(x-x')\,dx + \frac{1}{2W}\int_0^W \frac{\varrho-\bar{\varrho}}{\varepsilon\varepsilon_0}\,x^2\,dx\,. \tag{A11}$$

If we replace in (A11) x' by x, we obtain expression (24) of Section 2.

References

[1] J. C. ANDERSON, Thin Solid Films **37**, 127 (1976).
[2] J. C. ANDERSON, Thin Solid Films **38**, 151 (1976).
[3] A. DE VOS and D. VAN DHELSEN, The Temperature Dependence of the Electrical Properties of Thin Tellurium Films, to be published.
[4] A. DE VOS and J. AERTS, Thin Solid Films **46**, 223 (1977).
[5] K. L. CHOPRA, Thin Film Phenomena, McGraw-Hill Publ. Co., New York 1969.
[6] A. MANY, Y. GOLDSTEIN, and N. B. GROVER, Semiconductor Surfaces, North-Holland Publ. Co., Amsterdam 1965.
[7] H. F. MATARE, Defect Electronics in Semiconductors, Wiley, New York 1971.
[8] H. H. WIEDER, Intermetallic Semiconductor Films, Pergamon Press, Oxford 1970.
[9] A. H. MARSHAK and K. M. VAN VLIET, Solid State Electronics **21**, 429 (1978).
[10] A. H. MARSHAK and K. M. VAN VLIET, Solid State Electronics **21**, 417 (1978).
[11] A. VAN CALSTER and H. J. PAUWELS, Solid State Electronics **18**, 691 (1975).
[12] J. VOLGER, Phys. Rev. **79**, 1023 (1950).
[13] P. K. WEIMER, in: Physics of Thin Films, Vol. 2, Ed. G. HASS, Academic Press, New York 1964 (p. 147).
[14] S. M. SZE, Physics of Semiconductor Devices, Wiley, New York 1969.
[15] R. L. PETRITZ, Phys. Rev. **104**, 1508 (1956).

(Received January 12, 1979)

phys. stat. sol. (a) 54, 217 (1979)

Subject classification: 18.2; 21.4

Institut für Physik, Max-Planck-Institut für Metallforschung, Stuttgart and Institut für Theoretische und Angewandte Physik der Universität Stuttgart

The Parasusceptibility of Dysprosium

By

R. Herz and H. Kronmüller

The parasusceptibility, χ_p, of dysprosium single crystals is measured in the temperature range between 4.2 and 130 K. Above the Curie temperature of $T_C = 85$ K, the parasusceptibility can also be measured because under the action of an applied field in this temperature range the helical phase changes into a ferromagnetic spin structure. For the field dependence of χ_p a $H^{-0.47}$-law is obtained. The experimental results are discussed within the framework of the spin wave theory taking into account a quadratic, anisotropic dispersion law for the spin waves, as well as the effect of the energy gap, resulting from the anisotropy of the crystalline field.

Die Parasuszeptibilität χ_p von Dysprosium wird an Einkristallen im Temperaturbereich zwischen 4,2 und 130 K gemessen. Auch oberhalb der Curie-Temperatur $T_C = 85$ K kann die Parasuszeptibilität bestimmt werden, da unter dem Einfluß eines Magnetfeldes in diesem Temperaturbereich die helische Spinstruktur in eine ferromagnetische Spinstruktur übergeht. Für die Feldabhängigkeit von χ_p kann ein $H^{-0,47}$-Gesetz bestimmt werden. Die experimentellen Ergebnisse werden im Rahmen der Spinwellentheorie diskutiert, wobei eine quadratische Dispersionsrelation für die Spinwellenenergie und der Einfluß der Energielücke, die von der Anisotropie des Kristallfeldes herrührt, berücksichtigt werden.

1. Introduction

In a ferromagnetic material thermally excited spin waves cause a decrease of the saturation value of the magnetization, $M_s(0)$, usually described by the temperature-dependent spontaneous magnetization, $M_s(T)$. The thermal excitation of spin waves can be suppressed by application of a magnetic field. The resulting increase of the spontaneous magnetization is described by the paraeffect. In general this effect at lower and average field ranges ($H < 10^6$ A/m) is small in comparison to effects of structural defects. Therefore, only at higher fields the paraeffect is expected to be observable.

In this paper we present experimental results of the field dependence as well as the temperature dependence of the parasusceptibility of Dy. We try to explain these results with theoretical expressions for χ_p, derived under the assumption that the dispersion law of the spin waves shows a quadratic dependence on the wave vector. It will be shown that the energy gap, caused by the magnetocrystalline anisotropy, has a significant influence on the field and temperature dependence of the parasusceptibility.

2. Experimental Procedure

From a single-crystalline rod of Dy of 50 mm length and 5 mm diameter, and a purity of 99.8% (Metals Research, Ltd., Cambridge) we prepared a spherical specimen of about 2.4 mm diameter by spark-cutting. Subsequently this sample was cleaned by chemical polishing. The orientation of the sphere was determined by the Laue back-reflection method.

In the temperature range between 4.2 and 200 K measurements of the isothermal magnetization curves parallel to an easy *a*-axis were performed with a Foner vibrating sample magnetometer.

With a superconducting coil external fields up to 4×10^6 A/m could be applied. Because of the spherical specimen and its demagnetizing field of $H_d = -NM$, where $N = \frac{1}{3}$ and M means the net magnetization, the maximum value of the internal magnetic field was about 3×10^6 A/m.

Low temperatures were obtained by a conventional liquid helium cryostat, and the temperature could be electronically controlled to ± 0.1 K.

3. Experimental Results

3.1 *Field dependence*

In Fig. 1 we show the susceptibility, χ_E, in the field range of approach to ferromagnetic saturation in dependence on the internal magnetic field for temperatures between 4.2 and 140 K in a log–log representation. The susceptibility which is defined as

$$\chi(H) = \lim_{\Delta H \to 0} \frac{\Delta M}{\Delta H}\bigg|_H \tag{3.1}$$

was determined from the difference quotient

$$\chi(H) \approx \frac{\Delta M}{\Delta H}\bigg|_H = \frac{M_n - M_m}{H_n - H_m}\bigg|_H, \tag{3.2}$$

where

$$H = \frac{H_n + H_m}{2} \tag{3.3}$$

means the averaged internal magnetic field between successive measuring fields H_n and H_m.

In the temperature range between $T_C = 85$ K and $T_N = 178$ K, where Dy has a helical spin structure in zero magnetic field, the parasusceptibility was measured above the critical field of the transition from the helical into the ferromagnetic phase [1]. For temperatures above T_N Dy behaves paramagnetically.

Fig. 1 shows that there exist different field dependences of χ_E for temperatures $T < T_C$ and $T > T_C$.

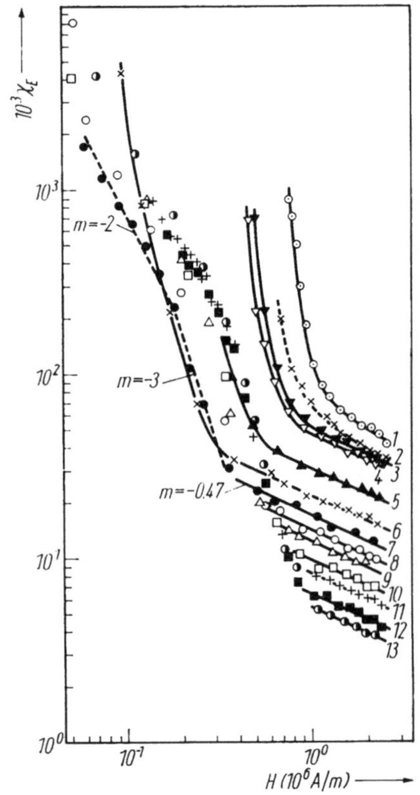

Fig. 1. Field dependence of the susceptibility in the field range of approach to saturation of Dy ($\boldsymbol{H} \parallel [11\bar{2}0]$) for different temperatures between 4.2 and 140 K. (1) $T = 138$, (2) 131.5, (3) 125, (4) 120.5, (5) 108, (6) 91, (7) 77.5, (8) 70.5, (9) 62, (10) 52.5, (11) 44, (12) 10, (13) 5 K

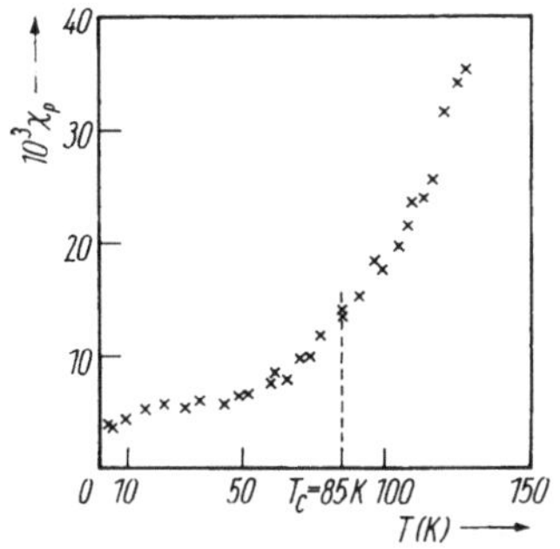

Fig. 2. Temperature dependence of the parasusceptibility of Dy measured at a field $H = 2.4 \times 10^6$ A/m

In the temperature range $T < T_C$ we observe three field regions where $\log \chi_E$ depends linearly on $\log H$ with different slopes. We find that χ_E obeys a H^{-2} law for low fields, which changes to a H^{-3} law for medium fields, and for high fields we obtained a $H^{-0.47}$ dependence of χ_E, which we will call χ_p. The positions of the different field ranges were found to be temperature-dependent, with the tendency to shift to lower fields for increasing temperatures.

For temperatures $T > T_C$ the $H^{-0.47}$ region is observed up to 130 K but for the low-field part of the $\chi(H)$ curves no simple field dependence could be analysed. The beginning of the paramagnetic region, where a $H^{-0.47}$ law was found, depends strongly on temperature and shifts to higher fields with increasing temperatures.

3.2 Temperature dependence

Fig. 2 represents the temperature dependence of χ_p in the temperature range between 4.5 and 130 K, and for a magnetic field of about 2.4×10^6 A/m. For temperatures 40 K $< T <$ 75 K we obtained a linear temperature dependence of χ_p. For temperatures below 40 K the linear dependence no longer holds, instead a flat maximum is observed. Also for $T >$ 75 K a deviation from the linearity is found: χ_p increases stronger than linear in T.

4. Theoretical Background

As derived by Holstein and Primakoff [2] the parasusceptibility χ_p can be derived from the partition sum Z:

$$\chi_p = \frac{k_B T}{V} \frac{\partial^2}{\partial H^2} (\ln Z) , \tag{4.1}$$

$$Z = \sum_n \exp\left(-\frac{E_n}{k_B T}\right), \tag{4.2}$$

where E_n are the energy eigenvalues of the Hamiltonian describing the ferromagnetically ordered spin system. The energy eigenvalues contain terms from the exchange interaction, from the dipole–dipole interaction, from the magnetostatic interaction with an applied magnetic field, and from the spin–orbit interactions. Neglecting interactions between spin waves, Holstein and Primakoff derived for $T < T_C/2$ the following result for the field and temperature dependence of χ_p:

$$\chi_p = \text{const}\ (T) \left[\left(\frac{M_s(0)}{H}\right)^{1/2} + \arcsin\left(\frac{M_s(0)}{H + M_s(0)}\right)^{1/2}\right]. \tag{4.3}$$

$$\chi_p = \alpha T . \tag{4.4}$$

For $H \gg M_s(0)$ the field dependence is given by

$$\chi_p = 2 \text{ const } (T) \left[\left(\frac{M_s(0)}{H} \right)^{1/2} \right]. \tag{4.5}$$

Theoretical investigations on the ferromagnetic rare-earth metal gadolinium (Gd) by Hof et al. [3] have shown that the dipole–dipole interaction modifies the field law (4.5) into

$$\chi_p \sim H^{-0.44} . \tag{4.6}$$

This theoretical result was confirmed by magnetization measurements for Gd [3].

In the case of the rare-earth metals the exchange integral is anisotropic and due to the indirect exchange interaction (RKKY). It has been shown previously [2], however, that this exchange anisotropy induces no anisotropy of χ_p.

The parasusceptibility χ_p can be determined quantitatively if we know the dispersion law of the spin waves. On the other hand, the wave vector dependence, $\varepsilon(\boldsymbol{q})$, of the spin wave spectrum is determined by the Fourier integral of the exchange integral which for Dy is not known explicitly. Therefore, we use in the following an expression for $\varepsilon(\boldsymbol{q})$ which accounts for the hexagonal symmetry and takes care of the existence of an energy gap:

$$\varepsilon(\boldsymbol{q}, H) = R q_z^2 + U q_\perp^2 + \Delta\varepsilon + g\mu_0 \frac{e\hbar}{2m} H \tag{4.7}$$

with

$$q_\perp^2 = q_x^2 + q_y^2 \tag{4.8}$$

and

$$R = k_B T_R \frac{c^2}{\pi^2}; \qquad T_R \approx 38 \text{K}, \tag{4.9}$$

$$U = k_B T_U \frac{a^2}{\pi^2}; \qquad T_U \approx 180 \text{ K}, \tag{4.10}$$

$$\frac{R}{U} \approx 0.5 \tag{4.11}$$

(a interatomic distance within the basal plane, c periodicity distance parallel to the c-axis). x, y, and z denote the coordinates, where x and y lie parallel to the basal planes and z is oriented parallel to the c-axis of the h.c.p. lattice of Dy. $\Delta\varepsilon$ describes the energy gap resulting from the crystalline anisotropy. R and U were determined by approximating $\varepsilon(\boldsymbol{q}, H)$ to the theoretical spin wave dispersion curve for low temperatures shown in Fig. 3 [4].

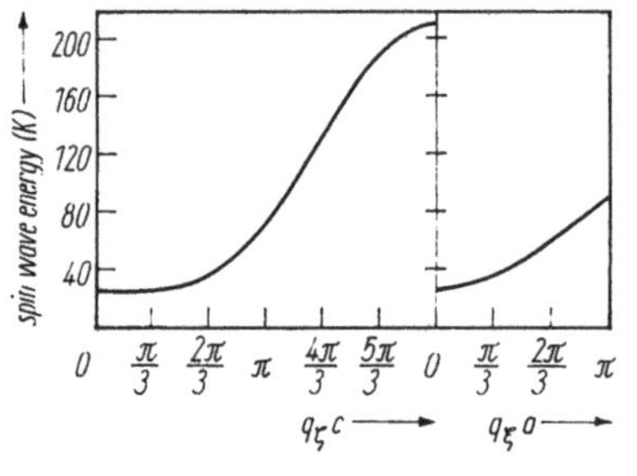

Fig. 3. Theoretical dispersion relation $\varepsilon(\boldsymbol{q})$ for spin waves of Dy [4]

Our starting expression for χ_p is given by

$$\chi_p = \mu_0 \frac{(geh/2m)^2}{(2\pi)^3} \frac{1}{k_B T} \int \frac{\exp(\varepsilon(\boldsymbol{q}, H)/k_B T)}{[\exp(\varepsilon(\boldsymbol{q}, H)/k_B T) - 1]^2} \, d^3q \,, \tag{4.12}$$

where the integration extends over the first Brillouin zone, which will be approximated by a cylinder with a radius of $2\pi/a\sqrt{3}$ and a length of $2\pi/c$.

In the following we distinguish between two cases:

$$y = \frac{\varepsilon(\boldsymbol{q}, H)}{k_B T} < 1 \,, \tag{4.13a}$$

$$y = \frac{\varepsilon(\boldsymbol{q}, H)}{k_B T} > 1 \,. \tag{4.13b}$$

In the first case the integrand can be developed into the following Taylor series:

$$\frac{e^y}{(e^y - 1)^2} = \frac{1}{y^2} - \frac{1}{12} + \frac{1}{240} y^2 \dots \,. \tag{4.14}$$

We shall only take into account the first and the second term, the other terms can be neglected. Then the resulting expression for χ_p is found to be

$$\chi_p = \mu_0 \frac{\pi(eh/2m)^2}{4a^2 c T_U \sqrt{k_B T_R}} T \left[\frac{\pi}{2\sqrt{\Delta\varepsilon + g\mu_0 (eh/2m) H}} - \frac{\sqrt{k_B T_R}}{\Delta\varepsilon + g\mu_0 (eh/2m) H + (4/3) k_B T} \right] - \frac{\mu_0 \pi (eh/2m)^2}{36 a^2 c k_B} \frac{1}{T} \,. \tag{4.15}$$

In the second case the integrand can be represented as a power series of e^{-ny}

$$\frac{e^y}{(e^y - 1)^2} = \frac{e^{-y}}{(1 - e^{-y})^2} = \sum_{n=1}^{\infty} n \, e^{-ny} \,. \tag{4.16}$$

For χ_p we find

$$\chi_p = \sum_{n=1}^{\infty} \mu_0 \frac{(eh/2m)^2 \pi^{3/2}}{8a^2 c k_B T_U \sqrt{T_R}} \frac{\sqrt{T}}{\sqrt{n}} \underbrace{\left(1 - e^{-(4/3) T_U (n/T)}\right)}_{\approx 1} \times \underbrace{\operatorname{erf}\left(\sqrt{T_R \frac{n}{T}}\right)}_{\approx 1} e^{[\Delta\varepsilon - g\mu_0 (eh/2m) H]/k_B T} \,. \tag{4.17}$$

Now we have to investigate under which conditions (4.15) or (4.17) holds.

In our measurements we used fields up to 2.4×10^6 A/m. The values for $\Delta\varepsilon$ can be determined from infrared resonance measurements [5, 6], or can be calculated by using the following formula [4]:

$$\Delta\varepsilon = \frac{6geh/2m}{M} \sqrt{2k_2 k_6^6} \,, \tag{4.18}$$

where k_2 and k_6^6 mean the macroscopic anisotropy constants. We find that (4.15) is valid for $T > 22$ K, whereas (4.17) has to be used for a discussion of χ_p in the temperature range below 22 K.

5. Discussion of Results

5.1 Field dependence

For the discussion of the field dependence of χ_p for constant temperature we rewrite (4.15) in a more comprehensive form neglecting the second and third terms:

$$\chi_p = \frac{A}{\sqrt{H_{eff}}} \tag{5.1}$$

with

$$H_{eff} = H_{an} + H = \frac{\Delta\varepsilon}{g\mu_0 e h/2m} + H \,. \tag{5.2}$$

If we compare the values of the anisotropy field H_{an} with the available field strengths, produced by our superconducting coil, we find that $H_{an} \gg H$. Therefore, from a theoretical point of view no significant field dependence of χ_p should be observed.

Let us now regard our susceptibility measurements. In Fig. 1 we have represented a log–log diagram for χ_E in dependence on the applied magnetic field, with the temperature as a parameter. We clearly can analyse that χ_E obeys a $H^{-0.47}$ law in the high-field region. From this result we may conclude that the spin waves only feel a fraction of the theoretical value $\Delta\varepsilon$. Consequently the applied field H has to be considered as the dominant term in the sum $(H_{an} + H)$. This could be possible if we assume that the spin wave excitations correspond to spins oscillating on elliptical cones around the a-axis, where the oscillation amplitude within the basal planes is much larger than in c-direction.

In Section 4 we found that (4.15) should be valid only for $T > 22$ K. If we now assume that the effective value of $\Delta\varepsilon$ is much smaller than predicted by (4.18), then condition (4.13a) will be valid also for lower temperatures than $T = 22$ K, and as a consequence the $H^{-0.5}$ dependence of χ_p should be measured also for $T < 22$ K. Our measurements confirm this expectation.

We also can interpret the temperature dependence of the beginning of the field range, where the paraeffect is dominating. As the energy gap shows a strong temperature dependence, $\Delta\varepsilon$ decreases with increasing temperature, H_{an} also becomes smaller for higher temperatures. Therefore, for higher temperatures the relation $\chi_p \sim H^{-0.47}$ becomes valid at lower fields. Our measurements reveal this behaviour for $T < T_C$, where our calculations should be valid.

5.2 Temperature dependence

In Fig. 4 our experimental results for the temperature dependence of χ_p, measured at a field of 2.4×10^6 A/m, are compared with our theoretical results as obtained

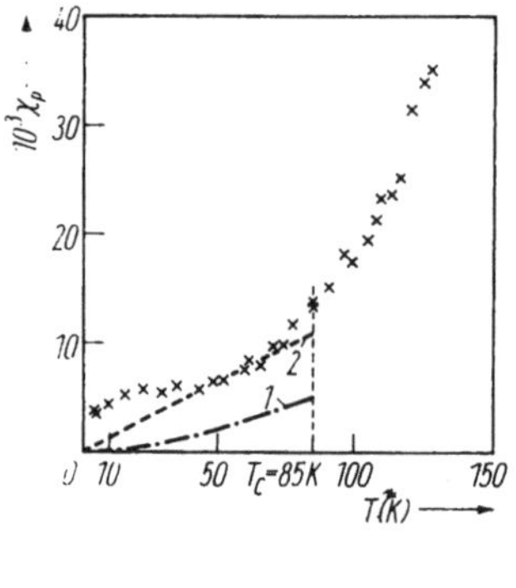

Fig. 4. Comparison between the theoretical and experimental temperature dependences of the parasusceptibility of Dy at a field $H = 2.4 \times 10^6$ A/m. × Experiment, measured values; (1) theoretical curve ($\Delta\varepsilon \neq 0$); (2) theoretical curve ($\Delta\varepsilon = 0$)

for the expressions (4.15) and (4.17). In one case (curve 1) the energy gap $\Delta\varepsilon$ as defined by (4.18) was taken fully into account. In the temperature range $T > 22$ K χ_p was determined by (4.15), and for $T < 22$ K equation (4.17) was applied considering only the first term ($n = 1$) of the sum. In the other case (curve 2) we assumed $\Delta\varepsilon$ to be negligible. In this case (4.15) was valid in the whole temperature range $T < T_C$.

If the two theoretical curves are compared with the experimental curve, it becomes obvious that the theoretical values of χ_p for $\Delta\varepsilon \neq 0$ are much smaller than the measured values. On the other hand, if we consider the calculated curve for $\Delta\varepsilon = 0$, we find a very good correspondence with our experimental curve in the temperature range between 40 and 75 K. Below $T = 40$ K, however, the experimental values deviate significantly from a linear temperature dependence of χ_p. At $T \to 0$ K the parasusceptibility extrapolates to a constant value of the order of magnitude $\chi_p(0) \approx \approx 3 \times 10^{-3}$. A similar behaviour previously was found for Gd single crystals [3]. An interpretation of this peculiar temperature dependence of χ_p could not be given so far. It will be the object of further work to investigate the effect of more realistic spin wave spectra on the temperature dependence of χ_p.

The deviation from the linearity for $T > 75$ K, i.e. in the vicinity of the Curie temperature $T_C = 85$ K is not astonishing, because in this temperature range spin wave theory no longer holds.

Acknowledgement

The authors gratefully acknowledge the interest and helpful remarks of Prof. Dr. A. Seeger during the preparation of this paper.

References

[1] R. Herz and H. Kronmüller, phys. stat. sol. (a) **47**, 451 (1978).
[2] T. Holstein and H. Primakoff, Phys. Rev. **58**, 1098 (1940).
[3] H.-J. Hof, H. Kronmüller, and B. Rothfuss, phys. stat. sol. (b) **71**, 585 (1975).
[4] B. R. Cooper, Solid State Phys. **21**, 393 (1968).
[5] H. S. Marsh and A. J. Sievers, J. appl. Phys. **40**, 1563 (1969).
[6] A. J. Sievers, J. appl. Phys. **41**, 980 (1970).

(Received May 10, 1979)

phys. stat. sol. (a) 54, 225 (1979)

Subject classification: 1.6; 10.2; 22.2.1

Matsushita Research Institute Tokyo, Inc., Ikuta, Tama-ku, Kawasaki [1])

Impurity Effect on the Formation of Terraces in GaAs LPE Growth

By

N. TOYODA, M. MIHARA, and T. HARA

Terraces on the surfaces of GaAs LPE layers grown from Sn or Ge doped solutions are studied. The concentration of Sn (X^l_{Sn}) or Ge (X^l_{Ge}) in the growth solution is varied from 0.01 to 15 at%. The terraces sensitively depend on the doping concentration. The width and height of terraces decrease with increasing X^l_{Sn} or X^l_{Ge}. Though the effects of Sn and Ge on the terrace formation differ when compared as a function of X^l_{Sn} or X^l_{Ge}, they agree well when compared as a function of the concentration of Sn (N_{Sn}) or Ge (N_{Ge}) in grown layers.

Es werden Terrassen auf den Oberflächen von GaAs-LPE-Schichten, die aus mit Sn oder Ge dotierten Lösungen gezüchtet wurden, beobachtet. Die Konzentration von Sn (X^l_{Sn}) oder Ge (X^l_{Ge}) in der Züchtungslösung wird von 0,01 bis 15 At% variiert. Die Terrassen hängen empfindlich von der Dotierungskonzentration ab. Die Breite und Höhe der Terrassen nehmen mit zunehmendem X^l_{Sn} oder X^l_{Ge} ab. Obwohl die Einflüsse von Se und Ge auf die Terrassenbildung als Funktion von X^l_{Sn} oder X^l_{Ge} verschieden sind, stimmen sie gut in Abhängigkeit von den Konzentrationen von Sn (N_{Sn}) oder Ge (N_{Ge}) in den gewachsenen Schichten überein.

1. Introduction

A characteristic pattern called terraces (waves) usually appears on the surfaces of GaAs [1 to 5], GaP [6 to 8], InP [9 to 12], and GaAlAs [13 to 17] LPE layers. These terraces lead to uneven surfaces and adversely affect the device performance.

The origin of terraces (waves) has been studied extensively. Previously terraces have been attributed to the instability of growth solution by constitutional supercooling [1 to 3]. Recently some workers [4, 6 to 8, 12] have shown that terraces disappear when the growth is made on substrates accurately oriented to the low-index planes, and have concluded that the origin of terraces is the substrate misorientation from low-index planes. Rode [18, 19] has suggested the correlation between terraces and surface reconstruction. Small and Potemski [15] have indicated that terraces should be distinguished from waves: Terraces are equilibrium forms, which would exist in an isothermal situation at zero growth rate, while waves are not present at zero growth rate.

Terraces sensitively depend on various growth conditions. It has been shown that terraces are affected by the supercooling of the growth solution [5, 9, 10] and disappear in the growth from the solution initially supercooled beyond a critical value [5]. Kobayashi and Furukawa [13] have reported that the temperature fluctuation of the furnace leads to terraces. Non-uniform wetting caused by residual surface oxide also affects the formation of terraces [11, 16, 17], especially in the growth of reactive materials. Furthermore, terraces can be affected by other growth conditions such as growth temperature, cooling rate, and impurity. However, few studies have examined their effects on the terrace formation.

[1]) Present name: Tokyo Research Laboratory, Matsushita Electronics Corp., 4896 Ikuta, Tama-ku, Kawasaki, Japan.

It is the purpose of this paper to examine how the formation of terraces is affected by the concentration of doping impurities in the growth solution.

2. Experimental Procedure

The growth was performed in a horizontal sliding boat made of high-purity graphite [20]. Before the growth runs the boat assembly was baked in vacuum at 1100 °C, and further under flowing Pd-diffused H_2 gas at 900 °C for 24 h.

A growth solution was prepared as follows: The high-purity Ga metal (99.9999%) and the GaAs source material (undoped polycrystals with $N_D - N_A = 1 \times 10^{16}$ cm^{-3}) were loaded in the boat, and baked under H_2 gas flow at 800 °C for 24 h to prepare a homogeneous solution. This baking also removed the residual oxides on the Ga metal, and led to a high-purity growth solution. The layers grown from the undoped solution were n-type and the net donor concentration was below 5×10^{14} cm^{-3}. In the doping experiment, Sn or Ge, which are representative donor and acceptor impurities, respectively, was added in the amount of 0.01 to 15 at%.

The substrates were Si-doped GaAs ($N_D - N_A = 1 \times 10^{18}$ cm^{-3}) oriented 0.1° off the (100) plane toward the ⟨110⟩ *B* direction. They were chemomechanically polished with NaOCl solution after mechanical lapping. The substrate was cleaned by organic solvents, etched with 5 H_2SO_4:1 H_2O:1 H_2O_2 and dipped in HCl to remove the residual oxides just prior to its introduction into the boat.

The boat assembly was positioned in a fused silica tube, and the LPE growth program was started. The silica tube was evacuated to 2×10^{-2} Torr, and back-filled with H_2 gas. After heating to 800 °C for 1 h, the substrate was brought into contact with the growth solution and the cooling was started with a constant rate 0.5 K/min. This time-temperature program corresponds to the equilibrium cooling technique [21].

The surface morphology of grown layers was examined with a Nomarski photomicroscope and also with a Talystep having a stylus of diameter 0.1 μm.

3. Results

First the surface morphology of undoped layers was examined. Fig. 1 shows the Nomarski photomicrograph and the Talystep height profile of the surface of an undoped layer. Typical terraces are observed. The height (*H*) and width (*W*) of terraces, which

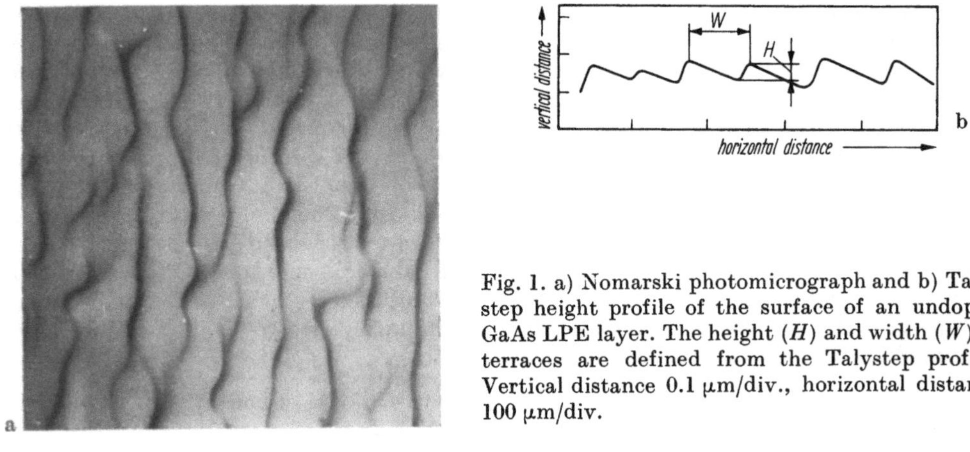

Fig. 1. a) Nomarski photomicrograph and b) Talystep height profile of the surface of an undoped GaAs LPE layer. The height (*H*) and width (*W*) of terraces are defined from the Talystep profile. Vertical distance 0.1 μm/div., horizontal distance 100 μm/div.

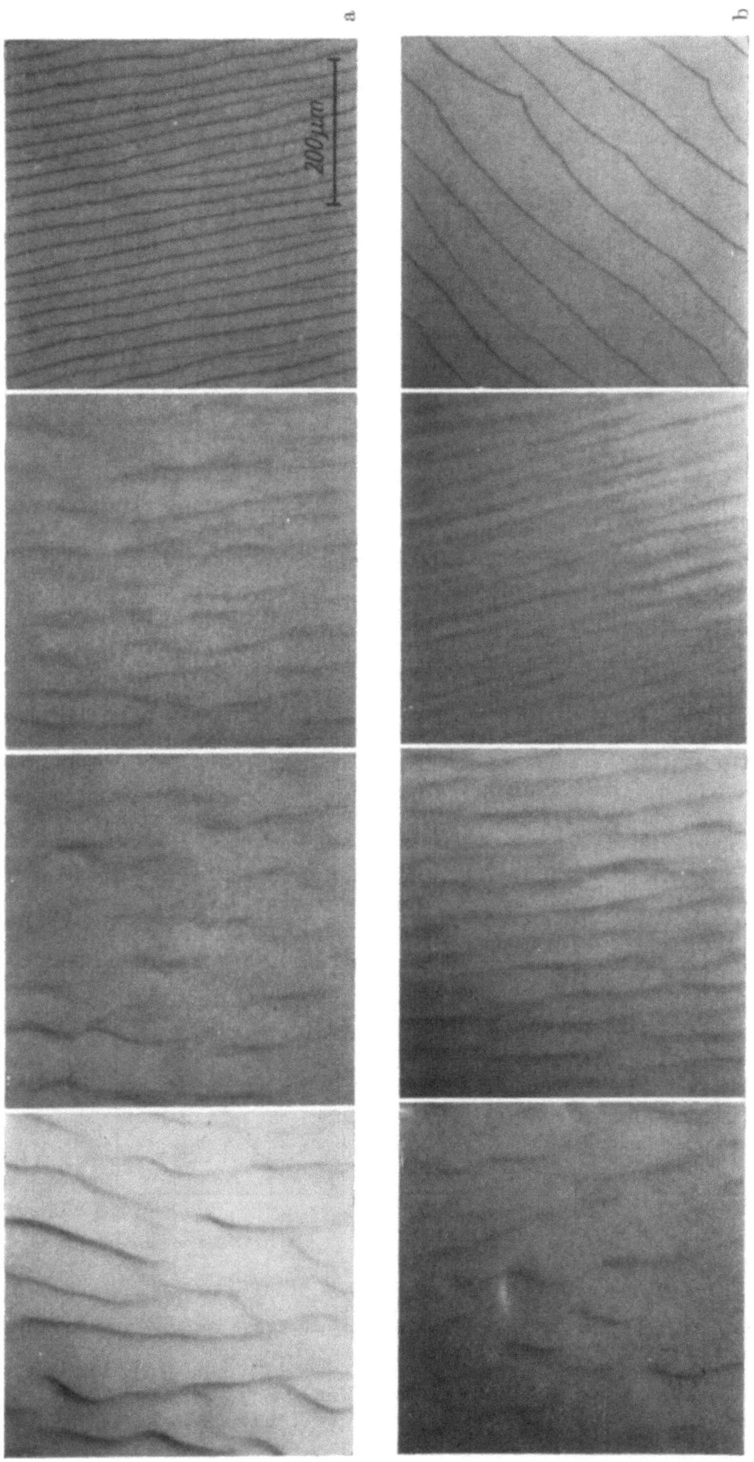

Fig. 2. Nomarski photomicrographs of GaAs LPE layers grown from Sn or Ge doped solutions. The concentrations of a) Sn (X^{l}_{Sn}) are 0.01, 0.1, 1, 15 at% (from left to right), and b) Ge (X^{l}_{Ge}) are 0.01, 0.1, 1, 7 at% (from left to right)

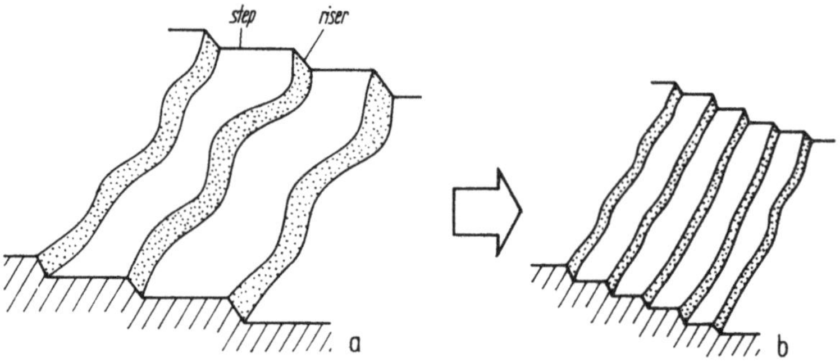

Fig. 3 Schematic drawings of the change of terrace structure with doping level. a) Low doping level, b) high doping level

are defined from the Tylastep profile shown in Fig. 1b, were determined to be 300 to 500 Å and 50 to 80 μm, respectively.

The doping experiments were made more than twice under the same growth conditions. The surface morphologies of grown layers were reproducible among runs. Fig. 2 shows the typical surface morphologies of layers grown from solutions with different Sn or Ge concentrations: (a) $X^l_{Sn} = 0.01$, 0.1, 1, 15 at%, (b) $X^l_{Ge} = 0.01$, 0.1, 1, 7 at%. Terraces clearly depend on the doping concentration. In Sn doping, terraces for $X^l_{Sn} \lessapprox 0.1$ at% are similar to those of the undoped layer shown in Fig. 1a. Thus terraces were not affected by Sn doping when $X^l_{Sn} \lessapprox 0.1$ at%. When X^l_{Sn} exceeds 1 at%, terraces rapidly change. In Ge doping, however, this effect becomes evident in the range $X^l_{Ge} > 0.1$ at%.

Fig. 3 shows schematically how the structure of terraces changes with the doping concentration. For low doping levels the risers of terraces are wavy. With the increase of doping level the risers become straight and parallel to one another. This trend applies except for $X^l_{Ge} = 7$ at%, where the terrace formation seems different from that for low doping levels.

To ascertain the doping effect quantitatively, the change of terrace width (W) with the doping concentration was examined. The terrace width (W) was measured at

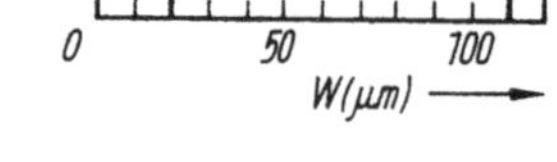

Fig. 4

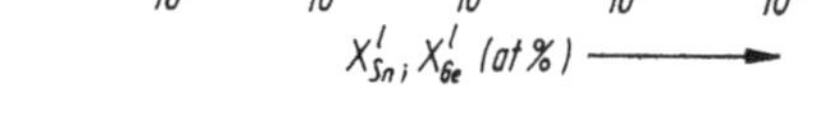

Fig. 5

Fig. 4. Distribution of terrace width (W) for $X^l_{Sn} = 0.01$ at%. The average value $\overline{W}$ and the standard deviation σ are 69 and 23 μm (33%), respectively

Fig. 5. Terrace width (W) as a function of the concentrations of Sn (X^l_{Sn}) or Ge (X^l_{Ge}) in the growth solution. The circles represent the average values of W. The length of the bars shows the magnitude of 2σ. ● Sn, ○ Ge

80 to 100 positions in individual wafers. Fig. 4 shows a typical distribution of W for $X^l_{Sn} = 0.01$ at%. Here the average value $\overline{W}$ and the standard deviation σ are 69 and 23 μm (33%), respectively. Fig. 5 shows the terrace width (W) as a function of Sn and Ge concentrations in the growth solution. The closed and open circles represent the average values of W for Sn and Ge doping, respectively. The length of the bars shows the magnitude of 2σ. It is found that there is a difference in the dependence of W on the doping concentration between Sn and Ge doping. In the Sn doping, W decreases monotonically with increasing X^l_{Sn}. In the Ge doping, W reaches a minimum at $X^l_{Ge} = 1$ at%. With further increasing X^l_{Ge}, W increases again.

4. Discussion

We have seen that the formation of terraces is affected by Sn and Ge doping, and that there is a difference between Sn and Ge doping in the dependence of terrace width (W) on the concentrations of Sn (X^l_{Sn}) and Ge (X^l_{Ge}) in the growth solution. Sn and Ge are incorporated into grown layers according to the Ga–As–Sn or Ga–As–Ge ternary phase diagram [22, 23]. The segregation coefficient differs between Sn and Ge. Thus we next examine the relation between the Sn or Ge concentration in grown layers and the formation of terraces.

To estimate the concentrations of Sn and Ge (N_{Sn}, N_{Ge}) in grown layers, we approximate that the Sn and Ge concentration are equal to the total ionized impurity concentration. For Ge doping, this approximation is valid within an accuracy ±10% for $X^l_{Ge} \lesssim 10$ at%, as shown by Rosztoczy and Wolfstein [24]. For Sn doping, Kressel et al. [25] have indicated that the approximation is valid within a factor of about 2 for $X^l_{Sn} \lesssim 16$ at%.

With this approximation, N_{Sn} and N_{Ge} can be derived from X^l_{Sn} and X^l_{Ge}, respectively, as follows. For Sn doping, the distribution coefficient k_{Sn}, determined by $(N_D - N_A)/X^l_{Sn}$, is (1.1 to 1.2) $\times 10^{-4}$ and the compensation ratio N_D/N_A is about 3 when the substrate orientation is (100) and the growth temperature (T_G) is 800 °C [22]. From these values, we obtain

$$N_{Sn} : N_D + N_A = 1.2 \times 10^{17} X^l_{Sn}\,. \tag{1}$$

For Ge doping, we can derive the values of k_{Ge} and N_D/N_A at $T_G = 800$ °C by interpolating the data by Rosztoczy and Wolfstein ($T_G = 900$ to 875 °C) [24] and Vilms and Garrett ($T_G = 760$ °C) [26], and obtain

$$N_{Ge} \doteqdot N_D + N_A = 4 \times 10^{18} X^l_{Ge}\,. \tag{2}$$

By using (1) and (2), the data in Fig. 5 are replotted in Fig. 6, which shows W as a function of N_{Sn} and N_{Ge}. The horizontal bars in Fig. 6 represent the accuracy of N_{Sn} estimated from the results by Kressel et al. [25]. It is noticed that the dependence of W on N_{Sn} or N_{Ge} agrees between Sn and Ge dopings, in contrast with the dependence of W on X^l_{Sn} or X^l_{Ge} shown in Fig. 5. It seems that Sn and Ge have the same effect on the formation of terraces.

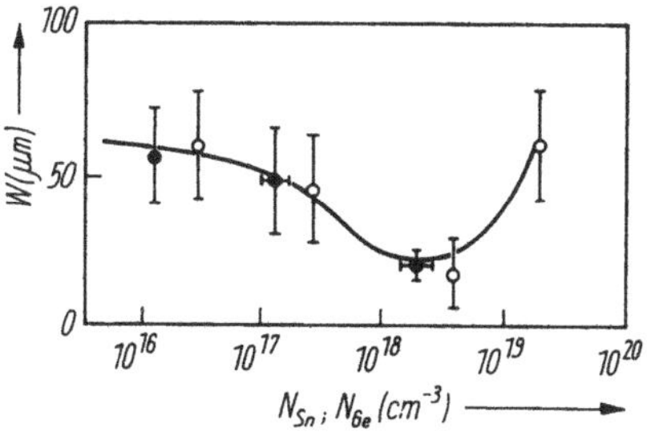

Fig. 6. Terrace width (W) as a function of the concentrations of Sn (N_{Sn}) or Ge (N_{Ge}) in grown layers. ● Sn, ○ Ge

These results suggest that (i) terrace formation is closely associated with the process of the impurity incorporation into grown layers, and (ii) the impurity effect does not depend sensitively on the species of impurity atom.

5. Summary

We have examined the effect of impurity on the terrace formation in GaAs LPE growth from Sn or Ge doped solutions. It is found that terraces sensitively depend on the doping concentration. Though the effects of Sn and Ge impurity on the terrace formation differ when compared as a function of the impurity concentration in the growth solution, they agree well when compared as a function of the total ionized impurity concentration, $N_D + N_A$, in grown layers.

Acknowledgements

The authors wish to thank I. Niikura, T. Hozuki, H. Sugibuchi, Y. Shimura, and S. Yokoyama for helpful discussions.

References

[1] E. Andre and J. M. LeDuc, Mater. Res. Bull. **3**, 1 (1968).
[2] B. L. Mattes and R. K. Route, J. Crystal Growth **16**, 219 (1972).
[3] J. T. Longo, J. S. Harris, Jr., E. R. Gertner, and J. C. Chu, J. Crystal Growth **15**, 107 (1972).
[4] E. Bauser, M. Frik, K. S. Leochner, L. Schmidt, and R. Ulrich, J. Crystal Growth **27**, 148 (1974).
[5] N. Toyoda, M. Mihara, and T. Hara, Appl. Phys. Letters **27**, 627 (1975).
[6] R. H. Saul and D. D. Roccasecca, J. appl. Phys. **44**, 1983 (1973).
[7] R. C. Peters and E. Andre, Acta electronica **17**, 9 (1974).
[8] A. Mottram and A. R. Peaker, J. Crystal Growth **27**, 193 (1974).
[9] A. Majerfeld and N. Moyes, Proc. 5th Biennial Cornell Electrical Engineering Conf., 1975 (p. 145).
[10] S. Guha, A. Majerfeld, N. Moyes, and P. N. Robson, Electronics Letters **11**, 303 (1975).
[11] V. Wrick, G. J. Scilla, L. F. Eastman, R. L. Henry, and E. M. Swiggard, Electronics Letters **12**, 394 (1976).
[12] K. Pak, T. Nishinaga, and S. Uchiyama, Japan. J. appl. Phys. **16**, 949 (1977).
[13] T. Kobayashi and Y. Furukawa, Japan. J. appl. Phys. **15**, 171 (1976).
[14] H. F. Lockwood, see [9] (p. 127).
[15] M. B. Small and R. M. Potemski, J. Crystal Growth **37**, 163 (1977).
[16] T. Fukui and T. Kobayashi, Japan. J. appl. Phys. **16**, 2081 (1977).
[17] T. Fukui, K. Wakita, Y. Horikoshi, and Y. Furukawa, Japan. J. appl. Phys. **15**, 2005 (1976).
[18] D. L. Rode, J. Crystal Growth **27**, 313 (1974).
[19] D. L. Rode, W. R. Wagner, and N. E. Schumaker, Appl. Phys. Letters **30**, 76 (1977).
[20] N. Toyoda, M. Mihara, and T. Hara, J. appl. Phys. **47**, 443 (1976).
[21] J. J. Hsieh, J. Crystal Growth **27**, 49 (1974).
[22] M. B. Panish, J. appl. Phys. **44**, 2659 (1973).
[23] M. B. Panish, J. appl. Phys. **44**, 2676 (1973).
[24] F. E. Rosztoczy and K. B. Wolfstein, J. appl. Phys. **42**, 426 (1971).
[25] H. Kressel, H. Nelson, and F. Z. Hawrylo, J. appl. Phys. **39**, 5647 (1968).
[26] J. Vilms and J. P. Garrett, Solid State Electronics **15**, 443 (1972).

(Received April 17, 1979)

phys. stat. sol. (a) 54, 231 (1979)

Subject classification: 16; 12.2; 22.8

Institute of Crystallography, Academy of Sciences of the USSR, Moscow (a), Institute of Crystallography, University Freiburg i. Br. (b), and Department of Physics, University Rostov (Don) (c)

Photoferroelastic Phenomena in Sb_5O_7I Crystals

By

V. M. FRIDKIN (a), R. NITSCHE (b), N. KORCHAGINA (c), N. A. KOSONOGOV (c), R. MAGOMADOV (b), A. I. RODIN (c), and K. A. VERKHOVSKAYA (b)

The influence of band-gap radiation on various properties of the two ferroelastic polytypes 2MC and 2MA of Sb_5O_7I is investigated. In addition to normal photoconduction, an influence of radiation on a) the spontaneous deformation (photodeformation), b) the double refraction (photorefractive effect), and c) the domain structure (photodomain effect) is noted. Free energy calculations indicate that these effects are caused by non-equilibrium carriers in concentrations of the order of 10^{16} cm^{-3}.

Es wird der Einfluß von Bandgap-Strahlung auf verschiedene Eigenschaften der beiden ferroelastischen Polytypen 2MC und 2MA von Sb_5O_7I untersucht. Zusätzlich zur normalen Photoleitung wird ein Einfluß der Strahlung auf a) die spontane Deformation (Photodeformation), b) die Doppelbrechung (photorefraktiver Effekt) und c) die Domänenstruktur (Photodomäneneffekt) bemerkt. Berechnungen der freien Energie zeigen, daß diese Effekte durch Nichtgleichgewichtsträger in Konzentrationen der Größenordnung 10^{16} cm^{-3} hervorgerufen werden.

1. Introduction

Pentaantimony-heptaoxide-iodide, Sb_5O_7I, can exist in at least eight polytypic modifications. Crystals of Sb_5O_7I, grown from the vapour phase [1], are either purely ferroelastic or ferroelastic + ferroelectric, depending on whether the polytypes have centric (C) or acentric (A) structures [2].

The simplest centric two-layer polytype, denoted 2MC¹) (M monoclinic) undergoes a phase transition at a Curie temperature of $T_0 = 481$ K with an accompanying change in symmetry: $P6_3/m$ (2HC-Sb_5O_7I) → $P2_1/c$ (2MC-Sb_5O_7I). In the monoclinic, low-temperature phase the material is ferroelastic. Due to its pseudotrigonal structure, a crystal can exist in three possible states of spontaneous deformation forming angles of 120° with each other in the *ac* plane. Mechanical switching is achieved by applying a uniaxial stress in the *ac* plane along the [20$\bar{1}$] direction of a domain. Loops of mechanical hysteresis have been observed by following the change in domain structure upon switching. The coercive stress at room temperature was estimated to 10^6 N/m^2 [2].

The simplest acentric two-layer polytype 2MA²) is ferroelastic + ferroelectric below $T_0 = 438$ K. The change in symmetry is $P\bar{6}$ (2HA-Sb_5O_7I) → Pc (2MA-Sb_5O_7I) and there exist three possible directions of spontaneous polarization in the *ac* plane, forming angles of 120° with each other. Switching of 2MA under electric fields and the types of domain structures have been studied in [2].

This paper is concerned with effects observed when light interacts with Sb_5O_7I crystals. In addition to being photoconductive (photocurrent: dark current $\approx 10^3$ in 2MC), Sb_5O_7I exhibits a photorefractive effect. The light-induced non-equilibrium

¹) Formerly called α-modification.

²) Formerly called β-modification.

electrons also influence the phase transitions, the ferroelastic properties, and the domain structure. These photoferroelastic phenomena are analogous to photoferroelectric phenomena [3], the mechanisms of which may be associeted with the interactions of electrons with a soft acoustic mode at the boundary of the Brillouin zone. In contrast to photoferroelectric phenomena, the mechanisms underlying photoferroelasticity cannot involve screening of the spontaneous polarization or the anomalous photovoltaic effect [4].

2. Photoconductivity and Intrinsic Optical Absorption

Fig. 1 shows the spectral dependence of the photocurrent at room temperature for 2MC (curve 2) and 2MA (curve 1). For both polytypes the maximum in photocurrent is close to the intrinsic optical absorption edge ($E_g = 3.3$ eV).

Fig. 2 is a plot of dark and photoconductivity (measured in a-direction) of 2MC versus temperature for illumination with band-gap light. Both curves reveal anomalies near the Curie temperature at 481 K.

In Fig. 3 the temperature dependence of the energy gap, $E_g(T)$, is depicted for the polytypes 2MC (curve 2) and 2MA (curve 1). In both cases, dE_g/dT shows a discontinuity, whereas E_g varies continuously. (Experimental accuracy for E_g was better than 10^{-3} eV.) This behaviour is typical for second-order phase transitions [5]. Numerical values for 2MC (f ferroeleastic, p paraelastic phase) are $(dE_g dT)_f \approx$ $\approx -1.46 \times 10^{-3}$ eVK^{-1}; $(dE_g/dT)_p \approx -1.43 \times 10^{-3}$ eVK^{-1}, i.e. $\Delta(dE_g/dT) \approx 3 \times$ $\times 10^{-5}$ eVK^{-1}.

In a pure ferroelastic like the polytype 2MC, the stress-induced change in optical absorption ΔE_g^σ should show a maximum at the Curie temperature T_0, in analogy to the electrical absorption in ferroelectrics [6].

In fact, the change in band gap ΔE_g^σ should vary linearly with the applied stress σ in the ferroelastic and quadratically in the paraelastic phase. To prove this, the change in absorption edge of a crystal, stressed in the ac plane, was measured (Fig. 4a). It can be seen that ΔE_g^σ has a maximum at $T_0 = 433$ K.[3]) In the ferroelastic phase,

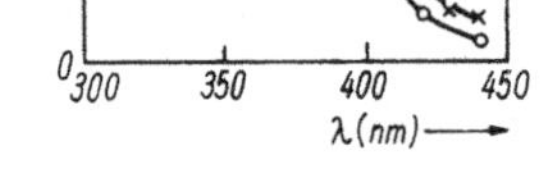

Fig. 1

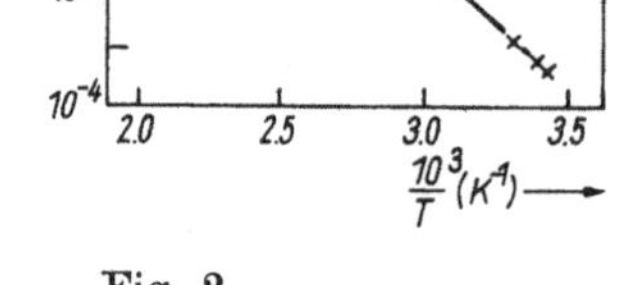

Fig. 2

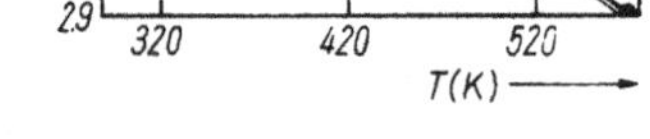

Fig. 3

Fig. 1. Spectral distribution of photoconductive response of (1) 2MA-Sb_5O_7I and (2) 2MC-Sb_5O_7I

Fig. 2. Temperature dependence of dark and photoconductivity (curves 1 and 2, respectively) in 2MC-Sb_5O_7I under band-gap illumination

Fig. 3. Temperature dependence of band gap for 2MA-Sb_5O_7I (curve 1) and 2MC-Sb_5O_7I (curve 2). Arrows indicate phase transition temperatures

[3]) This 2MC-Sb_5O_7I crystal was doped with arsenic and therefore had a reduced transition temperature of 433 K.

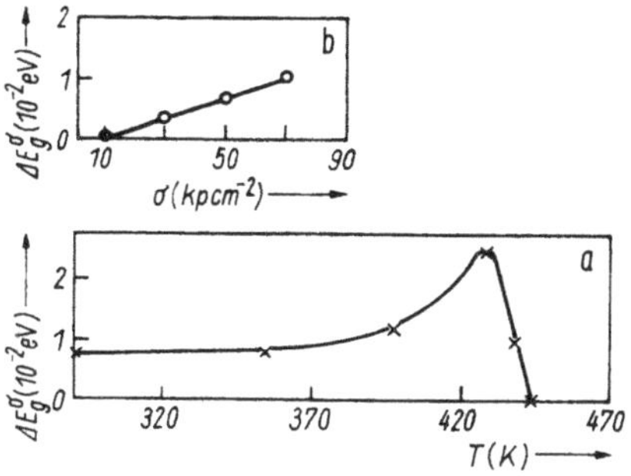

Fig. 4. Elastoabsorption in a 2MC-Sb_5O_7I crystal having a lowered transition temperature (433 K) because of a small As content. a) Temperature dependence of the stress-induced shift in band gap ΔE_g^σ. b) Stress-induced change of band gap ΔE_g^σ at 383 K

ΔE_g^σ varies linearly with σ (Fig. 4b). The slope at 383 K is $d\Delta E_g^\sigma/d\sigma = 2 \times 10^{-4}$ eV cm²kp⁻¹. In the paraelastic phase, ΔE_g^σ was too small to be measured.

3. Spontaneous Deformation and Spontaneous Birefringence

The temperature dependence of the spontaneous deformation $\eta(T)$ and of the spontaneous birefringence $\Delta n(T) = n_a(T) - n_c(T)$ are depicted in Fig. 5. The spontaneous deformation was determined by means of the dilatometric method described by Grekov et al. [7]. The spontaneous birefringence was measured by a compensation method at $\lambda = 633$ nm. Both quantities, $\eta(T)$ and $\Delta n(T)$ show a behaviour typical for second-order phase transitions including the transition temperature $T_0 = 481$ K. They can be fairly well described by the relations

$$\eta(T) = 4 \times 10^{-4} (T_0 - T)^{1/2} \tag{1}$$

and

$$\Delta n(T) = 1.5 \times 10^{-3} (T_0 - T)^{1/2} . \tag{2}$$

Thus, in ferroelastic 2MC-Sb_5O_7I the spontaneous deformation is a parameter determining the spontaneous birefringence according to $\Delta n \approx 4\eta$.

4. The Photorefractive Effect

In crystals of 2MC-Sb_5O_7I a photorefractive effect was detected. Similar to ferroelectrics, also in ferroelastics an influence of illumination on the spontaneous birefringence Δn (caused by non-equilibrium electrons) can be noted.

The effect was measured by illuminating a monodomain crystal with monochromatized light from a xenon lamp in the region of the photoconductivity maximum (Fig. 1) and simultaneously determining $\Delta n = n_a - n_c$ at $\lambda = 633$ nm (for details, see [8]). It was found that the sign of the photorefractive effect, $\delta\Delta n$, is identical with that of $\delta\Delta n$ caused by heating from the irradiation. Therefore, the true photorefractive effect and the "pyroeffect" could be separated by investigating spectral sensitivity and decay with time of $\delta\Delta n$.

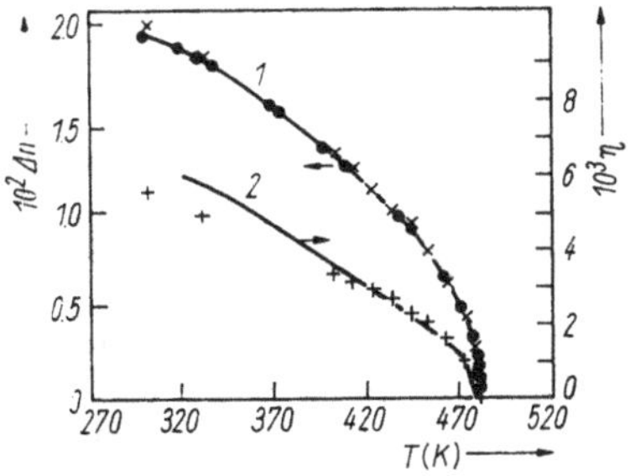

Fig. 5. Temperature dependence of (1) spontaneous birefringence Δn and (2) spontaneous deformation η in 2MC-Sb_5O_7I

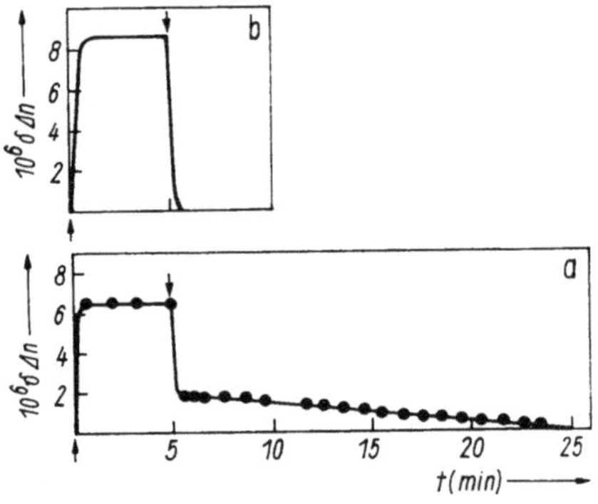

Fig. 6

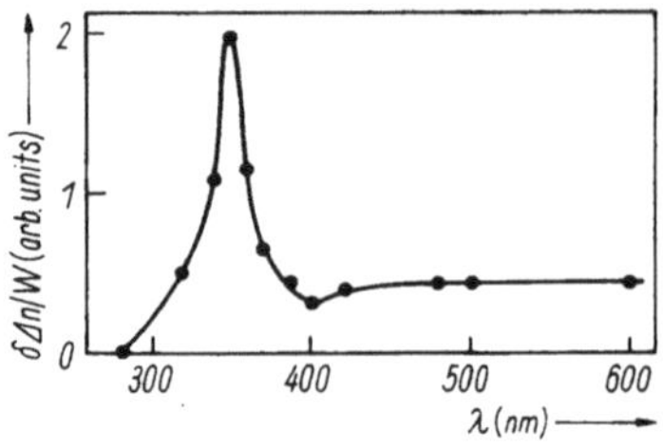

Fig. 7

Fig. 6. Kinetics of the photorefractive effect $\delta\Delta n$ at 98 K in 2MC-Sb_5O_7I. a) Excitation with intrinsic light, observation with red light ($\lambda = 633$ nm). b) Observation with red light only

Fig. 7. Spectral dependence of photorefractive effect at 98 K in 2MC-Sb_5O_7I. (Ordinate normalized to unit energy)

Fig. 6a shows $\delta\Delta n$ for a crystal at $T = 98$ K upon illumination with light of $\lambda = 350$ nm (corresponding to the photoconductivity maximum at that temperature). After switching off the exciting light, a residual photorefractive effect (magnitude $\delta\Delta n \approx 2 \times 10^{-6}$) persists, which decays slowly with time. This residual, true photorefractive effect is proportional to the intensity of the exciting (intrinsic) light. Its decay is accelerated with increasing temperature. In contrast, Fig. 6b is an experiment carried out under illumination with $\lambda = 480$ nm, a wavelength at which no appreciable photoconductivity was noted. Here the observed $\delta\Delta n$ was due only to a "pyroeffect", since $\delta\Delta n$ vanishes immediately after switching off the light.

Fig. 7 illustrates the spectral dependence of the photorefractive effect at $T = 98$ K. On the ordinate the magnitude of $\delta\Delta n$ is plotted (referred to unit energy of the exciting radiation), measured upon switching on the illumination. The maximum in Fig. 7 coincides with the maximum of photoconductivity at the given temperature.

5. The Photodomain Effect

A photodomain effect [9], i.e. changes in domain pattern under illumination with band-gap light when a crystal is cooled or heated in the vicinity of its Curie point, has been found in 2MA-Sb_5O_7I. The crystal was first observed in red light on the heating stage of a polarizing microscope. Then, additional illumination in the intrinsic region, where photoconduction takes place, was switched on and the domain structure photographed again. This is shown in Fig. 8, where the column "dark" refers to a heating (and cooling) cycle under red light only and the column "light" to a cycle with additional band-gap illumination. During the heating cycle no photodomain effect is found. However, upon cooling, during the transition from the high- to the low-temperature phase, much smaller domains are formed under illumination with band-gap light. These small domains persist down to room temperature and are preserved indefinitely.

The photodomain effect is found only in the absence of stress. With a uniaxial stress applied in the *ac* plane, the effect became unnoticeable. This indicates that the photodomain effect may be associated with the influence of the photogenerated non-

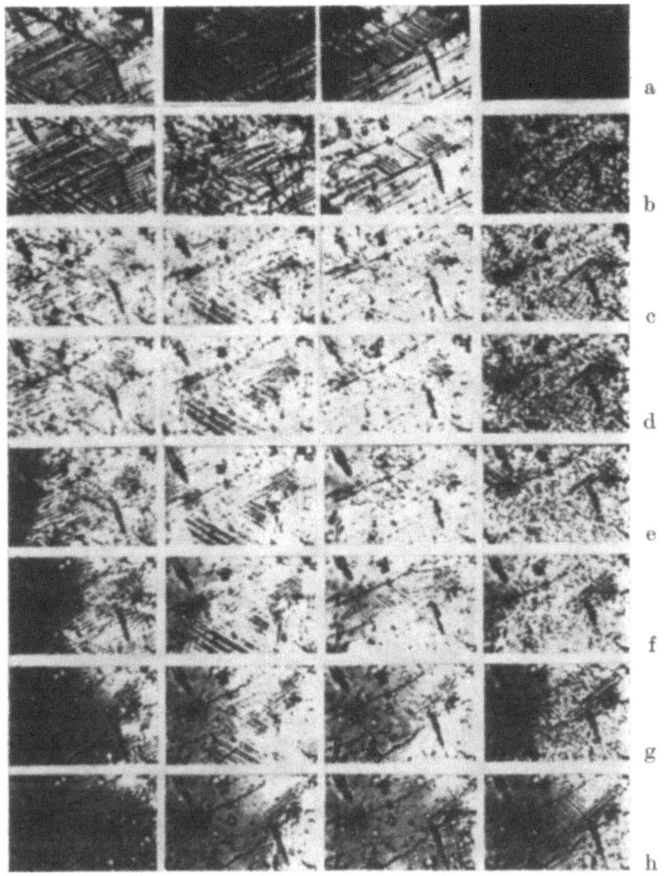

Fig. 8. The photodomain effect in 2MA-Sb_5O_7I; column "dark" is a heating and cooling cycle without, column "light" with additional band-gap illumination. a) $T = 373$, b) 426, c) 427, d) 427.2, e) 427.4, f) 427.6, g) 427.8, h) 428 K

equilibrium carriers on the elastic deformation near the Curie point (and not with screening of spontaneous polarization as in proper ferroelectrics [9]).

In 2MC-Sb_5O_7I no photodomain effect could be detected, possibly because of the high conductivity of this polytype near the Curie point (see Fig. 2).

The data of Fig. 8 also point to the absence of any appreciable photostimulated shift in Curie temperature. Such an influence of non-equilibrium carriers on T_C has previously been observed in photoferroelectrics [5].

6. Discussion

Let us consider the thermodynamics of photoferroelastic phenomena for an improper ferroelastic such as 2MC-Sb_5O_7I. Restricting ourselves to second-order phase transitions, the free energy F_f of a ferroelastic crystal can be expanded in terms of the spontaneous deformation [10]:

$$F_f(T, \eta, \sigma) = F_p + \alpha\eta^2 + \beta\eta^4 + \eta\sigma - \tfrac{1}{2} S_0\sigma^2 , \tag{3}$$

where

$$\alpha = \alpha'(T - T_0) . \tag{4}$$

F_p is the free energy of the paraelastic phase, β a constant, independent of temperature, σ a component of the mechanical stress tensor, and S_0 the non-critical part of the modulus of elasticity (far from the Curie point).

From the condition for the minimum of free energy, $\partial F_f/\partial\eta = 0$ we obtain for η_0, the spontaneous deformation at $\sigma = 0$,

$$\eta_0 = \left(\frac{\alpha'}{2\beta}\right)^{1/2} (T_0 - T)^{1/2} \qquad \text{for} \qquad T < T_0 . \tag{5}$$

Determining the modulus of elasticity $S = -(\partial^2F/\partial\sigma^2)_{\sigma=0}$ from $1/S = -(\partial^2F/\partial\eta^2)_{\sigma=0}$ we arrive at the Curie-Weiss law for a ferroelastic:

$$S = \frac{1}{4\,\alpha'(T_0 - T)} \qquad \text{for} \qquad T < T_0 . \tag{6}$$

We consider a photoferroelastic — like a photoferroelectric [5] — as an intrinsic semiconductor with an energy gap E_g, having a concentration of free (or localized in shallow traps) carriers N.

Then, at a sufficiently high concentration N of non-equilibrium carriers, the free energy of the photoferroelastic will be

$$F = F_f + F_{el} = F_f + E_gN . \tag{7}$$

Expanding E_g in terms of the phase transition parameter η gives

$$E_g^f(T, \eta, \sigma) = E_g^p(T) + a_1\eta^2 + a_2\eta^4 + E_g'\sigma + \tfrac{1}{2} E_g''\sigma^2 , \tag{8}$$

where

$$E_g' = \left(\frac{\partial E_g}{\partial\sigma}\right)_{\sigma=0} , \qquad E_g'' = \left(\frac{\partial^2 E_g}{\partial\sigma^2}\right)_{\sigma=0} . \tag{9}$$

By substituting (8) into (7) we arrive at an expression for the free energy of a photoferroelastic with terms arranged similarly as in (3):

$$F = F_{pN} + \alpha_N\eta^2 + \beta_N\eta^4 + \sigma(\eta + E_g'N) - \tfrac{1}{2}(S_0 + E_g''N)\,\sigma^2 , \tag{10}$$

where

$$F_{pN} = F_p + E_g^pN , \qquad \alpha_N = \alpha + a_1N , \qquad \beta_N = \beta + a_2N . \tag{11}$$

We shall now consider two phenomena in photoferroelastics on the basis of (10) and (11), namely (i) the photorefractive effect and (ii) the photostimulated shift in Curie temperature.

(i) The spontaneous birefringence in a proper ferroelastic is determined by the spontaneous deformation, i.e. $\Delta n = \Delta n(\eta)$. Therefore, it is natural to associate the photorefractive effect with photodeformation, i.e. with the influence of non-equilibrium carriers on spontaneous deformation. Applying the general equation of state

$$\left(\frac{\partial F}{\partial\sigma}\right) = \eta \tag{12}$$

to (10) and denoting with η_N the spontaneous deformation under illumination, it follows

$$\eta_N = \eta + E_g'N . \tag{13}$$

From the data of Fig. 4b, the magnitude of $E_g' = \partial\Delta E_g^\sigma/\partial\sigma$ is estimated to be of the order of 10^{-22} cm^3. Neglecting the temperature dependence of E_g' we have for the photodeformation

$$\Delta\eta = \eta_N - \eta \approx 10^{-22}N .$$

From (1) and (2) it had been shown that $\Delta n \approx 4\eta$. Therefore, for the photorefractive effect one obtains

$$\delta\Delta n \approx 4\eta = 4 \times 10^{-22}N . \tag{14}$$

The photorefractive effect actually observed in 2MC-Sb_5O_7I (see Fig. 6) is $\delta\Delta n \approx \approx 2 \times 10^{-6}$. It follows that the carrier concentration N in (14) should be of the order $N \approx 10^{16}$ cm^{-3}. This is a reasonable estimate, since in high-resistance semiconductors the trap concentration may reach 10^{17} to 10^{18} cm^{-3}.

(ii) The dependence of the shift in Curie temperature $\Delta T = T_0 - T_N$ under illumination on the concentration N of equilibrium carriers [5] is

$$\Delta T = \frac{a_1}{\alpha'} N . \tag{15}$$

Numerical values for a_1 and α' can be estimated from the temperature dependence of the spontaneous deformation (equation (5)) from the jump of $\Delta(\mathrm{d}E_g/\mathrm{d}T) = = a_1\alpha'/2\beta$ at the phase transition and from the expression for the elastic constant S (equation (6)) which is determined from the data on elastoabsorption in Fig. 4b.

These estimates lead to values of $\alpha' \approx 10^{10}$ cgs units and $a_1 \approx 10^{-10}$ cgs units, i.e. $\Delta T \approx 10^{-20}N$.

Thus, for a plausible trap concentration of $N = 10^{17}$ to 10^{18} cm^{-3}, the photostimulated shift of the Curie point would be of the order of $\Delta T \approx 10^{-2}$ to 10^{-3}, i.e. much lower than the accuracy of temperatuie measurement and in fact no such shift has been found.

It also should be noted that the peculiarities of elastoabsorption in a ferroelastic follow immediately from (8):

$$E_g^f \approx E_g^p + a_1(\eta_0 + S\sigma)^2 \approx E_g^p + a_1\eta_0^2 + 2a_1S\sigma \quad \text{for} \quad T < T_0 . \tag{16}$$

According to (16), E_g^f depends linearly on the stress in the ferroelastic phase (Fig. 4a, b), whereas $\mathrm{d}E_g^f/\mathrm{d}\sigma$ is determined by the Curie-Weiss law for S (equation (6)).

Acknowledgements

We are indebted to Mrs. M. Schuhmacher for growing the crystals. One of us (R. N.) is grateful for financial support by the Deutsche Forschungsgemeinschaft.

References

[1] V. Krämer, R. Nitsche, and M. Schuhmacher, J. Crystal Growth 24/25, 179 (1974).
[2] R. Nitsche, V. Krämer, M. Schuhmacher, and A. Bussmann, J. Crystal Growth 42, 549 (1977).
[3] V. M. Fridkin, Ferroelectrics–Semiconductors, Izd. Nauka, Moscow 1976 (in Russian).
[4] G. Chanussot, Ferroelectrics 20, 37 (1978).
[5] V. M. Fridkin, Soviet Phys. — J. exper. theor Phys., Letters 3, 252 (1966).
[6] G. Harbeke, J. Phys. Chem. Solids 24, 957 (1963).
[7] A. A. Grekov, E. D. Rogach, and A. G. Sukijazov, Soviet Phys. — Solid State 12, 3559 (1970).
[8] P. Günter and F. Micheron, Ferroelectrics 18, 27 (1978).
[9] V. M. Fridkin, A. A. Grekov, N. A. Kosonogov, and T. R. Volk, Ferroelectrics 4, 169 (1972).
[10] B. A. Strukov, Ferroelectrics, Izd. Nauka, Moscow 1979 (in Russian).

(Received March 7, 1979)

phys. stat. sol. (a) **54**, 239 (1979)

Subject classification: 1.3; 18.2; 21.1; 21.1.1

Institute of Metal Physics, Academy of Sciences of the USSR, Ural Scientific Centre, Sverdlovsk

Atomic Structure of the Invar Fe–Ni Alloys

By

S. F. DUBININ, S. G. TEPLOUCHOV, and S. K. SIDOROV

Quenched Ni–Fe monocrystals are studied in wide temperature and composition intervals by the method of elastic neutron scattering. Superstructural reflections, the wave-vector moduli of which are incommensurable with the basic vectors of the reciprocal lattice, are found in the $65.5 \leqq C \leqq 70$ interval (C is the atomic Fe concentration). The analysis of the dependence of the satellite intensities on the temperature and the external magnetic field allows to make conclusions on the nature of the incommensurable structure and to draw the phase diagram of the invar alloys.

Getemperte Ni–Fe-Einkristalle werden in einem großen Temperatur- und Zusammensetzungsbereich mit der Methode der elastischen Neutronenstreuung untersucht. Superstrukturreflexionen, deren Wellenvektormodulen inkommensurabel mit den Basisvektoren des reziproken Gitters sind, werden im Intervall $65{,}5 \leqq C \leqq 70$ beobachtet (C ist die atomare Fe-Konzentration). Die Analyse der Abhängigkeit der Satellitenintensitäten von der Temperatur und dem äußeren Magnetfeld erlaubt, Schlüsse über die Art der inkommensurablen Struktur zu ziehen und das Phasendiagramm der Invarlegierungen aufzustellen.

1. Introduction

Among the disordered ferromagnetic Ni–Fe alloys the alloys with 60 to 70 at% Fe may be chosen as a special group called invar alloys since their physical properties anomalously depend upon composition and temperature.

For a long time these alloys were considered to have the usual f.c.c. structure with Fm3m space group above the temperature of the martensite transformation. However, in [1] on the X-ray pattern of a monocrystal with 69 at% Fe at 300 K, there were found besides the structural reflections of the γ-phase the superstructural reflections giving evidence, strictly speaking, to the existence of an atomic structure with symmetry lower than the f.c.c. one. The neutronographical studies [2] carried out independently of the austenite phase at room temperature of a monocrystalline 67.7 at% Fe sample, cooled before to 4.2 K in a magnetic field perpendicular to the scattering vector, have shown the presence of such superstructural reflections whose moduli of the wave vectors $\boldsymbol{k}_i$ ($i = 1, 2, ...$) were incommensurable with the basic vectors of the reciprocal lattice. Detailed studies of the concentration dependences of different physical properties in the polycrystalline samples [3] also give evidence to the fact that far below the martensite transformation in Fe–Ni alloys a phase transition of unknown nature takes place.

All these experimental data tell that a new problem had appeared — *the problem of the atomic structure of the invar iron–nickel alloys*. Therefore, it is necessary to carry out thorough and systematic investigations of alloy monocrystals in an external magnetic field and without it in wide intervals of temperature and composition. Taking into account that the wave vectors of the unknown structure can lie in a non-symmetric direction [2] it is necessary to investigate a rather large volume of the Brillouin zone, which can be done in the most reliable way by the neutronographic method.

2. Samples and Experimental Method

Electrolytic nickel and carbonyl iron of 99.99% purity were the raw materials for monocrystal growth. The monocrystals in the concentration interval $50 \leqq C \leqq \leqq 70$ at% Fe were grown by Bridgman's method in alundum crucibles.

The alloys were subjected to a long homogenization and to quenching from 1100 °C into water. Then balls 12 mm in diameter were cut off from the obtained ingots. In addition samples of the alloys 50; 65.5; 67.7 at% Fe, 18 mm in diameter were prepared. In order to remove the stresses which appeared in preparing the monocrystals the surface layer was etched away. To determine the composition of the alloy a chemical analysis of the samples was done.

The studies of the elastic neutron scattering were carried out on a diffractometer with a double monochromator (two copper monocrystals were used in the experiment). The neutron wavelength λ was equal to 1.35 Å. The admixture of neutrons of second and third order was 1.5 and 3×10^{-2}%, respectively. The neutron diffraction patterns of the alloys were studied in the most informative part of the Brillouin zone, i.e. in its sections by the planes (100), (110), and (111). In these sections the scattering vector $\boldsymbol{q} = \boldsymbol{q}_2 - \boldsymbol{q}_1$ ($\boldsymbol{q}_2$, $\boldsymbol{q}_1$ are the wave vectors of the scattered and the incident waves, respectively, $|\boldsymbol{q}_2| = |\boldsymbol{q}_1| = 2\pi/\lambda$) varied within the interval $0 < \boldsymbol{q} < \boldsymbol{q}_m = \boldsymbol{k}_B$ ($\boldsymbol{k}_B$ is the wave vector on the Brillouin zone boundary).

3. Experimental Results and Discussion

3.1 Superstructural reflections

The study of the intensity of the elastic neutron scattering by a monocrystal of equiatomic composition in the interval 4.2 to 300 K in the external magnetic field $H = 1000$ Oe and without it did not reveal any peculiarities. With increasing iron concentration regions of diffuse (quasi-elastic) neutron scattering with distribution in the reciprocal lattice analogous to the earlier published one [4] appear in the alloy.

Finally, beginning approximately from 65.5 at% Fe the superstructural reflections (satellites) $\boldsymbol{k}_1$, $\boldsymbol{k}_2$, $\boldsymbol{k}_3$ in a wide interval of composition above the temperature of the

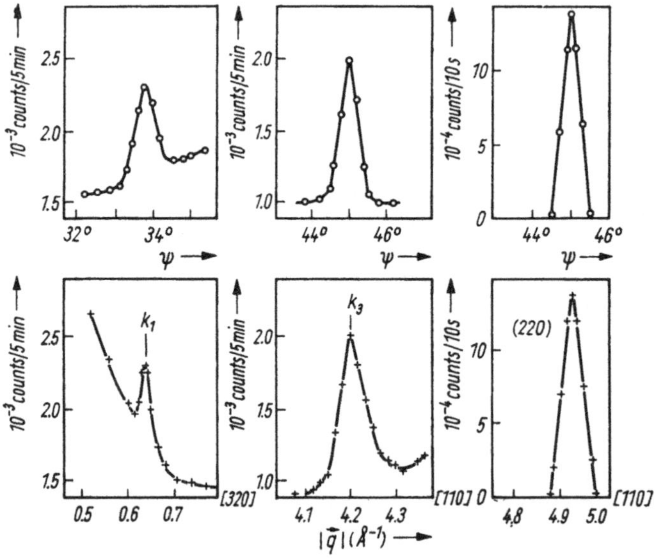

Fig. 1. Superstructural reflections (satellites) compared with the intensity of the structural peak (220)

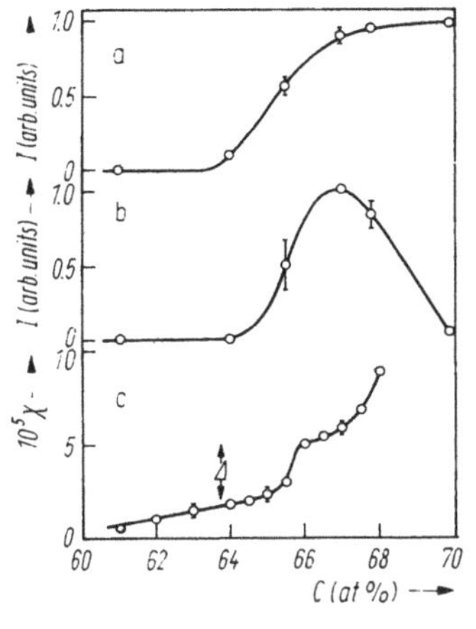

Fig. 2. a), b) Concentration dependence of the intensity of the satellites k_3 and k_1 at 300 K. c) Concentration dependence of the paraprocess susceptibility (in e.m.u./g), obtained in [3] at 225 K

martensite transformation, M_s, appear and remain. The reflections k_1 and k_2 depend upon H. Their intensities increase in the external magnetic field $H \perp q$ more than three times compared with their values in the non-magnetized state. The intensity of the satellite k_3 does not depend upon H. Fig. 1 shows the intensities of the satellites k_1 ($H \perp q$) and k_3 in the section (001) at room temperature in the alloy with 67.7 at% Fe in comparison with the structural reflection (220): in the lower part of the figure the dependence upon $|q| = 4\pi \sin \theta/\lambda$ is shown in the directions [320] and [110]: in the upper part the dependence upon ψ at fixed value of $|q|$ (2θ is the angle between the wave vectors of the incident and the scattered waves, ψ the angle between [100] and q). The coordinates of the reflections in the reciprocal lattice do not depend on the parameters C and T. The intensities J_{ki}, as it is seen in Fig. 2a and b depend upon the alloy composition. The satellites k_1, k_2 vanish with the heating of the alloys above the Curie temperature T_C. Fig. 3 shows the temperature dependence of the integral intensity of k_3 in the alloy with 67.7 at% Fe, it is seen that the coherent reflections vanish at $T_t \approx 880$ K, i.e. far above $T_C = 430$ K.

The coordinates of the superstructural reflections (in fractions of basic translations of the reciprocal lattice) are expressed as follows:

$$\left.\begin{aligned} k_1 &= (\pm 0.31;\ \pm 0.21;\ 0) \quad \text{and} \quad (\pm 0.21;\ \pm 0.31;\ 0)\,,\\ k_2 &= (\pm 0.488;\ \pm 0.464;\ 0) \quad \text{and} \quad (\pm 0.464;\ 0.488;\ 0)\,,\\ k_3 &= (1+\delta)\,(\pm \tfrac{5}{3};\ \pm \tfrac{5}{3};\ 0)\,, \end{aligned}\right\} \tag{1}$$

in accordance with [1] $\delta = 0.025$.

The thorough analysis of the coordinates (1) allows to consider that the satellites cannot be multiple diffraction harmonics of the f.c.c. structural reflections. The dependence of the intensity of the satellites upon the composition of the samples, having the same geometrical size, apparently excludes that the reflections are caused by the effects of multiple scattering. The superstructural reflections are also not connected with the effects of the surface scattering since their values are proportional to the number of atoms in the unit volume. All these follow from the comparison of the scattering pictures of the samples of the same composition but with different diameters.

Thus, the observed superstructure can be connected only with the change of the atomic structure of the invar alloys. Taking into account that the coordinates of the

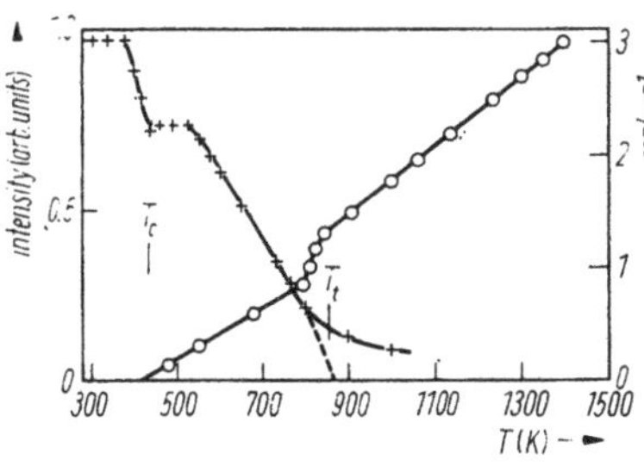

Fig. 3. + temperature dependence of the integrated intensity of the reflection k_3; ○ temperature dependence of the inverse paramagnetic susceptibility (in g/e.m.u.), obtained in [6]

wave vectors (1) are not rational fractions of the basic vectors of the reciprocal lattice, the formation of the incommensurable atomic structure takes place in alloys with $65.5 \leqq C \leqq 70$ at% Fe above the temperature of the martensite transformation, M_s, while at $C > 65.5$ at% Fe (see Fig. 2a and b) the structural phase transition takes place. Just in this region of composition the changes of the physical properties of the alloys have critical character [3]. As an example, Fig. 2c shows the concentration dependence of the paraprocess susceptibility, obtained in [3] at 225 K. It is seen from the figure that the jump of the susceptibility and the appearance of the satellites take place approximately at the same concentration of iron in the alloy.

3.2 *On the nature of the superstructural reflections*

As in this paper the structure of the quenched Fe–Ni alloys is studied, it is hard to assume that the observed superstructure can be connected with the development of static waves of composition or the order degree.

It is left to consider that the origin of the satellites is caused either by the modulated change of the scattering amplitude connected with the regular redistribution of the charge (spin) density of the atoms in the alloy, or by the modulation of the scattering phase due to the static displacements of the atoms or the spin density from the sites of the intitial f.c.c. lattice according, for example, to the sine law. The neutron scattering patterns in these two cases are different and they were described in detail in the review by Axe [5]. According to [5], in the first case the neutron pattern corresponds to the sinusoidal or helicoidal structure whose intensities of the satellites decrease proportionally to the square of the magnetic form factor with increasing value of $|\boldsymbol{q}|$. In the second case, the main properties of the diffraction picture are determined by the structural scattering amplitude

$$F(\boldsymbol{q}) = f\,\Delta R^n (\boldsymbol{e_k} \cdot \boldsymbol{q})^n \sum_{(\nu)} \exp\,(2\pi i \boldsymbol{k}_0 \cdot \boldsymbol{R})\, \delta(\boldsymbol{q} + 2\pi \boldsymbol{k}_0 + n\boldsymbol{k}_i)\,, \tag{2}$$

where f is the mean (per atom) nuclear-magnetic scattering amplitude, ΔR the displacement amplitude, $\boldsymbol{e_k}$ the polarization vector, ν the number of atoms in the elementary cell, $\boldsymbol{k}_0$ the commensurable vector of the structure, $\boldsymbol{R}$ the radius vector of the R lattice site, $n\boldsymbol{k}_i$ the incommensurable vector of order n of the i-th star.

From (2) three most peculiar features of the picture of neutron (X-ray) scattering by a crystal, in which the modulation of the scattering phase takes place, follow: the increase of the satellite intensity with increasing $|\boldsymbol{q}|$, the rules of cancellation of the satellites are determined by the factor $(\boldsymbol{e_k} \cdot \boldsymbol{q})$, the presence of the higher "diffraction harmonics" on the neutron (X-ray) pattern by $n\boldsymbol{k}_i$.

Before the classification of the satellites according to the listed features it is appropriate to study the diffraction pattern of the austenite phase at 300 K of a 67.7 at% Fe monocrystal sample, being cooled before to 4.2 K. In this case the satellite intensities

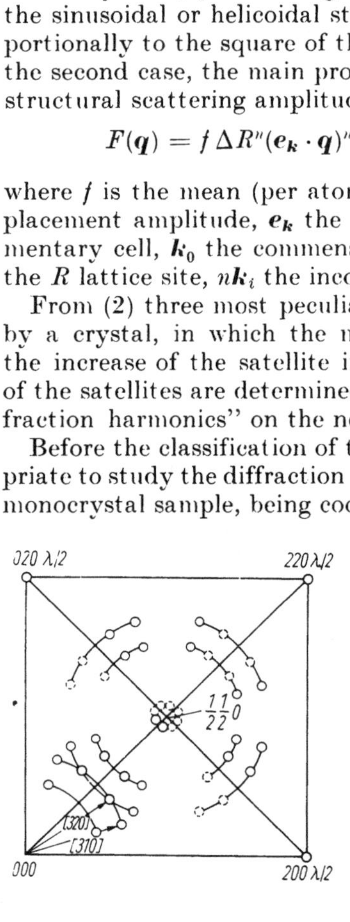

Fig. 4. Disposition of the satellites in the reciprocal lattice

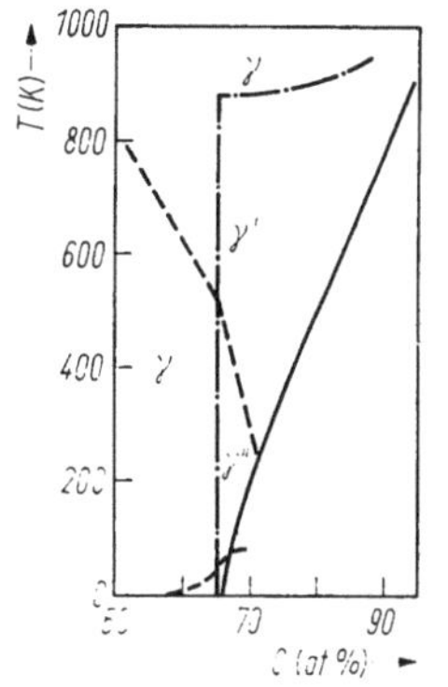

Fig. 5. State diagram of the invar alloys (solid line M_s)

should considerably increase due to the increase of stresses in the alloy because of the partial precipitation of the martensite. Really, a great number of strong coherent superstructural reflections were revealed on the neutron pattern of the alloy. It is convenient to show the distribution of the reflections in the reciprocal lattice. Fig. 4 shows the positions of the reflections in the reciprocal lattice by circles. The diameter of a circle corresponds to the angular size of the peak at half of its height. The satellites on the figure are roughly divided according to their sizes into three groups. The intensities of the peaks having dark outline are not less than the intensity $(220)_{\lambda/2}$. By the light outline there are drawn the reflections whose intensities are approximately three times less then $(220)_{\lambda/2}$. By the dashed line there are drawn the reflections one order weaker than $(220)_{\lambda/2}$.

The space distribution of the reflections in the reciprocal lattice is symmetrical to the (000) site and to the $(\frac{1}{2}, \frac{1}{2}, 0)$ position. The satellites are distributed in groups near the directions [110] and $[1\bar{1}0]$, connected by one another with the fourth-order axis. From the figure it is seen that the satellite $\boldsymbol{k}_2$ lies very close ($|\delta| = 0.04$) to the $(\frac{1}{2}; \frac{1}{2}; 0)$ position and hence it is necessary to be connected with long-wave ($\lambda \approx 100$ Å) modulations of the structure $\boldsymbol{k}_0 = (\frac{1}{2}, \frac{1}{2}, 0)$. With the precipitation of the martensite in a sample the coordinates of the $\boldsymbol{k}_1$ reflection did not change, however, due to the increase of the intensity as a whole the extra reflections $|\boldsymbol{k}_1^\alpha| = |\boldsymbol{k}_1|$ in the direction $\alpha = [320]$ appear to be noticeable on the neutron pattern. It is interesting that in the presence of the martensite in a sample the reflection $(1 + \delta)\,(\frac{5}{3}, \frac{5}{3}, 0)$ $\left((1 - \delta)\,(\frac{1}{3}, \frac{1}{3}, 0)\right.$ in the reduced zone) is split into four reflections located symmetrically to [110] in the direction $\boldsymbol{k}_1^\alpha$ (see Fig. 4). Let us denote them by $\boldsymbol{k}_3^{\alpha'}$.

The second diffraction harmonics of all the strong satellites are distinctly observed (not shown in the figure) on the neutron pattern. In [1] on the X-ray pattern of a 69 at% Fe monocrystal a very weak reflection has been revealed (0.94; 0.63; 0) which, as it follows from (1), corresponds to the third diffraction harmonic of the satellite $\boldsymbol{k}_1 = (0.31; 0.21; 0)$.

The dependences of the satellite intensities upon $\boldsymbol{q}$ are different. If $\boldsymbol{k}_1$ and $\boldsymbol{k}_2$ decrease with increasing scattering vector, the intensities $\boldsymbol{k}_3^{\alpha'}$ in the vicinity of $(\frac{5}{3}, \frac{5}{3}, 0)$ are greater than in $(\frac{1}{3}; \frac{1}{3}; 0)$.

The presence of the higher diffraction harmonics on the neutron pattern and the different character of the angular dependence of the satellite intensities can be understood if one assumes that the origin of the satellites $\boldsymbol{k}_3^{\alpha'}$ is connected with the existence of sinusoidal static waves of atomic displacements in the alloy and the origin of $\boldsymbol{k}_1^\alpha$ and $\boldsymbol{k}_2^\alpha$ is connected mainly with the modulated static redistribution of the spin (charge) density of atoms.

It is reasonable to consider now that the nature of the superstructure in the alloys without martensite in the region $M_s < T < T_C$ is analogous. This means that $\boldsymbol{k}_0 = (\frac{1}{2}; \frac{1}{2}; 0)$ is also the point of symmetry of the stars $\{\boldsymbol{k}_1\}$, $\{\boldsymbol{k}_2\}$, $\{\boldsymbol{k}_3\}$. As far as the satellites $\boldsymbol{k}_1$, $\boldsymbol{k}_2$ appear below T_C, it is apparently the spontaneous magnetostriction that gives rise to the redistribution of the spin density. Then the decrease of the intensity of $\boldsymbol{k}_1$ at room temperature, when $C > 67.7$ at% Fe, becomes understandable (Fig. 2). This is connected with the increase of the reduced temperature $\tau = 300/T_C$ (since T_C decreases) which approaches unity in the 70% Fe alloy. The dependence of $\boldsymbol{k}_1$ and $\boldsymbol{k}_2$ upon the external magnetic field also gives evidence to the striction model. Their intensities at $\boldsymbol{M} \perp \boldsymbol{q}$ increase three times since in the non-magnetized state only $\frac{1}{3}$ of the magnetic domains contain $\boldsymbol{k}_1$, $\boldsymbol{k}_2$ in the scattering plane.

Above T_C only the $\boldsymbol{k}_3$ reflection remains with the centre of symmetry $\boldsymbol{k}_0 = \boldsymbol{0}$. Its intensity depends upon $\boldsymbol{q}$ and becomes substantial only in the position $(1 + \delta)$ $(\frac{5}{3}; \frac{5}{3}; 0)$. The studies of the satellite intensities, belonging to all the rays of the $\{\boldsymbol{k}_3\}$ star, allowed to conclude that they are connected with the longitudinal modulated displacements of atoms from the sites of the f.c.c. lattice.

3.3 *State diagram*

The absence of hysteresis of the physical properties in the region of critical compositions and temperatures allows to state that the incommensurable phase transitions in the invar Fe–Ni alloys are of the second type. This fact allows to draw the phase diagram of the atomic structure of the alloys (Fig. 5). For the construction of the state diagram there were also used data from [3, 6], in which measurements of the magnetic susceptibility $\chi(C, T)$ have been carried out in a wide interval of compositions and temperatures. The validity of this follows from Fig. 3 where it is seen that the anomaly of $\chi(T)$ corresponds to T_t, the temperature where the superstructure vanishes.

The discovered incommensurable phases are denoted by γ' $(\boldsymbol{k}_0 = \boldsymbol{0}; \{\boldsymbol{k}_3\})$ and $\gamma''(\boldsymbol{k}_0 = \pm(\frac{1}{2}; \pm\frac{1}{2}; 0), \{\boldsymbol{k}_1\}, \{\boldsymbol{k}_2\}, \{\boldsymbol{k}_3\})$ and they are apparently connected with the loss of stability of the f.c.c. structure. The region of their existence is limited by the dash-dotted line in the figure. In the beginning the γ'-phase is created with decreasing temperature. Below the Curie temperature ($T_C(C, T)$ is shown by the dashed line) the incommensurable γ''-structure, the symmetry of which is lower than γ', is created. As far as the intensities of the satellites $\boldsymbol{k}_1$, $\boldsymbol{k}_2$, $\boldsymbol{k}_3$ are modulated by the magnetization, it is natural to consider that the invar property of the iron–nickel alloys is caused by the peculiarities of the temperature dependence of the γ''-phase.

In the lower part of Fig. 5 the Néel temperature T_N is shown by the dashed line. Below this temperature a non-collinear magnetic structure, which is determined by the superposition of the wave vectors $\boldsymbol{k} = \boldsymbol{0}$ and $\boldsymbol{k} = (\frac{1}{2}; \frac{1}{2}; 0)$, is created in the alloys. The magnetic structure of the invar Fe–Ni alloys will be published in detail elsewhere.

Acknowledgements

The authors wish to thank Yu. A. Izyumov and V. E. Naish for useful discussions and L. V. Smirnov for growing the excellent monocrystals.

References

[1] V. G. VEERARAGHAVAN, C. F. EAGEN, H. R. HARRISON, and P. G. WINCHELL, J. appl. Phys. 47, 4768 (1976).
[2] S. F. DUBININ, S. G. TEPLOUCHOV, and S. K. SIDOROV, Zh. eksper. teor. Fiz., Pisma 26, 177 (1977).
[3] G. T. DUBOVKA, phys. stat. sol. (a) 41, K63 (1977).
[4] YU. D. TYAPKIN, V. G. PUSHIN, R. R. ROMANOVA, and N. N. BUINOV, Fiz. Metallov i Metallovedenie 41, 1040 (1976).
[5] J. D. AXE, Proc. Conf. Neutron Scattering, Vol. 1, Gatlinburg (Tenn.) 1976 (p. 353).
[6] V. I. TCHECHERNIKOV, Zh. eksper. teor. Fiz. 42, 956 (1962).

(Received January 23, 1979)

phys. stat. sol. (a) 54, 245 (1979)

Subject classification: 20.3; 10.2; 20.1; 22.5.4

Solid State and Materials Science Laboratories,
Department of Physics, Osmania University, Hyderabad[1])

Thermoluminescence of BaFBr Crystals

By

M. Mutha Reddy, K. Somaiah, and H. Vari Babu

Thermoluminescence and optical absorption spectra of X- and γ-irradiated crystals are studied. Optical absorption studies reveal two absorption bands around 455 and 480 nm, a well defined peak at 580 nm, and a broad band at about 810 nm. The glow curves obtained are resolved by the thermal cleaning technique. Four glow peaks are obtained around 338, 369, 408, and 465 K, respectively. The peaks are analyzed using first-order kinetics. The trap depth and frequency factor of the above peaks are calculated using various methods. The 338 K peak is attributed to the detrapping of electrons from impurities and subsequent recombination with holes and whereas the other three peaks perhaps arise from the ionization of two types of F-centres and coagulation centres.

Thermolumineszenz und optische Absorptionsspektren von Röntgen- und γ-bestrahlten Kristallen werden untersucht. Die Untersuchungen der optischen Absorption ergeben zwei Absorptionsbanden um 455 und 480 nm, ein wohldefiniertes Maximum bei 580 nm und eine breite Bande bei etwa 810 nm. Die erhaltenen Glowkurven werden mit thermischer Ausheiztechnik aufgelöst. Vier Glowmaxima werden bei etwa 338, 369, 408 und 465 K erhalten. Die Maxima werden mit Kinetik erster Ordnung analysiert. Die Haftstellentiefe und der Frequenzfaktor der oben erwähnten Maxima werden mit verschiedenen Methoden berechnet. Das Maximum bei 338 K wird der Befreiung von Elektronen aus Störstellen und der nachfolgenden Rekombination mit Löchern zugeschrieben, während die anderen drei Maxima wahrscheinlich aus der Ionisation von zwei Arten von F-Zentren und Agglomerationszentren herrühren.

1. Introduction

Crystals of mixed dihalides of Ba, Sr, and Pb constitute an important class of partially ionic compounds having interesting optical properties [1]. All of them crystallize in the tetragonal space group P4/nmm [2 to 4]. F-centres in BaFCl were studied optically by Nicklaus and Fischer [5], and more recent work by Yuste et al. [6] has covered the optical absorption and electron spin resonance of both BaFCl and SrFCl. In alkali halides, the F-absorption band arises from the electronic transition from an s-like ground state to a triply degenerate p-like state. This degeneracy can be removed by replacing one of the neighbouring ions by a different alkali ion which is the well-known F_A-centre [7]. In the tetragonal BaFCl both anionic sites, fluorine and chlorine [5] have tetragonal symmetry. So two types of F-centres $F(F^-)$ and $F(Cl^-)$ are possible and their excited p-like state would be split into a singlet and a doublet state. Yuste et al. [6] have designated the two types of F-centres as I- and J-centres. By taking into account the analogy between the position of the I_2-band ($\lambda = 550$ nm, $\boldsymbol{E} \perp \boldsymbol{C}$) in BaFCl and that of the F-band in BaF_2, Nicklaus and Fischer [5] concluded that I- and J-centres might be attributed to $F(F^-)$ and $F(Cl^-)$ centres, respectively. On the other hand, a theoretical study of F-centres in BaFCl [8] has led to an attribution opposite to that of Nicklaus and Fischer.

[1]) Hyderabad-500007, India.

Thermoluminescence and optical absorption have been extensively used in the past to study colour centres in alkali halide crystals. In this paper we present the results of thermoluminescence and optical absorption of X- and γ-ray irradiated BaFBr crystals and also the values of the fundamental parameters like trap depth (E) and frequency factor (P_0) obtained from thermoluminescence data.

2. Experimental

Single crystals of BaFBr were grown by the flux method, the details of which were published elsewhere [9]. Transparent crystals free from traces of unreacted mass were used in the present study.

The crystals were X-irradiated at room temperature (RT). For this purpose an X-ray tube with copper target operated at 30 kV and 10 mA was used. The crystals were always kept at a distance of 2 cm from the target. The emitted light during heating of these samples was measured using an EMI 6256 Sphotomultiplier tube. The output of the photomultiplier tube was given to a dc amplifier and was recorded finally by a DIGILOG 200 XY recorder. The sample was heated at the rate of 0.5 K/s and the temperature of the sample was measured by a calibrated chromel–alumel thermocouple fixed very near the sample. The optical absorption spectra of γ-irradiated BaFBr crystals were recorded on a Cary-14(R) recording spectrophotometer, the dose rate γ-rays used being 0.325 MR/h.

3. Results

3.1 Glow curves

Fig. 1 a and b show the glow curves of BaFBr, X-irradiated for 40 min at RT and after thermal cleaning of the first peak, respectively. Four peaks at about 338, 369, 408, and 465 K were obtained. To resolve the peaks a thermal cleaning technique was employed [10]. In this method, the irradiated sample was heated up to about 340 K at the rate of 0.5 K/s and then rapidly cooled to RT. The sample was next heated up to about 370 K at the same rate and again rapidly cooled to RT. This process was repeated for the third peak, too. Finally the glow curve was recorded from RT to about 540 K and the curves obtained by this process are shown in Fig. 2. It was also observed that annealing at RT after γ- or X-irradiation for three or more days resulted in the elimination of the first peak (338 K). Fig. 3 shows the glow curve of a γ-irradiated BaFBr crystal annealed for 25 d.

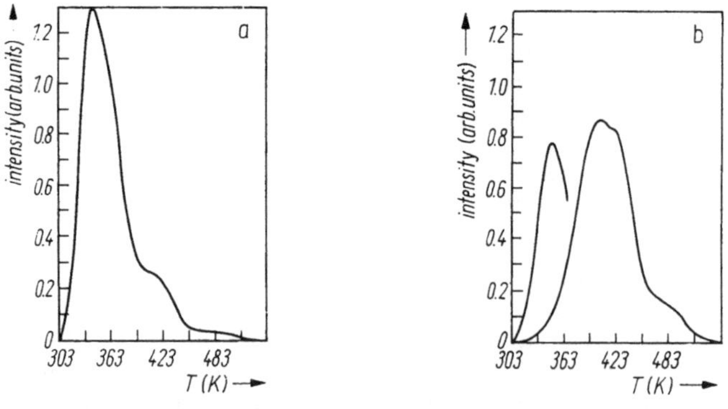

Fig. 1. a) Shows the glow curve of BaFBr crystal (X-irradiated for 40 min). b) Glow curve recorded by thermal cleaning technique (X-irradiated for 40 min). Thermal cleaning was applied to the peak at 338 K and afterwards the curve was recorded continuously from RT to about 540 K

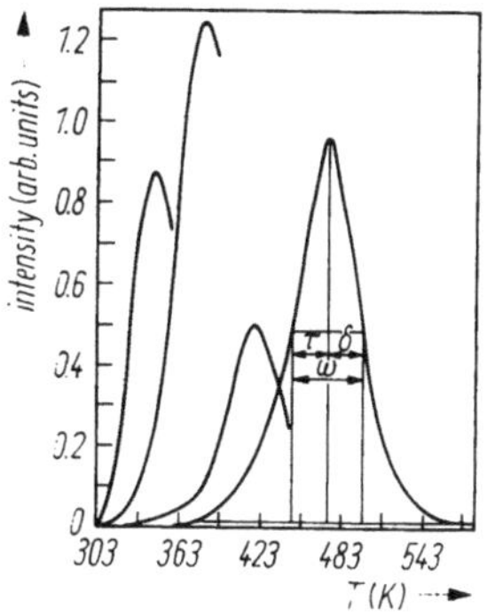

Fig. 2

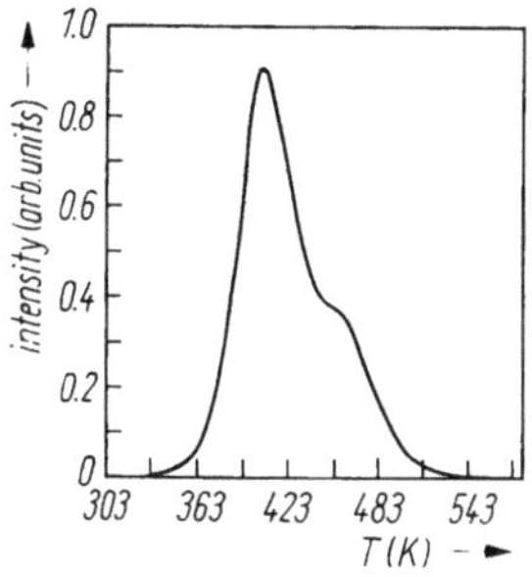

Fig. 3

Fig. 2. Glow curves of BaFBr crystal resolved into component peaks by the thermal cleaning technique

Fig. 3. Glow curve of BaFBr crystal (γ-irradiated for 10 h) annealed at RT for 25 d

The optical absorption spectra of BaFBr immediately after γ-irradiation at RT are shown in Fig. 4a. Two absorption bands around 455 and 480 nm, a well-defined peak at 580 nm, and a broad band at about 810 nm were observed. It is also observed that annealing at RT for 25 d after γ-irradiation resulted in the elimination of the 580 nn deak in the absorption spectra. Fig. 4b shows the absorption spectra BaFBr γ-irradiated and annealed for 25 d.

3.2 Analysis of glow curves

The glow curves obtained in the present investigation have been analyzed by the following methods. The calculated values of trap depth and frequency factor are tabulated in Table 1.

(i) The Halperin-Branner formula [11] refined by Chen [12] for first-order kinetics and constant P_0 reads as

$$E = 1.52 \frac{kT_g^2}{\tau} (1 - 1.58\Delta) - 3.16kT_g \tag{1}$$

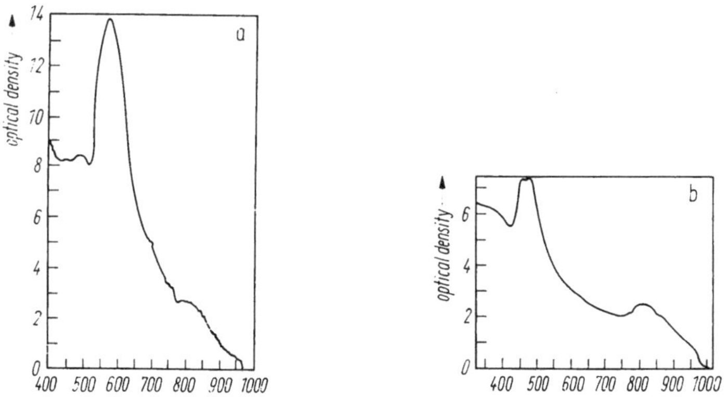

Fig. 4. a) Optical absorption spectra of BaFBr crystal (γ-irradiated) immediately after irradiation. b) Optical absorption spectra of BaFBr crystal (γ-irradiated) annealed at RT for 25 d

Table 1

Calculated values of E and P_0 by several methods for glow peaks in BaFBr

equation	values of E (eV) (1)	(2)	(3)	(4)	peak values of P_0 (s^{-1}) (1)	(2)	(3)	(4)
first-order kinetics								
a) $E = 1.52 \frac{kT_g^2}{\tau} (1 - 1.58\Delta) - 3.16kT_g$	0.67	0.70	0.78	0.88	2.95×10^8	1.25×10^8	1.14×10^8	0.92×10^8
$\frac{P_0}{\beta} = \frac{E}{kT_g^2} \exp(E/kT_g)$								
b) $E = 25kT_g$	0.72	0.79	0.89	1.00	2.66×10^9	2.44×10^9	2.21×10^9	1.88×10^9
c) $E = 2.29\,(kT_g^2/\omega)$	—	—	—	0.85	—	—	—	—

and the frequency factor P_0 is given by

$$\frac{P_0}{\beta} = \frac{E}{kT_g^2} \exp\left(\frac{E}{kT_g}\right), \tag{2}$$

where β is the heating rate of the sample, E the trap depth, k the Boltzmann constant, T_g the glow peak temperature, τ the half-width on the raising side, and $\Delta = 2kT_g/E$.

(ii) E is next calculated by the Randall-Wilkins [13] formula and is given by

$$E = 25kT_g\,. \tag{3}$$

(iii) From the total half-intensity width of the glow curve, E is also calculated by using equation (4) given by Chen [12] for first-order kinetics. It reads as

$$E = 2.29\,\frac{kT_g^2}{\omega}, \tag{4}$$

where ω is the total half-intensity width. Equation (4) is used only for the glow peak obtained at 465 K.

4. Discussion

Impurities and other imperfections give rise to traps at different depths in the forbidden energy region [14]. When the crystals are X- or γ-irradiated, the ionizing radiation produces many free electrons and these may be trapped at the impurities and also other imperfections like vacancies. Electrons trapped at anion vacancies give rise to F-centres. BaFBr crystals exhibit phosphorescence decay and it is generally expected that in the initial stages of decay, the comparatively shallow traps will be emptied giving rise to the luminescence light output, whereas the light output due to emptying of comparatively deeper traps will occur after a sufficiently long time. Fig. 3 shows the glow curve of a BaFBr crystal γ-irradiated for 10 h (0.325 MR/h) and annealed for 25 d. It is observed that annealing at RT resulted in the elimination of the peak obtained around 338 K. Therefore, the peak obtained at about 338 K perhaps arises due to the detrapping of electrons from impurities and subsequent recombination with holes.

As shown in Fig. 4b optical absorption spectra also exhibit the same behaviour. The absorption peak at about 580 nm was absent when the absorption spectra were recorded after the samples (γ-irradiated) were annealed at RT for 25 d. The other three peaks might be due to the bleaching of the two types of F-centres and some other coagulation centres. Until detailed investigations are made it is very difficult to say which glow peak corresponds to which type of centre. The geometrical factor $\mu_g = \delta/\omega$ is found to be less than 0.52 supporting the application of first-order kinetics.

Acknowledgement

The authors wish to thank Prof. K. V. Krishna Rao, Head, Department of Physics, Osmania University, for his encouragement in this work.

References

[1] J. L. Sommerdijk, J. M. P. J. Verstegen, and A. Bril, J. Lum. 8, 502 (1974).
[2] ASTM Card-Index 3-0304.
[3] L. K. Frevel, H. W. Rinn, and M. C. Anderson, Ind. engng. Chem. 18, 83 (1965).
[4] T. N. Bhat, H. L. Bhat, A. H. Rama Rao, M. R. Srinivasan, and P. S. Narayan, Curr. Sci. India 47, 204 (1978).
[5] E. Nicklaus and F. Fischer, phys. stat. sol. (b) 52, 453 (1972).

[6] M. YUSTE, L. TAUREL, M. RAHMANI, and D. LEMOYNE, J. Phys. Chem. Solids **37**, 961 (1976).
[7] H. OKHURA and T. UCHIDA, J. Phys. Soc. Japan **15**, 2114 (1960).
[8] S. LEFRANT and H. H. HARKER, Solid State Commun. **19**, 853 (1976).
[9] K. SOMAIAH and V. HARI BABU, Indian J. pure appl. Phys. **14**, 702 (1976).
[10] N. M. GUPTA, J. M. LUTHRA, and J. SHANKAR, Indian J. pure appl. Phys. **11**, 684 (1973).
[11] A. HALPERIN and A. A. BRANNER, Phys. Rev. **117**, 408 (1960).
[12] R. CHEN, J. appl. Phys. **40**, 570 (1969).
[13] J. J. RANDALL and M. H. F. WILKINS, Proc. Roy. Soc. **A184**, 366 (1945).
[14] K. V. RAO and SUJATA ROY, Solid State Commun. **20**, 941 (1976).

(Received March 7, 1979)

phys. stat. sol. (a) 54, 251 (1979)

Subject classification: 10.2 and 14.4; 22.5.2

Department of Electrical and Electronic Engineering, University of Nottingham[1])

The Relationship between the Defect Structure and the Intrinsic Ionic Conductivity of Monocrystalline Sodium Chloride

By

D. L. KIRK

A study is made of the relationship between the intrinsic ionic conductivity of monocrystalline sodium chloride and the defect structure of the material. Controlled deformation places various dislocation densities within the solid. Conductivity studies are made prior to and following the deformation. The annealing behaviour of the defect state and the intrinsic conductivity also is followed.

Es werden die Beziehungen zwischen der ionischen Intrinsicleitfähigkeit von einkristallinem Natriumchlorid und der Defektstruktur des Materials untersucht. Eine gesteuerte Deformation erzeugt verschiedene Versetzungsdichten innerhalb des Festkörpers. Vor und nach der Deformation werden Leitfähigkeitsuntersuchungen durchgeführt. Das Temperungsverhalten der Defekte und der Intrinsicleitfähigkeit wird ebenfalls verfolgt.

1. Introduction

The high-temperature ionic conductivity of the simple rock salt structures has been a topic inviting much experimental investigation over many years. The simple ionic structures of potassium and sodium chlorides present high-temperature Arrhenius conductivity plots that accord well with models involving the formation and movement of intrinsic and extrinsically created point defects. Such models and experimental data were critically reviewed in earlier years by Lidiard [1]. In the past decade even more attention has been focussed on the detailed behaviour of the intrinsic regions of the conductivity behaviour and in particular on the so called stage I of the Arrhenius plot. Much effort has been spent in examining the non-linearity of the region and the associated point defect mechanisms that control the high-temperature conductivity. The most thoroughly investigated and discussed materials have been the simple monocrystalline solids of sodium and potassium chloride. A definitive explanation of the curvature of stage I of these solids was originally put forward by Beaumont and Jacobs [2] who suggested that the ionic conduction, at high temperatures, was determined by the formation and motion of anion and cation vacancies. The model proposed was tested by computer fitting of observation to the Arrhenius plot predicted by the model, with optimization of the fitting to yield the critical parameters of the model. Since this work much attention has been focussed on this particular line of approach. However the method of fitting a model with up to eight variable parameters to experimental data imposes a severe test for that data and considerable variations in both the observations made and the results obtained from the fitting procedures have been reported. Recently an important article by Murthy and Pratt [3] has reviewed some of this previous work and has demonstrated the possible weaknesses of this method of approach.

---

[1]) Nottingham NG7 2RD, Great Britain.

In many of the previous studies little or no consideration has been made of the mechanical state of the solid or of the manner in which the mechanical state may interact with the point defect configuration to influence the resulting conductivity of the material. The aim of the present article is to report upon an examination of this particular point. The work demonstrates quite clearly that the dislocation content of an ionic solid can influence quite markedly the magnitude and temperature dependence of the intrinsic conductivity of monocrystalline sodium chloride.

2. Experimental Consideration

The measurements of ionic conductivity were performed upon nominally pure, monocrystalline samples of sodium chloride. They were grown from the melt using a modified Kyropoulos technique. The specimens utilized a conductivity rig that has been described previously [4]. This used an ac measurement of the ionic conductivity, in an atmosphere of spectroscopically pure argon, with platinium/colloidal graphite electrodes making intimate contact with the surface of the specimen. The conductivity and capacitance were measured at an angular frequency of 10000 rad s^{-1} (1592 Hz). Conductivity measurements were also taken over a range of frequencies from 50 Hz to 20 kHz to ensure that there was no significant dispersion of the conductivity values.

The mechanical state of the specimens was varied by increasing the dislocation content of the solid. This was achieved with controlled deformation of the specimens in an Instron testing machine. A typical geometry of the specimens investigated is displayed in Fig. 1, along with the load–compression characteristic of the material. The elastic, stage I easy glide and stage II hardening of the material are readily discernible. The slip planes and slip bands of this material are also shown along with an indication of the geometric axes along which the deformation was applied. Specimens which were utilized in the conductivity studies were deformed into stage I and II of the deformation characteristic. The appropriate percentage deformation associated with the defined geometry of specimen used for conductivity investigation are also indicated. The dislocation content of the material after the controlled deformation was studied with optical microscopy. This dislocation content was examined by etch pit studies. The etchants and methodology used to reveal the dislocations have been described previously in an article by Davidge and Pratt [5].

3. Experimental Observations

The high-temperature ionic conductivity of a nominally pure monocrystalline sample of sodium chloride is presented in Fig. 2. The data points were obtained from conductivity determinations performed upon four different specimens cleaved from the same transverse platelet of a crystalline boule of pure material. The specimens were unannealed after cleavage. There was excellent reproducibility which did not require

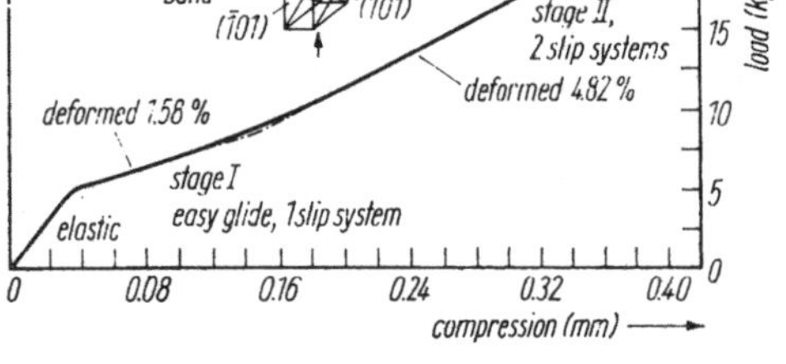

Fig. 1. A load–extension characteristic of the monocrystalline specimens of sodium chloride utilized in the present study. Compression was applied in the ⟨100⟩ direction. Dimensions of the inset: 4.18 × 4.62 × 7.75 mm^3

any high-temperature annealing of the crystal–electrode interface to attain consistent observation. Within this data, the intrinsic region of the ionic conductivity (stage I) showed a definite curvature towards an increased slope at the higher temperatures (stage I′). Q_{I}, the slope of the low-temperature portion of stage I, had a value of (1.78 ± 0.06) eV. This was the activation energy for the defect processes controlling this region. This activation energy is normally accepted as being related to the enthalpy of formation of a Schottky vacancy pair (h_s) and the enthalpy of motion of a cation vacancy (Δh_1). The relationship (1) takes the form

$$\left|\frac{h_s}{2} + \Delta h_1\right| = Q_{\mathrm{I}} \,. \tag{1}$$

Q_{II}, the slope of the extrinsic region of ionic conductivity of a pure crystal is steepened at low temperatures by the onset of an association reaction between divalent cations present as impurities in the material and the free cation vacancies. A line of minimum slope of (0.69 ± 0.03) eV could be drawn through the extrinsic data. This would represent the best value for the enthalpy of motion of the cation vacancy that can be obtained from the conductivity data of the nominally pure crystals that were studied. The relationship between Q_{II} and Δh_1 is

$$Q_{\mathrm{II}} = \Delta h_1 \,. \tag{2}$$

The values of Q_{I} and Q_{II} then yielded a value of (2.18 ± 0.09) eV for the energy of formation of the Schottky pair.

Utilizing a previously published value for the mobility of the cation vacancy of [4]

$$\mu_1 T = 1590 \exp\left(-\frac{0.67}{kT}\right) \frac{\mathrm{cm^2\,K}}{\mathrm{V\,s}}, \tag{3}$$

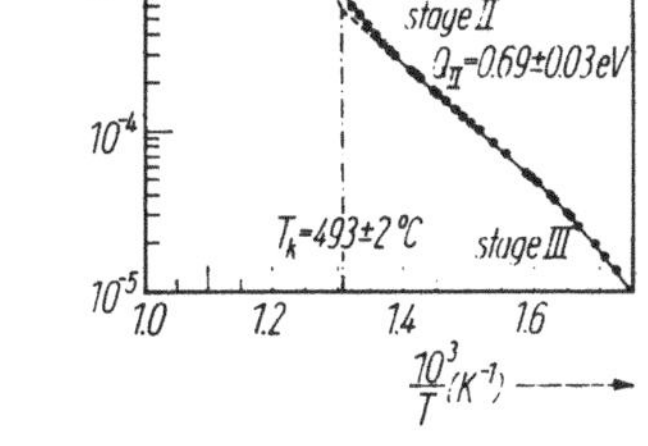

Fig. 2

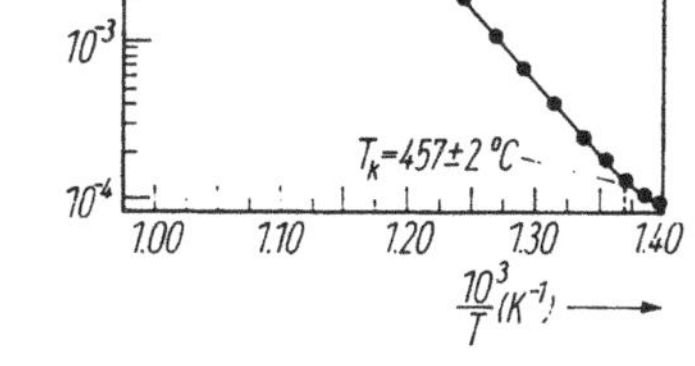

Fig. 3

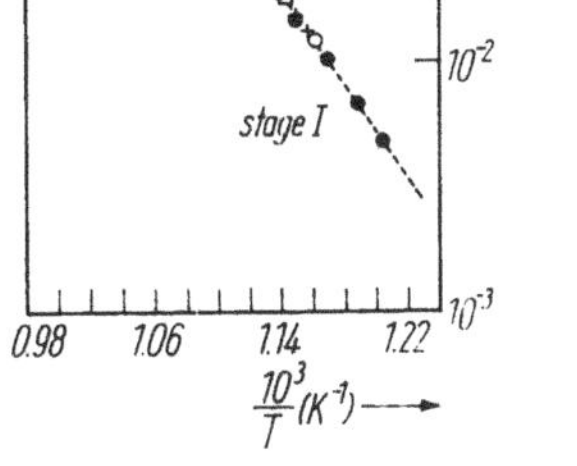

Fig. 4

Fig. 2. The intrinsic and extrinsic conductivity of a monocrystalline specimen of sodium chloride

Fig. 3. The ionic conductivity of monocrystalline sodium chloride intentionally doped with $(OH)^-$ ions

Fig. 4. The effect to deformation upon the intrinsic ionic conductivity of monocrystalline sodium chloride. + deformed 1.58%, ● deformed 4.82%, ○ annealed, □ annealed it was possible to calculate a value for the concentration of divalent cations present in the solid from the knee temperature existing between stage I and II of the Arrhenius conductivity plot. The knee temperature of (493 $\pm$ 2) °C indicated an effective divalent impurity content of (2.3 $\pm$ 0.1) ppm (expressed as a site fraction). This is to be contrasted with a total divalent impurity content of 28 ppm of divalent impurity ions as determined by analysis of the single-crystal material with mass spectrometry. Even accepting that large systematic errors are often present in analysis with mass

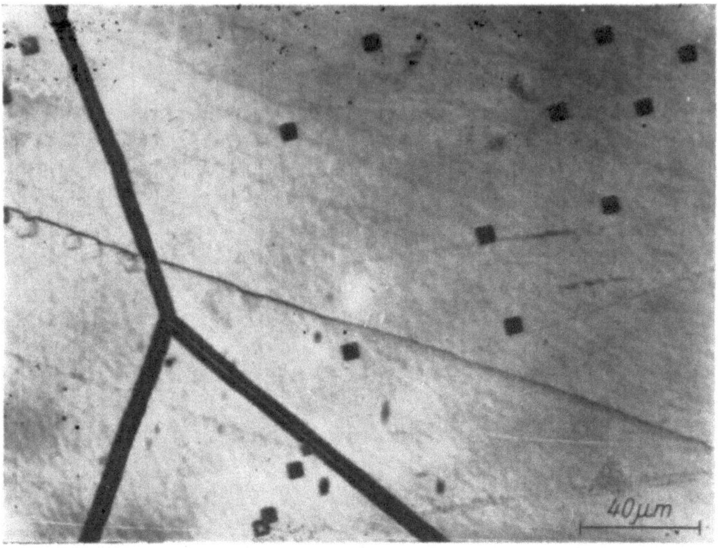

a

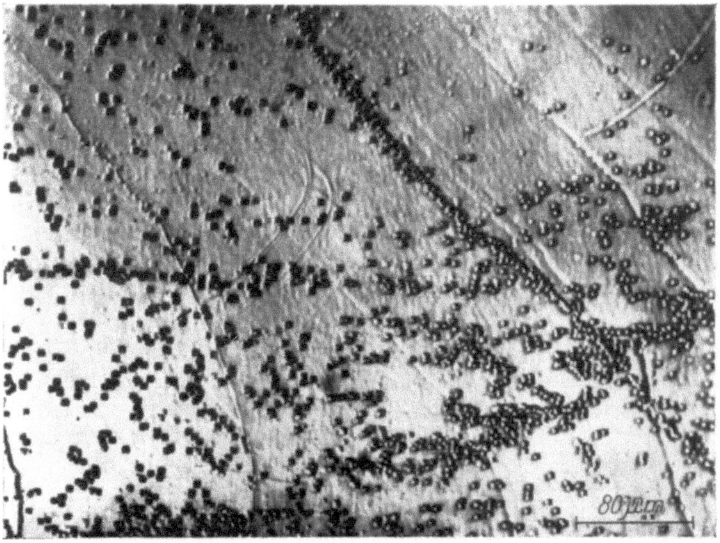

b

Fig. 5. a) Typical etch pit configuration to be found in undeformed and annealed specimens; b) etch pit configuration to be found in the deformed specimens prior to any conductivity determination; c) etch pit configuration present in the deformed specimens following a conductivity measurement; d) etch pit configuration following post deformation annealing spectrometry, the observation would suggest the presence of trace concentrations of anion impurities that are complexing with the unintentional cation impurities to yield an effective net, divalent cation content that is considerably reduced from that actually present.

The analysis with mass spectrometry showed the chief divalent cation impurity to be calcium. As this has a rejection ratio of 1, the knee temperature could not be reduced by zone refining. To resolve the variations in stage I′ from stage I more clearly, stage I was extended by incorporating $(OH)^-$ ions into the solid. Crystals were volume-doped by adding controlled amounts of sodium hydroxide to the crystal-

c

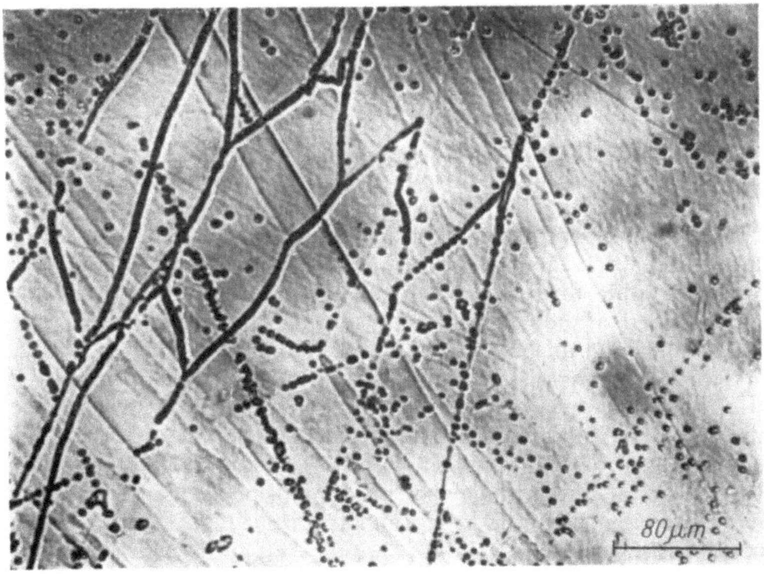

d

line melt. This reduced the knee temperature to 457 °C (Fig. 3) with an effective divalent impurity content calculated as being 8.9×10^{-1} ppm. Stage I was considerably extended and the activation energy Q_I for this enhanced region was found to have a value of (1.83 ± 0.01) eV. This yielded a more refined value for h_s of $(2.28 \pm \pm 0.7)$ eV.

The ionic conductivity of the crystals that had experienced the controlled deformation were then studied. Prior to deformation all the specimens were annealed for 24 h at 700 °C and then slowly cooled over a period of 36 h. This was done to anneal out any damage, defects, or dislocations introduced by the cleaving process.

Following annealing etch pit studies showed the dislocation content to be low and of the order of 10^6 cm^{-3}. After the annealing treatment, the specimens were deformed in the manner depicted in Fig. 1. The specimens were then placed in the conductivity rig and conductivity measurements were performed. It was observed that the deformation considerably enhanced the high-temperature conductivity with a much larger enhancement appearing in the crystal suffering the smaller percentage deformation (Fig. 4). Following the conductivity measurements the specimens were removed from the conductivity rig, after cooling, and then annealed for a period of 24 h at 700 °C. The specimens were then slowly cooled to room temperature and replaced in the conductivity cell with the intrinsic ionic conductivity being remeasured. The region of enhanced intrinsic conductivity, stage I′, was seen to be considerably reduced (Fig. 4). The region of ionic conductivity usually ascribed to the motion of the free cation vacancy (stage I) was found to be insensitive to the defect state of the crystal. The slope of this region $(Q_I = (1.84 \pm 0.01)$ eV) being in good agreement with that determined for the crystal doped with the hydroxyl anions.

When examined, after suitably etching, the etch pit configurations showed significant differences between the originally annealed, the deformed, and the deformed and subsequently annealed state. This is demonstrated in Fig. 5. The undeformed and annealed specimens showed a low dislocation content (Fig. 5a). The deformation increased markedly the dislocation density (Fig. 5b). The initial conductivity run performed upon the deformed samples transformed a mainly random dislocation array to that of parallel lines of dislocation walls (Fig. 5c) and the longer annealing treatment of 24 h at 700 °C resulted in the reduction of the dislocation content with the development of a grain and subgrain structure (Fig. 5d).

4. Discussion

A deformed crystal is a non-equilibrium structure that will revert to a configuration of lowest energy with prolonged annealing. Previous work [5] has shown that deformed crystals of sodium chloride recover principally within a temperature interval of between 400 and 500 °C. At 400 °C this previous study also showed that dislocations show a tendency to move and stack themselves into dislocation walls often aligned in certain preferred crystallographic directions. The density of the walls was greatly reduced in specimens annealed at 500 °C and maintaining the specimens at even higher temperatures developed the dislocation walls into subgrains. The recovery of the material, with the reduction of the dislocation density was seen to occur principally between 400 and 500 °C with an annealing time of 1 h. These developments are seen to be parallelled in the deformed and annealed conductivity specimens of Fig. 5. The initial conductivity observation, with the deformed specimen spending some time at an elevated temperature, did transform a somewhat random dislocation array (Fig. 5b) to a network of dislocation walls, with the 24 h anneal at 700 °C producing a grain and subgrain structure.

In addition to enhancing the dislocation structure of a solid, deformation increases the concentration of anion and cation vacancies [6] and if the deformation is particularly severe, the concentration of interstitials. Cation and anion vacancy concentrations readily revert to their equilibrium values by rapidly annealing out at temperatures below 100 °C; the annealing of the enhanced cation vacancy concentration proceeding with a reaction involving first-order kinetics [7]. Because the energies of motion of the interstitial defects are considerably smaller [8] than those of the vacancies, the interstitials would not be expected to remain stable at temperatures in excess of 100 °C. For such reasons, the only rational explanation of the structure-dependent enhanced ionic conduction would be in terms of the enhanced mobility of anion or cation vacancies along paths of preferred conduction.

Dislocations in ionic solids can act as paths of enhanced mobility for point defects and, in some cases, extrinsic impurities. Experimental work has shown that the diffusivity of the cationic species is not dependent upon defect state or the presence of grain boundaries [9]. A thorough investigation of the dependence of intrinsic ionic conductivity upon grain size in polycrystalline sodium chloride has been made by Graham [10]. The work showed that the contribution of the chlorine anion vacancy to the total conductivity was significant in the intrinsic regions even for single crystals and that the anion vacancy's contribution to the conduction process was reciprocally dependent upon grain size. The work also showed that the intergranular layer thickness providing the path of enhanced conduction was approximately 0.1 μm thick at 350 °C and that the contribution of the sodium ion vacancy to the conduction process was virtually independent of grain size. In so far as a grain or subgrain boundary can be considered to be made up of an ordered arrangement of dislocations, then the most likely explanation of the variable, enhanced, high-temperature conductivity (stage I′) is in terms of the preferred movement of anion vacancies through the grain and subgrain structure that is developing within the material during the high-temperature conductivity measurements and also during the subsequent annealing procedures. Charge flow experiments performed upon sodium chloride at elevated temperatures have suggested that there is an excess of chlorine ion vacancies present along the dislocation cores with a compensating Debye-Huckel cloud of sodium ion vacancies surrounding them [11]. The subgrains and grain boundaries would be in the same condition with the increased density of anion vacancies within the dislocation cores forming the proposed anion contribution to conduction.

Having clearly resolved the energy and magnitude of the cation vacancy's contribution to intrinsic conductivity, the anion contribution may be resolved by extrapolation. The form of the present enhancement is that of a sharp increase in the slope of the intrinsic conduction similar to that observed by Allnatt and Jacobs [12] in potassium chloride. This is in contrast to the data of Beaumont and Jacobs [2] whose intrinsic region appeared to present a continuous curvature with not such a significant enhancement associated with stage I′. The enhanced conduction when plotted as a function of reciprocal temperature (Fig. 6) yielded an activation energy of (2.7 ± ± 0.1) eV with a variable magnitude for the enhancement. If it is assumed that this enhanced anion conduction is compounded from enthalpies of formation and motion of the anion vacancy, then

$$Q_{\text{I}}' = \left| \frac{h_s}{2} + \Delta h_2 \right| , \tag{4}$$

where Δh_2 is the enthalpy of motion of the anion vacancy. Utilizing the value for h_s of (2.28 ± 0.07) eV this yielded an enthalpy of motion of the anion vacancy of (1.52 ± 0.08) eV. This is too high a value when compared with the generally accepted

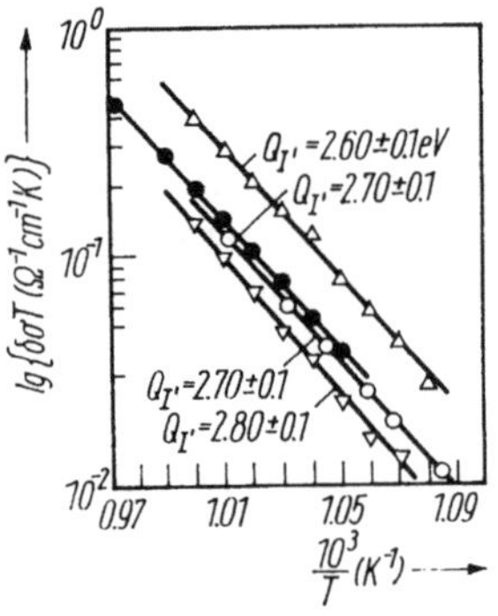

Fig. 6

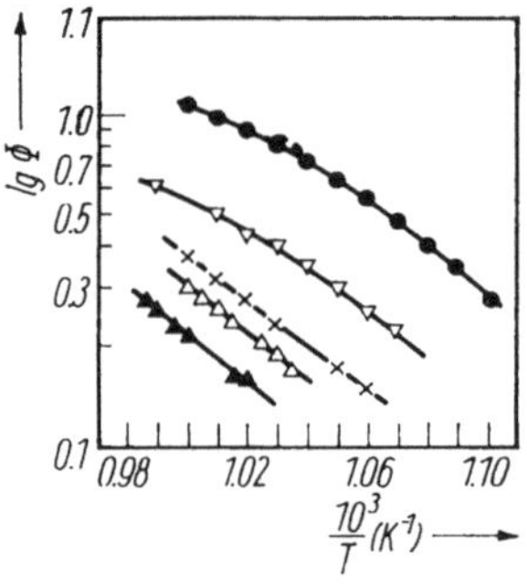

Fig. 7

Fig. 6. The enhanced conduction of stage I' presented as a function of reciprocal temperature. ○ pure crystal, ● OH-doped pure; pure crystal: △ deformed 1.58%, ▽ deformed 4.82%

Fig. 7. The dependence of the mobility ratio ($\Phi = \mu_2/\mu_1$) upon reciprocal temperature. ×, ● deformed crystal; ▲ crystal after annealing; △ unspecified deformation, pure; ▽ unspecified deformation, OH-doped

value of 0.95 eV for Δh_2 obtained from measurements of anion diffusion in sodium chloride [12].

With a knowledge of the anion and cation vacancy contributions to the ionic conductivity, the mobility ratio $\Phi(\mu_2/\mu_1)$ was determined as a function of temperature (Fig. 7). The curvature present in these mobility plots would not seem inconsistent with an annealing out of the paths of preferred conduction, particularly as the limiting slopes of these curves all possess the same values at the lower temperatures. From the slope of these plots and a knowledge of $\Delta h_1 = (0.68 \pm 0.02)$ eV and $\Delta s_1 = (0.19 \pm 0.05) \times 10^{-3}$ eV K^{-1} [4], the enthalpy and entropy of the postulated anion vacancy motion may also be determined. Applied to the data of Fig. 7, values of (2.0 ± 0.4), (2.0 ± 0.4), and (1.98 ± 0.4) eV and 1.53, 1.62, and 1.58×10^{-3} eV/K were obtained from the essentially linear plots of mobility ratio as a function of reciprocal temperature. It would be naive to assume that these figures are representative of motion through the bulk of the material. Again the enthalpy of motion is far too high and the large variation in Δs_2 is obviously interpretable in terms of a varying magnitude in the anion's contribution to the total conductivity.

5. Conclusions

An enhancement of the high-temperature intrinsic conductivity of monocrystalline sodium chlorid (stage I' of the Arrhenius plot) has been shown to be sensitive to the defect structure of the solid. Evidence of an anion contribution to the conductivity in undeformed crystals appears in the slight curvature of the intrinsic region. The data of Fig. 2 (the nominally pure, undeformed specimen) when fitted to a theoretical curve involving the eight variables associated with the intrinsic, extrinsic, and association regions of ionic conduction produced values for the enthalpy and entropy of anion motion of $\Delta h_2 = 0.89$ eV and $\Delta s_2 = 4.33 \times 10^{-4}$ eV K^{-1} [14]. The computer fitting involved a standard IBM least squares minimization routine. The values are in reasonable agreement with tracer measurements of anion diffusivity [13, 15] and they would suggest that two distinct types of ionic defect motion may be observed: one associated with anion vacancy movement through structurally perfect regions of the crystalline lattice and the other with an enhanced mobility along subgrain boundaries and possibly dislocations. The effects of deformation and annealing were consistent with this model. Care should then be exercised in the interpretation of the intrinsic ionic conduction, especially if the crystal growth, method of specimen preparation, or mode of measurement deform the material. It should be appreciated that most ionic solids become readily plastic at elevated temperatures. The present observations may go some way in explaining the variously reported behaviours of the intrinsic ionic conduction of monocrystalline sodium chloride.

References

[1] A. B. Lidiard, Hdb. Phys. **20**, 246 (1957).
[2] J. H. Beaumont and P. W. M. Jacobs, J. chem. Phys. **45**, 1496 (1966).
[3] C. S. N. Murthy and P. L. Pratt, Proc. II. Internat. Conf. Lattice Defects in Ionic Solids, Berlin (West) 1976 (pp. 307 to 315, Europ. Phys. J.).
[4] D. L. Kirk and P. L. Pratt, Proc. Brit. Ceram. Soc. **9**, 215 (1967).
[5] R. W. Davidge and P. L. Pratt, phys. stat. sol. **6**, 759 (1964).
[6] A. Taylor and P. L. Pratt, Phil. Mag. **4**, 665 (1959).
[7] R. W. Dreyfus and A. S. Nowick, Phys. Rev. **126**, 1367 (1962).
[8] K. Tharmalingham, J. Phys. Chem. Solids **25**, 225 (1964).
[9] J. F. Laurent and J. F. Benard, J. Phys. Chem. Solids **7**, 218 (1958).
[10] H. C. Graham, Dissertation, Ohio State University, 1965.
[11] R. L. Moment and R. B. Gordon, J. appl. Phys. **35**, 2489 (1964).
[12] A. R. Allnatt and P. W. M. Jacobs, Trans. Faraday Soc. **58**, 116 (1962).
[13] R. G. Fuller, Phys. Rev. **142**, 524 (1966).
[14] D. L. Kirk, Ph.D. Thesis, University of London, 1968.
[15] L. W. Barr, J. A. Morrison, and P. A. Schroeder, J. appl. Phys. **36**, 624 (1965).

(Received January 15, 1979)

phys. stat. sol. (a) 54, 261 (1979)

Subject classification: 13.3 and 14.3.4; 21.6; 22.6

School of Physical Sciences, The Flinders University of South Australia, Bedford Park[1])

Electrode Kinetics at the Pt/Yttria-Stabilized Zirconia Interface by Complex Impedance Dispersion Analysis

By

S. P. S. BADWAL and H. J. DE BRUIN

Low frequency impedance dispersion analysis, attributed to polarization at the electrode–electrolyte interface, exhibits a strong frequency dependence of the equivalent parallel resistor and capacitor. Such response for Pt/YSZ/Pt cells is analysed as a Cole-Cole complex impedance–admittance dispersion with a relaxation distribution. A physical interpretation based on the results obtained and observations of reactions between noble metals and ceramic oxides, suggests the creation of a complex polarized layer at the electrode–electrolyte interface involving a proliferation of cationic species as the possible cause of irreversible energy losses in the Faradaic reactions. Their kinetics is found to change with the oxygen partial pressure of the atmosphere.

Die Dispersionsanalyse der niederfrequenten Impedanz, die der Polarisation an der Elektroden–Elektrolyt-Grenzfläche zugeschrieben wird, zeigt eine starke Frequenzabhängigkeit des äquivalenten, parallelen Widerstands und der Kapazität. Ein derartiger Response für Pt/YSZ/Pt-Zellen wird als komplexe Cole-Cole-Impedanz–Admittanzdispersion mit einer Relaxationsverteilung analysiert. Eine physikalische Interpretation, die auf den erhaltenen Ergebnissen beruht sowie auf Beobachtungen der Reaktionen zwischen Edelmetallen und keramischen Oxiden, weist auf die Bildung einer komplexen Polarisationsschicht an der Grenzfläche Elektrode–Elektrolyt hin, die eine Anregung der Kationenarten als möglichen Grund für die irreversiblen Energieverluste in den Faraday-Reaktionen einschließt. Es wird gefunden, daß deren Kinetik sich mit dem Sauerstoffpartialdruck der Atmosphäre ändert.

1. Introduction

Interest in high-temperature electrolytic cells has concentrated largely on the study of the solid electrolytes with emphasis on transport numbers, concentrations of charge carrying species, electronic and ionic defects and their mobilities and the effects of oxygen partial pressure, dopant concentration, impurities, grain boundaries, and second phase precipitation on conductivity [1 to 3]. To evaluate the performance and application of high-temperature electrolytic cells, a knowledge of charge transfer phenomena at the electrode–electrolyte interface is essential. Yet a minority of authors [4 to 8] have addressed themselves to this problem. Our recent observations on reactions between noble metals and refractory oxides [9, 10] must have an important bearing on the interpretation of charge transfer data. Until then the chemical inertia of noble metal electrodes in cells operating at high temperatures in oxidizing atmospheres had been tacitly assumed. In this paper we apply complex impedance–admittance dispersion analysis to cells of the type Pt, $P_{O_2}/Y_{1-x}Zr_xO_{1.5+0.5x}/Pt$, P_{O_2} (for $x = 0.818$ and $x = 0.852$, to study the reaction between Pt and an ionically conducting ceramic oxide. As a basis we use our recent analysis [11] of charge transfer and polarization events at the electrode–electrolyte interface which gives rise to the non-ideal behaviour of such solid state galvanic cells.

If the only event at a noble metal electrode–solid electrolyte interface were the Faradaic reaction, $\frac{1}{2} O_2 + 2e^- = O^{2-}$, following a standard kinetic sequence in which

[1]) Bedford Park, 5042, South Australia.

a slow step dominates the reaction rate, then in the complex impedance plane one would expect an ideal semicircle whose radius is a measure of the rate constant for the reaction. In solid state cells at low frequencies (5 Hz to 5 kHz) and high temperatures (600 to 1000 °C) the observed semicircles are substantially depressed (15 to 45°) below the real axis. The actual kinetics of the charge transfer is little influenced by the properties of the grain boundaries in the electrolyte, but more so by the oxygen migration through the bulk [12]. Electrode pre-treatment to affect the reaction mechanism by changing the density of triple-phase-contacts [13, 14] has a minor influence on the depression of the arc. In general physical variables imposed on the interface cannot eliminate it [15 to 17] and chemical interference should be considered as a possible cause for the observed impedance behaviour. Faradaic side reactions can then be seen to create time dependent capacitors and resistors which give rise to the observed depressions of the semicircles as proposed by several authors [18 to 22]. It implies that MacDonald's condition of single species of mobile charge of either polarity will be difficult to achieve in such solid cells [15, 16].

Our earlier model to quantise the non-ideal behaviour of the impedance arcs [11, 18] suggests the existence of a complex capacitive layer in the solid electrolyte surface in which polarization involves a multiplicity of charge carriers of both polarities. The composition of the layer and hence its capacitance changes with frequency. Loss due to parasitic short-circuiting reactions also increases with frequency, as the thickness of the capacitor decreases. The multiplicity of charge carriers inherently causes a range of electron transfer reactions with different rates resonating at different frequencies. This together with the changes in loss phenomena gives rise to the variable resistor in the equivalent RC circuit.

2. Experimental

The impedance was measured for cells consisting of two identical parallel platinum electrodes in contact with discs of 8 or $10^{m}/_{0}$ Y_2O_3 stabilized ZrO_2 (YSZ). Measurements were made in pure oxygen, air, or argon between 450 and 1000 °C in the frequency interval 5 Hz to 10 MHz. YSZ discs approximately 10 mm ∅, between 1.5 and 4 mm thick and density 89% of the theoretical, were obtained from Ceramic Fabricators Ltd. Australia and polished with various grades of silicon carbide papers, diamond pastes (down to 1 to 2 μm), and alumina powder (0.1 to 0.3 μm). Electrodes were made

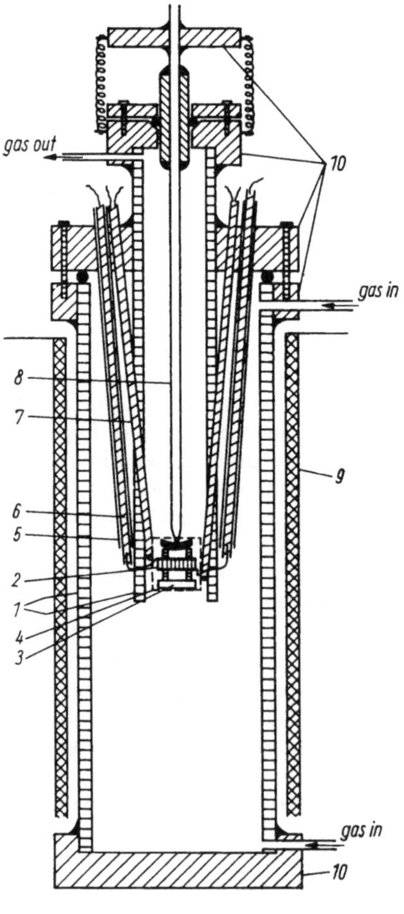

Fig. 1. Experimental cell assembly. 1 Alumina tube, 2 specimen, 3 alumina support, 4 square hole cut in alumina tube (1), 5 stainless steel shield, 6 alumina sheathing, 7 thermocouple, 8 alumina rod for applying slight compressive stress to cell components, 9 stainless steel cage, 10 brass structural members

from Johnson Matthey platinum paste N 758. A uniform thickness of the electrodes was obtained by repeated application of platinum paste followed by firing at 1000 °C. Experimental details have been described elsewhere [19]. In summary: oxygen partial pressures were measured using a CoO wire probe [23]. A Hewlett-Packard impedance meter, Model 4800A (frequency range 5 Hz to 500 kHz), and an rf vector impedance meter, Model 4815A (frequency range 500 kHz to 108 MHz) were used to measure amplitude and phase angle within $\pm 5\%$. The signal applied to the cell was usually 2.7 mV r.m.s. Impedance of various *RC* circuits made up of standard resistors and capacitors was compared with the simulated values to estimate the inductive effects in the shielded leads and to calibrate the impedance meters. All experimental equipment was operated inside an doubly shielded Faraday cage. The furnace assembly was modified for atmospheric control around the cell (Fig. 1). Initially the temperature of the sample was raised to 1000 °C and the system allowed to equilibrate with the surrounding gas atmosphere for several days. A dc current of 100 to 150 mA was passed for 1 h in either direction to stabilize the electrodes. The impedance was then measured at intervals of 50 to 60 °C after thermal equilibrium had been established. When the minimum temperature had been reached the experiments were repeated for an increasing temperature cycle. Most measurements were made between 5 Hz to 500 kHz during the off cycle of the temperature controller.

3. Results and Discussion

The impedance arc due to Faradaic dispersion at the electrode–electrolyte interface was generally observed between 5 Hz and 5 kHz. Its length varied substantially with oxygen pressure. Warburg impedance was only observed (Fig. 2 a and b) for experiments in argon ($P_{O_2} = 3 \times 10^{-6}$ atm), but may be operative in air and oxygen at fractional frequencies outside the range available to us.

The centres of the Faradaic arcs were invariably depressed substantially below the real axis; those for space charge phenomena at the grain boundaries in the electrolyte considerably less and those for the bulk conductivity hardly at all (Fig. 2c).

For thick electrolyte discs (> 2 mm) overlap of relaxation was observed, particularly at intermediate temperatures. Analysis of the data for such cells was difficult and led to large relative errors. Such data were generally not processed.

Using the Cole-Cole analysis for dispersion data [24], similar to that by Raistrick

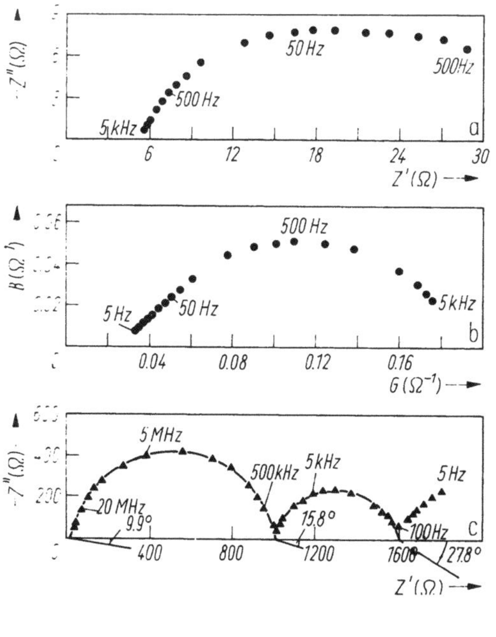

Fig. 2. a) Impedance and b) admittance response for a Pt/YSZ/Pt cell at 996 °C (in argon including a Warburg impedance). c) Impedance response of the same cell in O_2 at 457.4 °C showing the depressions for three relaxation phenomena: Faradaic reaction (right), space charge (middle), and bulk conductivity (left)

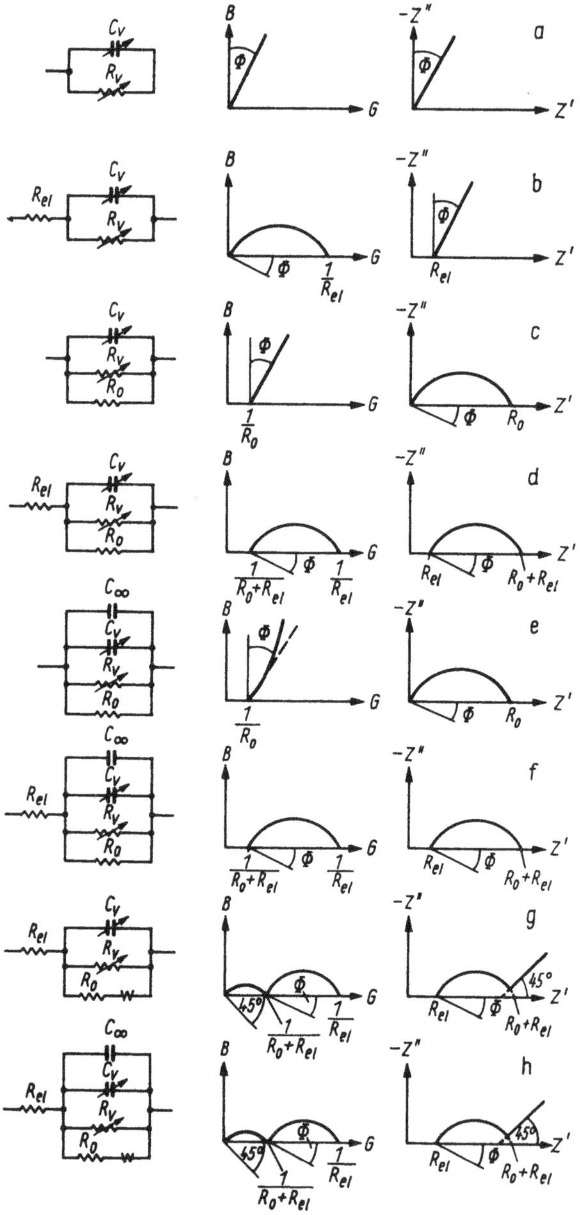

Fig. 3. Equivalent circuits and their dispersion in the impedance and admittance planes. a) Angle of depression $\Phi = f(a/b)$, relaxation τ_0; variable; b) $\Phi = f(a/b)$, relaxation variable; c) $\Phi = f(a/b)$, relaxation $R_0 C_{vm}/\cos\Phi$; d) $\Phi = f(a/b)$, relaxation $R_0 C_{vm}/\cos\Phi$; e) $\Phi = f(C_\infty, a/b)$, relaxation $R_0(C_{vm} + C_\infty)/\cos\Phi$; f) $\Phi = f(C_\infty, a/b)$, relaxation $R_0(C_{vm} + C_\infty)/\cos\Phi$; g) $\Phi = f(a/b)$; relaxation $R_0 C_{vm}/\cos\Phi$, h) $\Phi = f(C^\infty, a/b)$, relaxation $R_0(C_{vm} + C_\infty)/\cos\Phi$

et al. [21, 22], we suggested in an earlier paper [18] a physical interpretation of impedance data for Faradaic charge transfers with a frequency-sensitive relaxation. In Fig. 3 are shown equivalent circuits in an increasing order of complexity of which Fig. 3 d and g represented that earlier analysis. Those in Fig. 3f and h appear to be more appropriate for the current data. R_0 is the zero frequency resistance and is the steady state or equilibrium value of the resistance (η/I) for dc current densities smaller than the exchange current densities (η is the overpotential). C_∞ is some constant capacitance (at infinite frequency), representing the interface of two materials with substantially different band structures, which causes space charge, Helmholtz and other permanent capacitances. It is suggested that R_0 represents the charge transfer resistance of the rate determining step and C_∞ the double layer capacitance. The frequency dependent components $R_v(= 1/a\omega^n)$ and $C_v(= b\omega^{n-1})$ summarize the steps in a complex mechanism for the Faradaic reaction. R_{el} replaces the relatively low frequency limiting response for the electrolyte polarisations. This approximation can be justified in the analysis of Faradaic polarisation if the apparent relaxation times differ by more than two orders of magnitude. W is the Warburg impedance observable only in argon. For circuits 3a to d and 3g the depression of the center of the electrode arc, Φ, is simply related to the ratio a/b (= tan Φ) and the frequency exponent $n = (1 - 2\Phi/\pi)$.

However, for circuits 3e, 3f, and 3h $\Phi = f(a/b, C_\infty)$ (see below) and n is no longer directly related to Φ.

The presence of the frequency dependent components R_v and C_v does not provide a single valued relaxation time [18, 19] related to the reaction constant of the Faradaic reaction and the double layer capacitance. However, a relaxation time can be obtained in the conventional way from the maximum in the complex impedance plane:

$$\tau_0^{-1} = \omega_m = \cos \Phi/(C_{vm} + C_\infty)\, R_0 \tag{1}$$

in which C_{vm} is the value of C_v at $\omega = \omega_m$. This equation recognises the depression of the electrode arc as a rotation of the impedance axis through an angle Φ.

Two approaches were employed to analyse the experimental data. An extrapolation subtraction method was used to isolate the frequency independent components after which the response of non-ideal components were analysed. This is expressed in the equivalent circuit of Fig. 3d in oxygen and air and that of Fig. 3g in argon where Warburg impedance was clearly visible (Fig. 2a and b). Further refinement of parameters was obtained using a non-linear least squares analysis in the admittance and impedance modes based on equations (2) to (7).

$$G = Z_R/(Z_R^2 + Z_I^2)\,, \tag{2}$$

$$B = Z_I/(Z_R^2 + Z_I^2)\,, \tag{3}$$

$$Z_R = \frac{G_\omega}{G_\omega^2 + B_\omega^2} + R_{el}\,, \tag{4}$$

$$Z_I = \frac{B_\omega}{G_\omega^2 + B_\omega^2}\,, \tag{5}$$

$$G_\omega = \frac{R_{ct} + \sigma\omega^{-1/2}}{(R_{ct} + \sigma\omega^{-1/2})^2 + (\sigma\omega^{-1/2})^2} + a\omega^n\,, \tag{6}$$

$$B_\omega = \frac{\sigma\omega^{-1/2}}{(R_{ct} + \sigma\omega^{-1/2})^2 + (\sigma\omega^{-1/2})^2} + b\omega^n + \omega C_\infty\,. \tag{7}$$

This only slightly modified the values of the circuit parameters (Table 1) obtained above and values of C_∞ obtained were very small and lay between 0 to 10 μF. Furthermore the relaxation time $\tau_0^{-1} = \omega_m = \cos \Phi/C_{vm}R_0$.

In Fig. 4 are shown the simulated impedance responses for circuit 3f. For these curves R_0, n, R_{el}, C_∞, and a/b were identical, but the individual values of a and b were increased by an order of magnitude. This resulted in rotation of the axis by 2°, 16.8°, and 24°, respectively. Inversely non-linear regression analysis of these curves was in-

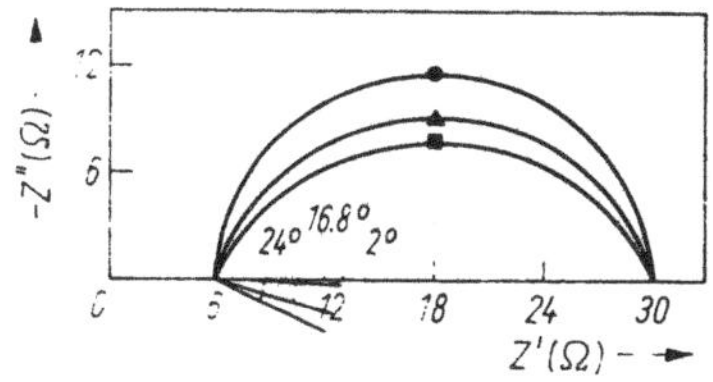

Fig. 4. Impedance response for the circuit 3f for the conditions:

	R_0 (Ω)	n	a	b	C_∞ (μF)	R_{el} (Ω)
●	24.0	0.72	9.33×10^{-6}	2.0×10^{-5}	20	6.0
▲	24.0	0.72	9.33×10^{-5}	2.0×10^{-4}	20	6.0
■	24.0	0.72	9.33×10^{-4}	2.0×10^{-3}	20	6.0

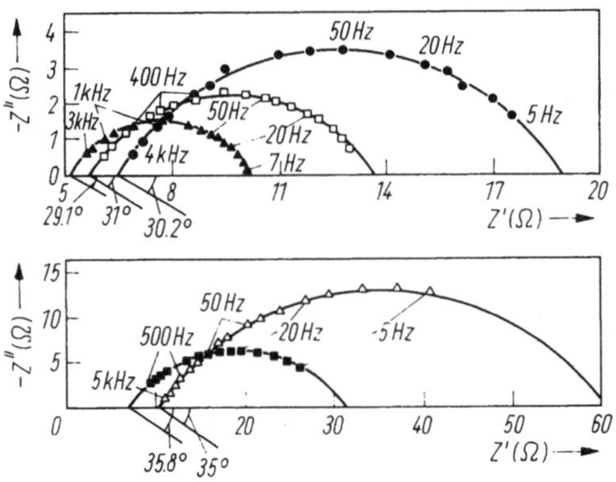

Fig. 5

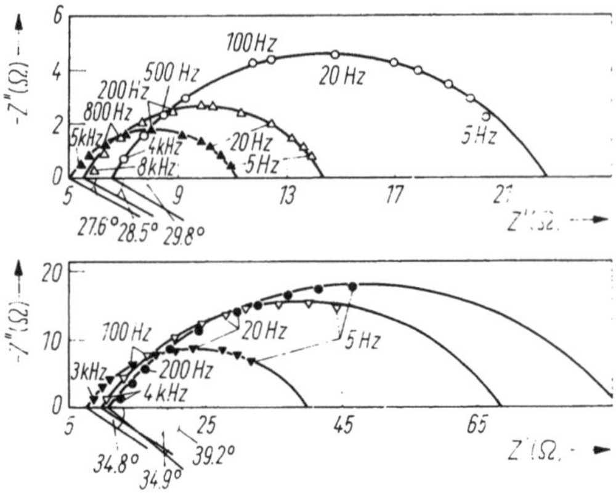

Fig. 6

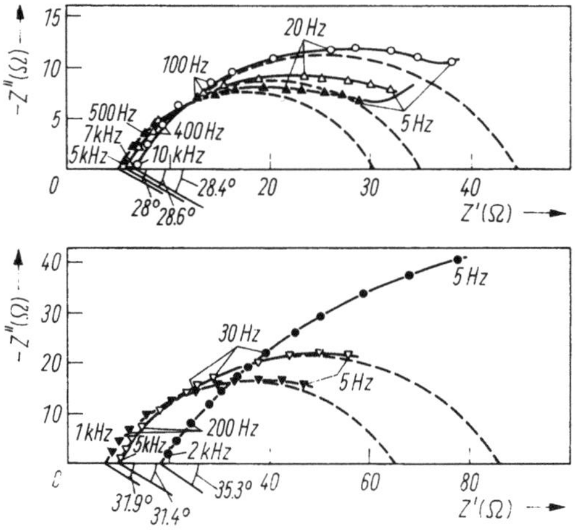

Fig. 7

Fig. 5. Cell's impedance in oxygen (symbols). The simulated curves (solid lines) have been calculated using circuit d in Fig. 3 for the values of circuit parameters in Table 1. ▲ T = 996.8, □ 947.0, ● 902.2, ■ 850.7, △ 799.2 °C

Fig. 6. Cell's impedance in air. ▲ T = 993.2, △ 946.1, ○ 900.2, ▼ 845,5. ▽ 813.3, ● 779.7 °C

Fig. 7. Cell's impedance response in argon. The dashed lines were simulated using circuit in Fig. 3d ($\sigma = 0$) and the solid lines that in 3g. ▲ T = 996.0, △ 949.5, ○ 902.0, ▼ 852.0, ▽ 800.2, ● 729.8 °C

sensitive to the difference between circuits d and f in Fig. 3. It did not provide an accurate estimate of C_∞ especially when values of b and a were large. However, the angle of depression of 24° in the last case is very close to the expected value of 25° for $C_\infty = 0$. The magnitude of C_∞ is too small to influence the Faradaic dispersion and can therefore not be derived from such low frequency data in contrast to the analysis of high frequency arcs.

The data in Fig. 5 and 6 are representative of the typical impedance response in pure oxygen and air, respectively. The simulated curves using the circuit of Fig. 3d are superimposed, showing that agreement is good within the experimental error. The results in argon under the influence of a Warburg impedance in series with R_0 required an analysis according to the circuit in Fig. 3g, as shown in Fig. 7.

Table 1
Various circuit parameters for Pt/YSZ/Pt cells

P_{O_2} (atm)	T (°C)	R_0 (Ω)	n	$a \times 10^4$	$b \times 10^4$	R_{el} (Ω)	σ ($\Omega rad^{0.5} s^{-0.5}$)	$\Phi = \tan^{-1}(a/b)'$	$\Phi = (1-n)\pi/2$	τ_0 (s)
1.0	996.8	5.0	0.69	7.0	12.5	5.2	0.0	29.1	27.8	7.8×10^{-4}
	947.0	7.9	0.63	11.9	20.4	5.7	0.0	30.4	33.8	1.7×10^{-3}
	902.2	12.0	0.64	9.4	16.9	6.5	0.4	29.1	32.4	2.8×10^{-3}
	850.7	24.1	0.60	11.8	16.4	7.1	0.0	35.8	36.0	6.5×10^{-3}
	799.2	48.9	0.61	10.4	15.2	10.2	0.0	34.4	35.3	1.9×10^{-2}
	718.4	—	—	—	—	21.5	—	—	—	—
	659.0	—	—	—	—	43.8	—	—	—	—
	602.2	—	—	—	—	93.8	—	—	—	—
	554.5	—	—	—	—	201.8	—	—	—	—
	504.7	—	—	—	—	541.7	—	—	—	—
	457.4	—	—	—	—	1596.0	—	—	—	—
0.21	993.2	5.8	0.65	8.2	15.6	5.1	1.8	27.6	31.7	8.5×10^{-4}
	946.1	8.8	0.66	10.1	18.6	5.6	—	28.5	31.6	2.3×10^{-3}
	900.2	15.9	0.64	8.7	15.2	6.6	—	29.8	32.6	3.7×10^{-3}
	845.5	32.5	0.62	9.3	13.4	7.5	—	34.8	34.1	8.8×10^{-3}
	813.3	59.1	0.63	8.5	12.2	9.6	—	34.8	33.5	2.1×10^{-2}
	779.7	74.8	0.57	12.7	15.6	10.8	—	39.2	38.3	3.7×10^{-2}
	722.7	—	—	—	—	20.5	—	—	—	—
	651.5	—	—	—	—	48.1	—	—	—	—
3×10^{-6}	996.0	25.3	0.69	4.2	7.8	5.2	17.0	28.0	28.2	4.0×10^{-3}
	949.5	29.5	0.69	4.2	7.7	5.7	15.9	28.5	28.2	4.9×10^{-3}
	902.0	38.0	0.69	4.2	7.8	6.6	15.2	28.4	28.2	7.2×10^{-3}
	852.0	58.5	0.65	5.6	9.1	7.5	4.4	31.9	31.5	1.4×10^{-2}
	800.2	75.7	0.65	5.3	8.8	10.3	6.3	31.4	31.5	2.0×10^{-2}
	729.8	166.3	0.62	6.5	9.2	18.3	3.1	35.3	34.3	6.7×10^{-2}
	657.3	—	—	—	—	44.1	—	—	—	—
	603.0	—	—	—	—	95.0	—	—	—	—
	555.4	—	—	—	—	209.2	—	—	—	—
	506.0	—	—	—	—	547.6	—	—	—	—
	457.4	—	—	—	—	1640.0	—	—	—	—

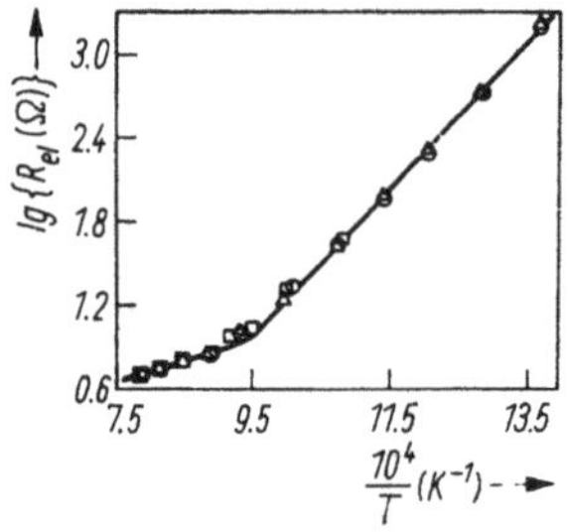

Fig. 8

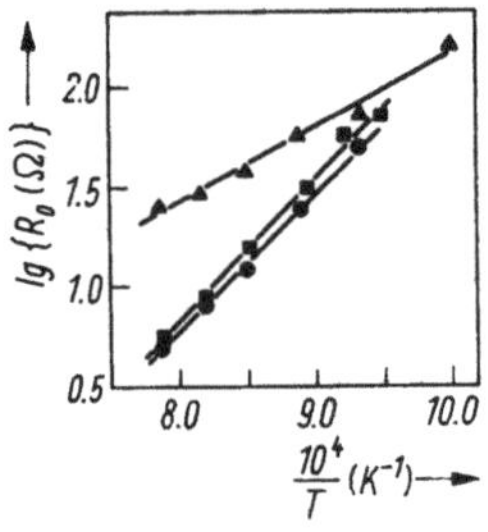

Fig. 9

Fig. 8. Arrhenius plots for electrolyte resistance R_{el}. Activation energy for $T > 775$ °C = (30.0 ± ± 1.3) kJ mol^{-1}. Activation energy for $T < 775$ °C = (98.7 ± 2.1) kJ mol^{-1}; ○ oxygen, □ air, △ argon

Fig. 9. Arrhenius plots for charge transfer resistance R_0. Activation energy in oxygen (●) is (131.4 ± 5.0) kJ mol^{-1}, in air (■) is (139.8 ± 6.3) kJ mol^{-1}, and in argon (▲) is (70.7 ± 4.4) kJ mol^{-1}

It is clear that extrapolation-subtraction and non-linear least squares analysis are insensitive to distinguish between circuits 3d and 3f or 3g and 3h for our experimental data although circuits 3f and 3h seem to be more appropriate.

The best estimates for the various parameters that determine the impedance response in the Pt/YSZ/Pt cells of the present study are shown in Table 1. The error in the parameters a and b is estimated to be less than ± 20%. This relatively large uncertainty is partly due to experimental error in the primary impedance data (± 5%) but more importantly to the calculation procedure described above in which they are secondary to the parameters R_{ct}, R_{el}, and n. The variance of a and b with temperature therefore does not show a significant trend, but their increase with oxygen partial pressure is real. General trends of the parameters with temperature and oxygen partial pressure are shown in Table 2. In Fig. 8 and 9 the Arrhenius diagrams for R_{el} and R_0 are drawn.

The activation energy for the electrolyte conductance is higher below 770 °C than above that temperature, but independent of oxygen partial pressure, as was expected. More importantly the overall Faradaic charge transfer reaction shows a clear change in mechanism when performed in argon and compared with that in oxygen and air. This is attributed to a different interfacial noble metal corrosion mechanism at high oxygen partial pressures. It agrees with the earlier proposed thermodynamic interpretation of the reaction welding process [10], which could contribute to the multiplicity of charge transfer mechanisms under the current experimental conditions.

Table 2

Trends of Faradaic dispersion parameters in Pt/YSZ/Pt cells

χ_i

	a	b	n	R_{ct}	R_{el}	τ_0
$(\partial \chi_i/\partial T)_{p_{O_2}}$	0	0	+	—	—	—
$(\partial \chi_i/\partial p_{O_2})_T$	+	+	0	—	0	—

The physical model proposed here, which in its simplest form is represented by the Faradaic equivalent circuit in Fig. 3f recognizes the presence of the ideal charge transfer resistance and double layer capacitance, as well as the frequency dependent components due to the complexity of events at the interface between the solid ionic conductor and the metal electrode. As suggested elsewhere [11, 18], the proliferation of charged species near the interface involves noble metal cations as well as those of the electrolyte in various stages of oxidation and reduction. The anionic species are qualitatively not so varied, but certainly not unique, as postulated in MacDonald's impedance dispersion theory [15]. The ionic mobilities in the solid electrolyte are slow compared with aqueous systems and consequently the composition and dimensions of the near-interface polarized region will vary with frequency. Not surprisingly, loss phenomena in this region are dependent on frequency as the simple charge transfer in the Faradaic reaction is complicated by parasitic electron transfers between the host of cationic and anionic species whose compositions and concentrations are a function of the frequency (and amplitude) of the applied potential.

The interface between two solids is not easily reproduced from specimen to specimen. Surface roughness, and the presence of impurities complicates reproducibility of results. So do segregation–combination of defect pairs in the surface layers, plastic deformation and residual stresses. Warburg impedance does not affect the non-ideal behaviour of electrode reactions [11] and is only significant at very low oxygen partial pressures and very low frequencies in fluorite related oxide electrolytes.

The distribution of relaxations expressed by the depression of the centre of arc in the Cole-Cole analysis is a mathematical model which relates only the product *RC* to frequency. Yet these quantities are physically distinct with a resistance related to a reaction rate and the concentrations and diffusion constants of reactants and products, while a capacitance is a measure of the degree of polarization at interfaces. Furthermore it is not difficult to fit a Cole-Cole arc to data generated on the basis of equivalent circuits involving a limited number of nested and/or series *RC* circuits each representing individual steps in competing Faradaic reactions. The analysis is as yet not adequate to unraffle the complexity of the reaction kinetics causing the low frequency impedance dispersions but it is based on a physical interpretation and supports earlier observations on the reactions taking place at the noble-metal–ceramic interface at elevated temperatures.

4. Conclusion

Polarization phenomena at the electrode–electrolyte interface, other than those due to the double layer and Faradaic charge transfers, are clearly significant. The variable capacitance is large compared with the constant value of the double layer and becomes a maximum under dc conditions. The data obtained here support our earlier suggestion that the proliferation of cationic species near the interface contributes to the creation of such a variable polarized layer with parasitic charge transfer losses. These can contribute to irreversible energy losses, such as in high temperature fuel cell operations.

Acknowledgements

We gratefully acknowledge the computational help of Mr. C. Klauber and Mr. T. D. Ainsworth. This project was supported by a grant from the Australian Research Grants Commission.

References

[1] S. H. Chu and M. A. Seitz, J. Solid State Chem. **23**, 297 (1978).
[2] T. H. Etsell and S. N. Flengas, Chem. Rev. **70**, 339 (1970).
[3] L. Heyne, Nat. Bur. Standards Spec. Publ. No. 296, 149 (1968).
[4] T. H. Etsell and S. N. Flengas, J. Electrochem. Soc. **118**, 1890 (1971).
[5] R. J. Brook, W. L. Pelzmann, and F. A. Kröger, J. Electrochem. Soc. **118**, 185 (1971).

[6] H. Yanagida, R. J. Brook, and F. A. Kröger, J. Electrochem. Soc., **117**, 593 (1970).
[7] R. L. Zahradnik, J. Electrochem. Soc. **117**, 1443 (1970).
[8] M. Kleitz, P. Fabry, and E. Schouler, in: Fast Ion Transport in Solids, Ed. W. van Gool, North-Holland, Amsterdam 1973 (p. 439).
[9] H. J. de Bruin, A. F. Moodie, and C. E. Warble, J. Mater. Sci. **7**, 909 (1972).
[10] H. J. de Bruin, Nature **272** (5655), 712 (1978).
[11] H. J. de Bruin and S. P. S. Badwal, J. Austral. Ceram. Soc. **14**, 20 (1979).
[12] P. Fabry and M. Kleitz, J. Electroanal. Chem. **57**, 165 (1974).
[13] D. O. Raleigh, Electroanalytical Chemistry, A Series of Advances, Vol. 6, Ed. A. J. Bard, Marcel Dekker, Inc., New York 1973.
[14] S. Pizzini, in: Fast Ion Transport in Solids, Ed. W. van Gool, North-Holland, Amsterdam 1973 (p. 461).
[15] J. R. Macdonald, in: Electrode Processes in Solid State Ionics, Ed. M. Kleitz and J. Dupuy, Reidel, Dordrecht 1976 (p. 149).
[16] D. R. Franceschetti and J. R. MacDonald, J. Electroanal. Chem. **82**, 271 (1977).
[17] R. J. Friauf, J. chem. Phys. **22**, 1329 (1954).
[18] H. J. de Bruin and S. P. S. Badwal, phys. stat. sol. (a) **49**, K181 (1978).
[19] S. P. S. Badwal and H. J. de Bruin, Austral. J. Chem. **31**, 2337 (1978).
[20] A. K. Jonscher, phys. stat. sol. (a) **32**, 665 (1975).
[21] I. D. Raistrick, Chun Ho, and R. A. Huggins, J. Electrochem. Soc. **123**, 1469 (1976).
[22] I. D. Raistrick, Chun Ho, Yaw Wen Hu, and R. A. Huggins, J. Electroanal. Chem. **77**, 319 (1977).
[23] A. Duquesnoy and F. Marion, Bull. Chim. Fr. 77 (1964).
[24] K. S. Cole and R. H. Cole, J. chem. Phys. **9**, 341 (1941).

(Received March 2, 1979)

phys. stat. sol. (a) 54, 271 (1979)

Subject classification: 1.4 and 10.2; 22.8.2

Centre de Mise en Forme des Matériaux, Ecole des Mines de Paris, Sophia Antipolis, Valbonne[1])

Coherent Precipitation near Dislocations

A Theoretical Analysis

By

F. MONTHEILLET and J. M. HAUDIN

A simple theoretical model is developed for a dislocation decorated by a cylindrical precipitate, which makes it possible to derive the energy change ΔW associated with the precipitation process versus radius R of the precipitate. The distance a between the cylinder axis and the dislocation line is taken so as to minimize ΔW. When the chemical and elastic energy terms are simultaneously considered, $\Delta W(R)$ is obtained by numerical computation. However, when the chemical energy terms are negligible, it is possible to calculate analytically the minimum of $\Delta W(a, R)$. In the former case, the precipitate grows until complete depletion of solute atoms in the matrix, after a very small critical radius is reached. In the latter case, growth leads to equilibrium radius and location. In both the cases, decoration of the dislocation is more likely to happen than homogeneous precipitation of coherent spheres.

Nous proposons un modèle théorique simple de dislocation décorée par un précipité cylindrique permettant de calculer la variation d'énergie ΔW associée au processus de précipitation en fonction du rayon R du précipité. La distance a entre son axe et la ligne de dislocation est choisie de façon à minimiser ΔW. Lorsque les termes d'énergie chimique et élastique interviennent simultanément, la courbe $\Delta W(R)$ est obtenue numériquement. Par contre, dans le cas où les termes d'énergie chimique sont négligeables, on peut calculer analytiquement le minimum de $\Delta W(a, R)$. Dans le premier cas, le précipité croît jusqu'à épuisement des atomes de soluté, après avoir franchi un rayon critique de germination voisin de zéro. Dans le second cas, la croissance aboutit à un rayon et une position d'équilibre. Dans les deux cas, la décoration de la dislocation est énergétiquement plus favorable que la précipitation homogène de sphères dans la matrice.

1. Introduction

In many alloys submitted to an ageing treatment after annealing, precipitation occurs on or near dislocations. This "decoration" induces specific diffraction contrasts in electron microscopy: images consist of several alternatively dark and bright fringes, similar to those of coherent precipitates discussed by Ashby and Brown [1]. Such precipitation of carbides [2, 3], nitrides [4, 10], or intermetallic compounds [5] on dislocations has been previously reported. Only a few attempts have been made to justify theoretically these phenomena: the free energy change associated with decoration has been calculated by Cahn [6] and Dollins [4]. The former assumed incoherent precipitation, and the latter considered spherical or disk-shaped coherent precipitates. However, in both cases some terms have been neglected in elastic energy calculations, and the results could be obtained only by numerical computation. Dundurs [11] has calculated the equilibrium positions of a screw or edge dislocation near or inside a cylindrical precipitate. This rather sophisticated derivation only takes into account the difference between elastic constants of matrix and second phase. Therefore, although precise results are given, practical inferences cannot be drawn concerning the decoration of dislocations in alloys.

---

[1]) 06560 Valbonne, France.

In this work, a simple model is developed from experimental observations, which is able to take into account all physical relevant parameters. It is further shown, in the case of intermetallic compounds, that the problem can be solved merely by analytical calculation.

2. Energy Change Associated with Decoration

2.1 The model

In most cases, precipitates on dislocations are likely to be coherent with the matrix, at least in the first stages of precipitation. Accordingly, the precipitation process consists in substituting a number of atoms in the matrix for an equal number of different atoms, which will form the second phase (Fig. 1).

The corresponding displacement field is assumed to be

$$\boldsymbol{U} = \boldsymbol{U}_{\mathrm{D}} + \boldsymbol{U}_{\mathrm{P}} , \tag{1}$$

where $\boldsymbol{U}_{\mathrm{D}}$ and $\boldsymbol{U}_{\mathrm{P}}$ are the displacement fields of a dislocation without precipitate and of a single coherent precipitate, respectively. These displacements will be calculated in isotropic linear elasticity theory. When the matrix and second phase have the same elastic constants, $\boldsymbol{U}$ is the exact solution of the problem as resulting from superposition of two equilibrium states. However, (1) is only expected to give a rather good approximation of the exact displacement field in general; corrections could then be brought by using image forces [8].

At the beginning of the precipitation process, the solute atoms diffuse in the crystal towards the region of minimum or maximum hydrostatic pressure, according to whether they are bigger or smaller than the matrix atoms (Cottrell effect). The map of hydrostatic pressure of an edge dislocation (Fig. 2) shows that precipitation is expected to start on the $0y$ symmetry axis. In agreement with numerous experimental observations [5], supported by computer simulations [5, 7], it is supposed that the precipitate at any given moment is a cylinder of radius R and of large length that can be taken as infinite. Its axis lies parallel to dislocation line D at an algebraic distance a. Thus the displacement field $\boldsymbol{U}$ remains unchanged by any translation parallel to D. The strains take place only in the $0xy$ plane normal to D. Furthermore, if the diffusion rate is high enough, a will minimize the total energy of the system for each radius R.

It would be desirable to calculate the change ΔG of free enthalpy in the precipitation process. However, the transformation takes place under constant temperature and pressure and furthermore changes of entropy and volume can be reasonably neglected. Thus we can merely use the energy change $\Delta W = W_{\mathrm{DD}} - W_{\mathrm{D}}$, where

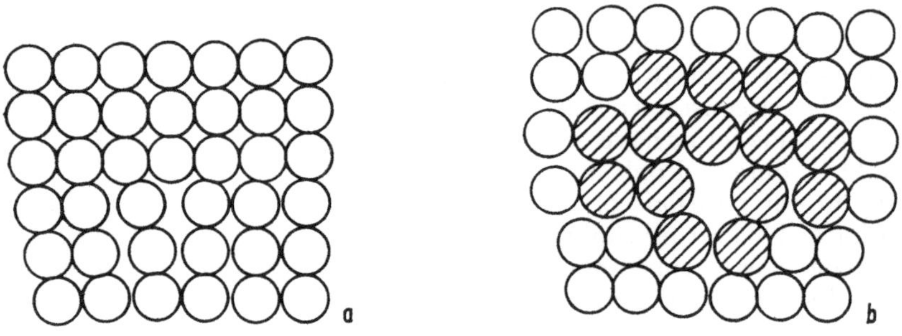

Fig. 1. Formation of a coherent precipitate near a dislocation. a) Dislocation without precipitate, b) decorated dislocation. The second phase is constituted here by only one atomic species

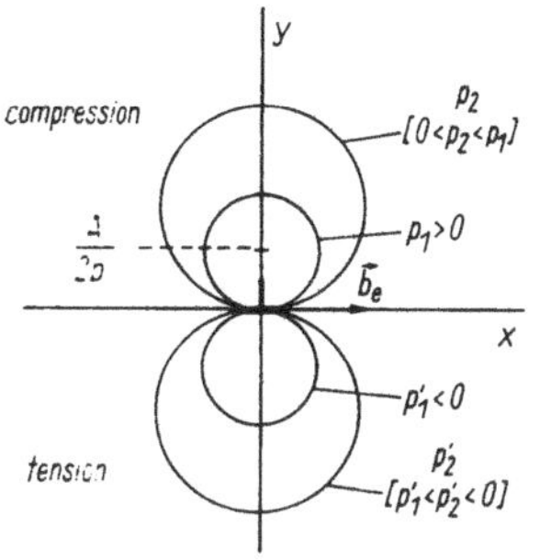

Fig. 2. Map of hydrostatic pressure of an edge dislocation. Isopressure curves are circles centred on $0y$ axis. A is a constant depending on material and Burgers vector $\boldsymbol{b}_e$

W_{DD} and W_D are the energy of the decorated dislocation and of the dislocation before precipitation, respectively. ΔW is the sum of both elastic and chemical terms. Irrespective of any assumption concerning the core structure, the corresponding energy is supposed to remain constant during the precipitation process.

2.2 Calculation of energy change ΔW

2.2.1 *Definition of elastic energy terms*

From (1) the strain tensor of the decorated dislocation is

$$\boldsymbol{\epsilon} = \boldsymbol{\epsilon}_D + \boldsymbol{\epsilon}_P . \tag{2}$$

Therefore, the elastic strain energy per unit volume is given by

$$w = \tfrac{1}{2}\lambda(\mathrm{Tr}\,\boldsymbol{\epsilon})^2 + \mu(\varepsilon_{xx}^2 + \dots + 2\varepsilon_{xy}^2 + \dots) ,$$

$$w = \tfrac{1}{2}\lambda(\mathrm{Tr}\,\boldsymbol{\epsilon}_D + \mathrm{Tr}\,\boldsymbol{\epsilon}_P)^2 + \mu[(\varepsilon_{xx}^D + \varepsilon_{xx}^P)^2 + \dots + 2(\varepsilon_{xy}^D + \varepsilon_{xy}^P)^2 + \dots] ,$$

that is

$$w = w_D + w_P + w_I , \tag{3}$$

where w_D and w_P are the elastic energy per unit volume of a dislocation without precipitate and of a single coherent precipitate, respectively.

$$w_I = \lambda\,\mathrm{Tr}\,\boldsymbol{\epsilon}_D\,\mathrm{Tr}\,\boldsymbol{\epsilon}_P + 2\mu(\varepsilon_{xx}^P\varepsilon_{xx}^D + \dots + 2\varepsilon_{xy}^P\varepsilon_{xy}^D)$$

can be considered as an interaction elastic energy between dislocation and precipitate.

The total elastic strain energy of the decorated dislocation per unit length is then

$$W_e = \iint_P w\,ds + \iint_M w\,ds ,$$

where both integrals are calculated in the $0xy$ plane inside the circular section of the cylindrical precipitate with the elastic constants of the second phase, and outside this circle with the elastic constants of the matrix, respectively. Then

$$W_e = \iint_P w_D\,ds + \iint_M w_D\,ds + \iint_P w_P\,ds + \iint_M w_P\,ds + \iint_P w_I\,ds + \iint_M w_I\,ds \tag{4}$$

or by grouping the terms according to their physical origin

$$W_e = W_D^* + W_P + W_I . \tag{5}$$

W_D^* is the elastic energy of the dislocation *in the presence of the precipitate*, W_P the elastic energy of the single coherent precipitate, and W_I the interaction elastic energy.

On the other hand, the growth of the precipitate involves a depletion of the solute atoms contained in the matrix, hence a decrease of solid solution elastic energy ΔW_{SS}.

We can thus write the total change of elastic energy

$$\Delta W_e = W_e - W_D + \Delta W_{SS}\,, \qquad \Delta W_e = W_D^* - W_D + W_P + W_I + \Delta W_{SS}\,,$$

or

$$\Delta W_e = W_{\delta'} + W_\delta + \Delta W_{SS}\,,$$

where

$$\left.\begin{aligned} W_{\delta'} &= W_D^* - W_D\,,\\ W_\delta &= W_P + W_I\,. \end{aligned}\right\} \tag{6}$$

It will be seen later that $W_{\delta'}$ and W_δ are the parts of the elastic energy correspondi ng to the modulus effect and to the size effect, respectively.

2.2.2 *Calculation of elastic energy terms*

The displacement field of a dislocation is given by classical formulae [8]. That of a cylindrical precipitate is purely radial and given as a function of the distance r from the axis

a) in the second phase ($0 \leqq r \leqq R$) by

$$U_P = -\frac{\omega\delta}{1+\omega}\, r\,, \tag{7a}$$

b) in the matrix ($r \geqq R$) by

$$U_P = \frac{\delta}{1+\omega}\,\frac{R^2}{r}\,, \tag{7b}$$

where

$$\omega = \frac{\mu_m}{\lambda_s + \mu_s} \quad \text{and} \quad \delta = \frac{1}{3}\,\frac{V_s - V_m}{V_m}\,.$$

μ_m, λ_s, μ_s are the elastic constants of matrix (m) and second phase (s), V_m and V_s the mean atomic volume in the matrix and second phase.

For an analytical calculation of W_e (4), three cases are to be considered, according to whether the dislocation core lies entirely inside the second phase ($|a| \leqq R - r_c$), entirely inside the matrix ($|a| \geqq R + r_c$), or partly in both media ($R - r_c \leqq |a| \leqq$ $\leqq R + r_c$). In the third case the analytical calculation is actually not possible.

We shall not present in detail the rather tedious calculations. The results are given below:

(i) dislocation core lying inside second phase $|a| \leqq R - r_c$

$$W_{\delta'} = \frac{\mu_m}{8\pi} B^2 \delta' \ln\left(\frac{R^2 - a^2}{r_c^2}\right), \tag{8}$$

$$W_P = \frac{2\mu_m}{1+\omega}\,\delta^2 \pi R^2\,, \tag{9}$$

$$W_I = \frac{2\mu_m}{1+\omega}\, b_e\,; \tag{10}$$

(ii) dislocation core lying inside matrix $|a| \geqq R + r_c$

$$W_{\delta'} = \frac{\mu_m}{8\pi}\,\delta' \left[B^2 \ln\left(\frac{a^2}{a^2 - R^2}\right) - \frac{3}{4}\, b_e^2\, \frac{R^2}{a^2}\right], \tag{8'}$$

$$W_P = \frac{2\mu_m}{1+\omega} \delta^2 \pi R^2 , \tag{9'}$$

$$W_I = \frac{2\mu_m}{1+\omega} \delta b_e \frac{R^2}{a} . \tag{10'}$$

δ and ω have been already defined and

$$\delta' = \frac{\mu_s - \mu_m}{\mu_m} , \qquad B^2 = \frac{3}{2} b_e^2 + b_s^2 ,$$

where b_e and b_s are the edge and screw components of Burgers vector, respectively.

The change of solid solution elastic energy can be calculated by means of the following formula:

$$\Delta W_{SS} = -\frac{\pi R^2}{V_s^2} w_a , \tag{11}$$

where w_a is the mean elastic energy of a solute atom in the matrix. It can be estimated very approximatively by a similar derivation to that of W_P, transposed to the case of a sphere [9].

2.2.3 *Some comments: size and modulus effects*

The term $W_{\delta'}$ vanishes when $\delta' = 0$, i.e. when $\mu_m = \mu_s$. It corresponds to the *modulus effect* of the precipitate, and has the same sign as δ', which shows that decoration will be favoured by the modulus effect when the second phase is softer than the matrix ($\delta' < 0$) and conversely.

The term W_δ vanishes when $\delta = 0$, i.e. when $V_m = V_s$. It corresponds to the *size effect* of the precipitate. W_P is always positive, but W_I has the same sign as the product δa. Therefore, when $\delta > 0$, if the precipitate can nucleate, it will be in the crystal region of negative hydrostatic pressure ($a < 0$, cf. Fig. 2), which involves a negative W_I, and conversely when $\delta < 0$. This result is in agreement with the well-known Cottrell effect. It is also noticeable that interaction energy W_I disappears for a pure screw dislocation.

2.2.4 *Chemical energy terms*

The formation of a second phase involves a change of chemical energy per unit length of precipitate

$$\Delta W_{CH} = \pi R^2 \, \Delta F , \tag{12}$$

where ΔF is the change of chemical energy per unit volume ($\Delta F < 0$).

Finally, the interfacial energy W_S between the matrix and second phase has to be considered

$$W_S = 2\pi R \gamma , \tag{13}$$

where γ is the interfacial energy density ($\gamma > 0$).

In conclusion, the formation of a cylindrical coherent precipitate near a dislocation induces a total energy change, taking into account both elastic and chemical energy changes

$$\Delta W = W_{DD} - W_D = W_{\delta'} + W_\delta + \Delta W_{SS} + \Delta W_{CH} + W_S . \tag{14}$$

3. Discussion

Equation (14) enables us to determine the position of the precipitate for each value of its radius R: if diffusion takes place fast enough, the actual value of a must minimize ΔW. It then becomes possible to compute the variations of ΔW versus R and therefore to forecast whether decoration is likely to take place.

3.1 General case

In most cases, the elastic and chemical energy terms are to be considered simultaneously. For instance, we have used in (14) physical parameters relevant to precipitation of niobium nitride Nb_2N in Nb, as given by Dollins [4]. Because of the small radius of

Table 1
Parameters used for numerical calculations

	precipitation Nb_2N in Nb [4]	R-phase in δ-ferrite [5]
a_m*) (Å)	3.30	2.88
	(b.c.c. structure)	(b.c.c. structure)
a_s**) (Å)	3.052	10.90
c_s**) (Å)	4.957	19.34
	(h.c.p. structure)	(hexagonal lattice with complex structure)
V_m (Å^3)	17.97	11.94
V_s (Å^3)	19.99	12.52
δ	0.036	0.016
ΔF (10^9 erg/cm^3)	− 4.57	0
γ (erg/cm^2)	20	0
μ_m (10^{11} dyn/cm^2)	3.75	8.3
μ_s (10^{11} dyn/cm^2)	19	8.3
δ'	4.07	0
$\nu_m = \nu_s$***)	0.38	0.29
b_e (Å)	2.86	2.49
b_s (Å)	0	0
r_c (Å)	2.86	2.49

*) a_m is the matrix parameter.
**) a_s and c_s are second-phase parameters.
***) ν_m, ν_s Poisson's ratios.

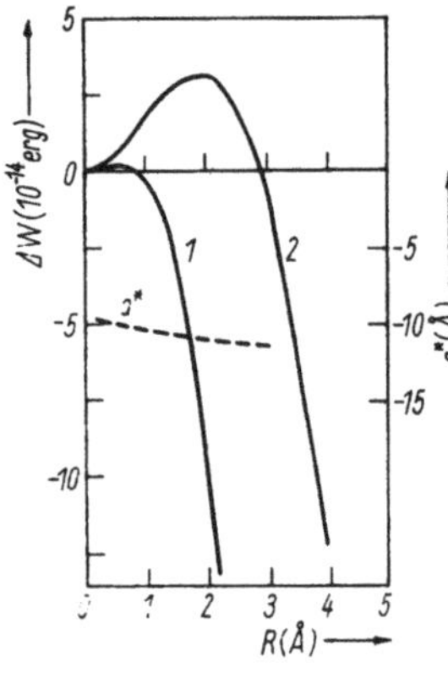

Fig. 3. Total energy changes associated with decoration of an edge dislocation (1) and with formation of a coherent spherical precipitate (2) vs. radius R. Volume of second phase is assumed to be the same in both cases. Equilibrium value a^* of a is also plotted vs. R. Dollins' parameters are used [4], corresponding to the precipitation of Nb_2N in Nb (see Table 1)

interstitial nitrogen atoms in the niobium matrix, W_{SS} was supposed here to be negligible. In any case, we must emphasize that a precise calculation of this term would be rather difficult. Curves plotted on Fig. 3 have been computed using Dollins' parameters given in Table 1. One notices that the precipitate axis remains roughly the same distance ($a \approx 11$ Å) from the dislocation line during the precipitation process. The curve $\Delta W(R)$ has a very low maximum for a critical radius $R_c \approx 0.5$ Å. Therefore, the activation energy for the precipitation process is almost zero.

It is interesting to compare this process with the formation of a coherent spherical precipitate in the matrix without dislocation (homogeneous precipitation). In such a case

$$\Delta W = W_e^{SPH} + W_{CH} + W_S ,$$

where W_e^{SPH} is the elastic energy of a spherical coherent precipitate. This term is derived in the same way as W_P [9]

$$W_e^{SPH} = \frac{6\mu_m}{1 + \omega'} \delta^2 \frac{4}{3} \pi R^3 ,$$

where

$$\omega' = \frac{4\mu_m}{3\lambda_s + 2\mu_s} .$$

To compare both types of precipitation, we have assumed that a given volume of the second phase is likely to form either a cylinder of infinite length having a radius R near the dislocation, or if there is no dislocation, coherent spheres of the same radius. The energy of one sphere must then be compared with the energy of a length of cylinder $L = \frac{4}{3} R$ to ensure that the volume of the second phase is the same in both cases. Fig. 3 shows that the homogeneous precipitation process (sphere) involves a much higher activation energy than the decoration of a dislocation (cylinder).

We have noted that Dollins [4], using the same values for physical parameters, gives a very different result: R_c is comprised between 7 and 15 Å according to the assumption made for the W_P-calculation and a is equal to 35 Å. Nevertheless, his derivation assumes a disk-shaped precipitate of thickness $t = R/10$, which of course induces larger strains in the matrix than an infinite cylinder for a given volume of the second phase. On the other hand, a careful examination of Dollins' derivation shows that the parts of W_P and W_I stored *within the precipitate* were neglected; it corresponds in the present work to

$$\iint_P w_P \, ds + \iint_P w_I \, ds .$$

3.2 Chemical energy terms are negligible

In a few cases, the chemical energy terms W_{CH} and W_S are likely to be small with respect to elastic energy terms. This involves that atomic bonds are of the same type and strength in matrix, second phase, and at the interface between both media. This is most probably true for instance when the second phase is an intermetallic compound, which has nearly the same composition as the matrix. When it is so, the elastic energy term ΔW_{SS} will be also negligible and ΔW reduces to $W_\delta + W_{\delta'}$.

This case is particularly interesting, because it makes it possible to analytically solve the system

$$\frac{\partial \Delta W}{\partial a} = 0 , \qquad \frac{\partial \Delta W}{\partial R} = 0 .$$

The derivation shows that in most cases $\Delta W(a, R)$ reaches an unique absolute minimum for values a^* and R^* of parameters given by the following equations:

$$a^* = -\frac{b_e}{2\pi\delta}, \tag{15}$$

$$R^* = \frac{\alpha}{2\pi\,|\delta|}, \tag{16}$$

where

$$\alpha = \left[b_e^2 - B^2\frac{f(\delta')}{12}\right]^{1/2} \quad \text{if} \quad \delta' \leqq 0\,,$$

$$\alpha = \left[b_e^2 - \frac{4}{3}\,B^2\,\frac{f(\delta')}{f(\delta') + 16}\right]^{1/2} \quad \text{if} \quad \delta' \geqq 0\,;$$

$$f(\delta') = \frac{\delta'(3\delta' + 4)}{\delta' + 1}.$$

Table 2
Equilibrium states for decorated dislocations, when chemical energy terms are negligible

		screw dislocation	edge or mixed dislocation
$\delta = 0$ modulus effect only	$\mu_s < \mu_m$	 $a^* = 0$, $R^* \to \infty$	 $a^* = 0$, $R^* \to \infty$
	$\mu_s \geqq \mu_m$	no decoration	no decoration
	$\mu_s < \mu_m$	 $a^* = 0$	 $a^* = -\frac{b_e}{2\pi\delta}$, $R^* > \|a^*\|$
$\delta \neq 0$ (schemes for $\delta > 0$)	$\mu_s = \mu_m$ (size effect only)	no decoration	 $a^* = -\frac{b_e}{2\pi\delta}$, $R^* = \|a^*\|$
	$\mu_m < \mu_s < \mu_s^M$	no decoration	 $a^* = -\frac{b_e}{2\pi\delta}$, $R^* < \|a^*\|$
	$\mu_s \geqq \mu_s^M$	no decoration	no decoration $\mu_s^M \approx 6\mu_m$ (edge dislocation) $\mu_m < \mu_s^M < 6\mu_m$ (mixed dislocation)

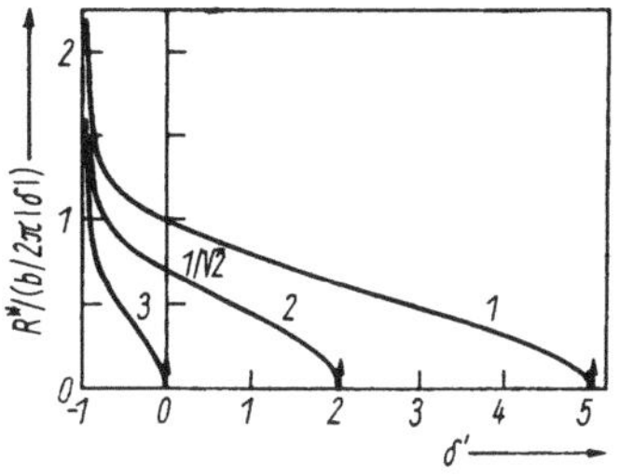

Fig. 4

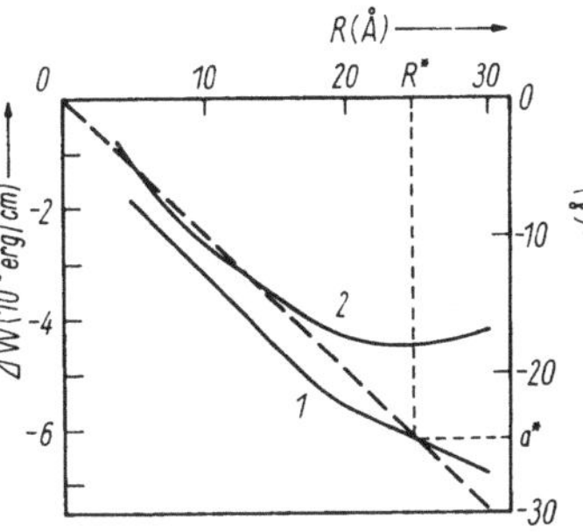

Fig. 5

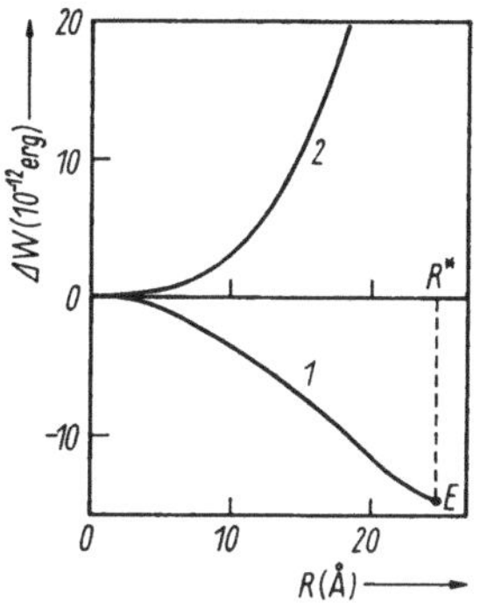

Fig. 6

Fig. 4. Variation of equilibrium radius R^* vs. δ' for an (1) edge, (2) mixed with $b_e = b_s$, and (3) screw dislocation

Fig. 5. Value of a minimizing ΔW (1) and corresponding energy change (2) vs. R. Parameters corresponding to precipitation of R-phase in δ-ferrite are used [5] (see Table 1). Chemical energy terms are neglected

Fig. 6. Total energy change associated with decoration of an edge dislocation (1) and with formation of a coherent spherical precipitate (2) vs. radius R. Volume of second phase is assumed to be the same in both cases. Decoration of the dislocation leads to equilibrium state (E) for a radiu R^*. Parameters corresponding to precipitation of R-phase in δ-ferrite are used [5] (see Table 1) Chemical energy terms are neglected

Therefore, decoration of the dislocation will generally take place, until an equilibrium state is reached, at which the cylindrical precipitate of radius R^* has its axis lying at an algebraic distance a^* from the dislocation. Table 2 summarizes all possible cases. It clearly appears that both following factors will favour decoration: the size effect of the second phase ($\delta \neq 0$) and the edge component of the dislocation ($b_e \neq 0$). When the modulus effect of the second phase is very large ($\mu_s \geqq \mu_s^M$), decoration no longer takes place. The critical value μ_s^M depends in a complex manner on the ratio b_s/b_e of screw and edge components of the Burgers vector. It increases from μ_m to nearly $6\mu_m$, when b_e grows from zero (pure screw dislocation) to b (pure edge dislocation). Fig. 4 shows the variation of the equilibrium radius R^* versus δ', when all other parameters are kept constant. The difference in behaviour of screw and edge dislocations can again be observed.

As an illustration of this case, we have used physical parameters relevant to precipitation of R-phase in δ-ferrite of a stainless steel (Table 1). The structure and chemical composition of this intermetallic phase, its growth near dislocations, and associated contrasts in electron microscopy have been studied elsewhere [5, 7]. Fig. 5 shows, for each value of the radius R of the precipitate, the value of a minimizing ΔW and the corresponding energy change, as obtained by numerical computation. ΔW reaches a minimum, when R is roughly 25 Å and $a = -R$. This result could be obtained of course directly from (15) and (16) with $f(\delta') = 0$ since here $\delta' = 0$.

Finally, in Fig. 6, the formation of a coherent sphere of R-phase without dislocation is compared to the decoration of a dislocation by a cylindrical precipitate of the same radius R and as previously of length $\frac{4}{3} R$. It becomes clearly apparent that the latter process, which involves an energy decrease for the system, will take place at the expense of the former.

4. Conclusions

A theoretical analysis has been developed, in which a decorated dislocation is considered as a superposition of a dislocation and a coherent cylindrical precipitate. The energy change involved by the decoration process has been analytically derived according to this model.

In the general case, both elastic and chemical energy terms are to be simultaneously considered, and a numerical computation of the $\Delta W(R)$ curve, where for each radius R of the precipitate a is determined in order to minimize ΔW, enables us to compare decoration of dislocations with other precipitation processes, and therefore to forecast whether decoration is likely to take place.

Furthermore in the particular case where chemical energy terms can be neglected when decoration takes place, the precipitate grows further until an equilibrium state is reached. The corresponding R^* and a^* parameters, defining size and location of the precipitate have been analytically calculated.

In both cases, the evaluated energy changes show that decoration of the dislocation is more likely to happen than homogeneous precipitation of coherent spheres.

References

[1] M. F. ASHBY and L. M. BROWN, Phil. Mag. **8**, 1083, 1649 (1963).
[2] M. H. LEWIS and B. HATTERSLEY, Acta metall. **13**, 1159 (1965).
[3] A. GELPI, Thesis, Paris 1971.
[4] C. C. DOLLINS, Acta metall. **18**, 1209 (1970).
[5] J. M. HAUDIN and F. MONTHEILLET, Metallography **11**, 391 (1978).
[6] J. W. CAHN, Acta metall. **5**, 169 (1957).
[7] F. MONTHEILLET, J. M. HAUDIN, and G. FRADE, phys. stat. sol. (a) **17**, 593 (1973).
[8] J. FRIEDEL, Dislocations, Pergamon Press, New York 1964.
[9] F. R. N. NABARRO, Proc. Roy. Soc. **A175**, 519 (1940).
[10] K. BORGGREEN and A. R. THÖLÉN, phys. stat. sol. (a) **39**, 223 (1977).
[11] J. DUNDURS, in: Mathematical Theory of Dislocations, Ed. T. MURA, ASME, Evanston 1969 (p. 70).

(Received March 22, 1979)

phys. stat. sol. (a) 54, 281 (1979)

Subject classification: 18.3; 22.8.2

Optisches Institut der Technischen Universität Berlin, Berlin (West)

Lichtbeugung an magnetischen Bubble-Bereichsgittern in substituiertem Yttrium–Eisengranat

Von

B. KUHLOW

Die in magnetischen Granatkristallen auftretenden Bubble-Bereiche können sich unter bestimmten Bedingungen in einem regelmäßigen hexagonalen Bereichsgitter anordnen. Das Verhältnis von Bubble-Durchmesser zu ihrem gegenseitigen Abstand im Gitter kann durch ein äußeres Magnetfeld verändert werden. Aufgrund des magneto-optischen Faraday-Effektes tritt eine Beugung des Lichtes an dem magnetischen Bereichsgitter auf. Im Fraunhoferschen Beugungsbild werden die Lagen, Polarisationszustände und Intensitäten der einzelnen Beugungsordnungen berechnet und die Intensitäten als Funktion des äußeren Magnetfeldes graphisch dargestellt. Die Übereinstimmung zwischen den beobachteten und den berechneten Beugungsbildern ist gut. Eine Anwendung der Beugungserscheinungen in der Datenverarbeitung wird diskutiert.

The bubble domains that appear in magnetic crystals are under certain circumstances arranged in a regular hexagonal grating. The ratio of diameter to distance of the domains in the grating is changed by an external magnetic field. Due to the magneto-optical Faraday effect light is diffracted by that domain grating. In the case of Fraunhofer diffraction the positions, polarization states, and intensities of the different diffraction orders are calculated. The intensities are represented graphically as a function of the external magnetic field. The agreement between the observed and calculated diffraction patterns is sufficient. An application of the diffraction for data processing is discussed.

1. Einleitung

In magnetisch uniaxialen ferro- und ferrimagnetischen Schichten können sogenannte Bubble-Bereiche auftreten. Das sind zylindrisch durch die Schicht hindurchgehende Bereiche einheitlicher Magnetisierung in Richtung einer Schichtnormalen, umgeben von einem „See" der entgegengesetzt gleichen Magnetisierung (Fig. 1). Diese Bereiche haben als Informationsträger für die Datenspeicherung große Bedeutung gewonnen.

Magnetische Bereichsstrukturen in optisch transparenten Schichten lassen sich mit Hilfe des magneto-optischen Faraday-Effektes abbilden. Wie bei jeder optischen Abbildung entsteht dabei in der hinteren Brennebene des abbildenden Objektivs ein der Objektstruktur entsprechendes Fraunhofersches Beugungsbild. Diese Beugungserscheinung wurde unabhängig voneinander von Dillon und Remeika [1] und Lambeck [2] entdeckt und von Boersch und Lambeck [3] und de Lang [4] für einfache Streifenbereichsstrukturen berechnet. Viele weitere Beobachtungen und Berechnungen dazu folgten [5 bis 13], sowie Vorschläge für technische Anwendungen [14 bis 16]. Über die Beugung an Bubble-Bereichsstrukturen wurde von Papworth [17], Woolhouse und Chaudhari [18] sowie Antonov et al. [19] berichtet. Während die in [17] durchgeführten Berechnungen der Beugung nur qualitativ sind, wurden in [19] die Phasen- bzw. Polarisationszustände der einzelnen Beugungsordnungen, die für die weitere Interferenz zur Bildentstehung wichtig sind, außer acht gelassen.

Hier werden für ein variables Bubble-Bereichsgitter die Orte der einzelnen Beugungsordnungen sowie ihre Intensitäten und Polarisationszustände jeweils angegeben und mit beobachteten Beugungsbildern verglichen.

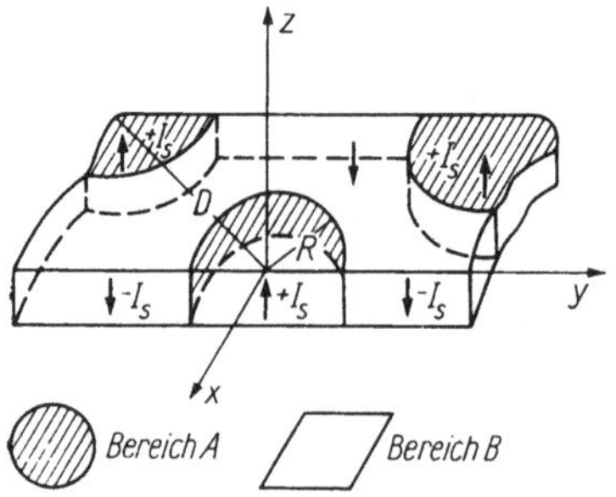

Fig. 1. Bubble-Bereiche (schraffiert) in einer Magnetschicht mit den Bereichsmagnetisierungen $+I_s$ und $-I_s$ in Richtung der uniaxialen Anisotropie

2. Versuchsdurchführung

Der verwendete ferrimagnetische Kristall der Zusammensetzung

$$Y_{1,92}Sm_{0,10}Ca_{0,98}Fe_{4,02}Ge_{0,98}O_{12}$$

war durch Epitaxie aus der flüssigen Phase (LPE) auf ein Gadolinium–Gallium–Granat-Kristallplättchen (GGG) erhalten worden. Die Dicke der einkristallinen Magnetschicht beträgt 5,3 μm, die Faradaydrehung für grünes Licht bei Zimmertemperatur $\beta = 1{,}6°$ und die Curietemperatur $T_C = 198$ °C.

Anstelle der im abmagnetisierten Zustand der Probe normalerweise auftretenden Streifenbreiche können auch Bubble-Bereiche erzeugt werden, die sich infolge magnetischer Wechselwirkung in einem regelmäßigen Bubble-Bereichsgitter anordnen [17, 18, 20]. Dazu wurden hier die vorhandenen, mäanderförmig verlaufenden Streifenbereiche zunächst durch ein starkers Magnetfeld in der Schichtebene ausgerichtet. Sodann wurde dieses Feld etwas gegen die Schichtebene geneigt und verringert. Dabei zerrissen die Streifenbereiche in jeweils mehrere Bubble-Bereiche, die sich in dem fehlerfreien Kristall in einem regelmäßigen, hexagonalen Bubble-Gitter anordneten. Dieses blieb als metastabiler Zustand auch erhalten, wenn das erzeugende Feld auf Null herabgeregelt wurde. Durch ein normal zur Schichtebene liegendes Feld können die Bereiche im Gitter in weiten Grenzen verändert werden.

Die Bereiche und die an ihnen auftretenden Beugungserscheinungen wurden mit Hilfe eines hochauflösenden Polarisationsmikroskops untersucht. Als Lichtquelle diente eine XBO 150 mit vorgeschaltetem Grünfilter. Als Polarisator wurde ein Prismenoperator und als Analysator ein drehbarer Folienpolarisator verwandt. Das abbildende Objektiv hatte eine Apertur von 0,65. Während für die Bereichsabbildung die Beleuchtungsapertur etwa 0,4 betrug, wurde diese für die Aufnahme des Fraunhoferschen Beugungsbildes jeweils beträchtlich herabgesetzt, um die einzelnen Beugungsordnungen getrennt erscheinen zu lassen.

3. Fraunhofer-Beugung am hexagonalen Bubble-Bereichsgitter

Im Polarisationsmikroskop wird linear polarisiertes Licht $E = E_0 \cos \omega t$ in den Kristall eingestrahlt. Bei verschwindendem Zirkulardichroismus, wie das für den vorliegenden Kristall als auch die meisten ähnlich zusammengesetzten, magnetischen Granatkristalle gilt, ist das Licht nach Durchlaufen der Magnetschicht unmittelbar hinter dem Kristall im Bereich A um $+\beta$ gedreht:

$$\left.\begin{array}{c} E(\mathrm{A}) = E_0 \cos \omega t \, \mathrm{e}^{i\beta}\,, \\ \text{und im Bereich B um } -\beta \text{ gedreht:} \\ E(\mathrm{B}) = E_0 \cos \omega t \, \mathrm{e}^{-i\beta}\,. \end{array}\right\} \qquad (1)$$

Ausgehend von der skalaren Beugungstheorie nach Kirchhoff ergibt sich das Fraunhofersche Beugungsbild in der hinteren Brennebene des abbildenden Objektivs (bis auf einen konstanten Faktor) als die Fouriertranformierte der Objektfunktion $E(x, y)$:

$$E(u, v) = \frac{\Phi}{\lambda f} \iint_{\text{Objekt}} E(x, y)\, e^{-i(ux+vy)}\, dx\, dy \tag{2}$$

mit $u = (2\pi/\lambda)\, x_f/f$ und $v = (2\pi/\lambda)\, y_f/f$. x, y sind die Koordinaten im Magnetisierungsobjekt und x_f, y_f die Koordinaten im Beugungsbild in der Brennweite f des Objektivs. λ ist die Lichtwellenlänge und Φ ein konstanter, geometrischer Phasenfaktor, so daß für jede monochromatische Lichtkomponente $\Phi/\lambda f = \text{const}$ ist.

Das vorliegende Bubble-Bereichsgitter hat als erzeugende Gittervektoren

$$\boldsymbol{C}(x_1 = D, y_1 = 0) \quad \text{und} \quad \boldsymbol{D}(x_2 = D/2, y_2 = \sqrt{3}\, D/2) \tag{3}$$

mit D als Gitterkonstanten (Fig. 2).

Dann ist die Lichterregung im Fraunhoferschen Beugungsbild

$$E(u, v) = \text{const} \underbrace{\sum_{\mu=0}^{N-1} \sum_{\nu=0}^{M-1} e^{-i[u(\mu x_1+\nu x_2)+v(\mu y_1+\nu y_2)]}}_{G(u, v)} \times$$

$$\times \underbrace{\iint_{\text{Grundzelle}} E(x, y)\, e^{-i(ux+vy)} dx\, dy}_{F(u, v)} . \tag{4}$$

Bei einem ausgedehnten Beugungsgitter mit einer großen Anzahl von Gitterperioden (z. B. N, $M \geqq 10$) liefert der Gitterfaktor $G(u, v)$ die Lagen der scharfen Gitterreflexe (Beugungsordnungen) im Beugungsbild. Dafür ergeben sich die zweidimensionalen Laueschen Gleichungen mit den Lösungsvektoren

$$\boldsymbol{C}^*\left(u_1 = 0, v_1 = \frac{2\pi}{D}\frac{2}{\sqrt{3}}\right) \quad \text{und} \quad \boldsymbol{D}^*\left(u_2 = \frac{2\pi}{D}, v_2 = -\frac{2\pi}{D}\frac{1}{\sqrt{3}}\right) \tag{5}$$

im reziproken Gitter und allen ganzzahligen vielfachen davon, an deren Endpunkten die Beugungsreflexe liegen, d.h.

$$m\boldsymbol{C}^* + n\boldsymbol{D}^* = \text{Beugungsreflex } (m, n) \quad \text{mit} \quad m, n = 0, \pm 1, \pm 2, \dots .$$

Das Beugungsbild hat damit ebenfalls hexagonale Symmetrie, jedoch sind dessen Symmetrieachsen um 90° gegen die des Bereichsgitters gedreht (Fig. 3).

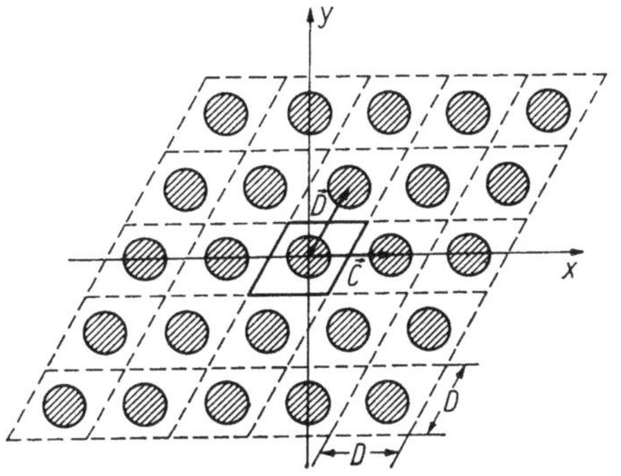

Fig. 2. Hexagonal angeordnetes Bereichsgitter, erzeugt durch mehrmaliges Aneinanderreihen einer rhombusförmigen Grundzelle (Kantenlänge D, spitzer Winkel 60°), mit zentral darinliegender Bubble, längs der Gittervektoren $\boldsymbol{C}$ und $\boldsymbol{D}$

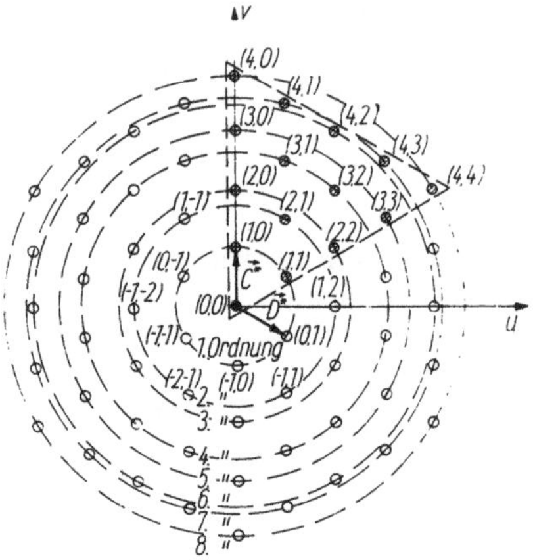

Fig. 3. Beugungsbild eines hexagonal angeordneten Bubble-Bereichsgitters in der Fraunhoferebene (u, v) mit Angabe der einzelnen Beugungsordnungen (m, n), die durch m-, n-maliges Aneinanderreihen der Vektoren $\boldsymbol{C}^*$, $\boldsymbol{D}^*$ erreicht werden. Die zueinandergehörigen Beugungsordnungen liegen auf konzentrischen Kreisen um den Mittelpunkt

Die Stärke der Lichterregungen in den einzelnen Beugungsordnungen ist proportional zu der Anzahl NM der an der Beugung beteiligten Gitterperioden.

Der Strukturfaktor $F(u, v)$ bestimmt die Verteilung der Lichterregung auf die einzelnen Beugungsordnungen. Das magnetische Bereichsgitter kann hier als sogenannte Stufenfunktion angenommen werden, da die Magnetisierung innerhalb der Bereiche jeweils stückweise konstant ist und die schmalen Blochwände optisch nicht auflösbar sind. Der Strukturfaktor zerfällt dann in zwei Summanden. Da der Bereich B aus der Fläche des Rhombus abzüglich der des Kreises besteht, ergibt sich

$$F(u, v) = E(\mathrm{B}) \iint\limits_{\text{Rhombus}} e^{-i(ux+vy)}\, dx\, dy + [E(\mathrm{A}) - E(\mathrm{B})] \iint\limits_{\text{Kreis}} e^{-i(ux+vy)} dx\, dy. \tag{6}$$

Der Strukturfaktor braucht nun nur an den festen Orten der Beugungsreflexe (m, n) berechnet zu werden. Dort ist er gleich den Fourierkoeffizienten der nach der Gitterperiodizität entwickelten Objektfunktion.

An der Stelle der nullten Beugungsordnung ($u = v = 0$) ist

$$F(0, 0) = E(\mathrm{B}) \frac{D^2 \sqrt{3}}{2} + [E(\mathrm{A}) - E(\mathrm{B})]\, \pi R^2 \tag{7}$$

mit R als Bubble-Radius.

Bei der Berechnung der Lichterregung in den höheren Beugungsordnungen (m oder $n \neq 0$) ist der Anteil des ersten Integrals in Gleichung (6) Null, da es sich über die Grundperiode der Fourierzerlegung erstreckt. Das Integral über den Kreis, abgekürzt durch $K(u, v)$, liefert zunächst allgemein für die Beugung an der Kreisscheibe

$$K(u, v) = 2\pi R^2 \frac{J_1\left(R\sqrt{u^2 + v^2}\right)}{R\sqrt{u^2 + v^2}} \tag{8}$$

mit J_1 als Besselfunktion 1. Ordnung. An den Stellen der höheren Beugungsordnungen mit den Koordinaten

$$u = n \frac{2\pi}{D} \quad \text{und} \quad v = (2m - n) \frac{2\pi}{D\sqrt{3}}; \qquad m \text{ oder } n \neq 0$$

ist damit die Lichterregung

$$F(m, n) = [E(\mathrm{A}) - E(\mathrm{B})]\, K(m, n) \tag{9}$$

mit

$$K(m, n) = 2\pi R^2 \frac{J_1\left(2\pi(R/D)\sqrt{n^2 + \frac{1}{3}(2m - 2)^2}\right)}{2\pi(R/D)\sqrt{n^2 + \frac{1}{3}(2m - n)^2}}.$$

Darin ist $r = (2\pi/D)\sqrt{n^2 + \frac{1}{3}(2m - n)^2}$ der Radius von der nullten zu den höheren Beugungsforderungen.

Nach Einsetzen der Lichterregungen $E(\mathrm{A})$ und $E(\mathrm{B})$ direkt hinter dem Beugungsgitter gemäß (1) in (7) und (9) erhält man für die relativen Lichterregungen in den Beugungsordnungen (bis auf den für alle Ordnungen gleichen Faktor $\Phi NM/\lambda f$)

$$F(0, 0) = E_0 \frac{D^2\sqrt{3}}{2} [\cos\beta + e^{i\pi/2} \sin\beta\,(2p - 1)] \cos\omega t\,, \tag{10}$$

$$F(m, n) = 2E_0 K(m, n)\, e^{i\pi/2} \sin\beta \cos\omega t \tag{11}$$

mit $p = (2\pi/\sqrt{3})\,(R/D)^2$ als Tastfaktor, der das Verhältnis der Flächen von gittererzeugendem Bubble-Bereich und Grundzelle beschreibt.

Das Licht in der nullten Beugungsordnung besteht aus zwei Anteilen, einem parallel zum einfallenden Licht polarisierten und einem senkrecht dazu polarisierten mit $e^{i\pi/2}$, die beide gleichphasig schwingen und sich zu einer linear polarisierten Schwingung zusammensetzen. Das Licht in den seitlichen Beugungsordnungen ist wegen des Faktors $e^{i\pi/2}$ stets linear und senkrecht zum einfallenden Licht polarisiert.

Im gesättigten Zustand der Schicht, in dem die gesamte Schicht entweder nur die Magnetisierung des Bereiches A oder die des Bereiches B trägt, d. h. $p = 1$ oder 0 ist, verschwinden die seitlichen Beugungsordnungen, und die verbleibenden beiden Komponenten in Gleichung (10) geben die Drehung der Polarisationsebene des Lichtes um die Faraday-Winkel $+\beta$ oder $-\beta$ an. Die visuell oder photometrisch erfaßten Intensitäten sind die Betragsquadrate der berechneten Lichterregungen in den einzelnen Beugungsordnugen. Hierbei mittelt sich der zeitliche Verlauf der Schwingung wegen der Trägheit der Beobachtungseinrichtung heraus.

Damit ergeben sich die folgenden relativen Intensitäten im Beugungsbild des magnetischen Bubble-Bereichsgitters:

1. parallel zum eingestrahlten Licht polarisiert:

$$I(0, 0)_{0^\circ} = E_0^2 \cos^2\beta \left(\frac{D^2\sqrt{3}}{2}\right)^2, \tag{12}$$

2. senkrecht zum eingestrahlten Licht polarisiert:

$$I(0, 0)_{90^\circ} = E_0^2 (2p - 1)^2 \sin^2\beta \left(\frac{D^2\sqrt{3}}{2}\right)^2 \tag{13}$$

$$\text{mit} \quad p = \frac{2\pi}{\sqrt{3}} \left(\frac{R}{D}\right)^2$$

und

$$I(m, n)_{90^\circ} = 4E_0^2 K^2(m, n) \sin^2\beta \tag{14}$$

mit $K(m, n)$ nach Gleichung (9).

Bei Kristallen mit geringer Faraday-Drehung β geht der überwiegende Lichtanteil in die ungedrehte Komponente der nullten Beugungsordnung. Dieser Anteil kann jedoch durch einen senkrecht dazu stehenden Analysator abgeblockt werden, hinter dem nur die Intensitäten $I(0, 0)_{90^\circ}$ und $I(m, n)_{90^\circ}$ erscheinen.

Im entmagnetisierten Zustand der Probe mit $p = \frac{1}{2}$, wo die Flächen der Bereiche A und B gleich groß sind, verschwindet der senkrecht zum einfallenden Licht polarisierte Anteil der nullten Ordnung. Das kann als empfindlicher Nachweis für die remanente Magnetisierung einer Probe benutzt werden [21]. Voraussetzung dafür ist jedoch, daß keine Mehrfachbeugung im Kristall auftritt [13].

4. Versuchsergebnisse

Durch ein in Richtung der Schichtnormalen liegendes Magnetfeld wurde das Bereichsgitter verändert. Die Orientierung des Feldes, in der sich die Bubbles als Umkehrbereiche zusammenziehen, wird positiv bezeichnet, die entgegengesetzte Orientierung negativ. Ausgehend vom remanenten Zustand im Felde Null nimmt im positiven, ansteigenden Feld bis ca. +62 A/cm der Durchmesser $2R$ der Bubbles linear mit dem Feld ab, während ihr gegenseitiger Abstand im weiterhin hexagonalen Gitter konstant bleibt. Oberhalb dieses Feldwertes weitet sich das Gitter stark auf, und die Bubbles kollabieren. Im negativen, ansteigenden Feld weiten die sich jetzt vergrößernden Bubbles das Gitter zunehmend auf, bis −28 A/cm um ca. 5%. Bei weiterer Felderhöhung deformieren die sich aneinanderdrängenden Bubbles zu sechseckigen Wabenbereichen, die zunehmend über den Kristallrand hinauslaufen. Die verbliebenen Streifenbereiche zwischen den Waben ziehen sich dann zu Bubbles zusammen, die schließlich kollabieren. Im Feldstärkebereich von −28 A/cm bis +62 A/cm wird das hexagonale Bubble-Gitter reversibel durchlaufen. Der Durchmesser $2R$ der Bubbles und ihr gegenseitiger Abstand D wurden als Funktion des Magnetfeldes H gemessen und sind in Fig. 4 aufgetragen. Ebenfalls aufgetragen ist das daraus ermittelte Verhältnis $2R/D$. Zwischen −28 und +62 A/cm ergibt sich für $2R/D$ eine nahzu lineare Abhängigkeit vom Feld H.

Die Größe $2R/D$ (bzw. R/D) des Bereichsgitters bestimmt wesentlich die Intensitätsverteilung in den einzelnen Beugungsordnungen, wobei die Beugungsreflexe gleicher Ordnung jeweils immer auf einem der um den Nullreflex konzentrisch angeordneten Kreise liegen (vgl. Fig. 3). Für die gemessenen Werte $2R$, D, und $2R/D$ des reversibel durchlaufenen, hexagonalen Bubble-Gitters wurden die relativen Intensitäten $I(0, 0)_{90°}$ und $I(m, n)_{90°}$ der Beugungsordnungen gemäß (13) und (14) berechnet. Diese sind in Fig. 5a und b über dem Magnetfeld dargestellt. Das Maximum der 1. Ordnung wurde zu 1 normiert. Bei Vorhandensein eines Bereichsgitters kann die Intensität der 1. Ordnung für $\beta \neq 0$ nie Null werden. Sämtliche anderen Beugungsordnungen können bei bestimmten vom Magnetfeld abhängigen Werten des Gitterparameters $2R/D$ ausfallen.

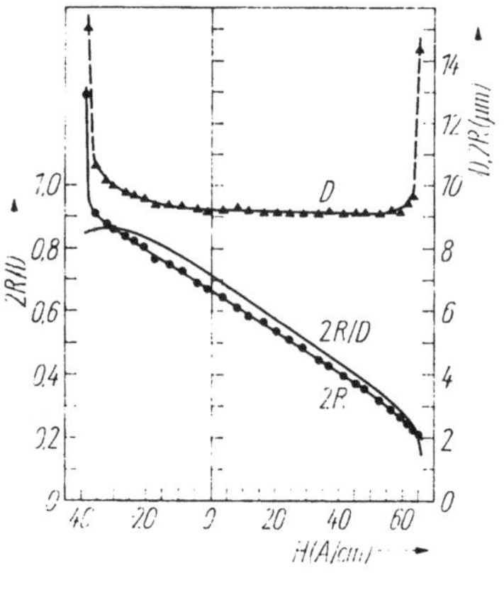

Fig. 4. Gemessener Durchmesser $2R$ und gegenseitiger Abstand D der Bubbles im hexagonalen Bereichsgitter sowie $2R/D$ als Funktion des äußeren Magnetfeldes H

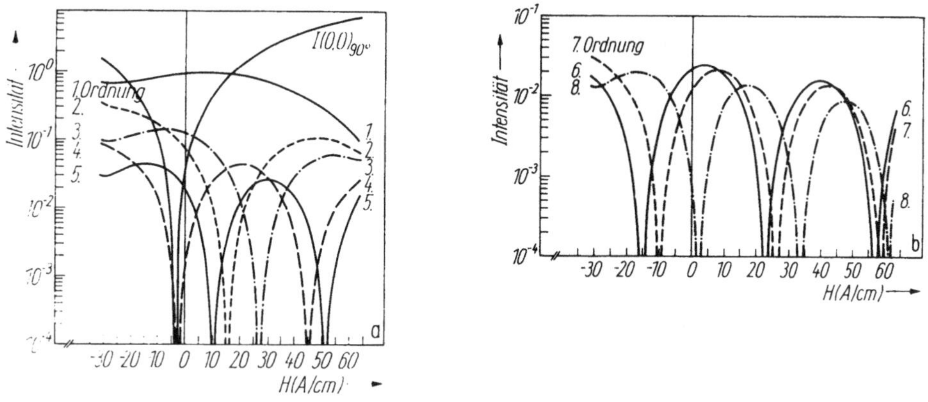

Fig. 5. Berechnete Intensitäten (in relativen Einheiten) des senkrecht zum einfallenden Licht polarisierten Anteils a) der nullten Beugungsordnung und der 1. bis 5. Beugungsordnungen und b) der 6. bis 8. Beugungsordnungen als Funktion des äußeren Magnetfeldes (vgl. Fig. 3)

Im vorliegenden Versuch wurde der entmagnetisierte Zustand der Probe mit $p = \frac{1}{2}$, d. h. $I(0, 0)_{90°} = 0$, nicht im Felde $H = 0$, sondern erst in einem negativen Feld von ≈ -3 A/cm erreicht.

Fig. 6 zeigt für fünf verschiedene Werte des Magnetfeldes das Bereichsgitter, kontrastiert in der Amplitudenstellung des Analysators $\alpha = 90° + \beta$, und darunter jeweils das dazugehörige Fraunhofersche Beugungsbild in der zum Polarisator gekreuzten Stellung des Analysators. In den Beugungsbildern sind also die Intensitäten $I(0, 0)_{90°}$ und $I(m, n)_{90°}$ zu sehen. Vergleicht man die Intensitäten der einzelnen Beugungsordnungen untereinander in den Beugungsbildern mit den berechneten Intensitäten in Fig. 5, jeweils an den betreffenden Feldwerten der Beugungsaufnahmen, nämlich bei -28, -18, 0, $+24$ und $+51$ A/cm, so findet man eine gute Übereinstimmung zwischen Theorie und Experiment.

Beachtenswert ist, daß innerhalb des reversibel durchlaufenen Bereichsgitters wegen des konstanten Bereichsabstandes auch die Lage der Beugungsordnungeen im Beugungsbild fest bleibt. Lediglich die Intensitäten der einzelnen Beugungsordnungen verändern sich bei Veränderung des Magnetfeldes gemäß der in Fig. 5 dargestelten Werte. Damit ist es möglich, das Beugungsbild durch ortsfeste Detektoren an den Stellen der einzelnen Beugungsordnungen zu registrieren. Zur Verarbeitung der nullten bis achten Ordnung sind neun Detektoren nötig. Bringt man z. B. ein Photodiodenfeld in Form des in Fig. 3 eingezeichneten Dreiecks mit 15 Dioden an den Stellen der Beugungsordnungen an, dann werden die 1., 3., 4., 5., 7. und 8. Ordnung je zweimal registriert und die 0., 2. und 6. Ordnung je einmal. Je nach dem an das Bereichsgitter angelegten Magnetfeld wird in den verschiedenen Dioden gemäß den „Nullstellen" in Fig. 5 kein Licht registriert werden. Mit Hilfe dieser Auslöschungsbedingung für die Beugungsordnungen können von den reversibel durchlaufenen Bereichsgitterzuständen mehrere durch ihre Nullstellen im Beugungsbild deutlich voneinander unterschieden werden. Bei Registrierung der ersten acht Ordnungen sind dies mindestens zwölf verschiedene Bereichsgitterzustände, deren Information im Beugungsbild in digitaler Form (Licht oder kein Licht an den einzelnen Dioden) vorliegt. Diese Lichtverteilung im Beugungsbild bleibt auch erhalten, wenn das zugehörige Bereichsgitter geringe Störungen aufweist, wie z. B. den Ausfall einiger Bubble-Bereiche. Damit kann die im Bereichsgitter jeweils abgespeicherte Information relativ unempfindlich gegen lokale Störungen, wie Kristallfehler, über das Beugungsbild ausgelesen werden. Da-

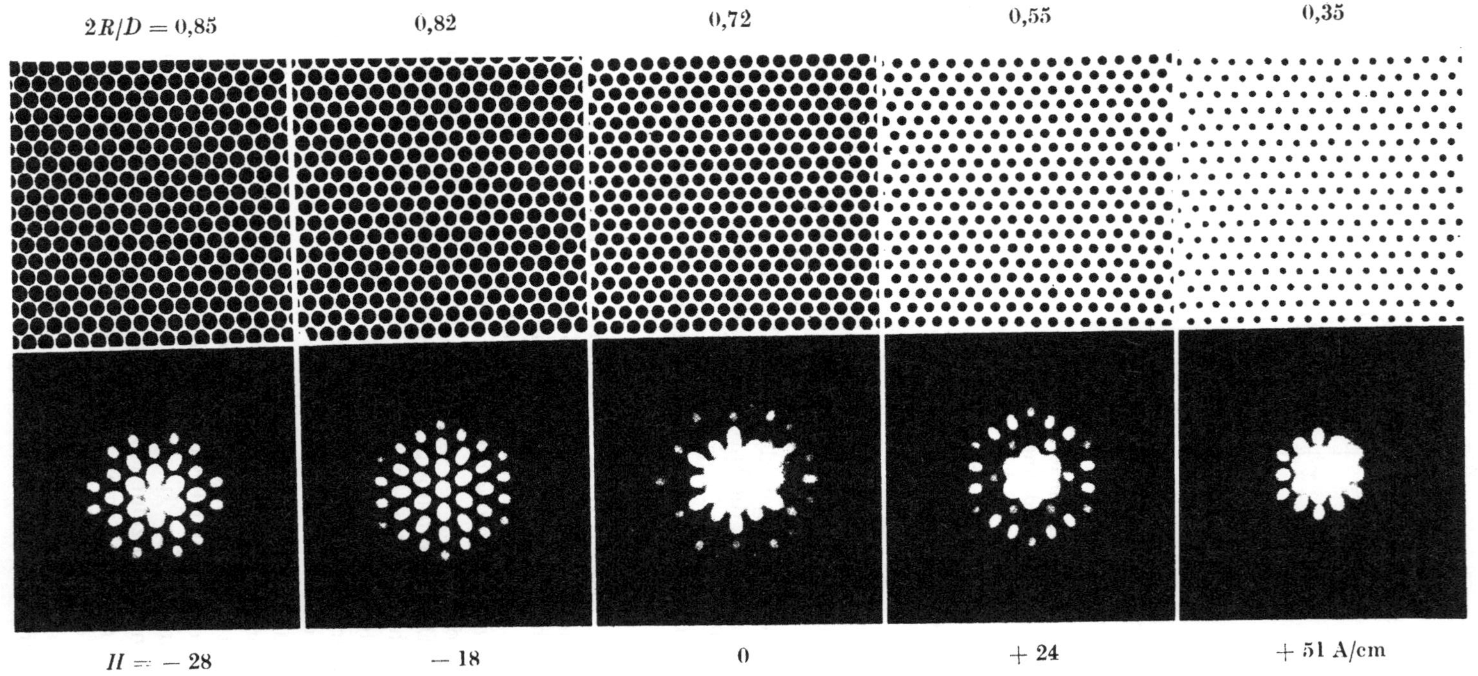

Fig. 6. In einer 5,3 μm dicken $Y_{1,92}Sm_{0,10}Ca_{0,98}Fe_{4,02}Ge_{0,98}O_{12}$-Einkristallschicht auftretende Bubble-Bereichsgitter (obere Bilder) und dazugehörige Fraunhofersche Beugungsbilder (Bilder jeweils darunter) für fünf verschiedene Werte des äußeren Magnetfeldes *H*. Die Kantenlänge der Bereichsbilder entspricht 180 μm im Kristall

neben ist der Auslesevorgang selbst für die abgespeicherte Information zerstörungsfrei. Da die einzelnen Zustände durch permanente Magnetfelder eingestellt werden, bietet sich diese Art der Informationsspeicherung im Bereichsgitter mit der Abfrage über das Beugungsbild zur zerstörungsfreien Speicherung und Verarbeitung von Daten an.

Danksagung

Herrn Prof. Dr. M. Lambeck danke ich für zahlreiche wertvolle Diskussionen.

Literatur

[1] J. F. Dillon, Jr. und J. P. Remeika, J. appl. Phys. **34**, 637 (1963).
[2] M. Lambeck, Z. angew. Phys. **15**, 272 (1963).
[3] H. Boersch und M. Lambeck, Z. Phys. **177**, 157 (1964).
[4] H. de Lang, Philips Res. Rep. **18**, 61 (1963).
[5] M. Lambeck, Dissertation, TU Berlin, I. Phys. Inst., 1964 (D 83).
[6] M. Lambeck, IEEE Trans. Magnetics **4**, 51 (1968).
[7] H. M. Haskal, IEEE Trans. Magnetics **6**, 542 (1970).
[8] R. S. Mezrich, IEEE Trans. Magnetics **6**, 537 (1970).
[9] R. V. Telesnin, A. G. Shishkov, E. N. Ilicheva, N. G. Kanavina und N. A. Etonomov, phys. stat. sol. (a) **12**, 303 (1972).
[10] G. R. Woolhouse und P. Chaudhari, phys. stat. sol. (a) **19**, K3 (1973).
[11] B. Kuhlow, Dissertation, TU Berlin, Opt. Inst., 1974 (D 83).
[12] B. Kuhlow und M. Lambeck, Physica (Utrecht) **80B**, 374 (1975).
[13] B. Kuhlow und M. Lambeck, J. Magnetism magnetic Mater. **4**, 337 (1977).
[14] T. R. Johansen, D. I. Norman und E. J. Torok, J. appl. Phys. **42**, 1715 (1971).
[15] H. W. Fuller, J. appl. Phys. **42**, 1816 (1971).
[16] B. E. Argyle und P. Chaudhari, IBM Tech. Discl. Bull. (USA) **18**, 603 (1975).
[17] K. R. Papworth, phys. stat. sol. (a) **22**, 373 (1974).
[18] G. R. Woolhouse und P. Chaudhari, Phil. Mag. **31**, 161 (1975).
[19] A. V. Antonov, I. A. Ignatev und K. O. Boltar, Soviet Phys. – Solid State **17**, 682 (1975).
[20] H. Dötsch, Mikrowellen Mag. **2**, 161 (1978).
[21] S. Kern und D. J. Craik, J. Phys. D **9**, L 81 (1976).

(Received April 9, 1979)

phys. stat. sol. (a) 54, 291 (1979)

Subject classification: 12.1; 21.1

Freiberg/Sa.[1])

On the Role of Grüneisen's Elastic Parameters in the Theory of Polycrystal Elasticity

A Critical Assessment

By

H. H. WAWRA†

Grüneisen's experimental elastic constants of polycrystalline aggregates are often used in order to judge the value of theories which aim at the calculation of effective elastic moduli, given the moduli of the constituent crystallites. The conditions of Grüneisen's experiments are reproduced in the experiments and it is found that his specimens were heavily *cold-worked*. For this reason Grüneisen's elastic constants should not be compared with theoretical polycrystal moduli which, as it is usually done, are calculated from the elastic moduli of *annealed* single crystals. A reliable assessment of the value of a theory can only be expected if also the experimental polycrystal moduli taken for comparison with the theoretical ones, refer to the annealed state. Good agreement with experiment is then found, if the so-called self-consistent elastic moduli are taken as theoretical polycrystal moduli.

Die experimentellen elastischen Grüneisenkonstanten von polykristallinen Aggregaten werden oft benutzt, um den Wert von Theorien abzuschätzen, die eine Berechnung der effektiven Elastizitätsmoduln aus den gegebenen Moduln der Ausgangskristallite vornehmen. Die Bedingungen der Grüneisenexperimente werden in den vorliegenden Experimenten reproduziert, und es wird gefunden, daß seine Proben stark *kalt gewalzt* waren. Aus diesem Grund sollten die elastischen Grüneisenkonstanten nicht mit theoretischen Polykristallmoduln verglichen werden, die, wie es üblicherweise getan wird, aus den Elastizitätsmoduln von *getemperten* Einkristallen berechnet werden. Eine vernünftige Zuordnung des Wertes einer Theorie läßt sich nur erwarten, wenn auch die experimentellen, polykristallinen Moduln, die für den Vergleich mit den theoretischen herangezogen werden, sich auf den getemperten Zustand beziehen. Gute Übereinstimmung mit dem Experiment wird dann gefunden, wenn die sogenannten selbstkonsistenten Elastizitätsmoduln als theoretische polykristalline Moduln benutzt werden.

1. Introduction

In the years 1907/08 Grüneisen [1, 2] has published experimental values of the shear modulus G, the Young's modulus E, and the Poisson ratio μ of numerous macro-isotropic polycrystals. Since Grüneisen has performed his experiments with great care his values have been accepted for a long time as the best values available for the concerned polycrystals. In [3 to 10] a number of authors are quoted who use Grüneisen's values as standard reference data in order to check theories which calculate the mentioned elastic parameters in terms of the elastic constants of the crystallites which build up the polycrystalline aggregate. It is then concluded that theories with results close to the Grüneisen values are good and others are not so good. It is the objective of this paper to reanalyze the experiments of Grüneisen by comparison with our experiments and to show on this ground that the problem of comparing theory and experiment in our field is much more subtle than it is apparently assumed by various authors. Since the most numerous and accurate experiments for the determination of elastic constants have probably been performed with copper we shall,

[1]) DDR-92 Freiberg, DDR.

for illustration, concentrate on this material. The results are very similar for many other materials.

2. Preliminaries

A basic fact which ought to be observed in all investigations on polycrystal elasticity is the following: A polycrystal can be macro-isotropic if, and only if, the distribution of the crystallographic axes of the crystallites is isotropic. This means that to all crystallographic orientations of the crystallites belong equal volume percentages of crystallites. If a polycrystal of a given material possesses such an isotropic orientation distribution, and if this is the only information available about it then all what can be predicted from this theory is that its elastic moduli lie anywhere between the bounds of Voigt and Reuss which in the systematic bound theory (see e.g. Kröner [11, 12]) are classified as first-order bounds. A polycrystal with isotropic orientation distribution is not necessarily macro-isotropic. In fact, anisotropic grain shape distribution can make such a crystal anisotropic. However, since this effect is not relevant for our present discussion we shall not go into its details.

The essential point we want to raise is the following: If, by some measurement, a polycrystal is found to be macro-isotropic then this information is so little that we cannot localize the values of its elastic moduli better than that they lie anywhere between the bounds of Voigt and Reuss. In the case of the shear modulus of a polycrystal of copper these bounds are near 5.40 and 4.00.[2]) The experimental values of Grüneisen lie at 4.51 and those of Bradfield at 4.83 [13]. Since both authors have measured with extreme care we conclude that the two specimens involved had, in fact, shear moduli of the given value, with an accuracy in the vicinity of 1‰. The above discussion gives a clear answer to this apparent discrepancy: Although both specimens were macro-isotropic polycrystals of copper they were in rather different structural states.

In order to find out more about these states we have performed experiments which reproduce Grüneisen's values as well as those of Bradfield. This enables us to state conditions on the internal state of the polycrystal in which we expect Grüneisen's and Bradfield's results. The numerical values listed are corrected with respect to a possible small texture. For this we have used the method described in [14]. Since the correction is much smaller than the effect to be discussed we shall not go into details here.

3. Experiments

Grüneisen chooses the coincidence of the Poisson number measured statically and dynamically as a criterion for the macro-isotropy of his specimens. Although this test is not very accurate let us assume that he succeeded in preparing reasonably macro-isotropic specimens. The isotropy, however, does not define sufficiently the internal state. Unfortunately, not much further information about this state is available from Grüneisen's papers. Hence we have tried to find out experimental conditions adequate to reproduce Grüneisen's elastic parameters, or, which means the same, to reproduce specimens with an internal state equal to that of Grüneisen's specimens.

We started with cold-drawn electrolytic copper (degree of reduction $\eta = 76\%$) which later was annealed for various time intervals at a certain fixed temperature and cooled to room temperature after each time interval. The following quantities were

---

[2]) Here and later, all numerical values of elastic moduli are given in units of 10^{10} N/m^2.

then measured at room temperature:

a) the Vickers hardness HV (30 N for 30 s),
b) the electrical conductivity $\varkappa$ (sigmatest of Förster),
c) the shear modulus G and Young's modulus E (elastomat of Förster).

The results are given in Fig. 1 to 4. The shear modulus G of the cold-drawn non-annealed specimen is about 4.545. It grows continuously with increasing annealing time and temperature. The value of the non-annealed specimen is very close to values measured by Grüneisen with two different specimens I and II. It is therefore suggestive to assume that the specimens used by Grüneisen were cold-drawn as well.

In order to understand the increase of the shear modulus with annealing time we recall that cold-working always implies the introduction of numerous structural defects such as dislocations, subgrain boundaries, etc. It is well understood in material sciences that these defects lower the apparent elastic moduli of a material. Of importance is, for instance, the bowing out of dislocations which is a kind of anelastic effect. Depending of the annealing temperature and time more or less of these defects annihilate so giving rise to the observed increase of the elastic moduli. Annealing at

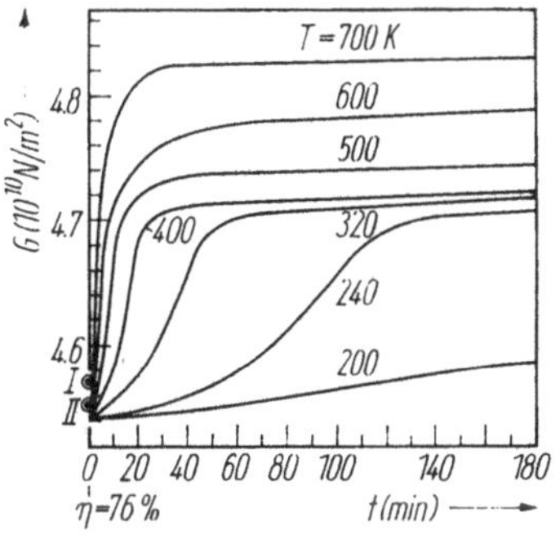

Fig. 1

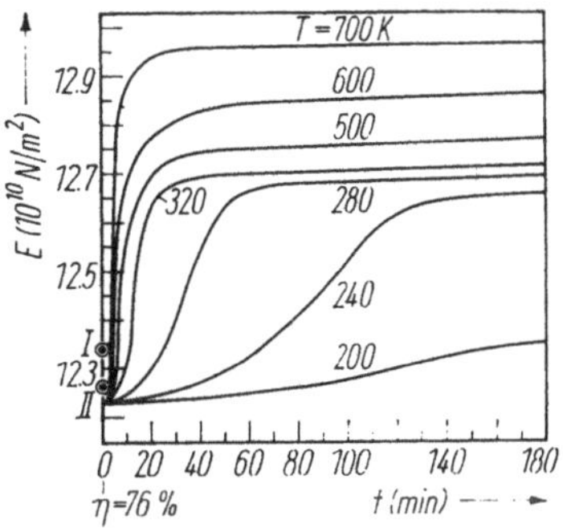

Fig. 2

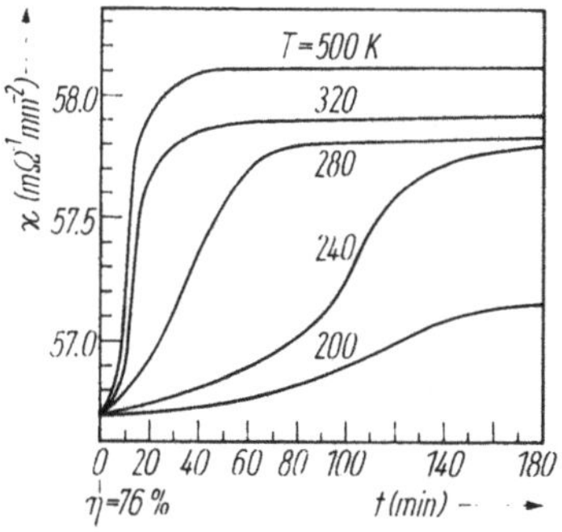

Fig. 3

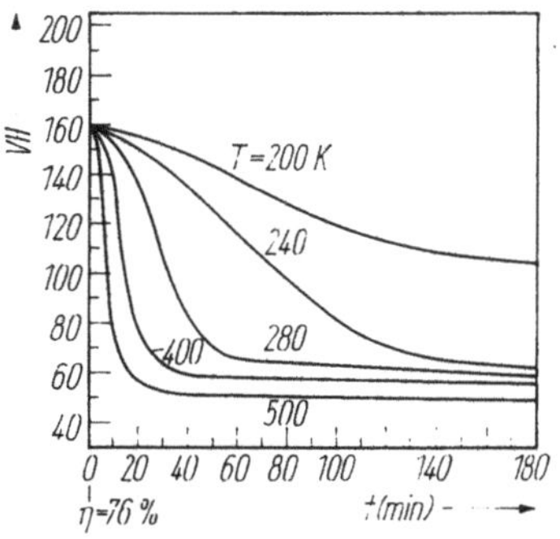

Fig. 4

Fig. 1. Shear modulus G of 76% cold-drawn electrolytic copper which subsequently has been annealed at temperatures from 200 K till 700 K. Measurements were performed at room temperature. Points I and II from Grüneisen. All values of G are corrected for small texture [14]

Fig. 2. Young's modulus E of the specimens of Fig. 1

Fig. 3. Electrical conductivities $\varkappa$ of the specimens of Fig. 1. $\varkappa = 53$: Grüneisen's specimen

Fig. 4. Vickers hardness VH (30/30) of the specimens of Fig. 1

700 K leads to a complete recrystallization of the specimen during which practically[3]) all structural defects and also some texture which may be present in the cold-drawn specimen, disappear. The shear modulus then assumes its maximum value which turns out to be that of Bradfield. In a certain sense, the polycrystal is now in an ideal state.

It is hard to say quantitatively how much the disappearing of the texture during the recrystallization contributes to the modulus effect. That this contribution is small can be seen as follows: A ⟨111⟩-texture in copper lowers G so that G increases when this texture is removed. On the other hand, a ⟨111⟩-texture in copper raises E so that E decreases if the texture is removed. This is opposite to the results shown in Fig. 2 so that we conclude that the texture effect is not decisive in our problem.

In order to specify the internal situation of the specimens somewhat better we have added measurements of the Vickers hardness (VH) and of the electrical conductivity. The decrease of VH is a measure of the disappearance of the sources of the hardness, i.e. of the structural defects. The electrical conductivity $\varkappa$, on the other hand, is sensitive not only to these defects but also to impurity atoms of another species. Grüneisen reports $\varkappa = 53$ and 32 m/Ωmm^2 for the specimens I and II, respectively, whereas our results on rather pure copper lie within 56.7 and 58.2 m/Ωmm^2. They provide evidence that Grüneisen's specimens were not of very pure copper. Grüneisen himself reports an impurity content of 0.15% As and 0.03% Fe for specimen I. The effect of these impurities on the moduli, however, is certainly much smaller than the effect under discussion in this paper so that we need not go into more details.

4. Conclusions

All theories known to us which calculate effective polycrystal moduli from the single crystal moduli (C_{ik}) assume that the latter moduli are those of the in situ crystallites. Thus if one wants to derive the cold-work (e.g. Grüneisen) moduli one has to use the moduli of single crystals which have the defect structure of the crystallite in situ. These moduli, however, are not available. Cold-working of a single crystal does not help because the defect structure developing in a plastically deformed single crystal is different from that obtained in a crystallite of a deformed polycrystal. On the other hand, both well-annealed single crystals and the crystallites of a well-annealed polycrystal are practically defect-free. This means that the quota of lattice defects (point defects and dislocations) influencing the elastic parameters in both species is negligible so that they have the same elastic moduli. This allows us to use the experimental moduli of well-annealed single crystals as if they were measured crystallite moduli of well-annealed polycrystals. Clearly the theoretical polycrystal moduli then obtained are to be compared with experimental moduli of well-annealed polycrystals, for instance those of Bradfield.

This discussion allows us to draw the following conclusion: Good agreement between theoretical predictions based upon the moduli of annealed single crystals and Grüneisen's polycrystal moduli (e.g. for copper $G = 4.552$, $E = 12.263$) shows that this theory is quantitatively not correct.

As examples which stand for many others we quote the theories of Laurent and Eudier [5], Huber and Schmid [15], and Shibuya [16]. Good theories based on defect-free single crystals give results close to those of Bradfield (e.g. for copper $G = 4.83$, $E = 12.96$). In order to calculate such values one needs the moduli of well-annealed single crystals. Such moduli have been provided in large numbers by the precise

[3]) The term "practically" indicates that the number of defects is not zero but so small that they have only a negligible influence on the elastic moduli.

measurements of the National Physical Laboratory in Teddington [17]. For copper the values are $C_{11} = 16.905$, $C_{12} = 12.193$, $C_{44} = 7.55$.

If the comparison between theory and experiment is performed along the indicated lines one finds that the so-called self-consistent elastic moduli, first discussed by Hershey [18] and Kröner [19], are very close to the experimental values of the annealed specimens in almost all cases. There are exceptions, for instance for graphite, where the texture- and defect-free state cannot be easily reached. The self-consistent moduli always lie between the second-order bounds of Hashin and Shtrikman [20] and the third-order bounds of Kröner [11]. For instance, the self-consistent shear modulus of an annealed polycrystal of copper is 4.826 when calculated with the single crystal moduli given above. The third-order bounds are 4.77 and 4.86.

The Voigt-Reuss-Hill (VRH) arithmetic average is 4.72, i.e. outside the third-order bounds. It has recently been explained by Kröner [12] that the self-consistent moduli are the true effective moduli of a macro-homogeneous and macro-isotropic polycrystal provided the arrangement of the grains is as irregular as possible, i.e. the polycrystal is in a state of "perfect disorder". The latter term has been given a quantitative meaning in Kröner's work. Apparently, the annealed polycrystal comes very close to the condition of perfect disorder. On the other hand, there is no theoretical basis at all that the VRH prescription has anything to do with this condition.

Our considerations certainly do not devaluate the great merit of Grüneisen's work whose aim was to provide the numerical values of the elastic moduli of materials of practical importance. Cold-worked materials are very important for many applications. The study of Grüneisen's work shows that his specimens were produced by cold-work after which "tiefgreifende Änderungen durch thermische oder mechanische Behandlung vermieden wurden" ([1], p. 846). At that time, nobody knew better than Grüneisen about the complicated nature of the internal state of solid materials. Hei writes: "Weder der Einfluß mechanischer oder thermischer Behandlung noch derjenge der chemischen Zusammensetzung scheint mir bisher hinreichend festgestellt zu sein" ([1], p. 802). The point is that thermal and mechanical treatment change the internal state of a material thereby also changing its elastic moduli, sometimes by a considerable amount.

References

[1] E. Grüneisen, Ann. Phys. **22**, 801 (1907).
[2] E. Grüneisen, Ann. Phys. **25**, 825 (1908).
[3] D. A. G. Bruggemann, Z. Phys. **92**, 570 (1934).
[4] H. Röhl, Ann. Phys. **16**, 904 (1933).
[5] P. Laurent and M. Eudier, Rev. Métall. **47**, 582 (1950).
[6] P. S. Rudman, Trans. MS AIME **233**, 864 (1965).
[7] W. Köster and H. Franz, Metallurg. Rev. **6**, 1 (1961).
[8] E. Vogt, Physikalische Eigenschaften der Metalle, Vol. I, Akademische Verlagsgesellschaft Geest & Portig, Leipzig 1958 (p. 125).
[9] J. K. Varma and M. D. Aggarval, J. appl. Phys. **46**, 2841 (1975).
[10] H. M. Ledbetter and E. R. Naimon, J. appl. Phys. **45**, 66 (1974).
[11] E. Kröner, J. Mech. Phys. Solids **25**, 137 (1977).
[12] E. Kröner, J. Phys. F **8**, 2261 (1978).
[13] G. Bradfield, Use in Industry of Elasticity Measurements in Metals with the Help of Mechanical Vibrations, Notes on Applied Science No. 30, Nat. Phys. Lab. London, 1960 (p. 4).
[14] H. H. Wawra, Z. Metallk. **65**, 584 (1974).
[15] A. Huber and E. Schmid, Helv. phys. Acta **7**, 620 (1934).
[16] Y. Shibuya, Sci. Rep. Res. Inst. Tôhoku Univ. **A1**, 161 (1949).

[17] H. PURSEY, Private Communication, National Physical Laboratory Teddington (Middlesex) 1974.
[18] A. V. HERSHEY, Trans. ASME, Ser. E (J. appl. Mech.) **21**, 127 (1963).
[19] E. KRÖNER, Z. Phys. **151**, 504 (1958).
[20] Z. HASHIN and S. SHTRIKMAN, J. Mech. Phys. Solids **11**, 127 (1963).

(Received January 3, 1979)

Obituary

Dr. H. H. Wawra lost his life in a car accident in June 1978. With Dr. Wawra the scientific world lost an eminent researcher who was always full of ideas; he has significantly advanced the field of the elasticity and other properties of polycrystalline materials. I personally lost a good friend and coworker.

E. Kröner

phys. stat. sol. (a) 54, 297 (1979)

Subject classification: 10

Institute of Metallography and Metal Physics,
Central Scientific Institute for Ferrous Metallurgy, Moscow

The "Superdislocation" Model of a Plastic Zone near a Wedge-Shaped Crack

By

V. A. SOLOVEV

A simple anisotropic dislocation theory of the plastic zone near a nonsymmetric wedge-shaped crack is developed. The zone is described by some superdislocations with multiple Burgers vectors. The main parameters of the equilibrium zone (the strength and spread of the relaxation shears) are determined. Simple expressions for a symmetrical quasi-brittle crack are derived both, in anisotropic and isotropic elasticity. Capabilities of the theory developed to describe the interaction of different shear-type planar defects are discussed. The critical pressure is calculated necessary for crack plastic closing as well as the parameters of cracks arising during deformation under pressure.

Построена анизотропная теория дислокационной пластической зоны около несимметричных клиновидных трещин. Зона описана несколькими эффективными сверхдислокациями с кратными векторами Бюргерса. Сформулированы условия равновесия трещины с зоной в нагруженном материале. Определены параметры равновесной зоны: характерные размеры и мощности релаксирующих сдвигов. Получены простые выражения, описывающие зону около симметрично нагруженной трещины и проведен предельный переход к изотропии. Обсуждается возможность использования развитой теории для описания взаимодействия различных плоских дефектов сдвигового типа. Вычислены критическое давление пластического Заплывания трещины, а также параметры трещин, образующихся при деформации под давлением.

1. Introduction

A dislocation model of a plastic zone arising near cracks in crystal materials was first proposed in [1, 2]. According to this model the zone forms on the crack continuation in the crack plane. There is a "dry friction" force acting on the dislocations of the zone. The formation of such a zone may relax plane and antiplane shears due to the crack; however, in the model of [1, 2] there is no opportunity to relax strains induced by the crack opening in the direction of the normal to the crack plane. As a consequence, the geometry of the zone formation accepted in [1] is too simplified (that is most clearly seen from the TEM observations [3]). The simple model of a plastic zone consisting of four noncoplanar symmetric "superdislocations" near the tips of symmetric, symmetrically loaded cracks has been developed in [4], using the solution for the stress field due to the interaction of a dislocation and a crack given in [5]. Nevertheless, in [4, 5] there was not taken into account the Burgers vector of the crack itself (i.e. wedging) which is quite important because crack nucleation in the internal parts of the crystal appear to be a result of the coalescence of dislocations piling up against obstacles. Moreover, only such cracks (Stroh cracks) are able to provide nucleation, storage, and development of microcracks followed by critical crack formation [6]. Contrary to the cracks considered in [4, 5] (Griffith cracks) the Stroh cracks are characterized by not only the critical configuration but also by the steady-state one which can pass into the critical one under proper conditions [7].

In [8] we had found stresses near a nonsymmetric Stroh crack in the presence of outer dislocations. In the present work the "superdislocation" model of a plastic zone is developed for the case of such a crack. The case of a quasi-brittle crack is analysed in more detail. The analysis is carried out in anisotropic elasticity theory, and then for symmetric loading the isotropic formulae are found by the limiting procedure developed in [9]. The important applications of the model worked out are discussed in the last part of the article.

2. The Model of the Stroh Crack with the Plastic Zone

The model under consideration of the plastic zone near the general slit-like crack is shown in Fig. 1. The crack whose wedging vector is $\boldsymbol{h}$ is influenced by the homogeneous stress $\sigma_{ij}^{(0)}$ and by the stress field of the zone which is described by a set of effective superdislocations in the vicinity of the crack termination. The superdislocations lie parallel to each of the crack edges in planes passing through these edges. Only the planes most favourable for plastic deformation are taken into account. The zone parameters are as follows: the magnitudes of effective dislocation Burgers vectors $m_s\boldsymbol{b}^{(s)}$ and their distances $r_s^{(\pm)}$ from the crack ends. They can all be simultaneously determined by the following conditions:

1. The shear stresses in slip planes (t) in the vicinity of the crack tips are equal to the dislocation emission stresses $\sigma_e^{(t)}$.

2. The total forces acting on superdislocations (s) in their slip planes are equal to "dry friction" forces $\sigma_p^{(s)}b^{(s)}$ in the material investigated multiplied by the magnitudes m_s of the Burgers vectors of the effective superdislocations with respect to the Burgers vectors of the elementary dislocations in corresponding slip planes.

The stress field of the s-th effective dislocation lying parallel to $0z$ is given by

$$\sigma_{kj}^{(s)} = \sum_{\nu=1}^{6} \frac{m_s U_{kj}^{(s\nu)}}{\zeta_\nu - \zeta_{s\nu}}, \tag{1}$$

where $\zeta_\nu = x + p_\nu y$, $\zeta_{s\nu} = x_s + p_\nu y_s$; x, y, z are rectangular coordinates of the observation point; x_s, y_s, $-\infty < z < +\infty$ designate the dislocation (s) position; p_ν are the complex roots of the six-order characteristic equation of anisotropic elasticity theory; $U_{kj}^{(s\nu)}$ are defined by p_ν, the Burgers vectors $\boldsymbol{b}^{(s)}$, the dislocation direction, and the elastic moduli of the media [10, 11].

The slit-like crack parallel to $0z$ and passing through the interval $[-\Delta, \Delta]$ of $0x$ is adequately described by the three dislocation arrays distributed with densities $\varrho_\alpha(x)$ $(\alpha = 1, 2, 3)$ and having linearly independent shear vectors $\boldsymbol{b}^{(\alpha)}$, the choice of which is governed by reasons of convenience (in particular we assume hereafter $\boldsymbol{b}^{(\alpha)} = \boldsymbol{n}^{(\alpha)}$, where $\boldsymbol{n}^{(1)}$ is the unit vector perpendicular to the crack edge and lying in the crack plane, $\boldsymbol{n}^{(2)}$ the unit vector perpendicular to the crack plane, $\boldsymbol{n}^{(3)}$ the unit vector parallel to the crack edge). If we define the stress field of a dislocation with

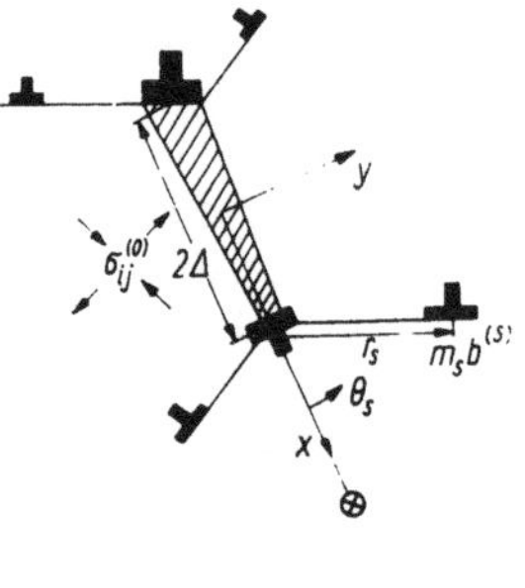

Fig. 1. The wedge-shaped crack formed by the dislocations piling-up against an obstacle and the superdislocations describing the plastic zone near the crack tips

Burgers vector $\boldsymbol{n}^{(\alpha)}$ passing along $0z$ through the point $(\xi, 0)$ as

$$\sigma_{kj}^{(\alpha)} = \sum_{\nu=1}^{6} \frac{Q_{kj}^{(\alpha\nu)}}{\zeta_\nu - \xi} \tag{2}$$

then the crack stresses are given [6] by the expression

$$\sigma_{kj}^{(\mathrm{cr})} = \sum_{\alpha=1}^{3} \sum_{\mu=1}^{6} Q_{kj}^{(\alpha\mu)} \left\{ \frac{h_\alpha + \omega_\alpha \zeta_\mu + \sum_{[s]} \sum_{\nu=1}^{6} m_s u_\alpha^{(s\nu)}}{\sqrt{\zeta_\mu^2 - \Delta^2}} + \right.$$

$$\left. + \sum_{[s]} \sum_{\nu=1}^{6} \frac{m_s u_\alpha^{(s\nu)}}{\zeta_{s\nu} - \zeta_\mu} \left(1 - \sqrt{\frac{\zeta_{s\nu}^2 - \Delta^2}{\zeta_\mu^2 - \Delta^2}}\right) \right\} \tag{3}$$

in which

$$h_\alpha = \int_{-\Delta}^{\Delta} \varrho_\alpha(\xi)\, \mathrm{d}\xi \,, \tag{4}$$

$$\omega_\alpha = \sum_{\beta=1}^{3} (A^{-1})_{\alpha\beta}\, \sigma_{2\beta}^{(0)} \,, \tag{5}$$

$$u_\alpha^{(s\nu)} = \sum_{\beta=1}^{3} (A^{-1})_{\alpha\beta}\, U_{2\beta}^{(s\nu)} \,, \tag{6}$$

and $(A^{-1})_{\alpha\beta}$ is the inverse matrix of

$$A_{\alpha\beta} = \sum_{\nu=1}^{6} Q_{2\beta}^{(\alpha\nu)} \,. \tag{7}$$

By use of the expressions (1) to (3) one could easily obtain the equations defining the parameters of the plastic zone. As a first step, the condition for the absence of emission from the wedged left-hand tip of the crack is the following:

$$n_i^{(t)} b_j^{(t)} \left\{ \sigma_{ij}^{(0)} + \sum_{[s]} \sigma_{ij}^{(s)} + \sigma_{ij}^{(\mathrm{cr})} \right\} \Big|_{\substack{x=-\Delta \\ y=\pm d_1}} = \sigma_{\mathrm{e}}^{(t)} b^{(t)} \,. \tag{8a}$$

Here t designates the slip system considered, $[s]$ labels the set of dislocation systems emitted (or seized)[1]),

$$d_1 = \tfrac{1}{2} h_2 + \sum_{[s_1]} m_s b_2^{(s)} \,, \tag{9a}$$

and $[s_1]$ is a set of systems emitted by the left tip only. Similarly, for the dislocation of the t-th system of the set $[s_2]$ of systems emitted from the right-hand crack tip

$$n_i^{(t)} b_j^{(t)} \left\{ \sigma_{ij}^{(0)} + \sum_{[s]} \sigma_{ij}^{(s)} + \sigma_{ij}^{(\mathrm{cr})} \right\} \Big|_{\substack{x=\Delta \\ y=\pm d_2}} = \sigma_{\mathrm{e}}^{(t)} b^{(t)} ; \tag{8b}$$

$$d_2 = \tfrac{1}{2} \sum_{[s_2]} m_s b_2^{(s)} \,. \tag{9b}$$

In just the same way the equilibrium conditions for each of the effective dislocations in their slip planes have the form

$$n_i^{(s)} b_j^{(s)} \left\{ \sigma_{ij}^{(0)} + \sum_{[t \neq s]} \sigma_{ij}^{(t)} + \sigma_{ij}^{(\mathrm{cr})} \right\} \Big|_{\substack{x=x_s \\ y=y_s}} = \sigma_p^{(s)} b^{(s)} \,. \tag{10}$$

---

[1]) The model presented describes the equilibrium state of the zone and is practically insensitive to the zone formation mechanism. If the zone forms by some dislocation source action it is necessary to use those source coordinates instead of d_1, d_2; however, such a difference does not bring appreciable changes in the zone parameters.

3. The Case of Quasi-Brittle Materials

Some simplification in equations (1) to (10) is possible if the zone in the material investigated obeys the quasi-brittle conditions, i.e. $r_s^{(\pm)} = \sqrt{(x_s \mp \Delta)^2 + y_s^2} \ll \Delta$. We could expect this is valid in materials with high Peierls stresses (Si, Ge, Fe_3Si etc.). In this case the crack tips could be considered independently. Thus (8b), (9b) may be written as

$$\sum_{\varkappa=1}^{3} \sum_{\mu=1}^{6} Q_{kj}^{(\varkappa\mu)} n_k^{(t)} b_j^{(t)} \left\{ \frac{h_\varkappa + \omega_\varkappa \Delta + \sum\limits_{[s]} \sum\limits_{\nu=1}^{6} m_s u_\varkappa^{(s\nu)}}{\sqrt{2\Delta p_\mu d_2}} + \sum_{[s_2]} \frac{m_s u_\varkappa^{(s\nu)}}{x_s - \Delta + p_\nu y_s} \times \right.$$
$$\left. \times \left[1 - \sqrt{\frac{(x_s - \Delta) + p_\nu y_s}{p_\mu d_2}}\right] \right\} - \sum_{[s_2]} \sum_{\nu=1}^{6} \frac{m_s U_{kj}^{(s\nu)} n_k^{(t)} b_j^{(t)}}{x_s - \Delta + p_\nu y_s} = \sigma_c b^{(t)} - \sigma_{ij}^{(0)} n_i^{(t)} b_j^{(t)} . \tag{11}$$

As a rule, the magnitude of the nonhomogeneous stress near the crack tips is much greater than $\sigma_c^{(t)}$ or $\sigma_{ij}^{(0)} n_i^{(t)} b_j^{(t)}$ and similarly d_1, d_2 are small compared the other characteristic distances of the problem. Therefore, in the case of full stress relaxation at the crack tips the conditions (11) turn to

$$\frac{h_\varkappa + \omega_\varkappa \Delta + \sum\limits_{[s]} \sum\limits_{\nu=1}^{6} m_s u_\varkappa^{(s\nu)}}{\sqrt{2\Delta}} - \sum_{[s_2]} \sum_{\nu=1}^{6} \frac{m_s u_\varkappa^{(s\nu)}}{\sqrt{r_s}\sqrt{\cos\theta_s + p_\nu \sin\theta_s}} = 0 \tag{12}$$

for any $\varkappa = 1, 2, 3$; and also to

$$\sum_{[s_2]} \sum_{\nu=1}^{6} \frac{m_s}{r_s(\cos\theta_s + p_\nu \sin\theta_s)} \left[\sum_{\varkappa=1}^{3} u_\varkappa^{(s\nu)} \sum_{\mu=1}^{6} Q_{kj}^{(\varkappa\mu)} n_k^{(t)} b_j^{(t)} - U_{kj}^{(s\nu)} n_k^{(t)} b_j^{(t)} \right] = 0 . \tag{13}$$

In these expressions

$$r_s = \sqrt{(x_s - \Delta)^2 + y_s^2} , \tag{14}$$
$$\cos\theta_s = \frac{x_s - \Delta}{r_s}, \qquad \sin\theta_s = \frac{y_s}{r_s}.$$

But if there is only partial relaxation at the crack tips, only some linear combinations of equations (12) hold.

The conditions (10) in a quasi-brittle approach take the form

$$\sigma_p^{(s)} b^{(s)} - \sigma_{ij}^{(0)} n_i^{(s)} b_j^{(s)} =$$
$$= \frac{1}{\sqrt{2\Delta r_s}} \sum_{\alpha=1}^{3} \left(h_\alpha + \omega_\varkappa \Delta + \sum_{[t]} \sum_{\nu=1}^{6} m_t u_\varkappa^{(t\nu)} \right) \sum_{\mu=1}^{6} \frac{Q_{kj}^{(\varkappa\mu)} n_k^{(s)} b_j^{(s)}}{\sqrt{\cos\theta_s + p_\mu \sin\theta_s}} +$$
$$+ \sum_{[t \neq s]} \frac{m_t}{r_s} \left\{ \sum_{\mu=1}^{6} \frac{U_{ij}^{(t\mu)} n_i^{(s)} b_j^{(s)}}{\cos\theta_s - \frac{r_t}{r_s}\cos\theta_t + p_\mu \left(\sin\theta_s - \frac{r_t}{r_s}\sin\theta_t\right)} + \right.$$
$$+ \sum_{\varkappa=1}^{3} \sum_{\mu,\nu=1}^{6} \frac{Q_{ij}^{(\varkappa\mu)} n_i^{(s)} b_j^{(s)} u_\varkappa^{(t\nu)}}{\cos\theta_s - \frac{r_t}{r_s}\cos\theta_t + p_\mu \sin\theta_s - p_\nu \frac{r_t}{r_s}\sin\theta_t} \times$$

$$\times \left[\sqrt{\frac{r_t(\cos\theta_t + p_\nu \sin\theta_t)}{r_s(\cos\theta_s + p_\mu \sin\theta_s)}} - 1\right]\Bigg\} + \frac{m_s}{r_s}\left[\sum_{\alpha=1}^{3} \sum_{\substack{\mu,\nu=1 \\ p_\mu \neq p_\nu}}^{6} \frac{Q_{ij}^{(\alpha\mu)} n_i^{(s)} b_j^{(s)} u_\alpha^{(s\nu)}}{(p_\mu - p_\nu)\sin\theta_s} \times \right.$$
$$\left. \times \left(\sqrt{\frac{\cos\theta_s + p_\nu \sin\theta_s}{\cos\theta_s + p_\mu \sin\theta_s}} - 1\right) - \sum_{\alpha=1}^{3} \sum_{\mu=1}^{6} \frac{Q_{ij}^{(\alpha\mu)} n_i^{(s)} b_j^{(s)} \sum_{p_\mu = p_\nu} u_\alpha^{(s\nu)}}{2(\cos\theta_s + p_\mu \sin\theta_s)}\right]. \tag{15}$$

In the right-hand part of this equation the following terms are consequently written: the interaction between the s-th superdislocation and the crack on the assumption that an effect of the superdislocations on the crack is absent; the interaction between the s-th dislocation and the other dislocations of the zone t (the first term in braces); the interaction between s and the images of t in the crack (the second term in braces); the interaction of s with the self-image in the crack.

Equations (12), (13), (15) (as well as analogous ones for the dislocation emission from the left tip of the crack) determine all the parameters m_s, r_s of the system considered.

It is clearly seen that equations (12), (15) become much simpler if external conditions are favourable only for one slip system to be excited, or if the zone is a symmetrical one. In these cases there is no need to consider other defects for determining the parameters m_1, r_1 of the primary system. In the following part we consider this case in some more detail.

4. The Case of a Symmetrically Loaded Crack

In terms of the quasi-brittle approach introduced above consider the case of a symmetrically loaded crack. The crack in this case is described by only one array of wedge dislocations with Burgers vectors in the $0y$ direction. Note that wedging and action of stresses play quite a different role in the stress distribution at the crack tips: under symmetrical tension $\sigma_{22}^{(0)} > 0$ their effects enhance each other near the sharp tip and relax each other near the base of the crack. So we assume the crack has a zone only near the right-hand (sharp) tip and it consists of two edge dislocations with Burgers vectors which lie in planes inclined to the crack plane at an angles $\pm\theta_1$. In such a case the parameters m_1 and r_1 of the equilibrium zone can be expressed in a closed form:

$$m_1 \approx \frac{A(h + \omega\Delta - 2mA_{12}/A)^2\,(q_1 q_0 + q_2)}{2\Delta q_0^2(\sigma_p b - \sigma_{ij}^{(0)} n_i b_j)}, \tag{16}$$

$$r_1 \approx \frac{A^2 b^2 (h + \omega\Delta - 2mA_{12}/A)^2\,(q_1 q_0 + q_2)^2}{2\Delta q_0^2(\sigma_p b - \sigma_{ij}^{(0)} n_i b_j)^2}. \tag{17}$$

In these formulae

$$A = \sum_{\mu=1}^{6} Q_{22}^{(2\mu)}, \qquad A_{12} = \sum_{\mu=1}^{6} U_{22}^{(t\mu)} \tag{18}$$

and all the anisotropy is localized in the nondimensional factors q_0, q_1, and q_2 all depending on the mutual orientation of the dislocations and the crack and being of the form

$$q_0 = (Ab)^{-1} \sum_{\nu=1}^{6} \frac{2U_{22}^{(+\nu)}}{\sqrt{\cos\theta_1 + p_\nu \sin\theta_1}}, \tag{19}$$

$$q_1 = \sum_{\mu=1}^{6} \frac{Q_{kj}^{(2\mu)} n_k^{(+)} b_j^{(+)}}{Ab\sqrt{\cos\theta_1 + p_\nu \sin\theta_1}}, \tag{20}$$

$$q_2 = \sum_{\mu=1}^{6} \frac{U_{kj}^{(-\mu)} n_k^{(+)} b_j^{(+)}}{2Ab^2 p_\mu \sin\theta_1} + \sum_{\mu=1}^{6} \left\{ \sum_{\substack{\nu=1 \\ (p_\mu \neq p_\nu)}}^{6} \frac{Q_{kj}^{(2\mu)} n_k^{(+)} b_j^{(+)} U_{22}^{(+\nu)}}{A^2 b^2 (p_\nu - p_\mu) \sin\theta_1} \times \right.$$
$$\left. \times \left(1 - \sqrt{\frac{\cos\theta_1 + p_\nu \sin\theta_1}{\cos\theta_1 + p_\mu \sin\theta_1}}\right) - \frac{Q_{kj}^{(2\mu)} n_k^{(+)} b_j^{(+)} \sum_{p_\nu = p_\mu} U_{22}^{(+\nu)}}{2A^2 b^2 (\cos\theta_1 + p_\mu \sin\theta_1)} \right\}. \tag{21}$$

Here $(+)$ denotes the upper and $(-)$ the lower superdislocation of the zone. Assuming superdislocations have Burgers vectors $m_1 \boldsymbol{b}^{(\pm)} = m_1 b(\pm \cos\theta_1, \sin\theta_1, 0)$, one could transform relations (16) to (21) to obtain the isotropic formulae by the rules developed in [9]. As a result we achieve the following relations (G being the shear modulus and ν being the Poisson ratio):

$$A = \frac{G}{2\pi(1-\nu)}, \qquad A_{12} = \frac{Gb}{2\pi(1-\nu)} \sin\theta_1, \tag{18a}$$

$$q_0 = 3 \sin\theta_1 \cos\frac{\theta_1}{2}, \tag{19a}$$

$$q_1 = \frac{1}{2} \sin\theta_1 \cos\frac{\theta_1}{2}, \tag{20a}$$

$$q_2 = -\tfrac{1}{2} \tag{21a}$$

which define the parameters of the equilibrium zone in isotropic media.

Quite analogous formulae could be written when only one important shear process occurs near the tip of the crack.

5. Some Applications of the Theory Developed

Up to this part we have treated just the model of the plastic zone near the cracks. Now let us point out that the theory developed above is useful to describe (in some situations) the interaction of the general pile-up-type shear defects such as twins, martensite crystals, slip bands, etc. In particular the formation and development of such defects near cracks and vice versa could be described. All the changes concern a number of pile-ups acting in the planes and the Burgers vectors of the dislocations providing plastic shears involved. In all cases it is convenient to describe the smaller defect by an effective superdislocation.

Below some phenomena will be discussed which include the interaction of cracks with other defects.

As an example, we discuss here the problem of the equilibrium parameters of two planar defects which turn into each other and are piled up at the line of their mutual crossing (see Fig. 2). Consider, e.g. a dislocation pile-up opening into a crack during plastic deformation under pressure. If the pressure is big enough not all the pile-up falls down into the crack. Each of $N-m$ dislocations of the pile-up (formula (3) and the next ones are now written for this pile-up) is influenced by shear stresses

$$f_1 = \sigma_{kj}^{(0)} n_k^{(\text{p-u})} b_j - \sigma_p b\,. \tag{22}$$

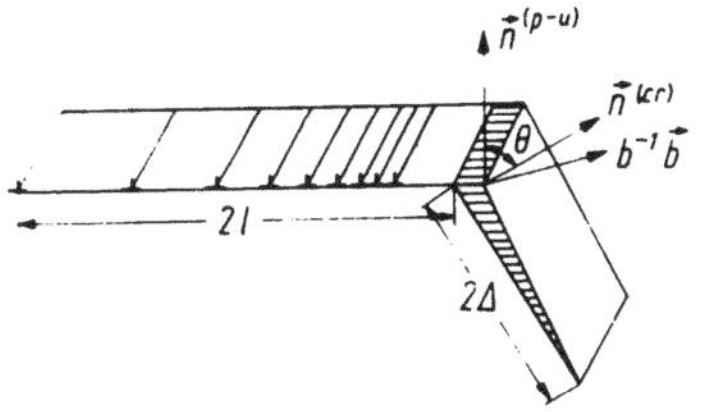

Fig. 2. The equilibrium state of the dislocation pile-up opening the crack in the presence of pressure p and shear stress σ_{12}

On the other hand, each of m dislocations of the crack is influenced by the force

$$f_2 = pb_k n_k^{(\mathrm{cr})} - \tilde{\sigma}_{kj}^{(0)} n_k^{(\mathrm{cr})} b_j + m^{-1}\gamma \approx pb_k n_k^{(\mathrm{cr})} - \tilde{\sigma}_{kj}^{(0)} n_k^{(\mathrm{cr})} b_j \tag{23}$$

(we single out the hydrostatic component of the stresses corresponding to $\sigma_{kj}^{(0)} = -p\delta_{kj} + \tilde{\sigma}_{kj}^{(0)}$). Making use of (16) and (17) immediately leads to the relation between the number of dislocations which contribute to the crack and the number of dislocations which remain in the pile-up and also to a relation between the lengths of these defects (assuming the pressure is high enough to allow the quasi-brittle approach to be used). It has only to be taken into account that the left-side end of the pile-up is free to move and the stress component $\sigma_{kj} n_k^{(\mathrm{p\text{-}u})} b_j$ equals zero at that end, so

$$0 = N - m + \sum_{\nu=1}^{6} m u^{(\nu *)} - \omega^* l \sim N - (A^*)^{-1}\, l(\sigma_{kj}^{(0)} n_k^{(\mathrm{p\text{-}u})} b_j - \sigma_p b)\,. \tag{24}$$

Here we keep the letter Δ for the crack length and use $2l$ for the pile-up length and an asterisk as a label of changes in A, ω, $u^{(\nu)}$, q_0, q_1, q_2 due to the pile-up and crack have interchanged their roles.

Hence the ratio of the number of dislocations forming the crack to the number of dislocations remaining in the pile-up is given by

$$\frac{m}{N} \approx \frac{2(q_0^* q_1^* + q_2^*)\,(\sigma_{ij}^{(0)} n_i^{(\mathrm{p\text{-}u})} b_j - \sigma_p b)}{(q_0^*)^2\,(pb - \sigma_{ij}^{(0)} n_i^{(\mathrm{cr})} b_j)} \tag{25}$$

and the ratio of the lengths of these defects is

$$\frac{\Delta}{2l} \approx \frac{(q_0^* q_1^* + q_2^*)^2}{(q_0^*)^2}\, \frac{(\sigma_{ij}^{(0)} n_i^{(\mathrm{p\text{-}u})} b_j - \sigma_p b)^2}{(pb - \sigma_{ij}^{(0)} n_i^{(\mathrm{cr})} b_j)^2}\,. \tag{26}$$

This is another proof of the estimation carried out in [12].

Formulae (25), (26) give also a way of evaluating the pressure necessary for a plastic closing of the crack. We have pointed out that under tension a plastic zone is mainly concentrated near the sharp tip of the crack. Application of pressure after unloading reverses the picture and leads to the zone formation near the crack base and therefore to a plastic closing of the crack. If we take into consideration that instead of $(\sigma_{ij}^{(0)} n_i^{(\mathrm{p\text{-}u})} b_j - \sigma_p b)$ the pile-up emitted by the crack is held in equilibrium by the "dry friction" force $\sigma_p b$ and that the pressure under consideration exceeds $\sigma_{ij}^{(0)}$ from (25) we derive the condition of pronounced crack plastic closing (at $m \approx N$) as follows:

$$p \sim p^{(\mathrm{c})} = 2(q_0^*)^{-2}\,(q_0^* q_1^* + q_2^*)\,\sigma_p \tag{27}$$

and the condition of full crack plastic closing down (at $m \lessapprox 1$):

$$p \gtreqqless p^{(\mathrm{c.d.})} = 2N(q_0^*)^{-2}\,(q_0^* q_1^* + q_2^*)\,\sigma_p\,. \tag{28}$$

These parameters, as is evident, play an important role in explaining the influence of pressure on the plastic behaviour of crystal materials.

The values of q_0^*, q_1^*, and q_2^* in the isotropic limit are given by

$$q_0^* = \frac{1}{2} \sin^2 \theta \cos \frac{\theta}{2}, \tag{19b}$$

$$q_1^* = \frac{3}{2} \sin^2 \theta \cos \frac{\theta}{2}, \tag{20b}$$

$$q_2^* = -\tfrac{1}{2} \sin^4 \theta . \tag{21b}$$

Here we assume that the Burgers vector of the effective superdislocation which describes the crack is $mb \sin \theta(\sin \theta, \cos \theta, 0)$ (on the assumption that shears run out of the crack because of the small friction force in the crack plane).

References

[1] D. S. DUGDALE, J. Mech. Phys. Solids 8, 100 (1960).
[2] B. A. BILBY, A. H. COTTRELL, and K. H. SWINDEN, Proc. Roy. Soc. London **A272**, 304 (1963).
[3] S. N. ZOLOTAREV, B. V. MOLOTILOV, and V. A. SOLOVEV, Fiz. tverd. Tela **18**, 3615 (1976).
[4] C. ATKINSON and D. L. CLEMENTS, Acta metall. **21**, 55 (1973).
[5] C. ATKINSON, Internat. J. Fract. Mech. **2**, 267 (1966).
[6] V. A. PETROV, V. I. VLADIMIROV, and A. N. ORLOV, phys. stat. sol. **42**, 197 (1970).
[7] V. A. SOLOVEV, Problemy Metallovedeniya i Fiziki Metallov, No. 3, Metallurgiya, Moscow 1976 (p. 258).
[8] V. A. SOLOVEV, phys. stat. sol. (b) **65**, 857 (1974).
[9] V. A. SOLOVEV and S. N. ZOLOTAREV, phys. stat. sol. (b) **91**, 319 (1978).
[10] A. N. STROH, Phil. Mag. **3**, 625 (1958).
[11] J. P. HIRTH and J. LOTHE, Theory of dislocations, McGraw Hill Publ. Co, New York 1968.
[12] V. A. SOLOVEV and V. A. STRELTSOV, phys. stat. sol. (a) **49**, K145 (1978).

(Received October 24, 1978)

Note added in proof:

We list here some corrections to [8]. Under accepted square root signs the signs before $\alpha\zeta_\nu$ and σ_0 in the formulas (15), (17), (19) have to be contrary to those used in [8]. Also, the sign change before $2C'_{66}$ in the denominator of (A.15), interchanging of captions to Fig. 3 and 4, and removing of a factor of 2 under the square roots in (A.16), (A.18) must be done.

phys. stat. sol. (a) 54, 305 (1979)

Subject classification: 10.1; 21.1

Sektion Physik der Pädagogischen Hochschule Dresden

A Quantitative Estimation of Specific Internal Stress Fields in Cyclically Deformed Nickel

By

C. HOLSTE, G. K. SCHMIDT, and R. TORBER

As a characteristic part of the internal stress field of fatigued f.c.c. metals the special stress field is considered which is correlated to the dislocations moving within a loading cycle in the dislocation-poor regions. For a quantitative estimation of this stress field (termed as polarization stress field) computer calculations are made for idealized dislocation arrangements. These arrangements are based on the idea of shuttling and separation (polarization) of dislocation groups of different signs. The results show that the polarization stress field can be divided into a half-macroscopic part strongly polarization-dependent (with a spatial distance of its maximum values in the order of the dimensions of the dislocation-poor regions) and into a nearly polarization-independent part (with a spatial distance of its maximum values in the order of the mean distance of the dislocations in a group).

Als charakteristischer Anteil des inneren Spannungsfeldes von ermüdeten k.f.z. Metallen wird das spezifische Spannungsfeld betrachtet, das mit den Versetzungen korreliert ist, die sich innerhalb des Lastzyklus in den versetzungsarmen Gebieten bewegen. Für eine quantitative Abschätzung dieses Spannungsfeldes (als Polarisationsspannungsfeld bezeichnet) werden Computerrechnungen für idealisierte Versetzungsanordnungen vorgenommen. Diese Anordnungen basieren auf der Vorstellung der Pendelbewegung und Trennung (Polarisation) von Versetzungsgruppen mit unterschiedlichem Vorzeichen. Die Ergebnisse zeigen, daß das Polarisationsspannungsfeld in einen stark polarisationsabhängigen halbmakroskopischen Anteil (mit einem räumlichen Abstand seiner Spitzenwerte in der Größenordnung der Abmessungen der versetzungsarmen Bereiche) und in einen nahezu polarisationsunabhängigen Anteil (mit einem räumlichen Abstand seiner Spitzenwerte in der Größenordnung des mittleren Abstandes der Versetzungen einer Gruppe) zerlegt werden kann.

1. Introduction

In recent years not only specific dislocation arrangements in fatigued f.c.c. metals have been investigated in detail (e.g. Woods [1], Mughrabi [2] on copper single crystals, Lukas and Klesnil [3] on copper polycrystals and Mecke [4] on nickel single crystals) but also the discussion about the dislocation behaviour within the single deformation cycle in the saturation stage of the stress amplitude has increased [5 to 10]. It is generally assumed that the dislocations in dislocation-poor regions may move in a more or less modified process of shutling to and fro between the dislocation-dense regions. So far there did not exist any experimentally and theoretically established results concerning the kind and degree of the accomodation of plastic strain in the dislocation-dense regions [7]. As yet along a single hysteresis loop only X-ray measurements [11, 12] on nickel polycrystals and magnetic measurements [13, 12] on nickel single crystals and nickel polycrystals are known. Especially the considerable, typical change of the coercive force at room temperature within a deformation cycle in the saturation stage [13] indicates that in a deformation cycle there occur characteristic changes of the mean amplitude and of the distribution of the amplitudes of long-range stress fields. The change of these stress fields should be correlated to a specific change of the density and arrangement of dislocations moving within the cycle.

The existence of long-range stresses in cyclically deformed metals has been shown experimentally for relatively large strain amplitudes by the TEM-investigations of Mughrabi [2] on neutron-irradiated copper single crystals. Considering the curvatures of bowed-out dislocation segments for a resolved shear strain amplitude $\gamma_{pa} = 5 \times 10^{-3}$ strain values of the internal stress field were determinedu p to the applied stress amplitude.

Up to the present there have not been any quantitative estimations of the long-range internal stresses which are correlated to the change of the density and arrangement of the movable dislocations within a cycle.

In the present paper several possible configurations of dislocations which have moved in the dislocation-poor regions are discussed and their calculated stress fields are given. The dislocation arrangements considered take into account the situation in nickel single crystals within an amplitude range from $\gamma_{pa} = 1 \times 10^{-3}$ to 5×10^{-3} and in nickel polycrystals within an amplitude range from $\gamma_{pa} = 1 \times 10^{-3}$ to $= 3 \times \times 10^{-3}$.

TEM-observations of the dislocation structures in fatigued nickel single crystals [4, 14] and in nickel polycrystals [14, 15] show that in the amplitude range considered here qualitatively the same amplitude-dependent structures occur as are found in fatigued copper single crystals [1] and copper polycrystals [3]. Therefore, the bulk of information obtained from copper investigations was also taken into consideration for the choice of the detailed structure of the dislocation arrangements, the stress field of which were calculated.

2. Choice and Discussion of the Basic Structure of the Treated Dislocation Arrangements

The experimental basis for the dislocation arrangements considered in the following are the TEM-observations by Mughrabi [2] on cyclically deformed, neutron-irradiated copper single crystals. As suggested by Mughrabi, the dislocation motion within a hysteresis cycle in the dislocation-poor regions is brought about mainly by the to- and fro-shuttling of groups of dislocations having predominantly screw character. These mobile groups of alternating sign are arranged in neighbouring glide planes separated by a certain distance. The action length, within which during cycling a relatively well-arranged shuttling of screw dislocations has developed is termed "range of coherency" in [7].

The basic mechanism of the dislocation shuttling suggested in [2] and [7], respectively, should be possible in all inhomogeneous dislocation arrangements (dislocation-poor regions along with dislocation-dense regions) for a large range of amplitudes in single crystals as well as also in polycrystals in spite of characteristic differences in the microstructures (e.g. wall structures in persistent slip bands [2, 4]). The assumed mechanism should be less efficient, when the glide paths of the screw dislocations are limited by dislocation-dense regions (e.g. cell walls) [7]. Regarding the results of observations in TEM [14, 15] this latter case should be excluded in the range of amplitudes considered here for nickel single crystals ($\gamma_{pa} \leqq 5 \times 10^{-3}$) but also even for nickel polycrystals ($\gamma_{pa} \lessapprox 3 \times 10^{-3}$).

Taking into account the above discussed mechanism idealized dislocation configurations with a basic structure according to Fig. 1 were chosen for the calculation of the stress fields (detail parameters see Section 3, Table 1). The specific choice of this idealized basic structure is connected with some further considerations of the cycle processes which will be referred to in the following. The shuttling of dislocation groups within the range of coherency (see Fig. 1) is connected with "bowing-out" the initial screw dislocations between the dislocation-dense regions. According to [9] in each half cycle

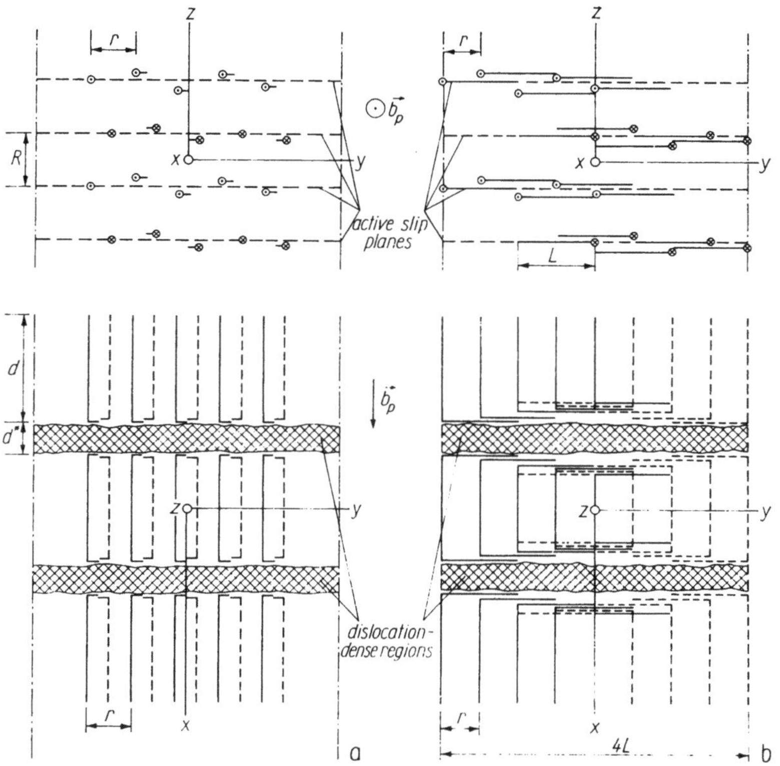

Fig. 1. Basic structure and detail parameters of the idealized dislocation configurations used for the numerical calculations (bottom: sections parallel to primary glide plane, top: sections perpendicular to primary Burgers vector $\boldsymbol{b}_p$). a) Configuration nearly unpolarized, b) configuration polarized to its maximum value within a half cycle (for explanation of the parameters see Section 3)

the screw dislocations will draw edge dislocations behind them which penetrate into the bordering regions of the multipole arrangements [2, 7] of the dislocation-dense regions and attain stable positions within the short-range multipole stress fields. In Fig. 1 the dislocations bowed out (small bowing in a half cycle (Fig. 1a) and maximum bowing (Fig. 1b)) are approximated by parts of pure screw dislocations and by straight edge dislocation segments. For clearness the dislocations of an active glide plane in Fig. 1 at the top are drawn in the z-direction and the edge dislocations in Fig. 1 at the bottom are shown in the x-direction side by side. For the numerical calculation constant x- and z-coordinates were used. If the possibility of accomodation of plastic strain in the dislocation-dense regions is supposed (this means the assumption of effective annihilation of the edge dislocation segments formed in a half cycle and taken into consideration in Fig. 1) the long-range stress field caused by the formed edge dislocations can be reduced or even vanish completely. According to observations in TEM [2] and to energetical considerations [9] it can be assumed that the dislocation motion at each side of the dislocations-dense regions is fairly well synchronized. In Fig. 1 this fact is taken into account, too.

For a more general consideration of the cycle processes a volume with a more or less locally limited dislocation motion during a hysteresis cycle is termed an active mesoscopic or half macroscopic volume (HMV). If the dislocations in the dislocation-dense regions do not take part, or only participate very insignificantly, in the dislocation motion within the cycle such an active HVM is bounded by the range of coherency, the distance between the dislocation-dense regions, and by the extension of the dislocation-dense regions perpendicular to the active glide planes. If the plastic strain can be accomodated in the multipole arrangements, the extension of an active HMV is possible across many dislocation-poor regions. Regardless of the dislocation motion, which is unknown in detail, it can be assumed that a relatively well-ordered dislocation motion takes place running from cycle to cycle nearly in the same way. In each half cycle, under the influence of a shear stress, the dislocation motion leads to a "polarization" of the dislocation arrangements. This polarization means that there is an increase in the concentration of dislocations of the same sign in the active glide planes in the boundary zones of a HMV. In opposite boundary zones of a HMV dislocations of opposite signs are accumulated. Therefore, the polarization is correlated to the formation of a local net Burgers vector surplus at the border of the HMV and this leads to characteristic long-range internal stress fields.

Fig. 1 in an idealized manner reflects the situation in a HMV for a dislocation arrangement almost unpolarized (Fig. 1a) and an arrangement polarized to its maximum value in a half cycle (Fig. 1b). In the numerical calculations the possibility of the effective annihilation of edge segments generated by the dislocation motion is not taken into account. Therefore, a HMV considered here is bounded in the direction of the x-axis by dislocation-dense regions. In Fig. 1 and also for the calculations only those dislocation arrangements in the dislocation-poor regions are paid attention to which will influence essentially the long-range stress field in a considered HMV.

The computer programme applied allows the determination of all components of the stress tensor in the coordinate system given in Fig. 1 for any line of points in the medium considered. However, the calculations were essentially limited to the ascertaintment of the course of the shear stress component τ_{13} along the x- and y-axis.

3. Parameters of the Dislocation Configurations Selected for Numerical Calculation

The detail parameters of all considered dislocation configurations are shown in Table 1. The values given in columns 2 to 5 for the distance d of the dislocation-dense regions, the extension d^* of the dislocation-dense regions along the x-axis, the density ϱ_s of the screw dislocations movable in the active slip planes, and the distance R of the active slip planes are varied in a larger range, regarding there by the special situation in single crystals and polycrystals. In spite of a great amount of experimental data it does not seem reasonable to be more exact. The values in columns 2 to 4 are taken from [14, 15, and 7]. The items in column 5 on the distance of the slip planes follow from considerations of the cross slip annihilation of screw dislocations [16, 14, 4] and from experimental observations [7].

The values in columns 6 to 10 were calculated from the parameters given in the first columns. The maximum glide path L in a half cycle in the sixth column results from $L = \gamma_{pa}/\varrho_s b(1 - m)$ if no accomodation of plastic strain in the dislocation-dense regions is permitted, where γ_{pa} is the resolved shear strain amplitude, b the absolute value of the primary Burgers vector, and m the volume fraction of the dislocation-dense regions. Only for the smallest amplitude a value different from zero ($m = 0.5$ [7]) was used. The range of coherence (see Fig. 1) is equal to $4L$ for all dislocation configurations considered. In column 7 the number N of the screw dislocation segments of an active slip plane within the total length $4L$ is given.

Table 1

Parameters of the dislocation configurations

experimental parameters				calculated parameters						configuration type
γ_{pa}	d (μm)	d^* (μm)	ϱ_s (cm^{-2})	R (μm)	L (μm)	N	r (μm)	n (μm^{-1})	B/b	
single crystals										
5.0×10^{-3}	1.5	0.3	1.5×10^9	0.1	1.33	9	0.33	1.5	4	5AA1
				0.25		1	0.133	3.75	10	5AA2
				0.5		1	0.067	7.55	20	5AA3
			1.0×10^9	0.1	2.0	9	0.5	1.0	4	5AB1
	1.2	0.3	1.5×10^9	0.1	1.33	9	0.33	1.5	4	5BA1
			1.0×10^9	0.1	2.0	9	0.5	1.0	4	5BB1
3.0×10^{-3}	1.5	0.3	1.5×10^9	0.1	0.8	5	0.4	1.5	2	3AA1
	1.2	0.3		0.1		5	0.4	1.5	2	3BA1
1.0×10^{-3}	1.2	0.8	0.5×10^9	0.1	1.6	4	1.1	0.5	1.5	1CC1
polycrystals										
3.0×10^{-3}	0.6	0.2	1.5×10^9	0.1	0.8	5	0.4	1.5	2	3DA1
1.0×10^{-3}	0.6	1.2	0.5×10^9	0.1	1.6	4	1.1	0.5	1.5	1EC1

With a given density of screw dislocations ϱ_s (related to the range of coherency) and with the slip path L corresponding to a given resolved shear strain amplitude, it follows $N \approx 4L\varrho_s R$. The distance of the screw dislocation segments in a dislocation group is determined by $r = 2L/(N - 1)$ (column 8). In column 9 the number per unit length of the screw dislocation segments of the same sign in an active slip plane is given by $n = \varrho_s R$. Finally in column 10 the quantity $B/b = L/r$ for the polarized state is a measure for the mean amount of the local net Burgers vector $\boldsymbol{B}$ of the edge dislocations lying in the active slip plane near the dislocation-dense regions. In addition, in the last column the different dislocation configurations are marked.

4. Quantitative Determination of the Stress Fields

The dislocation segments shown in Fig. 1 are thought of as parts of dislocation loops, the portions of which lying in the dislocation-dense regions do not yield a contribution to the long-range stress field. For the quantitative determination of the stress field of these segments we started from the Peach-Koehler formula in the form

$$\sigma_{\alpha\beta} = -\frac{G}{8\pi} \oint \sum_{i=1}^{3} \sum_{j=1}^{3} b_j e_{ij\alpha} \frac{\partial}{\partial x_j'} \nabla'^2 R \, dx_\beta' - \frac{G}{8\pi} \oint \sum_{i=1}^{3} \sum_{j=1}^{3} b_j e_{ij\beta} \frac{\partial}{\partial x_i'} \nabla'^2 R \, dx_\alpha' -$$

$$- \frac{G}{4\pi(1-\nu)} \oint \sum_{i=1}^{3} \sum_{j=1}^{3} \sum_{k=1}^{3} b_j e_{ijk} \left(\frac{\partial^3 R}{\partial x_i' \, \partial x_\alpha' \, \partial x_\beta'} - \delta_{\alpha\beta} \frac{\partial}{\partial x_i'} \nabla'^2 R \right) dx_k' \quad (1)$$

given in [17] taking the line integrals only over the dislocation segments of interest. In (1) α and β are the indices of the stress tensor, b_i the coordinates of the Burgers vector, x_i' and x_i the coordinates of any source and medium point, respectively, G and ν the shear modulus and Poisson's ratio, respectively $e_{ijk} = \boldsymbol{e}_i(\boldsymbol{e}_j \times \boldsymbol{e}_k)$ is the Einstein permutation operator, δ_{ij} the Kronecker delta, and $R = \left(\sum_{i=1}^{3} (x_i - x_i)^2 \right)^{1/2}$ the distance between any medium and source points. The constants $G = 8.66 \times 10^4$ MPa, $\nu = 0.3$ and $|\boldsymbol{b}| = 2.5$ Å were used.

When the partial differentiations in (1) are performed a dislocation segment can be decomposed into small pieces to be treated independent of each other. Thus it will be possible to solve the integral numerically. For the stress field originating from a small part of a dislocation segment with the coordinates $\Delta x_i'$ we obtained

$$\Delta\sigma_{\alpha\beta} = -\frac{G}{4\pi}\sum_{i=1}^{3}\sum_{j=1}^{3}\frac{x_i - x_i'}{R^3}\, b_j(e_{ij\alpha}\,\Delta x_\beta' + e_{ij\beta}\,\Delta x_\alpha') + \\ + \frac{G}{4\pi(1-\nu)}\sum_{i=1}^{3}\sum_{j=1}^{3}\sum_{k=1}^{3} b_j e_{ijk}\,\Delta x_k'\left[\frac{3}{R^5}(x_i - x_j')(x_\alpha - x_\alpha')(x_\beta - x_\beta') - \right. \\ \left. -\frac{\delta_{i\alpha}(x_\beta - x_\beta')}{R^3} - \frac{\delta_{i\beta}(x_\alpha - x_\alpha')}{R^3} + \frac{\delta_{\alpha\beta}(x_i - x_i')}{R^3}\right]. \qquad (2)$$

Taking into account the condition $\Delta x_i' \ll R$, a computer programme was written on the basis of expression (2) which allows to determine all components of the stress tensor for any arbitrary medium point.

5. Results of the Numerical Calculations

In Fig. 2 to 4 local distributions of the shear stress component τ_{13} are shown which are characteristic of the results of the numerical calculations. In Fig. 2 the changes of τ_{13} along the y-axis (Fig. 2a) and the x-axis (Fig. 2b) are represented for the dislocation arrangement 3AA1 polarized, as shown in Fig. 1b. In Fig. 3 for type 3AA1 those portions of the τ_{13} distribution can be seen which are caused by the edge dislocations and screw dislocations only. Fig. 4 shows the sourse of τ_{13} for the same type of configuration in the small polarized state according to Fig. 1a. For the polarized arrangements 5AA2 and 5AA3 characteristic deviations occur from the curve shown in Fig. 2b for the course of τ_{13} in the plane $z = 0$ in the direction of the y-axis. The deviations are caused by the relatively large distance R of the active slip planes compared with the distance r between the dislocations and are connected with the change of the sign of the maximum values (see below) given in column 6 of Table 2. However, the maximum values in the plane $z = 0$ of these types of configuration are not representative of the whole volume of a HMV.

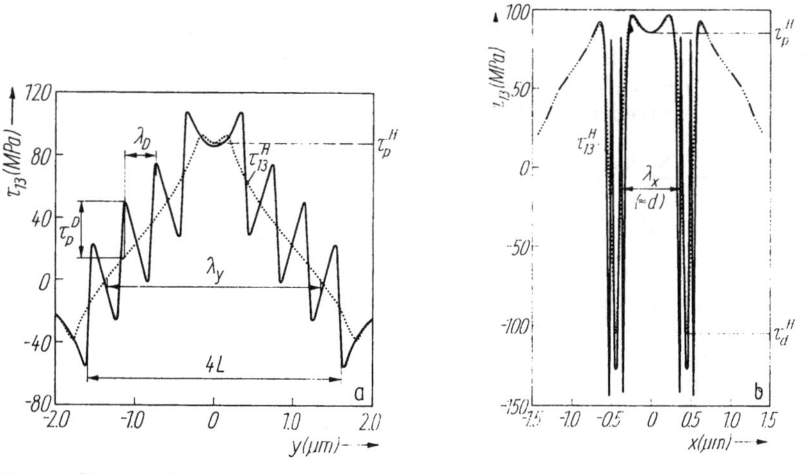

Fig. 2. Change of the stress component τ_{13} along the a) y-axis and b) x-axis of Fig. 1b for the polarized dislocation configuration 3AA1

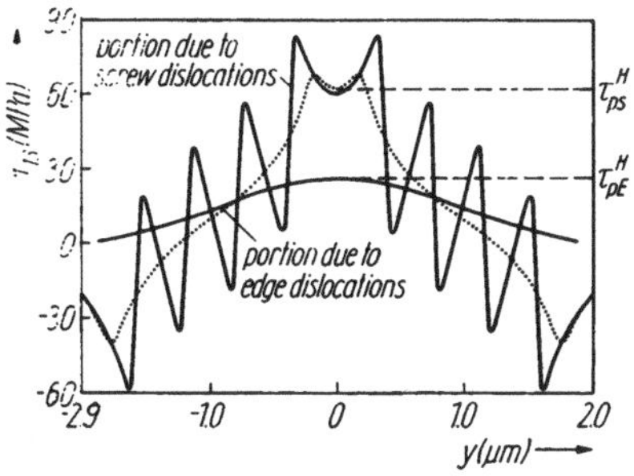

Fig. 3

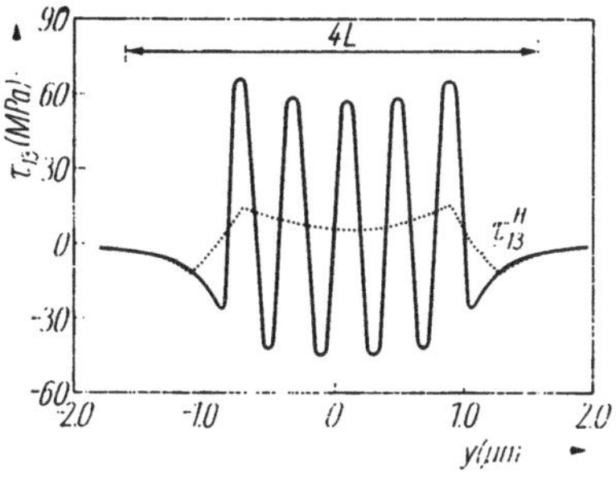

Fig. 4

Fig. 3. Portions of the function $\tau_{13}(0, y)$ represented in Fig. 2a which are caused by the edge dislocations and screw dislocations only

Fig. 4. Change of the stress component τ_{13} along the x-axis of Fig. 1a for the nearly unpolarized dislocation configuration 3AA1

From Fig. 2 it is obvious that $\tau_{13}(x, y)$ can be obtained approximately by the superposition of two functions $\tau_{13}^{H}(x, y)$ and $\tau_{13}^{D}(x, y)$, where τ_{13}^{H} at $x = 0$ and $y = 0$ is defined by

$$\tau_{13}^{H}(0, y) = \frac{1}{s_y} \int\limits_{y-S_y/2}^{y+S_y/2} \tau_{13}(0, y')\,dy' \quad \text{and} \quad \tau_{13}^{H}(x, 0) = \frac{1}{s_x} \int\limits_{x-S_x/2}^{x+S_x/2} \tau_{13}(x', 0)\,dx', \tag{3}$$

respectively. In the above expressions S_y is the mean distance between the peak values of the $\tau_{13}(0, y)$ curve, and S_x is the mean distance between the peak values of the oscillating $\tau_{13}(0, x)$ curve near the positions of the edge dislocations. The quantities λ_y and λ_x in Fig. 2 are of the same order of magnitude as the range of coherency $4L$ and the distance d of the dislocation-dense regions, respectively. λ_D corresponds to S_y and is approximately equal to the distance r of the dislocations in a group.

Table 2

Characteristic stress quantities of the calculated stress fields

dislocation configuration (see Table 1)	τ_p^H (MPa)	τ_p^D (MPa)	τ_{ps}^H (MPa)	τ_{pE}^H (MPa)	τ_d^H (MPa)	τ_{0s} (MPa) [14]
5AA1	142	34	90	52	−270	
5AA2	275	2,5	212	63	230	
5AA3	305	2,5	302	3	640	
5AB1	110	36	55	55	−300	} 55
5BA1	157	33	84	72	−270	
5BB1	125	35	51	74	−300	
3AA1	85	35	60	25	−100	} 52
3BA1	92	34	55	37	−100	
1CC1	62	40	12	50	−100	50
3DA1	98	35	46	52	− 80	120
1EC1	111	40	6	105	− 65	110

In Table 2, apart from the maximum values τ_p^H and τ_p^D of the functions defined above, also the maximum values of the portions of $\tau_{13}^{H}(0, y)$ caused by the screw dislocations and edge dislocations only τ_{ps}^{H} and τ_{pE}^{H}, respectively, are given for the dislocation-poor regions. Additionally the peak values τ_d^H of $\tau_{13}^H(x, 0)$ in the dislocation-dense regions are listed in Table 2.

6. Discussion

The stress field connected with the polarization of the dislocations movable in the dislocation-poor regions (in the following termed polarization stress field) can be considered as a characteristic part of the specific internal stress field τ_i in fatigued metals and should be strongly correlated to the cyclic processes. The numerical calculations show (Fig. 2 and 4) that generally the polarization stress field $\tau_i^{(P)}$ can be divided into a half-macroscopic (mesoscopic) part $\tau_i^{(H)}$ which changes considerably in a cycle (with a spatial distance of its maximum values of the order of the dimensions of the dislocation-poor regions) and into a nearly cycle-independent part $\tau_i^{(D)}$ (with a spatial distance of its maximum values of the order of the mean distance of the dislocations in a group).

A comparison of the calculated maximum values of the mesoscopic stress τ_p^H with the outer applied stress amplitude in the saturation stage (Table 2, column 6) and with qualitative and quantitative statements on the internal stress amplitudes [2, 6, 7, 10] shows that the suppositions assumed for the numerical calculations lead to an overestimation of the stress values, at least for some specific types of configuration. The unrealistically large values for configurations with distances of the glide planes $R > 0.1$ µm (5AA2, 5AA3) suggest that slip in a HMV is more homogeneous.

The stress values τ_{pE}^H and τ_d^H caused by the edge dislocation segments are, on the whole, too large compared with the outer stress. This indicates that it seems necessary to assume, at least in part, an effective annihilation of the edge dislocation segment formed in a cycle. This means, slip through the centre of the dislocation-dense region must be possible at least to a certain degree as mentioned in earlier papers [6, 7, 10] and the extension of a HMV in the x-direction of Fig. 1 should not be limited in general by the first neighbouring dislocation-dense regions. The relatively large peak values τ_{pS}^H corresponding to the greater amplitudes ($\gamma_{pa} > 2 \times 10^{-3}$), e.g. for the configurations 5AA1 and 3AA1, lead to the conclusion that for these amplitudes an undisturbed shuttling without annihilation of screw dislocations should not be possible any longer.

The whole polarization stress field for a macroscopic volume results from the superposition of the stress fields of the dislocation arrangements in the single active mesoscopic volumes. Caused by the locally different extensions of the active HMV and the different degree of polarization in the HMV, a spectrum of peak values of the mesoscopic stress field arises in the volume of a fatigued specimen. The characteristic frequency distribution of peak values changes within the deformation cycle from cycle to cycle in the same way. It should be the alteration of this frequency distribution and of the mean amplitude of the mesoscopic stress field which is responsible for the characteristic changes of stress-dependent measuring quantities throughout a cycle.

The possibility of dividing the macroscopic volume into active mesoscopic volumes, within which a dislocation motion takes place, repeated in the same way from cycle to cycle, leads to the idea that the mechanical behaviour in a deformation cycle (the hysteresis loop) of the saturation stage may be described to a first approximation by the mechanical behaviour of a system of elastic-ideally plastic mesoscopic volumes with different yield points. As a matter of fact, quantitative (e.g. [18 to 20]) and qualitative [22] investigations show that important properties of stabilized hysteresis loops (Bauschinger effect, shape of inner loops etc.) can be accounted for in a simple way on the basis of such a concept as was originally suggested by Masing [21].

On the assumption that the fatigued volume of a specimen can be interpreted on a mesoscopic level as a plastically heterogeneous body in the sense of Masing, the frequency distributions of the residual stresses resulting after unloading from different points on the hysteresis loop were estimated in [23]. The mesoscopic stress field discussed in the present paper is closely related to the residual stress field determined from hysteresis loops in [23] and [24], respectively.

7. Summary

The dislocations movable in the dislocation-poor regions within a stabilized deformation cycle (saturation of the outer flow stress) are connected with a specific internal stress field. For the quantitative estimation of this stress field the macroscopic volume of a fatigued specimen was thought of as to consist of active mesoscopic volumes, within which a shuttling of dislocation groups occurs along the dislocation-dense multipole arrangements repeated in the same way from cycle to cycle. This shuttling leads within each half cycle to a "polarization" of the arrangements of movable dislocations. For the characterization of the "polarization" stress field correlated to the movable dislocations, the stress fields of specific idealized dislocation arrangements were calculated numerically. As the results show, the polarization stress field can be divided into two parts with different spatial distances of their peak values. The mesoscopic part with a spatial distance of its maximum values of the order of the dimensions of the dislocation-poor regions changes strongly within a cycle and should be responsible for the characteristic change of stress-dependent quantities which are measured within a cycle. The other part remains relatively unchanged during the cycle and is correlated approximately to the mean density of movable dislocations.

Comparing the calculated maximum values of the mesoscopic stress field with values of the internal stress field resulting from experimental investigations it seems to be necessary to take into account, at least in part, the accomodation of plastic strain in the dislocation-dense regions. For large strain amplitudes an undisturbed shuttling of the srew dislocations should not be possible.

References

[1] P. J. Woods, Phil. Mag. **28**, 155 (1973).
[2] H. Mughrabi, Proc. 3rd. Internat. Conf. Strength of Metals and Alloys, Cambridge 1973 (p. 407).
[3] P. Lukas and M. Klesnil, Mater. Sci. Engng. **11**, 345 (1973).
[4] K. Mecke, phys. stat. sol. (a) **25**, K93 (1974).
[5] J. C. Grosskreutz, phys. stat. sol. (b) **47**, 11 (1971).
[6] S. J. Basinski, Z. S. Basinski, and A. Howie, Phil. Mag. **19**, 899 (1969).
[7] J. C. Grosskreutz and H. Mughrabi, Constitutive Equations in Plasticity, Ed. A. S. Argon, HIT Press, Cambridge (Mass.)/London 1975 (p. 251).
[8] J. M. Finney and C. Laird, Phil. Mag. **31**, 339 (1977).
[9] D. Kuhlmann-Wilsdorf and C. Laird, Mater. Sci. Engng. **27**, 137 (1977).
[10] H. Mughrabi, Mater. Sci. Engng. **33**, 207 (1978).
[11] K. H. Weile, G. K. Schmidt, and H.-D. Schumann, Proc. 5th Conf. on Dimensioning and Strength Calcul., Vol. II, Budapest 1974 (p. 164).
[12] K.-H. Weile and G. K. Schmidt, phys. stat. sol. (a) **31**, K155 (1975).
[13] H.-D. Schumann, phys. stat. sol. (a) **18**, K27 (1973).
[14] K. Mecke, Thesis, PH Dresden, 1975.
[15] Ch. Blochwitz, K. Mecke, and R. Stephan, Kristall und Technik **13**, 851 (1978).
[16] T. Takeuchi, Phys. Soc. Japan **27**, 436 (1969).
[17] J. P. Hirth and J. Lothe, Theory of Dislocations, Mc Graw-Hill, New York St. Louis/San Francisco/Toronto/London/Sydney 1968.
[18] J. F. Besseling, J. appl. Mech. **25**, 612 (1958).

[19] W. D. IVAN, Trans. ASME Ser. E 88, 893 (1966).
[20] C. HOLSTE, Wiss. Z. PH Dresden 1, Beiheft 2, 1 (1975).
[21] G. MASING, Wiss. Veröff. Siemens-Werke 3, 231 (1923).
[22] J. BURBACH, Techn. Mitt. Krupp Forsch.-Ber. 28, 55 (1970).
[23] C. HOLSTE, Wiss. Z. PH Dresden 3, Beiheft 3, 1 (1977).
[24] H.-J. BURMEISTER, C. HOLSTE, and F. LANGE, Kristall und Technik **14**, K31 (1979).

(Received May 9, 1979)

phys. stat. sol. (a) 54, 315 (1979)

Subject classification: 13.4 and 14.3; 22.8

Sektion Physik der Bergakademie Freiberg

Deep Levels in $ZnSiP_2$ Determined by Schottky TSC Measurements

By

G. KÜHNEL, W. SIEGEL, and E. ZIEGLER

Deep impurity levels in low-resistivity n-type $ZnSiP_2$ single crystals are investigated by thermally stimulated currents in the temperature range from 77 to 300 K. In undoped crystals traps are detected with activation energies 0.34, 0.48, 0.54 eV and concentrations 5.6×10^{16}, 2.0×10^{17}, and 1.1×10^{17} cm^{-3}, respectively. For the dominating level with $\Delta E_T = 0.48$ eV the thermal capture cross section is estimated to $\sigma = 4 \times 10^{-17}$ cm^2. From different facts it can be concluded, that the dominating level is a trapping centre for holes. For the determination of the concentration of deep impurities by TSC measurements on Schottky barriers an exact relation is given.

In niederohmigen n-$ZnSiP_2$-Kristallen werden tiefe Störstellen mit Hilfe thermisch stimulierter Ströme im Temperaturintervall von 77 K bis Raumtemperatur untersucht. In undotierten Kristallen werden Störniveaus zu 0.34; 0.48 und 0,54 eV mit zugehörigen Konzentrationen von $5,6 \times 10^{16}$; $2,0 \times 10^{17}$ bzw. $1,1 \times 10^{17}$ cm^{-3} nachgewiesen. Für das dominierende Niveau mit $\Delta E_T = 0.48$ eV wird der thermische Einfangquerschnitt zu 4×10^{-17} cm^2 abgeschätzt. Verschiedene Fakten lassen darauf schließen, daß es sich zumindest bei dem dominierenden Niveau um ein Haftzentrum für Löcher handelt. Zur Bestimmung der Konzentration tiefer Störstellen aus TSC-Messungen an Schottkybarrieren wird eine exakte Beziehung angegeben.

1. Introduction

Imperfections connected with deep energy levels play an important role for the recombination of charge carriers in semiconductors. These centres act either as recombination or as trapping centres and can influence both luminescence properties and photocurrent. Radiative transitions via these levels cause luminescence at energies far from the gap energy or nonradiative transitions reduce the luminescence intensity.

There is only little information about deep level centres in $ZnSiP_2$. This information results from Hall effect and conductivity measurements on high-resistivity samples [1, 2] as well as from photoconductivity measurements [3, 4]. Besides, from the large relaxation time of the photocurrent (10^{-3} to 10^{-1} s) a high concentration of trapping centres may be deduced.

In this work we report about investigations of deep levels by the method of thermostimulated currents (TSC) in low-resistivity n-type $ZnSiP_2$ crystals. TSC spectra give a survey about deep levels and allow the estimation of their parameters (activation energy ΔE_T, thermal capture cross-section σ, and concentration N_T). On low-resistivity samples the TSC measurements are disturbed by the high dark current. However, the dark current can be eliminated, if the measurements are performed on space charge structures. Because pn junctions of $ZnSiP_2$ could not be realized up to now, we have used Ni contacts with nearly ideal Schottky behaviour [5].

2. Experimental

The investigations were performed with $ZnSiP_2$ crystals, grown in the Zn or Sn melt [6]. The Schottky contacts were produced by electroless deposition of Ni on (112) growth faces in aqueous solution [5]. The electron concentration of the samples varied from 3×10^{16} to 3×10^{17} cm^{-3} at room temperature. Ohmic contacts were made by welding of Au wires by means of a capacitor discharge directly on the $ZnSiP_2$ samples [7]. The TSC apparatus used permits measurements under vacuum between liquid nitrogen temperature and room temperature with a maximum heating rate of 2 Ks^{-1}. The heating characteristic is approximately linear. The temperature is measured with a copper–constantan thermocouple. A small copper block with heater serves as sample holder. The specimen is fixed with silicon grease. The thermocouple is soldered directly on the copper block near the specimen. The deep levels were charged by excitation with light. Incandescent and monochromatic intrinsic light yielded no differences in the TSC spectra.

3. Results

Fig. 1 shows the measured TSC spectrum of a Ni/n-type $ZnSiP_2$ Schottky contact, obtained without an external voltage. One can see a broad band consisting of the superposition of three peaks. For the dominating peak at $T_M = 240$ K the determination of the deep level parameters is possible, but the two other peaks at about 190 and 270 K allow only a rough evaluation of the activation energy and the concentration. The TSC signal of the initial part ($T < 160$ K) differs from the zero line, but an analysis was not performed.

The general equation for bulk TSC studies deduced e.g. in [8] can be modified for Schottky junctions. On principle this equation is suitable for the determination of the activation energy, the capture cross-section, and the concentration by fitting of the experimental TSC curve, if all peaks are well separated. In practice, however, it is more favourable to use special experimental methods based on simplifying assumptions for the determination of these parameters [8 to 10]. Moreover, these methods are suitable even in the case of overlapping peaks. The agreement of the results obtained with different methods is often used as a confirmation of the assumptions.

With the heating rate method one gets the activation energy ΔE_T and the capture cross-section σ according to equation (1) deduced by neglecting retrapping [8]:

$$\Delta E_T = kT_M \left(\ln \frac{T_M^2}{\beta} + \ln \frac{N_{c,v} \sigma v_{th} k}{\Delta E_T} \right), \tag{1}$$

where T_M is the temperature at the maximum of the TSC current, k the Boltzmann constant, β the heating rate, $N_{c,v}$ the density of states in the conduction band and the

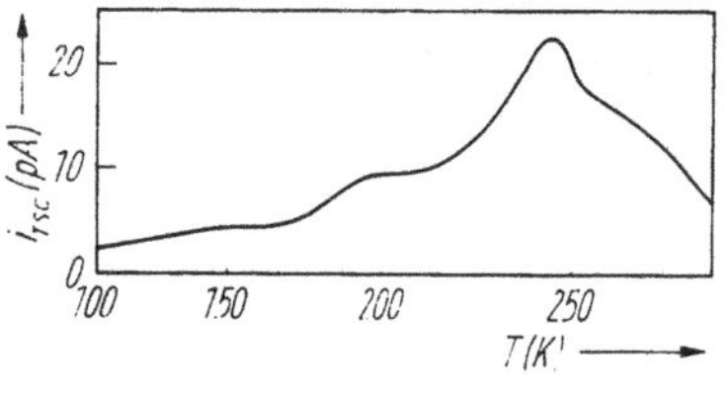

Fig. 1

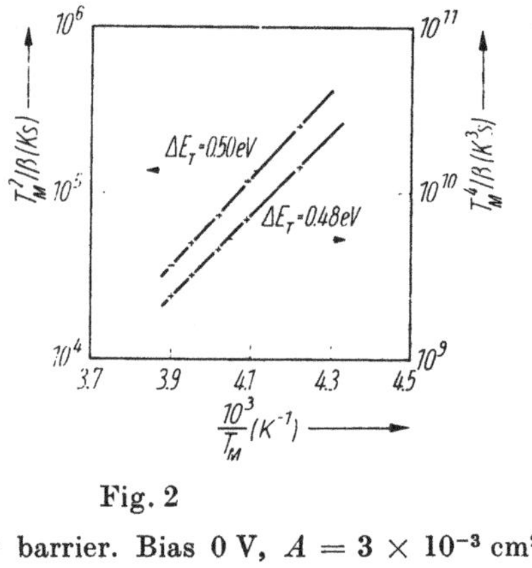

Fig. 2

Fig. 1. TSC spectrum of a Ni/n-type $ZnSiP_2$ Schottky barrier. Bias 0 V, $A = 3 \times 10^{-3}$ cm^2, $\beta = 0.85$ Ks^{-1}

Fig. 2. lg T_M^2/β and T_M^4/β, respectively, versus reciprocal temperature for the dominating trap valence band, respectively, and v_{th} the thermal velocity of the emitted carriers. The thermal capture cross-section σ is defined by the relation of the detailed balance given by Shockley and Read [11].

The heating rate was varied from 0.22 to 1.8 K s^{-1}. Only the level corresponding to $T_M = 240$ K was analysed, because the other peaks are not clearly defined. In Fig. 2 $\ln(T_M^2/\beta)$ is plotted against $1/T_M$. The activation energy of this level was determined as 0.50 eV. The temperature dependence of $N_{c,v}$, σ, and v_{th} was neglected in the first approximation because of the small change of T_M with heating rate. Regarding also the temperature dependence of $N_{c,v}$ and v_{th}, $\ln(T_M^4/\beta)$ must be plotted versus $1/T_M$ in order to determine the activation energy (see Fig. 2). The ΔE_T value deduced from this plot (0.48 eV) is nearly the same as that from the $\ln(T_M^2/\beta)-1/T$ plot. The difference of both ΔE_T values is of the order of the error of the activation energy.

The determination of the capture cross-section σ according to equation (1) is influenced strongly by errors of the activation energy. Therefore, the capture cross-section can be determined only approximately. We obtained $\sigma \approx 4 \times 10^{-17}$ cm^2 for the level at $T_M = 240$ K. This value is of the same order of magnitude as corresponding values in other semiconductors. The investigated temperature range is too small to conclude on the temperature dependence of σ.

The initial-rise method of Garlick and Gibson [12] gives ΔE_T values independent of retrapping (however, a more exact analysis [9] proves that this independence is not general, but is valid in most real cases):

$$I_{TSC} = \text{const} \cdot \exp\left(\frac{-\Delta E_T}{kT}\right) \quad \text{for} \quad I_{TSC} \ll I(T_M)\,. \tag{2}$$

In order to use the initial-rise method in our case of superimposed peaks we must make visible the initial increases of the different peaks in the TSC spectrum. This is possible with the cleaning procedure [13] (see Fig. 3a). From the $\ln i_{TSC}-1/T$ plot (see Fig. 3b) an activation energy $\Delta E_T = (0.46 \pm 0.04)$ eV was deduced for the peak at $T_M = 240$ K. In the same way the activation energy was determined for the level at $T_M = 190$ K: $\Delta E_T = (0.30 \pm 0.06)$ eV. For the peak at $T_M = 270$ K the cleaning procedure is not suitable as shown in Fig. 4.

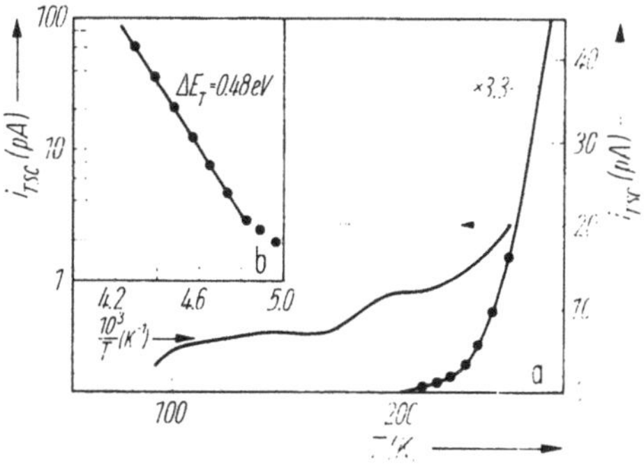

Fig. 3

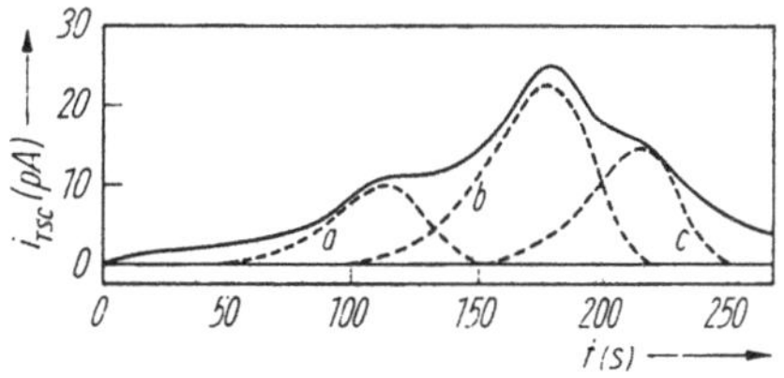

Fig. 4

Fig. 3. Determination of the activation energy of the 240 K peak by the initial-rise method. a) Cleaning of the initial part of TSC peak. b) lg $i_{TSC}-T^{-1}$ plot

Fig. 4. Decomposition of the measured i_{TSC}–time curve (full line) in separate curves (dashed line) for the three analysed traps. (a) $Q = 3.0 \times 10^9$, (b) 8×10^9, (c) 4.5×10^9 electrons

An approximation for the determination of ΔE_T is given by the $23kT_M$ rule [14, 15]. This rule is deduced assuming a standard capture cross-section and temperature independence of the other parameters. For the two energy levels determined with the other methods the $23kT_M$ relation yields $\Delta E_T = 0.47$ and 0.37 eV, respectively. Because these values agree reasonably with the energies calculated with the help of the heating rate and the initial-rise method, the energy of the level at $T_M = 270$ K, which is difficult to determine with other methods, is also estimated with this rule. One gets a value of 0.54 eV.

The concentration N_T of a deep level can be determined by TSC measurements on Schottky barriers in a simple way from the charge Q of the emitted carriers, but Q differs from the measured charge $Q_m = \int_0^\infty i_{TSC}\, dt$ because of the occurrence of the "displacement" current [16]. The measured charge Q_m is determined by the change of the total charge in the space charge region ($Q - Q_{displ}$) during the heating process. The Poisson equation was solved using the depletion approximation and a constant distance λ between the neutral bulk range and the crossing point of the Fermi energy with the deep level energy. This yields a relation between the measured charge Q_m and the concentration N_T for Schottky barriers under the condition that all levels are occupied

$$Q_m = Qa = AeN_T(\omega - \lambda)\, a\,, \tag{3}$$

where A is the junction area, e the electron charge, and ω the width of the space charge region in the unexcited state. The factor a has different values for majority and minority trapping centres.

For majority traps one obtains

$$a = \frac{1 + \frac{\lambda}{\omega} + \frac{\Delta}{\omega}}{2 + \frac{\Delta}{\omega}}\,, \tag{3a}$$

where Δ is the total change of the width of the space charge region during the thermal release of trapped carriers. Values between 0.5 and 1 result for a.

For minority traps

$$a = \frac{1 - \frac{\Delta}{\omega}}{2 - \frac{\Delta}{\omega}} \tag{3b}$$

is valid, which yields $a \leqq 0.5$. For most of the practical cases ($N_T/N_{Shallow} \leqq 1$) $a = 0.5$ is a good approximation (with an error $\leqq 20$ %).

Equation (3) can be used in an analogous manner for TSC measurements in pn junctions.

The retrapping of the thermally emitted carriers has no influence on the estimation of the concentration in space charge structures.

The concentration of the deep level centres was determined as follows: By variation of the excitation conditions a saturation of the TSC signal could be achieved. This confirms that all traps could be occupied. In order to avoid errors in the integration of the TSC curve due to a heating rate which is not exactly linear the i_{TSC}–time curve was plotted. Because the peaks are superimposed the measured TSC curve was decomposed into separate curves (Fig. 4). For this purpose TSC curves were calculated for the three levels (without retrapping) and fitted to the measured TSC curve. For this calculation the activation energies and capture cross-sections determined above have been used.

We assume, that our samples contain hole trapping centres (see Section 4). In this case $\lambda = 0$ is valid and a value 0.5 is used for a. The space charge width ω was obtained from $C-V$ measurements.

In this way concentrations of $N_T = 5.6 \times 10^{16}$, 2.0×10^{17}, 1.1×10^{17} cm^{-3} for the levels at $T_M = 190$, 240, 270 K, respectively, have been estimated.

4. Discussion

In TSC measurements on Schottky contacts or pn junctions retrapping is always neglected [17]: By reason of the high field strength in the space charge region ($\geqq 10^4$ V cm^{-1} at usual impurity concentrations) the emitted charge carriers are transported away in an extremely short time ($t < 10^{-11}$ s) and, therefore, the portion of retrapped carriers in space charge structures is negligibly small. This explains why the results obtained with the heating rate method which neglects retrapping agree sufficiently with those of the other methods.

TSC measurements yield no direct information whether electrons or holes are trapped by the deep levels. However, from the following we conclude that at least the level with the highest concentration ($T_M = 240$ K) is a trap centre for holes, because

(i) excitation with bias was not successful. This is clear if one takes into consideration that in forward biased Schottky contacts only majority carriers are injected. On the contrary the excitation with intrinsic illumination which created electron-hole pairs yields a TSC signal in our n-type samples;

(ii) traps of minority carriers with high concentration cause high relaxation times for the photoconductivity (see e.g. [18]) observed also in these $ZnSiP_2$ samples;

(iii) a clear frequency dispersion of the space charge capacity, connected with electron trapping centres was not detected.

It is remarkable, that in high-resistivity undoped $ZnSiP_2$ samples a deep acceptor with $E_A = 0.45$ eV was found by means of electrical measurements [1].

The step at $\hbar\omega = 1.6$ eV in the spectral response of photoconductivity [3] indicates also the existence of deep acceptors in low-resistivity n-type $ZnSiP_2$ samples with an energy similar to that of the 240 K peak. The concentration of compensating acceptors in low-resistivity samples [19], determined by Hall effect measurements, is of the same magnitude as the concentration of the deep levels, determined in this work. We assume, that the compensating acceptors in low-resistivity n-type $ZnSiP_2$ are deep acceptors and the same centres which cause the TSC signal. Hall effect measurements, however, do not provide information on the energy levels of the compensating acceptors.

The results of investigations of the recombination process in $ZnSiP_2$ [4] are not in full agreement with our results: imperfection levels with acceptor behaviour also determine essentially the recombination process of the non-steady-state charge carriers, however, the energetic levels differ from those found by us.

The deep level centre, according to $T_M = 240$ K, was observed in all investigated low-resistivity samples, and in high-resistivity crystals grown by different methods, too. The origin and the chemical nature of the investigated centres are unknown.

Our results show that deep impurity centres with large concentration exist in undoped $ZnSiP_2$ crystals in difference to GaP, the isoelectronic analogue. This fact can explain the low luminescence efficiency of $ZnSiP_2$ crystals [20] in contrast to GaP.

Acknowledgements

The single crystal growth of Dr. E. Buhrig and Dr. M. John, Sektion MWT der Bergakademie Freiberg and the helpful discussions of Dr. G. Richter, Zentralinstitut für Elektronenphysik Berlin are gratefully acknowledged.

References

[1] F. GÜNTHER, J. MONECKE, and H. A. SCHNEIDER, Freiberger Forschungsh. **B 207** (1979).
[2] I. S. GORBAN, G. A. GRISHCHENKO, A. P. SAKALAS, and I. I. TYCHINA, phys. stat. sol. (a) **35**, 571 (1976).
[3] G. KÜHNEL, W. SIEGEL, and E. ZIEGLER, phys. stat. sol. (a) **30**, K25 (1975).
[4] G. A. GRISHCHENKO, A. V. LJUBCHENKO, and I. I. TYCHINA, Ukr. fiz. Zh. **21**, 303 (1976).
[5] G. KÜHNEL, W. SIEGEL, and E. ZIEGLER, phys. stat. sol. (a) **49**, 705 (1978).
[6] H. BECHERER, E. BUHRIG, K. HEIN, M. JOHN, P. KIRSTEN, H. A. SCHNEIDER, W. SIEGEL, and K. WINKLER, Kristall und Technik **13**, 1053 (1978).
[7] W. SIEGEL, G. KÜHNEL, and E. ZIEGLER, Exper. Tech. Phys. **27**, 79 (1979).
[8] A. G. MILNES, Deep Impurities in Semiconductors, Chap. 9, J. Willey & Sons New York 1973.
[9] P. BRÄUNLICH, Invited paper presented at the Adv. Res. Inst. on Application of TL to Geological Problems, Spoleto (Italy), 5 to 16th Sept. 1966.
[10] R. STUCK, I. C. MULLER, J. P. PONPON, C. SCHARAGER, C. SCHWAB, and P. SIFFERT, J. appl. Phys. **47**, 1545 (1976).
[11] W. SHOCKLEY and W. T. READ, Phys. Rev. **87**, 835 (1952).
[12] G. F. J. GARLICK and A. F. GIBSON, Proc. Phys. Soc. **81**, 141 (1948).
[13] W. HOOGENSTRATEN, Philips Res. Rep. **13**, 515 (1958).
[14] H. SCHADE, J. appl. Phys. **40**, 2613 (1969).
[15] H.-J. DITTFELD and J. VOIGT, phys. stat. sol. **3**, 1941 (1963).
[16] C. T. SAH, L. FORBES, L. L. ROSIER, and A. F. TASCH, JR., Solid State Electronics **13**, 759 (1970).
[17] B. L. SMITH and M. A. CARTER, J. Phys. D 8, 254 (1975).
[18] J. AUTH, D. GENZOW, and K. H. HERRMANN, Photoelektrische Erscheinungen, Chap. 2, Akademie-Verlag, Berlin 1977.
[19] W. SIEGEL and E. ZIEGLER, phys. stat. sol. (a) **21**, 639 (1974).
[20] P. KIRSTEN, private communication.

(Received April 4, 1979)

phys. stat. sol. (a) 54, 321 (1979)

Subject classification: 14.2; 10.1; 21

Zentralinstitut für Festkörperphysik und Werkstofforschung der Akademie der Wissenschaften der DDR, Dresden

Pinning Mechanism in Superconducting NbTi Alloys at High Magnetic Fields

By

G. Fuchs

For the field dependence of the volume pinning force $F(H)$ of superconducting NbTi alloys a positive curvature is found at high magnetic fields. It is shown that the experimental $F(H)$ curve can be described in the high-field range by a model of plastic deformations of the flux line lattic proposed by Schmucker, when line pinning is assumed. The line pinning hypothesis is discussed.

Für die Feldabhängigkeit der Volumenpinningkraft $F(H)$ supraleitender NbTi-Proben wird bei hohen magnetischen Feldern eine positive Krümmung gefunden. Es wird gezeigt, daß die experimentellen $F(H)$-Kurven im Bereich hoher magnetischer Felder durch ein Modell plastischer Deformationen des Flußschlauchgitters das von Schmucker vorgeschlagen wurde, beschrieben werden können, wenn Linienpinning vorausgesetzt wird. Die Linienpinninghypothese wird diskutiert.

1. Introduction

The essential quantity of a high-field superconductor is the critical current density j_c in the transverse magnetic field H or the volume pinning force

$$F = j_c H \tag{1}$$

which is a measure for the interaction between the flux-line lattice (FLL) and the pinning centres of the material.

The pinning mechanism in type II superconductors is not clearly understood so far. Most pinning models are based on pin breaking, i.e., the pinning force is calculated by appropriate summation of the elementary pinning force f_p of the interaction between a single pinning centre and the FLL. Several authors [1 to 4] discussed the possibility of flux motion occurring by plastic flow of the FLL around strong pins rather than by pin breaking. This mechanism corresponds to the motion of flux-spots in type I superconductors which has been proved by Träuble and Essmann [5]. The question, whether this mechanism also occurs in high-field superconductors is of great practical interest because in this case the pinning force is independent of the elementary pinning force f_p.

In pinning models of plastic deformations of the FLL the field dependence of the pinning force $F(h)$ contains in the case of line pinning a term $(1 - h)^2$ [2, 4], where h is the reduced magnetic field $h = H/H_{c2}$. Therefore, the mechanism of plastic deformations of the FLL offers a possible explanation for the positive curvature of $F(h)$ at high magnetic fields found often for type-II superconductors [6 to 8], especially for high-field superconductors as Nb_3Sn [9] and Nb–50 wt% Ti [8].

In a preceding paper [10] NbTi alloys in the range of Ti concentrations between 44 and 66 wt% were investigated. Some experimental results in the field range close to H_{c2} which was not considered in this paper indicate a positive curvature of $F(h)$ at high magnetic fields. In order to decide this question unequivocally, in the present paper the critical current of these samples was measured thoroughly in the range of high magnetic fields varying the temperature between 1.5 and 4.2 K.

The experimental results will be compared with pinning models of plastic deformations of the FLL.

2. Measurements and Results

The measurements of the critical current density on NbTi single-wire samples of various composition reported in a preceding paper [10] were extended to the field range close to H_{c2}. The critical current density was measured in magnetic fields up to 15 T and in the temperature range from 1.5 to 4.2 K.

A criterion of 10^{-6} V across a sample length of about 1 mm was used to define the critical current.

The critical current was measured to low values increasing the magnetic field in steps of $\Delta H = 0.25$ T.

In Fig. 1 the pinning force is plotted as a function of magnetic field. The most important result of the measurements is the positive curvature of the $j_c(H)$ and $F(H)$ curves at high magnetic fields. This result is in agreement with measurements of the critical current density on Nb–50 wt% Ti by an ac technique [8].

By plotting of $F^{1/2}$ versus H straight lines result in the range of high magnetic fields as shown in Fig. 2. The upper critical field H_{c2} was defined by extrapolation of the $F^{1/2}(H)$ straight lines to $F = 0$. This definition of H_{c2} is similar to that used by Montgomery and Sampson [11] and Shapira and Neuringer [12].

Fig. 3 shows that most values of H_{c2} of the present paper are close to those reported by Suenaga and Ralls [13] for homogeneous samples.

It is obvious that the curvature of $F(h)$ at high magnetic fields and the determined value of H_{c2} depend on the choice of the extrapolation procedure when the extrapolation is started from relatively high values of the critical current. Hampshire and Taylor [14], for example, found by linear extrapolation of $j_c(H)$ to $j_c = 0$ leading to a negative curvature of $F(h)$ at high magnetic fields, a value of $\mu_0 H_{c2} \approx 10.5$ T for Nb–44 wt% Ti samples. This value is by about 1 T smaller than that of the homogeneous sample (see Fig. 3). In [10] the experimental $F(H)$ curves were extrapolated by using simple functions $f^{\text{eff}}(H)$ which describe the $F(H)$ curves in the range of medium fields. The field obtained by extrapolation of the experimental $F(H)$ curves to $F = 0$ with these functions was denoted as H_{c2}^{eff}. The experimental data have been written

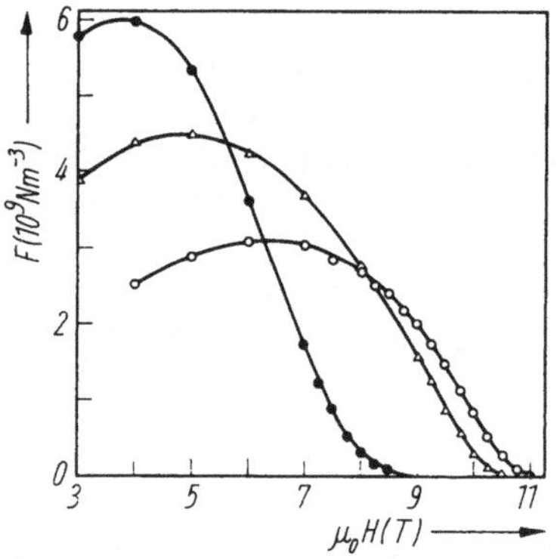

Fig. 1

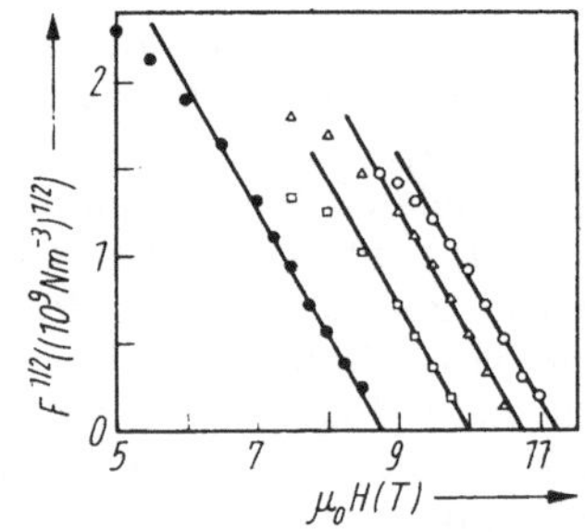

Fig. 2

Fig. 1. Volume pinning force as a function of magnetic field $T = 4.2$ K. Samples No. 1 (△), 14 (●), and 17 (○)

Fig. 2. $F^{1/2}$ as a function of H at $T = 4.2$ K. Samples No. 1 (△), 14 (●), 17 (○), and 27 (□)

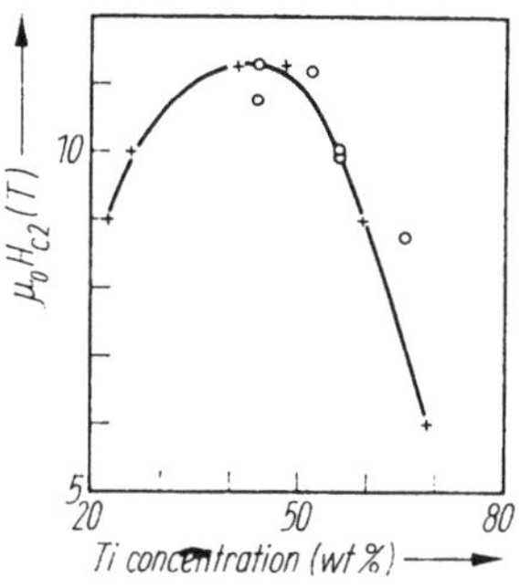

Fig. 3

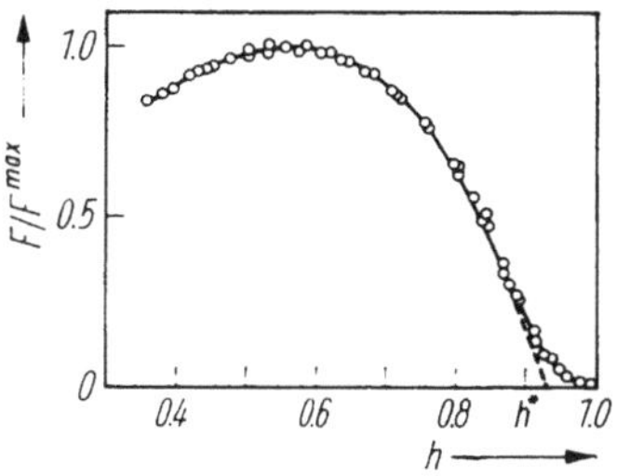

Fig. 4

Fig. 3. Upper critical field as a function of Ti concentration at $T = 4.2$ K. + data from [14] for homogeneous samples, ○ samples from this paper

Fig. 4. Reduced volume pinning force as a function of reduced magnetic field in the temperature range 1.5 K $\leqq T \leqq$ 4.2 K for sample 17

in the form of a scaling law

$$F(h^{\text{eff}}, T) = f^{\text{eff}}(h^{\text{eff}}) \, [H_{c2}^{\text{eff}}(T)]^n \tag{2}$$

with $h^{\text{eff}} = H/H_{c2}^{\text{eff}}$.

Fig. 4 shows that by plotting $F/F^{\max}$ versus $h = H/H_{c2}$ a universal curve results in the measured temperature range. The corresponding scaling law

$$F(h, T) = f(h) \, [H_{c2}(T)]^n \tag{3}$$

has the same exponent n as (3), since the ratio $h^* = H_{c2}^{\text{eff}}/H_{c2}$ is temperature independent (see Fig. 4).

In Table 1 the values of n, h^* and H_{c2} are given.

Table 1

sample No.	Ti content (wt %)	H_{c2} (4.2 K) (T)	$H_{c2}^{\text{eff}}/H_{c2}$	n	ϱ (10^{14} m^{-2})	$\varrho^{-1/2}$ (10^{-8} m)
1	44	10.75	0.92	2.0	5.33	4.33
4	44	11.3	0.94	2.21	6.07	4.06
17	52	11.2	0.93	2.33	5.29	4.35
16	56	9.95	0.91	2.15	4.45	4.74
27	56	10.0	0.90	2.18	4.86	4.54
14	66	8.75	0.85	2.0	4.68	4.62

3. Interpretation of the Experimental Results

A possible explanation for the positive curvature of the experimental $F(h)$ curves at high magnetic fields would be that in NbTi plastic deformations of the FLL and line pinning occur [2, 4].

In this section the applicability of two models of plastic deformations of the FLL on the measured NbTi samples will be tested.

3.1 Kramer's model of synchronous shearing of the FLL

Kramer [2] examined a model of unbreakable line pins lying in planes separated by a distance $\varrho^{-1/2}$ parallel to the Lorentz force. Replacing the expression [15]

$$\tau_{\max} = \frac{c_{66}}{2\pi}$$

used by Kramer for the theoretical shear strength of the FLL by the correct expression [16]

$$\tau_{\max} = \frac{c_{66}}{30},$$

c_{66} being the shear modulus of the FLL [17], the model yields

$$F(h) = 1.38 \times 10^{-4} \frac{\mu_0^{3/2} H_{c2}^{5/2}}{\Phi_0^{1/2} \varkappa^2} \frac{1}{(1 - a_0 \varrho^{1/2})^2} h^{1/2} (1 - h)^2 . \quad (4)$$

In (3) $\varkappa$ is the Ginzburg-Landau parameter and a_0 the flux line spacing.

The magnitude of the pinning force (4) in the upper limit, that is for $\varrho^{-1/2} = 2a_0$ with $\varkappa = 40$ [14] and $\mu_0 H_{c2} = 11$ T, is by one order of magnitude too small compared with the experimental values.

3.2 Schmucker's model of plastic flow of the FLL

Schmucker [4] calculated the pinning force in the case of plastic flow of the FLL around model pinning forces

$$f = f_0 \, e^{-(x^2+y^2)^{1/2}/w - |z|/l} \quad (5)$$

acting in x-direction, while the flux lines are parallel to the z-axis. f_0 is the amplitude of the pinning force and l and w are the extensions of the model pinning centres parallel and perpendicular to the flux lines, respectively.

Schmucker approximately obtained

$$F \approx \varrho_v \tau_F w l \left[0.93 + \left(\frac{c_{44}}{c_{66}} \right)^{1/2} \frac{w}{l} \right], \quad (6)$$

where ϱ_v is the volume density of model pinning centres, τ_F the flow stress, and c_{44} the tilting modulus of the FLL. Assuming line pins of the sample diameter length and of the flux line core radius, i.e. $l = 10^{-4}$ m and $w = \xi$ and using the expressions

$$c_{66} = 8.9 \times 10^{-2} \mu_0 \left(\frac{H_{c2}}{\varkappa} \right)^2 (1 - h)^2 \quad (7)$$

and

$$c_{44} = \mu_0 H_{c2}^2 h^2 \quad (8)$$

for the elastic moduli of the FLL [17] one gets

$$\left(\frac{c_{44}}{c_{66}} \right)^{1/2} \frac{w}{l} \approx 0.01$$

and (6) reduces to

$$F \approx 2.14 \times 10^2 \varrho \xi \tau_F . \quad (9)$$

Here ϱ is the number of line pins per m².

According to Schmucker [4, 18] τ_F is approximately given by

$$\tau_F \approx \frac{c_{66}}{50}. \tag{10}$$

Using (7) and (10) one gets from (9)

$$F(h) \approx 0.38\varrho\xi\mu_0 \left(\frac{H_{c2}}{\varkappa}\right)^2 (1-h)^2 . \tag{11}$$

The field dependence of (11) corresponds to the straight lines of Fig. 2, where $F^{1/2}$ is plotted versus H. Values of ϱ resulting from the slope of these straight lines are given in Table 1. It is interesting to note that the density of line pins is similar in all the measured samples.

It can be concluded that Schmucker's model of plastic flow of the FLL in the line pinning limit offers a possible explanation for the $F(h)$ results measured on NbTi samples at high magnetic fields. The assumption of line pinning in NbTi will be discussed in detail in the following section.

4. Line Pinning in NbTi

Discussing the line pinning hypothesis it must be taken into consideration that the question of point or line pinning cannot be reduced to the geometry of a single pin. Line pinning, that is a parallel displacement of the flux lines from their undisturbed positions, is not only produced by line pins but also by pins of any geometry when the space d between two net pins on a flux line is sufficiently small. This has been shown by Kramer [2] who approximately calculated the shape of a flux line pinned by two point forces. Neglecting compressional deformations of the FLL Kramer found as criterion for line pinning

$$d \ll l^* \quad \text{with} \quad l^* = a_0 \left(\frac{c_{44}}{c_{66}}\right)^{1/2}. \tag{12}$$

For NbTi one gets, for example, $l^* = 2 \times 10^{-6}$ m at $h = 0.4$ and $l^* = 8.7 \times 10^{-6}$ m at $h = 0.8$ using (7) and (8) and the values $\mu_0 H_{c2} = 11$ T and $\varkappa = 40$.

In Nb–44 wt% Ti pinning is caused by a dislocation cell structure [14, 19, 20] and in the Ti-rich samples ($\geqq 50$ wt% Ti) the pinning effect is attributed mainly to normally conducting α-Ti precipitations [21 to 23]. Typical values of the cell size and the number of α-Ti precipitations per m² are $D = 5 \times 10^{-8}$ m [20] and $\varrho = 10^{14}$ to 10^{15} m^{-2} [22, 23], respectively. As in both cases the spacing between neighbouring pins is about 5×10^{-8} m, the line pinning criterion (12) would be satisfied even for relatively weakly pinned flux lines with a small number of net pins.

A more realistic criterion for line pinning which includes compressional deformations of the FLL can be derived on the basis of the calculation of the displacement field $\mathbf{s}$ of the FLL near a model pinning centre (5) reported by Schmucker [4].

In the case of a single pin which has been considered by Schmucker line pinning results for $l \gg w$. In order to estimate the critical space of point-shaped pins ($l = w$) leading to line pinning, a flux line with the initial coordinates (0, 0) is considered which is pinned at $z = 0$ and $z = d$. The shape of the deformed flux line results by superposition of the components s_x of the displacement fields $\mathbf{s}$ of the FLL due to the two pins. The bowing of the flux line can be expressed by

$$\frac{\delta s}{s} = 1 - \frac{2s_x(d/2)}{s_x(0) + s_x(d)}, \tag{13}$$

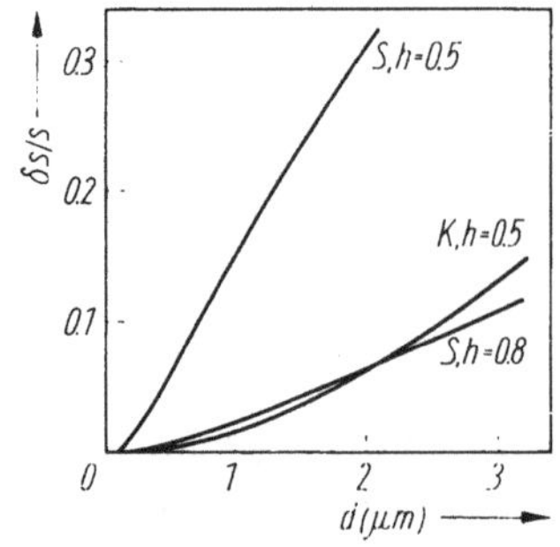

Fig. 5. Parameter $\delta s/s$ of bowing of a flux-line pinned at $z = 0$ and $z = d$ as a function of d at constant reduced magnetic field according to Kramer (K) and Schmucker (S), respectively

where $s_x(z)$ according to Schmucker is given by

$$s_x(z) = \frac{f_0}{2c_{44}lw} \int_0^\infty \frac{1/a_1\, e^{-a_1k|z|} - kl\, e^{-|z|/l}}{(1/w^2 + k^2)\,(1/l^2 - a_1^2k^2)}\, dk \tag{14}$$

with $a_1 = (c_{66}/c_{44})^{1/2}$.

In Fig. 5 $\delta s/s$ is plotted as a function of the pin spacing for point-shaped model pins with $l = w = \xi = 5 \times 10^{-9}$ m. Fig. 5 shows that the bowing of the flux line is greater than expected from the model considered by Kramer. Nevertheless, line pinning is expected at least in the range of high magnetic fields, where the line pinning criterion

$$\frac{\delta s}{s} \ll 1$$

is satisfied for most flux lines. At medium magnetic fields only relatively strong line pins are expected to be in the line pinning limit. Assuming a spacing between neighbouring pins of $d_0 = 5 \times 10^{-8}$ m and setting $\delta s/s \leq 0.01$ line pinning would result for flux lines for which the spacing d between net pins is given by $d \lessapprox 3d_0$.

5. Conclusions

Measuring the pinning force close to H_{c2} a positive curvature of the field dependence of the pinning force $F(h)$ at high magnetic fields was found for NbTi samples. It is shown that the experimental $F(h)$ curves in the field range $0.85 \lessapprox h \leq 1$ can be described by a model of plastic flow of the FLL proposed by Schmucker, when line pinning is assumed. A criterion for line pinning is given which is based on the calculation of the displacement field of the FLL due to a model pinning centre reported by Schmucker. Applying this criterion to NbTi it is shown that the assumption of line pinning is justified for NbTi at least at high magnetic fields.

The density of line pins resulting from a comparison of the experimental data with Schmucker's model corresponds to a mean space of line pins of $\varrho^{-1/2} \approx 4.5 \times 10^{-8}$ m and is of the order of magnitude of the cell size and the spacing between two α-Ti precipitations, respectively. Since the density of line pins increases with the density of pins in the material, it may be concluded that in NbTi alloys the pinning force at high magnetic fields would increase by a refinement of the cell structure and a higher density of α-Ti precipitations, respectively.

Acknowledgements

The measurements were performed in the International Laboratory of High Magnetic Fields and Low Temperatures in Wrocław. The author wishes to thank Dr. C. Bazan and the colleagues from the International Laboratory for their help in the measurements.

I am grateful to Dr. A. Gladun for his reading and comments on the manuscript and to Dipl.-Phys. A. Möbius for help in the numerical calculations.

References

[1] H. T. Coffey, Phys. Rev. **166**, 447 (1968).
[2] E. J. Kramer, J. appl. Phys. **44**, 1360 (1973).
[3] M. I. Turchinskaya and A. L. Roitburd, Soviet Phys. — J. exper. theor. Phys., Letters **20**, 79 (1974).
[4] R. Schmucker, phys. stat. sol. (b) **80**, 89 (1977).
[5] H. Träuble and U. Essmann, phys. stat. sol. **25**, 395 (1968).
[6] W. A. Fietz and W. W. Webb, Phys. Rev. **178**, 657 (1969).
[7] R. E. Enstrom and J. R. Appert, J. appl. Phys. **43**, 1915 (1972).
[8] H. Küpfer, Phil. Mag. **36**, 859 (1977).
[9] T. Luhman, C. S. Pande, and D. Dew-Hughes, J. appl. Phys. **47**, 1459 (1976).
[10] G. Fuchs, phys. stat. sol. (a) **45**, 239 (1978).
[11] D. B. Montgomery and W. Sampson, Appl. Phys. Letters **6**, 108 (1965).
[12] Y. Shapira and L. J. Neuringer, Phys. Rev. **140**, A 163 (1965).
[13] M. Suenaga and K. M. Ralls, J. appl. Phys. **40**, 4457 (1969).
[14] R. G. Hampshire and M. T. Taylor, J. Phys. F **2**, 89 (1972).
[15] J. Frenkel, Z. Phys. **37**, 572 (1926).
[16] A. Seeger, Hdb. Phys. Vol. VII/2, Springer-Verlag, 1958 (p. 1).
[17] R. Labusch, phys. stat. sol. **32**, 439 (1969).
[18] R. Schmucker, Phil. Mag. **35**, 431 (1977).
[19] C. Baker, J. Mat. Sci. **5**, 40 (1970).
[20] D. F. Neal, A. C. Barber, A. Woolcock, and J. A. F. Gidley, Acta metall. **19**, 143 (1971).
[21] I. Pfeiffer and H. Hillmann, Acta metall. **16**, 1429 (1968).
[22] H. Hillmann, Siemens Forsch.- u. Entwickl.-Ber. **3**, 197 (1974).
[23] J. Willbrand and W. Schlump, Z. Metall. **66**, 714 (1975).

(Received April 9, 1979)

phys. stat. sol. (a) 54, 477 (1979)

Subject classification: 13.4 and 20.3; 22.8.1

Solid State Department, Physical Laboratory, Utrecht[1])

The Luminescence of Calcium Molybdate

By

J. A. GROENINK, C. HAKFOORT, and G. BLASSE

The luminescence properties of calcium molybdate are reported. Two emission bands are observed, one in the green and another in the orange spectral region. The shorter wavelength emission band is assigned to the $^3T_1 \rightarrow {}^1A_1$ transition in the tetrahedral molybdate group. The longer wavelength band is ascribed to a similar transition in a molybdate group lacking an oxygen ion, i.e. a MoO_3 group. Similar observations are made in the case of calcium tungstate.

Es werden die Lumineszenzeigenschaften von Kalziummolybdat mitgeteilt. Zwei Emissionsbanden werden beobachtet, eine im grünen, die andere im orangen Spektralbereich. Die Emissionsbande mit der kürzeren Wellenlänge wird dem $^3T_1 \rightarrow {}^1A_1$-Übergang in der tetraedrischen Molybdatgruppe zugeordnet. Die Bande mit der längeren Wellenlänge wird einem ähnlichen Übergang in einer Molybdatgruppe, in der ein Sauerstoffion fehlt, d. h. einer MoO_3-Gruppe, zugeschrieben. Ähnliche Beobachtungen werden für den Fall des Kalziumwolframats gemacht.

1. Introduction

After the early work of Kröger [1] and Botden [2] there has been a renewed interest in the molybdate and tungstate luminescence of scheelites during recent years. Beard et al. [3] reported on the temperature dependence of the quantum efficiency of $CaWO_4$ and $CdWO_4$. Van Loo [4, 5] discussed the luminescence of $PbWO_4$ and $PbMoO_4$. Treadaway and Powell [6, 7] and Grasser et al. [8, 9] have studied the $CaWO_4$ luminescence in detail. Blasse and Schipper [10] reported on the luminescence of $SrWO_4$ and $BaWO_4$. Recently Tyner and Drickamer [11] presented a study on the pressure and temperature dependence of the decay time and the intensity of the luminescence of several molybdates and tungstates.

Usually one emission band is found upon excitation into the optical band gap. In $CaWO_4$ [7, 9], $PbWO_4$ [4, 5], $SrWO_4$ [10], and $BaWO_4$ [10] an additional emission band at longer wavelengths was found upon excitation into the tail of the band gap. Such an additional emission band has not been reported for the molybdates, as far as we know.

Although the tungstates and the vanadates have been studied extensively, there has not been great interest in the molybdates. This work was undertaken to elucidate the luminescence of $CaMoO_4$, and especially the decay time of the emission, for this has been reported to be very long [12]. Further we report some additional observations on the luminescence of $CaWO_4$.

2. Experimental

Samples of doped and undoped $CaMoO_4$ were prepared by firing intimate mixtures of $CaCO_3$, MoO_3, Na_2MoO_4, and Y_2O_3 for 4 h at 850 °C in air, followed by milling and firing for 4 h at 1000 °C in air. The compositions of the starting mixtures for the Y- and Na-doped samples were $CaCO_3 + (1 + \delta)\, MoO_3 + \delta\, Y_2O_3$ ($\delta = 0.0025$) and $CaCO_3 + (1 + \delta)\, MoO_3 + \delta\, Na_2MoO_4$ ($\delta = 0.0025$), respectively. Samples of doped

[1]) P.O. Box 80000, 3508 TA Utrecht, The Netherlands.

and undoped $CaWO_4$ were prepared by firing intimate mixtures of $CaCO_3$, WO_3, Na_2CO_3, and Y_2O_3 for 12 h at 950 °C in air, followed by milling and firing for 12 h at 1000 °C in air. Starting mixtures for the Y- and Na-doped samples were $CaCO_3 + (1 + 3\delta)\, WO_3 + \delta Y_2O_3$ ($\delta = 0.01$ and 0.0025) and $CaCO_3 + (1 + 2\delta)\, WO_3 + \delta Na_2CO_3$ ($\delta = 0.00375$, 0.0125, and 0.025), respectively.

The compounds were checked by X-ray analysis on a Philips diffractometer using the CuKα radiation. All samples appeared to be single-phased.

Emission and excitation spectra of their luminescence were recorded either on a Perkin-Elmer MPF-2A spectrofluorimeter, equipped with an Oxford Instruments CF 100 continuous flow He cryostat or on a Perkin-Elmer MPF-3L spectrofluorimeter, equipped with a cryoson XL2 liquid-nitrogen cryostat. Excitation spectra were taken with a combination of a Schott UG5 filter and a cuvet with a $NiSO_4$ solution in the excitation beam; in the emission beam Balzers K_1 to K_6 interference filters were used together with the standard mounted set of UV cut-off filters. Emission spectra were obtained using Baird Atomic broad-band interference filters with a maximum transmission at the excitation wavelength; in the emission beam the standard mounted set of UV cut-off filters was used.

The excitation spectra were corrected for the lamp intensity and the UG5-$NiSO_4$ filter using lumogen as a standard. The emission spectra were corrected for the sensitivity of the photomultiplier according to the curve given by the manufacturer.

The performance of the luminescence decay measurements has been described in detail by van Loo and Wolterink [13]. As a light source we used an EG & G Xe flash lamp; excitation and emission wavelengths were selected with the set of filters mentioned above and a filter which cuts off radiation above 2.03 eV.

The temperature dependence of the emission intensity was measured on the MPF-2A spectrofluorimeter from LHeT to 300 K and on the MPF-3L spectrofluorimeter from 100 to 430 K.

3. Results

All the samples under examination were white powders. The $CaMoO_4$ samples show a green luminescence under short wavelength (4.0 to 5.0 eV) UV excitation. Under 3.9 eV excitation, undoped $CaMoO_4$ and $CaMoO_4$ doped with Na show an orange emission, while in the case of $CaMoO_4$ doped with Y the maximum of the emission band is in between these two (Fig. 1). Doping with Na or Y does not introduce new emission bands, but increases or decreases the intensity of the orange emission, respectively. The emission spectra consist of a broad band, the halfwidth of which is about 0.45 eV in the case of the green emission, and about 0.35 eV in the case of the orange emission.

Excitation spectra of the luminescence of undoped $CaMoO_4$ are presented in Fig. 2. Recording the excitation spectrum of the green emission (at 2.58 eV) yields a broad

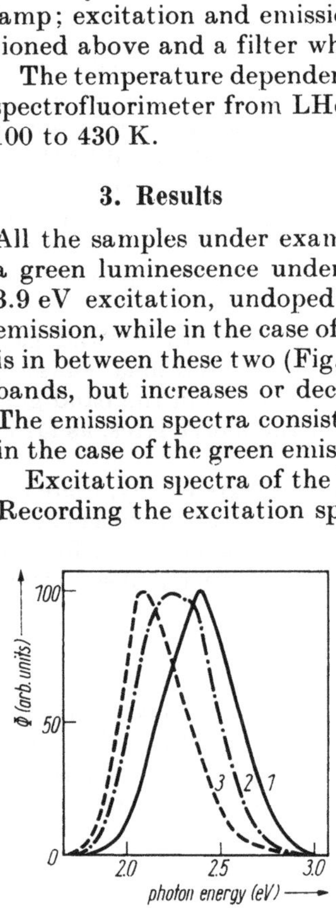

Fig. 1. Spectral energy distribution of the emission of several $CaMoO_4$ samples at 77 K. Φ denotes the spectral radiant power per unit energy. (1) Undoped, Na- and Y-doped, $E_{exc} = 4.51$ eV; (2) Y-doped, $E_{exc} = 3.85$ eV; (3) undoped and Na-doped, $E_{exc} = 3.85$ eV

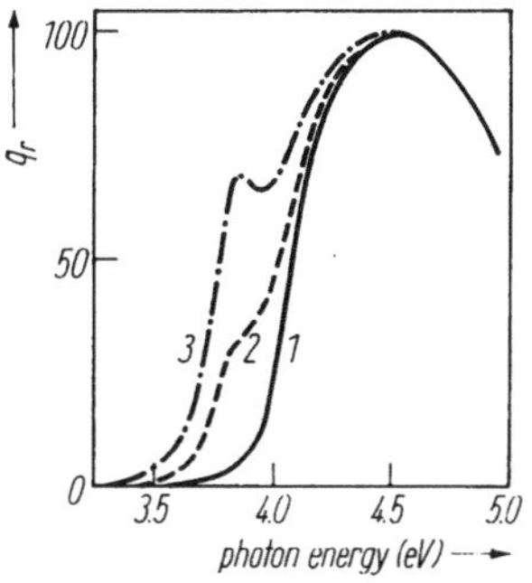

Fig. 2

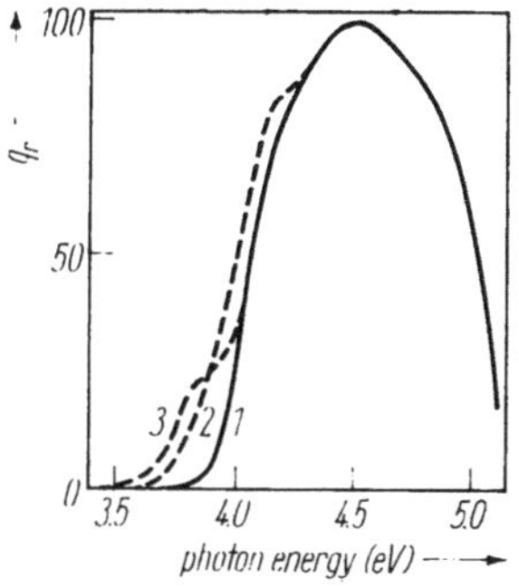

Fig. 3

Fig. 2. Relative excitation spectra of the luminescence of $CaMoO_4$ at several emission energies at 77 K. q_r denotes the relative quantum output. (1) $E_{em} = 2.21$, (2) 1.94, (3) 1.82 eV

Fig. 3. Relative excitation spectra of the luminescence of several $CaMoO_4$ samples at 77 K. q_r denotes the relative quantum output. (1) Undoped, Na- and Y-doped, $E_{em} = 1.94$ eV; (2) Y-doped, $E_{me} = 1.94$ eV; (3) undoped and Na-doped, $E_{em} = 1.94$ eV

excitation band. At lower emission energies an additional excitation band appears at lower energy. The considerable spectral overlap of the green and the orange emission (Fig. 1) causes the simultaneous presence of the excitation band of the green emission, when recording the excitation spectrum of the orange emission.

Excitation spectra of the luminescence of undoped, Na- and Y-doped $CaMoO_4$ samples are presented in Fig. 3. As can be seen from Fig. 1 to 3 we observed two emission bands, which can be excited separately. The present results suggest that the green emission which has been reported before [1, 9] is the intrinsic molybdate emission. The orange emission, which has not been reported earlier, must be due to a defect centre the concentration of which changes by doping with aliovalent ions.

For both emissions the decay time and the intensity of the emission were recorded as a function of temperature. The decay curves appeared to be single exponentials at

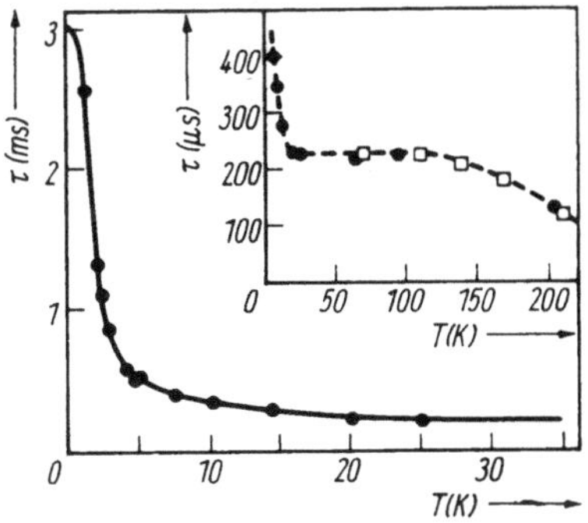

Fig. 4

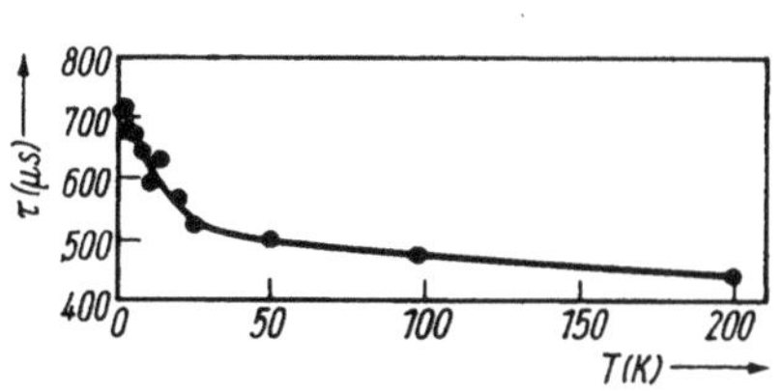

Fig. 5

Fig. 4. Temperature dependence of the decay time of the green emission of $CaMoO_4$. $E_{exc} = 4.47$ eV; circles this work, squares [11]. The solid line represents a fitted curve (see text)

Fig. 5. Temperature dependence of the decay time of the orange emission of $CaMoO_4$. $E_{exc} =$ $= 3.85$ eV; the solid line represents a fitted curve (see text)

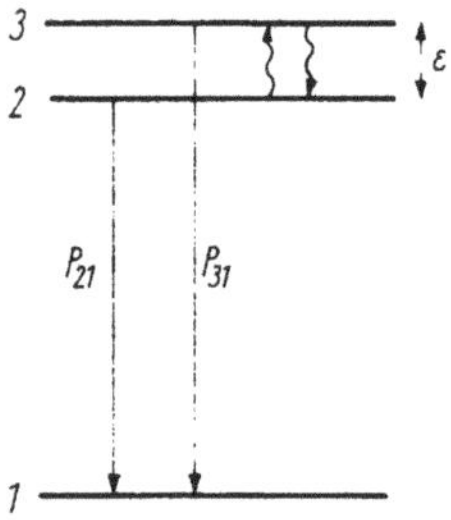

Fig. 6. Three-level energy scheme

all temperatures. The experimental points are given in Fig. 4 and 5. The decay time of both emissions decreases with increasing temperature in the low-temperature region. This points to an energy level scheme as given in Fig. 6, with $p_{21} < p_{31}$, so that level 2 acts as an optical trap [14].

The solid line in Fig. 4 and 5 represents a fit of the measured decay times to $\tau = [1 + \exp(\varepsilon/kT)]/[p_{31} + p_{21} \exp(\varepsilon/kT)]$. This equation gives the temperature dependence of the decay time of the three-level scheme in Fig. 6, under the assumption of thermal equilibrium between the levels 2 and 3 [14]. Further we assume that energy transfer is absent. This is fairly reasonable because excitation into the green centre does not yield the orange emission. The parameters, p_{21}, p_{31}, and ε for the two emissions are presented in Table 1.

Table 1
Parameters used for fitting the experimental decay curves of $CaMoO_4$

	ε (meV)	p_{31} (10^3 s^{-1})	p_{21} (10^3 s^{-1})
green emission	0.6 ± 0.1	8.0 ± 0.2	0.33 ± 0.03
red emission	2.0 ± 0.5	3.1 ± 0.3	1.43 ± 0.04

The temperature dependence of the intensity of both emissions is given in Fig. 7. The increase of the emission intensity of the orange emission in the low-temperature region is probably due to an increasing absorption strength. Because excitation took place in the low-energy side (at 3.69 eV) of the spectrum, a thermal shift of the absorption edge effects the emission intensity strongly. Excitation at higher energies yields also the green emission and can, therefore, not be applied.

In the case of $CaWO_4$ also excitation and emission spectra were recorded. We found in all samples two emission bands, a blue emission band resulting from excitation at energies higher than about 4.8 eV, and a green emission band, resulting from excitation at energies of about 4.6 eV. These two emission bands have been observed earlier by Grasser and Scharmann [9] and Treadaway and Powell [7].

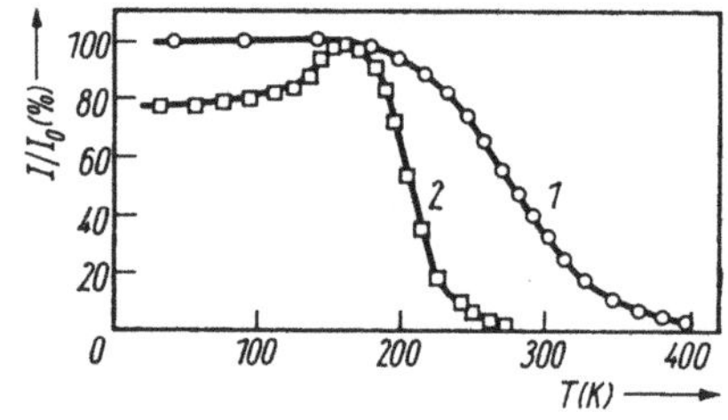

Fig. 7. Temperature dependence of the emission intensity of the green and the orange emission of $CaMoO_4$. (1) Green emission, $E_{exc} = 4.27$ eV; (2) orange emission, $E_{exc} = 3.69$ eV

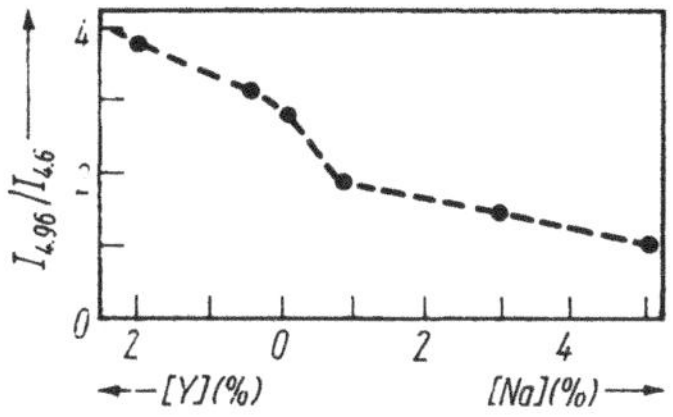

Fig. 8. Ratio of the emission intensity of $CaWO_4$ at 2.48 eV excited at 4.96 and 4.6 eV as a function of the nominal Na and Y-concentration, measured at 77 K

The blue emission band consists of a broad band with a halfwidth of about 0.7 eV and a maximum at 2.75 eV. The green emission band has its maximum at 2.38 eV and a halfwidth of about 0.44 eV. The green emission was quenched at room temperature.

In the case of the Na-doped samples, the green emission was stronger than in the case of the undoped and the Y-doped samples. This is shown in Fig. 8 where the ratio of the intensity of the 2.48 eV emission (which is in between the blue and the green emission) at excitation energies of 4.96 and 4.6 eV is plotted versus the nominal Na or Y content. The higher this ratio, the weaker the green emission will be and the lower the concentration of green emitting centres. We realize that this is a rough and artificial method to estimate the concentration of the green emitting centres, but in view of the overlap of the emission and excitation bands involved we have no other possibility.

4. Discussion

4.1 The temperature dependence of the emissions

The temperature dependence of the green emission of $CaMoO_4$ is given in Fig. 7. Fig. 9 presents a plot of ln $[(I/I_0) - 1]$ versus $1/T$. In this figure the results of Botden [2] on $CaMoO_4$ are presented, too. In both cases one observes two regions, with a higher slope in the higher temperature region. In addition, we made the following observations: a) Excitation into the green centre yields never the orange emission, independent of temperature. b) When doped with Sm^{3+} (1%), the Sm^{3+} emission starts to increase in the high-temperature region in Fig. 9, which has also been observed by Botden [2]. c) Experiments on mixed crystals of $CaWO_4$ and $CaMoO_4$ show no increase of the quenching temperature of the MoO_4^{2-} emission at low MoO_4^{2-} concentrations [2].

In view of the observations mentioned above we conclude, in agreement with [2], that the decrease of the emission intensity in the low-temperature region (150 to 320 K) is due to thermal quenching of the molybdate luminescence. At a certain temperature (about 320 K) the probability of energy transfer between the molybdate groups exceeds the probability of the non-radiative transitions in the centre itself. This is concluded from the increase of the Sm^{3+} emission intensity and the higher slope

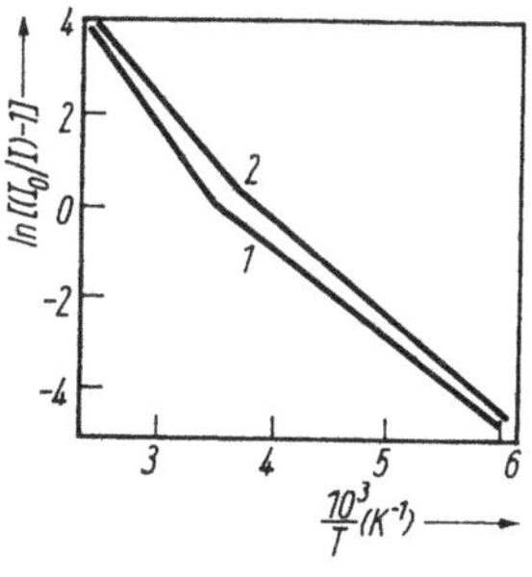

Fig. 9. Temperature dependence of the intensity of the green emission of $CaMoO_4$, plotted as ln $[(I_0/I) - 1]$ vs. $1/T$. (1) This work, (2) Botden [2]

in the high-temperature region in Fig. 9. This model implies that the temperature dependence of the energy transfer probability must be more pronounced than the temperature dependence of the non-radiative transitions. This is in line with a recent theoretical analysis by Fonger and Struck [15].

The temperature dependence of the orange emission of $CaMoO_4$ is shown in Fig. 7. The slight temperature dependence at low temperatures was discussed in Section 3. At high temperatures the quenching can be described by $I_0/I = [1 + c \exp(\Delta E/kT)]^{-1}$ which represents thermal quenching in the centre itself.

Botden [2] and Treadaway and Powell [6, 7] have reported on the temperature dependence of the luminescence of $CaWO_4$ and $CaWO_4{:}\,Sm^{3+}$. Their studies reveal that the decrease of the intensity of the blue tungstate emission occurs simultaneously with the increase of the Sm^{3+} emission. This implies energy transfer between the tungstate groups and finally to the samarium ions. The model for the energy transfer processes proposed in [6] can, however, not completely account for the observed temperature dependence of the emission. It was supposed, therefore, that thermal quenching processes in the tungstate group itself are playing a role, too. The situation in $CaWO_4$ is, therefore, different from that in $CaMoO_4$, because there is no temperature region in which the thermal quenching processes dominate the temperature dependence of the emission.

It is interesting to compare $YVO_4{:}Eu^{3+}$ and $CaMoO_4{:}Sm^{3+}$ in this respect. Whereas in the case of YVO_4 [16] energy migration starts at much lower temperatures than thermal quenching, we have the reversed situation in $CaMoO_4$. The experiments mentioned above suggest that $CaWO_4$ is in between these extremes.

4.2 *The nature of the shorter wavelength emission*

It is well accepted that the shorter wavelength emission in these compounds is due to optical transitions within the MoO_4^{2-} or WO_4^{2-} group. In the known closed-shell tetrahedral tetroxo-transition metal complexes the highest completely filled MO, and the lowest unfilled MO are supposed to be t_1 and $2e_1$, respectively. This gives rise to a $^1A_1$ ground state. The first excited configuration t_12e_1 gives rise to four states viz. $^1T_1$, $^1T_2$, $^3T_1$, and $^3T_2$. The most probable ordering of these states is $^3T_1 \approx {}^3T_2 < < {}^1T_1 < {}^1T_2$ [17]. The emission in these compounds is ascribed to transitions from the lower lying triplets $^3T_{1,2}$ to the $^1A_1$ ground state [9, 12].

In T_d-symmetry the $^3T_1 \to {}^1A_1$ transition is orbitally forbidden as an electric dipole transition, while $^3T_2 \to {}^1A_1$ is an orbitally allowed transition. Both $^3T_1 \to {}^1A_1$ and $^3T_2 \to {}^1A_1$ are spin forbidden. The actual site symmetry in $CaMoO_4$, however, is S_4. Due to the symmetry lowering from T_d to S_4, T_1 will split into an A- and an E-state, and T_2 into a B- and an E-state. The ground state repesentation is now A. Under the selection rules in S_4, $A \to A$ is forbidden, whereas $E \to A$ and $B \to A$ are allowed transitions. Note that the $^3T_1 \to {}^1A_1$ transition contains under S_4 a forbidden and an allowed component.

The decay parameters (Table 1) show that the emission transition from the higher emitting level has a roughly 25 times higher probability than the emission from the lower lying level. From this we conclude that the green emission of $CaMoO_4$ originates from the split $^3T_1$ level, because this level gives rise to an allowed and a forbidden transition. On the basis of similar arguments Ronde and Blasse [18] concluded that the emission of the isoelectronic VO_4^{3-} complex originates from the $^3T_1$ level, too. It is not very surprising that the splitting of the $^3T_1$ level is about 4.5 cm^{-1}, because Ronde and Blasse found for the splitting of this level also a few cm^{-1}. Furthermore, Ballhausen and Trabjerg [19] suggest a splitting of the $^1T_1$ level in the case of $LiClO_4$ ·

$\cdot$ 3 $H_2O/LiMnO_4 \cdot 3\, H_2O$ of less than 10 cm^{-1}, while Butowiez [20] reports 6 cm^{-1} for this splitting in the case of K_2CrO_4.

In view of the similarities between the decay times as a function of temperature of the shorter wavelength emissions in $CaWO_4$ and $CaMoO_4$, and in view of the "isoelectronic" configuration of the tetrahedral groups, we suppose that in $CaWO_4$, too, the intrinsic blue emission corresponds to a transition from the split $^3T_1$ level to the $^1A_1$ level.

A comparison of the decay times, measured in the temperature-independent region, of the "isoelectronic" VO_4^{3-}, MoO_4^{2-}, and WO_4^{2-} complexes reveals that in that sequence the decay time decreases: for VO_4^{3-} about 10^{-3} s [18], for MoO_4^{2-} about 10^{-4} s (this work, [11]), and for WO_4^{2-} about 10^{-5} s [7, 9, 11]. The emission corresponds in all three cases to a transition which is orbitally allowed, but spin forbidden. The decrease of the decay time in the sequence vanadate, molybdate, and tungstate can be ascribed to the increase of the spin–orbit coupling in this sequence.

4.3 The nature of the longer wavelength emission

Our experiments on Na- and Y-doped samples of $CaMoO_4$ and $CaWO_4$ reveal that the concentration of the centre, responsible for the longer wavelength emission increases in the sequence: Y-doped, undoped, and Na-doped samples. Doping with Na will increase, doping with Y will decrease the oxygen ion vacancy concentration, and, therefore, the concentration of MoO_3 (WO_3) groups.

On the basis of quenching experiments, Grasser and Scharmann suggest that the green emission band in $CaWO_4$ originates from WO_3 groups. Our doping experiments give additional chemical evidence for this model. It is obvious, therefore, to ascribe the green emission in $CaWO_4$ and the orange emission in $CaMoO_4$ to a transition on a WO_3 and a MoO_3 group, respectively.

We suppose that the symmetry of such a MoO_3 (WO_3) group is C_{3v}. Relaxation to a planar MoO_3 (WO_3) group is very unlikely, because this would lead to very immobile oxygen ion vacancies. This is contradicted by various experiments which show that oxygen is a rather mobile species in these compounds [21 to 23].

If we consider now the nine 2p orbitals on the three O^{2-} ions, the group theoretical representations of the combinations of these orbitals are $\Gamma_\sigma = A_1 + E$ and $\Gamma_\pi = A_1 + A_2 + E$. None of the s-, p-, or d-orbitals of the metal ions have the A_2-representation. The combination with representation A_2 is, therefore, non-bonding. We assume that it will be the highest filled orbital in the ground state of the $MoO_3(WO_3)$ complex. It seems also reasonable to assume that the d_{z^2} orbital (directed along the trigonal axis) will be the lowest unoccupied d-orbital, since it points to the effectively positive oxygen ion vacancy. The d_{z^2} orbital transforms as a_1.

It is this simple picture which is able to show that at low temperatures the emission transition is orbitally forbidden. The ground state will be $^1A_1$, while the first excited configuration a_2a_1 gives rise to $^1A_2$ and $^3A_2$. The latter one will be the lower. The selection rules in C_{3v} show that transitions between A_2 and A_1 are orbitally forbidden and between A_1 or E and A_1 allowed. All possible transitions at higher energy are, therefore, orbitally allowed, since only A_1- and E-representations are involved. The decay parameters show that the probability of the transition from the lowest excited level to the ground state is smaller than that from the higher level (Table 1). This is in accordance with the model given above.

The difference between the transition probabilities is not as great as may be expected for an allowed and a forbidden transition. It should be realized, however, that we considered the isolated $MoO_3(WO_3)$ complex by assuming C_{3v} symmetry. The actual site symmetry is only C_1 without any symmetry selection rules at all. Our results show, nevertheless, that C_{3v} symmetry is a reasonable first-order approximation.

Acknowledgements

The authors wish to thank Mr. J. Kamphorst who performed the measurements on $CaWO_4$. The investigations were performed as a part of the research programme of the Stichting voor Fundamenteel Onderzoek der Materie (FOM) with financial support from the Nederlandse Organisatie voor Zuiver Wetenschappelijk Onderzoek (ZWO).

References

[1] F. A. KRÖGER, Some Aspects of the Luminescence of Solids, Elsevier Publ. Co., Amsterdam 1948.
[2] TH. P. J. BOTDEN, Philips Res. Rep. **6**, 425 (1951).
[3] G. B. BEARD, W. H. KELLEY, and M. L. MALLOY, J. appl. Phys. **33**, 144 (1962).
[4] W. VAN LOO, phys. stat. sol. (a) **27**, 565 (1975).
[5] W. VAN LOO, phys. stat. sol. (a) **28**, 227 (1975).
[6] M. J. TREADAWAY and R. C. POWELL, Phys. Rev. B **11**, 862 (1975).
[7] M. J. TREADAWAY and R. C. POWELL, J. chem. Phys. **61**, 4003 (1974).
[8] R. GRASSER, E. PITT, A. SCHARMANN, and G. ZIMMERER, phys. stat. sol. (b) **69**, 359 (1975).
[9] R. GRASSER and A. SCHARMANN, J. Lum. **12**, 473 (1976).
[10] G. BLASSE and W. J. SCHIPPER, phys. stat. sol. (a) **25**, K163 (1974).
[11] C. E. TYNER and H. G. DRICKAMER, J. chem. Phys. **67**, 4103 (1977).
[12] G. BLASSE and A. BRIL, J. Lum. **3**, 109 (1970).
[13] W. VAN LOO and D. J. WOLTERINK, Phys. Letters A **47**, 83 (1974).
[14] W. VAN LOO, J. Lum. **10**, 221 (1975).
[15] W. H. FONGER and C. W. STRUCK, J. Lum. **17**, 241 (1978).
[16] G. BLASSE, Philips Res. Rep. **23**, 344 (1968).
[17] C. J. BALLHAUSEN, Theor. chim. Acta (Berlin) **1**, 285 (1963).
[18] H. RONDE and G. BLASSE, J. inorg. nuclear Chem. **40**, 215 (1978).
[19] C. J. BALLHAUSEN and I. B. TRABJERG, Mol. Phys. **24**, 689 (1972).
[20] B. BUTOWIEZ, C.R. Acad. Sci. (France) **B267**, 1234 (1968).
[21] M. A. RIGDON and R. E. GRACE, J. Amer. Ceram. Soc. **56**, 475 (1973).
[22] K. NASSAU and G. M. LOIACONO, J. Phys. Chem. Solids **24**, 1503 (1963).
[23] J. A. GROENINK and H. BINSMA, J. Solid State Chem., accepted.

(Received April 5, 1979)

phys. stat. sol. (a) **54**, 337 (1979)

Subject classification: 14.3.3; 14.3.4; 22

Elektronska Industrija, Fabrika Poluprovodnika, Niš[1])

Analytical Approximations of Non-Degenerate Schottky Diode Current–Voltage Characteristics

By

R. S. Popović

The current–voltage characteristics of non-degenerate Schottky diodes are analysed using the thermionic theory and allowing for quantum effects. A general expression for the current–voltage characteristic is derived, containing one analytically unsolvable integral. An evaluation of this integral in the form of the sum of three alternating series is presented. Each of these series has a clear physical meaning, corresponding to thermionic emission, thermal-field emission, and field emission, respectively. The solution of the integral in the form of alternating series makes the numerical evaluation of the integral easy, and/or provides a method to derive simple analytical expressions approximating the integral, along with an easy estimation of truncation errors. Based on such approximations, analytical expressions for the voltage–current Schottky-diode characteristics are derived.

Рассмотрена вольт–амперная характеристика невырожденного диода Шоттки. Использован метод обобщенной термоэлектронной эмиссии, учитывающий квантовые эффекты. Обобщенное выражение для вольт–амперной характеристики содержит один аналитически нерешаемый интеграл. Найдено решение этого интеграла в форме трех рядов с чередующимися знаками. У каждого из этих рядов определенное физическое значение: один представляет собой термоэлектронную эмиссию, туннелирование под влиянием температуры и туннельную эмиссию. Решение интеграла в форме рядов с чередующимися знаками облегчает вычисление этого интеграла численным способом. Кроме того, благодаря такому подходу удалось найти относительно простые аналитические выражения, аппроксимирующие вольт–амперные характеристики диода Шоттки.

1. Introduction

There are two basic theories describing the carrier transport phenomenon through the metal–semiconductor contact: the diffusion theory developed by Schottky [1] and the thermionic theory proposed by Bethe [2]. Crowell and Sze [3] have condensed these two theories and developed a unified thermionic-diffusion theory. The two basic theories can be derived from of the unified one if certain conditions are met. Thus, for higher impurity concentrations the thermionic theory is valid [4]. Because the range of "higher" impurity concentrations covers almost all concentrations of practical interest (for n-Si, $N > 2.5 \times 10^{14}\ \text{cm}^{-3}$ at $T = 300\ \text{K}$ [5]), the thermionic theory approach has been mainly used and developed lately.

The original form of the thermionic theory did not include the quantum-mechanical reflection and tunnelling of carriers through the barrier. Further investigations have pointed out the importance of these quantum effects. Initially, they have been analysed assuming that the barrier had the shape of a simple contact depletion region potential, [6 to 8]. Later on, the form of the barrier top was modified. Two models were proposed and exploited to allow for this modification. The first one, the image charge force effect (see [4], p. 364), was implemented into the transport theory by

[1]) Bul. V. Vlahovića bb, 18000 Niš, Yugoslavia.

Rideout and Crowell [9]. The second one is based on the influence of the electric space charge due to the quantum-mechanical tunnelling of electrons from the metal into the forbidden band of the semiconductor. It was proposed by Heine [10] and developed by Pellegrini [11, 12]. Both models lead essentially to the same result — lowering the barrier height and rounding-off its top. In the present paper we shall use the first model because of its mathematical simplicity. Additionally, we shall assume an intimate metal–semiconductor contact, with no interface dielectric layer.

The implementation of the quantum effects into the thermionic theory increased greatly the complexity of the mathematical treatment of the contact current–voltage characteristics. Even with the use of WKB approximation and replacement of the real barrier shape by a parabola, only rough mathematical solutions of the problem have been found [5, 9]. The shortcomings of such results are a limited parameter range of their validity and an undefined approximation error.

In this paper the approximate analytical expressions for the $J(V)$ characteristics of the non-degenerate Schottky diodes have been derived for various temperature intervals. The solutions have the form of alternating series convenient either for numerical computations or analytical approximations. The properties of the alternating series make an easy estimation of the approximation error possible.

In the next section an analysis and transformation of Fonash's [13] "analytical" expression (up to integration) for the $J(V)$ characteristics of a Schottky diode is made. The derivation of the high-temperature approximation is presented in Section 3. Separate expressions for the thermionic emission and thermal-field emission contributions to the current have been found. The known Rideout and Crowell approximate expression [9] can be deduced from the sum of these two components. Section 4 is devoted to the low-temperature approximation of the characteristics. We present a few analytical expressions which have not yet been reported in the literature. In Section 5 we discuss our approach in comparison with other ones.

2. Current–Voltage Characteristics

The general expression of the current–voltage characteristics of a metal–semiconductor contact is of the following form [13]:

$$J = n_{\rm m} \int\limits_0^\infty P(E_x, m_x)\, N_{\rm i}(E_x)\, {\rm d}E_x\,, \tag{1}$$

where $n_{\rm m}$ is the number of equivalent minima of the semiconductor conduction band, $N_{\rm i}(E_x)$ the supply function, E_x the electron energy corresponding to the motion normal to the metal–semiconductor interface, m_x is the electron effective mass for tunnelling along the x-axis.

This expression includes, through the carrier penetration probability $P(E_x, m_x)$, both important quantum effects: reflection for those electrons passing over the barrier and tunnelling for electrons having energies below the barrier top.

Approximate analytical expressions of the penetration probability, defined in such a way, can be obtained by applying the generalized WKB approximation [14] to a few simple potential barrier shapes.

We shall use here the expression

$$P(E_x, m_x) = \left[1 + \exp\left(\frac{qV_{\rm m} - E_x}{E_{\rm t}}\right)\right]^{-1} \tag{2}$$

derived by Rideout and Crowell [9] for the parabola fitting best the top of the real metal–semiconductor contact barrier (Fig. 1). In defining the real potential barrier shape, they made use of the image charge force barrier lowering and rounding-off of

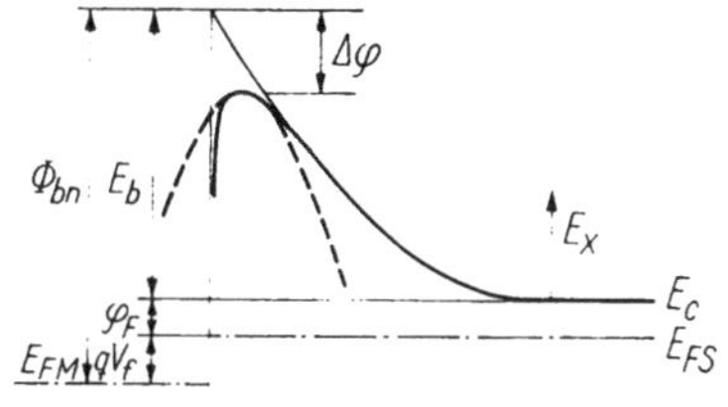

Fig. 1. The contact potential diagram. Full line real barrier; dashed line best fitting parabola. (E_b built-in barrier potential; E_c energy of the bottom of the semiconductor conduction band; E_F Fermi level; V_f forward voltage drop; Φ_{bn} metal-semiconductor barrier height, neglecting barrier lowering)

its top. The energy of the potential barrier top, qV_m, in (2) is measured relative to the conduction band bottom of the semiconductor substrate and is given by (see Fig. 1)

$$qV_m = \Phi_{bn} - \varphi_F - \Delta\varphi - qV_f\,. \tag{3}$$

The image charge barrier lowering is given by [9]

$$\Delta\varphi \approx E_b \left[\pi \left(\frac{E_b}{E_{11}}\right)^{3/2}\right]^{-1/2}, \tag{4}$$

where

$$E_b = \Phi_{bn} - \varphi_F - qV_f\,, \tag{5}$$

$$E_{11} = \frac{q^2}{2\varepsilon_0}\left(\frac{N}{K_s K_d^2}\right)^{1/3} \tag{6}$$

(N is the donor impurity concentration in the semiconductor, K_s and K_d are the semiconductor relative static and dynamic dielectric constants, respectively).

The characteristic energy E_t in (2) is defined as

$$E_t \approx \frac{E_{00}}{\pi}\left(8\,\frac{E_b}{\Delta\varphi}\right)^{1/2}, \tag{7}$$

$$E_{00} = \frac{qh}{4\pi}\left(\frac{N}{m_x K_s K_d}\right)^{1/2}. \tag{8}$$

According to the results cited in [9], by using the parabola of best fit instead of the correct barrier shape, we exaggerate the value of the tunnelling probability. However, there are two strong arguments approving such an approximation:

(i) The disagreement between the tunnelling probability of the parabolic barrier and that of the correct one is noticeable only for carriers tunnelling far below the barrier top [9]. But the disagreement is smaller for higher semiconductor impurity concentrations. Therefore, in spite of using the parabolic barrier approximation, we do not make too big a mistake; because if there exists any tunnelling in a low impurity concentration diode, it takes place mainly just below the barrier top where the disagreement of the parabola and the real barrier is negligible; at higher impurity concentration contacts, the majority of the tunnelling carriers may go further below the barrier top, but having in mind the above said, the approximation error still cannot be too large.

(ii) Note that we are using a WKB-type approximation to describe the penetration probability. According to the results of Gundlach and Simmons [15], the WKB approximation gives a correct functional dependence of the tunnelling probability, but its numerical value is up to ten times smaller than the one obtained by a direct solution of the Schrödinger equation. Consequently, if a parabolic barrier approximation gives rise to a slightly greater tunnelling probability, this can only make a benefit to the precision of the final results, because the two errors partially compensate each other.

Let us go back to (1): The second term of the integrand is the supply function. For the case of a non-degenerate semiconductor, it is given by [13]

$$N_i(E_x) = (A_i^* T k^{-1}) \exp\left(-\frac{\varphi_F}{kT}\right)\left[1 - \exp\left(-\frac{qV_f}{kT}\right)\right]\exp\left(-\frac{E_x}{kT}\right), \tag{9}$$

where

$$A_i^* = A_0 \frac{m_t}{m_0}\left(\frac{m_l}{m_x}\right)^{1/2}. \tag{10}$$

Here A_i^* is the effective Richardson constant, A_0 the Richardson constant, m_0 the electron mass in vacuum, and m_t and m_l the transverse and longitudinal effective electron mass, respectively. Using (2) and (9), expression (1) becomes

$$J = A \int_0^\infty \left\{\left[1 + \exp\left(\frac{qV_m - E_x}{E_t}\right)\right]\exp\left(\frac{E_x}{kT}\right)\right\}^{-1} dE_x, \tag{11}$$

where we introduced the function

$$A = n_m (A_i^* T k^{-1}) \exp\left(-\frac{\varphi_F}{kT}\right)\left[1 - \exp\left(-\frac{qV_f}{kT}\right)\right]. \tag{12}$$

The integral in (11) is analytically unsolvable in general [9]. It is inconvenient even for a numerical integration because of its infinite upper limit. However, making the change of variables

$$y = \exp\left(-\frac{E_x}{kT}\right), \tag{13}$$

expression (11) is reduced to

$$J = kTAB \int_0^1 (B + y^{kT/E_t})^{-1} \, dy, \tag{14}$$

where

$$B = \exp\left(-\frac{qV_m}{E_t}\right). \tag{15}$$

Generally, the integral (14) could be evaluated analytically if the parameters kT and E_t were rational numbers [16]. But this is quite inconvenient because, for each parameter value a different analytical expression should be obtained. For this reason we shall derive approximative expressions holding for wide ranges of parameter values. Moreover, this approach allowed a clear distinction of the various current components in Schottky diodes.

3. High-Temperature Approximation

We are proceeding with the derivation of an analytical expression of $J(V)$ characteristics for the case when the value of the carrier thermal energy kT is larger than the characteristic energy E_t. In order to do this, we shall write the integral

$$I = \int_0^1 (B + y^c)^{-1} \, dy \tag{16}$$

in (14) in the form of the sum

$$I = I_{TE} + I'_{FE}\,, \tag{17}$$

where

$$I_{TE} = \int_0^{y_1} B^{-1}(1 + B^{-1}y^c)^{-1}\,dy\,, \tag{18}$$

$$I'_{FE} = \int_{y_1}^{1} y^{-c}(1 + By^{-c})^{-1}\,dy\,, \tag{19}$$

$$c = \frac{kT}{E_t}, \tag{20}$$

$$y_1 = B^{1/c}\,. \tag{21}$$

Note the fact that we divided the integration range $y \in [0, 1]$ into two intervals at the point $y = y_1$. According to (13), (15), and (21), these intervals correspond to the following ranges of the original variable:

$$y \in [0, y_1] \Rightarrow E_x \in (+\infty, qV_m]\,, \tag{22}$$

$$y \in [y_1, 1] \Rightarrow E_x \in [qV_m, 0]\,. \tag{23}$$

This means that I_{TE}, equation (18), expresses the flux of electrons passing over the barrier top, e.g. thermionic emission, while I'_{FE}, equation (19), refers to the flux of those passing through the barrier, e.g. field emission.

In mathematical terms, the value y_1 is chosen so that the second terms in the brackets of (18) and (19) are less than unity in the whole integration intervals except just at $y = y_1$. Therefore, we may treat the expression in the brackets as geometrical series and write

$$I_{TE} = \lim_{\varepsilon \to 0} \int_0^{y_1-\varepsilon} B^{-1}[1 - B^{-1}y^c + (B^{-1}y^c)^2 \ldots]\,dy\,, \tag{24}$$

$$I'_{FE} = \lim_{\varepsilon \to 0} \int_{y_1+\varepsilon}^{1} y^{-c}[1 - By^{-c} + (By^{-c})^2 \ldots]\,dy\,, \tag{25}$$

where $\varepsilon \gtrapprox 0$. By integrating the series term by term and taking the limit, we obtain

$$I_{TE} = B^{(1/c)-1} \sum_{n=0}^{+\infty} (-1)^n \frac{1}{nc+1}, \tag{26}$$

$$I'_{FE} = \sum_{n=0}^{+\infty} (-1)^n B^n \frac{1}{1-(n+1)\,c} - B^{(1/c)-1} \sum_{n=0}^{+\infty} (-1)^n \frac{1}{1-(n+1)\,c}\,. \tag{27}$$

The sequences of terms belonging to these series converge to zero. In the cases of the first and the third series, it is obvious; in the second series, this depends upon the value of the parameter B. But by definition (15), $B < 1$. Hence, the corresponding sequence of terms converges, too. In addition to that, the series are alternating, so according to Leibniz' criterion, they must be convergent.

The expressions (26) and (27) provide an easy way of evaluating the numerical value of the integral (14) or (11). Because of the properties of alternating series, the maximum value and sign of the truncation error can be easily established. Moreover, it turned out that it was possible to find simple analytical expressions approximating the sums of these series.

First of all, note that parameter B is practically always in the range

$$0 < B \ll 1 . \tag{28}$$

Its typical value lies between 10^{-200} and 10^{-10} for an unbiased metal–silicon contact with impurity concentration between 10^{15} and 10^{18} cm^{-3} and $\Phi_{bn} = 0.7$ eV. Therefore, for $c > 1$ the sum of the second series in (26) and (27) is negligible compared to the other two:

$$I \approx I_{TE} + I_{TFE} , \tag{29}$$

where

$$I_{TFE} = B^{(1/c)-1} \sum_{n=0}^{+\infty} (-1)^n \frac{1}{(n+1)c - 1} . \tag{30}$$

The relative error is of the order of $B^{2(1-1/c)}$. Since we are dealing with higher temperatures, we may assume that the last series (30) refers to the thermal-field emission fraction of the total current.

Next, by adding the corresponding terms of series (26) and (30), we have

$$I \approx B^{(1/c)-1} \left(1 + 2 \sum_{n=1}^{+\infty} (-1)^{n+1} \frac{1}{n^2c^2 - 1}\right). \tag{31}$$

If $c \gg 1$, we may retain only the first term of series (31) and neglect 1 compared to c^2. Equation (31) becomes

$$I \approx B^{(1/c)-1} \left[1 + 2 \left(\frac{E_t}{kT}\right)^2\right]. \tag{32}$$

We recognize the expression in the square brackets to be the same as the corresponding one derived by Rideout and Crowell [9] for the high-temperature approximation of $J(V)$ Schottky diode characteristics. However, our approach makes an easy truncation error estimation possible.

An even better approximation can be obtained by making use of the similarity between our series and the known numerical series of ln 2 [17].

Here we present only the results, the details of the derivation being shown in Appendix A:

$$I_{TE} = B^{(1/c)-1} \left[1 - \frac{1}{c} \ln 2 + \frac{\theta_1}{c(c+1)}\right]; \qquad 0 < \theta_1 < 1 ; \tag{33}$$

$$I_{TFE} = B^{(1/c)-1} \frac{1}{c} \left(\ln 2 + \frac{\theta_2}{c-1}\right); \qquad 0 < \theta_2 < 1 . \tag{34}$$

Now we have

$$I \approx B^{(1/c)-1} \left[1 + \frac{1}{c} \frac{c(\theta_1 + \theta_2) + \theta_2 - \theta_1}{c^2 - 1}\right]. \tag{35}$$

Putting the most probable values

$$\theta_1 \approx \theta_2 = \tfrac{1}{2}$$

we obtain

$$I \approx B^{(1/c)-1} \left(1 + \frac{1}{c^2 - 1}\right). \tag{36}$$

We now turn our attention to the $J(V)$ characteristics itself. Substituting (36) in (14), we have

$$J \approx kTAB^{1/c} \left(1 + \frac{1}{c^2 - 1}\right). \tag{37}$$

Using again the complete relations instead of the symbols A and B and (3) for qV_{m}, we finally come to

$$J = J_{\mathrm{s}}\left[\exp\left(\frac{qV_{\mathrm{f}}}{kT}\right) - 1\right]; \tag{38}$$

$$J_{\mathrm{s}} = n_{\mathrm{m}} A_{\mathrm{i}}^{*} T^2 \exp\left(-\frac{\Phi_{\mathrm{bn}} - \Delta\varphi}{kT}\right) \frac{(kT)^2}{(kT)^2 - E_{\mathrm{t}}^2} \tag{39}$$

provided $kT > E_{\mathrm{t}}$.

Therefore, our result for the high-temperature approximation of the Schottky diode current–voltage characteristics is similar to the simple well-know formula, the only difference being in the form of the saturation current.

4. Low-Temperature Approximation

Expressions (26) and (27) for I_{TE} and I_{FE} are still valid if $c < 1$, the only difference being that the magnitude of the first sum of (27) dominates. However, it is rather inconvenient to use these expressions because, for sufficiently small n, the terms of the series in (27) may be increasing. In addition to this, for some particular values of the parameter c ($c = 1/(n + 1)$), the series in (27) are not convergent if considered separately, for their n-th terms tend to infinity. For this reason we shall specially derive an expression for I when $c < 1$.

We shall approximate the parameter E_{t} by a rational number. Then we may write

$$\frac{kT}{E_{\mathrm{t}}} = c \approx \frac{p}{q'}, \tag{40}$$

where p and q' are integers. Since $c < 1$, it must be $p < q'$. Now let us make the change of variable in the integral (16)

$$y = Z^{q'} . \tag{41}$$

This integral becomes

$$I = q' \int_0^1 \frac{Z^{q'} - 1}{B + Z^p} \mathrm{d}Z . \tag{42}$$

Suppose $q' - 1 > p$. Having divided the polynomials under the integral sign, we obtain

$$I = q' \left[\sum_{n=0}^{m-1} (-1)^n B^n \int_0^1 Z^{l-(n+1)q} \,\mathrm{d}Z + (-1)^m B^m \int_0^1 \frac{Z^{l-mp}}{Z^p + B} \mathrm{d}Z\right], \tag{43}$$

where

$$l = q' - 1 , \tag{44}$$

$$m < \left[\frac{1}{p}\right]. \tag{45}$$

Evaluating the integrals in (43), we have

$$I = q' \left[\sum_{n=0}^{m-1} (-1)^n B^n \frac{1}{q' - (n + 1)\, p} + (-1)^m B^m \int_0^1 \frac{Z^{l-mp}\, \mathrm{d}Z}{Z^p + B}\right]. \tag{46}$$

As it is shown in Appendix B, by choosing a sufficiently small number of terms in the sum (46), we can make the magnitudes of terms decreasing, and the value of the integral to be smaller than the last term of the sum. Therefore,

$$I \approx I_{\mathrm{FE}} = \sum_{n=0}^{m-1} (-1)^n B^n \frac{1}{1-(n+1)\,c}\,, \tag{47}$$

where m is subject to the condition

$$m \leqq \left[\frac{1}{c} - B\right]. \tag{48}$$

The approximation error we made is of the order of B^{m-1}. The smaller the parameters B and c are, the smaller is the error made.

We have denoted the sum (47) by I_{FE}. Since it is valid for $c < 1$, it evidently expresses the tunnelling flux of carriers through the barrier. It is interesting to note that the sum (47) is identical to the sum of the first few terms in the expression for I_{FE} (27). This enables us to write, instead of (17),

$$I = I_{\mathrm{TE}} + I_{\mathrm{TFE}} + \dddot{I}_{\mathrm{FE}}\,. \tag{49}$$

Each of the terms in (49) becomes dominant in its "own" interval of the parameter c and negligible in the "foreign" intervals, in which it can even change the sign (I_{TFE} and I_{FE}). Naturally, this has no physical meaning.

For sufficiently small values of c and B, we can approximate the sum (47) by its first term, see Appendix B. Then we have

$$I_{\mathrm{FE}} \approx \frac{1}{1-c} \quad \text{for} \quad c \ll 1\,, \qquad B \ll 1\,. \tag{50}$$

We have not yet an analytical expression for I if $c = 1$. However, it can be found in a fairly straightforward fashion

$$I = \int_0^1 (B + y)^{-1}\, \mathrm{d}y = \ln \frac{B+1}{B}\,. \tag{51}$$

Since $B \ll 1$, it is

$$I \approx -\ln B = \frac{qV_{\mathrm{m}}}{E_{\mathrm{t}}}\,, \quad \text{if} \quad kT = E_{\mathrm{t}}\,. \tag{52}$$

Now we can finally write the relation for the $J(V)$ characteristics. Starting from the general expression (14), we obtain

$$J = kTABI_{\mathrm{FE}}\,, \tag{53}$$

or, in the complete form,

$$J = J_{\mathrm{s}} \left\{ \exp\left(\frac{qV_{\mathrm{f}}}{E_{\mathrm{t}}}\right) - \exp\left[qV_{\mathrm{f}}\left(\frac{1}{E_{\mathrm{t}}} - \frac{1}{kT}\right)\right]\right\}, \tag{54}$$

$$J_{\mathrm{s}} = n_{\mathrm{m}} A_{\mathrm{i}}^{*} T^2 \exp\left(-\frac{\Phi_{\mathrm{bn}} - \Delta\varphi}{E_{\mathrm{t}}}\right) \left(\frac{N_{\mathrm{c}}}{N}\right)^{(kT/E_{\mathrm{t}})-1} I_{\mathrm{FE}}\,, \tag{55}$$

where N_{c} is the effective density of states in the semiconductor conduction band and I_{FE} is a function not differing much from unity,

$$I_{\mathrm{FE}} \approx 1\,. \tag{56}$$

Its more precise value is given by the sum (47) or (51), and the approximate one by

$$I_{\mathrm{FE}} \approx \frac{E_\mathrm{t}}{E_\mathrm{t} - kT}, \quad \text{if} \quad kT \ll E_\mathrm{t}, \tag{57}$$

$$I_{\mathrm{FE}} \approx \frac{qV_\mathrm{m}}{E_\mathrm{t}}, \quad \text{if} \quad kT = E_\mathrm{t}. \tag{58}$$

To the author's knowledge, expressions like (54) and (55) have not yet been reported in the literature. Without going into the detailed analysis of these relations now, we shall only emphasize a few of their features:

Equation (54) shows at least a qualitatively correct current versus voltage dependence observed in Schottky diodes with higher impurity concentrations: (i) the forward voltage drop can be larger than that in reverse direction — similar to the p–n junction backward diodes; (ii) at forward bias, the characteristics displays a maximum of the dynamic resistivity [6, 18].

If $kT \ll E_\mathrm{t}$, (55) gives the known dependence $J_\mathrm{s} \sim N^{1/2}$ [19], but only as a particular case, at $kT/E_\mathrm{t} = \frac{1}{2}$.

5. Discussion and Conclusions

In the majority of papers considering the current–voltage characteristics of the metal–semiconductor contact, the analysis is made in such a way that the whole parameter domain is divided into three principal regions. In each of them, only the dominant mechanism of the carrier transport is considered out of the possible three ones: thermionic emission, thermal-field emission, and field emission. This approach is physically very instructive, but it suffers from lacking mathematical consistency.

On the other hand, a few papers have been published [9, 11, 13] where, using the generalized WKB approximation, the carrier flux has been evaluated totally, without any partition into current fractions. This way of the problem treatment is mathematically correct, but physically less evident.

In the present paper a clear relationship between these two approaches has been developed. We have started the analysis according to the exact, total current approach, equation (1). Then, by purely mathematical transformations, we have derived relations (26) and (27), or (49), the physical meaning of which is quite in agreement with the results of the classical phenomenological approach.

According to our results, the current of the Schottky diode can be represented by the sum of components J_{TE}, J_{TFE}, and J_{FE} in the entire range of the parameter c. Although each of them is dominant in its "own" range of the parameter c, this does not mean we may always easily replace the total current by its dominant fraction.

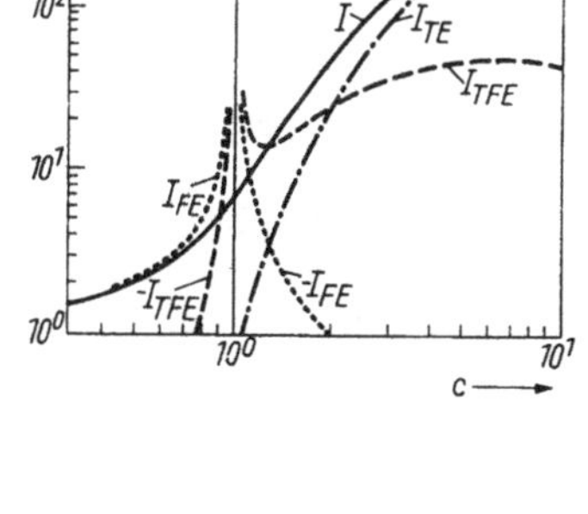

Fig. 2. Dependence of the integral I and of its parts upon the parameter c, for $B = 10^{-3}$

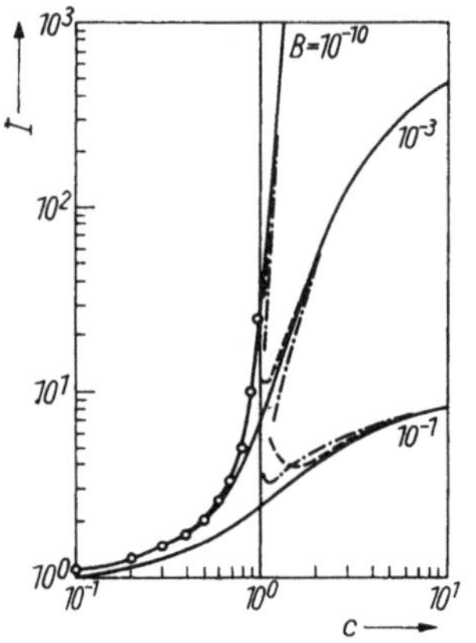

Fig. 3. Exact and approximate values of the integral I versus parameter c. Full lines exact values; dashed lines approximations by (36); dash-dotted lines approximations by (32); acircles low-temperature approximation by (50)

This is particularly true for values of the parameter c in the vicinity of 1 and 2, as indicated in Fig. 2. This figure shows the exact numerical values of the integral I (equation (16)), together with its "components", the sums of series in (26) and (27). We see that an especially bad approximation is obtained if I is replaced by I_{TFE} in the interval $1 < c < 2$. The approximation is not so incorrect in terms of the absolute values, but it shows a totally false functional dependence upon the parameter c in the whole interval. On the other hand, the approximations of I by I_{FE} (for $c < 1$) and by I_{TE} (for $c > 2$) are fully justified.

As far as the approximations of the correct I-values are concerned, it is better to use the analytical approximate expressions (32), (36), and (50) instead of the complete series representing the partial components of the integral I. Although one can easily estimate the maximum errors for the alternating properties of the series approximated by their first few terms, it is interesting to verify the quality of approximation in the whole domain of c. Therefore, the exact numerical values of the integral I (16) have been computed for a few values of the parameter B. Of course, the calculations have been done indirectly, using the series (26) and (27). The results are shown in Fig. 3, together with the appropriate approximate values. It can be seen that

(i) If $c < 1$, the agreement between the exact curves and the dots representing the approximate values calculated by (50) is fairly good, provided the magnitude of c is sufficiently small. What is "sufficiently small" depends upon the magnitude of the constant B. If $B = 10^{-1}$, it should be, say, $c < 0.1$; if $B = 10^{-10}$, $c < 0.95$ is sufficient.

(ii) If $c > 1$, the quality of approximation depends again upon the magnitude of B. The approximation by our expression (36) is slightly better than that obtained by the use of expression (32) of Ridout and Crowell's type. However, the application of the latter is fully justified for higher values of c. It is significant to note that expression (32) evidently represents the total current of the contact, rather than only its thermionic part, as it was claimed in [9]. In addition and connection to that, expression (32) is still valid if $1.5 < c < 2$.

(iii) Neither of these approximations is valid in the close proximity of the point $c = 1$. However, we can see from Fig. 3 that each of the three curves $I(c)$ can be approximated very well in the vicinity of this point by a straight line (in log–log coordinates). Therefore, since we know the exact values of I for $c = 1$ (equation (51) or (52)), and for any other (close) c (equations (26) and (27)), we can always write the equation of this line. This equation will be a good approximation of the curve $I(c)$ if $c \approx 1$.

Appendix A

Consider expression (26) for I_{TE}. We can rewrite it in the form

$$I_{TE} = B^{(1/c)-1}\left(1 - \frac{1}{c}\,S_{TE}\right), \tag{A1}$$

where

$$S_{TE} = \sum_{n=1}^{+\infty} (-1)^{n+1} \frac{1}{n + 1/c}. \tag{A2}$$

If $c \gg 1$, the latter series does not differ so much from the numerical series expansion of ln 2 [17],

$$\ln 2 = \sum_{n=1}^{+\infty} (-1)^{n+1} \frac{1}{n}. \tag{A3}$$

To make use of this similarity, we shall form the difference

$$R_{TE} = S_{TE} - \ln 2\,, \tag{A4}$$

$$R_{TE} = \sum_{n=1}^{+\infty} (-1)^{n} \frac{1}{(nc + 1)\,n}. \tag{A5}$$

Note the series (A5) converges much faster than that in (A2). It is alternating and therefore the magnitude of its sum is smaller than its first term, the sign of the sum being the same as that of the first term. Thus we can write

$$R_{TE} = -\frac{\theta_1}{c + 1}; \tag{A6}$$

$$0 < \theta_1 < 1\,. \tag{A7}$$

Now, from (A4) and (A6)

$$S_{TE} = \ln 2 - \frac{\theta_1}{c + 1}. \tag{A8}$$

Substituting (A8) into (A1), we obtain

$$I_{TE} = B^{(1/c)-1}\left[1 - \frac{1}{c}\ln 2 + \frac{\theta_1}{c(c + 1)}\right]. \tag{A9}$$

Likewise, starting from

$$I_{TFE} = B^{(1/c)-1} \frac{1}{c} \sum_{n=1}^{+\infty} (-1)^{n+1} \frac{1}{n - 1/c} \tag{A10}$$

we come to

$$I_{TFE} = B^{(1/c)-1} \frac{1}{c}\left(\ln 2 + \frac{\theta_2}{c - 1}\right); \tag{A11}$$

$$0 < \theta_2 < 1\,. \tag{A12}$$

Appendix B

The sufficient condition for the sequence of terms of the sum (46) to be decreasing is that the ratio of the magnitudes of the two last subsequent terms is less than unity,

$$\left|\frac{a_{m-1}}{a_{m-2}}\right| = \frac{B^{m-1}(q' - mp)^{-1}}{B^{m-2}[q' - (m - 1)\,p]^{-1}} < 1\,, \tag{B1}$$

or, after some rearrangements,

$$m < \left(\frac{1}{c} - \frac{B}{1-B}\right) \approx \left(\frac{1}{c} - B\right). \tag{B2}$$

Let us estimate now the magnitude of the last term (with the integral) in (46):

$$|a_m| = q' B^m \int_0^1 \frac{Z^{l-mp}}{Z^p + B} \mathrm{d}Z\,. \tag{B3}$$

Since $Z^p > 0$ in the whole integration interval, we have

$$|a_m| < q' B^m \int_0^1 \frac{Z^{l-mp}}{B} \mathrm{d}Z\,, \tag{B4}$$

$$|a_m| < q' B^{m-1} \frac{1}{q' - mp} < B^{m-1} \frac{1}{1 - mc}\,, \tag{B5}$$

where we made use of (40) and (44). Comparing the relation (B5) with the last term of the sum (47), we conclude that

$$|a_m| < |a_{m-1}| \tag{B6}$$

which justifies the approximation made in (47). Therefore, the integral (16) can be approximated with a pretty good precision by the sum (47), provided the condition (B2) is fulfilled.

References

[1] W. Schottky, Naturwissenschaften **26**, 843 (1938).
[2] H. A. Bethe, MIT Radiation Laboratory, Report No. 43-12, 1942.
[3] C. R. Crowell and S. M. Sze, Solid State Electronics **9**, 1035 (1966).
[4] S. M. Sze, Physics of Semiconductor Devices, Wiley–Interscience, New York 1969 (p. 382).
[5] V. I. Strikha, Teoreticheskie osnovy raboty kontakta metall–poluprovodnik, Naukova Dumka, Kiev 1974.
[6] J. W. Conley, C. B. Duke, G. D. Mahan, and J. J. Tiemann, Phys. Rev. **150**, 466 (1966).
[7] F. A. Padovani and R. Stratton, Solid State Electronics **9**, 695 (1966).
[8] C. R. Crowell and V. L. Rideout, Solid State Electronics **12**, 89 (1969).
[9] V. L. Rideout and C. R. Crowell, Solid State Electronics **13**, 993 (1970).
[10] V. Heine, Phys. Rev. **138**, A1689 (1965).
[11] B. Pellegrini, Phys. Rev. B **7**, 5299 (1973).
[12] B. Pellegrini, Solid State Electronics **17**, 217 (1974).
[13] S. J. Fonash, Solid State Electronics **15**, 783 (1972).
[14] E. C. Kemble, Fundamental Principles of Quantum Mechanics, Dover Publ., Inc., New York 1958 (p. 109).
[15] K. H. Gundlach and J. G. Simmons, Thin Solid Films **4**, 61 (1969).
[16] N. S. Piskunov, Differentsialnoe i integralnoe ischisleniya, Vol. 1, Izd. Nauka, Moskva 1964 (p. 351).
[17] T. Pejović, Matematička analiza, Vol. 3, Gradjevinska knjiga, Beograd 1967 (p. 109).
[18] R. Stratton and F. A. Padovani, Solid State Electronics **10**, 813 (1967).
[19] C. Y. Chang, Y. K. Fang, and S. M. Sze, Solid State Electronics **14**, 541 (1971).

(Received December 6, 1978)

phys. stat. sol. (a) 54, 349 (1979)

Subject classification: 7; 12

Technical University of Advanced Transport Engineering, KTF-VŠD, Žilina[1])

Some Invariance Relations for Acoustic Waves in Crystals Deformed by Hydrostatic Pressure

By

P. HEGEDÜŠ and S. KOLNÍK

Third order elastic coefficients can be determined by using experimental values of ultrasonic velocities and their changes by hydrostatic pressure. These elastic coefficients are correct only in the case when the experimental data satisfy some invariance relations. Propagation and polarization directions of crystals of Laue groups CI, CII, TI and HI fulfilling mentioned relations are given.

Значения констант упругости третьего порядка можно получить из экспериментально измеренных значений скоростей звука и их зависимостей от внешнего гидростатического и одноосного сжатия. Коректные значения упругих констант можно определить только в случаях, когда экспериментальные значения удовлетворяют инвариантам. В настоящей работе найдены направления, распространения, в которых существуют инварианты для кристаллов, принадлежающих Лауе группам CI, CII, TI и HI, подвергнутых гидростатическим давлением.

1. Introduction

Higher-order elastic coefficients are important in the calculation of many anharmonic effects such as thermal expansion and the interaction of thermal and acoustic phonons [1]. Brugger [2] defines the third-order elastic (TOE) coefficients using a thermodynamic method. In this definition TOE coefficients are given by the derivatives of the isentropic second-order elastic (ISOE) coefficients with respect to the Lagrangian strains at constant temperature. The TOE coefficients can be determined from velocity measurements and from their changes by hydrostatic pressure or uniaxial compression [3]. The invariance relations as important criteria enable one to judge the accuracy and correctness of the measured values of the ultrasonic velocities and their changes by hydrostatic pressure for the different propagation and polarization directions.

In the present paper they are analysed using Thurston and Brugger's formulae [3] for propagation and polarization directions for which the invariance relations are fulfilled in the case of crystals deformed by hydrostatic pressure and belonging to Laue groups CI, CII, TI, and HI.

2. The Basic Formulae

Let us consider an acoustic wave in natural state of the crystal in the form

$$U_i = U_{0i} \exp \left[j\omega \left(t - \frac{a_k N_k}{W} \right) \right], \tag{1}$$

where t is the time, U_{0i} the amplitude, ω the angular frequency, a_k are coordinates, N_k are components of the unit vector at propagation direction $\boldsymbol{N}$, and W is the ultrasonic natural phase velocity. The summation over repeated indices is implied from 1 to 3.

---

[1]) 010 88 Žilina, Veľký Diel, Czechoslovakia.

Substituting (1) into the equation of motion one obtaines the well-known Christoffel's equation in the form

$$|w_{ik} - \varrho_0 W^2 \delta_{ik}|\ U_{0k} = 0\ , \tag{2}$$

where δ_{ik} is Kronecker's tensor,

$$w_{ik} = C^s_{ijkl} N_j N_l \tag{3}$$

is a symmetric tensor of second rank, ϱ_0 the mass density in the natural state, and C^s_{ijkl} are ISOE coefficients. Three eigenvalues $\varrho_0^{\mathrm{L}} W^2$, $\varrho_0^{\mathrm{T1}} W^2$, and $\varrho_0^{\mathrm{T2}} W^2$ follow from (2) for arbitrary propagation direction $\boldsymbol{N}$, where ${}^{\mathrm{L}}W$, ${}^{\mathrm{T1}}W$, and ${}^{\mathrm{T2}}W$ are quasi-longitudinal and two quasi-transverse velocities of acoustic waves.

The first basic invariant of the symmetric tensor (3) in the direction $\boldsymbol{N}$ is

$$\{w_{ii}\}_{\boldsymbol{N}} = \varrho_0 \{{}^{\mathrm{L}}W^2 + {}^{\mathrm{T1}}W^2 + {}^{\mathrm{T2}}W^2\}_{\boldsymbol{N}}\ . \tag{4}$$

We can show by using (3) and (4) that in crystals of cubic, tetragonal, and hexagonal symmetry [4] there exist propagation directions satisfying the equation

$$\{w_{ii}\}_{\boldsymbol{N}_1} = \{w_{ii}\}_{\boldsymbol{N}_2} \tag{5}$$

although in general for $\boldsymbol{N}_1 \neq \boldsymbol{N}_2$

$$\{{}^A W\}_{\boldsymbol{N}_1} \neq \{{}^A W\}_{\boldsymbol{N}_2}\ ,$$

where $A = 1 = \mathrm{L}$, $A = 2 = \mathrm{T1}$, and $A = 3 = \mathrm{T2}$.

Equation (5) corresponds to the invariance of SOE coefficients. Thurston and Brugger in [3] conveniently express TOE coefficients by the stress derivatives (noted by a prime) of the $\varrho_0 W^2$ eigenvalues at zero static stress (noted by zero). Their general formula for the cases of hydrostatic pressure is

$$-\ (\varrho_0^A W^2)'_0 = 1 + 2\,{}^A w\ {}^A F + {}^A G\ , \tag{6}$$

where

$${}^A F = S^T_{aars}\,{}^A U_r\,{}^A U_s\ , \tag{7}$$

$${}^A G = S^T_{aauv} C_{uvprqs} N_r N_s\,{}^A U_p\,{}^A U_q\ , \tag{8}$$

$${}^A w = (\varrho_0\,{}^A W^2)_0 = C_{prqs} N_r N_s\,{}^A U_p\,{}^A U_q\ . \tag{9}$$

S^T_{aars} are the isothermal second-order compliances [5], C_{uvprqs} are TOE coefficients, N_r and ${}^A U_p$ are unit vectors components along propagation and polarization directions in the absence of a static stress, i.e. in the natural state.

By using (6) to (9) we get the result that in the crystals deformed by hydrostatic pressure there exist directions of propagation $\boldsymbol{N}_1 \neq \boldsymbol{N}_2$ for which

$$\left\{\sum_A (\varrho_0\,{}^A W^2)'_0\right\}_{\boldsymbol{N}_1} = \left\{\sum_A (\varrho_0\,{}^A W^2)'_0\right\}_{\boldsymbol{N}_2} \tag{10}$$

although in general for $\boldsymbol{N}_1 \neq \boldsymbol{N}_2$

$$\{(\varrho_0\,{}^A W^2)'_0\}_{\boldsymbol{N}_1} \neq \{(\varrho_0\,{}^A W^2)'_0\}_{\boldsymbol{N}_2}\ .$$

Relation (10) hereafter is called invariance relation of TOE coefficients.

3. Invariance Relations for Second-Order Elastic Coefficients

(i) The first invariant of the symmetric tensor (3) in crystals of the Laue groups CI and CII according to [4] is

$$\{w_{ii}\}_{\boldsymbol{N}} = C^s_{11} + 2C^s_{44}\ , \tag{11}$$

where C^s_{11} and C^s_{44} are ISOE coefficients in Voigt notation.

It follows from (11) that (5) is fulfilled for arbitrary directions in a cubic crystal.

(ii) The first invariant of the symmetric tensor (3) for crystals of the Laue group TI according to [4] is

$$\{w_{ii}\}_{N\perp[001]} = C^{s}_{11} + C^{s}_{44} + C^{s}_{66} \,. \tag{12}$$

It follows from (12) that (5) in crystals of the Laue group TI is fulfilled for arbitrary directions perpendicular to the tetragonal axis.

(iii) The first invariant of the symmetric tensor (3) for crystals of the Laue group HI according to [4] is

$$\{w_{ii}\}_{N\perp[001]} = \tfrac{1}{2}\,(3C^{s}_{11} - C^{s}_{12} + 2C^{s}_{44}) \,. \tag{13}$$

It follows from (13) that (5) in crystals of the Laue group HI is fulfilled for arbitrary directions perpendicular to the hexagonal axis.

4. Invariance Relations for Third-Order Elastic Coefficients

4.1 Laue groups CI and CII of the cubic system

In the case of cubic crystals belonging to the Laue groups CI and CII homogeneously deformed by hydrostatic pressure one gets from (7)

$$F = S^{T}_{jkaa} U_j U_k = (S^{T}_{11} + 2S^{T}_{12})\, U_j U_k \delta_{jk} = \frac{1}{C^{T}_{11} + 2C^{T}_{12}} = \frac{1}{3B^{T}_{0}}\,, \tag{14}$$

where

$$3B^{T}_{0} = C^{T}_{11} + 2C^{T}_{12}$$

is the isothermal bulk modulus at zero pressure. Inserting (8), (9), and (14) into (6) we obtain

$$-\{(\varrho_0\,{}^{A}W^2)'\}_N = 1 + \frac{1}{3B^{T}_{0}}\,(2\varrho_0\,{}^{A}W^2 + C_{jrksaa}\,{}^{A}U_j\,{}^{A}U_k N_r N_s) \,. \tag{15}$$

Relation (15) is fulfilled for each mode A in the propagation direction N. By summation of (15) we get

$$-\sum_{A} \{(\varrho_0\,{}^{A}W^2)'\}_N = 3 + \frac{2}{3B^{T}_{0}} \sum \{\varrho_0\,{}^{A}W^2\}_N +$$

$$+ \frac{1}{3B^{T}_{0}} \left\{\sum_{A} C_{jrksaa}\,{}^{A}U_j\,{}^{A}U_k N_r N_s\right\}_N = 3 + \frac{1}{3B^{T}_{0}}\,\{2(C^{s}_{11} + 2C^{s}_{44}) + C_{irisaa} N_r N_s\}_N \,. \tag{16}$$

Relation (16) is correct for all crystals of CI and CII symmetry homogeneously deformed by hydrostatic pressure in arbitrary propagation directions for three mutually perpendicular modes.

Relation (16) for the Laue group CII has the form

$$-\sum_{A} \{(\varrho_0\,{}^{A}W^2)'\}_N = 3 + \frac{1}{3B^{T}_{0}}\,[2(C^{s}_{11} + 2C^{s}_{44}) +$$

$$+ C_{111} + C_{112} + C_{113} + 2(C_{144} + C_{155} + C_{166})] \tag{17}$$

and for the Laue group CI

$$-\sum_{A} \{(\varrho_0\,{}^{A}W^2)'\}_N = 3 + \frac{1}{3B^{T}_{0}}\,[2(C^{s}_{11} + 2C^{s}_{44}) +$$

$$+ C_{111} + 2C_{112} + 2C_{144} + 4C_{166}] \,. \tag{18}$$

On the right-hand side of (17) and (18) there are all quantities constant and these do not depend on the propagation direction.

It follows from (17) and (18) that in cubic crystals homogeneously deformed by hydrostatic pressure relation (10) is fulfilled for arbitrary propagation directions.

4.2 Laue group TI of the tetragonal system

In the case of tetragonal crystals of the Laue group TI homogeneously deformed by hydrostatic pressure we have for all propagation directions perpendicular to the hexagonal axis

$$\left.\begin{aligned} &\boldsymbol{N} \equiv [N_1 N_2 0]; && N_1^2 + N_2^2 = 1\,, \\ &{}^{\mathrm{L}}\boldsymbol{U} \equiv [{}^{L}U_1 {}^{L}U_2 0]; && {}^{\mathrm{L}}U_1^2 + {}^{L}U_2^2 = 1\,, \\ &{}^{\mathrm{T1}}\boldsymbol{U} \equiv [{}^{\mathrm{T1}}U_1 {}^{\mathrm{T1}}U_2 0]; && {}^{\mathrm{T1}}U_1^2 + {}^{\mathrm{T1}}U_2^2 = 1\,, \\ &{}^{\mathrm{T2}}\boldsymbol{U} \equiv [001]; && {}^{\mathrm{L}}U_i\,{}^{\mathrm{L}}U_j + {}^{\mathrm{T1}}U_i\,{}^{\mathrm{T1}}U_j = \delta_{ij}\,, \quad {}^{\mathrm{L}}U_1\,{}^{\mathrm{T1}}U_1 + {}^{\mathrm{L}}U_1\,{}^{\mathrm{T1}}U_2 = 0\,. \end{aligned}\right\} \tag{19}$$

For arbitrary propagation direction $\boldsymbol{N} \perp [001]$ from (7) and (19) we get

$$\left.\begin{aligned} {}^{\mathrm{L}}F &= {}^{\mathrm{T1}}F = S_{11}^{\mathrm{T}} + S_{12}^{\mathrm{T}} + S_{13}^{\mathrm{T}}\,, \\ {}^{\mathrm{T2}}F &= 2S_{13}^{\mathrm{T}} + S_{33}^{\mathrm{T}}\,. \end{aligned}\right\} \tag{20}$$

In the case of the Laue group TI by using (8), (19), and (20) we get

$$\left.\begin{aligned} {}^{A}G = {} & S_{11}^{\mathrm{T}}[(C_{111} + C_{112})\,(N_1^2\,{}^{A}U_1^2 + N_2^2\,{}^{A}U_2^2) + \\ & + 2C_{166}(N_1^2\,{}^{A}U_2^2 + N_2^2\,{}^{A}U_1^2) + 4(C_{112} + C_{116})\,N_1 N_2\,{}^{A}U_1\,{}^{A}U_2 + \\ & + (C_{144} + C_{155})\,{}^{A}U_3^2] + \\ & + S_{12}^{\mathrm{T}}[(C_{111} + C_{112})\,(N_1^2\,{}^{A}U_1^2 + N_2^2\,{}^{A}U_2^2) + \\ & + 2C_{166}(N_1^2\,{}^{A}U_2^2 + N_2^2\,{}^{A}U_1^2) + 4(C_{112} + C_{116})\,N_1 N_2\,{}^{A}U_1\,{}^{A}U_2 + \\ & + (C_{144} + C_{155})\,{}^{A}U_3^2] + \\ & + S_{13}^{\mathrm{T}}[(2C_{113} + C_{111} + C_{112})\,(N_1^2\,{}^{A}U_1^2 + N_2^2\,{}^{A}U_2^2) + \\ & + 2(C_{366} + C_{166})\,(N_1^2\,{}^{A}U_2^2 + N_2^2\,{}^{A}U_1^2) + \\ & + 4(C_{123} + C_{112} + C_{166} + C_{366})\,N_1 N_2\,{}^{A}U_1\,{}^{A}U_2 + \\ & + 2(C_{144} + C_{344})\,{}^{A}U_3^2] + \\ & + S_{33}^{\mathrm{T}}[C_{113}(N_1^2\,{}^{A}U_1^2 + N_2^2\,{}^{A}U_2^2) + C_{366}(N_1^2\,{}^{A}U_2^2 + N_2^2\,{}^{A}U_1^2) + \\ & + 2(C_{123} + C_{366})\,N_1 N_2\,{}^{A}U_1\,{}^{A}U_2 + C_{344}\,{}^{A}U_3^2]\,. \end{aligned}\right\} \tag{21}$$

By using (6), (12), (19), and (21) we get

$$\begin{aligned} -\sum_A \{(\varrho_0\,{}^{A}W^2)'\}_{N\perp[001]} &= 3 + \sum_A (2\,{}^{A}F\,{}^{A}w + {}^{A}G) \\ &= 3 + (S_{11}^{\mathrm{T}} + S_{12}^{\mathrm{T}} + S_{13}^{\mathrm{T}})\,(2C_{11}^{\mathrm{s}} + 2C_{66}^{\mathrm{s}} + C_{111} + \\ &\quad + C_{112} + C_{144} + C_{155} + 2C_{166}) + \\ &\quad + (2S_{13}^{\mathrm{T}} + S_{33}^{\mathrm{T}})\,(2C_{44}^{\mathrm{s}} + C_{113} + C_{344} + C_{366})\,. \end{aligned} \tag{22}$$

It follows from (22) that in tetragonal crystals homogeneously deformed by hydrostatic pressure relation (10) is fulfilled for arbitrary direction perpendicular to the tetragonal axis.

4.3 Laue group HI of the hexagonal system

In the case of hexagonal crystals of the Laue group HI homogeneously deformed by hydrostatic pressure for all propagation directions perpendicular to the hexagonal axis the conditions (19) are fulfilled, too. For arbitrary directions $\boldsymbol{N} \perp [001]$ from (7) we get

$$\left.\begin{aligned} {}^{\mathrm{L}}F &= {}^{\mathrm{T1}}F = S^{\mathrm{T}}_{11} + S^{\mathrm{T}}_{12} + S^{\mathrm{T}}_{13},\\ {}^{\mathrm{T2}}F &= 2S^{\mathrm{T}}_{13} + S^{\mathrm{T}}_{33}\,. \end{aligned}\right\} \tag{23}$$

By using (6), (13), and (23) in a similar way as in the case of tetragonal crystals we obtain

$$-\sum_{A} \{(\varrho_0^A W^2)'\}_{N\perp[001]} = 3 + \sum_{A} (2\, {}^AF\, {}^Aw + {}^AG)]$$

$$= 3 + (S^{\mathrm{T}}_{11} + S^{\mathrm{T}}_{12} + S^{\mathrm{T}}_{13})\,[3C^{\mathrm{s}}_{11} - C^{\mathrm{s}}_{12} + C_{111} + \tfrac{1}{2}(C_{112} + C_{222}) + C_{144} + C_{155}] -$$

$$- (2S^{\mathrm{T}}_{13} + S^{\mathrm{T}}_{33})\,[2C^{\mathrm{s}}_{44} + C_{344} + \tfrac{1}{2}(3C_{113} - C_{123})]\,. \tag{24}$$

It follows from (24) that in hexagonal crystals homogeneously deformed by hydrostatic pressure relation (10) is satisfied for arbitrary directions perpendicular to the hexagonal axis.

5. Conclusion

The results reported in this paper give a possibility to classify experimental data of ultrasonic velocities and their changes with hydrostatic pressure with respect to their applicability in calculations of TOE coefficients. The experimental data of ultrasonic velocities and their changes with pressure must satisfy the following relations: for Laue group CII (11) and (17), for CI (11) and (18), for TI (12) and (22) and for HI (13) and (24). If this does not occur the measured values of the velocity and its changes must be considered as distorted by experimental errors. If different authors reported different results for the invariance relations for the same crystals measured at comparable experimental conditions, then their results indicate different elastic properties of the investigated crystals.

References

[1] W. P. Mason and T. B. Bateman, J. Acoust. Soc. Amer. **36**, 644 (1964).
[2] K. Brugger, Phys. Rev. **133**, A 1611 (1964).
[3] R. N. Thurston and K. Brugger, Phys. Rev. **133**, A 1604 (1964).
[4] P. Hegedüs and S. Kolník, Acustica to be published.
[5] Yu. Sirotin and M. P. Saskolskaya, Osnovy kiristallofiziki, Izd. Nauka, Moskva 1975.

(Received March 20, 1979)

phys. stat. sol. (a) **54**, 355 (1979)

Subject classification: 10.2 and 12.2; 22.6

Department of Mechanical Engineering, Faculty of Engineering, Ibaraki University, Hitachi[1]) *(a)*,
Départment de Physique des Matériaux, associè au C.N.R.S. Equipe Recherche n° 172, Université Claude Bernard de Lyon 1, Villeurbanne[2]) *(b), and*
Department of Physics, Faculty of Science, Sciencé University of Tokyo[3]) *(c)*

Elastic Parameters, Elastic Energy, and Stress Fields of Dislocations in TiO_2 Rutile Crystals

By

Y. MOTOHASHI (a), M. G. BLANCHIN (b), E. VICARIO (b), G. FONTAINE (b), and S. OTAKE (c)

The elastic parameters of dislocations in rutile crystals in the temperature range from 100 to 298 K are determined by numerical calculations in terms of an anisotropic elasticity theory. The results are applied to possible dislocation reactions and elastic stress fields of dislocations. The energy factor of edge dislocations in planes of important zone axes and the dependence of the energy factor on the edge-screw composition angle θ for verious dislocations are analysed. Moreover, the relation between the dislocation parameters and the choice of slip system is considered based on the criterion of minimum dislocation energy. According to the analysis of dislocation reactions, it appears that the $\langle\bar{1}01\rangle$ $\{010\}$, $\langle\bar{1}\bar{1}1\rangle$ $\{101\}$, and $\langle 100\rangle$ $\{101\}$ type dislocations are energetically unstable and there is a chance of their dissociating into other perfect or partial dislocations. It is believed that the $\langle\bar{1}01\rangle$ $\{101\}$ type dislocation is possible to dissociate into $1/2\langle\bar{1}01\rangle$ $\{101\}$ type collinear partial dislocations.

Les paramètres élastiques de différentes dislocations dans des monocristaux de rutile ont été calculés numériquement en fonction de la temperature entre 100 et 298 K en utilisant la théorie de l'élasticité anisotrope. Les résultats obtenus sont utilisés pour étudier les réactions de dislocations possibles et le champ de contraintes élastiques autour de certaines dislocations. On analyse également les valeurs du coefficient d'energie K pour des dislocations coins appartenant à des plans correspondant à des axes de zone importants et la variation de ce coefficient en fonction de l'angle d'orientation de la ligne pour différentes dislocations. On discute enfin la relation entre les valeurs des paramètres élastiques des dislocations et le choix du système de glissement par rapport au critère d'énergie de dislocation minimale. L'analyse des réactions de dislocations montre que les dislocations $\langle\bar{1}01\rangle$ $\{010\}$, $\langle\bar{1}\bar{1}1\rangle$ $\{101\}$ et $\langle 100\rangle$ $\{101\}$ sont énergétiquement instables et doivent donc se dissocier en d'autres dislocations parfaites ou partielles. Il apparait d'autre part que la dissociation d'une dislocation $\langle\bar{1}01\rangle$ $\{101\}$ en deux partielles colinéaires $\frac{1}{2}\langle\bar{1}01\rangle$ $\{101\}$ est toujours énergétiquement favorable.

1. Introduction

The thermodynamically stable form of titanium dioxide TiO_2 is rutile, with a tetragonal crystalline structure. Rutile has been extensively studied owing to its interesting electrical and optical properties, which are strongly correlated with the ability of this compound to lose stoichiometry under reduced oxygen pressure. All these properties are obviously dependent on the interaction between dislocations and point defects [1]. Therefore, the knowledge of plastic properties of rutile single crystals appeared as a necessary purpose of study. The earliest studies in this field [2 to 4]

[1]) Nakanarúsawa, Hitachi, Ibaraki 316, Japan.
[2]) 69621 Villeurbanne, France.
[3]) Shinjuku-ku, Tokyo 162, Japan.

were made using slip-line and etch-pit observations in order to identify the slip systems in specimens generally deformed by flexion: the active glide planes were found to be {101}, {110}, and {100}. Ashbee and Smallman [5, 6] first performed uniaxial compression tests in air on synthetic bulk single crystals with different crystallographic orientations in order to successively activate several possible glide systems. It was concluded that slip systems observed were only {101} ⟨10$\bar{1}$⟩ and {110} ⟨001⟩ and the both systems were active above 850 K, but the former was much easier than the latter. Similar experiments on synthetic single crystals were recently performed by Blanchin et al. [7] under controlled oxygen pressure to maintain the initial stoichiometry of the crystals as far as possible when the temperature is increased. In this study the active slip systems were found again to be {101} ⟨10$\bar{1}$⟩ and {110} ⟨001⟩ and the latter one did not provide any ductility of the deformed samples [1, 7]. Foils directly cut from the bulk deformed crystals and thinned by ion bombardment were observed by transmission electron-microscopy. These observations revealed high densities of edge dislocations with a ⟨10$\bar{1}$⟩ Burgers vector in the active {101} planes but without any obvious dissociation of these dislocations.

In terms of the dislocation theory both the slip plane and the slip direction may be governed by the energy required to create the needed dislocations for continuous slip. Recently, Steeds [8] treated some analytic cases of the elastic parameters for the dislocations belonging to the {110} ⟨001⟩ and the {101} ⟨10$\bar{1}$⟩ systems. However, so far as the authors are aware, to date the elastic properties of the interesting dislocations in rutile have not been investigated; at least no result has been given in the literature.

In the present paper, the dislocation energies and the stress fields around some edge dislocations for various slip systems are presented in terms of anisotropic elasticity theory and the relations between the dislocation parameters and the choice of slip system are discussed.

2. Results and Discussion

2.1 Dislocation parameters

2.1.1 Experimental data

The unit cell corresponding to the tetragonal structure of rutile (space group P4/mnm) is shown in Fig. 1 with the commonly used axes. The values of the cell parameters are taken from the most recent crystallographic study by Straumanis et al. [9], and $c/a = 0.6439$.

The elastic constants used in the present paper are those obtained by Fritz [10]. Dislocation parameters and stress fields around dislocations are calculated in terms of anisotropic elasticity theory following the calculation procedure developed by Eshelby et al. [11].

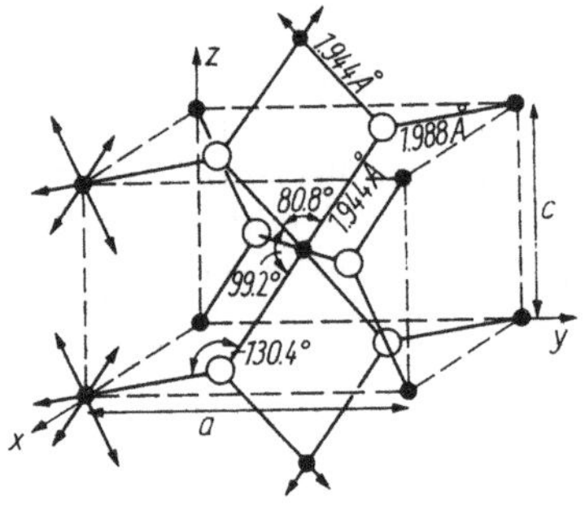

Fig. 1. The rutile unit cell: $a = 4.5937$ Å and $c = 2.9581$ Å. ○ O, ● Ti

2.1.2 *Calculation data*

According to Eshelby [12], the elastic energy per unit length of a long straight dislocation of orientation θ is given by

$$E(\theta) = \frac{K(\theta)\, b^2}{4\pi} \lg \left(\frac{R}{r_0}\right), \tag{1}$$

where b is the magnitude of the Burgers vector, r_0 and R are the inner and outer cut-off radii, respectively, θ is the angle between the orientation of the dislocation line and its Burgers vector, and K is the energy factor which is given by

$$Kb^2 = \operatorname{Im} b_i \sum_{l=1}^{3} (\pm)_{(l)} B_{i2(l)} D_{(l)}\,, \tag{2}$$

where Im denotes imaginary part of, and the quantities $B_{i2(l)}$ and $D_{(l)}$ are functions of the elastic constants referred to the dislocation axis. $B_{i2(l)}$ and $D_{(l)}$ can be calculated from the formulas of Foreman [13].

Usually, the elastic constants of rutile crystals have been given for a coordinate system x_i in which x_1, x_2, and x_3 are parallel to the $[a00]$, $[0a0]$, and $[00c]$ lattice vectors, respectively (Fig. 1). Therefore, it is necessary to find the elastic constants which are referred to the dislocation axes, i.e. to a set of orthogonal axes with the x_3-axis along the dislocation line. Transformation of the elastic constants referred to the crystal axis into those referred to the dislocation axis is given by

$$C_{ijkl} = \alpha_{ip}\alpha_{jq}\alpha_{kr}\alpha_{ls}C_{pqrs}\,, \tag{3}$$

where α_{ij} is the cosine of the angle between the x_i dislocation axis and x_j crystal axis.

2.1.3 *Results*

First of all, the energy factors of dislocations in planes of $\langle 10\bar{1}\rangle$, $\langle 001\rangle$, and $\langle 100\rangle$ zone axes are calculated since important slip vectors in rutile are $\langle a0c\rangle$, $\langle 00c\rangle$, and $\langle a00\rangle$. Fig. 2 shows variations of the K-values of pure edge dislocations in the planes of these axes. Among the K-values for the $\langle 10\bar{1}\rangle$ zone axis, the K for a (100) is smallest and the K for a (101) is also very small, with the same order of magnitude. On the other hand, the K-value for the (101) plane of the $\langle 001\rangle$ and the $\langle 100\rangle$ zone axes is very large as compared with K's for other planes of these zone axes. Fig. 3 shows

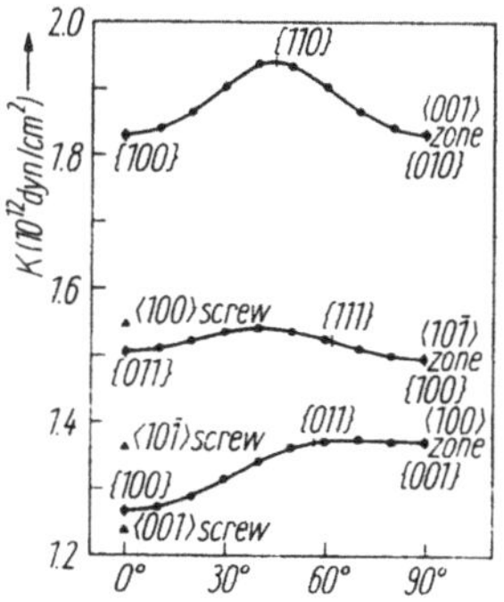

Fig. 2

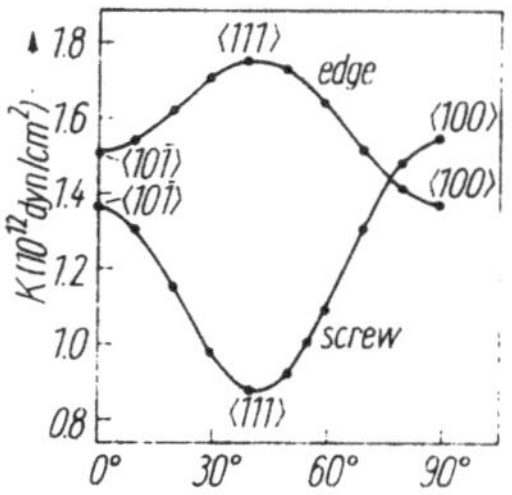

Fig. 3

Fig. 2. Variation of the energy factor, K, at 298 K for pure edge dislocations belonging to the $\{pqr\}$ $\langle 001\rangle$, $\{p'q'r'\}$ $\langle 10\bar{1}\rangle$, and $\{p''q''r''\}$ $\langle 100\rangle$ slip systems

Fig. 3. Variation of K for pure edge and pure screw dislocations in the $\{101\}$ $\langle lmn\rangle$ slip system at 298 K

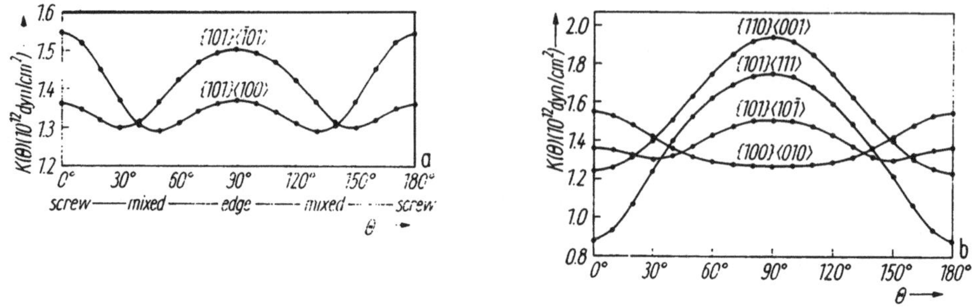

Fig. 4. Dependence of edge–screw composition angle θ on the K-value of various dislocations at 298 K

variations of the K-values of pure edge and pure screw dislocations with various Burgers vectors in the {101} slip plane. The dependence of θ on the K-value at 298 K of important slip dislocations is shown in Fig. 4. The K's of the {101} ⟨10$\bar{1}$⟩ and {101} ⟨100⟩ dislocations become the smallest when their dislocation lines lie along a mixed orientation. In cases of the {100} ⟨100⟩ and the {101} ⟨100⟩ dislocations, the K for the screw orientation is largest. These results clearly indicate that the elastic properties of dislocations in rutile are quite anisotropic.

Table 1

Dislocation parameters for a rutile crystal at 298 K

(a)

slip system	energy factor, K (10^{12} dyn/cm^2)	energy, Kb^2 (10^{-4} erg/cm)	magnitude of Burgers vector b (Å)	interplanar spacing d (Å)	shear modulus, C (10^{12} dyn/cm^2)	dislocation width, ζ ($= Kd/2C$)
{101}⟨10$\bar{1}$⟩ edge	1.503	44.86	5.4637	2.487	1.727	1.082
{101}$\frac{1}{2}$⟨10$\bar{1}$⟩ edge	1.503	11.22	2.7319	2.487	1.727	1.082
{101}⟨010⟩ edge	1.368	28.87	4.5937	2.487	1.727	0.985
{101}⟨$\bar{1}$11⟩ edge	1.748	89.09	7.1382	2.487	0.7159	3.036
{110}⟨001⟩ edge	1.938	16.96	2.9581	3.248	1.239	2.540
{100}⟨010⟩ edge	1.265	26.70	4.5937	4.593	1.930	1.505
{100}⟨001⟩ edge	1.827	15.99	2.9581	4.593	1.239	3.386
{001}⟨100⟩ edge	1.368	28.87	4.5937	2.958	1.239	1.633
{111}⟨$\bar{1}$01⟩ edge	1.519	45.34	5.4637	2.187	1.598	1.039
{111}$\frac{1}{2}$⟨$\bar{1}$01⟩ edge	1.519	11.34	2.7319	2.187	1.598	1.039
⟨10$\bar{1}$⟩ screw	1.361	40.63	5.4637			
$\frac{1}{2}$⟨10$\bar{1}$⟩ screw	1.361	10.16	2.7319			
⟨010⟩ screw	1.546	32.63	4.5937			
⟨$\bar{1}$11⟩ screw	0.8773	44.70	7.1382			
⟨001⟩ screw	1.239	10.84	2.9581			

(b)

slip system	d/b	ζ (Å)	S	τ_P/C
{101}⟨10$\bar{1}$⟩	0.455	1.08	0.717	0.15
{110}⟨001⟩	1.098	2.54	0.05	6×10^{-5}
{100}⟨001⟩	1.55	3.39	0.01	2×10^{-6}

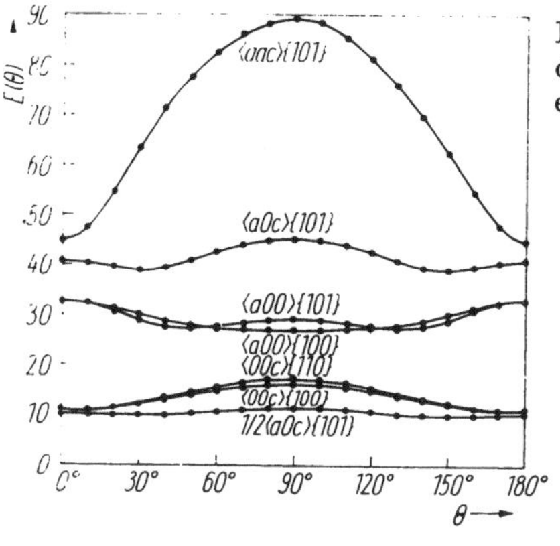

Fig. 5. Dependence of θ on the elastic energy E of various dislocations at 298 K. Unit of energy $4\pi 10^4/\log(R/r_0)$ erg s/cm

In terms of dislocation theory, it is usually assumed that the direction and plane of slip are determined by the criterion of minimum dislocation energy. The dislocation energies and the dislocation parameters of pure edge and pure screw dislocations for various slip systems are summarized in Table 1a, in which the interplanar spacing d and the width ζ of the dislocations are also listed. ζ is calculated from the simple formula $\zeta = Kd/2C$, where C is the shear modulus on the slip plane in the direction of the Burgers vector. Since the nature of the bonds in rutile is rather complex, partly ionic and partly covalent, we performed the calculation of d in a simple way, using the clasical formulae for the tetragonal structures.

The elastic energy of the edge dislocation for the {100} ⟨001⟩ system is smallest among those of the important dislocations. This slip dislocation, however, has not been observed in rutile.

A consideration for the choice of slip system in rutile, based on a comparison among only the energies of pure edge dislocations for each slip system, is thorough enough because the orientation dependence of the dislocation energy is different for each dislocation system as shown in Fig. 4. Therefore, it is assumed, to compare more precisely the self-energies of each dislocation, that (i) the dislocations belonging to the $\{pqr\}$ $\langle lmn\rangle$ slip systems exist as circular loops with the same radius r_l ($r_l \gg R$), on the $\{pqr\}$ planes, and (ii) the energy necessary to make the circular loop from the straight shape and the interaction energy are negligibly small. Then, the elastic energy of this dislocation loop is given by[4])

$$E_l = \frac{b^2 r_l}{4\pi} \log\left(\frac{R}{r_0}\right) \int_0^{2\pi} K(\theta)\, d\theta\,. \tag{4}$$

Fig. 5 shows $E(\theta)$, the elastic energy of dislocation as a function of θ for various slip systems at 298 K. The E_l-value corresponds with the area surrounded by the x-axis and the $E(\theta)$ curve from $\theta = 0$ to 2π.

2.2 *Discussion on the choice of slip system*

It is found from Fig. 5 that the E_l of a dislocation in the {100} ⟨001⟩ slip system is again smallest among those of perfect dislocations, yet this slip system is hardly observed. The E_l-value for the {110} ⟨001⟩ system is very small too, very close to the

[4]) When the relation $r_l \ll R$ does not hold, the approximation of (4) may be bad. A precise analysis of this problem has been presented in another paper [14].

other one. As explained above, the {110} ⟨001⟩ system can effectively be activated above 850 K under rather high stresses. However, systematic experiments showed that the behaviour of specimens deformed by this glide system remains essentially brittle even at very low strain rates [1, 7]. For the same values of temperature, stress, and strain-rate, the activation of the {101} ⟨10$\bar{1}$⟩ slip system is much easier than the {110} ⟨001⟩ system and provides a substantial ductility of the crystals, although the corresponding E_l-value is somewhat higher than that for the {110} ⟨001⟩ system.

Therefore, let us make a comparison of other dislocation parameters for these systems. These reviewed parameters are:

1. the ratio d/b (Table 1a),
2. the width ζ of the dislocations (Table 1a),
3. the ratio S of the stress needed to move the dislocation to the rigid planar shear stress [12],

$$S = \frac{4\pi\zeta}{b} \exp\left(-\frac{2\pi\zeta}{b}\right), \tag{5}$$

4. the Peierls stress τ_p given by [12]

$$\frac{\tau_p}{C} = \frac{2K}{C} \exp\left(-\frac{4\pi\zeta}{b}\right). \tag{6}$$

The corresponding values are listed in Table 1b and the conclusions are the same as those concerning the E_l-value: the activation of the {100} ⟨001⟩ and {110} ⟨001⟩ should be easy. Then the question remains, why is it not?

For a crystalline oxide structure like that of rutile, the bond and coordination problems do have a prominent importance. Therefore, the activation of the easiest slip plane can be essentially explained by structural considerations.

It is well known that the rutile structure can be described as an almost hexagonal close-packed arrangement of the oxygen ions (where titanium ions fill alternate rows of octahedral interstices [1]). And we can see that the {101} ⟨10$\bar{1}$⟩ system corresponds effectively to prismatic slip of this h.c.p. arrangement but the {100} ⟨001⟩ and the {110} ⟨001⟩ ones do not correspond to any known slip system of the h.c.p. structure. Furthermore, the easiest glide system in the h.c.p. structures with $c/a < 1.6$, as rutile, is precisely the prismatic one. Another cause will be that the effect of covalent and/or ionic bonds contained in the rutile crystal on the choice of slip system; in particular the choice of slip plane is quite important, because it is known [15] that, for a bismuth-type crystal, which is a rhombohedral crystal structure and contains both covalent and metallic bond layers in it, the choice of slip plane is mainly determined by the covalent bond density of the interplanar spacings. At present, this problem for rutile crystals remains unsolved.

On the other hand the fairly large E_l-value of a {101} ⟨10$\bar{1}$⟩ type perfect dislocation could imply that this dislocation should be dissociated into {101} $\frac{1}{2}$⟨10$\bar{1}$⟩ type collinear partial dislocations. The E_l-value of the {101} $\frac{1}{2}$⟨10$\bar{1}$⟩ partial dislocation is effectively the smallest one among the others, as shown in Fig. 5. The concept of this dissociation is also supported by structural considerations, because the fault resulting from such a dissociation occurs only on the titanium ion arrangements while the h.c.p. arrangement of the oxygen ions remains approximately unaffected [1, 16]. In the same view, a similar dissociation of the perfect {110} ⟨001⟩ dislocation seems very difficult because both the anion and cation sublattices are disturbed [1, 4].

Up to date our electron microscopy observations using weak-beam techniques could not, unfortunately, allow a choice between the ⟨10$\bar{1}$⟩ and the $\frac{1}{2}$⟨10$\bar{1}$⟩ Burgers vectors.

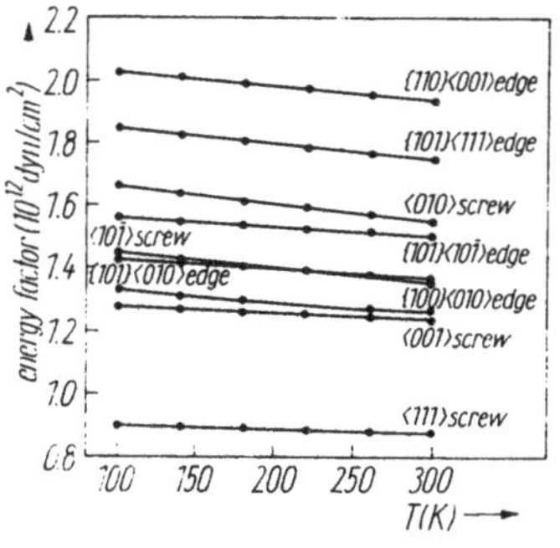

Fig. 6. Temperature dependence of the K-value for various dislocations in rutile

Some experimental difficulties arise from parasitic contrasts due to the segregation of impurity atoms on the glide dislocations. If the dissociation exists, the width should be less than 50 Å, as shown by the observations. However, the present results about the elastic energies appear as a good support for the dissociation of the $\langle 10\bar{1}\rangle$ dislocations.

The arguments stated above about the values of E_l for the choice of the slip system are not sufficient, because the E_l's treated here are the values at 298 K but the plastic deformation of the crystals usually occurs at temperatures above 850 K. The temperature dependence of the K's for various dislocations is shown in Fig. 6. The values of K decrease monotonically as the temperature is increased for the temperature range from 100 to 298 K. There is no phase change or singularity points for temperatures from 298 up to 850 K, and so it is believed that the relative magnitude of the energies among various dislocations at temperatures below 298 K will be preserved above 850 K (see Fig. 6). Therefore, the comparison of the self-energies discussed here may be valid. Measurements of the elastic constants above 850 K are desirable.

2.3 *Dislocation reactions*

In the same view we examined also different possible reactions in order to determine those which are energetically possible.

The possible Burgers vectors are the lattice vectors $\langle 101\rangle$, $\langle 001\rangle$, $\langle 100\rangle$, and $\langle 111\rangle$ and possible slip planes are the low-index planes $\{101\}$, $\{100\}$, $\{001\}$, $\{110\}$, and $\{111\}$. The plane $\{101\}$ is the easiest slip plane and contains the important slip vectors $\langle 101\rangle$, $\langle 111\rangle$, and $\langle 100\rangle$. In this section, dissociations of dislocations in the $\{101\}$ plane into other dislocations are principally studied. The reactions are as follows:

$$\left.\begin{aligned} &\text{a)}\ \langle\bar{1}01\rangle \to \langle\bar{1}00\rangle + \langle 001\rangle\,, \\ &\text{b)}\ \langle\bar{1}\bar{1}1\rangle \to \langle\bar{1}01\rangle + \langle 0\bar{1}0\rangle\,, \\ &\text{c)}\ \langle 100\rangle \to \tfrac{1}{2}\langle 101\rangle + \tfrac{1}{2}\langle 10\bar{1}\rangle\,. \end{aligned}\right\} \qquad (7)$$

The criterion for reaction will be that the energy of the product dislocations is less than the energy of the reactant dislocation, or neglecting core energy, $\Delta E > 0$, where

$$\Delta E = \frac{1}{4\pi}\,(k_1 b_1^2 - k_2 b_2^2 - k_3 b_3^2)\log\left(\frac{R}{r_0}\right), \qquad (8)$$

where the subscript 1 refers to the reactant dislocation, the subscripts 2 and 3 refer to the product dislocations. It is also assumed that the orientation of the dislocations do not change after the dissociation. When $\Delta E > 0$, the dislocation dissociation is energetically possible.

Tabelle 2

Detailed analyses for each dislocation reaction and its energy for a rutile crystal at 298 K

		reactant dislocation		product dislocations				
(a)		$\langle \bar{a}0c \rangle$	→	$\langle \bar{a}00 \rangle$	+	$\langle 00c \rangle$	ΔE	$\Delta E_{\text{ratio}} = \Delta E / K_1 b_1^2 \times 100(\%)$
	d.l.	s.p. e.s.c. (θ)		s.p. e.s.c. (θ)		s.p. e.s.c. (θ)		
	[010]	(101) edge (90°)		(001) edge (90°)		(100) edge (90°)	0.0	0.0
	$[\bar{1}11]$	(101) mixed (40.06°)		$(0\bar{1}1)$ mixed (49.95°)		(110) mixed (65.52°)	− 3.85	− 9.8
	$[\bar{1}01]$	— screw (0°)		(010) mixed (32.78°)		(010) mixed (57.22°)	− 3.29	−8.0
	[001]	(010) mixed (57.22°)		(010) edge (90°)		— screw (0°)	+ 5.87	+13.5
	$[\bar{1}00]$	(010) mixed (32.78°)		— screw (0°)		(010) edge (90°)	− 6.82	−16.3
(b)		$\langle \bar{a}\bar{a}c \rangle$	→	$\langle \bar{a}0c \rangle$	+	$\langle 0\bar{a}0 \rangle$		
near	[575]	(101) edge (90°)		(101) mixed (49.95°)		(101) mixed (40.06°)	+20.8	+23.4
	[010]	(101) mixed (49.94°)		(101) edge (90°)		— screw (0°)	+ 0.11	+ 0.1
	$[\bar{1}01]$	(101) mixed (40.06°)		— screw (0°)		(101) edge (90°)	+ 1.84	+ 2.6
	$[11\bar{1}]$	— screw (0°)		(101) mixed (40.06°)		(101) mixed (49.95°)	−21.69	−48.5
(c)		$\langle a00 \rangle$	→	$1/2 \langle a0c \rangle$	+	$1/2 \langle a0\bar{c} \rangle$		
	$[01\bar{1}]$	(011) edge (90°)		$(\bar{1}11)$ mixed (36.48°)		(111) mixed (36.48°)	+ 7.72	+26.7
	$[11\bar{1}]$	(011) mixed (49.95°)		$(\bar{1}21)$ mixed		(101) mixed (40.06°)	+ 6.24	+22.9
	[100]	— screw (0°)		(010) mixed (32.78°)		(010) mixed (32.78°)	+11.73	+35.9
	[001]	(010) edge (90°)		(010) mixed (57.22°)		(010) mixed (122.78°)	+ 4.99	+18.7
	$[10\bar{1}]$	(010) mixed (32.78°)		(010) mixed (65.56°)		— screw (0°)	+ 8.31	+28.2

d.l. The orientation of the dislocation line, s.p. the slip plane, e.s.c. the edge–screw composition, θ the angle between the Burgers vector and the d.l.

Analysis shows there are many physically non-equivalent and important dislocation dissociations, and they are summarized in Table 2. According to the analysis of reaction energy, ΔE, the following results can be concluded:

(i) The dissociation of a $\langle\bar{1}01\rangle$ $\{101\}$ type dislocation into a $\langle\bar{1}00\rangle$ and a $\langle 001\rangle$ type dislocations is energetically impossible and the $\langle\bar{1}01\rangle$ dislocation may be stable in the $\{101\}$ plane. A $\langle\bar{1}01\rangle$ $\{010\}$ type dislocation, however, may be unstable and dissociates into $\langle\bar{1}00\rangle$ and $\langle 001\rangle$ dislocations.

(ii) A $\langle\bar{1}\bar{1}1\rangle$ dislocation in the $\{101\}$ plane except for a $\langle\bar{1}\bar{1}1\rangle$ pure screw dislocation may be unstable, because there is an energy gain when the $\langle\bar{1}\bar{1}1\rangle$ dislocation dissociates into $\langle\bar{1}01\rangle$ and $\langle 0\bar{1}0\rangle$ dislocations.

(iii) The dissociation of a $\langle 100\rangle$ dislocation into two $\frac{1}{2}\langle 101\rangle$ type partial dislocations is always possible.

(iv) In the following reaction

$$\langle 001\rangle \rightarrow \tfrac{1}{2}\langle 101\rangle + \tfrac{1}{2}\langle\bar{1}01\rangle \tag{9}$$

the reaction energy, ΔE, is always negative; then it may be impossible that the $\langle 001\rangle$ dislocation dissociates into $\frac{1}{2}\langle 101\rangle$ type partial dislocations.

Table 3

The angle φ_i between the x-axis and the zero-stress line, defined positive in the counterclockwise sense and measured from the x-axis

dislocation	φ_1	φ_2	φ_3
$\{101\}\langle 10\bar{1}\rangle$ edge	$+54.29°$	$+82.10°$	$-42.89°$
$\{101\}\langle 111\rangle$ edge	$+43.82°$	$+87.73°$	$-46.45°$
$\{101\}\langle 010\rangle$ edge	$+47.86$	$90°$	$-47.86°$
$\{110\}\langle 001\rangle$ edge	$+43.95°$	$90°$	$-43.95°$
$\{100\}\langle 010\rangle$ edge	$+45.0°$	$90°$	$-45.0°$
$\{100\}\langle 001\rangle$ edge	$+40.87°$	$90°$	$-40.87°$
$\{001\}\langle 100\rangle$ edge	$+49.14°$	$90°$	$-49.17°$
$\{111\}\langle 10\bar{1}\rangle$ edge	$+41.28°$	$+89.86°$	$-47.75°$

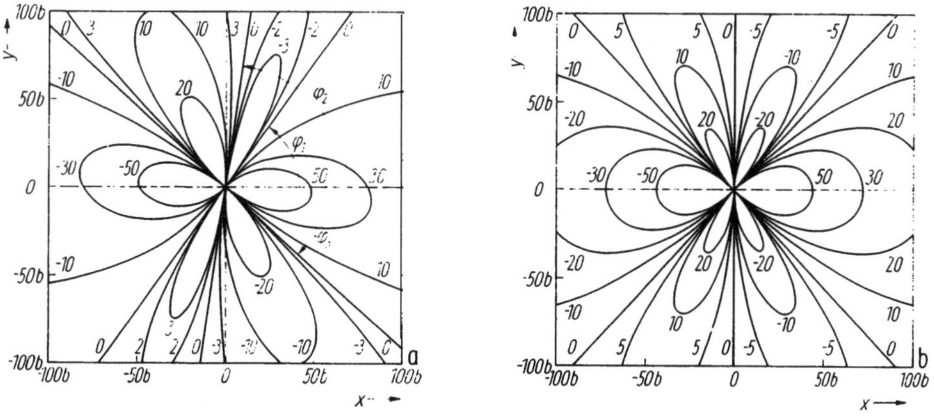

Fig. 7. The elastic stress field of a single edge dislocation, τ_{xy}, at 298 K in rutile. Unit of stress 10^8 dyn/cm^2. Unit of distance b. a) $\{101\}$ $\langle 10\bar{1}\rangle$ system, b) $\{101\}$ $\langle 010\rangle$ system

2.4 Dislocation stress field

The stress field, τ_{xy}, around a {101} ⟨10$\bar{1}$⟩ type edge dislocation is shown in Fig. 7a. The stress field around the dislocation is asymmetrical with respect to the x- and y-axes, but it has a two-fold symmetry about the z-axis (except for changes in sign).

In general, the stress fields of dislocations are characterized by positions of zero-stress contours in the field. The zero-stress contours around an isotropic dislocation are along straight lines, $x = 0y$ and $x = \pm y$, whereas the zero-stress lines in the stress field of the {101} ⟨10$\bar{1}$⟩ edge dislocation in rutile are $x = 0.7188y$, $x = 0.1388y$, and $x = -1.077y$. Fig. 7b shows the stress field, τ_{xy}, around a {101} ⟨010⟩ edge dislocation, in which the zero-stress lines are $x = \pm 0.9048y$ and $x = 0y$.

The angle φ_i between the zero-stress lines and the x-axis in the stress field, τ_{xy}, around various edge dislocations at 298 K are summarized in Table 3. The stress fields of edge dislocations with the Burgers vectors ⟨10$\bar{1}$⟩ and ⟨$\bar{1}\bar{1}$1⟩, such as the {101} ⟨10$\bar{1}$⟩ dislocation show a distinct anisotropic effect. On the other hand, the stress fields of dislocations with the Burgers vectors ⟨100⟩ and ⟨001⟩ are symmetrical with respect to the x- and y-axes and show rather isotropic features.

3. Conclusion

According to the experimental results, {101} ⟨10$\bar{1}$⟩ is the easiest slip system in rutile. The present calculations suggest that such a ⟨10$\bar{1}$⟩ dislocation should be dissociated into collinear partial dislocations with a $\frac{1}{2}$⟨10$\bar{1}$⟩ Burgers vector. This assumption is strongly supported by structural considerations. {110} ⟨001⟩ is the easiest slip system after {101} $\frac{1}{2}$⟨10$\bar{1}$⟩, but its activation requires a much higher stress because no suitable dissociation of the ⟨001⟩ Burgers vector can be expected. The stress field of a ⟨10$\bar{1}$⟩ edge dislocation shows a distinct anisotropic effect.

References

[1] M. G. Blanchin, Thése Doctorat d'Etat Es-Sciences, Lyon 1978.
[2] J. B. Wachtman and L. H. Maxwell, J. Amer. Ceram. Soc. **37**, 291 (1954).
[3] W. M. Hirthe, N. R. Adsit, and J. O. Brittain, in: Direct Observations of Imperfections in Crystals, Interscience Publishers, New York 1962 (p. 135).
[4] W. H. Hirthe and J. O. Brittain, J. Amer. Ceram. Soc. **45**, 546 (1962).
[5] K. H. G. Ashbee and R. E. Smallman, J. Amer. Ceram. Soc. **46**, 211 (1963).
[6] K. H. G. Ashbee and R. E. Smallman, Proc. Roy. Soc. **A274**, 195 (1963).
[7] M. G. Blanchin, L. Kubin, and G. Fontaine, to be published.
[8] J. W. Steeds, in: Introduction to Anisotropic Elasticity Theory of Dislocations, Clarendon Press, Oxford 1973 (p. 201).
[9] M. E. Straumanis, T. Ejima, and W. J. James, Acta cryst. **14**, 493 (1961).
[10] I. J. Fritz, J. Phys. Chem. Solids **35**, 817 (1974).
[11] J. D. Eshelby, W. T. Read, and W. Shockley, Acta metall. **1**, 251 (1953).
[12] J. D. Eshelby, Phil. Mag. **40**, 903 (1949).
[13] A. J. E. Foreman, Acta metall. **3**, 322 (1955).
[14] Y. Motohashi and S. Ohtake, phys. stat. sol. (a) **50**, 449 (1978).
[15] Y. Motohashi and S. Ohtake, phys. stat. sol. (a) **50**, 335 (1978).
[16] K. H. G. Ashbee, R. E. Smallman, and G. K. Williamson, Proc. Roy. Soc. **A276**, 542 (1964).

(Received March 19, 1979)

phys. stat. sol. (a) **54**, 365 (1979)

Subject classification: 18.1 and 19; 21.1; 21.4

Physics Department, Cluj-Napoca University[1])

Knight Shift and Magnetic Susceptibilities of Intermetallic Compounds $PrCu_4$ and $PrCu_5$

By

I. Pop, E. Rus, M. Coldea, and O. Pop

The magnetic susceptibility and Knight shift of $PrCu_4$ and $PrCu_5$ intermetallic compounds are investigated over the temperature range 80 to 800 K and 140 to 400 K, respectively. The most important contributions to the magnetic susceptibility are the Langevin term, expressing the paramagnetism of the localized f-electrons, and a temperature independent term for which all the contributions are estimated. The phenomenological exchange integral J_{sf} between conduction electron spins and localized 4f-electron spins is derived to be -3.65×10^{-3} eV for $PrCu_4$ and 2.21×10^{-3} eV for $PrCu_5$. As in the case of $CeCu_5$ compound the s–f coupling for $PrCu_5$ is inversed in respect of that reported so far for all other compounds of rare-earth elements with nonmagnetic ones.

Die magnetische Suszeptibilität und die Knight-Verschiebung der intermetallischen Verbindungen $PrCu_4$ und $PrCu_5$ werden im Temperaturbereich von 80 bis 800 K bzw. 140 bis 400 K untersucht. Die wesentlichsten Beiträge zur magnetischen Suszeptibilität sind der Langevin-Term, der den Paramagnetismus der lokalisierten f-Elektronen ausdrückt, und ein Temperaturunabhängiger Term, für den alle Beiträge berechnet werden. Das phänomenologische Austauschintegral J_{sf} zwischen den Spins der Leitungselektronen und den Spins der lokalisierten 4f-Elektronen wird zu $-3{,}65 \times 10^{-3}$ eV für $PrCu_4$ und $2{,}21 \times 10^{-3}$ eV für $PrCu_5$ abgeleitet. Wie im Fall der $CeCu_5$-Verbindung ist die s–f-Kopplung für $PrCu_5$ invertiert bezüglich derjenigen, die bisher für alle anderen Verbindungen von Seltenen-Erd- mit unmagnetischen Elementen berichtet wurden.

1. Introduction

The phase diagram of the Pr–Cu system indicates the existence of different types of intermetallic compounds as PrCu [1], $PrCu_2$ [1], $PrCu_4$ [3], $PrCu_5$ [2], and $PrCu_6$ [1]. Concerning the existence and the structures of the compounds $PrCu_4$ and $PrCu_5$ there are some contradictory opinions. So, Dwight [2] reports the preparation of the compound $PrCu_5$ of hexagonal structure with the lattice parameters $a = 5.122$ Å and $c = 4.108$ Å, while Buschow and van der Goot [4] show that at this concentration of the components there is no single phase. No magnetic investigation on these compounds has been reported so far.

The purpose of this paper is to investigate the magnetic behaviour of these compounds in a large temperature range. It was interesting to follow the influence of the crystal structure on the magnetic interaction between the localized 4f electrons of Pr ions and the conduction electrons.

In a previous paper [5] we have reported the magnetic characteristics of $CeCu_4$ and $CeCu_5$ intermetallic compounds, so we can draw conclusions concerning the magnetic behaviour of these four compounds.

2. Experimental

The investigated intermetallic compounds $PrCu_4$ and $PrCu_5$ were prepared from stoichiometric amounts of the elements in a cold-boat high-frequency melting system under high-purity argon atmosphere.

[1]) 3400 Cluj-Napoca, Romania.

The X-ray and metallographic studies pointed to the existence of a single phase in $PrCu_4$ and $PrCu_5$.

The magnetic susceptibility was measured in the temperature range 80 to 800 K with a Faraday-type magnetic balance, having 10^{-8} e.m.u./g sensitivity.

The absorption NMR spectra of ^{63}Cu nuclei were recorded on a JEOL spectrometer in the temperature range 140 to 400 K. The Knight shift was determined with reference to the ^{63}Cu resonance in CuCl powder.

3. Results and Discussion

Our X-ray studies have confirmed the orthorhombic crystal structure of the $PrCu_4$ compound with lattice parameters $a = 4.54$ Å; $b = 8.08$ Å; $c = 9.22$ Å. The density calculated from X-ray data and the experimental one are given in Table 1. One can

Table 1

compound	M	density (g/cm³) exp.	density (g/cm³) calc.	θ_P (K)	$\mu_{eff}(\mu_B)$	μ_{eff} ion Pr^{3+} (μ_B)
$PrCu_4$	395.08	7.76	7.78	−6	3.68	3.62
$PrCu_5$	458.61	8.13	8.21	−8	3.52	3.62
$PrCu_6$	522.14	8.38		−12	3.59	3.62

see the very good agreement between both data.

Concerning the $PrCu_5$ intermetallic compound, our results support the hexagonal symmetry reported by Dwight [2]. The lattice parameters determined by us are as follows: $a = 5.125$ Å and $c = 4.113$ Å which are close to the above data [2].

The thermal variation of the reciprocal magnetic susceptibility for the investigated samples is shown in Fig. 1.

The experimental data fit a Curie-Weiss law with an additional temperature independent term

$$\chi = \frac{C}{T - \theta_P} + \chi_0 .$$

From the Curie constants the effective magnetic moments of Pr ions for the two compounds have been determined. The obtained values (Table 1) are very close to the free Pr^{3+} values. This result confirms that the compounds were realized and represent only a single phase. These results show that the 4f electronic state of Pr^{3+} ions is unaffected by alloying and by the different kinds of crystal symmetry. It should be noted that the same conclusion was drawn for the compounds $CeCu_4$ and $CeCu_5$ [5].

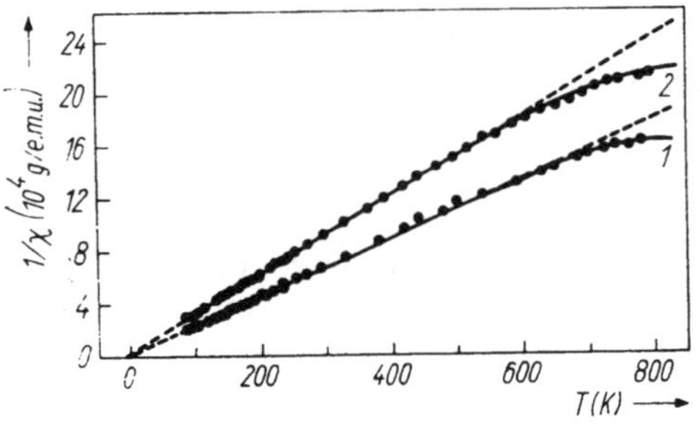

Fig. 1. Temperature dependence of the reciprocal susceptibility for (1) $PrCu_4$ and (2) $PrCu_5$

Table 2

compound	calculated values (10^{-6} e.m.u./mol)					experimental values (10^{-6} e.m.u./mol)		
	χ_P^0	$\chi_{orb}^{5d^1}$	χ_{VV}^{4f}	χ_{dia}	χ_L	f	χ_P	χ_0
$PrCu_4$	71.89	38.8	110	−68	−24	1.70	122.44	179.28
$PrCu_5$	79.78	38.8	110	−80	−26.59	2.35	187.1	229.3
$PrCu_6$	87.22	38.8	110	−92	−29.24	2.5	219.30	246.86

The magnetic susceptibilities plotted against $(T - \theta_P)^{-1}$ provide the temperature independent term χ_0 (Table 2).

The temperature independent term in the magnetic susceptibility may be expressed as

$$\chi_0 = \chi_p + \chi_{orb}^{5d^1} + \chi_{VV}^{3f} + \chi_L + \chi_{dia} ,$$

where χ_P is the Pauli susceptibility, $\chi_{orb}^{5d^1}$ the orbital contribution of the $5d^1$ electrons, χ_{VV}^{4f} the van Vleck paramagnetism of the 4f electrons, χ_L the Landau diamagnetism of the conduction electrons, and χ_{dia} the diamagnetism of the Pr and Cu atomic cores.

The values of these contributions, calculated in a similar way as for $PrCu_6$ [6] are listed in Table 2, which contains also the enhancement factor of the Pauli susceptibility $f = \chi_P/\chi_P^0$,where χ_P^0 is the free-electron Pauli susceptibility.

The Knight shift of the two compounds is positive and temperature dependent. The dependences of the Knight shift on temperature and f electron magnetic susceptibility are given in Fig. 2 a and b.

In the case of a uniform polarization of the conduction electrons by the 4f spins, the Knight shift and magnetic susceptibility are linearly related through [7, 8]

$$K = K_0 \left[1 + \frac{J_{sf}(g_f - 1)\, \chi_f}{2 g_f \mu_B^2}\right],$$

where J_{sf} is the phenomenological exchange constant, g_f the 4f electron g-value, and K_0 the Knight shift due to the Pauli paramagnetism.

The Knight shift versus the susceptibility χ_f intersects the Knight shift axis at $K_0 = 0.209\%$ for $PrCu_4$ and at $K_0 = 0.235\%$ for $PrCu_5$. From the slope of this curve we have determined the phenomenological exchange constants $J_{sf} = -3.65 \times 10^{-3}$ eV

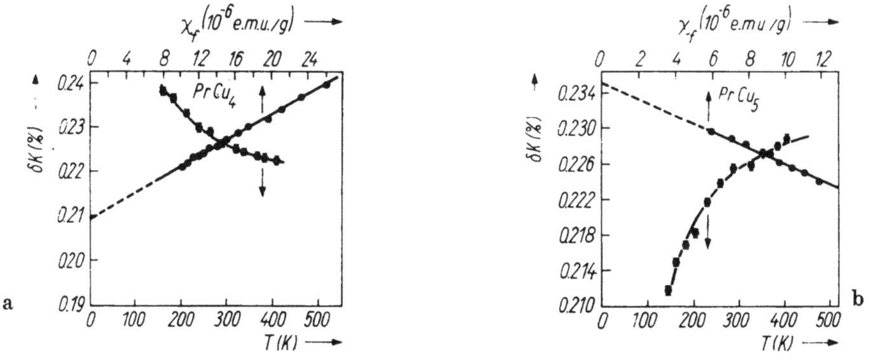

Fig. 2. Temperature and magnetic susceptibility dependence of the Knight shift for a) $PrCu_4$ and b) $PrCu_5$

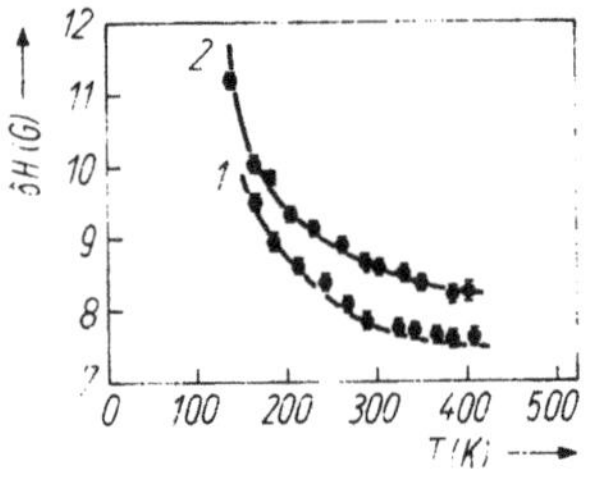

Fig. 3. Temperature dependence of the ^{63}Cu NMR linewidth for compounds (1) $PrCu_4$ and (2) $PrCu_5$

for $PrCu_4$ and $J_{sf} = 2.21 \times 10^{-3}$ eV for $PrCu_5$. The obtained K_0-values are in good agreement with those reported for the isostructural compounds $LaCu_4$, $CeCu_4$, and $CeCu_5$, respectively. These values of the phenomenological exchange integrals are of the same order of magnitude as for compound $PrCu_6$, namely $J_{sf} = 7.29 \times 10^{-3}$ eV [6].

It is interesting to see that the phenomenological exchange constant for $PrCu_4$ is negative, while for $PrCu_5$ it is positive. This means that an inverse coupling between conduction electron spins and 4f spins in $PrCu_5$ in comparison with that in $PrCu_4$ was obtained. The early results obtained by different authors indicate that in the intermetallic compounds of rare-earth elements from the first half series with nonmagnetic elements the exchange integrals J_{sf} are negative. So, our results obtained for the $PrCu_5$ and $CeCu_5$ [5] compounds point out the possibility of reversal s–f coupling. These results are for the first time reported in the literature [5].

It should be mentioned that the compounds $PrCu_4$ and $PrCu_6$ are orthorhombic, while $PrCu_5$ is hexagonal and that in this series the conduction electron density is changing from one compound to another.

We assume that this reversal coupling may be attributed to changes in the interatomic distances and in the conduction electron density. It is well known that the spin polarization and the exchange interaction as a function of distance from the magnetic ion have oscillatory character [9].

The values of the exchange integrals for all three compounds are small, so our case is situated in the neighbourhood of a point where the sign of spin polarization changes. A small difference in the distances may lead to a change in the sign of the phenomenological exchange integral.

The temperature dependence of the NMR half-width in $PrCu_4$ and $PrCu_5$ compounds is shown in Fig. 3. Since the temperature part of the Knight shift is very small, demagnetizing fields and anisotropic Knight shifts are partly responsible for the increase in linewidth towards lower temperatures.

References

[1] M. Hansen and K. Anderko, Contribution of Binary Alloys, McGraw-Hill Publ. Co., New York/Toronto/London 1958.
[2] A. E. Dwight, Trans. ASM **53**, 479 (1961).
[3] A. Bystrom, Acta chem. Scand. **6**, 709 (1952).
[4] K. W. Buschow and A. S. van der Goot, J. less-common Metals **19**, 433 (1969).
[5] I. Pop, E. Rus, M. Coldea, and O. Pop, J. Phys. Chem. Solids, in the press.
[6] M. Coldea and I. Pop, Phil. Mag. **28**, 881 (1973).
[7] V. Jaccarino, J. appl. Phys. Suppl. **32**, 102 (1961).
[8] V. Jaccarino, B. T. Mathias, M. Peter, H. Suhl, and M. J. Wernick, Phys. Rev. Letters **5**, 251 (1960).
[9] W. E. Wallace, Rare-Earth Intermetallics, Academic Press, New York/London 1973.

(Received February 1, 1979)

phys. stat. sol. (a) 54, 369 (1979)

Subject classification: 11 and 20.3; 20.1; 22.8.1

Department of Physics, University of Western Ontario, London[1])

Thermoluminescence and Optical Absorption of X-Ray Irradiated $NaBrO_3$[2])

By

V. S. Sivasankar and P. W. Whippey

Single crystals of sodium bromate grown by slow cooling of a saturated aqueous solution, are subjected to X-irradiation at 77 K. Thermoluminescence glow curves of the irradiated crystals recorded in the temperature range from 77 to 400 K show four glow peaks. Optical absorption spectra of the damaged crystals also are recorded as a function of radiation dose and at different stages of thermal annealing between 77 and 400 K. On the basis of the results of optical measurements, the glow peaks at 115, 159, 229, and 300 K are identified as due to O^{2-} in the vicinity of BrO_3, BrO_3 itself, O^{2-} in the neighbourhood of BrO_2, and BrO_2 itself, respectively. The variations in the thermoluminescence spectrum with the dose of irradiation are consistent with the assignment of second-order kinetics for all four glow peaks. The trapping parameters also are determined for these centers.

Natriumbromateinkristalle, die durch langsame Abkühlung aus gesättigter wäßriger Lösung gezüchtet wurden, werden einer Röntgenstrahlung bei 77 K ausgesetzt. Die im Temperaturbereich von 77 bis 400 K aufgezeichneten Thermolumineszenzkurven zeigen vier Glowmaxima. Die optischen Absorptionsspektren der strahlengeschädigten Kristalle werden in Abhängigkeit von der Bestrahlungsdosis und bei verschiedenen Stufen der thermischen Ausheilung zwischen 77 und 400 K aufgezeichnet. Auf der Grundlage der Ergebnisse der optischen Messungen werden die Glowmaxima bei 115, 159, 229 und 300 K als O^{2-} in der Nähe von BrO_3, als BrO_3 selbst, als O^{2-} in der Nachbarschaft von BrO_2 bzw. als BrO_2 selbst identifiziert. Die Änderungen des Thermolumineszenzspektrums mit der Bestrahlungsdosis sind konsistent mit der Annahme einer Kinetik zweiter Ordnung für alle vier Glowmaxima. Die Haftstellenparameter werden ebenfalls für diese Zentren bestimmt.

1. Introduction

Sodium bromate and sodium chlorate crystals are isomorphous. On this basis, the properties of radiation damage products in these two crystalline compounds are expected to be similar. In an earlier paper we have reported the results of our investigations on the thermal stability of radiation damage products in $NaClO_3$ [1]. The nature of specific products in $NaBrO_3$ crystals has previously been studied by other workers who have used different experimental techniques [2 to 7]. In particular, thermoluminescence and optical absorption measurements on $NaBrO_3$ irradiated at room temperature by γ-rays have been reported by Andersen et al. [8]. It has been established long since that the damage products formed in these crystals are governed by the type of radiation employed. Also, by their very nature, some of the products formed either decompose or are otherwise unstable when the samples are irradiated at room temperature. Therefore, it is necessary to subject the crystals to radiation damage at much lower temperatures so as to better understand some of the primary processes and the subsequent behavior of these primary products under thermal treatment. In this paper we report results of our studies on $NaBrO_3$ irradiated by

[1]) London, Ontario N6A 3K7, Canada.

[2]) Work supported by NRC Canada, and Centre for Chemical Physics UWO.

X-rays at 77 K. Correlation of the thermoluminescence spectrum with the optical spectra recorded at different temperatures has enabled a positive identification of the damage products and also their thermal stability.

Crystalline $NaBrO_3$ possesses a cubic structure with its space group represented by $T^4(P2_13)$ [9]. Four molecules constitute a unit cell whose edge length is 6.705 Å at 300 K. The bromate ion is a flat trigonal pyramid having the three oxygen atoms of its base 2.95 Å apart. The bromine atom is at a height of 0.58 Å above this base and at a distance of 1.78 Å from each of the oxygen atoms. The metal-to-oxygen distance is 2.38 Å. The symmetry axis of each of the four molecules in a unit cell is parallel to a body diagonal of the unit cell. Sodium bromate is also an optically active crystal.

2. Experimental

Single crystals of $NaBrO_3$ were grown by slow cooling of a saturated aqueous solution. $5 \times 5 \times 2$ mm³ samples were cut from larger single crystals and were polished to a window glass finish. These samples were exposed to X-rays at 30 kV and 20 mA for a duration which varied from 30 min to 10 h, at an approximate dose rate of 1.3×10^6 R/h. Thermoluminescence measurements were carried out using an EMI9558 photomultiplier tube. Optical absorption spectra were recorded on a Cary 14 spectrophotometer.

3. Results

Thermoluminescence and optical absorption spectra have been recorded using samples irradiated for different durations. Within the dose range of investigation, none of the species formed due to irradiation has exhibited saturation. Four glow peaks appear in the thermoluminescence spectrum recorded from 77 to 400 K. With a linear heating rate of 10 K/min, the peak temperatures of samples irradiated for 10 h are at 115, 159, 229 and 300 K as shown in Fig. 1. The peak temperatures are found

Table 1

Parameters of the glow peaks observed with X-irradiated $NaBrO_3$

T_m	order	E	s	E_m	species
115	II	0.122	2.199×10^3	4.2	O^{2-} (A)
159	II	0.268	4.063×10^6	2.5	BrO_3
229	II	0.326	8.284×10^4	3.1	O^{2-} (B)
300	II	0.507	1.081×10^6	3.7	BrO_2

$T_m(K)$ position of the glow peak, order means order of kinetics, E (eV) trap depth, s (cm³/s) frequency factor, E_m (eV) position of the associated optical absorption band.

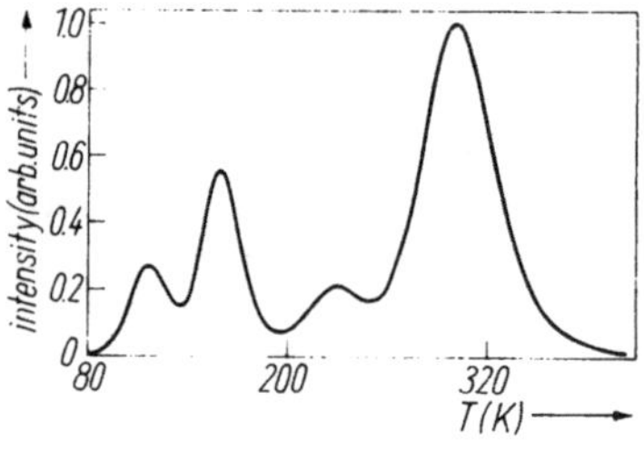

Fig. 1. Thermoluminescence spectrum of $NaBrO_3$ irradiated for 10 h by X-rays at 77 K. Linear heating rate of 10 K/min was employed

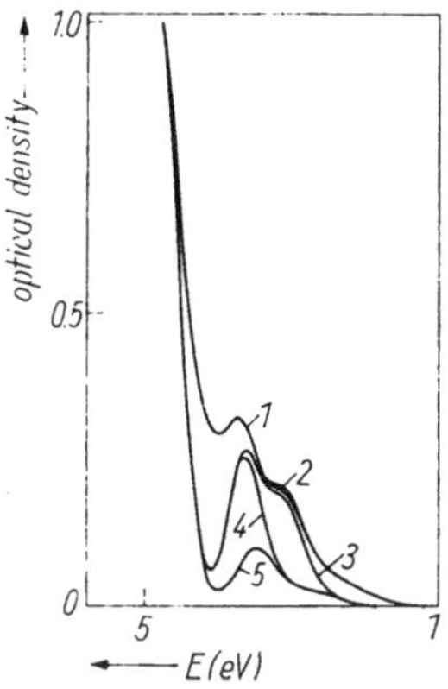

Fig. 2

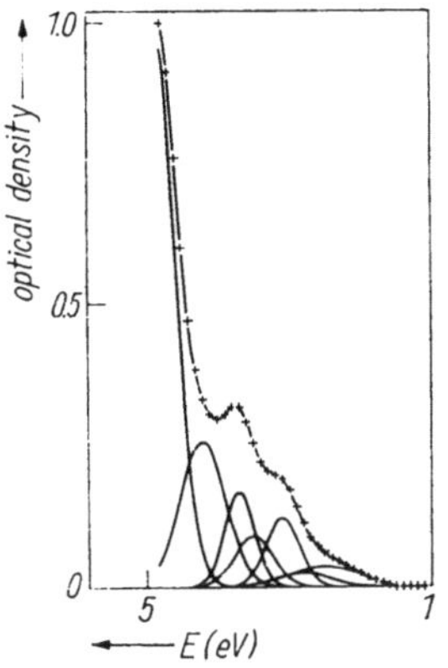

Fig. 3

Fig. 2. Optical absorption spectra of $NaBrO_3$ irradiated for 10 h by X-rays at 77 K. The temperature at which these spectra are recorded are: (1) 77, (2) 140, (3) 200, (4) 260, (5) 360 K

Fig. 3. Resolution of the optical absorption spectrum recorded at 77 K, of $NaBrO_3$ irradiated for 10 h by X-rays at 77 K. The component bands, the total spectrum, and the data points (marked +) are shown

to decrease with increasing dose rate. Therefore, the glow peaks are considered to follow second-order kinetics. Starting with the familiar rate equations for second-order processes [10], a nonlinear least-squares fit has been carried out. The frequency factor and trap depth of the centers responsible for these peaks are obtained by the curve fitting procedure. These results are presented in Table 1. The computed glow curves using the parameters given in Table 1 resemble the observed glow curves so closely that for clarity only the latter are shown in Fig. 1. The identification of the centers is based on the following results of optical absorption measurements.

Optical absorption spectra were recorded at regular intervals of temperature between 77 and 400 K. Spectra at five different temperatures which are significant for the interpretation of our results are shown in Fig. 2. Unlike the spectra in $NaClO_3$, all these are quite complex. In order to obtain definite, quantitative information,

Table 2

Parameters of optical absorption bands and their behaviour under thermal treatment

E_m	W	T_i	T_f	species	reference
4.85	0.65	—	—	edge	[2, 11]
4.20	0.74	80	140	O^{2-} (A)	present work
3.70	0.49	240	360	BrO_2	present work
3.50	0.63	—	—	BrO^-	[3]
3.10	0.55	190	260	O^{2-} (B)	present work
2.80	0.91	—	—	Br_2	[3]
2.50	1.11	125	200	BrO_3	present work

E_m (eV), W (eV) peak position and width, respectively, T_i (K) temperature at which the band begins to decrease in intensity, T_f (K) temperature at which the band completely disappears.

Note: The blanks in place of the values of T_i and T_f for certain bands indicate that the corresponding band is not affected by temperature in the range 77 to 400 K.

each of them has been resolved into component Gaussian-shaped absorption bands. Fig. 3 shows the components, the total spectrum, and the data points for the spectrum obtained at 77 K. In all of the optical measurements the data points were obtained at regular intervals of 5 Å. However, in Fig. 3 only a few of these which essentially define the shape of the observed absorption spectrum are shown explicitly. Since our spectra are similar to those reported by Levy et al. [2], initial estimates for the parameters as well as the number of component bands have been taken from their results. In all, there are seven components with maximum absorption at 4.85, 4.2, 3.7, 3.5, 3.1, 2.8, and 2.5 eV. In the spectrum at 400 K, the highest temperature of investigation, three components centered at 2.8, 3.5, and 4.85 eV are still present. In going from 77 to 400 K the order in which the remaining four components disappear are as follows: 4.2, 2.5, 3.1, and 3.7 eV. Table 2 summarizes the results of the optical measurements.

4. Discussion

Before attempting to interpret the thermoluminescence spectrum, it is essential to identify the centers which give rise to the optical absorption. The very intense absorption band at 4.85 eV is attributed by Levy et al. [2] to either the true band edge absorption or an apparent edge attributable to a very intense absorption band due to an unknown entity. However, the work of Sastry et al. [11] on the indirect band-to-band transitions in alkali halates demonstrates unambiguously that the fundamental absorption in sodium bromate is at 4.85 eV. Since this band does not anneal up to 400 K, it is not of any further interest to the present investigation.

Fischer and Gummer [12, 13] have reported several optical absorption bands due to O^{2-} centers trapped in alkali halides. From a careful comparison with their data we have deduced that the band at 4.2 eV is due to O^{2-} trapped in the neighborhood of a BrO_3 center (designated as O^{2-} (A) center), whereas the band at 3.1 eV is due to a O^{2-} center with a BrO_2 radical in its vicinity (designated as O^{2-} (B) center).

We propose that the band at 4.2 eV is due to BrO_3 with a nearby O^{2-} ion. Boyd and Larsen [6] originally proposed a primary reaction involving BrO_3. However, since their measurements were made at room temperature, the presence of this unstable radical could not be directly established by them. Andersen et al. [8] have observed a band at 4.3 eV when the crystals were irradiated at −78 °C and have attributed it to a BrO_3 center influenced by a BrO^- center. Optical absorption and electron spin resonance studies [1, 14] on NaCl irradiated at 77 K by X-rays, have demonstrated very clearly the presence of ClO_3 radical perturbed by an O^{2-} ion in its environment. This interpretation is consistent with the behavior of these centers under thermal treatment.

It is difficult to provide direct evidence to show that the band at 3.7 eV is due to a BrO_3 radical influenced by an O^{2-} center. This band has been observed by Levy et al. [2] but no interpretation has been offered. The counterpart radical, namely ClO_2, has been observed in $NaClO_3$ [1, 14] and in $KClO_3$ [15]. The annealing characteristics of this band together with the results of thermoluminescence measurements strongly support our explanation.

Ramasastry and Murti [3] have identified the bands at 3.5 and 2.8 eV as due to a BrO^- center and a Br_2 center, respectively. Similar bands have been observed in other alkali halates also [4 to 6].

It remains to explain the origin of the four glow peaks appearing at 115, 159, 229, and 300 K. Out of the seven components into which the optical absorption spectrum at 77 K could be resolved, only three are stable up to 400 K. These are the bands at 2.8, 3.5, and 4.85 eV, and are due to Br_2 centers, BrO^- centers, and the edge absorption, respectively. In the order in which they disappear on heating the crystals from 77 K, the four unstable centers are O^{2-} (A) (4.3 eV), BrO_3 (2.5 eV), O^{2-} (B) (3.1 eV), and BrO_2 (3.7 eV). Specifically the 4.2 eV band decreases above 80 K and disappears at 140 K whereas the 2.5 eV band begins to decrease at 125 K and disappears at 200 K. Therefore, the glow peaks at 115 and 159 K are attributed to O^{2-} (A) centers and BrO_3 centers, respectively.

Similarly, the glow peaks at 229 and 300 K are attributed to O^{2-} (B) and BrO_2 centers, respectively, because the intensity of the 3.1 eV band decreases above 190 K, finally disappearing at 260 K whereas the 3.7 eV band is reduced in intensity above 240 K and disappears completely at 360 K. The disappearance of O^{2-} (A) followed by BrO_3 and of O^{2-} (B) followed by BrO_2 indicates an interesting causal relation. A BrO_3 or BrO_2 radical located at an anion site possesses an effective positive charge whereas an O^{2-} center at a similar site bears an effective negative charge. It is likely that the Coulomb attraction between these opposite effective charges is the dominant factor for the O^{2-} centers to be trapped adjacent to BrO_3 or BrO_2 centers. It therefore follows that if and when an O^{2-} center is rendered unstable, the "associated" BrO_3 or BrO_2 center also becomes unstable. These results are similar to those reported in $NaClO_3$ [1] except that in sodium chlorate the glow peak due to ClO_2 centers has not been observed. At this stage it is difficult to specify the mechanisms by which these centers decay.

References

[1] V. S. Sivasankar and P. W. Whippey, Canad. J. Phys. **57**, 128 (1979).
[2] P. W. Levy, M. Goldberg, and P. J. Herley, J. phys. Chem. **76**, 3751 (1972).
[3] C. Ramasastry and Y. V. G. S. Murti, Indian J. pure appl. Phys. **2**, 35 (1964).
[4] T. Andersen, J. R. Byberg, and K. L. Olsen, J. phys. Chem. **71**, 4129 (1967).
[5] T. Andersen, Ph. D. Thesis, Univ. Aarhus, 1968.
[6] G. E. Boyd and Q. V. Larson, J. phys. Chem. **68**, 2627 (1964).
[7] F. T. Gamble, J. chem. Phys. **47**, 1193 (1967).
[8] T. Andersen, H. E. L. Madsen, and K. Olesen, Trans. Faraday Soc. **62**, 2409 (1966).
[9] R. W. G. Wyckoff, Crystal Structure, Interscience Publishers, New York 1951.
[10] R. Chen, J. appl. Phys. **40**, 570 (1969).
[11] S. B. S. Sastry, R. B. Tripathi, and C. Ramasastry, J. Phys. Chem. Solids **34**, 481 (1973).
[12] F. Fischer and G. Gummer, Z. Phys. **189**, 97 (1966).
[13] G. Gummer, Z. Phys. **215**, 256 (1968).
[14] O. Vinther, J. chem. Phys. **57**, 183 (1972).
[15] H. G. Heal, Canad. J. Chem. **37**, 979 (1959).

(Received March 19, 1979)

phys. stat. sol. (a) 54, 375 (1979)

Subject classification: 1; 22.1.2

I. V. Kurchatov Institute of Atomic Energy, Moscow[1])

On the Theory of the Bragg Reflection in the Case of Multiple X-Ray Diffraction

By

V. G. KOHN

A general theory of multiple X-ray diffraction in the Bragg geometry is formulated. Special attention is paid to obtain simple and compact formulae which are convenient for computer programming. Two cases are considered: the incident beam is either a plane-wave or a spherical-wave, and the reflection occurs either at one surface or at several surfaces of the same crystal. The theory allows to calculate the angular and spectral characteristics of the many-beam X-ray monochromators and interferometers in a general case. Results of the numerical calculation of the reflection coefficients are described for the three-beam (111/11$\bar{1}$) case of the CuK$_\alpha$-radiation diffraction in silicon.

Es wird eine allgemeine Theorie der Mehrfach-Röntgenbeugung in Bragg-Geometrie formuliert. Insbesondere wird eine einfache und geschlossene Formel erhalten, die eine bequeme Computerprogrammierung gestattet. Zwei Fälle werden betrachtet: der einfallende Strahl ist entweder eine ebene Welle oder eine Kugelwelle, und die Reflexion tritt entweder an einer Oberfläche oder an einigen Oberflächen desselben Kristalls auf. Die Theorie erlaubt, die Winkel- und Spektralcharakteristiken der Mehrstrahl-Röntgenmonochromatoren und Interferometer im allgemeinen Fall zu berechnen. Die Ergebnisse der numerischen Berechnung der Reflexionskoeffizienten werden für den Dreistrahl (111/11$\bar{1}$)-Fall der CuK$_\alpha$-Strahlenbeugung in Silizium beschrieben.

1. Introduction

Since 1965 when Borrmann and Hartwig observed the enhancement of the anomalous transmission in the many-beam region of angles, the interest in the multiple X-ray diffraction has continuously increased. However, many papers which have been published during these years, both experimental and theoretical, have been devoted mainly to the case of the X-ray diffraction in the Laue geometry when all radiation passes through the crystal. Among the few investigations on the many-beam Bragg reflection we want to mention the papers on the Renninger effect and the possibility connected with it to measure precisely the lattice parameter of a crystal [1, 2]. But in these works the knowledge of the exact character of the angular dependence of the many-beam reflection does not play an essential role. Therefore, the theoretical study of this phenomenon was not stimulated by these works.

Quite another situation arose in recent time when papers were published in which devices, using the many-beam Bragg reflection for obtaining a closed trajectory of X-rays were suggested. One can mention, for example, the three-beam X-ray interferometer by Graeff and Bonse [3] and also the X-ray resonator [4]. It is essential that in the many-beam region of angles energy can be transferred from any beam to any other one of the many-beam set by choosing the corresponding reflection surface. On the one hand one can solve the problem of beam splitting and following recombination and on the other hand, reduce the angular divergency and carry out the mono-

[1]) 123182 Moscow, USSR.

chromatization of the incident beam by choosing the proper transition with the use of the reflection at various surfaces of the same crystal.

For finding the optimum variants solving these problems the theoretical investigation of the phenomenon of the many-beam Bragg reflection becomes very important. The present paper is devoted just to this question. As it is known, the consecutive calculation and analysis of both the dispersion surface and the reflection coefficients in the many-beam region of angles can be carried out only by numerical methods. Therefore, such a formulation of the dynamical theory of many-beam scattering is necessary that is convenient for a computer program to be worked out. For the Laue case this question has been analysed in [5] in which the scattering matrix has been determined and the problem has been reduced to the eigenvalue problem of this matrix.

In Section 2 the results of [5] are generalized for the Bragg-reflection case. The results of the numerical calculation of both reflection and transmission coefficients for the three-beam $(111/11\bar{1})$ case are described in Section 3. In Section 4 a general formula is obtained which describes the case of reflection at several surfaces of the same crystal. This formula is a basic one for the calculation of many-beam monochromators. The incident-spherical-wave approach was shown [6, 7] to be a basis for the calculation of the diffraction pattern when investigating the coherent phenomena which arise in X-ray diffraction with fixing the diffraction pattern on a film. The spherical-wave theory of the Bragg reflection is developed in Section 5.

2. Reflection of a Plane Wave at One Crystal Surface

Let a monochromatic plane wave with the wave vector $\boldsymbol{k}_0$ and the frequency ω fall on a crystal plate of thickness t. Let $\boldsymbol{n}$ be the normal to the plate surface directed towards the plate. It is necessary for observing multiple diffraction that the vector $\boldsymbol{k}_0$ is approximately equal to the vector $\boldsymbol{\varkappa}_0$ with the length $\varkappa_0 = \omega_0/c$ (c is the light velocity) and direction satisfying the Bragg condition for any set of reciprocal lattice vectors $\boldsymbol{h}_m$ simultaneously, namely

$$\boldsymbol{\varkappa}_m = \boldsymbol{\varkappa}_0 + \boldsymbol{h}_m\,, \qquad \varkappa_m^2 = \varkappa_0^2\,. \tag{2.1}$$

We assume that the incident radiation is slightly nonmonochromatic and has some angular divergence. Then ω_0 is the frequency corresponding to the intensity maximum of the radiation spectrum and the vector $\boldsymbol{k}_0$ can be expressed in the form

$$\boldsymbol{k}_0 = \boldsymbol{\varkappa}_0 + \boldsymbol{q}_0\,, \qquad \boldsymbol{q}_0 = \varkappa_0(\theta_1\boldsymbol{e}_1 + \theta_2\boldsymbol{e}_2 + \theta_\omega\boldsymbol{s}_0)\,. \tag{2.2}$$

Here $\boldsymbol{s}_0$ is the unit vector directed along $\boldsymbol{\varkappa}_0$, $\boldsymbol{e}_1$ and $\boldsymbol{e}_2$ are the mutually perpendicular unit vectors lying in the plane perpendicular to $\boldsymbol{s}_0$. The parameters θ_1 and θ_2 characterize the angular deviation of the wave vector of the incident beam from the Bragg direction $\boldsymbol{s}_0$, and $\theta_\omega = (\omega - \omega_0)/\omega_0$ characterizes the relative frequency shift.

To describe the electromagnetic field of X-rays inside a crystal, we use Maxwell's equations for Fourier components of an electrical field vector $\boldsymbol{E}(\boldsymbol{k}, \omega)$. As usual, expressing the current density induced by the electromagnetic wave through the electric field vector and preserving only strong waves, we obtain a set of linear homogeneous equations for scalar amplitudes of the strong wave (for more detail see [5]). The complete X-ray radiation field can be written in the form

$$\boldsymbol{E}_s(r) = \mathrm{e}^{i\boldsymbol{k}_0\boldsymbol{r}}\, A_s(\omega) \sum_j \lambda_s(j, t)\, \mathrm{e}^{i\varepsilon_j z/2} \sum_{ms'} \sqrt{\frac{\gamma_0}{\gamma_m}}\, \mathrm{e}^{i\boldsymbol{h}_m(\boldsymbol{r}-\boldsymbol{r}_1)}\, B_{ms'}(j)\, \boldsymbol{e}_{ms'}\,. \tag{2.3}$$

Here $z = (\boldsymbol{r} - \boldsymbol{r}_1)\,\boldsymbol{n}$, $\boldsymbol{r}_1$ is the coordinate of a point lying on the entrance surface of the crystal, $\gamma_m = \boldsymbol{s}_m\boldsymbol{n}$, $\boldsymbol{s}_m$ is a unit vector along $\boldsymbol{\varkappa}_m$. The quantities $B_{ms}(j)$ and ε_j are the eigenvectors and eigenvalues of the scattering matrix $G_{mn}^{ss'}$,

$$\sum_{ns'} G_{mn}^{ss'} B_{ns'}(j) = \varepsilon_j B_{ms}(j) . \tag{2.4}$$

$A_s(\omega)$ is the amplitude of the incident plane-polarized wave in the polarization state s ($s = \pi, \sigma$), $\boldsymbol{e}_{ms}$ are the unit vectors lying in the plane perpendicular to $\boldsymbol{s}_m$ and determining the polarization state for the wave m.

The polarization vectors $\boldsymbol{e}_{ms}$ can be chosen in an arbitrary way. We are interested in the systematic multiple diffraction [8] when all the reciprocal lattice vectors lie in the same plane. In this case the vectors $\boldsymbol{e}_{ms}$ suggested in [9] are most convenient, namely, all the vectors $\boldsymbol{e}_{m\sigma}$ lie in the same plane as the vectors $\boldsymbol{h}_m$ and $\boldsymbol{e}_{m\pi}$ are determined from the condition that the vectors $\boldsymbol{e}_{m\pi}$, $\boldsymbol{e}_{m\sigma}$, and $\boldsymbol{s}_m$ form a right-handed system of orthogonal directions.

The scattering matrix G is determined through the Fourier components of a complex polarizability tensor $\boldsymbol{\chi}$. The expression for $\boldsymbol{\chi}$ with account of all possible interactions between the X-rays and the crystal, has been obtained in [10]. Taking into consideration only the main contributions to $\boldsymbol{\chi}$, namely, Thomson scattering and the dipole part of a photoelectric absorption, we have [5, 7], according to (2.3)

$$G_{mn}^{ss'} = \frac{1}{\sqrt{\gamma_m \gamma_n}} \varkappa_0 \chi_{m-n} (\boldsymbol{e}_{ms} \boldsymbol{e}_{ns'}) - \alpha_m \delta_{mn}^{ss'} , \tag{2.5}$$

where

$$\alpha_m = \frac{2\boldsymbol{h}_m \boldsymbol{q}_0}{\varkappa_0 \gamma_m} \tag{2.6}$$

is a parameter characterizing the deviation from the Bragg condition for the beam m.

In the present work, we consider the cases in which at least one of the quantities γ_m is negative, i.e. the plane-wave with the wave vector $\boldsymbol{k}_m$ goes out of the crystal through the entrance surface (reflection). It is natural to refer all these cases to the Bragg reflection even if several of the diffracting waves pass through the crystal. Indeed, due to the presence of a multiplier $(\gamma_m \gamma_n)^{-1/2}$ the matrix G_r describing the scattering in a nonabsorbing crystal becomes nonhermitian. Therefore, it can have a complex eigenvalue for which the field diminishes exponentially when it penetrates into the crystal. This phenomenon is known in the two-wave case and arises just in the case of Bragg reflection.

The quantities $\lambda_s(j, t)$ determine the value of an excitation of the j-th mode of Bloch waves. They are defined by boundary conditions on both entrance and exit crystal surfaces. After averaging over a region with linear dimensions much greater than an interatomic distance a, we obtain, as usual, the boundary conditions for each m-th beam in the superposition (2.3) separately. Then the amplitude of the Bragg-reflected beams has to vanish on the exit surface of the crystal plate $z = t$ and the amplitude of Laue diffracted beams has to be equal to zero on the entrance surface $z = 0$. These conditions are conveniently written in the form

$$\sum_j c_m(j, t)\, B_{ms'}(j)\, \lambda_s(j, t) = \delta_{m0} \delta_{ss'} , \tag{2.7}$$

where

$$c_m(j, t) = \begin{cases} 1; & \gamma_m > 0 , \\ \exp{[i\varepsilon_j t/2]}; & \gamma_m < 0 . \end{cases} \tag{2.8}$$

The set (2.4) does not define the length of the vectors B_{ms}, but the general field (2.3) according to (2.7) does not depend on the choice of the length of the vectors. The vectors are conveniently normalized to unity. The eigenvalue problem (2.4) in the many-beam case with any value of the vector $\boldsymbol{q}_0$ can be solved only by numerical methods using a computer. We note that, contrary to the Laue case, the methods of diagonalization of nonhermitian matrices have to be used here.

If the crystal has an inversion centre in the point $\boldsymbol{\varrho}_0$ then it is easy to verify directly that the substitution $B_{ms} = \tilde{B}_{ms} \exp(-i\boldsymbol{h}_m\boldsymbol{\varrho}_0)$ gives the symmetrical complex matrix $\tilde{G}$. For the numerical diagonalization of this matrix the method of complex rotations was found to be very convenient, because it provides an acceptable calculation time and also stability and simplicity of the computer procedure. This method is a direct generalization of the rotation method (or Jakobi method) [11] for real symmetrical matrices in complex arithmetics. The eigenvectors of the complex symmetrical matrix are mutually orthogonal

$$\sum_{ms} \tilde{B}_{ms}(j)\, \tilde{B}_{ms}(j') = \delta_{jj'} \,. \tag{2.9}$$

But unlike the Laue case, equation (2.9) does not give the possibility to determine the quantities λ_j in an analytical form because of the complicated character of the set (2.7). Therefore, the set (2.7) has also to be solved numerically. Here, following the two-beam case, it is natural to consider separately the case of an infinitely thick plate (practically $t \gg L_{\text{ex}}$, where L_{ex} is the extinction length).

Indeed, the imaginary part of the quantity ε_j, can have any sign and a value as large as $1/L_{\text{ex}}$ in the complete reflection region. Therefore, some of the quantities $c_m(j)$ are exponentially large at $t \gg L_{\text{ex}}$. In the two-beam case such modes were known to be not excited in a crystal. The analogous situation arises also in the case of multiple diffraction. In the thick crystal only such modes are excited for which $\varepsilon_j'' > 0$ ($\varepsilon = \varepsilon' + i\varepsilon''$). The corresponding coefficients $\lambda(j)$ are defined from the truncated set (2.7) which consists of the equation with multipliers $c_m = 1$. As can be shown, the number of eigenvalues ε_j with negative imaginary part is exactly equal to the double number of Bragg reflecting beams. Consequently, the truncated set has only one solution.

From the experimental point of view the amplitude of the X-ray electric field leaving the crystal is of interest. Let us consider the radiation part going out of the crystal along the $\boldsymbol{k}_m$ direction. If the polarization state of the radiation is s' and one of the incident wave is s, then the field amplitude of this part reads

$$\boldsymbol{E}_m^{s's}(r) = \exp\left[i\boldsymbol{k}_0\boldsymbol{r}_1 + i\boldsymbol{k}_m(\boldsymbol{r} - \boldsymbol{r}_1)\right] \left(\frac{\gamma_0}{\gamma_m}\right)^{1/2} \boldsymbol{e}_{ms'} P_m^{s's}(\boldsymbol{q}_0, t)\, A_s \,, \tag{2.10}$$

$$P_m^{s's}(\boldsymbol{q}_0, t) = \sum_j B_{ms'}(j)\, \lambda_s(j, t)\, D_m(j, t) \,. \tag{2.11}$$

Here

$$\left.\begin{aligned} \boldsymbol{k}_m &= \boldsymbol{k}_0 + \boldsymbol{h}_m - \alpha_m(\boldsymbol{q}_0)\, \boldsymbol{n}/2 \,, \\ D_m(j, t) &= \begin{cases} \exp\left[i(\varepsilon_j + \alpha_m)\, t/2\right]; & \gamma_m > 0 \,, \\ 1; & \gamma_m < 0 \,. \end{cases} \end{aligned}\right\} \tag{2.12}$$

The parameter α_m in (2.12) is determined by (2.6). It arises in the expression for the vacuum wave vector because of the dispersion law $k_m^2 = k_0^2$. Formula (2.10) is valid for the Bragg-reflected beams in the space outside the entrance crystal surface and for the Laue-diffracted beams in that outside the exit surface.

The coefficients of transmission R_0 and reflection R_m, $m \neq 0$, are defined simply in terms of the amplitude $P_m^{s's}$. Let the incident radiation be plane-polarized, for example synchrotron radiation. If φ is the angle between the polarization plane and the vector $\boldsymbol{e}_{0\pi}$ then

$$R_m(\varphi) = \frac{|\gamma_m|}{\gamma_0} \frac{|\boldsymbol{E}_m|^2}{|A|^2} = \bar{R}_m + X_m \cos(2\varphi) + Y_m \sin(2\varphi) \,, \tag{2.13}$$

where

$$\left.\begin{aligned} \bar{R}_m &= \tfrac{1}{2} \sum_{ss'} |P_m^{ss'}|^2 , & |A|^2 &= \sum_s |A_s|^2 , \\ X_m &= \bar{R}_m - \sum_s |P_m^{s\sigma}|^2 , & Y_m &= \operatorname{Re}\left\{\sum_s P_m^{s\pi} P_m^{s\sigma *}\right\} . \end{aligned}\right\} \tag{2.14}$$

In the case of unpolarized radiation we have to average (2.13) over the angle φ. As a result, $R_m = \bar{R}_m$.

The dependence of the vector $\boldsymbol{q}_0$ on the three parameter projections is of most interest. These parameters define direction and frequency of the incident beam. However, in the case of the systematic diffraction, the quantities R_m depend effectively on only two parameters. Indeed, due to a nonresonance interaction between X-rays and crystal, we can neglect the slight frequency dependence of the quantities χ_m (see equation (2.5)) in the frequency region corresponding to the Bragg diffraction. Then the matrix G depends on $\boldsymbol{q}_0$ only through the parameter α_m. Let us introduce the unit vector $\boldsymbol{n}_{0R}$ [7] along the direction connecting the centre of a circle described by the reciprocal-lattice-vector polygon and the point 0. After choosing the vector $\boldsymbol{e}_2$ in (2.2) in the reciprocal-lattice-vector plane $\boldsymbol{e}_2 = \boldsymbol{e}_{0\sigma}$, we have

$$\alpha_m = \frac{2}{\gamma_m} \left[(\boldsymbol{h}_m \boldsymbol{n}_{0R})\, \theta + (\boldsymbol{h}_m \boldsymbol{e}_{0\sigma})\, \theta_2\right] , \tag{2.15}$$

$$\theta = \theta_1 \cos \theta_0 + \theta_\omega \sin \theta_0 , \qquad \theta_\omega = (\omega - \omega_0)/\omega_0 , \tag{2.16}$$

where θ_0 is the angle between $\boldsymbol{s}_0$ and the unit vector $\boldsymbol{n}_h$ of a normal to the reciprocal-lattice-vector plane.

According to (2.16) the change of the incident radiation frequency is compensated by a corresponding change of the beam direction. At small scattering angles ($\theta_0 < \pi/4$) realized in the Laue case, the necessary angular shift does not exceed, as a rule, the relative frequency change. On the contrary, a coplanar diffraction (when $\theta_0 = \pi/2$) is realized easily and is of more interest in the cases of Bragg reflection. In the coplanar case the parameter θ describes purely a frequency dependence, and the dependence on the angle in the plane perpendicular to the reciprocal-lattice-vector plane vanishes in an approximation linear in $\boldsymbol{q}_0$.

3. Concrete Example: The Bragg Reflection of CuK$_\alpha$ Radiation in the Three-Beam (111/11$\bar{1}$) Case in Silicon

We worked out the FORTRAN computer program for the calculation of the coefficient $R_m(\theta, \theta_2, t, \varphi)$ along a scheme described in the previous section. As an example, we consider the three-beam (111/11$\bar{1}$) case of CuK$_\alpha$ radiation diffraction ($\lambda = 1.54$ Å) in a silicon crystal plate with the entrance surface perpendicular to the [1$\bar{1}$0] direction. This case is of interest because it permits an analytical solution of the problem on the $\theta_2 = 0$ line that is situated in the symmetry plane. The solution for the Laue geometry was obtained for the first time in [12, 13]. A generalization to the Bragg reflection case for an infinitely thick crystal is trivial, therefore, we write the final result only. Let the polarization state of the incident beam be standard (π or σ), then

$$R_{1s}(\theta) = R_{2s}(\theta) = \frac{0.5}{|y_s \pm \sqrt{y_s^2 - 1}|^2} , \tag{3.1}$$

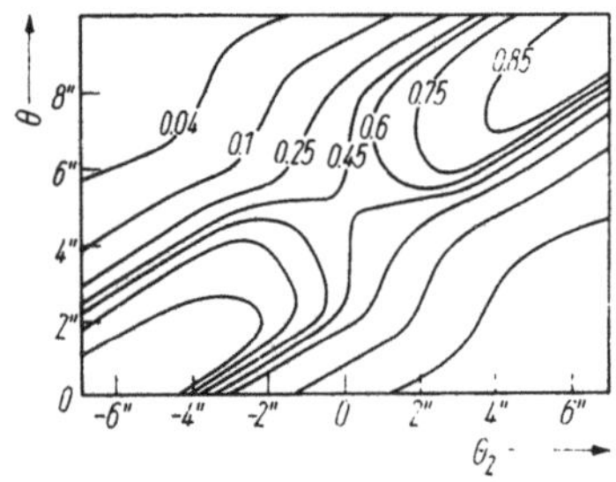

Fig. 1

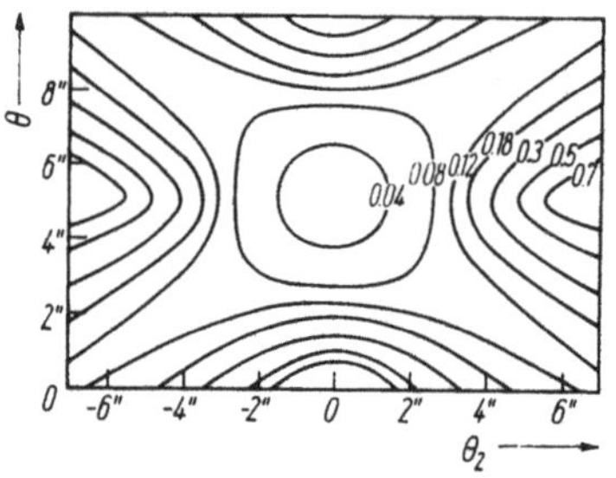

Fig. 2

Fig. 1. The coefficient $\overline{R}_1$ of a reflection of nonpolarized radiation from a thick silicon plate in three-beam $(111/11\overline{1})$ case. The entrance surface is perpendicular to $(1\overline{1}0)$

Fig. 2. The transmission coefficient $\overline{R}_0$ for a plate with at thickness 2 μm and the same experimental conditions as in Fig. 1

where

$$\left.\begin{aligned} y_s &= \frac{\theta - \theta_0}{\theta_{1s}}, \qquad \theta_0 = \frac{2\chi_0}{A}, \qquad \theta_{1s} = -\frac{2\chi_1 C_s}{A}, \\ A &= \frac{2}{\varkappa_0}(\boldsymbol{h}_1 \boldsymbol{n}_{0R}), \qquad C_s^2 = \begin{cases} 1 + 2(\boldsymbol{s}_0\boldsymbol{s}_1)^2 - (\boldsymbol{s}_1\boldsymbol{s}_2); & s = \pi, \\ 1 + (\boldsymbol{s}_1\boldsymbol{s}_2); & s = \sigma. \end{cases} \end{aligned}\right\} \tag{3.2}$$

Formula (3.1) differs from the corresponding expression for the two-beam reflection coefficient (see, for example, [14]) only by a factor 0.5 and the polarization factor C_s. The real part of the quantity θ_{1s} is easily seen to determine the angular region of total reflection, with θ_{1s} being proportional to C_s. In the present case $C_s^2 \lessapprox 2$, contrary to the two-beam case where $C_s^2 < 1$. Therefore, the three-beam angular region of total reflection along the line $\theta_2 = 0$ is broader than the two-beam one, but the value of the maximum is only 1/2 because of the energy conservation law.

It is easy to understand, taking into account the symmetry of the considered case that the relation $\overline{R}_2(\theta, \theta_2) = \overline{R}_1(\theta, -\theta_2)$ holds. Consequently it is sufficient to consider one of the diffraction beams only. Fig. 1 shows the reflection coefficient $\overline{R}_1(\theta, \theta_2)$ for nonpolarized radiation in the angular region $0 \leqq \theta \leqq 10''$, $-7'' \leqq \theta_2 \leqq 7''$ in the thick crystal. The calculations have been carried out with account of the true absorption of X-rays in the crystal, this fact being the reason for a maximum asymmetry in the two-beam region. As can be seen from the figure, the analytical solution (3.1) on the $\theta_2 = 0$ line does not give the right sense of the three-beam total reflection region which is a gentle pass with a smooth minimum along the two-beam line and with a smooth maximum in the perpendicular direction.

Fig. 2 shows the transmission coefficient $\overline{R}_0(\theta, \theta_2)$ for the same angular region and for a plate with thickness $t = 2$ μm. It follows from the calculation that the extinction length is reduced when penetrating in the many-beam region. The reflection coefficient $\overline{R}_1$ for the plate of such a thickness differs slightly from that presented on Fig. 1.

4. Reflection of a Plane Wave at Several Surfaces of the Same Crystal

The calculation of the coefficient of many-beam reflection at several surfaces of the same crystal in the general case is a somewhat more complicated problem. In the present section we restrict our treatment to the general formulation of this problem in a form which is convenient for working out the computer program.

Let the plane wave with the wave vector $\boldsymbol{k}_0$ fall on the first surface of a crystal with the inner normal vector $\boldsymbol{n}_1$. Several beams can go out of the crystal. Let us consider the beam with the wave vector $\boldsymbol{k}_1$ falling on the second surface of the crystal with the inner normal $\boldsymbol{n}_2$ from which, in particular, the beam with the wave vector $\boldsymbol{k}_2$ goes out, and so on. After n reflections the amplitude of the electric field can be written in the form

$$\boldsymbol{E}_n^{s's}(r) = \mathrm{e}^{i\varphi(r)}\, \boldsymbol{e}_{ns'} \Gamma P_{12\ldots n}^{s's}(\boldsymbol{q}_0)\, A_s\,, \tag{4.1}$$

where

$$\varphi(r) = \sum_{m=0}^{n} \boldsymbol{k}_m(\boldsymbol{r}_{m+1} - \boldsymbol{r}_m)\,, \qquad \boldsymbol{k}_m = \boldsymbol{\varkappa}_m + \boldsymbol{q}_m\,, \tag{4.2}$$

$$\boldsymbol{q}_m = \boldsymbol{q}_{m-1} - \alpha_m \boldsymbol{n}_m/2\,, \qquad \alpha_m = 2\,\frac{(\boldsymbol{h}_m \boldsymbol{q}_{m-1})}{\varkappa_0 \gamma_{mm}}\,, \tag{4.3}$$

$$\Gamma = \left\{\frac{\gamma_{01}}{\gamma_{11}}\,\frac{\gamma_{12}}{\gamma_{22}}\ldots\frac{\gamma_{n-1,n}}{\gamma_{nn}}\right\}^{1/2}, \qquad \gamma_{km} = (\boldsymbol{s}_k \boldsymbol{n}_m)\,, \tag{4.4}$$

$$P_{12\ldots n}^{s's}(\boldsymbol{q}_0) = \sum_{s_1\ldots s_{n-1}} P_n^{s's_{n-1}}(\boldsymbol{q}_{n-1}) \ldots P_2^{s_2s_1}(\boldsymbol{q}_1)\, P_1^{s_1s}(\boldsymbol{q}_0)\,. \tag{4.5}$$

Here $\boldsymbol{h}_m = \boldsymbol{\varkappa}_m - \boldsymbol{\varkappa}_{m-1}$, $\boldsymbol{r}_0$ is the point on the source surface, $\boldsymbol{r}_{n+1} = \boldsymbol{r}$ the detection point, $\boldsymbol{r}_m$ the point on the m-th surface of the crystal.

The vector $\boldsymbol{q}_m$ determines the degree of difference between the vectors $\boldsymbol{k}_m$ and $\boldsymbol{\varkappa}_m$, i.e. in fact the extent to which the Bragg conditions in the m-th reflection is not satisfied. It is essential that only the vector $\boldsymbol{q}_0$ is independent among the vectors $\boldsymbol{q}_m$. The remaining vectors are determined by $\boldsymbol{q}_0$ through (4.3). From the physical point of view formula (4.3) describes the angular divergence change of the incident beam, which arises from the condition of X-ray frequency conservation at reflection. In the approximation linear in $q/\varkappa$ this condition is reduced to the equality $\boldsymbol{\varkappa}_{m-1}\boldsymbol{q}_{m-1} = \boldsymbol{\varkappa}_m \boldsymbol{q}_m$. We note that the vectors $\boldsymbol{\varkappa}_{m-1}$, $\boldsymbol{q}_{m-1}$ at the m-th reflection play the same role as the vectors $\boldsymbol{\varkappa}_0$, $\boldsymbol{q}_0$ play at the first reflection, i.e. one many-beam set of vectors is numbered differently at every reflection.

The amplitude $P^{s's}$, according to (2.15), depends only on the two parameters θ and θ_2. In the case of several reflections, the formulae describing the direct transition from $\theta^{(m-1)}$, $\theta_2^{(m-1)}$ to $\theta^{(m)}$, $\theta_2^{(m)}$ are thought to be useful besides (4.3). Let us introduce the orthonormalized basis $\boldsymbol{n}_{mR}$, $\boldsymbol{e}_{m\sigma}$, $\boldsymbol{n}_h$ tied with point m and mark them for the sake of brevity through $\boldsymbol{a}_s^{(m)}$, $s = 1, 2, 3$. Then

$$\boldsymbol{q}_m = \sum_s \xi_s^{(m)} \boldsymbol{a}_s^{(m)}\,. \tag{4.6}$$

After substitution of (3.6) into (3.3), we have

$$\xi_s^{(m)} = \sum_{s'=1,2} D_m^{ss'} \xi_{s'}^{(m-1)}\,, \tag{4.7}$$

where

$$D_m^{ss'} = (\boldsymbol{a}_s^{(m)} \boldsymbol{a}_{s'}^{(m-1)}) - \frac{1}{\varkappa_0 \gamma_{mm}}\, (\boldsymbol{a}_s^{(m)} \boldsymbol{n}_m)\, (\boldsymbol{h}_m \boldsymbol{a}_{s'}^{(m-1)})\,. \tag{4.8}$$

The matrix D_m is three-dimensional, but we are interested only in the two-dimensional part with $s, s' = 1, 2$. In this case, (4.7) gives the necessary connection because $\xi_1 = \theta$, $\xi_2 = \theta_2$. Thus the two-dimensional transition matrices can be calculated easily in every concrete case by (4.8). This matrix makes it possible to calculate the two-beam regions of total reflection for all surfaces and to write the diagram analogous to the Du Mond diagram of the theory of two-beam spectrometers [14].

5. Reflection of the Spherical Wave

As it is known, an atom is the source of X-ray radiation, and coherence is preserved only for the radiation of a separate atom. Therefore, the incident-spherical-wave approach is the most natural one for the calculation of interference phenomena. The theory of multiple spherical-wave X-ray diffraction in the Laue case has been formulated in [7]. In this section we present the generalization of this theory to the Bragg reflection case.

Let the radiation of a source placed at the point $\boldsymbol{r}_0$ fall on a crystal. We present the electric field amplitude as a twofold Fourier integral and consider the little angular region near the Bragg direction. Within this angular region the amplitude in the space before the crystal can be written in the form

$$\boldsymbol{E}_s^{(in)}(\boldsymbol{r}, \omega) = \frac{2\pi i}{\varkappa_0} \int \frac{\mathrm{d}\boldsymbol{q}_{0\perp}}{(2\pi)^2} \, \mathrm{e}^{i\boldsymbol{Q}_0(\boldsymbol{r}-\boldsymbol{r}_0)} \sum_s A_s(\omega) \, \boldsymbol{e}_{0s} \,, \tag{5.1}$$

where

$$\boldsymbol{Q}_0 = \boldsymbol{k}_0 - \frac{q_{0\perp}^2}{2\varkappa_0} \boldsymbol{s}_0 \,, \qquad \boldsymbol{q}_{0\perp} = \varkappa_0(\theta_1 \boldsymbol{e}_{0\pi} + \theta_2 \boldsymbol{e}_{0\sigma}) \,. \tag{5.2}$$

Here we preserve, as in [7], the term quadratic in $\boldsymbol{q}_{0\perp}$ in the wave vector expression, and assume that the amplitude A_s and the polarization vectors $\boldsymbol{e}_{0s}$ do not depend on $\boldsymbol{q}_{0\perp}$. Although the characteristic values of the angles θ_1 and θ_2 are small ($\approx 10^{-5}$), the phase connected with the second-order term is of order $\varkappa_0 \theta^2 L$, where L is the source–crystal distance. Consequently, it may be essential when L enlarges. As it has been shown in [6] this term becomes essential when L is equal to ≈ 1 m, a situation that is experimentally achievable.

As a result of the reflection at the first crystal surface the vector $\boldsymbol{Q}_0$ is replaced by $\boldsymbol{Q}_1$. With account of terms to second power in $q/\varkappa$, we have

$$\boldsymbol{Q}_1 = \boldsymbol{k}_1 - \frac{q_{0\perp}^2}{2\varkappa_0} \boldsymbol{s}_0 - \alpha_1^{(2)} \boldsymbol{n}_1/2 \,, \tag{5.3}$$

where the vector $\boldsymbol{k}_1$ is determined by (4.2), (4.3) and the quantity $\alpha_1^{(2)}$ is determined by formula (8) in [7] which can be written in another form as

$$\alpha_1^{(2)} = C_1 \,, \qquad C_m = \frac{1}{\varkappa_0 \gamma_{mm}} [q_{0\perp}^2(1 - \boldsymbol{s}_0 \boldsymbol{s}_m) + q_m^2 - q_0^2] \,. \tag{5.4}$$

Considering n reflections consistently, we obtain easily the following expression for $\boldsymbol{Q}_n$:

$$\boldsymbol{Q}_n = \boldsymbol{k}_n - \frac{q_{0\perp}^2}{2\varkappa_0} \boldsymbol{s}_0 - \sum_{m=1}^{n} \alpha_m^{(2)} \boldsymbol{n}_m/2 \tag{5.5}$$

in which the quantity $\alpha_m^{(2)}$ is defined by the recurrence formula

$$\alpha_m^{(2)} = C_m - \sum_{k=1}^{m-1} \alpha_k^{(2)} \frac{\gamma_{mk}}{\gamma_{mm}} \,. \tag{5.6}$$

We note that the second-order corrections must be considered only where the corresponding vector is multiplied by a long distance of the order of 1 m. Because all the reflecting surfaces belong in our case to the same crystal, the distances between them are small. Therefore, we can replace all $\boldsymbol{Q}_m$, $m = 1, \dots, n-1$ by $\boldsymbol{k}_m$ with a good accuracy. We can also neglect the second-order terms in α_m when calculating the reflection amplitude and use the formulae (4.3), (4.5). However, these terms must be considered in $\boldsymbol{Q}_0$, if the source–crystal distance is long, and also in $\boldsymbol{Q}_n$, if there is a long crystal–film distance. Although the formula for $\boldsymbol{Q}_n$ is a rather cumbersome one, in concrete examples considerable simplifications arise.

Considering the above, we can write the amplitude of a field passed through the crystal in the form

$$\boldsymbol{E}_n^{s's}(\boldsymbol{r}, \omega) = \frac{2\pi i}{\varkappa_0} \boldsymbol{e}_{ns'} \int \frac{d\boldsymbol{q}_{0\perp}}{(2\pi)^2} \Gamma \, e^{i\varphi(\boldsymbol{q}_0, \boldsymbol{r})} P_{12\ldots n}^{s's}(\boldsymbol{q}_0) \, A_s(\omega) , \tag{5.7}$$

where

$$\varphi(\boldsymbol{q}_0, \boldsymbol{r}) = \boldsymbol{Q}_0(\boldsymbol{r}_1 - \boldsymbol{r}_0) + \sum_{m=1}^{n-1} \boldsymbol{k}_m(\boldsymbol{r}_{m+1} - \boldsymbol{r}_m) + \boldsymbol{Q}_n(\boldsymbol{r} - \boldsymbol{r}_n) . \tag{5.8}$$

If the crystal is cut in such a way that the last surface reflects into the same direction two or more beams which pass on differrent ways (interferometer), then the term $\Gamma e^{i\varphi} P$ has to be replaced by the sum of several terms of the same type.

In accordance with the method of geometrical optics, the trajectory of rays is defined from the condition of equality to zero of the terms linear in $\boldsymbol{q}_{0\perp}$ in the phase (5.8). Because the component of $\boldsymbol{q}_m$ parallel $\boldsymbol{\varkappa}_m$ does not depend on $\boldsymbol{q}_{0\perp}$, this condition is satisfied when the vectors $(\boldsymbol{r}_{m+1} - \boldsymbol{r}_m)$ are parallel $\boldsymbol{\varkappa}_m$. The diffraction pattern arises on the plane perpendicular $\boldsymbol{\varkappa}_n$. In the nonpolarized radiation case the intensity of the n-th beam is equal to

$$I_n(\boldsymbol{r}) = \frac{\gamma}{2\pi} \int \frac{d\theta_\omega}{(\theta_\omega^2 + \gamma^2)} \sum_{ss'} |F_n^{ss'}(\boldsymbol{r}, \theta_\omega)|^2 , \tag{5.9}$$

$$F_n^{ss'}(\boldsymbol{r}, \theta_\omega) = \frac{\varkappa_0}{2\pi} \Gamma \int d\theta_1 \, d\theta_2 \, e^{i\varphi} P_{12\ldots n}^{ss'}(\boldsymbol{q}_0) . \tag{5.10}$$

Here we also take into account the nonmonochromaticity of the radiation which is of order of the natural width of a characteristic line of a X-ray tube spectrum; γ is a relative half-width of the line.

According to (2.15), the amplitude $P^{ss'}$ depends on θ_1 and θ_ω only through a definite combination θ. In the nonplanar case we can proceed to a new variable θ instead θ_1 in (5.10), then the role of θ_ω reduces only to the shift of the whole pattern by some vector proportional θ_ω [7]. In the cases close to coplanar, however, the shift vector becomes very long leading to a strong averaging of the diffraction pattern along the direction of the shift. In the coplanar case, the dependence on the distance along this direction vanishes (in an approximation linear in θ), and the integration over θ_ω needs to be carried out independently. In this case the spectrum characteristics of the reflected radiation, in particular, the coherence length, are determined completely by the diffraction process, not by the source.

References

[1] B. Post, J. appl. Cryst. 8, 452 (1975).
[2] T. Hom, W. Kiszenick, and B. Post, J. appl. Cryst. 8, 457 (1975).
[3] W. Graeff and U. Bonse, Z. Phys. **B27**, 19 (1977).
[4] A. G. Rostomyan and P. A. Besirganyan, Rep. Acad. Sci. USSR **238**, 73 (1978).
[5] V. G. Kohn, Fiz. tverd. Tela **18**, 2538 (1976).
[6] A. M. Afanasev and V. G. Kohn, Fiz. tverd. Tela **19**, 1775 (1977).
[7] V. G. Kohn, Fiz. tverd. Tela **19**, 3567 (1977).
[8] B. J. Isherwood and C. A. Wallace, Acta cryst. **A27**, 119 (1971).
[9] T. Joko and A. Fukuhara, J. Phys. Soc. Japan **22**, 597 (1967).

[10] A. M. AFANASEV and YU. KAGAN, Acta cryst. A24, 163 (1968).
[11] N. S. BACHVALOV, Numerical Methods, Vol. I, Izd. Nauka, Moscow 1973 (p. 400) (in Russian).
[12] G. HILDEBRANDT, phys. stat. sol. 24, 245 (1967).
[13] P. P. EWALD and Y. HENO, Acta cryst. A24, 5 (1968).
[14] Z. G. PINSKER, Dynamical Scattering of X-Rays in Crystals, Springer-Verlag, 1978.

(Received March 2, 1979)

phys. stat. sol. (a) 54, 385 (1979)

Subject classification: 2; 20.1; 22

Central Research Institute for Physics, Budapest[1]) *(a) and*
Department of Electrical Engineering, University of Edinburgh (b)

Natural Optical Activity and Related Phenomena in As_2S_3 Glasses

By

J. HAJTÓ (a) and P. J. S. EWEN (b)

Natural optical activity is observed in As_2S_3 glasses. The value of the rotation angle depends on the incident polarisation direction relative to the sample i.e. optical anisotropy is found. The existence of optical activity is due to a structural pecularity of individual asymmetric molecules. To explain the observed anisotropy it is assumed that the layer structure of the crystalline As_2S_3 is conserved to some extent in the glass i.e. a proportion of the total number of molecules are oriented in the corresponding planes.

Нами наблюдалась естественная оптическая активность в As_2S_3-стекле. Величина угла вращения зависела от направления поляризации падающего луча, т.е. была обнаружена оптическая анизотропия. Существование оптической активности является следствием особенностей структуры индивидуальных молекул. Наблюденная оптическая активность обьясняется в предположении, что слоистая структура кристалла As_2S_3 частично остается в стекле, т.е. часть полного числа молекул ориентируется вдоль плоскости соответствующих слоев.

1. Introduction

A variety of photo-induced effects have been observed in amorphous chalcogenide materials, particularly in a-As_2S_3: photodarkening [1 to 3], photopolymerisation [4, 5], photodissociation [6], volume [7] and refractive index [8] changes have all been reported for this glass. Optical isotropy was not considered in these earlier studies but recent results [9] on evaporated a-As_2S_3 films show that illumination of this material leads to its becoming optically anisotropic. In the case of bulk a-As_2S_3 it has been shown that optical anisotropy can be induced by the application of an electric field [10] (electro-optic effect) or a magnetic field [11] (Faraday effect).

An amorphous solid may be optically anisotropic even in the absence of applied fields: birefringence can be produced in glasses by stress, the presence of laminations, structural anisotropy, or the presence of "particles" (e.g. phase-separated regions) smaller in size than the wavelength of the light used in examining the material [12]. The purpose of the present work was to investigate the optical isotropy of bulk a-As_2S_3 before and after illumination and to determine the origin of any anisotropy observed. The optical isotropy was investigated using polarimetry and transmission measurements.

2. Experimental

The samples were cut (of dimensions $1.5 \times 1.5 \times 0.7$ and $2.5 \times 1.5 \times 1.2$ cm^3) from a block of a-As_2S_3 obtained from the American Optical Co. and were polished on all six faces. The material was of good optical quality yielding an ellipticity ratio of 27 dB or less with polarised radiation from a He–Ne laser ($\lambda = 6328$ Å).

[1]) P.O.B. 49, H-1525 Budapest, Hungary.

The light sources used in the three types of experiment were either a dye laser (Spectra-Physics Model 375) tunable over the range $\lambda = 5700$ to 6500 Å or a He–Ne laser (Spectra-Physics Model 125) operating at $\lambda = 6328$ Å. In the polarimetry measurements a polarisation rotator (Spectra-Physics Model 310-21) was used to vary the plane of polarisation of the incident beam and a polaroid sheet mounted on a divided circle was used to determine the plane of polarisation of the beam emerging from the sample. The transmission was monitored with a laser power meter (Spectra-Physics Model 404).

In a number of experiments transmission and emergent beam polarisation were monitored simultaneously.

3. Results

3.1 Optical activity and related oscillation of transmission

When low-intensity polarised laser beam (0.2 mW power with diameter of 0.5 mm) of wavelength 6300 to 6500 Å was passed through the samples it was found that the plane of polarisation was rotated. The experimental arrangement can be seen in Fig. 1. The value of the optical rotation angle is given as $\varrho = (\gamma - \beta)$ where γ is the angle of the incident polarisation plane relative to plane A fixed to the sample and β is the angle of the emergent light polarisation plane relative to the same reference plane A. The value of the rotation angle depended on γ; ϱ was zero for four orthogonal directions and showed maxima for polarisations at 45° to these directions (Fig. 2).

The maxima values of the optical rotation ϱ_{max} varied from one region to another. There were two main types of regions, those which gave rise to a larger value for ϱ_{max} (for a thickness of 1.2 cm ϱ_{max} was typically 85 to 90° in these regions) and those for which ϱ_{max} was smaller (for a thickness of 1.2 cm ϱ_{max} was typically 10 to 12° in these regions). In the border regions between these principal types intermediate values of ϱ_{max} were obtained (Fig. 1).

It was found that the rotation angle ϱ could be altered by increasing the incident laser power density above a few W/cm^2. In the case of the regions yielding a smaller value of ϱ_{max}, $\varrho(t)$ (t being the time of light exposure) increased during illumination and reached a saturation value ϱ_∞, if the incident polarisation was such that $\varrho(t)_{t=0} = 0$. In Fig. 3 we plotted the values of ϱ_∞ obtained for various incident power densities; the points can be fitted by a straight line. Those regions which yielded a large value of ϱ_{max} showed the reverse behaviour, $\varrho(t)$ decreased during illumination when the incident polarisation angle was one for which $\varrho(t)_{t=0} = \varrho_{max}$. If the position of the incident beam was the same as before, namely $\varrho(t)_{t=0} = 0$, no optical rotation was observed, $\varrho(t)$ remained zero. This is the position, where oscillation in the transmission can be observed (Fig. 4) and the value of the transmission oscillated between two characteristic values under the influence of continuous laser radiation at a power density of $P = 54$ to 91 W/cm^2. The experimental results are collected in Table 1.

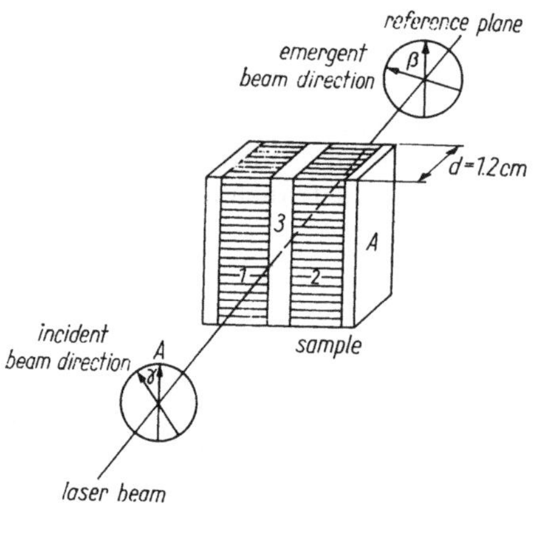

Fig. 1. Experimental arrangement for observation of optical rotation. For region 1: $\varrho \approx 85°$, region 2: $\varrho \approx 12°$, region 3: $\varrho \approx 12°$ to 85°

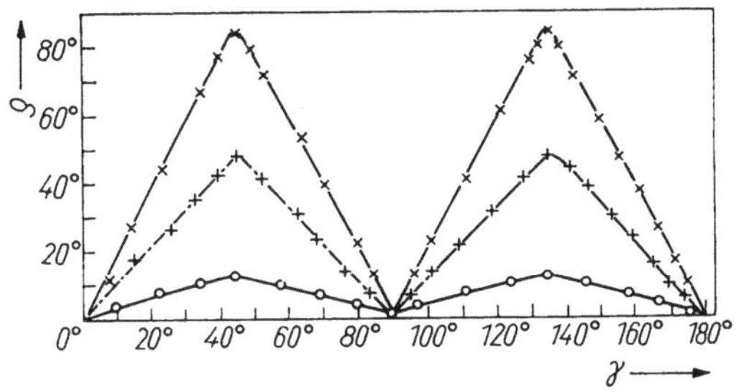

Fig. 2

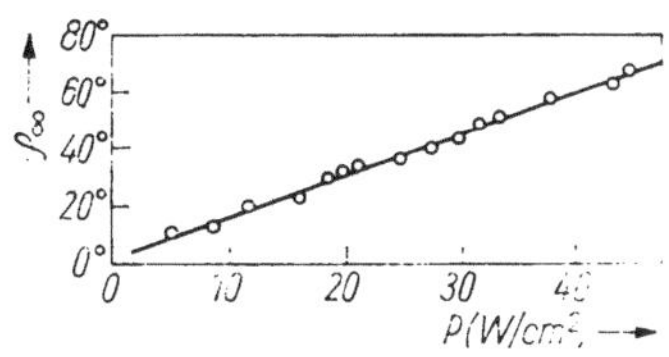

Fig. 3

Fig. 2. Orientation anisotropy in a-As_2S_3. –×–× region 1, –○–○– region 2, –+–+– region 3

Fig. 3. Laser-induced optical rotation. d = 1.2 cm, λ = 6328 Å

Table 1

Transmission oscillation in As_2S_3 glass. Laser orientation: non-rotation state

laser power density (W/cm^2)	magnitude of oscillation (percentage of transmitted light) (%)	frequency of oscillation (s^{-1})
60.1	4.1	4.78
63.7	5.4	4.87
70.8	10.6	4.93
80.8	13.5	4.97
90.7	17.1	4.98

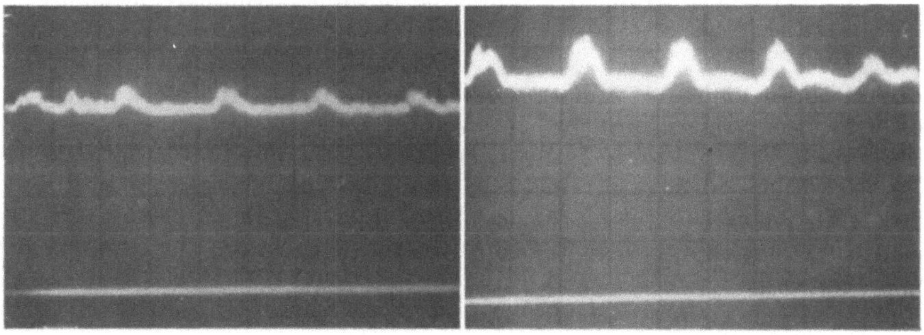

a b

Fig. 4. Oscillation of the transmission under the influence of continuous laser irradiation at a) P = 63.7, b) 90.7 W/cm^2 power density. Ordinate: transmission, abscissa: time (0.2 s/divis)

3.2 Transmission changes

Numerous reports of photodarkening in sputtered and evaporated films of a-As_2S_3 exist [1 to 3] and it has also been observed in melt-quenched bulk specimens [13]. In the present investigation a decrease in the transmission through the bulk a-As_2S_3 samples as a function of irradiation was always observed, the rate of decrease depended on the incident power density. Typical plots of the fractional change in the transmitted laser intensity $I(t)$, as a function of exposure time, t, are shown in Fig. 5. It is noteworthy that in the present experiments the wavelengths of the light inducing the decrease of the transmission (6300—6500 Å) correspond to $\alpha \approx 1\ cm^{-1}$, α being the absorption coefficient of the unilluminated material (Fig. 5).

It was also found that the kinetics of the transmission change depended on the orientation of the sample relative to the plane of polarisation of the incident light.

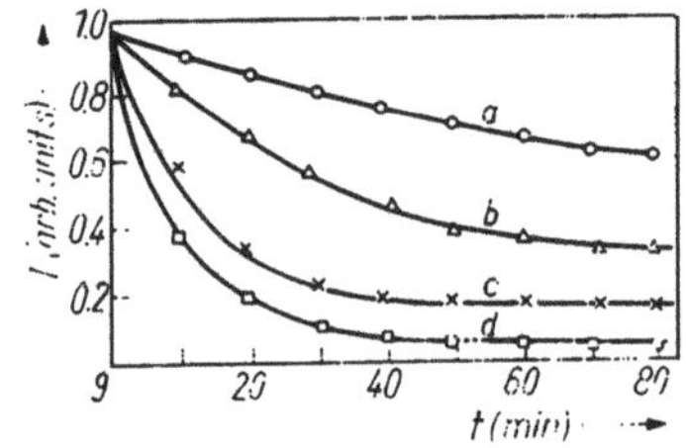

Fig. 5. Changes of the intensity of transmitted light in a-As_2S_3. (a) $P = 10$, (b) 25, (c) 45, (d) 45 W/cm^2

Curves a and b of Fig. 5 were obtained for two different orientations of the sample; the frequency and power density of the incident beam were identical in each case and both curves were obtained from a region of the sample for which ϱ_{max} was large. In the case of curve c the sample was oriented so that no rotation of the incident polarisation occurred before illumination (for this orientation no optical induced rotation occurred during illumination), while for curve d the sample was oriented to yield the maximum rotation of the incident polarisation before illumination (a rotation of 75° occurred during this run). Comparison of the two curves shows that the transmission change is faster when it is accompanied by rotation of the incident polarisation.

If in the course of illumination the plane of polarisation of the incident beam was altered, the transmission increased discontinuously and then started to decrease in the usual way. When the polarisation was changed back to the original direction it was found that the transmission for this polarisation was larger than it would have been if the incident polarisation had not altered, i.e. the original transmission change can be partly reversed by changing the plane of polarisation of the incident light. A recovery of 6% in transmittance was observed in some of these experiments. This effect may be of technological importance as it is the basis for a new type of optical memory.

When one of the samples was photographed in ordinary light after it had been illuminated with laser irradiation ($P = 20$ W/cm^2) no changes in any optical properties were observable (Fig. 5, curve *a*). However when it was illuminated with polarised light and photographed through an analyser, changes in the laser-irradiated regions were observed; the path of the beam through the sample was clearly visible (Fig. 6b). This confirms that laser-induced changes are taking place inside the sample during irradiation.

To our knowledge this is the first observation of laser-induced polarisation phenomenon in chalcogenide glasses.

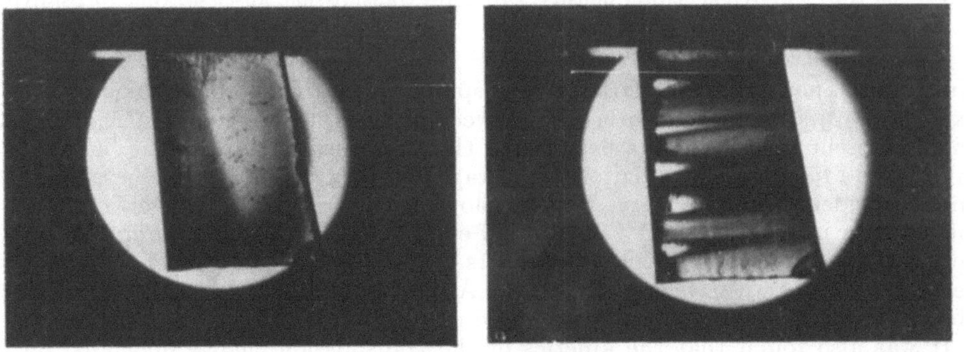

a b

Fig. 6. Laser-induced polarisation phenomenon in As_2S_3 glass

4. Discussion

If an electric field E is applied to a medium having polarisability α, a polarisation P will be induced obeying the relation $\boldsymbol{P} = \boldsymbol{\alpha}\boldsymbol{E}$, where $\boldsymbol{\alpha}$ is a tensor. In our study E is the field associated with an electromagnetic radiation and it can be expressed as $E = E_0 \cos \omega_L t$ where ω_L is the laser frequency. Since the molecular dimension is of the order of Angstroms and the laser wavelength is of the order of thousands of Ångstroms the field is constant over many molecular units (Born approximation). Therefore the average polarizability of the sample can be approximated by averaging the product of the volume V, density of the sample, and the polarizability of each molecule $(\boldsymbol{\alpha}(i))$. Further evidence to this is that the value of the molar refraction $(4/3\pi\alpha_P)$ determined from the Lorentz-Lorentz equation is in excellent agreement with the measured value for a-As_2S_3 (9.42 and 9.44, respectively) [14].

The observed optical activity is due to a structural peculiarity of individual molecules. The As_2S_3 molecule is asymmetric therefore the polarizability tensor is anisotropic, i.e. the glass contains optically anisotropic elements in molecular scale. In this case the optical activity is a natural consequence of the asymmetric molecular structure. If all the molecular units were distributed randomly according to the classical model of glassy state no anisotropy in the optical activity would be observed. The recent experiments show that this is not the case for the glassy As_2S_3.

Two kinds of optical anisotropy have to be explained: a) where the value of the optical rotation angle depended on γ (see Section 3.1) at a fixed position of the sample (orientation anisotropy), b) where the value of the optical rotation angle changes in different regions of the glass (bulk optical anisotropy).

Ewans and Young reported [15] that there is a marked dichroism between light polarised parallel and perpendicular to the c-axis of crystalline As_2S_3 as a result of its layer structure. X-ray diffraction studies of As_2S_3 glass [16] have shown that the nearest-neighbour and next-nearest-neighbour coordinations are the same as in crystal. Thus the layer structure is expected to be conserved to some extent in a-As_2S_3. Therefore the absorption and propagation of linearly polarised light is strongly affected by the relative orientation of the polarisability tensor (which is tied to the symmetry axis of the molecule) to the polarisation plane of the light. This explains the orientational anisotropy. If the number of plane-oriented molecular units varies in different regions of the glass, the value of the optical rotation angle varies, too. This is the cause of the bulk anisotropy.

To discuss the observed laser-induced polarisation phenomenon in the glass, the chemical bond approach will be helpful. In the present case a structural change in the As–S network might not be expected since the photon energies used in these experiments (1.91 to 1.95 eV) are smaller than the energy required to rupture an As–S bond (2.12 eV). Indeed simultaneous Raman experiments have confirmed that no significant structural change occurs during laser irradiation [17]. Most of the polarisability of As_2S_3 molecules arises from the lone-pair electrons of S atoms [18]. Lone pairs are probably more localized than bonding electrons because of the small nearest-neighbour overlap. Due to the large bond-free solid angle [19] a rearrangement of the lone-pair charge density of S atoms occurs corresponding to the direction of the electric field E (field of applied polarised light). A similar mechanism is possible for group V elements, too, namely for As atoms that are threefold coordinated but the bond-free solid angle is smaller. This rearrangement of the charge density results in energetically more stable state so the light-induced polarisation will be permanent (see Fig. 6).

The linearly polarised light is commonly regarded as a result of the superposition of two circularly polarised waves. The optical activity is produced as an indirect effect of the difference in velocities of propagation of right and left circularly polarised waves. The angle of optical rotation can be expressed directly in terms of the difference of the two refraction indices [20]

$$\varrho = \frac{\pi d}{\lambda} (n_l - n_r) , \tag{1}$$

where λ is the wavelength of light and d the thickness of sample. Using the measured values of $\lambda = 0.63\ \mu m$ and $\varrho = 80°$ for a thickness of 1.2 cm glass, the calculated difference in the refractive indices is $(n_l - n_r) = 4.7 \times 10^{-5}$, so considerable rotations can be produced in spite of $(n_l - n_r)$ being very small. Or, in other words, a small change in polarisability ($\varepsilon = n^2$) connected with the charge density rearrangement of lone-pair electrons causes considerable rotation of the polarised light.

The oscillatory behaviour of the transmission is not a unique feature of As_2S_3 glass only. Similar oscillation of the transmission was observed recently in amorphous $GeSe_2$ films, too [21]. In both cases the oscillation was induced by a continuous laser irradiation ($h\nu = 1.96$ eV) in the low-absorption region of these materials. At present time the mechanism responsible for the phenomena is not clear. The observed oscillation of transmission in $GeSe_2$ was explained by a sudden but reversible change in the refractive index (n) of the material [22].

We suppose that the recent observed oscillation indicates the existence of metastable states excited by nearly band gap radiation [23]. A more quantitative analysis, especially on a change in polarisation direction is required for providing more insight into these phenomena.

References

[1] K. TANAKA, Appl. Phys. Letters **26**, 243 (1975).
[2] I. P. DE NEUFVILLE, S. C. MOSS, and S. R. OVSHINSKY, J. non-crystall. Solids **13**, 191 (1974).
[3] T. IGO and Y. TOYOSHIMA, J. non-crystall. Solids **11**, 304 (1973).
[4] I. P. DE NEUFVILLE R. SEGUIN, S. C. MOSS, and S. R. OVSHINSKY, Proc. V. Internat. Conf. Amorphous Liquid Semicond., Vol. 2, Ed. J. STUKE and W. BERING, Taylor & Francis, London 1974 (p. 737).
[5] S. A. SOLIN, Phys. Rev. B **15**, 2084 (1977).
[6] Y. S. BERKES, S. W. ING, and W. J. HILLEGAS, J. appl. Phys. **42**, 4908 (1971).
[7] T. IGO, Y. NOGUCHI, and H. NAGAI, Appl. Phys. Letters **25**, 193 (1974).
[8] I. P. DE NEUFVILLE, R. SEGUIN, S. C. MOSS, and S. R. OVSHINSKY, see [4] (p. 740).
[9] V. G. ZHDANOV and V. K. MALINOVSKII, Soviet tech. Phys. Letters **3**, 387 (1977).
[10] N. F. BORELLI, Phys. Chem. Glasses **12**, 93 (1971).
[11] G. GREVENDONK, P. RUYMBEEK, P. VAN DEN KEYBUS, and A. DE BOCK, phys. stat. sol. (b) **78**, K89 (1976).
[12] I. O. ISARD, S. I. DESAI, and K. MUNILA, J. Phys. D. **11**, 433 (1978).
[13] H. HAMANAKA, K. TANAKA, and S. IZIMA, Solid State Commun. **23**, 63 (1977).
[14] A. R. HILTON, C. E. JONES, and M. BRAU, Phys. Chem. Glasses **7**, 105 (1966).
[15] B. L. EWANS and P. A. YOUNG, Proc. Roy. Soc. **A297**, 230 (1967).
[16] A. A. VAIPOLIN and E. A. POVAI-KOSHITS, Fiz. tverd. Tela **5**, 246 (1963).
[17] J. HAJTÓ and P. J. S. EWEN, Rep. Centr. Res. Inst. Phys., Budapest, KFKI-1978–97.
[18] M. KASTNER, Phys. Rev. Letters **28**, 355 (1972).
[19] M. KASTNER, Phys. Rev. B **7**, 5237 (1973).
[20] E. U. CONDON, Rev. mod. Phys. **9**, 432 (1937).
[21] J. HAJTÓ, G. ZENTAI, and I. KÓSA SOMOGYI, Solid State Commun. **23**, 401 (1977).
[22] J. GAZSÓ and J. HAJTÓ, phys. stat. sol. (a) **45**, 181 (1978).
[23] R. A. STREET, Proc. VII. Internat. Conf. Amorphous Liquid Semicond., Edinburgh, Ed. W. E. SPEAR, Centre Ind. Consultancy and Liaison,, Edinburgh 1977 (p. 509).

(Received March 2, 1979)

phys. stat. sol. (a) 54, 391 (1979)

Subject classification: 14.3.1; 22.2.3

Department of Physics and Astrophysics, University of Delhi[1])

The Effect of Non-Stoichiometric Excess of In on the Electrical Transport Properties of Polycrystalline p-Type InSb Thin Films

By

P. C. MATHUR, O. P. TANEJA[2]), K. V. KRISHNA, and A. L. DAWAR[3])

Polycrystalline p-type InSb films with different indium contents are fabricated by co-evaporating InSb and In on a glass substrate. The electrical transport properties (Hall coefficient and dc conductivity) are studied in the temperature range 77 to 500 K. The grain boundary barrier potential is calculated as a function of concentration of In in the films. It is observed that the effect of excess In is to reduce the value of the grain boundary barrier potential, thereby increasing the mobility. The ratio of the electron-to-hole mobility b also is calculated.

Durch gleichzeitiges Aufdampfen von InSb und In auf Glassubstrate werden polykristalline p-leitende InSb-Schichten mit unterschiedlichem Indiumanteil hergestellt. Die elektrischen Transporteigenschaften (Hallkoeffizient und Gleichstromleitfähigkeit) werden im Temperaturbereich von 77 bis 500 K untersucht. Das Korngrenzenbarrierenpotential wird als Funktion der In-Konzentration in diesen Schichten berechnet. Es wird beobachtet, daß der Einfluß von überschüssigem In in einer Reduzierung der Höhe des Korngrenzenbarrierenpotentials besteht, wobei die Beweglichkeit erhöht wird. Das Verhältnis der Elektronen- zur Löcherbeweglichkeit b wird ebenfalls berechnet.

1. Introduction

The inhomogeneities incorporated during the growth of polycrystalline as well as single crystal thin films of a compound semiconductor materially effect the transport properties [1, 2]. While the incorporated inhomogeneities are in general undesirable, the excess of either constituent may, in some cases, short circuit the grain boundary potential Φ of a polycrystalline film and may thereby considerably improve the mobility of the charge carriers. Indium, in excess of the stoichiometric ratio in the dendritic films of InSb, is known [3, 4] to cause electrostatic shortening of the transverse Hall potential and an increase in the conductivity σ. However, the Hall mobility μ_H ($\mu_H = R_H\sigma$, where R_H is the Hall coefficient) of the dendritic films was not appreciably affected.

The present paper reports the study of the effect of excess In on the electrical transport properties of two-phase (InSb + In) p-type polycrystalline films. It has been found that the excess of In causes a reduction in the value of Φ thereby considerably increasing the carrier mobility.

2. Experimental Details

The films (21×4 mm^2) rectangular in shape were grown from p-type bulk InSb by the vacuum evaporation technique under a vacuum of $\approx 5 \times 10^{-6}$ Torr on a glass substrate preheated to 350 °C. The substrate was then cooled to room temperature in about 4 h. The excess of In in the films was introduced by co-evaporating it from a separate boat. The measurements were made on three films having different In concentrations as shown in Table 1. All the films were grown to the same thickness (0.33 μm) which

[1]) Delhi-110007, India.

[2]) Present address: Lecturer, Government College, Gurgaon, Haryana, India.

[3]) Present address: Defence Science Laboratory, Metcalfe House, Delhi-110054, India.

Table 1

film	b	Φ (meV)	μ_{pm} (cm^2/Vs)	inversion temperature (K)
I (InSb)	18	7.7	97	242
II (InSb + In 1.0%)	8	1.4	401	256
III (InSb + In 1.9%)	3	1.1	541	363

was measured with the help of a quartz crystal thickness monitor [5]. Electron micrographs of the films showed the existence of some voids and the average grain diameter was found to be about 150 Å. High-purity silver dots were evaporated on to the films under high vacuum. The size of the voltage contacts was ≈ 0.5 mm and the current contacts were spread across the entire width of the films. For the measurements of R_H and σ, the five-probe technique as described by Putley [6] was used. The ohmic nature of the contacts was verified throughout the temperature range by the I–U characteristics. The current and the magnetic field ($B \approx 6$ kG) directions were reversed so as to eliminate the errors due to the thermomagnetic effects. The geometrical influence of the films on Hall measurements was avoided as the length-to-width ratio was greater than five. Hall measurements, as a function of temperature, were reproducible over two to three runs indicating that the structure of the films was fixed by annealing. The Hall coefficient was found to be independent of the magnetic field and the sample current. The overall maximum error in the mobility is estimated to be 7%. The sample was mounted on a copper block with electrical insulation and is kept in a Dewar flask containing liquid nitrogen. The temperature was measured with the help of a copper–constantan thermocouple soldered to the copper block.

3. Results and Discussion

The observed variation of R_H with temperature (lg R_H versus $1/T$) for all the films is shown in Fig. 1. The slow temperature variation of R_H in the low-temperature region is characteristic of a nearly degenerate extrinsic semiconductor. The decrease of the Hall coefficient with the increase of In content in the InSb films at any temperature can be due to either

a) an increase in the free carrier concentration as a result of In atoms going substitutional at vacant In sites in the InSb lattice, or

b) the electrostatic shortening of the transverse Hall field as a result of In incorporating itself in the form of metallic filaments in between the InSb grains.

However, the first possibility is unlikely because in a degenerate semiconductor the low-temperature mobility should decrease with the increase of free carrier concentration. The experimental results shown in Fig. 2 demonstrate that the Hall mobility increases appreciably with the increase of In content in the films.

The Hall coefficient falls rapidly at temperatures above 175 K and approaches zero. This shows that at higher temperature, besides the holes, the intrinsic electrons are also contributing to the conduction process. The value of R_H is given by

$$R_H = \frac{(1 - Xb)^2}{ep\,(1 + Xb)^2}, \tag{1}$$

where b is the ratio of the electron-to-hole mobility and X is the ratio of the concentration of electrons (n) to holes (p). The Hall coefficient rises again in magnitude as the temperature is further increased, thereby showing the predominance of the free

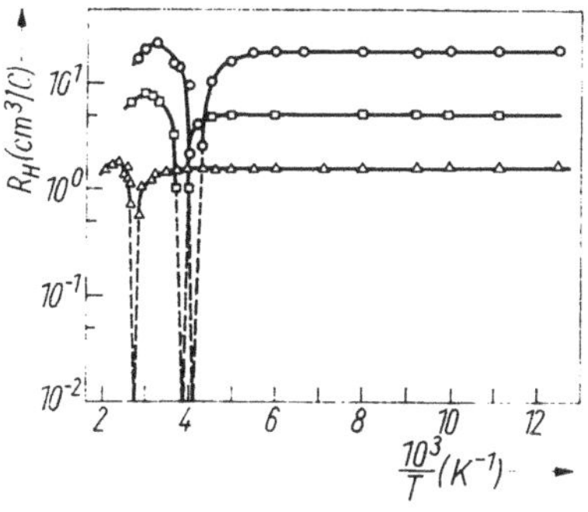

Fig. 1

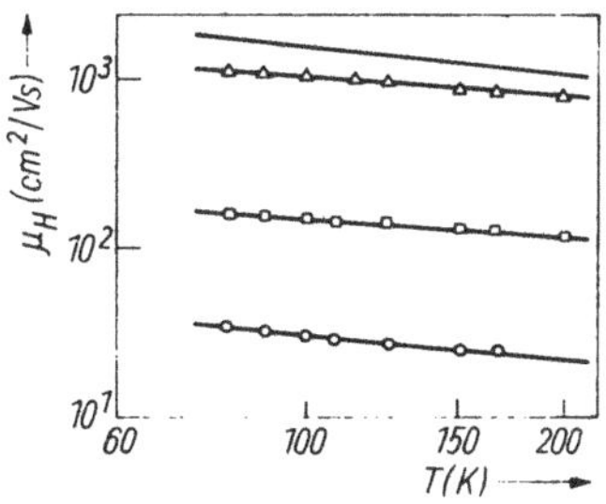

Fig. 2

Fig. 1. The variation of Hall coefficient as a function of temperature for different films; ○ film I, □ II, △ III

Fig. 2. The temperature variation of Hall mobility for different films; ○ film I, □ II, △ III. The solid line is for bulk

electron conduction process. This is due to the high mobility of electrons as compared to that of holes, inspite of the fact that $n < p$. At temperatures where R_H goes through zero (inversion temperature) the contribution of electrons and holes to the conduction process cancels and $p = nb$. Above the inversion temperature R_H rises again, reaches a peak, and at still higher temperatures it corresponds to the intrinsic behaviour. The shift of the inversion temperature towards higher temperatures with the increase of In content shows that the mobility of the holes increases with the increase of concentration of In and therefore more intrinsic electrons are needed to compensate for the increase in the conductivity due to the holes. The value of b can be calculated by the expression

$$\frac{R_{H_1}}{R_{H_2}} = \frac{(b-1)^2}{4b}, \tag{2}$$

where R_{H_1} and R_{H_2} are the minimum (algebraically) and extrinsic values of the Hall coefficient, respectively. The calculated values of b for all the three films are shown in Table 1. It can be observed that the value of b decreases with the increase of In content. This result is consistent with the observed shift of the inversion temperature towards higher temperatures. The values of b obtained in the present experiment are in agreement with the values reported by Ling et al. [7].

The observed variation of μ_H as a function of temperature ($\lg \mu_H$ versus $\lg T$) is shown in Fig. 2. The solid line represents the experimental bulk mobility data for a degenerate ($p = 5 \times 10^{17}$ cm^{-3}) single crystal. The mobility for all the three films is found to decrease with the increase of temperature. However, the temperature variation of mobility becomes steeper with the increase of In concentration. This shows that the lattice mobility μ_l becomes more important as compared to the grain barrier limited mobility μ_B with the increase of In content. The value of μ_B has been calculated from the observed Hall mobility data in the following manner:

$$\frac{1}{\mu_H} = \frac{1}{\mu_d} + \frac{1}{\mu_B}, \tag{3}$$

where μ_d is the diffused scattering mobility given by [8]

$$\mu_d = \frac{\mu_l}{1 + 3/8\gamma}, \tag{4}$$

γ being the ratio of the film thickness to the mean free path.

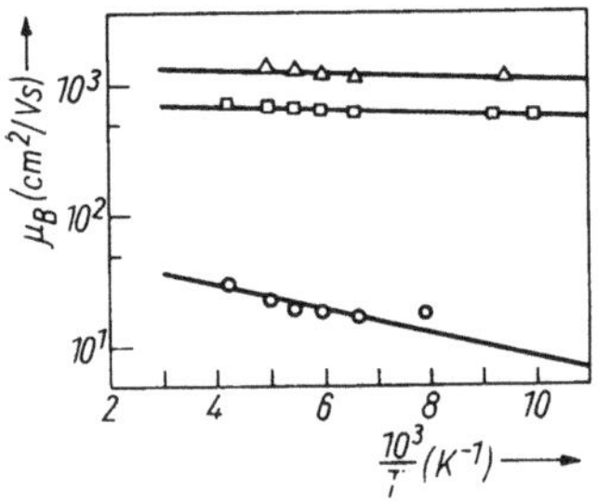

Fig. 3. The temperature variation of μ_B for different films; ○ film I, □ II, △ III

The temperature variation of μ_B calculated from (3) is shown in Fig. 3. The value of μ_B is found to increase exponentially with temperature in all the three films. However, it is observed that the rate at which μ_B increases with temperature, decreases as the In content is increased. The dependence of μ_B on the grain barrier potential can be expressed as [9]

$$\mu_B = \frac{e v_{th}}{4NkT} \exp\left(-\frac{e\Phi}{kT}\right), \tag{5}$$

where N is the number of crystallites per unit length and v_{th} is the average thermal velocity. The values of the barrier potential Φ for the three films calculated using (5) are shown in Table 1. It is found that Φ decreases with the increase of In content in the films.

The values of the effective hole mobility (μ_{pm}) at temperatures corresponding to the peak Hall coefficient R_H are shown in Table 1 and are calculated as follows: The electron density (n_m) at this temperature is given by

$$n_m = \frac{N_a - N_d}{b - 1}, \tag{6}$$

where

$$N_a - N_d = \frac{1}{eR_{H_2}} \tag{7}$$

is the free carrier concentration, and e the charge of an electron. The hole concentration p_m at this temperature will be

$$p_m = (N_a - N_d) + n_m . \tag{8}$$

The effective hole mobility at this temperature can be calculated by the expression

$$\mu_{pm} = \frac{\sigma_m}{e[n_m(b + 1) + N_a - N_d]}, \tag{9}$$

where σ_m is the effective dc conductivity. It is found that the value of μ_{pm} increases with the increase of In content in the film.

References

[1] H. WEISS, J. appl. Phys. **32**, 2064 (1961).
[2] A. MULLER and M. WILHELM, Z. Naturf. **19a**, 254 (1964).
[3] H. H. WIEDER and J. A. CORROLL, J. appl. Phys. **36**, 2321 (1965).
[4] H. H. WIEDER, Solid State Electronics **9**, 373 (1966).
[5] JUH IZENG LUE, J. Phys. E **10**, 161 (1977).
[6] E. H. PUTLEY, Hall Coefficient and Related Phenomenon, Butterworths, London 1960 (p. 42).
[7] C. H. LING, J. H. FISHER, and J. C. ANDERSON, Thin Solid Films **14**, 267 (1972).
[8] E. H. SONDHEIMER, Adv. Phys. **1**, 1 (1952).
[9] M. S. TYAGI, phys. stat. sol. (a) **30**, 609 (1975).

(Received November 2, 1978)

phys. stat. sol. (a) 54, 395 (1979)

Subject classification: 10.1; 21

Department of Metallurgy and Science of Materials, Oxford University[1])

The Extension of Jogs on Dissociated Dislocations in F.C.C. Metals

By

C. B. CARTER[2])

Experimental observations of jogs on dissociated dislocations are discussed. It is shown that using the weak-beam technique of electron microscopy both extended and apparently contracted jogs can exist on the same dislocation: the extension of jogs depends on several factors including the height of the jog and the character of the dislocation. It is found that complex jogs which may be combinations of ⟨110⟩ and ⟨112⟩ jog lines, also exist, and the occurrence of these may be related to the mechanism of the dislocation intersection process.

Es werden experimentelle Beobachtungen von Versetzungssprüngen an aufgespaltenen Versetzungen diskutiert. Mit Hilfe der elektronenmikroskopischen "weak-beam"-Technik wird gezeigt, daß sowohl aufgespaltene als anscheinend auch zusammengezogene Sprünge an derselben Versetzung existieren können. Die Aufspaltung der Versetzungssprünge hängt von mehreren Faktoren ab, u. a. von der Höhe des Sprunges und dem Charakter der Versetzung. Es wurden komplexe Versetzungssprünge gefunden, welche eine Kombination von ⟨110⟩- und ⟨112⟩-Sprunglinien sein können. Ihr Auftreten kann mit den Überschneidungsprozessen von Versetzungen in Zusammenhang gebracht werden.

1. Introduction

The configuration of jogs on dissociated dislocations in f.c.c. materials is expected to be a major factor in determining the ease with which jogs move [1, 2]. For example, if jogs on screw dislocations are extended, the jogs may pin the dislocation resulting in the formation of a dislocation dipole or of point defects if the dislocation is made to move. Similarly if jogs on edge dislocations are extended, absorption or emission of point defects at elevated temperatures or during irradiation may be more difficult, thus reducing the ease of dislocation climb under such conditions [3, 4].

It has been argued that jogs may lower their energy by adopting a dissociated configuration and that the amount of dissociation will depend on the height and sign of the jog and on the width and orientation of the dislocation [5 to 8]. The geometry of extended jogs has been examined in detail either by considering the geometry of extended super-jogs and extrapolating the results to the case of a unit jog [2, 9], or by considering ab initio the atomic configuration of a unit jog [10]. Because dislocations in these materials are extended on {111} planes, the former approach predicted unit jogs composed of ⟨110⟩ jog lines [11] which are referred to as vacancy- or interstitial-type because they correspond to rows of 1/3-vacancy or 1/3-interstitial, respectively.

The analysis of the atomic configuration of jog lines in different orientations indicated that ⟨112⟩ jog lines should have similar low energies; they are of particular interest because their formation involves only shear [10, 12].

[1]) Parks Road, Oxford, England.

[2]) Present address: Department of Materials Science and Engineering, Cornell University, Bard Hall, Ithaca, N.Y. 14853, USA.

In spite of the obvious importance of these defects, extended jogs have until recently only been reported for the very low stacking-fault energy (SFE) layer material SnS_2 [13]. This was due to the fact that prior to the development of the weak-beam method of electron microscopy [14], Shockley partial dislocations, formed by the dissociation of glide dislocations in f.c.c. materials, could not be studied in sufficient detail to distinguish between cusps in the partials caused by solute interaction and those due to jogs. Using the weak-beam technique, jogs on dissociated dislocations can be shown to cause cusps in both Shockley partial dislocations, although additional information is desirable to unambiguously distinguish jogs from effects due to solute pinning [15 to 17]. Observations on jogs have shown that most jogs appear to be extended by less than the ≈ 30 Å necessary to unambiguously indicate an extension. However, a preliminary report [16] has shown that extended jogs can be detected providing the SFE of the material is sufficiently low; in this case the extension appeared to depend on the orientation of the dislocation and the jog line appeared to be along a $\langle 110 \rangle$ direction. Extended jogs on other extended defects (double ribbons and P-type nodes) have also been observed and are discussed in greater detail elsewhere [18 to 20].

In this paper experimental observations of the extension of jogs in an alloy of copper – 15.8 at% aluminium will be discussed. These observations provide a method for obtaining a direct estimate of the energy of a jog line, the results of which may be extrapolated to materials of higher SFE where direct observation of the extension is not possible.

2. Experimental Details

The material used in this study was copper – 15.8 at% aluminium, kindly supplied by Mr. A. Y. Stathopoulos. The alloy has a SFE of (4 ± 1) mJ m^{-2} such that the separation of Shockley partial dislocations is ≈ 40 and ≈ 11 nm at edge and screw respectively [21]. Jet-polished specimens were prepared in the usual way after deforming the discs by bending to introduce dislocations both on the (111) planes parallel to the surface and on inclined $\{111\}$ glide planes. The weak-beam method of electron microscopy was used throughout; theoretical and experimental aspects of the technique have been discussed elsewhere [22, 23]. The dislocations were imaged using the different $\langle 220 \rangle$ reflections with those for which $|\boldsymbol{g} \cdot \boldsymbol{b}| = 2$, being particularly important because the two Shockley partials are then both in contrast, having $|\boldsymbol{g} \cdot \boldsymbol{b}| = 1$.

Two principal methods were used to identify the jogs. If constrictions on a number of neighbouring dislocations with the same (or opposite) Burgers vector $\boldsymbol{b}_T$, each lying on a (111) plane, are observed to be aligned close to the trace of the cross-slip plane (i.e. the $\langle 110 \rangle$ direction parallel to $\boldsymbol{b}_T$ hereafter referred to as $\langle 110 \rangle \parallel \boldsymbol{b}_T$), then it can be assumed that the constrictions occur at jogs and were created by intersection of dislocations on the cross-slip plane [17]. The same observation might be made for other $\langle 110 \rangle$ directions, but in such cases subsequent glide of the dislocations would almost certainly destroy this alignment. When constrictions on isolated dislocations were observed, identification of the jog was achieved by annealing the specimen at temperatures sufficiently high to remove cusps caused by solute pinning [15].

The extension of jogs considered here involves the formation of jog lines or stair-rod partial dislocations. The former will show no contrast under weak-beam imaging conditions as they represent a dislocation dipole of unit height (see Fig. 1), and the stair-rod partials will almost invariably also show no contrast either because the Burgers vector ($\boldsymbol{b}_{SR}$), and thus $\boldsymbol{g} \cdot \boldsymbol{b}$, is too small or because $\boldsymbol{b}_{SR}$ is perpendicular to $\boldsymbol{g}$. It will thus be necessary to deduce the line direction of the jog-line or stair-rod partial dislocations from the position of the cusps in the Shockley partial dislocations.

3. Experimental Results

As stated above, the extension of a jog is expected to depend on several parameters including the height of the jog. This can be understood from the schematic diagrams shown in Fig. 1a to c. The amount of extension of a unit jog (Fig. 1a) will be determined principally by the balance between the effective line tension of the Shockley partial dislocations at A and B (i.e. a self-energy term [24] which tends to straighten the Shockley partials) and a repulsion term due to the interaction between partials 1 and 2, which together tend to lengthen the jog line, and the SFE and line energy of the jog line (effectively a jog-line tension) which act to reduce the extension. The case of the large super-jog (Fig. 1c) is similar, but there are now two separate constrictions CD and EF with the Shockley partials on two planes having to extend stair-rod partial dislocations in order to reduce their self- and interaction energies; the line tension of the stair-rod partial will in general differ from that of a jog line. In the intermediate case (Fig. 1b) the two constrictions are sufficiently close that the inclined Shockley partials may have little effect and the other Shockleys must extend two stair-rod dislocations instead of either one or a jog line. Thus, it may be that unit jogs and large super-jogs are extended, but jogs at some intermediate height are effectively constricted [7]. Experimentally it is unlikely that the height of a jog can be determined to an accuracy of greater than ≈ 2.5 nm or $\approx 10d_{111}$, and therefore the height of jogs in the area of interest is indeterminable unless the formation of the particular jog has been observed [17].

The difference between the images of small contracted and extended jogs is illustrated in Fig. 2. Both $\boldsymbol{g}$- and $\bar{\boldsymbol{g}}$-images are shown because the contrast from the Shockley partials is alternately inside and outside the stacking-fault area, and this affects the apparent shape of the cusps at the jog. The orientation of the dislocation at B is 60° with partial dislocations 1 and 2 having edge and 30° character, respectively; at A the dislocation has curved into the edge orientation so that both partials have 60° character. This observation therefore suggests that the extension of a jog is strongly dependent on the orientation of the dislocation on which it lies. It has been shown [25] that Shockley partial dislocations in this material are theoretically unstable within $\approx 11°$ of the edge orientation, and recent experimental observations on isolated Shockley partial dislocations have confirmed that such dislocations are extremely flexible [19]. Futhermore, it has recently been shown [26 to 28] that Shockley partials will be particularly resistant to bending when close to the 30° orientation, and these two factors lead to the jog at A being very asymmetric and constricted. The latter point follows from the low line tension of the edge partial causing the dislocation to bend rather than pull out a jog line.

The constrictions shown in Fig. 3 also lie on dislocations which curve through the edge orientation and most are measurably extended. The width of dissociation of the dislocations varies significantly, indicating the influence of strong solute interactions [19, 21], and in several positions more localized pinning can be detected (e.g. at A). It is suggested that the effect of solute pinning may lead to the extension of jogs on 60° dislocations (e.g. at B), which would otherwise be contracted. It can also

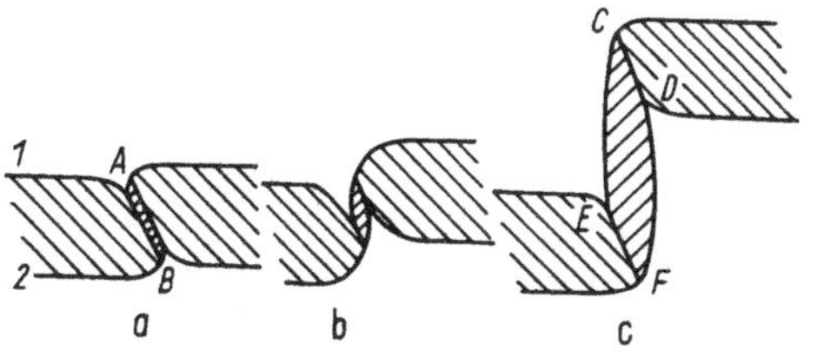

Fig. 1. Schematic diagrams to illustrate how the height of a jog may affect its extension: a) a unit jog or jog line; b) intermediate height $\approx 5d_{111}$; c) a large super-jog, height $\approx 20d_{111}$

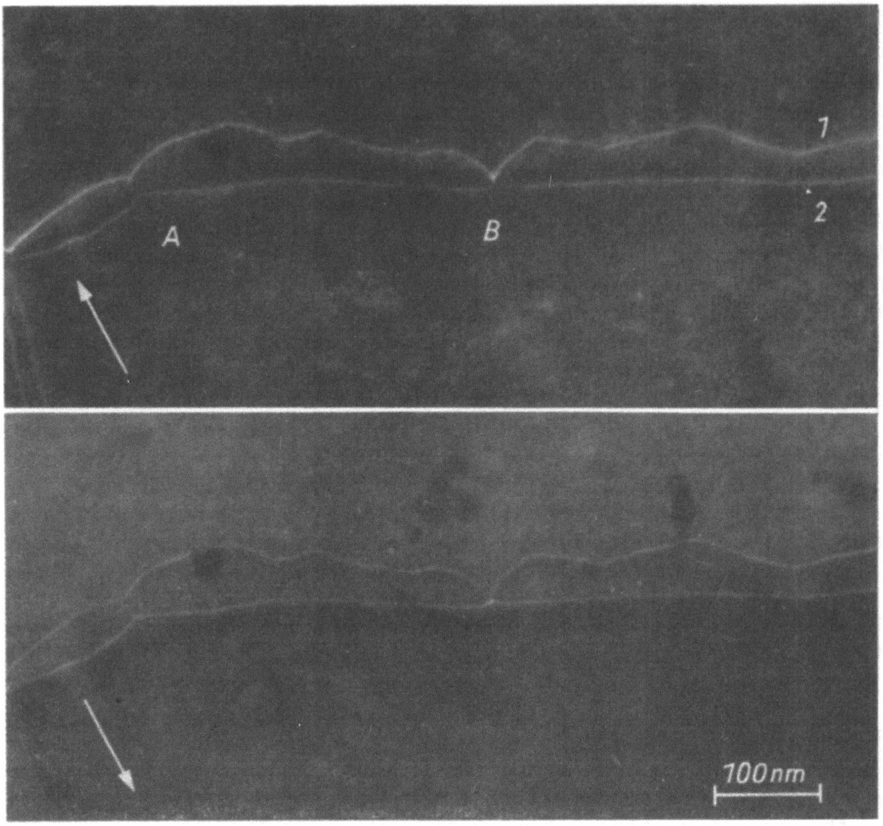

Fig. 2. Two jogs A and B on one dissociated dislocation; the jogs are respectively extended and apparently totally contracted. It is suggested that the difference is mainly due to the change in character of the dislocation

be seen that jogs aligned close to the same ⟨110⟩ || $\boldsymbol{b}_T$ may also differ in the amount of extension although the dislocations have the same character.

When a pair of jogs is relatively close together, as, e.g., those pairs along the line EF in Fig. 3, the jogs are almost invariably constricted. This situation is similar to that shown in Fig. 1 b for a single jog: i.e., when two jogs are very close to one another, the Shockley partial between the jogs cannot bow out as far as would otherwise be the case, and therefore cannot exert the same tension on the jog line (assuming a unit jog), and thus the jogs appear contracted.

Careful examination of Fig. 3 also reveals that the sets of jogs lying close to lines ⟨110⟩ || $\boldsymbol{b}_T$ are never exactly collinear and that the displacements of the jogs from these lines are not constant for a particular dislocation. Thus, for the lines 1, 2, and 3 drawn parallel to $\boldsymbol{b}_T$ passing through jogs on the far right dislocation, line 3 passes close to all the jogs, while lines 1 and 2 show large deviations. This is also clear from the pairs of jogs along EF, where the separation of the jogs varies in a random manner. Possible explanations for these observations are:

(i) The dislocations may be at different heights and the projected positions (onto (111)) of the jogs will differ. However, the observed deviation can be as large as 20 nm, which would imply differences in the height of the dislocations of ≈ 50 nm. Obser-

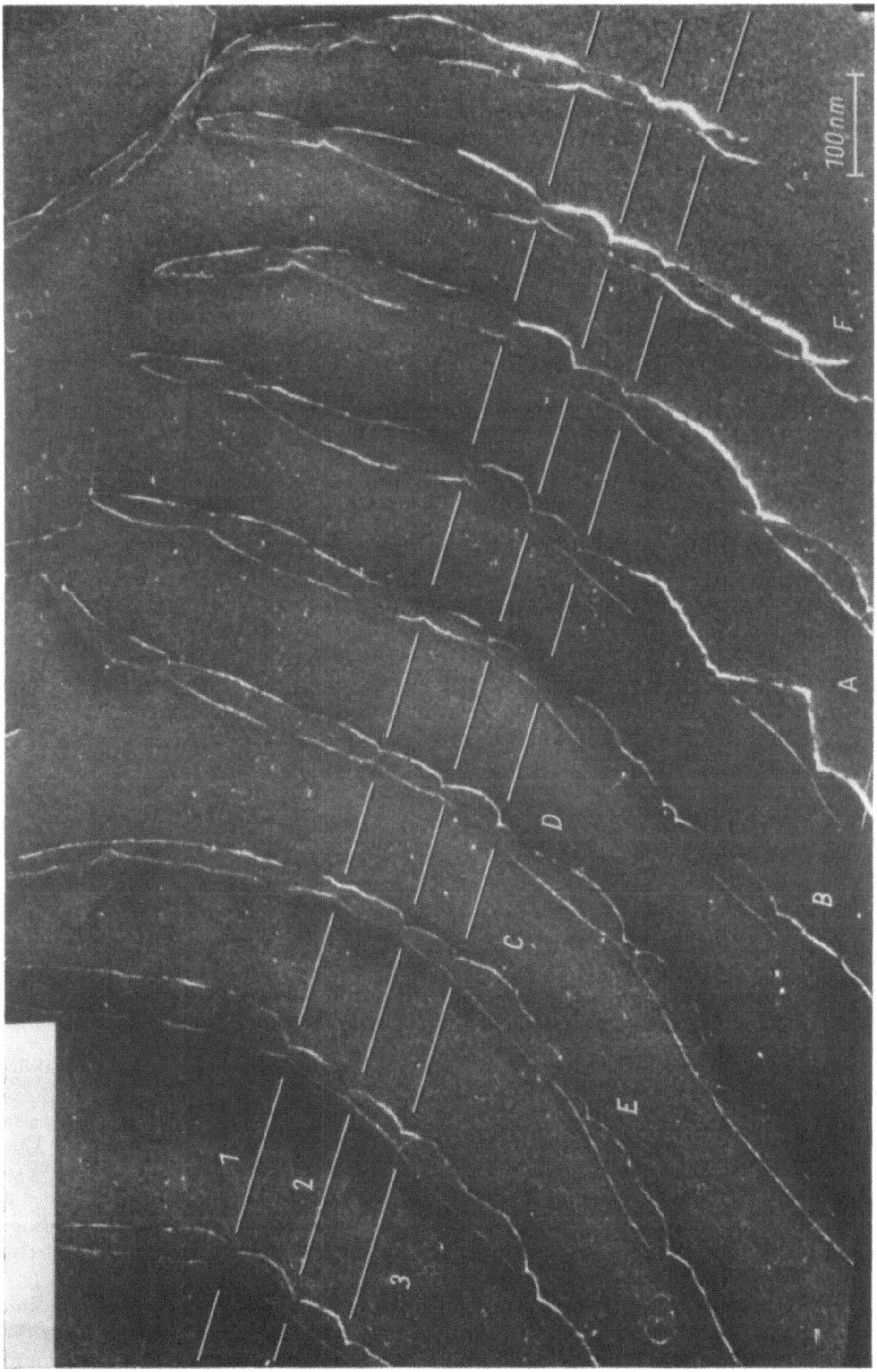

Fig. 3. Rows of jogs lying parallel to $\boldsymbol{b}_\mathrm{T}$ and $\boldsymbol{g}$. The displacement of the jogs from the lines 1, 2, and 3 and the pairs of jogs, e.g. E and F, are discussed in the text

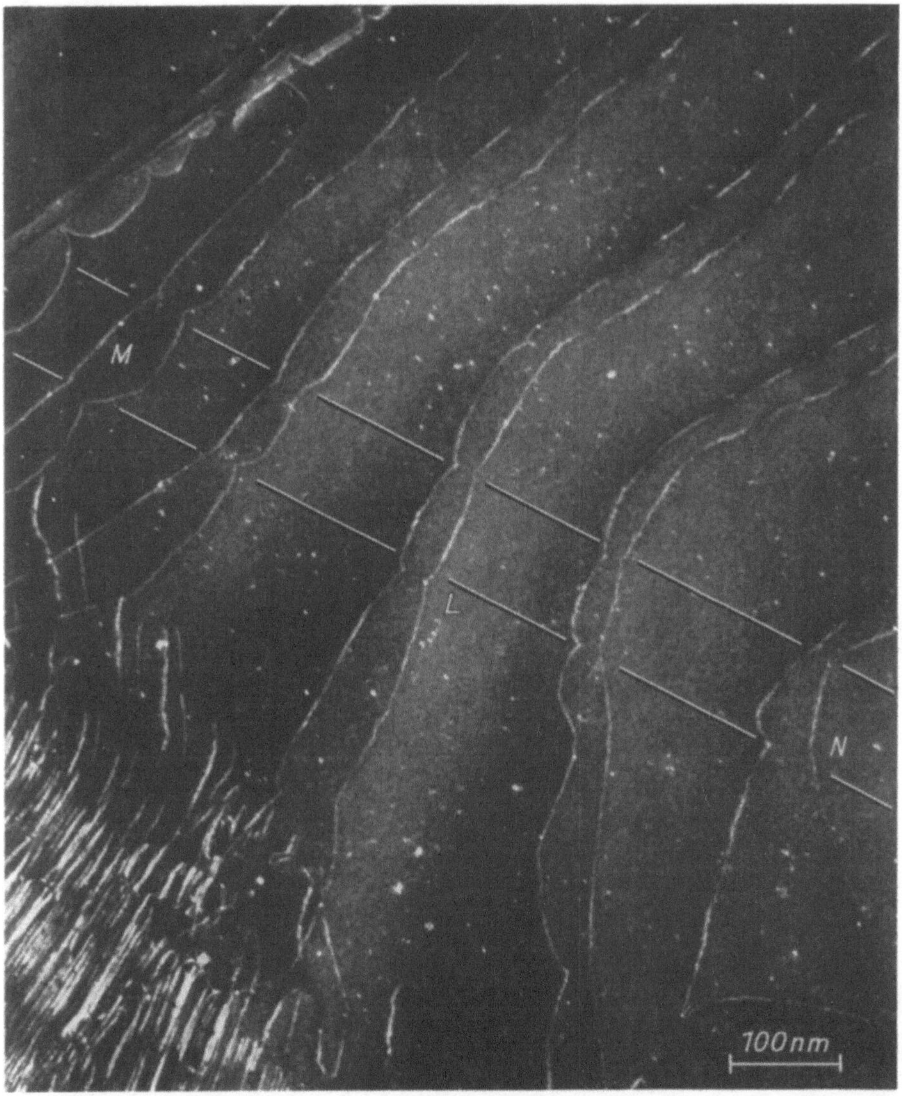

Fig. 4. Rows of jogs lying parallel to $\boldsymbol{b}_T$ and $\boldsymbol{g}$. The lines connecting the cusps in the partial dislocations make angles with $\boldsymbol{b}_T$ of up to 30° and may signify the presence of ⟨112⟩ jog lines

vations on the intersection of these dislocations with the foil surface indicate that this is not the case.

(ii) The inclined dislocation may cross-slip between intersecting the different dislocations. However, assuming it continues to move in the same direction, the inclined dislocation would not be able to cross-slip back onto its original glide plane, and this mechanism cannot thus explain the observation.

(iii) The Burgers vector of the inclined dislocation must itself have been inclined to the (111) plane in order to create the observed jogs. Thus, if the inclined dislocations contained large super-jogs, these could glide along the dislocations and pass through the (111) planes containing the observed dislocations, thus effectively moving the glide plane of the inclined dislocation by an amount equal to the height of the large super-jogs. This mechanism would require a large number of relatively closely spaced superjogs, and although such dislocations were not observed in these materials, the mechanism cannot be entirely discounted.

(iv) Dislocation climb could allow the jogs to move along the near-edge dislocations [15], but the large number of vacancies required for a unit jog to climb a distance of 20 nm suggests that this would be unlikely to occur during room temperature deformation unless large numbers of point defects were produced during the deformation.

(v) It has been noted [29] that cooperative glide resulting in cross-slip may occur at a node during the intersection process. If the dislocations in Fig. 3 are defined to be $\overrightarrow{BC}(d)$ (i.e., $\boldsymbol{b}_T = \overrightarrow{BC}$ and the glide plane is d [30]), then the intersecting dislocations must be $\pm\overrightarrow{BD}(a)$ or $\pm\overrightarrow{CD}(a)$ to give the observed result. The inclined dislocations will then create jogs in $\overrightarrow{BC}(d)$ as observed, but will not themselves be jogged and could not emit or absorb point defects. Thus a net generation of point defects would still be necessary.

It therefore appears that none of the above explanations is entirely satisfactory, but it appears that jog-climb must be involved in the intersection process.

In the observations discussed above, the line connecting the cusps on the Shockley partial dislocations has always been along a $\langle 110 \rangle$ direction, as would be expected for dissociated super-jogs or the $\frac{1}{3}$-interstitial/vacancy jog lines. A similar result has been found for the extended jogs on double ribbons [18, 20].

In Fig. 4 the jogs are again aligned close to $\langle 110 \rangle \parallel \boldsymbol{b}_T$, but in this example the lines connecting the cusps in the Shockley partial dislocations often are at an angle Φ to the direction of $\boldsymbol{b}_T$ such that $|\Phi| \leqq 30°$. It can be seen that the angle varies for jogs along a particular $\langle 110 \rangle \parallel \boldsymbol{b}_T$, e.g. MN, where Φ takes the values $+30°$ at M, $0°$ at L, and $-15°$ at N. Jogs for which $\Phi = 30°$, could be the $\langle 112 \rangle$ jog lines as predicted [10] and as shown schematically in Fig. 5, but because the jog lines are not themselves visible, the configuration may consist of two $\langle 110 \rangle$ jog lines [2] as also shown in Fig. 5. The presence of jogs similar to N where $0° < \Phi < 30°$ suggests that jogs composed of two jog lines are present.

The observation that the jog lines are not the same for each jog along a particular $\langle 110 \rangle \parallel \boldsymbol{b}_T$, is a further indication that the intersection process is a complicated one.

The extension of small or unit jogs is stongly dependent on the character of the partial dislocations due to the orientation dependence of the effective line tension. In the case of super-jogs this can result in a strong asymmetry of the jog, which was not considered in Fig. 1, but which is shown experimentally in Fig. 6. The dislocations are in each case widely extended at one end of the jog and contracted at the other. It should be noted that both bends will be either acute or obtuse and the stair-rod involved would be the same at both ends of the jog if both were extended. The reason for the asymmetry is that the partial dislocation can lower its line energy by rotating towards the screw or 30° orientation, and as shown schematically in Fig. 7a, this

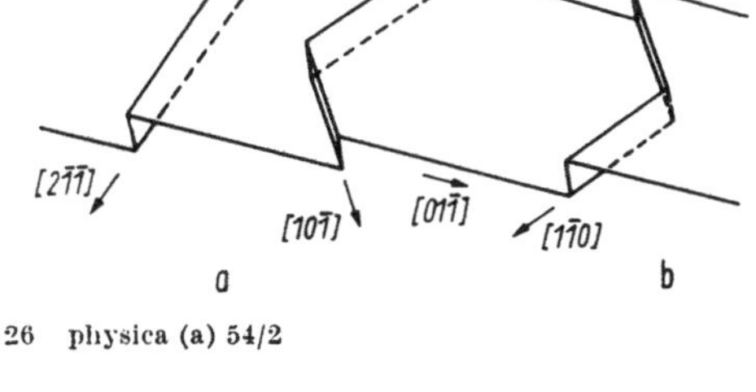

Fig. 5. Schematic diagrams to illustrate three possible configurations for jog lines connecting cusps (not shown) in the partial dislocations when the connecting line is along a $\langle 112 \rangle$ direction

will only occur at one end of the jog. If the partial dislocation is rotated into the edge orientation, it will become very flexible and continue to rotate towards the 60° orientation. The three examples shown in Fig. 6 illustrate both the acute and the obtuse cases and differ considerably in the length of dislocation lying on the cross-slip plane. They could in principle all be unit jogs, but it is more likely that they are super-jogs with the displacement of the two ends possibly occurring during the formation process [17]. Extended jogs on dissociated screw dislocations would by the same arguments

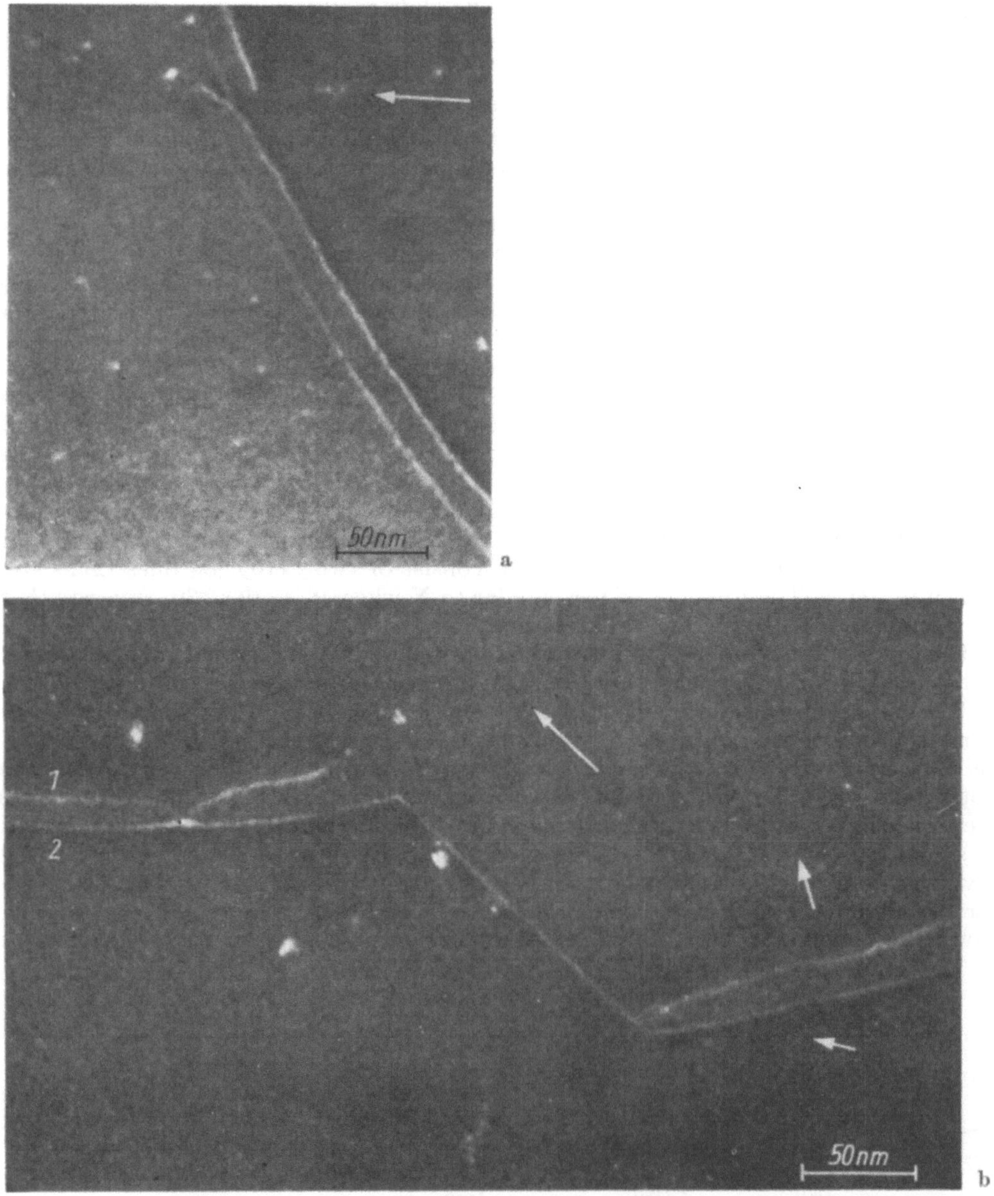

Fig. 6

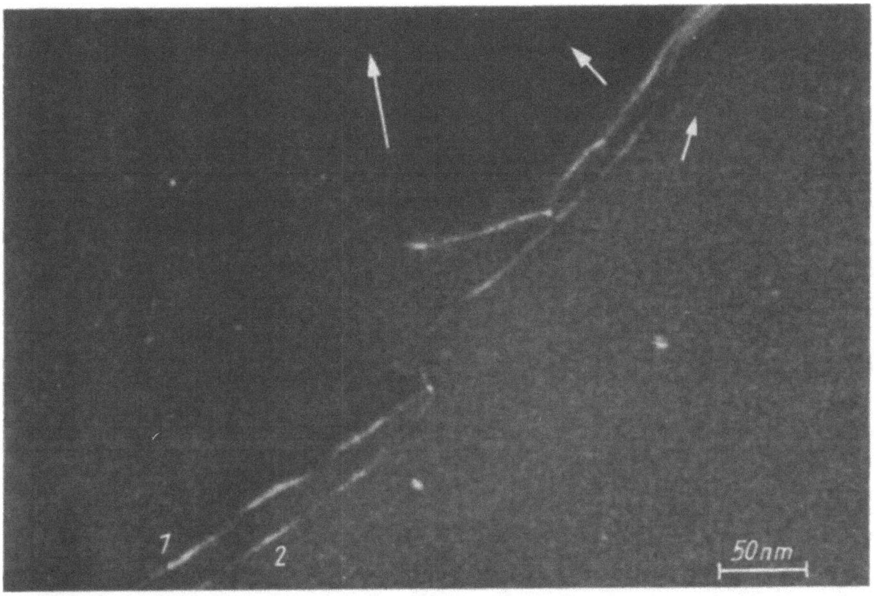

c

Fig. 6. Illustration of the asymmetry of jogs when the two ends of the jogs are displaced from one another. The large arrows indicate $\boldsymbol{g}$; the small arrows define $\boldsymbol{b}_\text{p}$

also be asymmetric as shown in Fig. 7 b, although it may not be possible to detect such fine detail. This asymmetry of jogs on dissociated dislocations is analogous to that observed when dissociated dislocations intersect small precipitates [19], inclined stacking-faults [19], and the foil surface [31].

As shown previously [31], jogs can also be studied on dislocations which are inclined to the foil plane, but because of the strong influence of the foil surfaces on the extension of such dislocations in low SFE materials neither the extension nor the shape of the jog can be directly related to the corresponding features of jogs not influenced by surfaces. These jogs can appear very asymmetric as shown in Fig. 8, even when the dislocations are close to the edge orientation (i.e., their Burgers vector is in the plane of the foil: foil plane (111)). These inclined dislocations form part of a pile-up as shown in Fig. 8a, the leading dislocations interacting with dislocations of opposite sign on a nearby parallel glide plane. It is suggested that the asymmetry of the constrictions

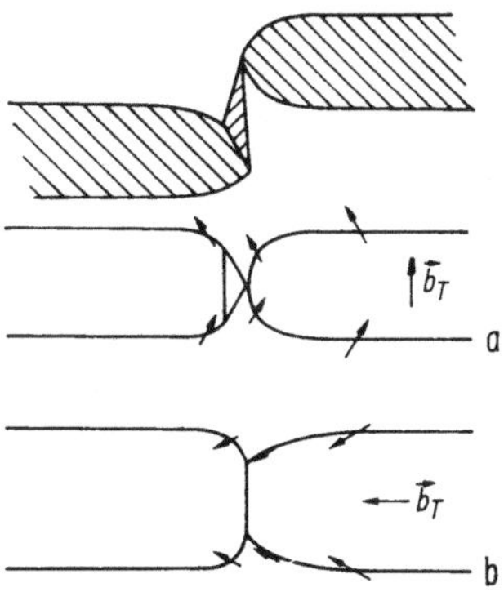

Fig. 7. Schematic diagrams to illustrate the intrinsic asymmetry of extended jogs on a) pure edge dislocations and b) pure screw dislocations. The arrows identify the Burgers vectors of the partial dislocations (on the dislocation line) and of the perfect dislocation $\boldsymbol{b}_\text{T}$

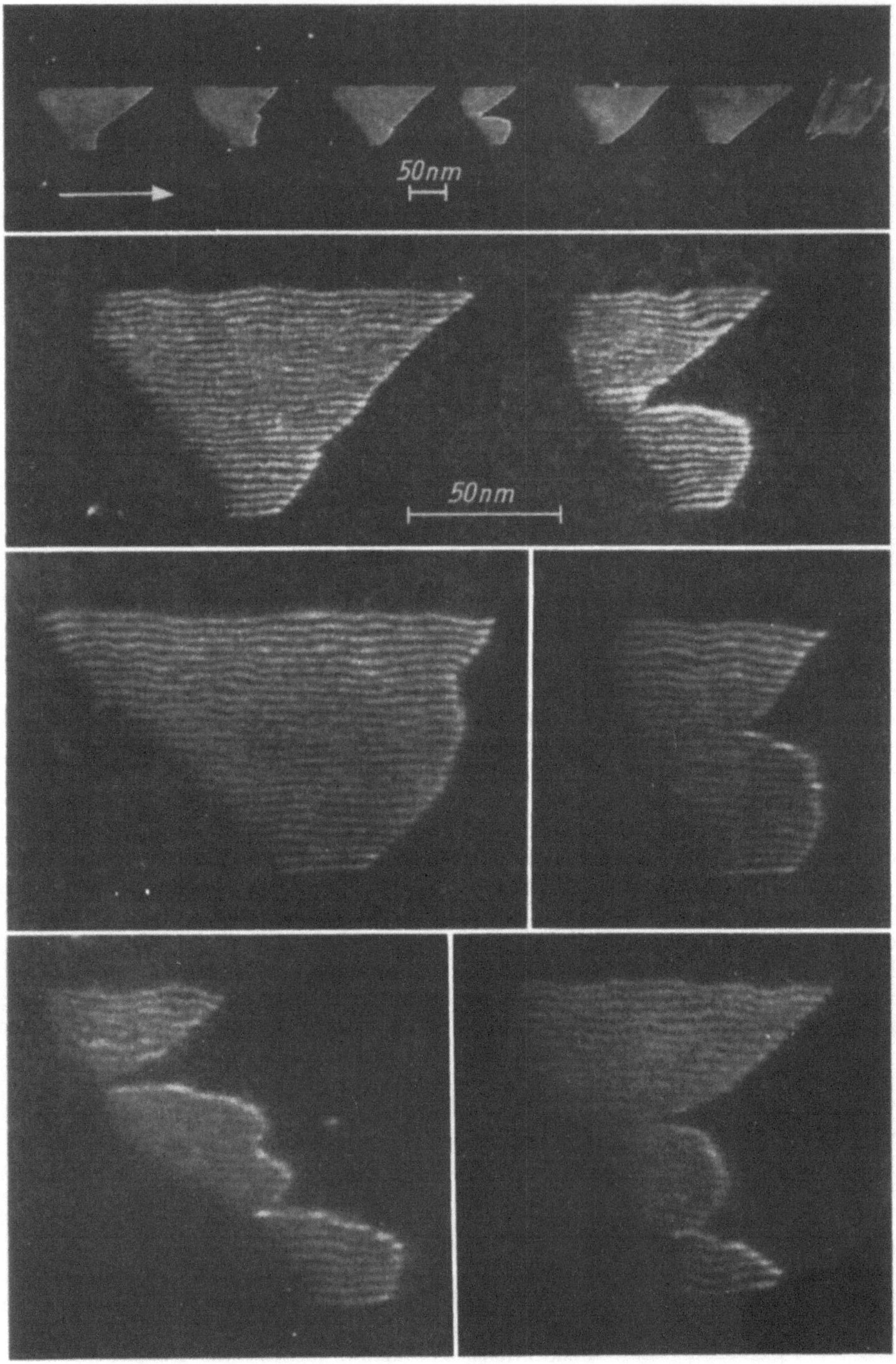

a b c d e f

Fig. 8. Jogs on inclined dislocations. Those shown in b) are from an enlarged area of a), while the other examples are from further to the left of the pile-up in a). The arrow indicates $\boldsymbol{b}_T$

in Fig. 8b to f is due to the dislocation having moved in a direction left to right under the influence of a stress causing the right-hand partial to move first. The fact that not all the dislocations contain jogs while some contain two, can be explained by the intersecting dislocations gliding on a {111} plane other than the cross-slip plane; each jog would then have been created by a different intersecting dislocation.

If the intersecting dislocation had been gliding on the cross-slip plane (i.e. the (111) plane in this example), then one dislocation could create a jog in each inclined dislocation. These jogs would then lie close to the ⟨110⟩ || $\boldsymbol{b}_T$ (cf. Fig. 3 and 4), and if extended would appear as cusps on the partial dislocations, which particularly with strong-beam imaging conditions could be confused with wavyness or zig-zagging due to the instability predicted by anisotropic elasticity [25].

4. Conclusions

The extension of jogs on dissociated dislocations varies according to a number of factors:

(i) The orientation of the dislocation; when one partial dislocation is close to pure edge, the jog is likely to be contracted.

(ii) The height of the dislocation; jogs of intermediate height may be more constricted than either unit jogs or large super-jogs. The extension of super-jogs may also vary along their length.

(iii) The applied stresses acting on the dislocation may distort the jogs because different forces act on the two partial dislocations, and this distortion may be frozen-in by the strong solute pinning forces when the applied stress is removed.

Observations strongly suggest that ⟨110⟩ jog lines do occur, and when the jogs are not distorted, such observations provide a direct method for the determination of jog-line energies. Such energies may also be appropriate to higher stacking-fault energy materials, where the extension of jogs cannot be directly determined or even detected.

Evidence is also presented to support the suggestion [10] that ⟨112⟩ jog lines will be energetically possible, although it is not possible to distinguish these from jogs consisting of two ⟨110⟩ jog lines.

Finally, the displacement of jogs from their position expected from a consideration of their mechanism of formation indicates that the intersection process is more complicated than expected and may include jog-climb.

Acknowledgements

The author would like to thank Prof. Sir P. Hirsch for his interest in this work and for provision of laboratory facilities. He is also grateful to Prof. S. L. Sass for provision of photographic facilities.

References

[1] N. F. Mott, Trans. MS AIME **218**, 962 (1960).
[2] P. B. Hirsch, Phil. Mag. **7**, 67 (1962).
[3] R. M. Thomson and R. W. Balluffi, J. appl. Phys. **33**, 803 (1962).
[4] B. Escaig, Acta metall. **11**, 595 (1963).
[5] A. N. Stroh, Phil. Mag. **3**, 625 (1958).
[6] J. Friedel, Dislocations, Pergamon Press, London/Oxford 1964.
[7] T. Jøssang, Canad. J. Phys. **45**, 3357 (1967).
[8] C. B. Carter, Ph. D. Thesis, Oxford Univ., 1975.
[9] J. Weertman, Phil. Mag. **8**, 967 (1963).
[10] K.-H. Pfeffer, P. Schiller, and A. Seeger, phys. stat. sol. **8**, 517 (1965).

[11] N. THOMPSON, Rep. Bristol Conf. Defects in Crystalline Solids, Phys. Soc., London 1955 (p. 153).
[12] J. P. HIRTH and J. LOTHE, Theory of Dislocations, McGraw-Hill Publ. Co., New York 1968.
[13] R. SIEMS, P. DELAVIGNETTE, and S. AMELINCKX, Phil. Mag. **9**, 121 (1964).
[14] D. J. H. COCKAYNE, I. L. F. RAY, and M. J. WHELAN, Phil. Mag. **20**, 1265 (1969).
[15] C. B. CARTER and I. L. F. RAY, Phil. Mag. **29**, 1231 (1974).
[16] C. B. CARTER and P. B. HIRSCH, Proc. IV. Internat. Conf. Strength of Metals and Alloys, Nancy 1976, Vol. 1 (p. 79).
[17] C. B. CARTER and P. B. HIRSCH, Phil. Mag. **35**, 1509 (1977).
[18] C. B. CARTER, Proc. Emag 77, Glasgow, Inst. Phys., London 1977 (p. 173).
[19] C. B. CARTER, to be published (1979).
[20] C. B. CARTER, to be published (1979).
[21] C. B. CARTER and P. M. HAZZLEDINE, unpublished research (1978).
[22] D. J. H. COCKAYNE, Z. Naturf. **27a**, 452 (1972).
[23] I. L. F. RAY and D. J. H. COCKAYNE, Proc. Roy. Soc. **A325**, 534 (1971).
[24] L. M. BROWN, Phil. Mag. **10**, 441 (1964).
[25] L. M. CLAREBROUGH and A. K. HEAD, phys. stat. sol. **33**, 431 (1969).
[26] R. O. SCATTERGOOD and D. J. BACON, Acta metall. **24**, 707 (1976).
[27] P. M. HAZZLEDINE, private communication, 1976.
[28] S. M. HOLMES, private communication, 1977.
[29] J. WASHBURN, Appl. Phys. Letters **7**, 183 (1965).
[30] N. THOMPSON, Proc. Phys. Soc. **B66**, 481 (1953).
[31] P. M. HAZZLEDINE, H. P. KARNTHALER, and E. WINTNER, Phil. Mag. **32**, 81 (1975).

(Received February 9, 1979)

phys. stat. sol. (a) 54, 407 (1979)

Subject classification: 10.2; 22.2

Departments of Physics (a) and Applied Mathematics (b), University of Port Elizabeth

The Nucleation and Generation of Dislocation Half-Loops in III–V Compounds

By

J. H. BASSON (a), H. BOOYENS (a), and C. A. B. BALL (b)

Misfit dislocation generation by means of the nucleation of dislocation half-loops is considered. The theory developed for semicircular dislocation loops is extended to half-loops lying in the Peierls troughs of the crystal. Contour plots of dislocation energy versus dislocation dimensions are produced from which the non-equlibrium behaviour of the loops can be deduced.

Es wird die Bildung von Misfit-Versetzungen durch die Keimbildung von Versetzungshalbschleifen untersucht. Die für halb-zirkulare Versetzungsschleifen entwickelte Theorie wird auf Halb-Versetzungen, die in den Peierlströgen des Kristalls liegen, angewendet. Die Konturbilder der Versetzungsenergie über der Versetzungsdimension werden aufgezeichnet, aus denen sich das Nichtgleichgewichtsverhalten der Schleifen ableiten läßt.

1. Introduction

Among the mechanisms by which misfit dislocations can be generated during growth of an epilayer on a substrate is the nucleation of half-loops at sources on the surface of the crystal and subsequent expansion of the half-loops towards the interface under the influence of the force which arises from the misfit strain [1 to 4]. However, before the dislocation is free to expand, an energy barrier must be overcome, and this activation energy must be provided at the source. On reaching the interface the leading portion of the loop extends laterally into the interfacial plane, leaving a length of misfit-relieving dislocation line.

Previously [1 to 4] only semicircular dislocation loops have been treated in this connection. However, in many crystals and, in particular, in group III–V crystals dislocations tend to lie in Peierls troughs owing to the high Peierls stresses associated with these materials [5, 6]. Thus, it is expected that dislocation half-loops in these structures would tend to have a more angular shape.

The energy associated with such an angular dislocation half-loop in the presence of a coherency strain is derived. Owing to the fact that, in general, the energy is a function of two dimensions of the dislocation loop, contours of energy versus these two dimensions are plotted so as to determine the conditions most favourable for its nucleation and expansion, as well as its behaviour under any given conditions.

The theory derived is applied to a typical case, that of (001)-oriented GaAs–$Ga_{0.7}Al_{0.3}As$ bicrystal structures with a misfit of 5.3×10^{-4}.

2. The Energy of an Angular Dislocation Half-Loop

The half-loop to be considered has two parallel sides in screw orientation, and its leading edge in 60° orientation. Other dislocation loops can be treated in a similar way. In general such an angular dislocation half-loop will assume the shape as sketched in Fig. 1. Every angular half-loop will be characterized by means of the angles α and ω, where α is the angle between the screw parts of the loop and the interfacial plane

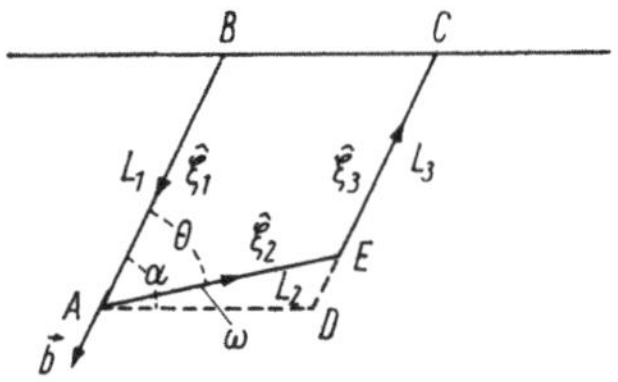

Fig. 1. Diagrammatical representation of a general angular dislocation half-loop

and ω is the angle between the 60° part and the interfacial plane. The loop will be known as the α, ω loop.

The self-energy of the dislocation loop is determined according to the method laid out by Hirth and Lothe [6]. Firstly, the self-energy of each segment of the loop is determined. Then the interaction energy between each pair of dislocation segments is determined separately. The total self-energy of the dislocation loop is thus a sum of six terms:

$$\omega = \omega_1 + \omega_2 + \omega_3 + \omega_{12} + \omega_{23} + \omega_{31}\,, \tag{1}$$

where ω_i ($i = 1, 2, 3$) is the self-energy of segment i and ω_{ij} ($i, j = 1, 2, 3$; $i \neq j$) is the ineraction energy between segments i and j.

The self-eneries ω_1, ω_2, and ω_3 are respectively given by

$$\omega_1 = \frac{\mu b^2}{4\pi} L_1 \ln\left(\frac{L_1}{e\varrho}\right), \tag{2}$$

$$\omega_2 = \frac{\mu b^2(1 - \nu \cos^2\theta)}{4\pi(1 - \nu)} L_2 \ln\left(\frac{L_2}{e\varrho}\right), \tag{3}$$

and

$$\omega_3 = \frac{\mu b^2}{4\pi} L_3 \ln\left(\frac{L_3}{e\varrho}\right), \tag{4}$$

where L_1, L_2, and L_3 are the lengths of segments 1, 2, and 3 respectively, as shown in Fig. 1, and $\varrho = b/8$ is a cut-off parameter. The angle θ is the angle between the 60° and the screw segments as shown in Fig. 1.

The interaction energies are given by

$$\omega_{12} = \frac{-\mu b^2 \cos\theta}{4\pi}\left[L_2 \ln\left(\frac{\sqrt{L_1^2 + L_2^2 - 2L_1L_2\cos\theta} + L_1 - L_2\cos\theta}{L_2}\right) + \right.$$
$$+ L_1 \ln\left(\frac{\sqrt{L_1^2 + L_2^2 - 2L_1L_2\cos\theta} + L_2 - L_1\cos\theta}{L_1}\right) -$$
$$\left. - (L_1 + L_2)\ln(1 - \cos\theta)\right], \tag{5}$$

$$\omega_{23} = \frac{\mu b^2 \cos\theta}{4\pi}\left[L_2 \ln\left(\frac{\sqrt{L_2^2 + L_3^2 + 2L_2L_3\cos\theta} + L_3 + L_2\cos\theta}{L_2}\right) + \right.$$
$$+ L_3 \ln\left(\frac{\sqrt{L_2^2 + L_3^2 + 2L_2L_3\cos\theta} + L_2 + L_3\cos\theta}{L_3}\right) -$$
$$\left. - (L_2 + L_3)\ln(1 + \cos\theta)\right], \tag{6}$$

and

$$\omega_{31} = \frac{-\mu b^2}{4\pi} \Bigg\{ \sqrt{(L_3 + L_2 \cos\theta - L_1)^2 + \eta^2} + \sqrt{L_2^2 \cos^2\theta + \eta^2} -$$

$$- \sqrt{(L_3 + L_2 \cos\theta)^2 + \eta^2} - \sqrt{(L_2 \cos\theta - L_1)^2 + \eta^2} -$$

$$- \frac{1}{2} (L_1 - L_3 - L_2 \cos\theta) \times$$

$$\times \ln \left[\frac{\sqrt{(L_3 + L_2 \cos\theta - L_1)^2 + \eta^2} + L_1 - L_3 - L_2 \cos\theta}{\sqrt{(L_3 + L_2 \cos\theta - L_1)^2 + \eta^2} - L_1 + L_3 + L_2 \cos\theta} \right] -$$

$$- \frac{1}{2} L_2 \cos \ln \left(\frac{\sqrt{L_2^2 \cos^2\theta + \eta^2} + L_2 \cos\theta}{\sqrt{L_2^2 \cos^2\theta + \eta^2} - L_2 \cos\theta} \right) +$$

$$+ \frac{1}{2} (L_1 - L_2 \cos\theta) \ln \left[\frac{\sqrt{(L_1 - L_2 \cos\theta)^2 + \eta^2} + L_1 - L_2 \cos\theta}{\sqrt{(L_1 - L_2 \cos\theta)^2 + \eta^2} - L_1 + L_2 \cos\theta} \right] +$$

$$+ \frac{1}{2} (L_3 + L_2 \cos\theta) \ln \left[\frac{\sqrt{(L_3 + L_2 \cos\theta)^2 + \eta^2} + L_3 + L_2 \cos\theta}{\sqrt{(L_3 + L_2 \cos\theta)^2 + \eta^2} - L_3 - L_2 \cos\theta} \right] \Bigg\}, \tag{7}$$

where η is the perpendicular distance between segments 1 and 3.

The energy released as the dislocation loop expands, is given by [6]

$$\omega_e = - \frac{2\mu(1 + \nu)}{1 - \nu} f\gamma A , \tag{8}$$

where μ and ν are the shear modulus and Poisson's ratio, respectively, and are assumed to be the same for the two crystals, f is the misfit between the two crystals, and γ is the factor relating the resolved shear stress acting on the dislocation loop in the slip plane and parallel to the direction of the Burgers vector to the misfit stress parallel to the interfacial plane. A is the area of the loop and is given by

$$A = \tfrac{1}{2} (L_1 + L_3)\, \eta \,. \tag{9}$$

The total change in energy as the dislocation expands, is given by a sum of equations (2) to (8). It contains only two independent parameters since

$$L_3 = L_1 - L_2 \frac{\sin\omega}{\sin\alpha} \tag{10}$$

and

$$\eta = L_2 \sin\theta \,, \tag{11}$$

as can be deduced from Fig. 1. Consequently, it is possible to plot contours of this energy as a function of the two parameters L_1 and L_2, and thus to predict the behaviour of the expanding loop from these contours.

Note that in the above treatment the energy of the surface step created by the dislocation loop has been negelected. This can be done provided that the critical thicknesses of formation of the loops concerned are greater than an order of 50 Å [4].

3. Angular Dislocation Half-Loops in an (001)-Oriented $GaAs-Ga_{0.7}Al_{0.3}As$ Structure

Two different types of angular half-loops can exist on the slip planes of an (001)-oriented system, namely a 60,0 and a 120,60 dislocation loop. The 60,0 loop has its screw edges inclined at 60° to the interface and its leading edge parallel to the interface, while the 120,60 loop has its screw edges inclined at 120° and its leading edge at 60° to the interfacial plane. In Fig. 2a and b energy contours derived from the formulae given in Section 2 are plotted for the 60,0 and 120,60 loops, respectively. Note that, whereas in the case of the 60,0 loop $L_1 = L_3$, in the case of the 120,60 loop $L_3 = L_1 - L_2$ from equation (10). Thus, in the latter case the energy contours are confined to the region below the $L_1 = L_2$ line.

The solid line in Fig. 2a and b represents a ridge of maximum energy which any dislocation loop expanding from the source has to overcome before it can extend in size. The activation energy required by the loop to surmount this ridge is supplied at the source. On the ridge of maximum energy there exists a point corresponding to the minimum of all these maximum energy points — a saddle point —, which is represented in Fig. 2a and b by a filled-in circle. This point corresponds to the minimum possible activation energy required by the loop to be able to nucleate and expand. The values of the saddle-point energy, E_m, together with the corresponding values of L_1 and L_2, as read off from Fig. 2a and b, are given in Table 1 for dislocation half-loops in an (001)-

Table 1
The critical values L_{1c}, L_{2c}, E_m, and h_c for 60,0 and 120,60 dislocation half-loops in an (001)-oriented $GaAs-Ga_{0.7}Al_{0.3}$ As heteroepitaxial structure

α	ω	L_{1c} (μm)	L_{2c} (μm)	E_m (10^{19} erg)	h_c (μm)
60	0	0.55	0.75	1.02μ	0.39
120	60	1.00	0.80	1.02μ	0.71

oriented $GaAs-Ga_{0.7}Al_{0.3}As$ structure. The critical epilayer thickness at which it becomes possible for such a dislocation loop to form, is that thickness of the epilayer enabling it to accommodate the dislocation loop. The critical thickness can thus be

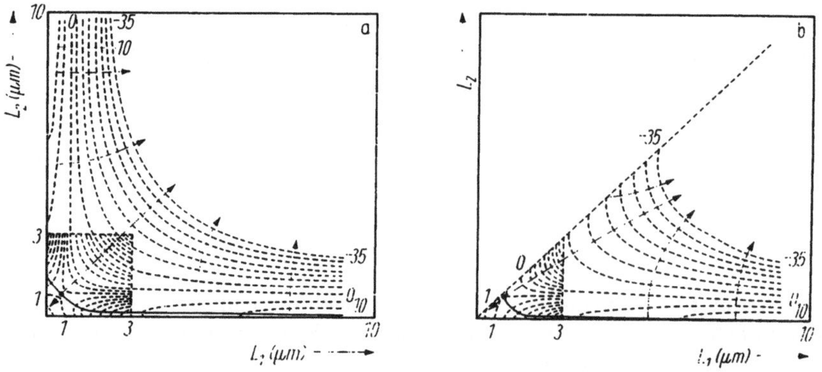

Fig. 2. Contour plot of the quantity $(E/\mu) \times 10^{19}$ versus L_1 and L_2 for the a) 60,0 and b) 120,60 half-loops. Neighbouring contours within the 3 μm block differ by 0.5 cm^3, while those outside the 3 μm block differ by 5 cm^3

calculated from

$$h_c = L_{1c} \cos \varphi \sin \alpha \,, \tag{12}$$

where L_{1c} is the value of L_1 corresponding to the saddle point, and φ is the angle between the direction normal to the slip plane and the interfacial plane. The values of h_c calculated from (12) are also given in Table 1. Note that the 120,0 and 60,0 and the 120,60 and 60,120 dislocation loops are pairwise equivalent in that they obey the same energy expressions.

It can be seen from Table 1 that, although the activation energy required to nucleate the 60,0 and 120,60 loops is the same, the critical thickness at which the 60,0 loop can form, is lower. Thus, it is expected that the 60,0 loops will be favoured for low epilayer thicknesses.

4. Dislocation

Much useful information can be gained by a careful study of the contour plots in Fig. 2a and b. Predictions concerning the manner in which a dislocation half-loop which nucleates at a given size and shape, continues to expand can be made. Every loop will grow in such a way as to follow that route on the contour plot along which the contours are closest together, i.e. along which the energy decrease is most rapid.

In Fig. 2a five different points representing five different sizes of nucleation have been chosen as points of departure for the loops. A loop nucleating on the saddle point expands in such a way as to keep L_1 and L_2 approximately equal. Dislocations which are initially long and narrow (L_1 large and L_2 small), tend to increase only slightly in length while widening rapidly, while wide flat loops (L_1 small and L_2 large) remain at the same width while lengthening.

From the contour plot in Fig. 2b it can be seen that a 120,60 loop nucleating at the saddle point follows a route which deviates considerably from the $L_1 = L_2$ line. Further, if the loop nucleates at a point on the $L_1 = L_2$ line, it expands along a route which leads away from this line. Hence, it is energetically unfavourable for the loop to assume a triangular shape with $L_3 = 0$. In addition, a loop which is very long and narrow, lengthens slightly as it widens, while one of intermediate length initially remains constants in length and then increases considerably in length on the loop's becoming wider.

Note that in the case of the 60,0 loop (Fig. 2a), the ridge of maximum energy runs approximately parallel to the $L_2 = 0$ axis, while there is no such energy ridge parallel to the $L_1 = 0$ axis. This indicates that long, narrow loops have a minimum width of nucleation, while the same restriction does not apply to wide, shallow loops.

Consider the situation when a 60,0 loop has reached the interface. It extends itself into the interfacial plane, leaving a length of misfit-relieving dislocation line there. The length L_1 is now constrained to increase with the thickness of the epilayer, while there is no such constraint on the length L_2. This might force the dislocation loop to expand along a route which deviates from the route types shown in Fig. 2a. Hence, for example, a long narrow loop, which would otherwise have tended to remain constant in length, would now be constrained to increase in length. The route drawn in on the contour plot in Fig. 2a for the case where L_1 is very large, would deviate from a straight line parallel to the L_2 axis to a line curving towards the right. On the other hand, the width of a wide, flat loop is unaffected by the fact that the loop is constrained to fit into the epilayer, and remains constant while its length increases. This seems to indicate that, after the loop has attained a certain width, it does not bow out any further into the interfacial plane, provided that minimal misfit is relieved by the dislocation.

Comparison of the two contour plots in Fig. 2a and b indicates that point for point the 60,0 loop has a lower energy configuration than the 120,60 loop. This seems to indicate that, if a 120,60 loop should form, it would be energetically favourable for its leading edge, which is inclined at 60° to the interfacial plane, to bow out into a direction parallel to the interfacial plane in order to assume the form of a 120,0 loop.

Finally, it must be noted that misfit dislocation half-loops do not form unless a high misfit or localized high stresses are present [1 to 4]. Thus, in the special case which has been considered in this work, high localized stresses have to be present to act as sources of activation energy for the formation of the dislocation loops. This theory can be applied to any other III–V compound heteroepitaxial structure.

Acknowledgement

Financial assistance from the South African Council for Scientific and Industrial Research is greatly appreciated.

References

[1] J. W. MATTHEWS, in: Epitaxial Growth, Part B, Ed. J. W. MATTHEWS, Academic Press, New York 1975 (p. 590).
[2] J. W. MATTHEWS, A. E. BLAKESLEE, and S. MADER, Thin Solid Films **33**, 253 (1976).
[3] J. W. MATTHEWS, J. Vacuum Sci. Technol. **12**, 126 (1975).
[4] J. W. MATTHEWS, S. MADER, and T. B. LIGHT, J. appl. Phys. **41**, 3800 (1970).
[5] H. G. VAN BUEREN, Imperfections in Crystals, North-Holland Publ. Co., Amsterdam 1961.
[6] J. P. HIRTH and J. LOTHE, Theory of Dislocations, McGraw-Hill Publ. Co., New York 1968.

(Received April 2, 1979)

phys. stat. sol. (a) 54, 413 (1979)

Subject classification: 13.4 and 14.3; 22.8.1

Department of Physics, Indian Institute of Technology, Madras[1])

Electrical Conductivity and Thermal Depolarisation Studies of CrO_4^{2-} and $Cr_2O_7^{2-}$ Doped KDP and ADP Crystals

By

B. V. R. CHOWDARI and Y. RAVI SEKHAR

The electrical conductivity and thermal depolarisation of both undoped and anionic impurity doped KDP and ADP crystals are studied. The proton transport in both the $H_2PO_4^-$ and NH_4^+ sublattices seems to be responsible for the conductivity and ITC of ADP crystals doped with CrO_4^{2-} and $Cr_2O_7^{2-}$ ions. In the case of KDP crystals doped with these ions, the proton transport in $H_2PO_4^-$ sublattice itself can account for the experimental results. The activation energies for the proton migration in these lattices are determined.

Es werden die elektrische Leitfähigkeit und die thermische Depolarisation von undotierten und mit Anionenstörstellen dotierten KDP- und ADP-Kristallen untersucht. Der Protonentransport sowohl im $H_2PO_4^-$- als auch im NH^+-Untergitter scheint für die Leitfähigkeit und ITC von mit CrO_4^{2-}- und $Cr_2O_7^{2-}$-Ionen dotierten ADP-Kristallen verantwortlich zu sein. Im Falle von KDP Kristallen, die mit diesen Ionen dotiert werden, lassen sich die experimentellen Ergebnisse mit dem Protonentransport im $H_2PO_4^-$-Untergitter allein erklären. Es werden die Aktivierungsenergien für die Protonenwanderung in diesen Gittern bestimmt.

1. Introduction

Studies on KH_2PO_4 (KDP) and $NH_4H_2PO_4$ (ADP) crystals have been of great interest in recent years in view of their dielectric, ferroelectric, and thermal properties. Hydrogen bonds in these crystals play a prominent role in determining these properties [1]. Unlike in simple alkali halides of NaCl-type, these crystals can have two types of cationic vacancies which may govern the electrical conductivity in these crystals. Although a large number of reports are available in the literature [2 to 7] on electrical conductivity studies of KDP and ADP crystals, there seems to be no unique mechanism which explains the electrical conductvity and related properties of these crystals. For example, the earlier studies [2 to 5] on the electrical conductivity of KDP and ADP crystals have been explained in terms of proton defects named as L- and D-defects in the $H_2PO_4^-$ sublattice. Later studies indicate that the protonic conduction in NH_4^+ sublattice could also be considered for explaining the electrical conductivity in ADP crystals [6]. However, recent studies on these crystals in a wider temperature region by Sharon and Kalia [7] indicate that the proton migration occurs through a synchronous phosphate rotation mechanism. Further, most of the electrical conductivity studies on these crystals doped with anionic impurity have been confined to SO_4^{2-} only [2 to 5]. These experiments have shown that conductivity is an activated process and is not altered much on deuteration but increases proportionally to the concentration of sulphate impurity doped into these crystals.

In view of the above-said discrepancies in the data, it appears to be desirable to carry out conductivity studies in a wider temperature region, particularly with different aliovalent anionic impurities. Therefore, for our study we have chosen CrO_4^{2-}

[1]) Madras-600036, India.

and $Cr_2O_7^{2-}$ as dopants in view of their characteristic optical and magnetic properties [8, 9], which unlike in the case of SO_4^{2-} can be readily used for identifying the presence of the impurity in the lattice.

The study of thermal depolarisation currents, which is also referred to as ionic thermo-currents (ITC) is of considerable interest in view of the information one can draw about the relaxation of dipolar complexes such as a impurity–vacancy complex [10]. No such data seems to be available on ITC measurements in these crystals except our preliminary report on $Cr_2O_7^{2-}$ doped KDP and ADP crystals [11]. In order to correlate electrical conductivity and ITC data, ITC measurements have been made on CrO_4^{2-} doped KDP and ADP crystals and the results are reported here along with those for $Cr_2O_7^{2-}$ doped crystals.

2. Experimental

Single crystals of pure and doped KDP and ADP have been grown by slow evaporation of the respective saturated solutions. K_2CrO_4 and $K_2Cr_2O_7$ have been added as impurity. In order to confirm the presence of the impurity in the lattice, optical absorption spectra have been recorded using a Cary-14 spectrophotometer. The resistance of the measuring sample was calculated from the current for a known voltage. The current in turn was obtained from the potential drop across the input resistance of a GR 1230A electrometer in series with the crystals. Temperature control and measurement were accurate up to ± 0.5 K.

For the thermal depolarisation measurements good electrical contact between the crystal and the electrodes was obtained by coating the crystal faces with silver point. The sample was first cooled to the required temperature using a small quantity of liquid nitrogen as coolant and then polarised by means of an applied electric field of strength 3000 V/cm. Electric field was applied for about 10 to 15 min and then the crystal was cooled to liquid nitrogen temperature after which the field was switched off. The specimen was warmed to room temperature at a constant heating rate, typically 5 K/min. The depolarisation current was measured using a Keithley 610 C electrometer.

3. Results and Discussion

3.1 Electrical conductivity studies

Before the electrical conductivity studies were made, the optical absorption spectra of KDP and ADP crystals doped with CrO_4^{2-} and $Cr_2O_7^{2-}$ have been studied. Two characteristic bands at about 2700 and 3700 Å have been observed in both the doped crystals. This observation indeed confirms the presence of the impurity unequivocally [8].

Fig. 1 shows the lg σ versus $1000/T$ plots for CrO_4^{2-} and $Cr_2O_7^{2-}$ doped KDP crystals and that for undoped KDP crystals. It can be noticed that there is a knee around 180°C

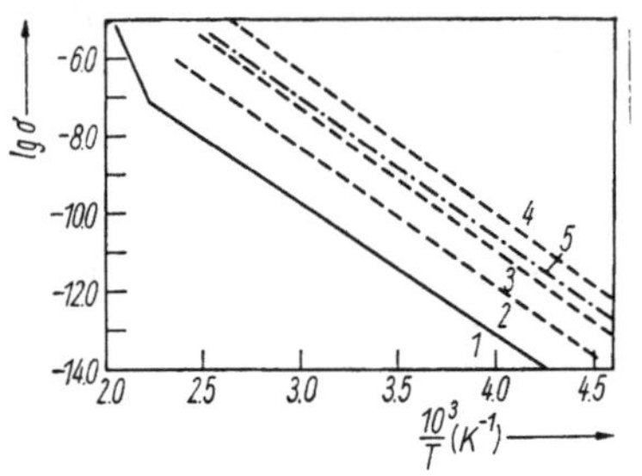

Fig. 1. The temperature dependence of the conductivity of (1) pure KDP and CrO_4^{2-} and $Cr_2O_7^{2-}$ doped KDP crystals. Curves 2, 3, and 4 refer to CrO_4^{2-} doped KDP in the increasing order of concentration of impurity, curve 5 to KDP:$Cr_2O_7^{2-}$

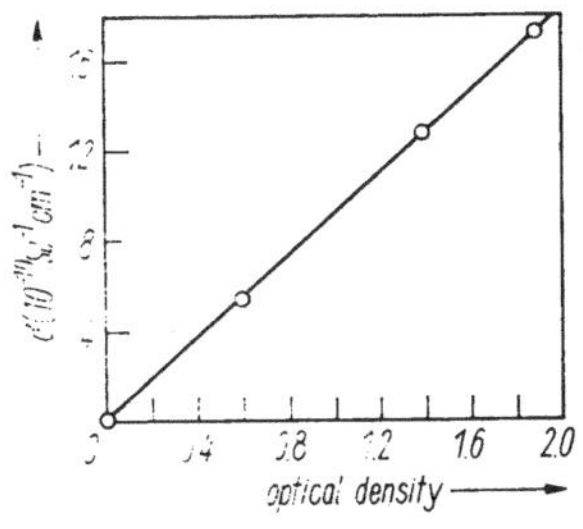

Fig. 2. The conductivity vs. concentration data for CrO_4^{2-} doped KDP at 80 °C (optical density of the 3700 Å band of CrO_4^{2-})

with a difference in the slope between high-temperature and low-temperature regions. Fig. 2 shows the conductivity (σ) at 80 °C plotted as a function of the optical density of 3700 Å band of CrO_4^{2-} in KDP. Since the optical density of the impurity is proportional to the concentration of the impurity, the straight line plot given in Fig. 2 shows the conductivity as a function of concentration of the impurity. It can be noticed that the conductivity (σ) at 80 °C increases proportionally with concentration of the CrO_4^{2-} impurity. This linear relation holds well for the conductivity at other temperatures in the range −120 to 180 °C. Similar observation has been made for $Cr_2O_7^{2-}$ doped KDP. The activation energies are determined by noting the slopes of the $\lg \sigma$ versus $1000/T$ plots. These along with $\lg \sigma_0$ values are given in Table 1 and compared with those obtained earlier for pure and SO_4^{2-} doped crystals.

Table 1

Activation energies E and $\lg \sigma_0$ factors for KDP and ADP crystals doped with molecular impurities

	KDP		ADP		
	E (eV)	$\lg \sigma_0$	E (eV)	$\lg \sigma_0$	references
pure	0.55 (< 180 °C)	3.7	0.61	3.5	present work
	0.80 (> 180 °C)	−1.0			
CrO_4^{2-} doped	0.54	3.2	0.59	3.2	present work
$Cr_2O_7^{2-}$ doped	0.53	2.9	0.60	3.0	present work
pure	0.53 (< 180 °C)	4.25	0.68	1.23	[6]
	0.99 (> 180 °C)	−1.73			
SO_4^{2-} doped	0.52		0.47		[2, 3]

In order to discuss the possible mechanism for conduction in these crystals, a brief review of the available information on these crystals is given. Earlier measurements on the conductivity in KDP and ADP crystals [2, 3] indicate that the conduction in these crystals is due to the proton transfer by means of L- and D-defects. In KDP, every phosphate tetrahedron is tetrahedrally surrounded by four phosphate groups, each linked by a hydrogen bond, with one proton in each hydrogen bond and two protons associated with each PO_4 group so that the anion may be considered as $H_2PO_4^-$ [12]. Thus each hydrogen bond of the phosphate lattice has two positions for its proton, each located towards the end of the bond. In the high-temperature paraelectric phase, a proton may jump from one phosphate to another giving rise to ionisation defects: HPO_4^{2-}, H_3PO_4. In addition it is possible for protons to jump from one bond to another on the same phsophate group. Such an interbond jump leaves a hydrogen bond without a proton (called an L-defect) and produces a bond with a proton in each of its two proton positions. Such a doubly occupied bond is termed as D-defect. Thus the proton migration occurs through the formation of L- and D-defects. On the other hand, Sharon and Kalia have concluded in their recent studies that the proton migration occurs through a synchronous phosphate rotation mechanism.

It has been observed in the present studies that the activation energy for conductivity is higher at high temperatures (above 180 °C) than at low temperatures. The presence of the knee around 180 °C could be due to a phase transition at which the hydrogen bond network breaks up as proposed by O'Keeffe and Perrino [13].

It has also been observed in the present case that the conductivity increases with increased concentration of the dopant as could be seen in Fig. 2, although the activation energy is not much different from that of the pure KDP crystals. The substitution of divalent CrO_4^{2-} or $Cr_2O_7^{2-}$ for trivalent PO_4^{3-} results in a proton vacancy for maintaining the charge neutrality of the crystal. Earlier ESR studies on Fe^{3+} doped KDP [14] also indicate the formation of proton vacancies rather than potassium vacancies for maintaining the charge neutrality of the crystal. The formation of such proton vacancies in turn results in the enhancement of the L-defect concentration. Thus the increase in conductivity in doped crystals may be attributed to increase in the L-defect concentration in KDP crystals. However, according to Sharon and Kalia the conduction process is controlled by the frequency of rotation of $H_2PO_4^-$. Thus, the increase in the conductivity in doped crystals is attributed to an increase in the frequency of rotation of $H_2PO_4^-$.

Fig. 3 shows lg σ versus $1000/T$ for pure and doped ADP crystals. No knee has been observed in this case unlike in KDP. Just as in the case of KDP, in this case also the conductivity has been found to increase with the concentration of the impurity. Activation energies obtained by analysing the plots given in Fig. 3 are presented in Table 1.

Since KDP and ADP are isomorphic in structure, the L-defect conduction mechanism is expected to hold well here also. However, as pointed out by Harris and Vella [6], the hydrogen bond network in the NH_4^+ sublattice may also contribute to the conductivity. This suggestion has not yet been adequately supported. In order to verify this, we have employed a very useful and different technique (ITC) the results and discussion of which are presented below.

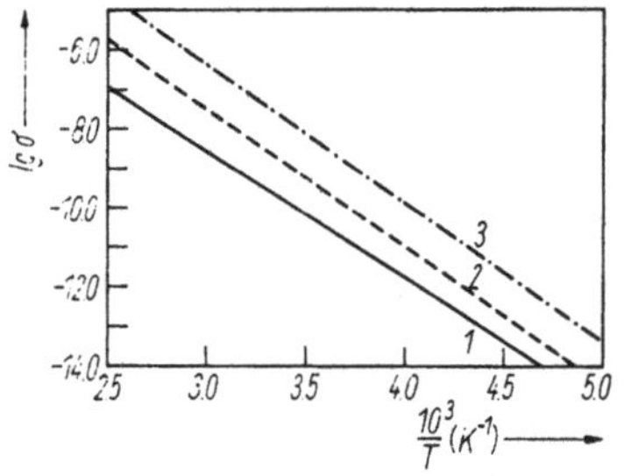

Fig. 3

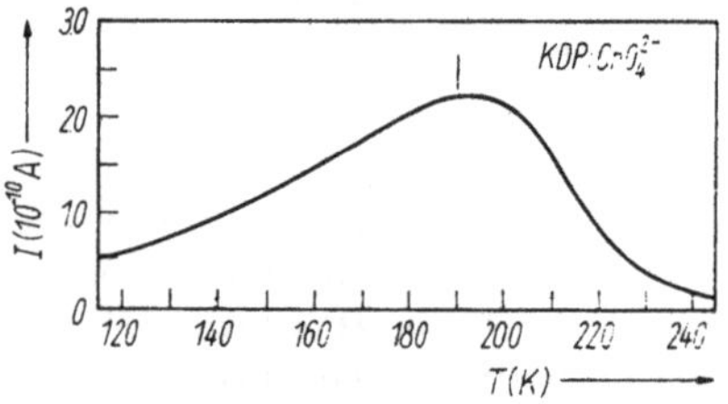

Fig. 4

Fig. 3. The temperature dependence of the conductivity of pure, CrO_4^{2-} and $Cr_2O_7^{2-}$ doped ADP crystals. (1) Pure ADP, (2) ADP:CrO_4^{2-}, (3) ADP:$Cr_2O_7^{2-}$

Fig. 4. ITC spectrum of KDP crystal doped with CrO_4^{2-}. $T_p = 200$ K

3.2 Thermal depolarisation studies (ITC)

Fig. 4 shows the ITC spectrum of CrO_4^{2-} doped KDP crystals in the range of 100 to 300 K. A peak around 190 K has been observed while no such peak could be seen in pure KDP. Various polarising temperatures and heating rates were tried, the result obtained most prominently is shown in the figure. The spectrum was analysed using the available methods [10] to calculate the activation energy (E) and the pre-exponential factor (τ_0). A programme written for IBM 370/155 computer has been used to find best fitting values. Analysis reveals that the dipoles responsible for the depolarisation current have a characteristic pre-exponential factor (τ_0) of the order 10^{-14} s and the activation energy (E) 0.51 eV.

When the host PO_4^{3-} ion is substituted by a aliovalent impurity ion, the charge compensation is usually achieved by the creation of proton vacancies. This is so in the present case also as is evident from conductivity studies discussed in Section 3.1. If the charge compensating species is in the local environment of the impurity ion, then a dipolar complex is formed which may reorient by a thermally activated jump of one of the constituents of the complex. Thus the detection of dipoles with relaxation parameters of the appropriate magnitude would support a vacancy transport mechanism. From the present ITC studies, the activation energy of the dipolar complex is found to be 0.51 eV. Since the mechanism for conductivity and thermal depolarisation involves identical proton jumps, one can correlate the conductivity and ITC results. From the electrical conductivity studies of the CrO_4^{2-} doped KDP discussed in Section 3.1, the activation energy in the low-temperature paraelectric phase was shown to be 0.54 eV. Thus by correlating these two studies the ITC peak occurring at 190 K with an activation energy of 0.51 eV in KDP may be attributed to impurity–proton vacancy dipoles. As reported earlier [11] only one ITC peak at 200 K with an activation energy of 0.55 eV has been observed in $Cr_2O_7^{2-}$ doped KDP crystals. By correlating those studies with the present electrical studies in which the activation energy has been observed to be 0.53 eV, we can conclude the formation of impurity–proton vacancy dipoles in $Cr_2O_7^{2-}$ doped KDP crystals, too.

Fig. 5 shows the ITC spectrum for CrO_4^{2-} doped ADP crystals when the polarising temperature was 250 K. Two peaks seen in the figure at 225 and 245 K are better resolved at this temperature. No such peaks have been observed in undoped crystals. Analysis of the two peaks has been carried out and relevant parameters are given in Table 2. Two peaks that are observed in pressent ITC studies could be explained by considering the charge separation mechanism [6, 15]. According to this mechanism, formation of molecular defects such as NH_3 and H_3PO_4 is accomplished by a proton transfer from the NH_4^+ sublattice to the $H_2PO_4^-$ sublattice. Since NH_3 has formed at the NH_4^+ site by loss of a proton, it has an effective negative charge. Thus there is proton conduction in the NH_4^+ sublattice in addition to the one in the $H_2PO_4^-$ sublattice. The proton conduction in the $H_2PO_4^-$ sublattice occurs by the same mechanism as that takes place in KDP. Thus the simultaneous proton conduction in both NH^+ and $H_2PO_4^-$ sublattices give rise to two ITC peaks. Similar observation has been made

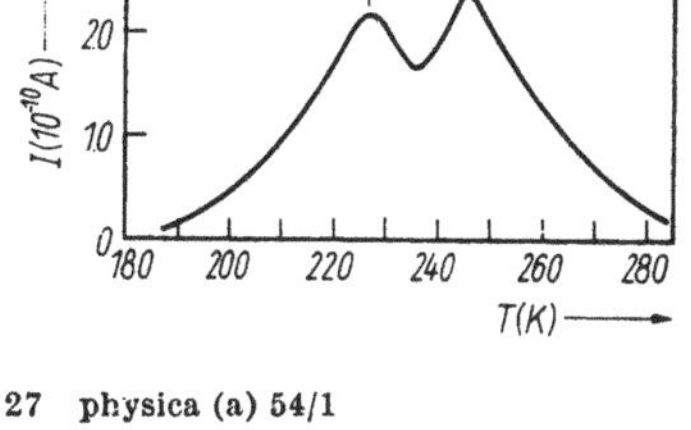

Fig. 5. ITC spectrum of ADP crystal doped with CrO_4^{2-}. $T_p = 250$ K

Table 2
ITC relaxation parameters for CrO_4^{2-}, $Cr_2O_7^{2-}$ doped KDP and ADP crystals

	KDP			ADP		
	T_m (K)	E (eV)	τ_0 (s)	T_m (K)	E (eV)	τ_0 (s)
CrO_4^{2-}	190	0.51	10^{-14}	225	0.58	10^{-16}
				245	0.24	10^{-6}
$Cr_2O_7^{2-}$	200	0.55	10^{-14}	210	0.44	10^{-16}
				235	0.28	10^{-6}

with $Cr_2O_7^{2-}$ doped ADP crystals, too [11]. The two peaks are at 210 and 235 K with the corresponding activation energies 0.44 and 0.28 eV, respectively. The activation energy for protonic conduction in the NH_4^+ sublattice of ADP was earlier reported erroneously to be 0.74 eV. A similar mechanism was proposed by Ah Mee Horr and Jacobs in order to explain their ITC data on ammonium perchlorate crystals doped with SO_4^{2-} and CrO_4^{2-} ions [15].

As the acitvation energy (0.59 eV) obtained from electrical conductivity studies is very close to the activation energy (0.58 eV) obtained by analysing the 225 K peak in the ITC spectrum, one can attribute the 225 K peak to protonic conduction in the $H_2PO_4^-$ sublattice. The activation energy for protonic conduction in the $H_2PO_4^-$ sublattice of KDP was found to be 0.54 eV which further supports the above-said assignment. The other peak at 245 K with activation energy 0.24 eV may be ascribed to protonic conduction in the NH_4^+ sublattice. The activation energy obtained from the present studies for protonic conduction in the NH_4^+ sublattice compares very well with the value obtained in the cases of NH_4ClO_4 [15] and NH_4Cl [16]. Since no such protonic conduction in cationic sublattice is possible in KDP, only one ITC peak is indeed expected.

4. Conclusion

The electrical conductivity studies on both undoped KDP and ADP crystals agree well with the earlier studies. Our studies confirm the existence of a knee in a conductivity versus $1000/T$ plot for pure KDP around 180 °C while no such knee seems to be present in similar plots for ADP. Proton motion appears to be the main mechanism for conductivity in these crystals. The CrO_4^{2-} and $Cr_2O_7^{2-}$ impurity effects on the conductivity are seen in both KDP and ADP crystals. The enhancement of conductivity in the doped crystals could be explained by considering the creation of proton vacancies in both crystals. ITC spectra of CrO_4^{2-} doped KDP and ADP crystals reveal the existence of one peak at 190 K in KDP and two at 225 and 245 K in ADP. These peaks are due to the impurity–proton vacancy dipole complexes. Thus while the proton transport in both $H_2PO_4^-$ and NH_4^+ sublattices seems to be responsible for conduction in ADP crystals, proton conduction in the $H_2PO_4^-$ sublattice only accounts for the conductivity in KDP crystals. This conclusion is based on a correlative study of ITC and electrical conductivity.

Acknowledgements

Thanks are due to Dr. S. Radhakrishna for his helpful discussions and comments on the manuscript. Thanks are also due to Prof. C. Ramasastry for his interest in the work. One of us (Y.R.S.) wishes to thank CSIR, New Delhi, for financial assistance in the form of JRF.

References

[1] L. GLASSER, Chem. Rev. **75**, 21 (1975).
[2] M. O'KEEFFE and C. T. PERRINO, J. Phys. Chem. Solids **28**, 211 (1967).
[3] E. J. MURPHY, J. appl. Phys. **35**, 2609 (1964).
[4] J. M. POLLOCK and M. SHARAN, J. chem. Phys. **51**, 3604 (1969).
[5] C. T. PERRINO and P. WENTRCEK, J. Solid State Chem. **10**, 36 (1974).
[6] L. B. HARRIS and G. J. VELLA, J. chem. Phys. **58**, 4550 (1973).
[7] M. SHARON and A. K. KALIA, J. Solid State Chem. **21**, 171 (1977).
[8] B. V. R. CHOWDARI and Y. RAVI SEKHAR, Chem. Phys. Letters **59**, 311 (1978).
[9] B. V. R. CHOWDARI and Y. RAVI SEKHAR, Solid State Commun., **31** (1979).
[10] S. RADHAKRISHNA and S. HARIDOSS, Crystal Lattice Defects **7**, 191 (1978).
[11] B. V. R. CHOWDARI and Y. RAVI SEKHAR, Solid State Commun. **29**, 687 (1979).
[12] T. PLESSER and H. STILLER, Solid State Commun. **7**, 323 (1969).
[13] M. O'KEEFFE and C. T. PERRINO, J. Phys. Chem. Solids **28**, 1086 (1967).
[14] K. TSUCHIDA and R. ABE, J. Phys. Soc. Japan **40**, 204 (1976).
[15] AH MEE HOR and P. W. M. JACOBS, J. Solid State Chem. **24**, 33 (1978).
[16] A. KESSLER, J. Electrochem. Soc. **123**, 1239 (1976).

(Received April 2, 1979)

phys. stat. sol. (a) 54, 421 (1979)

Subject classification: 14.3.2; 14.3.3; 22.2.2

Laboratoire d'Electronique et de Physique des Interfaces, Institut des Sciences de l'Ingenieur, Vandoeuvre-Les-Nancy[1])

A Method for Determining Interfacial Parameters of MIS Schottky Barriers

By

P. H. NGUYEN, C. BOUTRIT, B. LEPLEY, and S. RAVELET

A brief description of the electrical characteristics of MIS Schottky barriers is given. This electrical model of MIS Schottky barriers is applied to determine interfacial parameters: the width of the insulating layer, the average doping density, and the additional capacitances due to localized interfacial states. The control of these parameters is important for the optimization of MIS Schottky solar cells. In particular, various contributions relating to the "excess capacitance", as introduced by Vasudev et al. are discussed. Applications to Au/n-InP structures are proposed. An apparent zero bias band bending U_{bi}^* is obtained that is greater than the built-in potential U_{bi} of the intimate metal–semiconductor contact. A number of different possible interpretations are examined in order to explain this discrepancy.

La caractérisation électrique des barrières Schottky-MIS est brièvement présentée. On utilise un modèle électrique de ces structures pour déterminer les paramètres interfaciaux tels que l'épaisseur de la couche isolante, la densité moyenne de dopage et les capacités additionnelles dues aux états d'interface localisés. En particulier, les différentes contributions possibles à la „capacité d'excès" de Vasudev et al. ont été discutées. Le controle de ces paramètres est important, notamment dans l'optimisation des cellules solaires Schottky-MIS. Des applications aux structures Au/n-InP sont présentées. Un potentiel de diffusion apparent U_{bi}^*, plus élevé que le potentiel U_{bi} du contact intime métal-semiconducteur, est obtenu; différentes interprétations sont envisagées pour expliquer cet écart.

1. Introduction

There is considerable interest in MIS Schottky barrier (MIS-SB) structures nowadays because of their uses in microelectronics and for their potential application as photovoltaic devices. Some of the advantages offered by MIS-SB solar cells are the following: (i) low temperature and low manufacturing cost, (ii) adaptability to amorphous or polycrystalline thin films, and (iii) high current output and high radiation resistance in the case of direct band gap semiconductors. Following the theory of Card and Rhoderick [1] we describe a simple method for determining critical interface parameters from electrical measurements such as: (i) thickness of the insulating layer δ, (ii) uniform semiconductor doping concentration N_0, and (iii) additional capacitances C_{sa} and C_{sb} due to interface states which are in equilibrium with the metal and the semiconductor, respectively.

2. Theory

2.1 *Analysis of* C_m^{-2}–U_m *characteristics*

Schottky barrier heights are usually obtained by means of standard I–U, C_m^{-2}–U_m curves and photoemission threshold measurements [2]. For moderately doped semiconductors with chemically etch-polished surfaces, the three methods for determining

[1]) Parc Robert Bentz, 54500 Vandoeuvre, France.

the barrier height are compatible with each other when the interfacial layer is sufficiently thin. However, it has been noted that these values usually differ in the relevant publications [3]. For etch-polished interfaces the discrepancies between the photoemission values Φ_B (Ph), $I-U$ values $\Phi_B(I-U)$, and $C_m^{-2}-U_m$ values $\Phi_B(C-U)$ are explained in terms of the thickness of the interfacial layer [4] and the presence of interface states [5]. Thus, for MIS-SB structures the most reliable and accurate value of Φ_B would be obtained from the photoelectric measurements using monochromatic light with photon energy that is less than the band gap of the semiconductor. The theory of Card and Rhoderick for electrical characteristics of these contacts (in the absence of illumination) has been used to explain the increases observed in the $I-U$ barrier height of the cell due to the oxide layer.

The reverse bias capacitance measurements C_m-U_m are made at a very high frequency (100 kHz or 1 MHz) so that the interface states are generally unable to respond to the ac signal. Cowley [5] has discussed the deviation of the intercept on the U_m axis from the zero bias band bending U_{bi}. Two limiting cases are of interest: (i) in the case where the occupation of the interface states is governed entirely by the metal Fermi level the slope of the graph C_m^{-2} as a function of U_m gives the semiconductor doping concentration N_0, (ii) in the case where the occupation of the interface states is governed entirely by the semiconductor Fermi level, an apparent semiconductor doping density is deduced from the slope of the $C_m^{-2}-U_m$ curve. As regards the intermediate case where the interface states communicate with the two reservoirs, straightforward algebra shows that the $C_m^{-2}-U_m$ characteristic remains linear.

If the interface states are able to respond partially to the ac signal, a much more complicated situation arises because the interface states are now filled from the metal and from the semiconductor. Hence the differential capacitance is given by $C_m = \partial(Q_{sc} + Q_{sb})/\partial U_m$.

In Fig. 1, Q_{sc} is the charge in the depletion region, Q_{sb} the charge in the interface states in equilibrium with the semiconductor, and U_m the applied voltage in reverse bias. If we assume a simple model with a charge-free interfacial layer, the G_m-C_m parallel network measured with a capacitance bridge assembly (ignoring the effect of the series resistance) is equivalent to the electrical model shown in Fig. 2 where G_m is the parallel conductance of the junction. The differential capacitance can be written as

$$C_m^{-1} = (C_i + C_{sa})^{-1} + (C_{sb} + C_{sc})^{-1}, \tag{1}$$

where the additional capacitances are defined by $C_{sa} = -\partial Q_{sa}/\partial\Delta$, $C_{sb} = \partial Q_{sb}/\partial U_B$, and the oxide capacitance $C_i = \varepsilon_i/\delta = -\partial Q_m/\partial\Delta$. The terms ε_i and δ are the permittivity and the thickness of the oxide layer, respectively. The conservation of charges is given by

$$Q_m + Q_{sa} + Q_{sb} + Q_{sc} = 0, \tag{2}$$

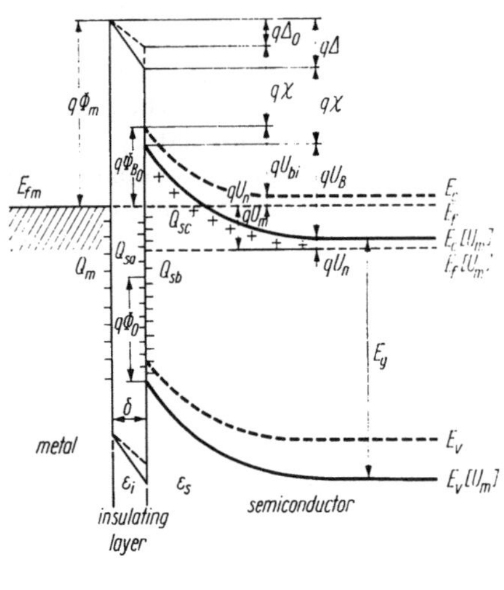

Fig. 1. Energy band diagram for MIS Schottky contact with a charge-free interfacial layer under reverse bias (continuous line) and under thermal equilibrium (dashed line)

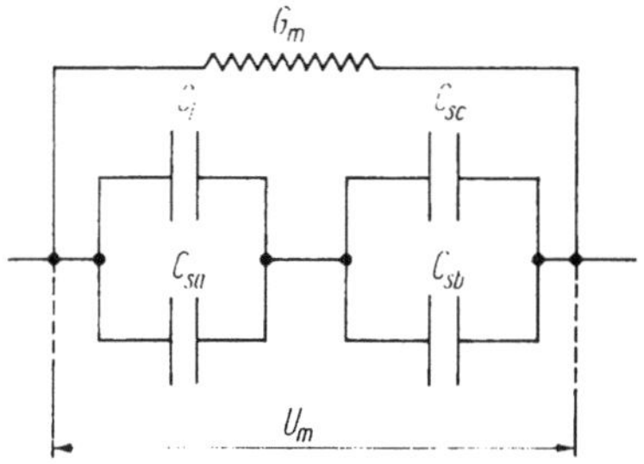

Fig. 2. Equivalent circuit of a MIS Schottky contact including interface state effect

Q_m is the charge in the surface of the metal and Q_{sa} the charge in the interface states in equilibrium with the metal. The surface potential U_B in the semiconductor and the potential drop Δ across the oxide are defined by $U_m = (U_B - U_{bi}) + (\Delta - \Delta_0)$ where U_{bi} and Δ_0 are zero bias values, respectively. From the depletion approximation and for a non-uniform donor distribution, the space charge capacitance C_{sc} is always defined by $C_{sc} = \partial Q_{sc}/\partial U_B = \varepsilon_s/W$ where ε_s is the permittivity of the semiconductor and W the thickness of the depletion region. In deriving equation (1) we have assumed that all the capacitances (except C_{sc}) remain constant as U_m changes.

According to the models of Kar and Dahlke [6] and using the analysis of Nicollian and Goetzberger [7] for a continuum of states we can write the additional capacitances C_{sa}, C_{sb} in the following forms:

$$C_{sa} = q^2 \lambda D_{ss} = q^2 D_{sa} , \tag{3}$$

$$C_{sb} = q^2(1 - \lambda) D_{ss} \frac{\text{tg}^{-1}(\omega\tau)}{\omega\tau} = q^2 D_{sb} \frac{\text{tg}^{-1}(\omega\tau)}{\omega\tau}, \tag{4}$$

where ω is the angular frequency in s^{-1}, τ the interface time constant in s, D_{ss} the total density of interface states in $eV^{-1}m^{-2}$. The parameter λ has been introduced to constitute a continuous transition from the interface recombination controlled case, $\lambda = 1$ and $\delta = 0$, to the oxide tunnelling controlled case where $\lambda = 0$ when $\delta \gg \delta_0$. The characteristic length δ_0 can be defined by

$$\lambda^{-1} = 1 + \exp\left((\delta - \delta_0)/\delta_1\right) . \tag{5}$$

The rapidity of the transition is determined by the value of δ_1. Several authors [8, 9] have suggested that δ_0 is about 25 to 30 Å for intermediate MOS silicon structures. But it seems that the characteristic length δ_0 would be a logarithmic function of δ if one were to express the tunnel current through the oxide into interface states as a function of λ, in accordance with the analysis of Freeman and Dahlke [10]. In fact this tunnel current governs the transition between the two previous limiting cases.

Equation (1) can be written in the following form [11]:[2])

$$(C_m - C_{00})^{-2} = \frac{2}{q \varepsilon_s N_p} (U^*_{int} - U_m) , \tag{6}$$

where $N_p = N_0 \left((1 + \alpha\lambda^*)/(1 + \alpha)\right)^3$ is the apparent doping concentration, and the intercept voltage U^*_{int} is a function of the built-in voltage U_{bi}:

$$U^*_{int} = \frac{1 + \alpha}{1 + \alpha\lambda^*} (U_{bi} - U_2) + \frac{U_1^{1/2}(U_{bi} - U_2)^{1/2}}{1 + \alpha\lambda^*} + \frac{U_1}{4(1 + \alpha)(1 + \alpha\lambda^*)} . \tag{7}$$

[2]) A somewhat similar approach has been proposed where C_m is defined by $C_m = - \partial Q_m/\partial U_m$ and with the assumption $\omega\tau \ll 1$.

The parameters α, λ^* are defined by $\alpha = (C_{sa} + C_{sb})/C_i$ and $\lambda^* = C_{sa}/(C_{sa} + C_{sb})$, respectively. The potentials U_1, U_2 are defined by $U_1 = 2q\varepsilon_s N_0/C_i^2$ and $U_2 = kT/q$, respectively, if the donors are completely ionized at room temperature [5].

2.2 Analysis of "excess capacitance"

The "excess capacitance" C_{00} in (6) is defined by

$$C_{00} = C_{sb}(C_{sa} + C_i)/(C_{sb} + C_{sa} + C_i)\,, \tag{8}$$

Vasudev et al. [12] have introduced a graphical method to determine C_{00}. They assumed that the intercept voltage U_{int}^* is identical to $(U_{bi} - U_2)$ and N_p to N_0. These assumptions are correct when the junction is an intimate Schottky contact ($\delta < 10$ Å).

Consequently, Pellegrini and Salardi [13] have explained the excess capitance as an effect of non-linear dielectric susceptibility in intimate metal–GaAs junctions. This capacitance is defined by

$$C_p = \frac{2}{3} \frac{qN_0}{\varepsilon_s E_\chi}\,, \tag{9}$$

where E_χ is the ratio of the electric field E in the semiconductor on the electric susceptibility $\chi(E)$. Another contribution that deals with the excess capacitance is the edge-effect capacitance [14] defined by

$$C_e = \frac{\varepsilon_s}{2\pi S_c^{1/2}}\,, \tag{10}$$

where S_c is the area of the circular contact. Since C_p and C_e are the additional capacitances for the depletion capacitance C_{sc}, the excess capacitance C_{00} may be written as

$$\begin{aligned} C_{00} &= (C_{sb} + C_p + C_e)\,(C_{sa} + C_i)/(C_{sb} + C_p + C_e + C_{sa} + C_i) = \\ &= C_{sb} + C_p + C_e\,, \end{aligned} \tag{11}$$

if

$$C_{sa} + C_i \gg C_{sb} + C_p + C_e\,.$$

This supplementary capacitance has two components, one frequency-dependent C_{sb} deriving from the interface states in equilibrium with the semiconductor, the other the doping concentration-dependent C_p. We can ignore C_e for diodes with an area in excess of 1 mm². In order to calculate the values of the additional capacitances C_{sb}, C_{sa} arising from the interface states, we assumed the electric susceptibility in InP to be a slowly varying function in relation to the electric field since this semiconductor is a "semicovalent" one [15]. In this case, C_{00} may be determined by (8).

3. Experimental Procedure

N-type indium phosphide MIS-SB structures were produced by evaporating the top contact (Au) onto the thin oxide layer which had grown under ambient conditions. The junction area was typically 2 mm². Experimental details on chemically etch-polished surface diodes, ohmic contacts, and technical measurements are given in an earlier publication [16].

3.1 Numerical techniques

Non-linear least squares fits are performed on C_m–U_m characteristics using different numerical techniques such as the simplex minimization method or the method of conjugate gradients. Several numerical techniques must be used to find a real optimum vector with physically acceptable values. We adopted a two stage procedure as follows:

an initial optimization to determine U^*_{int}, N_p, and C_{00},
a second optimization to determine U^*_{bi}, N_0, δ, C_{sb}, and C_{sa}.

The new values of U^*_{int}, N_p, and C_{00} are then deduced for comparison with the previous results.

3.2 Results and discussion

Table 1 shows the results obtained from the first optimization. The values of the intercept voltage are too large in comparison with the value 0.49 V of the Au/n-InP

Table 1
Results on Au/n-InP MIS Schottky barrier diodes from the first optimization

devices	surface preparation	U^*_{int} (V)	N_p (10^{23} m^{-3})	C_{00} (pF/mm^2)
I_1 (111)	0.2% Br–CH_4O	1.20	2.4	165
I_2 (111)	no	1.10	4.3	46
I_3 (111)	no	1.17	4.2	90
I_4 (111)	9 N HCl	0.88	4.7	61

intimate Schottky barrier height [17]. Consequently, the assumption that excess capacitance can be ascribed only to the effect of non-linear electrical susceptibility for Au/n-InP cannot hold. The apparent doping concentration N_p is only marginally reduced in comparison with the values determined from the Schottky model in assuming $C_{00} = 0$. In qualitative terms, the excess capacitance is largely dependent on the surface preparation.

Table 2 shows the results obtained from the second optimization using the same C_m–U_m characteristics. The parameters U^*_{int}, C_{00}, and N_p computed from these

Table 2
Results on Au/n-InP MIS Schottky barrier diodes from the second optimization

devices	U^*_{bi} (V)	N_0 (10^{23} m^{-3})	C_{sb} (pF/mm^2)	C_{sa} (pF/mm^2)	δ ($\varepsilon_i = 4\varepsilon_0$) (Å)
I_1 (111)	1.02	2.4	160	3000	24
I_2 (111)	0.93	4.4	39	830	18
I_3 (111)	1.00	4.3	86	1630	18
I_4 (111)	0.78	4.8	52	1100	13

values are consistent with the previous ones. Hitherto, we have been unable to determine the densities D_{ss} of the interface states, the interface time constant τ, and the parameter λ using equations (3) and (4). The width of the oxide layer is clearly dependent on the surface treatment. The dielectric constant is taken to be equal to 4 if the oxide consists mainly of In_2O_3 [18].

The optimized values of the built-in voltage are large in comparison with the results obtained by Mead and Spitzer [17] using Au/n-InP ideal Schottky diodes. To explain this discrepancy we assumed the existence either of slow states at the interface or of negative fixed charges in the oxide [19] or a combination of the two possibilities.

In the first case, where the states fail to respond to an ac signal, we can write

$$-Q'_{ss} = \varepsilon_i(U^*_{bi} - U_{bi})/\delta + (2q\varepsilon_s N_0)^{1/2} (U^{*1/2}_{bi} - U^{1/2}_{bi}), \tag{12}$$

where Q'_{ss} is the total charge due to slow interface states, U^*_{bi} is the apparent built-in voltage listed in Table 2.

In the alternative, where the discrepancy is ascribed only to the presence of fixed charges in the interfacial layer, we have

$$-\frac{\delta_{eff}}{\varepsilon_i} Q_{fix} = (U^*_{bi} - U_{bi}) + \left(2q\varepsilon_s N_0 \frac{\delta^2}{\varepsilon_i^2}\right)^{1/2} (U^{*1/2}_{bi} - U^{1/2}_{bi}). \tag{13}$$

Equations (12) and (13) are obtained on the basis of the following assumptions: (i) the difference $(\varphi_m - \chi)$ between the work function of the metal and the affinity of the semiconductor is taken to be equal to the free surface value in vacuum, (ii) the contribution of majority carriers, by the potential U_2, is ignored. If the oxide charges uniformly occupy all the interfacial layer, the effective oxide thickness δ_{eff} introduced by Fonash [20] is calculated as $\delta/2$ in accordance with the analysis of Landsberg and Klimpke [21]. With $U_{bi} = 0.35$ V with Au/n-InP Schottky barriers, in the first case we find $Q'_{ss} = -10^{-2}$ C/m² whereby $D'_{ss} = 5 \times 10^{12}$ eV^{-1} cm^{-2} if these interface states uniformly occupy all the band gap of the semiconductor. In the second case, we find $Q_{fix} = -2 \times 10^{-2}$ C/m², whereby the density N_{fix} of fixed charges is computed to be 1.3×10^{13} states/cm². Therefore, if the discrepancy is due to a combination of both fixed charges and slow interface states, the previous values are overestimated. The last assumption seems to be more realistic since the density of fixed charges is large and the thickness of the insulating layer is low.

The analysis of Cowley and Sze [22] and the experimental results of Kurtin et al. [15] are used to evaluate the total density D_{ss} for InP. Other authors [23] have used a similar approach to obtain the interface density of states in terms of the experimentally determined "index of interface behaviour" S. Ignoring the screening effect, the total density D_{ss} can be approximated as

$$q^2 D_{ss} = \frac{\varepsilon_i}{\delta}\left(\frac{2.3}{S} - 1\right), \tag{14}$$

where typically $\delta = 20$ Å, $\varepsilon_i = 4\varepsilon_0$, and $S = 0.10$ for InP, D_{ss} is calculated to be 2.4×10^{14} eV^{-1} cm^{-2}. From Table 2, we deduced 5×10^{11} to 2×10^{12} eV^{-1} cm^{-2} for D_{sa} only. This discrepancy between the expected value, which is of the order of 10^{14} cm^{-2} eV^{-1}, and the calculated results could be due either to the difficulties encountered in non-linear least squares optimizations, or to the fact that $(C_i + C_{sa}) \gg (C_{sc} + C_{sb})$. The results for C_{sb} are consistent with the values of C_{00} obtained in the first optimization. If we assume $D_{sb} = 10^{14}$ eV^{-1} cm^{-2} and $C_{sb} = 80$ pF/mm², the typical interface time constant is estimated to be 5×10^{-3} s. This value is in the vicinity of the majority carrier lifetime of 2×10^{-3} s in n-InP [24].

4. Conclusion

In this paper, we have proposed an electrical model for C_m–U_m characteristics in reverse bias in the absence of illumination. This model represents a continuous transition from the Schottky model ($\delta \ll \delta_0$) to the MIS model ($\delta \gg \delta_0$). It seems that the characteristic transition length δ_0 would be a logarithmic function of the width δ of the interfacial layer. We have shown that the excess capacitance may be explained in terms of an effect of interface states which communicate with both the metal and the semiconductor. Certain applications for Au/n-InP structures are proposed. Accurate results were obtained in respect of average doping density and the additional capacitance arising from the effect of interface states in equilibrium with the semiconductor.

The discrepancy between the apparent built-in voltage obtained by curve fitting and the "real" built-in voltage of Au/n-InP Schottky barriers is ascribed to the existence of fixed charges in the insulating layer or of slow interface states which are unable to respond to ac signals.

Acknowledgements

The authors are grateful to N. T. Linh (LCR Thomson CSF) for supplying the InP crystals used in the investigations, and to J. M. Hess for cooperation in this work. We would also like to thank C. Michel and J. C. Georges for constructive discussions.

References

[1] H. C. Card and E. H. Rhoderick, J. Phys. D 4, 1589 (1971).
[2] E. H. Rhoderick, Metal–Semiconductor Contacts, Univ. Press, Oxford 1978.
[3] S. Ashok, J. M. Borrego, and R. J. Gutman, Electronics Letters 14, 332 (1978).
[4] B. Lepley and S. Ravelet, phys. stat. sol. (a) 33, 517 (1976).
[5] A. M. Cowley, J. appl. Phys. 38, 3024 (1967).
[6] S. Kar and W. E. Dahlke, Solid State Electronics 15, 221 (1972).
[7] E. H. Nicollian and A. Goetzberger, Bell Syst. tech. J. 46, 1055 (1967).
[8] H. C. Card and E. H. Rhoderick, J. Phys. D 4, 1602 (1971).
[9] G. Sarrabayrouse and J. Buxo, phys. stat. sol. (a) 49, 603 (1978).
[10] L. B. Freeman and W. E. Dahlke, Solid State Electronics 13, 1483 (1970).
[11] P. H. Nguyen, Thèse de Docteur Ingénieur, Université de Nancy I, 1977.
[12] P. K. Vasudev, B. L. Mattes, E. Petras, and R. H. Bube, Solid State Electronics 19, 557 (1976).
[13] B. Pellegrini and G. Salardi, Solid State Electronics 21, 465 (1978).
[14] A. M. Goodman, J. appl. Phys. 34, 329 (1963).
[15] S. Kurtin, J. C. McGill, and C. A. Mead, Phys. Rev. Letters 22, 1433 (1969).
[16] J. M. Hess, P. H. Nguyen, B. Lepley, and S. Ravelet, phys. stat. sol. (a) 46, K55 (1978).
[17] C. A. Mead and W. G. Spitzer, Phys. Rev. 134, A713 (1964).
[18] D. L. Lile and D. A. Collins, Appl. Phys. Letters 28, 554 (1976).
[19] D. L. Pulfrey, IEEE Trans. Electron Devices 25, 1308 (1978).
[20] S. J. Fonash, J. appl. Phys. 47, 3597 (1976).
[21] P. T. Landsberg and C. Klimpke, Proc. Roy. Soc. A354, 101 (1977).
[22] A. M. Cowley and S. M. Sze, J. appl. Phys. 36, 3212 (1966).
[23] S. G. Louis, J. R. Chelikowsky, and M. L. Cohen, Phys. Rev. B 15, 2154 (1977).
[24] T. S. Moss, in: Semiconductors and Semimetals, Vol. 2, Physics of III–V Compounds, Ed. R. K. Willardson and A. C. Beer, Academic Press, New York 1966 (Chap. 9, p. 234).

(Received April 4, 1979)

phys. stat. sol. (a) 54, 429 (1979)

Subject classification: 18.2.1

Department of Electrical Engineering, Imperial College, London[1])

Domain Wall Mass in Magnetic Bubble Films under Large In-Plane Fields

By

T. H. O'DELL

The concept of effective mass for a magnetic domain wall is reviewed and then the particular problem of a straight wall in a magnetic bubble film is considered. The effect of an externally applied magnetic field which is parallel to the plane of the film and either parallel or perpendicular to the plane of the wall is then discussed and the two possibilities are shown to lead to very different results.

Der Begriff der effektiven Masse der magnetischen Domänenwände wird analysiert und das Problem der flachen Wand in einer magnetischen Blase besprochen. Der Einfluß eines externen magnetischen Feldes, das parallel zu der Fläche der magnetischen Schicht und entweder senkrecht oder parallel zur Fläche der Wand ist, wird dann untersucht. Es wird gezeigt, daß diese zwei Möglichkeiten sehr verschiedene Resultate ergeben.

1. Introduction

The idea that a magnetic domain wall should have an effective mass was first introduced by Döring [1] who considered the problem of a plane wall in an infinite medium with uniaxial anisotropy. The statics, and the motion at constant velocity of such a wall, had been considered previously by Landau and Lifshitz [2]. Döring's well known result

$$m_D = 2/\mu_0 \gamma^2 \Delta \tag{1}$$

for the effective mass per unit area of wall, kg/m², where γ is the magnetomechanical ratio and Δ is the wall width parameter $(A/K_u)^{1/2}$, was explained from a more simple point of view by Becker [3].

More recent theoretical work [4 to 10] has been concerned with the mass of domain walls which have a more realistic structure than the simple one-dimensional Landau-Lifshitz solution for the infinite planar wall, particular attention being paid to the kind of wall structure which is expected in magnetic bubble domain films [11 to 15].

In this paper, a very special case is considered which has been the situation in a number of dynamic experiments. This is the case in which a straight domain wall is stabilised in a magnetic bubble domain film, either by means of an externally applied field gradient [16 to 22] or by making the wall one of a set of parallel domain walls [23 to 28]. The interesting feature of these two particular experimental configurations is that a large in-plane field may be applied, either perpendicular or parallel to the wall plane, and this greatly extends the range of applied drive fields and wall velocities which may be investigated [29].

The initial conditions, that is the static structures of domain walls of this kind, have been considered in a previous paper [30]. The intention here is to obtain the equation of motion for such walls and understand the origin of the effective mass which may be attributed to the walls under these special conditions.

---

[1]) London SW 7 2AZ, Great Britain.

2. The Wall Mass from a Field Point of View

Both Döring [1] and Becker [3] obtained the expression (1) for the effective wall mass by considering the energy of the moving wall, as a function of its velocity, and then using a mechanical analogy by introducing a kinetic energy $\frac{1}{2} mv^2$.

When in-plane fields are present, it is more straightforward to use the field equations of the problem and these are the components of the well known Landau-Lifshitz equation

$$\frac{-1}{|\gamma|} \frac{d\boldsymbol{M}}{dt} = (\boldsymbol{M} \times \boldsymbol{F}) - \lambda(\boldsymbol{F} - (\boldsymbol{F} \cdot \boldsymbol{M}) \boldsymbol{M}/M_s^2) \tag{2}$$

which is written here in its original form [2] and in which $\boldsymbol{F}$ is the total effective magnetic field made up from the exchange, anisotropy, magnetostatic and applied fields. The material constant $\lambda \ll M_s$ gives a measure of the magnetomechanical damping in the Landau-Lifshitz model, λ/M_s being identical to the well established Gilbert damping parameter α [31]. This field model was used in a previous paper [29] on constant velocity motion of an isolated domain wall having a large in-plane field applied perpendicular to its plane.

As an introduction to the problem being considered here, let us see how the expression for the Döring mass (1) may be obtained from (2) by considering the simple Landau-Lifshitz wall structure in an infinite uniaxial medium with an easy direction of magnetisation along z. The wall structure will be given by

$$M_z = -M_s \tanh(y/\Delta) \tag{3}$$

and

$$M_x = \pm M_s \operatorname{sech}(y/\Delta) . \tag{4}$$

$\boldsymbol{M}$ rotating in the plane of the wall (x, z) as we pass across the wall width from one domain to the other. The choice of sign in (4) indicates that this kind of wall may have two possible screw senses.

The usual rigid wall approximation is now made which means that we assume that the dynamic wall structure is given by

$$M_z = -M_s \tanh\left((y' - v_y t)/\Delta\right) , \tag{5}$$

$$M_x = M_s \cos(\psi) \operatorname{sech}\left((y' - v_y t)/\Delta\right) , \tag{6}$$

and

$$M_y = M_s \sin(\psi) \operatorname{sech}\left((y' - v_y t)/\Delta\right) \tag{7}$$

when the wall moves along the y axis under the influence of a small drive field B_z. Equations (5), (6), and (7) show that the magnetisation within the moving wall is assumed to cant out of the wall plane by a small angle ψ, which is assumed to be independent of y.

The components of the vector $\boldsymbol{F}$, discussed in [29, 30, 32], will be given by

$$F_z = (2A/M_s^2)\, \partial^2 M_z/\partial y^2 + \mu_0 M_z(Q + 1) + B_z , \tag{8}$$

$$F_x = (2A/\mu_0 M_s^2)\, \partial^2 M_x/\partial y^2 + \mu_0 M_x , \tag{9}$$

and

$$F_y = (2A/\mu_0 M_s^2)\, \partial^2 M_x/\partial^2 y^2 , \tag{10}$$

where we are using the parameter

$$Q = 2K_u/\mu_0 M_s^2 . \tag{11}$$

Equations (5) to (10) may now be substituted into (2) to obtain the equation of motion. Because the space derivatives of $\boldsymbol{M}$ have been defined by (5), (6), and (7), equation (2) reduces to a set of ordinary differential equations, time being the only independent variable when we choose to consider any single point on the moving wall. Once the rigid wall approximation has been made, any point on the wall may be chosen but the algebra is simpler if the centre of the moving wall, $y' = v_y t$, is taken. This method is still valid when v_y is a function of time.

The result of substituting equations (5) to (10) into (2) is that the x and y components of (2) both yield the simple result

$$\frac{1}{|\gamma|} \frac{d\psi}{dt} = B_z - \mu_0 \lambda \cos(\psi) \sin(\psi) \tag{12}$$

while the z component of (2) establishes the same relationship,

$$v_y = (|\gamma| \Delta/M_s) (\mu_0 M_s^2 \sin(\psi) \cos(\psi) + \lambda B_z) \tag{13}$$

which applies when v_y is a constant [29]. We must now restrict our argument to the special case of low drive so that ψ may be assumed small and (12) and (13) linearised by setting $\sin(\psi) \approx \psi$ and $\cos(\psi) \approx 1$. It is only under such assumptions of linearity that the concept of an effective mass may be introduced in any case.

Eliminating ψ from the linearised forms of (12) and (13) yields a first order linear differential equation for v_y

$$\frac{2}{\mu_0 \gamma^2 \Delta} \frac{dv_y}{dt} + \frac{2\lambda}{|\gamma| \Delta} v_y = 2M_s B_z + \frac{2\lambda}{\mu_0 M_s |\gamma|} \frac{dB_z}{dt} \tag{14}$$

when the term involving λ^2/M_s^2, that is α^2, is neglected. The last term in (14), involving dB_z/dt, may also be neglected in all practical circumstances because the rise time of the drive field would have to be less than 10 ps before this term could become comparable with $2M_s B_z$.

Omitting the last term in (14), in view of the remarks above, we immediately recognise that the pressure, or force per unit area, on the wall due to the drive field, $2M_s B_z$ on the right-hand side, is in equilibrium with the sum of the pressures due to inertia, $(2/\mu_0 \gamma^2 \Delta)\,(dv_y/dt)$, which is $m\, dv_y/dt$ when we refer back to (1), and the pressure due to the viscous damping drag on the wall, $(2\lambda/|\gamma| \Delta)\, v_y$. In the steady state, $dv_y/dt = 0$, equation (14) yields the well known Landau-Lifshitz mobility relationship

$$v_y = \frac{|\gamma| \Delta M_s}{\lambda} B_z = \frac{|\gamma| \Delta}{\alpha} B_z . \tag{15}$$

3. Wall Mass when an In-Plane Field is Applied

In the previous section, wall dynamics in an infinite medium were reviewed. We now wish to extend this to the case of a thin layer of material, having its easy direction of magnetisation normal to the plane of the layer (x, y) and having a large in-plane field applied, either along x or y. This was the situation in the experiments referred to above [16 to 28].

The transition from an infinite medium to a thin layer would normally have a considerable effect upon the wall dynamics [12, 14, 33 to 35] because the wall structure can no longer be assumed one-dimensional, as described by (3) and (4). The problem is no easier when a small in-plane field is applied, below $\mu_0 M_s$ [36], but for large in-plane fields, between $\mu_0 M_s$ and $\mu_0 M_s Q$, it is possible to assume a one dimensional wall structure [29, 30] and proceed with the problem in a fairly simple way. This is fortunate because the experimental data was obtained under large in-plane fields too and these fields were directed either along y [16 to 22] or along x [23 to 28]. We consider these two possible orientations separately as they lead to very different results.

3.1 A large in-plane field applied parallel to the wall plane

When a magnetic field B_x is applied to the domain wall which was considered in Section 2, the static wall structure is no longer given by (3) and (4). If $\mu_0 M_s < B_x < \mu_0 M_s Q$, however, the new static wall structure may be found quite simply [30]. The result shows that there is no longer a choice of two possible screw-senses for the wall, represented here by the choice of sign in (4). The wall must now have a screw-sense of such a sign that $\boldsymbol{M}$ and B_x are parallel at the wall centre. The functions tanh (y/Δ) and sech (y/Δ), in (3) and (4), are replaced by functions which have a very similar form [30]. In detail, $\boldsymbol{M}$ rotates in (x, z) so that we may write

$$M_z = M_s \cos(\theta) \tag{16}$$

and

$$M_x = M_s \sin(\theta) \tag{17}$$

and the relationship between θ and the independent variable y is given by

$$y/\Delta = (1 - b^2)^{-1/2} \log_\varepsilon \left| \frac{\left(1 - b - (1 - b^2)^{1/2}\right) + \left(1 - b + (1 - b^2)^{1/2}\right) \tan(\theta/2)}{\left(1 - b + (1 - b^2)^{1/2}\right) + \left(1 - b - (1 - b^2)^{1/2}\right) \tan(\theta/2)} \right|, \tag{18}$$

[30] where

$$b = B_x/\mu_0 M_s Q \tag{19}$$

expresses the effect of the in-plane field in a dimensionless way. It may also be shown [30] that this one-dimensional wall structure, strictly true for the infinite uniaxial medium, also applies as a good approximation to the case of a thin layer provided $\mu_0 M_s < B_x < \mu_0 M_s Q$. That is, as stated above, the in-plane field must be large.

The new static wall structure, described by (16), (17), and (18), while complicated algebraically, may be treated in just the same way as (3) and (4) in that the rigid wall approximation can now be made, introducing a small canting angle ψ and replacing y by $(y' - v_y t)$. We then proceed in just the same way as in Section 2, adding B_x to (9) to include the in-plane field, and find that the x and y components of (2) now give

$$\frac{1}{|\gamma|} \frac{\mathrm{d}\psi}{\mathrm{d}t} = B_z - \mu_0 \lambda \left(\cos(\psi) + B_x/\mu_0 M_s\right) \sin(\psi), \tag{20}$$

in place of (12) while the z component of (2) yields,

$$v_y = \frac{|\gamma|\, \Delta}{M_s(1 - b)} \left(\mu_0 M_s^2 \sin(\psi) \cos(\psi) + M_s B_x \sin(\psi) + \lambda B_z\right) \tag{21}$$

which should be compared with (13). The term $(1 - b)$ which occurs in the denominator of (21) comes from differentiating (18) to find the value of $\mathrm{d}\theta/\mathrm{d}y'$ at the wall centre needed to form $\mathrm{d}M_z/\mathrm{d}t = (\mathrm{d}M_z/\mathrm{d}\theta)\,(\mathrm{d}\theta/\mathrm{d}y')\,(\mathrm{d}y'/\mathrm{d}t)$.

Equations (20) and (21) may be linearised by considering only small values of ψ and then ψ eliminated by combining the two equations to give

$$\frac{2(1 - b)}{\mu_0 \gamma^2 \Delta(1 + Qb)} \frac{\mathrm{d}v_y}{\mathrm{d}t} + \frac{2\lambda(1 - b)}{|\gamma|\, \Delta} v_y = 2 M_s B_z, \tag{22}$$

where the terms involving λ^2/M_s^2 and $\mathrm{d}B_z/\mathrm{d}t$ have been neglected. Comparing (22) and (14) it is clear that the effective mass per unit area of the wall has been reduced by the application of the in-plane field B_x and, using (1), we may write

$$m^{-1} = m_{\mathrm{D}}^{-1}(1 + Qb)/(1 - b). \tag{23}$$

For $Q \gg 1$, equation (23) predicts a linear relationship between m^{-1} and the in-plane field B_x. This result has been deduced from the general theory of wall motion [7, 37] and a linear relationship has been observed experimentally [28] but the rate of increase of m^{-1} with B_x was much greater than predicted by (23). This was probably due to the use of too large a drive field in these experiments, causing ψ to develop a large value. Our assumption that ψ is small is essential in the derivation of (23), making (20) and (21) linear. The concept of an effective mass is only valid when a linear equation of motion (22) may be assumed. The transient response shown in [28] was, in fact, a triangular oscillation in the position of the wall, not sinusoidal, and this oscillation decayed linearly with time, not exponentially. Such a transient response is clearly non-linear and cannot be modelled by means of a linear second order equation, as (22) would be when B_z depends linearly upon y.

Finally, we note that the steady-state solution of (22) predicts an increase in the wall mobility as a result of applying the in-plane field B_x. Comparing (14) and (22) it is clear that the mobility is increased by a factor $(1 - b)^{-1}$. Again, this result has little practical interest because it applies to the very low drive fields which allow ψ, in (20), to have a small value.

3.2 A large in-plane field applied perpendicular to the wall plane

When an in-plane magnetic field B_y is applied to the domain wall which was considered in Section 2, far more dramatic changes occur in the problem compared to the case which was considered in Section 3.1. Quite a small drive field, combined with a large in-plane field, $\mu_0 M_s < B_y < \mu_0 M_s Q$, can cause wall motion at very high velocities and the dynamic wall structure [29] can then take quite a different form to the static wall structure [30].

To see the reason for this, we should refer back to (12) which shows that steady-state motion with the simple Landau-Lifshitz model, $d\psi/dt = 0$, is impossible for $B_z > \mu_0 \lambda/2$. If B_z is increased towards $\mu_0\lambda/2$, ψ increases towards its maximum possible steady value $\pi/4$ and the velocity reaches a maximum

$$v_W = \mu_0 |\gamma| M_s \Delta/2 \tag{24}$$

which has become known as the Walker velocity [38]. This is rather an artificial result, however, in that $\psi = \pi/4$ is by no means a small deviation away from the initial static wall structure (3) and (4).

A very different situation applies when there is an in-plane field perpendicular to the wall plane because this can be arranged to give a high wall velocity when $\boldsymbol{M}$ makes a very small angle with the wall plane [29] and we may safely assume that (5), (6), and (7) can be used again. Equations (8) and (9) still apply while we must add B_y to (10) and also add the small field B_y/Q to allow for the tilted magnetisation within the domains on either side of the wall. With these assumptions the x and y components of (2) both yield

$$\frac{1}{|\gamma|} \frac{d\psi}{dt} = B_z - \mu_0 \lambda \big(\sin(\psi) - (Q+1)\, c\big) \cos(\psi)\,, \tag{25}$$

the z component gives

$$v_y = \frac{|\gamma|\, \Delta}{M_s} \left[\mu_0 M_s^2 \cos(\psi) \big(\sin(\psi) - (Q+1)\, c\big) + \lambda B_z\right], \tag{26}$$

where

$$c = B_y/\mu_0 M_s Q\,. \tag{27}$$

Equations (25) and (26) may be compared with (12) or (20) and (13) or (21). No term corresponding to the $(1 - b)$ in (21) occurs in (26) because we have neglected the small tilt of $\boldsymbol{M}$ within the domains by using (5), (6), and (7).

We now restrict our discussion to motion near the peak velocity, which was the subject of [29], where it was shown that, as $B_y/\mu_0 M_s$ tends towards its maximum possible value Q, the domain wall may move at a peak velocity for which ψ tends to be closer and closer to π. We may then write $\psi = \psi_0 + \psi_1$ in (25) and (26) where $\psi_0 \approx \pi$ and ψ_1 is a small angle for which $\sin \psi_1 \approx \psi_1$ and $\cos \psi_1 \approx 1$. This makes (25) and (26) linear in ψ_1, it may be eliminated and the equation of motion

$$\frac{2}{\mu_0 \gamma^2 \Delta(1 + c(Q + 1) \sin(\psi_0))} \frac{dv_y}{dt} + \frac{2\lambda}{|\gamma| \Delta} v_y = 2M_s B_z , \tag{28}$$

obtained for comparison with (14) or (22). Terms involving λ^2/M_s^2 and dB_z/dt have been neglected.

Equation (28) shows that the way in which the effective wall mass, the coefficient of dv_y/dt, depends upon the in-plane field B_y is much more complicated than in the previous case involving B_x. This is because the angle ψ_0 is a function of both B_y and B_z [29]. For example, at the peak velocity itself, ψ_0 has a value $\hat{\psi}_0$ given by

$$4 \sin(\hat{\psi}_0) = [c(Q + 1) - (c^2(Q + 1)^2 + 8)^{1/2}] \tag{29}$$

[29] and as B_y increases, $\sin(\hat{\psi}_0)$ tends to $-[c(Q + 1)]^{-1}$ and the mass which is

$$m = \frac{2}{\mu_0 \gamma^2 \Delta[1 + c(Q + 1) \sin(\psi_0)]} \tag{30}$$

from (28) tends to become infinite. Just below the peak velocity, where $\psi_0 = \pi$ [29], the mass can be identical to m_D (1) and at a lower velocity still we would expect the mass to fall below m_D as ψ_0 falls below π.

The very large increase in the mass which a domain wall should have as it approaches its peak velocity, when an in-plane field is applied perpendicular to the wall plane, could well explain the triangular waveform of the wall position–time transients recorded by de Leeuw et al. [22]. A triangular $y(t)$ response suggests almost constant velocity, and thus a large inertia because the drive field is actually changing sign while the velocity remains almost constant. The velocity then changes sign by means of a complete switch in the wall screw-sense, as discussed in [29].

4. Conclusions

The reasons which lie behind the effective mass of the domain wall, and why this mass should depend upon the in-plane field and so strongly upon its orientation with respect to the wall plane, may be understood by adopting Becker's point of view [3] for the simple wall discussed in Section 2.

In the simple Landau-Lifshitz wall, described by (3) and (4), $\boldsymbol{M}$ lies in the plane of the wall and there will be a magnetic field $\mu_0 M_s$ along x at the wall centre. The energy density associated with this field is $-\frac{1}{2} \mu_0 M_s^2$, and the fact that this is a negative energy, representing a minimum energy or equilibrium condition, is the key to this problem. The magnetic moments at the wall centre lie parallel to the magnetic field there, when the wall is stationary.

When the wall begins to move, $\boldsymbol{M}$ cants out of the wall plane by the small angle ψ and the energy density at the wall centre increases by $\frac{1}{2} \mu_0 M_s^2 \sin^2 \psi$. For steady-state motion, (12) and (13) tell us that ψ increases linearly with velocity when $v_y \ll v_W$ or $B_z \ll \mu_0 \lambda$, that is

$$\psi = v_y/\mu_0 |\gamma| M_s \Delta , \tag{31}$$

so that the increase in wall energy density $\frac{1}{2}\mu_0 M_s^2 \sin^2 \psi$ is

$$\delta\varepsilon = v_y^2/2\mu_0\gamma^2\Delta^2 \tag{32}$$

per unit volume of wall. Equation (32) results in the usual expression for m_D, (1), when $\delta\varepsilon$ is multiplied by an effective wall width 2Δ and equated to $\frac{1}{2} m_D v_y^2$.

When a field, B_x, is applied along the direction of the wall the angle ψ increases by a far smaller amount for a given wall velocity than when there is no in-plane field. The reason for this is that the magnetic moments at the wall centre have a large value of dM_z/dt because of the torque due to the small y component of $\boldsymbol{M}$ and the x component of the magnetic field. This field is normally $\mu_0 M_s$ and when we add the field B_x to this, $\boldsymbol{M}$ does not need to develop such a large y component in order to produce the same dM_z/dt as before. In detail, (20) and (22) give the steady-state relationship between ψ and v_y as

$$\psi = [(1 - b)/\mu_0 |\gamma| M_s(1 + Qb) \Delta] v_y \tag{33}$$

in place of (31). The mass given by (23) may be obtained by the same simple argument given above for the Döring mass, m_D.

When a field B_y is applied to a domain wall which is moving close to its peak velocity along y, the large value of dM_z/dt at the wall centre is mainly due to the torque developed by M_x and B_y. If this was the only factor involved the peak velocity would be observed when $\boldsymbol{M}$ lay in the plane of the wall. However, the torque due to M_y and the magnetic field $\mu_0 M_x$ is still available and this may add or subtract from the larger $M_x B_y$ term. This is why the peak velocity actually occurs when ψ exceeds π [29] and at the peak velocity $d\psi/dv_y$ approaches infinity, compared to the finite slopes given by (31) and (33). A more general picture of the way in which the wall mass depends upon B_y and B_z may be obtained from Fig. 1 [29] where the relationship between ψ and B_z is shown for a range of values of B_y. Because v_y and B_z are linearly related in this model, $d\psi/dv_y$ is proportional to the slope of these curves. The wall mass is thus proportional to the same slope.

Acknowledgements

This paper was written during a short visit to the Magnetism Laboratory of the KFKI, Budapest, Hungary, and the author would like to thank Dr. L. Gál there for many useful discussions. The financial support of the British Council and the Hungarian Academy of Sciences during this visit is also gratefully acknowledged.

References

[1] W. Döring, Z. Naturf. **3a**, 373 (1948).
[2] L. Landau and E. Lifshitz, Phys. Z. SU. 8, 153 (1935); in: L. D. Landau, Collected Papers, Ed. D. ter Haar, Pergamon Press, Oxford 1965 (p. 101 to 114); in L. D. Landau, Men of Physics, D. ter Haar (Ed.), Vol. 1, Pergamon Press, Oxford 1965 (p. 178 to 194).
[3] R. Becker, Z. Phys. **133**, 134 (1952).
[4] M. Prutton and K. D. Leaver, Phys. Letters **6**, 15 (1963).
[5] U. Enz, Helv. phys. Acta **37**, 245 (1964).
[6] E. Feldtkeller, phys. stat. sol. **27**, 161 (1968).
[7] J. C. Slonczewski, Internat. J. Magnetism **2**, 85 (1971).
[8] E. Schlömann, Appl. Phys. Letters **20**, 190 (1972).
[9] A. Hubert, Theorie der Domänenwände in geordneten Medien, Springer Verlag, Berlin 1974 (p. 134).
[10] A. Aharoni, J. appl. Phys. **47**, 3329 (1976).
[11] E. Schlömann, AIP Conf. Proc. **10**, 478 (1972).
[12] J. C. Slonczewski, J. appl. Phys. **45**, 2705 (1974).

[13] A. Hubert, J. appl. Phys. 46, 2276 (1975).
[14] G. M. Nedlin and R. Kh. Shapiro, Soviet Phys. — Solid State 17, 1357 (1975).
[15] A. Aharoni, Phil. Mag. B 38, 393 (1978).
[16] F. H. de Leeuw, IEEE Trans. Magnetics 9, 614 (1973).
[17] P. J. Rijnierse and F. H. de Leeuw, AIP Conf. Proc. 18, 199 (1973).
[18] F. H. de Leeuw, J. appl. Phys. 45, 3106 (1974).
[19] F. H. dee Leeuw and J. M. Robertson, J. appl. Phys. 46, 3182 (1975).
[20] F. H. de Leeuw, IEEE Trans. Magnetics 13, 1172 (1977).
[21] F. H. de Leeuw, Physica (Utrecht) 86/88 B+C, 1320 (1977).
[22] F. H. de Leeuw, R. van den Doel, and J. M. Robertson, J. appl. Phys. 49, 768 (1978).
[23] J. A. Seitchik, W. D. Doyle, and G. K. Goldberg, J. appl. Phys. 42, 1272 (1971).
[24] B. E. Argyle and A. P. Malozemoff, AIP Conf. Proc. 10, 344 (1972).
[25] H. Callen, R. M. Josephs, J. A. Seitchik, and B. F. Stein, Appl. Phys. Letters 21, 366 (1972).
[26] B. E. Argyle and A. Halperin, IEEE Trans. Magnetics 9, 238 (1973).
[27] J. W. Moody, R. W. Shaw, R. M. Sandfort, and R. L. Stermer, IEEE Trans. Magnetics 9, 377 (1973).
[28] R. W. Shaw, J. W. Moody, and R. M. Sandfort, J. appl. Phys. 45, 2672 (1974).
[29] T. H. O'Dell, phys. stat. sol. (a) 50, 179 (1978).
[30] T. H. O'Dell, phys. stat. sol. (a) 48, 59 (1978).
[31] T. L. Gilbert and J. M. Kelly, Proc. Conf. Magnetism and Magnetic Mater., Pittsburgh 1955. AIEE Special Publication T-78 (pp. 253—263).
[32] H. Kronmüller, Z. angew. Phys. 32, 49 (1971).
[33] J. C. Slonczewski, J. appl. Phys. 44, 1759 (1973).
[34] A. A. Thiele, J. appl. Phys. 45, 377 (1974).
[35] E. Schlömann, J. appl. Phys. 47, 1142 (1976).
[36] G. M. Nedlin and R. Kh. Shapiro, Soviet Phys.-Solid State 19, 1707 (1977).
[37] B. E. Argyle and A. P. Malozemoff, AIP Conf. Proc. 10, 344 (1972).
[38] N. L. Schryer and L. R. Walker, J. appl. Phys. 45, 5406 (1974).

(Received March 2, 1979)

See Short Note by V. V. Golubkov et al.

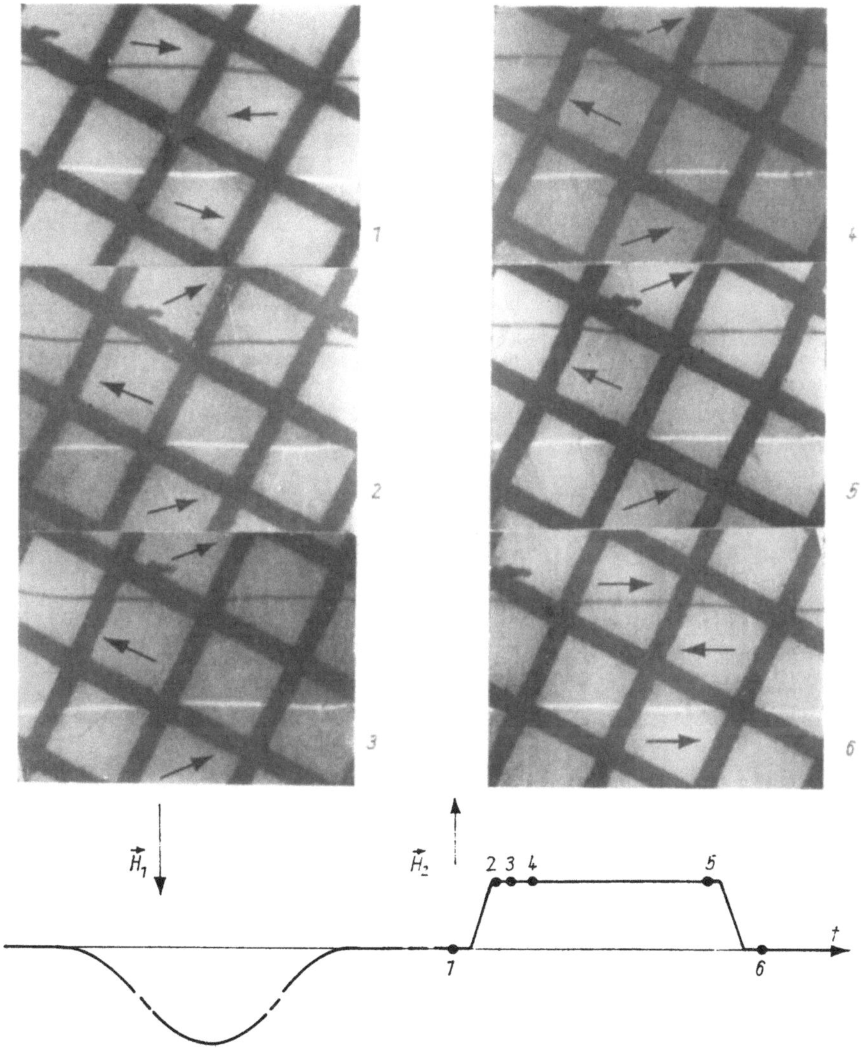

Fig. 1. Stroboscopic micrographs showing Néel wall transitions into cross-tie walls and temporal diagram of the applied fields. The hard axis is vertical, the period of the grid is 33 μm, $\boldsymbol{H}_1$ saturating pulse field, $\boldsymbol{H}_2$ writing pulse field. Figures near the micrographs correspond to the moments of observation shown in the diagram

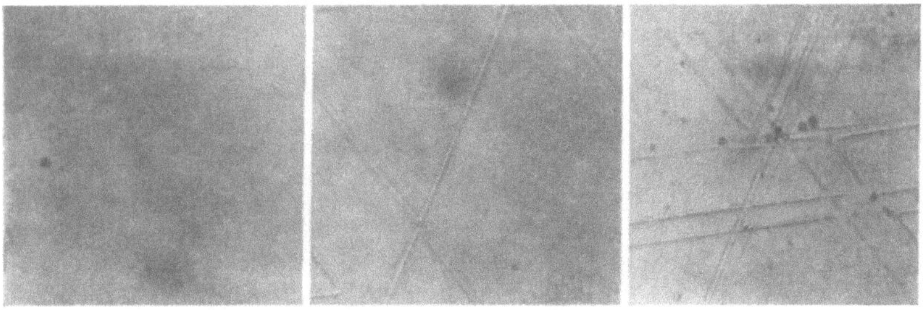

a b c

Fig. 1. Typical Sirtl etch patterns of wafers from different lots; a) lot No. 1, b) lot No. 2, c) lot No. 3; (320 ×)

See Short Note by J. KOSHY and C. C. OGBONNA

a

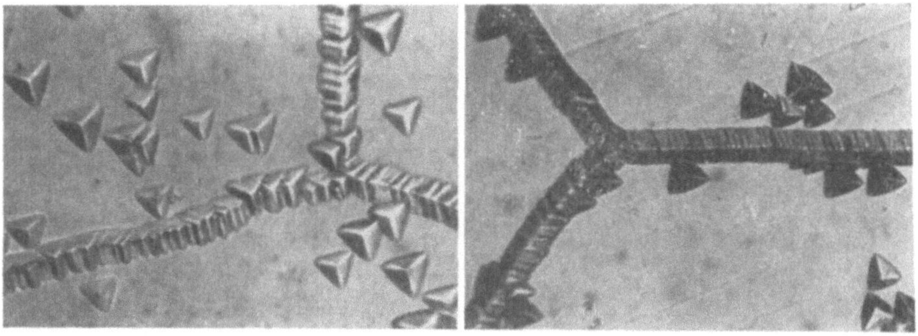

b

Fig. 1. Etch patterns produced on (111) cleavages of BaF_2; a) etched in sodium carbonate, b) etched in potassium bicarbonate; (275 ×)

a b

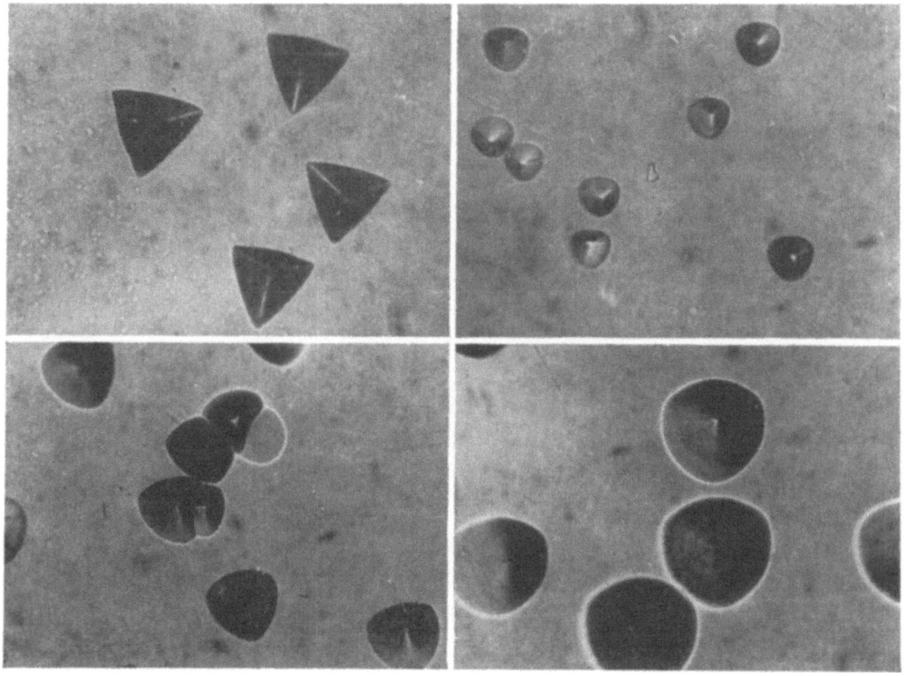

c d

Fig. 2. Etch pattern on (111) cleavage of BaF_2; a) etched in calcium carbonate solution, b) etched in calcium carbonate solution with 300 ppm of 1M oxalic acid, c) etched in calcium carbonate solution with 400 ppm of 1M oxalic acid, d) etched in calcium carbonate solution with 600 ppm of 1M oxalic acid; (250 ×)

phys. stat. sol. (a) 54, K1 (1979)

Subject classification: 18.2.1; 21.1.1

Electron Optics Laboratory, Department of Physics, Moscow State University[1)]

Dynamics of Néel Wall Transitions into Cross-Tie Walls

By

V.V. GOLUBKOV, V.I. PETROV, and G.V. SPIVAK

The process of Néel wall transitions into cross-tie walls in thin permalloy films while a pulse magnetic field is applied has been investigated. This is a process of writing information in magnetic memory devices that utilize the domain walls as channels for propagation of information in the form of Bloch pairs /1/. In investigations of cross-tie wall creation carried out earlier /1, 2/ either the powder technique or the Kerr effect were used and one could only register the situation after the action of the pulse magnetic field of a given width and amplitude. Using these techniques, however, it is impossible to determine how the process of the transition is going on and answer correctly the following questions: 1) whether all the Bloch pairs nucleate at the same moment, 2) whether cross-ties move after nucleation or only Bloch lines move, and 3) what are Bloch lines and cross-tie velocities. To answer these questions one needs to observe the process directly.

The problem has been solved by means of stroboscopic transmission electron microscopy /3/. We used a microscope with a time resolution of 0.5 ns /4/.

A set of two magnetic field pulses H_1 and H_2 was applied along the hard axis of a thin 300 Å permalloy film. The pulse field $H_1 > H_k$ of width 1 μs, causes the formation of Néel walls. The pulse field $H_2 = 0.45\ H_k$ acting in the opposite direction results in Néel wall transitions into cross-tie walls. The micrographs corresponding to different moments of this process and a temporal diagram of the applied pulse fields are shown in Fig. 1 (see on the photo pages before the Short Notes part). Because of the fact that the picture

1) 117234 Moscow, USSR.

formation time (0.5 ns) was much shorter than the repetition period one could not see the image on the screen and the exposure time of each picture was as long as 18 min.

In order to use a stroboscopic mode of operation the process under investigation has to be reproducible. Such a reproducibility was pointed out earlier only for domain walls and ripples /3, 5/. From the micrographs in Fig. 1 it is clearly seen that Bloch pair nucleation and movement along the domain wall are also reproducible. After the action of the saturating pulse field H_1 strip domains with nearly 180^0 Néel walls along the easy axis appear (moment 1). The magnetization direction inside the domains is marked by arrows, the magnetization direction inside the Néel walls is directed downward. The next (writing) pulse H_2 is acting in a direction that is opposite to that in the Néel walls. Just after the rising time of the pulse H_2 (2.5 ns) magnetization inside the domains turns a little in the direction of the applied field, but no Bloch pairs have arisen yet (moment 2). At the moment 3 when the field H_2 has already been for 2.5 ns at its maximum value several Bloch pairs have nucleated. In spite of the fact that the value of the field H_2 does not change, which can be deduced by the invariability of the magnetization direction inside the domains, some pairs appear with still longer delay - only by the moment 4, e.g. 2.5 ns later. Experiments carried out at higher fields H_2 show that a pair nucleation delay exists there too, it decreases with the increase of the field H_2. After nucleation both the Bloch line and the cross-tie move but in opposite directions and the section of the wall between them where magnetization follows the external field H_2 is growing up (moments 4 and 5). The velocities of Bloch line and cross-tie obtained directly from the micrographs are $2x10^4$ cm/s and $1.1x10^4$ cm/s, respectively. If at the moment 5 we stop the Bloch line and the cross-tie movement by switching off the field H_2 then the cross-tie walls remain (moment 6). When the pulse is long enough so that the Bloch lines and the cross-tie have time for reaching each other and annihilate, then after switching off the field H_2 the walls become Néel ones, but the magnetization direction inside them will be opposite to that at the beginning of the cycle (moment 1).

From the experiments described, we can summarize the following:

1. The process of Bloch pair nucleation and movement is quite well reproducible and it is possible to use a stroboscopic technique to investigate them.

2. There exists a time delay of Bloch pair nucleation that decreases with the increase of the amplitude of the writing pulse. It allows us to explain the fact /2/ that at a given amplitude of the writing pulse there exists a critical value of its width when the pair density reaches its maximum value after the pulse, the increase of the width turns to the annihilation of some pairs and the decrease - to a situation when some pairs have not yet nucleated.

3. After nucleation both Bloch lines and cross-ties move, but the latter have lower speed. The interruption of the process by switching off the external field results in stabilization of the cross-tie walls.

References

/1/ L.J. SCHWEE, N.K. IRONS, and W.E. ANDERSON, IEEE Trans. Magnetics MAG-10, 564 (1974).

/2/ A.G. SHISHKOV, E.N. ILYCHEVA, N.G. KANAVINA, O.S. KOLOTOV, R.V. TELESNIN, and N.B. SHIROKOVA, Fiz. Metallov i Metallovedenie 45, 59 (1978).

/3/ V.I. PETROV and G.V. SPIVAK, Z. angew. Phys. 27, 188 (1969).

/4/ V.V. GOLUBKOV, O.S. KOLOTOV, V.I. PETROV, G.V. SPIVAK, and R.V. TELESNIN, Prib. i Tekh. Eksper. No. 1, 185 (1976).

/5/ U.A. DURASOVA, I.S. KOLOTOV, O.S. KOLOTOV, V.I. PETROV, G.V. SPIVAK, and R.V. TELESNIN, Czech. J. Phys. B21, 451 (1971).

(Received March 28, 1979)

phys. stat. sol. (a) 54, K5 (1979)
Subject classification: 9 and 10.2; 22.1.2

Faculty of Electronic Engineering, University of Niš[1)]

Conditions for the Creation of Dislocations by Diffusion of Phosphorus into Silicon

By
N.D. STOJADINOVIĆ

Introduction Phosphorus diffusion into silicon has been widely studied in the past because of its great importance for fabrication of silicon devices. It has been well established that high-temperature diffusion of large concentrations of phosphorus into shallow surface layers of silicon results in lattice strain which is often relieved through the creation of dislocations /1 to 4/. Also, it has been established that dislocations created by phosphorus diffusion into silicon usually are arranged in regular patterns along common slip planes /4/. Principally, the strain and dislocations are believed to result from the mismatch between the ionic radius of the phosphorus atom that occupies a substitutional site and the covalent radius of the silicon /1/.

On the other hand, it has been established that dislocations have a very deleterious effect on electrical characteristics, yield, and reliability of silicon devices /5/. In order to avoid dislocations and to ensure satisfactory electrical characteristics, yield, and reliability of silicon devices, one must know the conditions under which they begin to appear. Unfortunately, despite much work on the creation of dislocations by phosphorus diffusion into silicon there exists a confusion concerning the conditions under which they occur /1, 4, 6 to 8/.

Because of that, we have further investigated theoretically as well as experimentally the creation of dislocations by phosphorus diffusion into silicon, with emphasis on the conditions under which they begin to appear. This paper describes our investigations.

Theory The Quesser-Shockley model /9, 10/ predicts the creation of dislocations when the total amount of phosphorus diffused into silicon, Q, exceeds the critical value, Q_c, defined by the following expression:

$$Q_c = \frac{b}{4\pi \sqrt{2}\, \beta\, (1 + \nu)} \ln \left(2 \sqrt{2} \frac{x_0}{r_0} \right) , \qquad (1)$$

1) Beogradska 14, 18000 Niš, Yugoslavia.

where b is the magnitude of Burgers vector, β is the lattice contraction coefficient, ν is Poisson's ratio, r_o is the effective dislocation core radius, and x_o is the depth where dislocations reach. Recently, we have established that x_o can be expressed as follows /11/:

$$x_o \approx 2\sqrt{Dt_c} , \qquad (2)$$

where D is the diffusion coefficient of phosphorus at the diffusion temperature and t_c is the critical diffusion time (the diffusion time for which dislocations begin to appear).

Assuming that the phosphorus profile conforms to the complementary error-function[2)], the critical total amount of phosphorus diffused into silicon can be expressed as follows /13/:

$$Q_c = \frac{2}{\sqrt{\pi}} \sqrt{Dt_c} N_o , \qquad (3)$$

where N_o is the solubility of phosphorus at the diffusion temperature, while the other symbols are as before.

Using (1), (2), and (3), we obtain the equation which describes a boundary condition for the creation of dislocations by phosphorus diffusion into silicon:

$$t_c^{1/2} = p \ln(q\, t_c^{1/2}) , \qquad (4)$$

where we have put

$$p = \frac{b}{8\sqrt{2\pi D}\,\beta N_o(1+\nu)} ; \qquad q = \frac{4\sqrt{2D}}{r_o} . \qquad (5)$$

The critical diffusion time can be easily estimated solving (4) numerically, by any available method /14/. Using the estimated critical diffusion time, the depth where dislocations reach and the critical total amount of phosphorus diffused can be estimated from (2) and (3), respectively.

Finally, we remark that the analysis presented here is quite general and can be applied to diffusion of other impurities into silicon.

Experimental procedure To check the accuracy of the theoretical considerations, the experiments have been carried out as follows: the silicon wafers were cut from commercially available dislocation-free Czochralski

2) It is a quite reasonable assumption for case of phosphorus diffusion with a relatively low phosphorus glass deposition rate /4, 12/.

Table 1

No. of lot	diffusion time t (min)	total amount of phosphorus diffused $Q^{*)}$ (10^{16} cm^{-2})
1	10	1.48
2	20	1.88
3	30	2.41

*) The values of Q were calculated from the electron mobility and the measured sheet resistivity.

single crystals (p-type, orientation [111], and resistivity 3 Ω cm), lapped and chemically polished to 300 μm thickness to remove the surface damage and strain. The wafers were cleaned by standard procedure and then they were separated into three lots.

After that, each of the wafers was subjected to phosphorus diffusion from $POCl_3$ /4/ at 1100 °C for different times, as shown in Table 1. Gas flow rates in the diffusion system were adjusted to provide conditions for a relatively low phosphorus glass deposition rate, i.e. the phosphorus profile reasonably conforming to the complementary error-function /12/: N_2 2000 cm^3/min and O_2 30 cm^3/min (by pass) and N_2 80 cm^3/min (through bubbler).

On the completion of diffusion processes the wafers from each of the three lots were thinly etched in Sirtl's etch /5/ to expose dislocations. After that, the control wafers were taken from each of the three lots and were subjected to successive thin etching in Sirtl's etch to ensure the measuring of the depth where dislocations reach.

Results and discussion Dislocations in the wafers from each of three lots were observed by a microscope and the typical results are shown in Fig. 1 (see on the photo pages before the Short Notes part). It is obvious that the wafers No. 1 are free of dislocations, in the wafers No. 2 the dislocations just begin to appear, and finally in the wafers No. 3 a relatively large number of dislocations is created.

Consequently, the experimental values of the critical diffusion time and the total amount of phosphorus diffused for the creation of dislocations are: $t_c \approx 20$ min and $Q_c \approx 1.88 \times 10^{16}$ cm^{-2}. Observation of the dislocations in the control wafers No. 2 by a microscope gives the corresponding value of the

T a b l e 2

	critical diffusion time t_c (min)	critical total amount of phosphorus diffused Q_c (10^{16} cm^{-2})	depth where dislocations reach x_o (μm)
theor.	22.5	1.58	0.72
exper.	20.0	1.88	0.85

depth where dislocations reach: $x_o \approx 0.85$ μm. This result is in accordance with the previously reported result /3/, stating that dislocations created by phosphorus diffusion into silicon penetrate approximately one third of the diffusion depth.

The experimental as well as theoretical values of the critical diffusion time, the critical amount of phosphorus diffused, and the depth where dislocations reach are given in Table 2. Equation (4) was solved numerically by Newton-Raphson's method /14/ to give the theoretical value of t_c, and then theoretical values of Q_c and x_o were calculated from (3) and (2), respectively.

Numerical values of constants used in these calculations are given in Table 3.

T a b l e 3

magnitude of Burgers vector b (cm)	3.84×10^{-8}	/16/
lattice contraction coefficient β (cm^3)	10^{-24}	/8, 16/
Poisson's ratio ν	0.27	/16/
effective dislocation core radius r_o (cm)	1.94×10^{-8}	/8, 9/
diffusion coefficient of phosphorus D (cm^2/s)	9.5×10^{-13}	/8, 17/
solubility of phosphorus N_o (cm^{-3})	3.9×10^{20}	/4, 18/

From these results it follows that the agreement between the theoretically and experimentally obtained conditions for the creation of dislocations by phosphorus diffusion into silicon is remarkably good.

Conclusion In this paper a very simple numerical method for the determination of the conditions for the creation of dislocations during phosphorus diffusion into silicon has been developed starting from the Quesser-Shockley model and using the reasonable assumption of a complementary error-function phosphorus profile. From the results given it follows that the conditions for the creation of dislocations calculated using the mentioned method are in remarkably good agreement with those obtained experimentally. Finally, we emphasize that using these results one can optimize the technological parameters of phosphorus diffusion in order to avoid dislocations and to ensure satisfactory electrical characteristics, yield, and reliability of silicon devices.

The author wishes to thank Mrs. V. Džunić of the Faculty of Electronic Engineering, University of Niš, for language help, and Mr. M. Stojanović of Ei-Niš, Semiconductor Factory, for assisting with the experiments.

References

/1/ S. PRUSSIN, J. appl. Phys. 32, 1876 (1961).

/2/ J. WASHBURN, G. THOMAS, and H.J. QUESSER, J. appl. Phys. 35, 1909 (1964).

/3/ M.L. JOSHI and F. WILHELM, J. Electrochem. Soc. 112, 185 (1965).

/4/ R.A. MC DONALD, G.G. EHLENBERGER, and T.R. HUFFMAN, Solid State Electronics 9, 807 (1966).

/5/ G.H. SCHWUTTKE, Microelectronics and Reliability 9, 397 (1970).

/6/ W. CZAJA, J. appl. Phys. 37, 3441 (1966).

/7/ H. MURAOKA, M. NAKAMURA, and T. YONEZAWA, Toshiba Rev. 27, 1118 (1972).

/8/ M. WATANBLE, H. MURAOKA, and T. YONEZAWA, Proc. 6th Conf. Solid State Devices, Tokyo 1973 (p. 269).

/9/ H.J. QUESSER, J. appl. Phys. 32, 1776 (1961).

/10/ S.M. HU, Atomic Diffusion in Semiconductors, Plenum Press, New York 1973 (p. 311).

/11/ N.D. STOJADINOVIĆ, M.Sc. Thesis, Fac. Electronic Engn. Univ. Niš, 1977.

/12/ M. NISHIDA, IEEE Trans. Electron Devices ED-20, 221 (1973).

/13/ S.K. GHANDHI, The Theory and Practice of Microelectronics, J. Willey and Sons, New York 1968 (p. 455).

/14/ R. BECKETT and J. HURT, Numerical Calculations and Algorithms, Mc Graw Hill Book Co., New York 1968 (p. 178).

/15/ E. SIRTL and A. ADLER, Z. Metallk. 52, 527 (1961).

/16/ R.K. JAIN and R.J. VAN OVERSTRAETEN, phys. stat. sol. (a) 25, 125 (1974).

/17/ S. MATSUMOTO, Y. AKAO, K. KOHIYAMA, and T. NIIMI, J. Electrochem. Soc. 125, 1840 (1978).

/18/ G. MASETTI, D. NOBILI, and S. SOLMI, Proc. 3th Internat. Symp. Silicon Materials Science and Technology, Philadelphia 1977 (p. 648).

(Received April 12, 1979)

phys. stat. sol. (a) 54, K11 (1979)
Subject classification: 1.6; 22.5.3

Department of Physics, University of Ibadan

Etching of Synthetic Barium Fluoride Single Crystals

By
J. KOSHY and C.C. OGBONNA

By etching cleavage faces of barium fluoride in 0.5 M oxalic acid for 6 min Patel and Singh /1/ obtained conical circular pits on the octahedral cleavages. In the present work attempts have been made to obtain other etchants and etching conditions which produce symmetrical and reliable etch pits on (111) BaF_2 cleavages. The etching formulae and etching results thus obtained are described in this short note.

Optical grade BaF_2 monocrystals obtained from Harshaw Chemical Co., USA, were cleaved along (111) cleavage faces. The following new etchants and etching conditions have been found for freshly cleaved BaF_2:

etchants	concentration (aqueous solution)	time of etching	shape of pits
sodium carbonate	0.1 M	3 min	triangular
sodium bicarbonate	0.1 M	3 min	triangular
potassium carbonate	0.15 M	2 min	triangular
potassium bicarbonate	0.15 M	2 min	triangular
calcium carbonate	saturated	10 min	triangular

In all the cases crystals were etched without agitation at room temperature. In the case of alkali metal carbonate and bicarbonate etchants increasing the concentration may result in the dehydration of the specimen. No such effect has been observed in the case of calcium carbonate etchant. The crystals after etching were cleaned ultrasonically in distilled water. The etch patterns were examined optically after depositing thin silver films to enhance contrast.

Fig. 1 (see on the photo pages before the Short Notes part) shows typical examples of etch patterns produced on (111) cleavage faces etched in sodium carbonate and potassium bicarbonate, respectively. Sharply defined triangular pits and intersecting grain boundaries are clearly seen on the figure. Pits on prolonged etching did not change the shape or orientation but only the size of the pits. Matched pair, one piece etched in one etchant and the other in another etchant produced excellent matching in shape and orientation of the pits and hence the pits are produced at the sites of dislocations. The average density of pits was found to be $4x10^4$ pits/cm^2.

It was found that an addition of a minute quantity of oxalic acid as impurity in the etchant has a remarkable influence on the etching of these crystals. In a saturated calcium carbonate aqueous solution impurities of 1 M oxalic acid have been added in the order of 300 ppm, 400 ppm, and 600 ppm. (111) crystal cleavages were then etched simultaneously in these etchants for 10 min. The etch patterns thus produced are shown in Fig. 2 (see on the photo pages before the Short Notes part).

The following points merit discussion:

(1) The pits produced in pure saturated $CaCO_3$ solution are triangular in shape as shown in Fig. 2a.

(2) The shape of the pits changed completely by an addition of 300 ppm of 1 M oxalic acid. The size of the pits became noticeably smaller for the same etching time.

(3) As the impurity content increases the size of the pits started increasing and became appreciably bigger as the impurity content reaches 600 ppm for the same etching time.

On increasing the oxalic acid impurity it was found that the etching rate decreases till a few ppm. Once the oxalic acid impurity has reached 300 ppm the etching rate and the shape of the pits are controlled by oxalic acid impurity which appears to have inhibited the etching action of $CaCO_3$ solution.

References

/1/ A.R. PATEL and R.P. SINGH, Indian J. pure appl. Phys. 7, 335 (1969).

(Received May 29, 1979)

phys. stat. sol. (a) 54, K13 (1979)
Subject classification: 10.1; 21; 21.1

Sektion Physik der Martin-Luther-Universität Halle/Wittenberg[1]

The Composition Dependence of the Effective Vacancy Formation Enthalpy in Copper-Rich Brass by Means of Positron Annihilation

By
G. DLUBEK and O. BRÜMMER

Positron annihilation has been developed in the last decade to a method most frequently used for the study of vacancies in metals in thermal equilibrium /1/. This method can be applied with success also to concentrated alloys /2, 3/. The purpose of the present work is the estimation of the effective vacancy formation enthalpy H^F_{eff} in brass and its dependence on composition varying from 0 to 48 wt% Zn by measurements of the peak height of the 2 γ-angular correlation curve.

For the measurement a conventional equipment /4/ having a resolution of 5 mrad x 230 mrad was used. Positrons of an external 30 mCi ^{22}Na source were focused on the sample with the help of a 5 kG magnetic field. This results in a coincidence counting rate of 2.5×10^4 min; per measured point 10^6 counts were accumulated. The sample material was prepared by melting appropriate amounts of Cu and Zn in an induction furnace, its homogeneity was checked by electron microprobe analysis. The purity of samples with 5.4, 15, and 37 wt% Zn was 99.8%, all other samples were of 99.999% purity. To minimise Zn evaporation all samples were covered with a 1 μm thin Ni foil. Further, most of the samples were measured at temperatures lower than 600 $^{\circ}$C (see Fig. 1).

Fig. 1 shows the measured peak heights for different brass alloys, Cu and Zn were included. The curves which show the usual behaviour known for pure metals can be subdivided into two different regions. Starting from room temperature the peak height shows a slow increase. This region is followed by a second one exhibiting a rapid growth of the peak height caused by thermal vacancies. Only for brass with 48 and 42 wt% Zn a deviation of the peak height from the normal S-like behaviour is observed. This is caused by the long range

1) Friedemann-Bach-Platz 6, DDR-402 Halle (Saale), DDR.

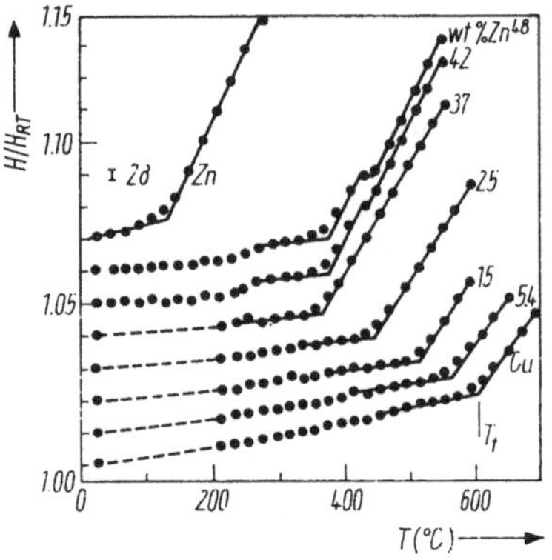

Fig. 1. Peak height H of the 2 γ angular correlation curve normalized to the initial value at room temperature H_{RT} as a function of the sample temperature T for copper-rich brass. T_t indicates the positron annihilation threshold temperature. For clarity the curves were shifted vertically by one unit with respect to the previous ones

order-disorder transformation of the β phase at 465 °C and will be discussed in detail in a separate paper.

By the intersection of straight line approximations to the prevacancy and the vacancy regions of the peak height curves the so-called threshold temperature T_t is defined (see Fig. 1) which characterizes the onset of positron trapping by thermal vacancies. For pure metals it is well confirmed that T_t is linearly related to the monovacancy formation enthalpy H^F_{1v}, $H^F_{1v} = \alpha T_t$ /5, 6/. This method is also very attractive for the estimation of the effective vacancy formation enthalpy H^F_{eff} in alloys from positron annihilation data. The state of sample, processes of ordering or clustering, and phase transformations could be of influence not only on vacancy concentration in alloy /7/ but also on the behaviour of the positron itself /1/. The method of threshold temperature, however, seems to be rather insensitive for these effects as well as for undesirable experimental effects, such as limited statistical accuracy and limited temperature range of measurement, electronic drift, thermal expansion of sample holder, etc. For the constant in the relation $H^F_{eff} = \alpha T_t$ we use a value of $\alpha = 15.4\ k = 1.33 \times 10^{-3}\ eVK^{-1}$ (k Boltzmann constant) which is based on the empirical relations $Q^{SD}_{1v} = 28\ kT_t$ /5/ and $H^F_{1v} = 0.55\ Q^{SD}_{1v}$ /8/ (Q^{SD}_{1v} monovacancy self-diffusion enthalpy). Both of the relations are confirmed for f.c.c., h.c.p., and b.c.c. metals.

Fig. 2 presents the threshold temperature T_t as well as the effective formation enthalpy H^F_{eff} for the investigated alloys. The results for the pure metals Cu and Zn agree with those estimated by a complete trapping model analysis (Cu : 1.16 eV /9/, Zn : 0.54 eV /10/). This supports the reliability of our results for the alloy. As Fig. 2 shows H^F_{eff} varies linearly with Zn concentration throughout the f.c.c. α phase. It remains virtually constant

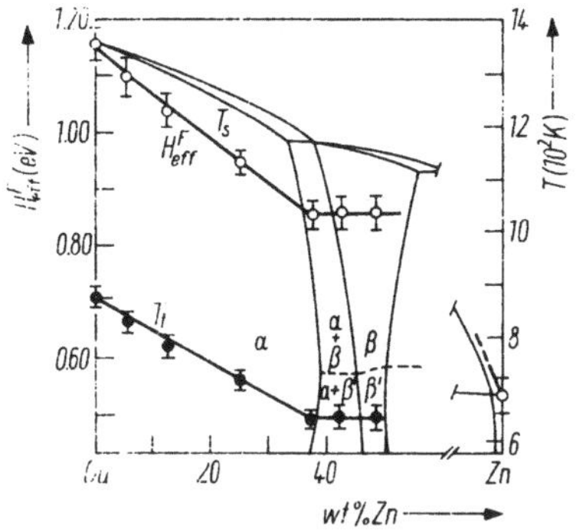

Fig. 2. Composition dependence of the effective vacancy formation enthalpy H^F_{eff} for copper-rich brass. T_t indicates the positron annihilation threshold temperature from which H^F_{eff} was estimated using the relation $H^F_{eff} = 1.33x$ $x10^{-3}$ T_t (K). The corresponding part of the phase diagram /11/ is shown also where T_s means the solidus temperature of the alloy

within the α + β and the b.c.c. β phase. Our results for α brass agree very well with those obtained by Schultz et al. /3/ using the same method but disagree with their constant value of 0.54 eV for α + β brass. Since their curves measured for brass containing 40, 45, and 48 at% Zn exhibit an unusual behaviour (see Fig. 1 in /3/) we believe that their value of 0.54 eV is not correct.

For pure metals an empirical relationship between the vacancy formation enthalpy and the melting temperature T_m of the type H^F_{1v}/T_m = const is well known /8/. In the literature (e.g. in /7/) it was suggested that a correlation between H^F_{eff} and the solidus temperature T_s of the alloy should exist likewise. In order to check this suggestion we have included in Fig. 2 the corresponding part of the phase diagram of brass /11/. The temperature scale was adopted in such a way that H^F_{eff} and T_s coincide for the pure metals. As can be seen H^F_{eff}/T_s is not constant but there is a clear semi-quantitative relationship between H^F_{eff} and T_s.

The results of this work demonstrate that the positron annihilation is very useful for the study of vacancies in thermal equilibrium. By the method of threshold temperature the effective vacancy formation enthalpy H^F_{eff} and its composition dependence in alloys can be estimated. In copper-rich brass H^F_{eff} follows in a semi-quantitative manner the solidus temperature of the alloy.

References

/1/ R.N. WEST, in: Topics in Current Physics, Positrons in Solids, Ed. P. HAUTOJÄRVI, Springer-Verlag, Heidelberg 1978.

/2/ H. FUKUSHIMA and M. DOYAMA, J. Phys. F 6, 677 (1976).

/3/ P.J. SCHULTZ, T.E. JACKMAN, J. FABIAN, E.A. WILLIAMS,

J.R. MAC DONALD, and I.K. MACKENZIE, to be published.

/4/ G. DLUBEK, C. BALZEROWSKI, W. PRESTING, and H. THOMAS, Exper. Tech. Phys. 26, 357 (1978).

/5/ I.K. MACKENZIE and P.C. LICHTENBERGER, Appl. Phys. 9, 3311 (1976).

/6/ S.M. KIM and W.J.L. BUYERS, J. Phys. F 8, L103 (1978).

/7/ C. KINOSHITA and T. EGUCHI, Acta metall. 20, 45 (1972).

/8/ M. DOYAMA and J.S. KOEHLER, Acta metall. 24, 871 (1976).

/9/ G. DLUBEK, O. BRÜMMER, and N. MEYENDORF, Appl. Phys. 13, 67 (1977).

/10/ B.T.A. MCKEE, W. TRIFTSHÄUSER, and A. STEWART, Phys. Rev. Letters 28, 358 (1972).

/11/ M. HANSEN, Constitution of Binary Alloys, McGraw-Hill, New-York 1958 (p. 650).

(Received March 30, 1979)

phys. stat. sol. (a) 54, K17 (1979)
Subject classification: 1.3; 21; 21.1

Laboratoire de Métallurgie Physique, Faculté des Sciences, Poitiers

Etude par diffusion centrale des rayons X de la démixtion d'un alliage Al-Zn 22 at%

Par
C. TEMPLIER et J. DELAFOND

Introduction La théorie statistique de Langer /1, 2/ généralise toutes les théories précédentes de Hillert /3/, Cagn /4 à 6/, Cook /7/, elle décrit le comportement non linéaire de la décomposition spinodale. L'utilisation des lois d'échelles permet d'obtenir une seule séquence de courbes S(k, t) (facteur de structure) présentant un pic qui croît et se déplace vers les petites valeurs de k. Le déplacement du max obéit aux lois suivantes:

$$k_{max} = C_1 t^{a'}, \tag{1}$$

$$S(k_{max}, t) = C_2 t^{a''}. \tag{2}$$

Parallèlement, des résultats obtenus par simulation sur un modèle d'Ising Cinétique avec interactions entre proches voisins /8/ ou bien à longue distance /9, 10/ permettent de prévoir des lois du même type.

Afin de vérifier ces lois nous avons étudié par diffusion centrale des rayons X la démixtion d'un alliage Al-Zn de concentration critique C = 22 at% (valeur obtenue d'après les travaux de Anatharaman et Satyanarayana /11/).

Tableau 1

Résultats théoriques

Langer /1, 2/	Marro et al. /8/			Binder /10/
$a' = -0,212$	T/T_C	a'	a''	$a' = -1/6$
$a'' \approx 0,87$	0,6	-0,21	0,69	
	0,8	-0,25	0,74	
	0,9	-0,25	0,65	$a'' = -3\,a'$

Etude d' une cinétique à 100 °C d' un alliage Al-Zn 22 at% trempé à l' eau

Dès la mise en température du four on observe une augmentation de la diffusion tandis que le maximum se déplace vers les petites valeurs de k.

Nous avons tracé les courbes (1) et (2) ln $S(k_{max}, t)$ et ln $k_{max}(t)$ en fonction de ln t. Ces deux courbes possèdent une partie linéaire pour t > 300 s avec des pentes $a' \approx -0,17$ et $a'' \approx 0,43$.

Les deux premiers relevés correspondent à la montée en température.

Les valeurs expérimentales de a' et a'' sont proches de celles données par Binder et Stauffer /9/. Ces derniers les ont obtenues à partir d' une théorie de diffusion d' amas contenant à basse température une forte proportion d' atomes de soluté, ce qui est le cas de notre alliage si la démixtion est très avancée. Par contre, nos résultats sont différents de ceux calculés par Marro et al. /8/.

On peut donc en conclure dès maintenant que la décomposition est pratiquement terminée après une trempe à l' eau.

Le critère que nous avons retenu pour évaluer la fin de la réaction spinodale est l' intensité intégrée constante, nous avons donc calculé l' intensité intégrée due aux zones seules à différents instants de l' évolution.

La diminution de Q_0 s' explique par une réversion de l' état obtenu après trempe.

En effet, si on calcule Q_0 à l' aide de la formule de Gérold /12/

$$Q_0 = \frac{(m_1 - m_A)(m_A - m_2)}{V_{at}} \, p \, (Z_B - Z_A)^2 ,$$

m_A concentration initiale de l' alliage, m_1 concentration des zones enrichies, m_2 concentration de la matrice appauvrie, Z_A et Z_B sont les numéros atomiques du zinc et de l' aluminium, p est la fraction du volume de l' alliage initial ségrégé sous forme de zones.

Tableau 2

t(s)	$\Theta = \Theta_a$ 0	$\Theta = 100\,^{\circ}C$ 1200	 2400	 5400
Q_0 (10^{24} electrons/cm^3)	$2 \pm 0,2$	$1,5 \pm 0,2$	$1,4 \pm 0,2$	$1,2 \pm 0,2$

D' après les travaux de Anatharaman et Satyaranayana, on obtient

à $\Theta_a \approx 17\ ^oC$ $Q_o \approx 1,6x10^{24}$ electrons/cm^3

et $\Theta \approx 100\ ^oC$ $Q_o \approx 1,04x10^{24}$ electrons/cm^3.

Comme nous nous intéressons aux premiers instants de la démixtion, il est nécessaire soit de ralentir le phénomène, soit d' opérer une trempe plus rapide.

Rundmann et Bartel /13/ ont étudié le phénomène de décomposition spinodale sur un alliage aluminium-zinc 22% contenant 0,1% magnésium, le magnésium ayant pour effet de ralentir la formation des zones d' équilibre.

Cependant, la décomposition est là aussi quasiment terminée après la trempe.

C' est pourquoi nous étudierons la démixtion d' un alliage Al-Zn 22at%, hypertrempé.

Etude d' un alliage Al-Zn 22 at% hypertrempé Dans une enceinte sous vide, l' alliage est porté à l' état liquide puis projeté et écrasé entre deux plaques de cuivre.

Immédiatement après hypertrempé, on devrait obtenir une diffusion constante quel que soit l' angle et égale à I_L, la diffusion de Laue de la solution solide. (I_L (théorique) = 39 electrons/atom , I_L (expérimentale) = 43 electrons/ atom). C' est ce que l' on observe pour k ≈ 0,9 Å^{-1} mais il existe près du centre une diffusion relativement importante.

Nous obtenons cette diffusion au centre pour tous les alliages hypertrempés, quelles que soient leurs concentrations en zinc.

La microscopie électronique montre que ces alliages sont formés de petits grains dont certains sont suffisamment petits pour être responsables de la diffusion après hypertrempé.

Après une trempe à partir de l' état liquide, la démixtion est très lente à la température ambiante, ce qui peut s' expliquer si les lacunes sont piégées aux joints de grains. Au cours d' un revenu à hautes températures (127 ^oC, 140 ^oC, 150 ^oC), ces grains grossissent, ce qui s' accompagne d' un resserrement de la diffusion au centre. Cette structure des alliages hypertrempés nous conduit à choisir les conditions expérimentales suivantes:

- A la diffusion enregistrée au cours d' une cinétique isotherme, nous soustrayons la diffusion obtenue après hypertrempé. Nous éliminons ainsi

Tableau 3

T (°C)	80	100	108	127	140	150
a'	-0,22 ± 0,02	-0,23 ± 0,02	-0,30 ± 0,03	-0,31 ± 0,03	-0,32 ± 0,03	-0,39 ± 0,04
a''	0,81 ± 0,08	0,72 ± 0,07	0,71 ± 0,07	0,69 ± 0,07	0,63 ± 0,07	0,72 ± 0,07

la diffusion due aux grains, la diffusion de Laue de la solution solide, la diffusion géométrique due à l'appareillage; pour ne garder que la diffusion caractéristique de la démixtion. Mais il existe quelques inconvénients à cette manière d'opérer:

- la diffusion due aux grains se resserre au cours du revenu,
- la diffusion de Laue diminue puisque la matrice s'appauvrie en zinc.

Ladiffusion retranchée est donc trop importante, et ceci d'autant plus que l'angle est grand et la température élevée.

Nous avons opéré des cinétiques isothermes à 80, 100, 108, 127, 140 et 150 °C. Pour chacune des ces températures, nous avons tracé ln k_{max} et ln S(k_{max}, t) en fonction de ln t. Chacune de ces courbes possèdent une partie linéaire mais les pentes a' et a'' varient avec la température.

Conclusion Pour un alliage aussi concentré que Al-Zn 22 at% la technique d'hypertrempé permet l'étude des premiers instants de la démixtion.

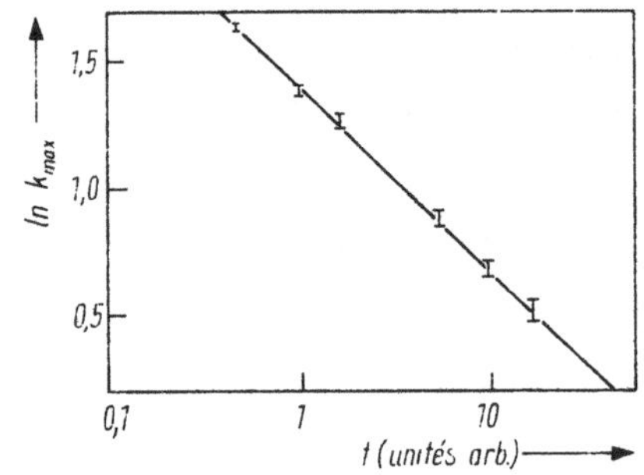

Fig.1

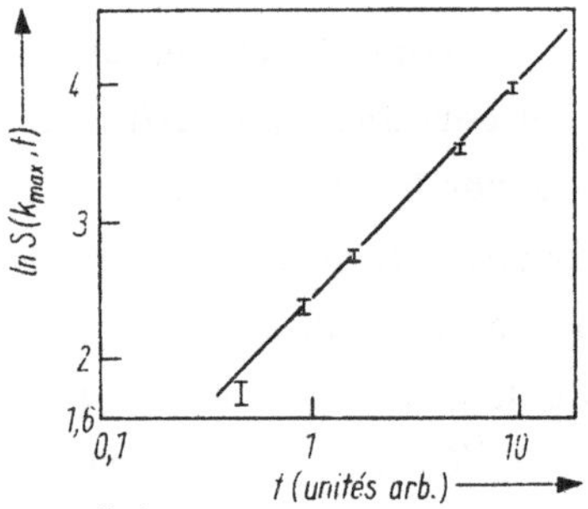

Fig.2

Fig. 1. Courbe expérimentale ln k_{max} = f(ln t) pour Θ = 127 °C, a' ≈ 0,31

Fig. 2. Courbe expérimentale ln S(k_{max}, t) = g(ln t) pour Θ = 127 °C, a'' ≈ 0,69

L' allure des courbes expérimentales ln k_{max} = f(ln t) (Fig. 1) et ln S(k_{max}, t) = = g(ln t) (Fig. 2) montrent que le déplacement et la croissance du maximum de diffusion au cours d'un revenu sont assez bien décrits par les lois (1) et (2).

Si, à basse température, les valeurs a' et a'' sont voisines des valeurs théoriques données par Langer, Marro et al. on constate des écarts aux températures plus élevées. Pour ces températures, ces divergences peuvent s'expliquer par un mécanisme autre que la diffusion, lié au grossissement des grains qui change la concentration en lacunes.

Bibliographie

/1/ J.S. LANGER, Ann. Phys. 65, 53 (1971).
/2/ J.S. LANGER et M. BAR-ON, Ann. Phys. 78, 421 (1973).
/3/ M. HILLERT, Acta metall. 9, 525 (1961).
/4/ J.W. CAHN, Acta metall. 9, 795 (1961).
/5/ J.W. CAHN, Acta metall. 10, 179 (1962).
/6/ J.W. CAHN, Trans. Metall. Soc. AIME 242, 166 (1968).
/7/ H.E. COOK, Acta metall. 14, 1685 (1966).
/8/ J. MARRO, A.B. BORTZ, M.H. KALOS et J.L. LEBOWITZ, Phys. Rev. B 12, 2000 (1975).
/9/ K. BINDER et D. STAUFFER, Phys. Rev. Letters 33, 17 (1974).
/10/ K. BINDER, Phys. Rev. B 15, 4425 (1977).
/11/ T.R. ANATHARAMAN et SATYANARAYANA, Scripta Met. 7, 189 (1973).
/12/ V. GEROLD, Small Angle X-Ray Scattering, Gordon and Breach Science Publishers, 1966.
/13/ K.B. RUNDMANN et T.L. BARTEL, Metall. Trans. 6A, 1887 (1975).

(Received April 5, 1979)

phys. stat. sol. (a) 54, K23 (1979)
Subject classification: 3; 1.6; 22.1.2

Zentralinstitut für Festkörperphysik und Werkstofforschung
der Akademie der Wissenschaften der DDR,
Institutsteil für magnetische Werkstoffe, Jena

Current-Voltage Characteristic of Reactive Sputtering with Element Targets

By
J. SALM, K. STEENBECK, and E. STEINBEISS

Introduction Reactive sputtering, particularly with element targets, has received considerable attention for depositing thin films of chemical compounds. One characteristic of this technique is the influence of gettering processes on the reactive gas partial pressure as the most important parameter determining the film composition. This partial pressure is substantially affected by interactions between the sputtering process and the reactions of the gas with the target surface and with the condensing film material. Many authors have investigated the dependence of the sputtering rate on the reactive gas concentration /1, 2/. Some qualitative aspects of current-voltage characteristics have been described in /2/. This paper presents current voltage (I-U) measurements of reactive sputtering using Si targets in a dc plasmatron and also a model for calculating the I-U characteristics.

Experimental We used a dc planar plasmatron with Si target, the ring-shaped erosion zone of which amounts to 10 cm^2. An Ar-O_2 mixture was led in by a needle valve. The oxygen partial pressure, being p_O when the discharge was switched off, could be adjusted by the O_2 percentage in the discharge gas and independently by the total gas pressure, which was about 5 Pa. Fig. 1 shows the I-U characteristics, measured in pure argon ($p_O = 0$, curve 1) and at high oxygen content, respectively ($p_O = 3.3$ Pa, curve 2). Curve 1 follows the relation /3/

$$I_O = g\, U^m \qquad (1)$$

with $m = 4.5$ and $g = 3.4\times10^{-13}$ AV^{-m}. At medium oxygen pressures p_O the measured I-U curves are n-shaped (Fig. 2), moving to higher currents I with increasing p_O. Using a simplified model we will show in the following section how the I-U characteristic relates to the process of target oxidation.

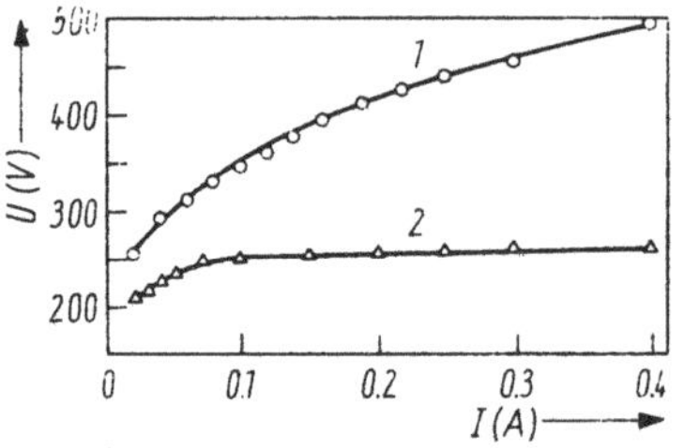

Fig.1

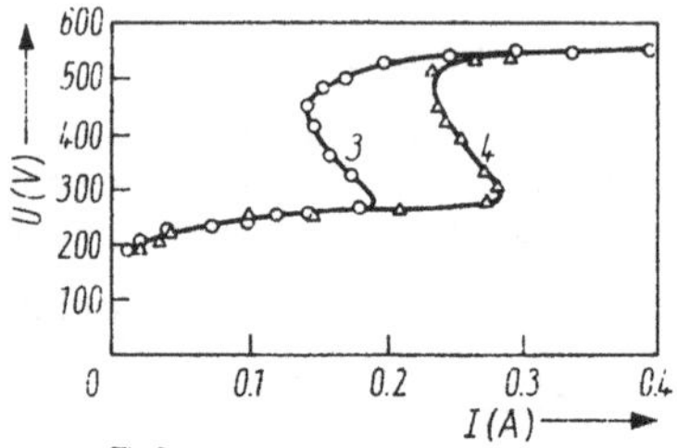

Fig.2

Fig. 1. Measured I-U characteristic at $p_o = 0$ (1) and $p_o = 3.3$ Pa (2)

Fig. 2. Measured I-U characteristic at $p_o = 0.27$ Pa (3) and $p_o = 0.53$ Pa (4)

Theoretical description of I-U characteristics The oxidized target surface compared with the non-oxidized one causes a lower resistance of the gas discharge due to increased electron emission. Therefore we assume the linear relationship

$$I = I_o\,(1 + \alpha\Theta\,)\,, \tag{2}$$

where I_o denotes the current of the oxygen free target and Θ the fraction of the target coated with oxygen ($0 < \Theta < 1$). The increased electron emission is expressed by $\alpha(U)$.

The stationary equilibrium between oxygen adsorption on the target and removal by sputtering can be described by

$$\varkappa\, p_{eff}\, \sigma(1 - \Theta) = j\,\eta_o \Theta / e \tag{3}$$

with the collision number $\varkappa = 3.81 \times 10^{18}$ cm^{-2} s^{-1} Pa^{-1} for O atoms at 300 K, the sticking coefficient $0 < \sigma < 1$, current density j, sputter yield for oxygen η_o, and electron charge e. The effective oxygen partial pressure p_{eff} is affected by the quoted gettering process, which can be approximated by the relation /1/

$$p_{eff} = p_o - A\,\gamma\, j\, \eta_m (1 - \Theta)/e\,S\ , \tag{4}$$

where A is the effective target area, S is the speed of the external pump, η_m is the sputter yield for Si, and γ is a conversion factor, which equals kT in /1/. We modify (4) taking into consideration that the efficiency of the gettering surface F equals the degree of oxydation ϑ of the film deposited upon it:

$$\gamma = \gamma_o\, \vartheta \qquad (0 < \vartheta < 1)\ ; \tag{5}$$

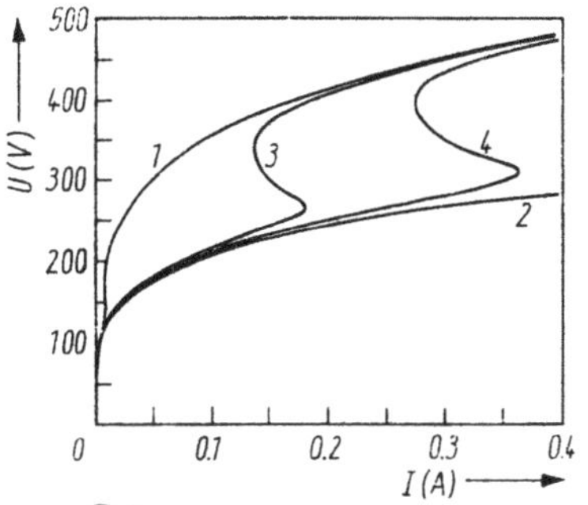

Fig.3

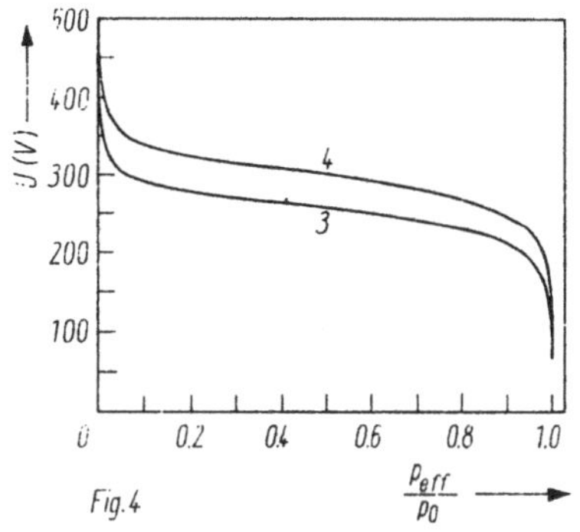

Fig.4

Fig. 3. Calculated I-U characteristic for $p_o = 1.3\times10^{-2}$ Pa (1), 5.3 Pa (2), 0.27 Pa (3), and 0.53 Pa (4)

Fig. 4. Effective oxygen partial pressure p_{eff}/p_o, calculated for $p_o = 0.27$ Pa (3) and $p_o = 0.53$ Pa (4)

ϑ depends on p_{eff} and on the Si deposition rate $n_{Si} = \eta_m Aj(1-\Theta)/eF$. For ϑ we use a relation similar to /4/:

$$2\vartheta = \varkappa p_{eff}(1-\vartheta)/n_{Si} \; . \tag{6}$$

Using (1) to (6) we get the following expression for the I-U characteristic as a function of the parameter Θ, $0<\Theta<1$:

$$I = p_o \sigma \varkappa e A(1-\Theta)\left[1 + w\frac{(1-\Theta)^2}{\Theta}\right] \Big/ \eta_o \left[\Theta + (z+w)(1-\Theta)^2\right] , \tag{7}$$

$$U = \left[I/g(1+\alpha\Theta)\right]^{1/m} \tag{8}$$

with the abbreviations $z = \sigma\varkappa\gamma_o A\, \eta_m/\eta_o S$ and $w = 2\sigma A\, \eta_m/\eta_o F$. The effective oxygen partial pressure p_{eff} results from (4) and (7) to

$$p_{eff}/p_o = \Theta + w(1-\Theta)^2/\Theta + (z+w)(1-\Theta)^2 \; . \tag{9}$$

Substituting the following estimated orders of magnitude: $\eta_m = 0.5$; $\eta_o/\eta_m = 5$; $\sigma = 1$; $\alpha = 10$, $F = 10^3$ cm^2, $A = 10$ cm^2 as well as the applied pump speed $S = 6$ l/s and $\gamma_o = 4.1\times10^{-18}$ Pa l at T = 300 K we get the I-U curves shown in Fig. 3. Curve 1 refers to small p_o, curve 2 to high, and curves 3 and 4 to medium values for p_o, respectively. The calculated oxygen partial pressure p_{eff}/p_o as a function of the discharge voltage is presented in Fig. 4.

The above calculations qualitatively agree with the measured behaviour of the I-U characteristics, even at constant values of η_m, η_o, α, and F assumed. A more detailed analysis taking into account the locus dependences and the thermal desorption in (3) yields similar results.

References

/1/ F. SHINOKI and A. ITOH, J. appl. Phys. 46, 3381 (1975).

/2/ B. GORANCHEV, V. ORLINOV, and V. POPOVA, Thin Solid Films 33, 173 (1976).

/3/ J.A. THORNTON, J. Vacuum Sci. Technol. 15, 171 (1978).

/4/ E. RITTER, Mh. Chemie 95, 795 (1964).

(Received April 2, 1979)

phys. stat. sol. (a) 54, K27 (1979)

Subject classification: 14.3.3; 22.4.2; 22.4.3

Istituto di Fisica CNR-GNSM, Parma (a),
Departamento de Semiconductores, Facultad de Fisica-Matematica,
Universidad de La Habana (b), and
Institute of Physics, Nicholas Copernicus University Torun (c)

Properties of the ZnSe/ZnTe Heterojunction Prepared by a Multi-Source Evaporation of ZnTe:Sb on ZnSe Single Crystals

By

N. ROMEO (a), D. SEURET (b), and F. FIRST (c)

ZnSe and ZnTe with their direct band gap of 2.6 and 2.2 eV, respectively, are interesting materials for visible light emitting devices.

Unfortunately, ZnSe can be made only n-type, while ZnTe can be made only p-type, so that it is impossible to make homojunctions with such materials. For this reason, several atttempts /1 to 5/ have been made to obtain efficient light emission by making a heterojunction between p-ZnTe and n-ZnSe. Two methods have been principally used such as liquid phase epitaxy /1 to 4/ and evaporation of a ZnSe thin film on a ZnTe single crystal /5/. In this paper we describe a new method of preparation of this heterojunction in which Sb-doped ZnTe is deposited by a multi-source apparatus on ZnSe single crystals. The ZnSe substrates were prepared by the high pressure Bridgman method, they were oriented by chemical etching and cut parallel to (111) crystallographic planes. Then, the crystals were purified in molten zinc by the Aven and Woodbury method /6/. Before placing the ZnSe crystals in the vaccum apparatus, they were mechanically polished and etched in 50% NaOH hot solution. The resistivity and electron mobility of these crystals were 10^{-1} Ωcm and $5x10^2$ cm^2 V^{-1} s^{-1}, respectively. Ohmic contacts to ZnSe were made by alloying an indium dot in an atmosphere of H_2 and N_2.

The ZnTe films were prepared by a multi-source method in which Zn, Te, and Sb as a dopant, were evaporated from three separated sources /7/. The evaporation rates of the sources were measured by a quartz monitor and the temperatures of the sources were controlled by Pt/Pt-Rh thermocouples connected to temperature controllers.

The substrate temperature was kept at 250 $^{\circ}$C. By regulating the ratio between the evaporation rates of Zn and Te to the value of 2 with a Sb doping

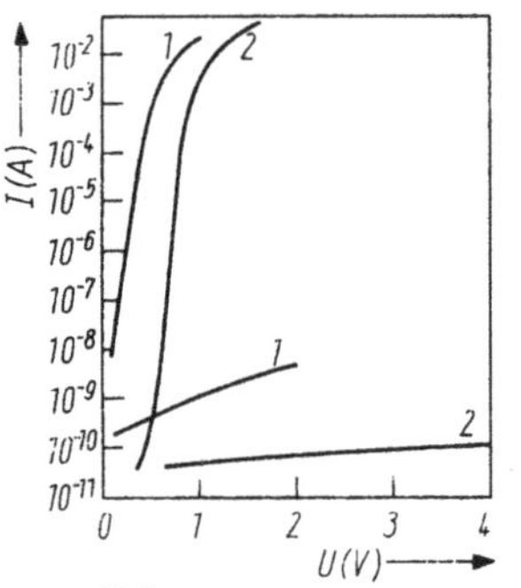

Fig.1

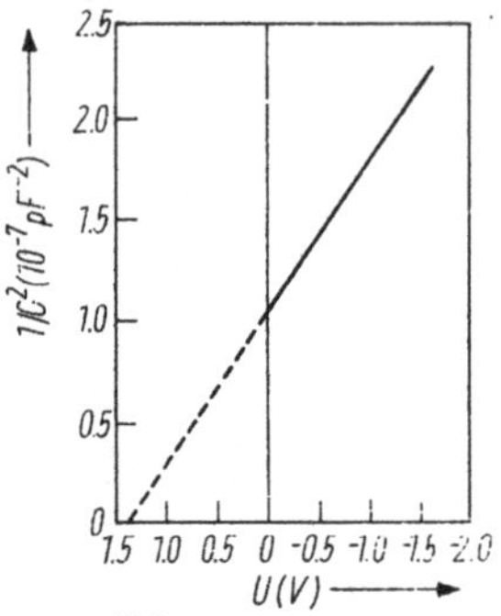

Fig.2

Fig. 1. Room temperature (curves 1) and liquid nitrogen temperature (curves 2) I-U characteristic of a ZnSe/ZnTe heterojunction

Fig. 2. Plot of $1/C^2$, where C is the capacity, as a function of the applied voltage in a ZnSe/ZnTe heterojunction at a frequency of 100 kHz

of about 2% in respect to the overall flux of Zn+Te, p-type films whose resistivity was 10^{-1} to 10^{-2} Ωcm were easily obtained. Ohmic contacts to ZnTe have been made by evaporating gold in high vacuum. The absorption spectrum of such films indicates a forbidden gap of 2.2 eV very near to the value of the single crystal. The I-U characteristic of the ZnSe/ZnTe heterojunctions obtained by the above described method is displayed in Fig. 1, in which a high rectification factor of the order of 10^8 can be noticed. Several I-U characteristics have been taken by varying the temperature from 300 to 77 K. It has been observed that the slopes of the I-U curves do not vary with temperature suggesting that the transport process through the junction is tunneling, probably through the interface states. The $1/C^2$ plot as a function of the reverse voltage, which is reported in Fig. 2, gives a diffusion voltage of about 1.4 V very near to the value reported in /3/ for an abrupt heterojunction grown by the liquid-phase epitaxy. When we exposed the heterojunction to solar light of about 70 mW/cm^2, we observed an open circuit photovoltage of 0.45 V and a short circuit photocurrent of 10^{-2} mA/cm^2. The photocurrent spectrum is displayed in Fig. 3 in which three peaks are present at energies of 2.6, 2.4, and 2.2 eV corresponding to the energy gap value of ZnSe, to an acceptor level in ZnSe, and to the energy gap value of ZnTe, respectively. At liquid nitrogen temperature, the hetero-

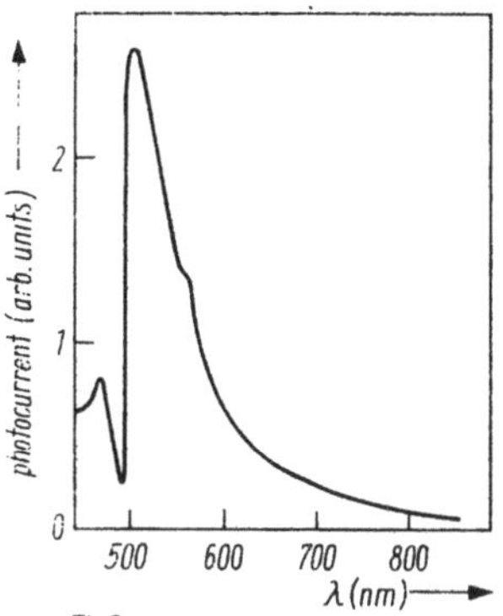

Fig.3

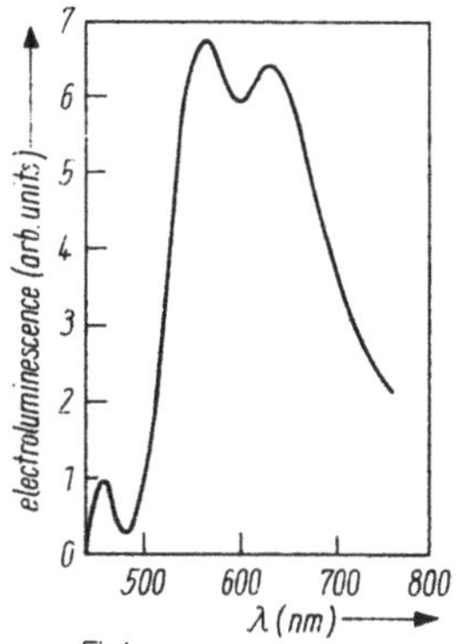

Fig.4

Fig. 3. Room temperature photocurrent spectrum of a ZnSe/ZnTe heterojunction

Fig. 4. Liquid nitrogen temperature electroluminescent spectrum of a ZnSe/ZnTe heterojunction

junctions, when forward biased, emit light which is visible by naked eye. The spectrum of such an emission, which is displayed in Fig. 4, presents three peaks at energies of 2.6, 2.17, 1.96 eV. A similar spectrum was reported in /3/ for a heterojunction prepared by liquid-phase epitaxy and was attributed to hole injection in ZnSe. The external efficiency of this heterojunction is below $10^{-3}\%$ also at liquid nitrogen temperature. This is probably due to a high density of states at the interface which are determined by the high mismatch between the lattice constants of the two materials. A band model, which could explain our experimental results, could be that suggested in /3/ for an abrupt heterojunction.

References

/1/ Y. TSUJIMOTO and M. FUKAI, Japan. J. appl. Phys. 6, 1024 (1967).

/2/ K.K. DUBENSKII, A.V. RUMYANTSEVA, YU.S. RYZHKIN, and V.I. TRAPEZNIKOV, Soviet Phys. - Semicond. 4, 845 (1970).

/3/ S. FUJITA, S. ARAI, F. ITOH, and T. SAKAGUCHI, J. appl. Phys. 46, 3070 (1975).

/4/ H.J. LOZYKOWSKI, H.L. OCZKOWSKI, and F. FIRST, J. Lum. 11, 75 (1975).

/5/ Y. LE FLOCH and H. ARNOULD, Solid State Electronics 16, 941 (1973).
/6/ H. AVEN and H.H. WOODBURY, Appl. Phys. Letters 1, 53 (1962).
/7/ N. ROMEO, G. SBERVEGLIERI, L. TARRICONE, J. VIDAL, and A. WOJTOWICZ, phys. stat. sol. (a) 47, 371 (1978).

(Received April 17, 1979)

phys. stat. sol. (a) 54, K31 (1979)
Subject classification: 12.1; 21

Metal Physics Laboratory, National Research Center, Dokki[1)]

Effect of the Type of Pre-Torsional Deformation on the Creep Rates of an Aluminium-Silicon-Zinc Alloy

By
G.A. HASSAN

The effect of extraneous deformation on the creep curves has been extensively investigated. It seems that extraneous deformation, under certain conditions, results in a decrease in the creep rates, which is due, in general, to the increase in the induced internal stresses /1 to 5/.

The aim of the present work, in which the literature seems lacking, is to compare the effect of the type of pre-torsional deformation given at room temperature on the tensile creep rate. The torsional deformation is carried out either unidirectionally or bidirectionally.

The material used is an aluminium dilute alloy containing 0.25 wt% silicon and 0.26 wt% zinc, in the form of wire samples (1 mm in diameter and of length 100 mm).

The effect of pre-torsional deformation on the tensile creep rate is investigated at 300 °C and under an applied tensile stress of 3 kp/mm^2. The examined alloy samples are subjected to tensile creep tests under three different conditions:

a) samples annealed at 350 °C for 3 h,

b) one set of the annealed samples are subjected (while in place) to unidirectional type of torsion and given five twists (N = +5) before the creep test,

c) another set of the annealed samples are subjected to bidirectional type of torsion before the creep test. The samples are given five turns of twist in the forward clockwise direction followed by an equal number of twists in the anti-clockwise direction ($N = \pm 5$).

The results of observations of the creep strain-time relation are illustrated in Fig. 1. It can be seen that the creep strain at any instance of time of predeformed samples is relatively low compared with those of annealed samples.

---

1) Dokki, Cairo, Egypt.

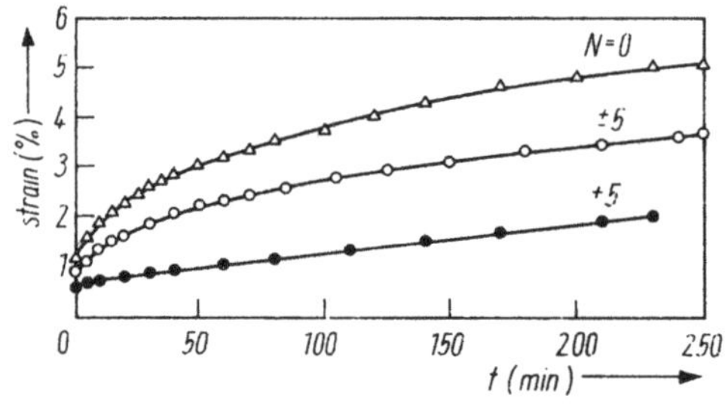

Fig. 1. Effect of the type of pre-torsional deformation given at room temperature on the strain-time relation of aluminum-silicon-zinc alloy. Test temperature 300 °C; stress 3 kp/mm^2; turns of twist N = 0 (annealed, △), ±5 (bidirectional twisting, ○), +5 (unidirectional twisting, ●)

Furthermore, the type of pre-torsional deformation affects the creep rate of the tested samples. The creep strain at any instance of time is higher for samples c) than those of samples b), though it is lower than that of annealed samples a).

The higher creep rate of bidirectionally twisted samples as compared to that found in unidirectionally twisted samples may be a consequence of the increased probability for mutual dislocation annihilation of opposite signs, and due to the reduction of internal stresses upon the reversal of the direction of twisting /6 to 9/.

References

/1/ T. HASEGAWA, S. KARASHIMA, and Y. IKEUCHI, Acta metall. 21, 887 (1973).

/2/ T. HASEGAWA, Y. IKEUCHI, and S. KARASHIMA, Metal Sci. J. 6, 78 (1972).

/3/ G.A. HASSAN, phys. stat. sol. (a) 42, K115 (1977).

/4/ K.H. GEORGY, M.A.H. ELLISY, and G.A. HASSAN, Conf. Solid State Phys., April 1977; Proc. Math. Phys. Soc. Egypt No. 445 (1977).

/5/ G.A. HASSAN, K.H. GEORGY, F.H. HAMMAD, and J. CADEK, phys. stat. sol. (a) 37, K213 (1976).

/6/ G.A. HASSAN, phys. stat. sol. (a) 41, 541 (1977).

/7/ G.A. HASSAN, phys. stat. sol. (a) 40, 43 (1977).

/8/ G.A. HASSAN, Aluminium 53, 436 (1977).

/9/ F.H. HAMMAD and G.A. HASSAN, J. Mater. Sci. 12, 1279 (1977).

(Received March 26, 1979)

phys. stat. sol. (a) 54, K33 (1979)
Subject classification: 10.1; 21

Department of Metal Physics, Charles University, Prague[1]

The Critical Resolved Shear Stress of Mg-Cd Single Crystals

By
P. LUKÁČ

Introduction The effect of foreign atoms in solid solution on the critical resolved shear stress of single crystals has been the subject of several theoretical studies as reviewed e.g. by Haasen /1/ and Labusch /2/. The hardening effect has been considered to be due to the interaction between solute atoms and dislocations. Solid solution hardening can be measured by the increase of the critical resolved shear stress (CRSS) τ_0 with solute concentration, c, in a constant strain rate tensile test.

This work presents the results of an experimental investigation of the influence of cadmium additions on the CRSS of magnesium single crystals.

Results and discussion Polycrystalline rods were prepared from Mg of 99.99% purity and Cd of 99.99%. Single crystals were grown from the melt in a graphite mould using a modified Bridgman method /3/. The mould was lowered under an argon atmosphere in a vertical resistance furnace at the rate of 40 mm/h. The crystals were then annealed (in an argon atmosphere) for 5 h at 450 $^{\circ}$C and cooled to room temperature. Specimens with the orientation allowing only basal slip were used.

The single crystals were deformed on an Instron machine at a constant cross-head speed which gave an initial shear strain rate of 1.0 to 1.2×10^{-4} s^{-1}. The tensile tests were performed at temperatures between 77 and 295 K. Resolved shear stress and shear strain were calculated by the usual formulae for single glide. The CRSS were estimated in the usual way of back extrapolation of stage A of the work hardening curve.

The temperature dependence of the CRSS for Mg-Cd single crystals is shown in Fig. 1. It is obvious that the CRSS decreases with increasing temperature up to 240 K. Beyond 240 K the CRSS is independent of temperature.

---

1) Ke Karlovu 5, 121 16 Prague, CSSR.

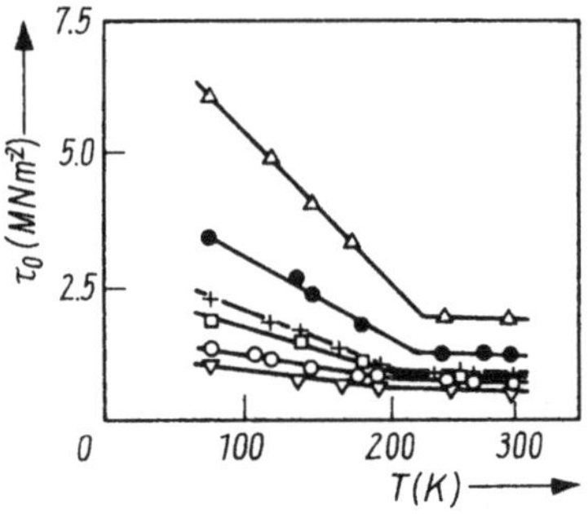

Fig.1

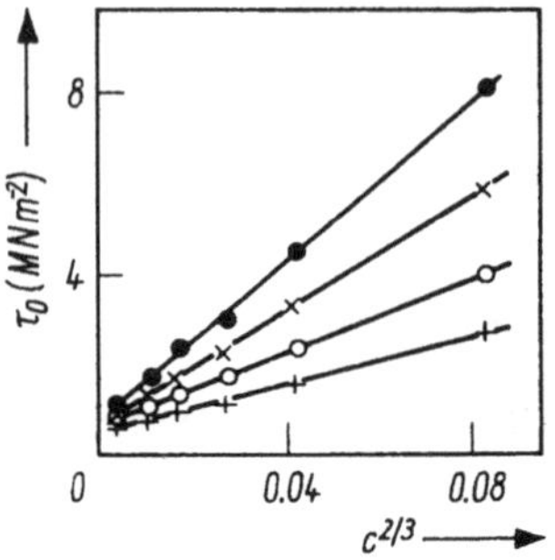

Fig.2

Fig. 1. Temperature dependence of the CRSS for Mg-Cd alloy single crystals. ▽ 0.02, ○ 0.11, □ 0.22, + 0.44, ● 0.90, △ 2.4 at% Cd

Fig. 2. Critical resolved shear stress dependent on $c^{2/3}$. ● 0, × 77, ○ 150, + 200 K

Below 220 K the temperature dependence of the CRSS can be given in the form

$$\tau_0(T, c) = \tau_0(0, c) - BT \ , \tag{1}$$

where the CRSS at 0 K, $\tau_0(0, c)$, and B depend on the concentration of cadmium. A similar temperature dependence of the CRSS has been also observed for Mg-Zn /4/, Mg-Al /5/, Mg-In /5/, and Cd-Zn /6/ single crystals.

Fig. 2 shows the CRSS plotted against $c^{2/3}$ where c is the concentration (in atomic fraction) of cadmium as the solute. The values of the CRSS at 0 K have been obtained by extrapolation of the curves in Fig. 1. It is seen that at all temperatures and for the whole concentration range studied the concentration dependence of the CRSS can be described by the empirical relation

$$\tau_0(T, c) = \tau_0(T, 0) + Sc^{2/3} \ , \tag{2}$$

where the slope S is a function of temperature.

The observed concentration dependence of the CRSS at 0 K can be accounted for by the theory of Labusch /2, 7/ according to which the increase of the CRSS at 0 K with the concentration is expressed in the form

$$\tau_0(0,\ c) = ZG\ \varepsilon^{4/3}\ c^{2/3} + \tau_0(0,\ 0)\ , \tag{3}$$

where Z is a numerical constant, G is the shear modulus, and ε is a parameter describing the interaction between solute atoms and dislocations.

Using experimental data /8, 9/ the value of ε can be calculated according to

$$\varepsilon = \left[(\delta)^2 + \alpha^2(\eta')\right]^{1/2}\ , \tag{4}$$

where the relative change of the lattice parameter b with alloying

$$\delta = \frac{1}{n}\ \frac{db}{dc} \tag{5}$$

and

$$\eta' = \frac{\frac{1}{G}\ \frac{dG}{dc}}{1 + \frac{1}{2}\left|\frac{1}{G}\ \frac{dG}{dc}\right|} \tag{6}$$

If interaction with screw dislocations is used (α = 3) then Z becomes equal to 1/60 whereas for interaction with edge dislocations (α = 16) Z comes out to be 1/250. The value of Z for interactions with edge dislocations is closer to 1/550 as estimated theoretically by Labusch /7/.

References

/1/ P. HAASEN, in: Physical Metallurgy, Ed. R.W. CAHN, North-Holland, Amsterdam 1970 (p. 1011).

/2/ R. LABUSCH, Acta metall. 20, 917 (1972).

/3/ H. SCHARF, P. LUKÁČ, M. BŎCEK, and P. HAASEN, Z. Metallk. 59, 799 (1968).

/4/ A. AKHTAR and E. TEGHTSOONIAN, Acta metall. 17, 1339 (1969).

/5/ A. AKHTAR and E. TEGHTSOONIAN, Phil. Mag. 25, 897 (1972).

/6/ M. ROJKO and P. LUKÁČ, J. Mater. Sci. 8, 1065 (1973).

/7/ R. LABUSCH, phys. stat. sol. 41, 659 (1970).

/8/ W.B. PEARSON, Handbook of Lattice Spacing and Structure of Metals and Alloys, Pergamon Press, London 1958.

/9/ H. SCHARF and H.G. FRIEDEL, Z. Metallk. 58, 729 (1967).

(Received April 7, 1979)

phys. stat. sol. (a) 54, K37 (1979)
Subject classification: 14.2; 21

V.I. Ulyanov (Lenin) Electrical Engineering Institute, Leningrad (a) and Ingenieurhochschule Dresden (b)

Superconducting Properties of Nb Films Deposited in a Quasi-Closed Volume

By

V. P. LEDOVSKOI (a), N. A. POTSAR (a), V. L. LASKA (a), and E. SEIDOWSKI (b)

Two basic methods are used at present for obtaining superconducting Nb films, namely ion-sputtering and electron-beam evaporation. For the improvement of the superconducting properties of the films ultra-high vacuum systems are normally used /1, 2/. Pre-evaporation of a Nb target is also applied during a period of some ten minutes to several hours /3, 4/. The attempts to obtain films in a vacuum of 10^{-5} to 10^{-6} Torr lead to significant deterioration of the superconducting properties owing to dissolved gases, even if one uses liquid nitrogen traps /5/. However, it is this range of pressures of residual gases that is most tempting from the point of view of manufacturing effectiveness, therefore in the present paper we give experimental results on the superconducting properties of Nb films deposited at pressures of residual gases of about 10^{-4} to 10^{-5} Torr by electron-beam evaporation in a quasi-closed volume. Evaporation and deposition of Nb take place in a quasi-closed volume which is an additional chamber manufactured of Nb sheets of 0.2 to 0.5 mm thickness, and maintained inside the main technological volume. The dimensions of the main volume are 500x500x500 cm^3 while the dimensions of the quasi-closed volume are 10x10x10 cm^3. The vacuum system is pumped by means of a conventional oil pump to a vacuum of about 10^{-4} to 10^{-5} Torr /6/. Nb films on "sitall" (SiO_2 + Al_2O_3 + Ta) and "polycor" (Al_2O_3 + MgO) substrates were deposited with the rates 0.35 to 1.8 μm/min. The substrate temperature was 650 to 900 ^oC. The films were examined without vacuum annealing. The power of the electron beam varied from 1 to 2 kW.

The superconducting properties of the deposited films were investigated in a conventional 10 l cryostat by means of the four-probe method. The temperature was measured by a germanium transducer with maximum total error not greater than 0.1 K. Investigating the superconducting properties, the tran-

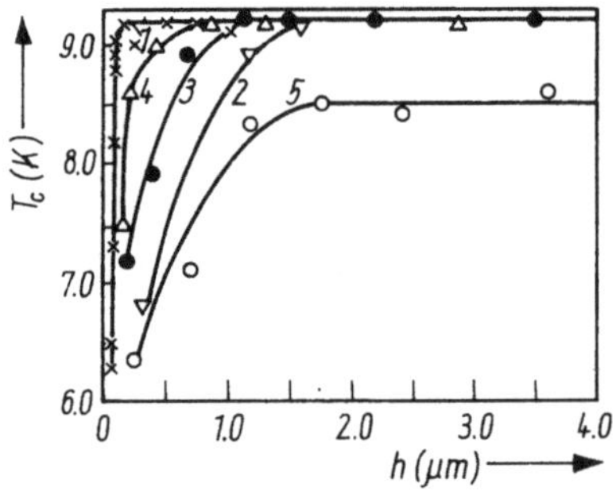

Fig. 1. Transition temperature of Nb films versus film thickness; (1) data of Mayadas et al. /7/ for single crystal films; (2) data of Alekseevskii et al. /5/ for films deposited in vacuum of 10^{-5} to 10^{-6} Torr; (3) and (4) films obtained by electron-beam evaporation in quasi-closed volume, deposition rates on polycor substrates 0.35 and 1.4 μm/min, respectively; (5) films deposited in "open" vacuum

sition temperature T_c, the resistance ratio $R_{300\ K}/R_{10\ K}$, and the critical current density at a temperature of 0.5 K below the transition temperature were registered.

Simultaneously with the measurements of the superconducting properties structural and morphological characteristics of the films were determined by means of X-ray diffraction analysis, electron diffraction examination, and transmission electron microscopy.

Fig. 1 illustrates the transition temperature versus film thickness which allows an estimation of the purity of the superconducting films. Curve 1 presents the data of Mayadas et al. /7/ for single crystal films deposited in ultra-high vacuum on oriented substrate. Curve 2 characterises properties of films deposited in a vacuum of 10^{-5} to 10^{-6} Torr with a system of liquid-nitrogen traps /5/. Curves 3 and 4 are given for films obtained by electron-beam evaporation in a quasi-closed volume. The deposition rates on polycor substrate were 0.35 and 1.4 μm/min, respectively. The substrate temperature was 750 °C and the pressure of residual gases in the main chamber was 10^{-4} to 10^{-5} Torr. Increasing the substrate temperature up to 1200 °C does not influence the superconducting properties of the obtained films. Temperatures lower than 750 °C could not be obtained owing to the radiation from the target.

Fig. 1 demonstrates that the method of quasi-closed volume utilises the gettering properties of the evaporated material to a large extent and provides a purity of the films comparable with that of films deposited in ultra-high vacuum. Curve 5 corresponding to films deposited in "open" vacuum (without quasi-closed volume) is given for comparison. The transition temperature versus film thickness for films deposited on polycor (curve 1) and sitall (curve 2)

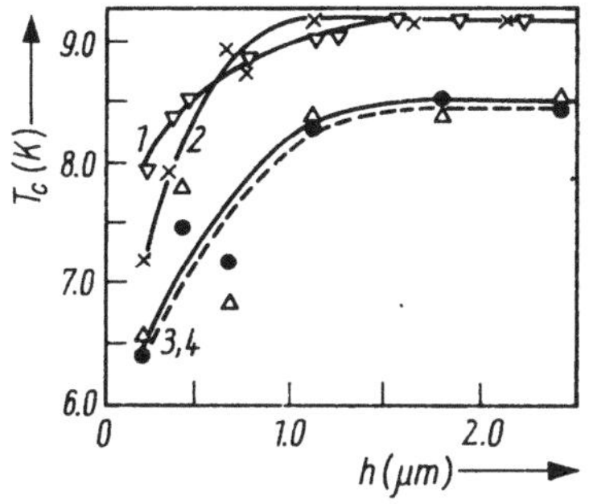

Fig. 2. Transition temperature of Nb films versus film thickness; (1) and (2) films deposited at a rate of 0.35 μm/min in quasi-closed volume on polycor and sitall substrates, respectively; (3) and (4) films deposited under the same conditions, but in "open" vacuum

with a rate of 0.35 μm/min is given in Fig. 2. The substrate temperature was 750 $^{\circ}$C. Curves 3 and 4 are both for films deposited under the same conditions but in "open" vacuum. The comparison of the diagrams leads to the assumption that the influence of the substrate material develops only in high-purity films. It is interesting to note that the characteristic T_c versus film thickness curves correspond to the grain-size versus film thickness dependences. According to X-ray diffraction analysis and electron microscopy data the dimensions of crystallites increase with film thickness. In films of small thickness the grain size is less on sitall substrate. Beginning from a thickness of about 0.6 μm the films on polycor became more fine-grained.

The resistance ratio $R_{300\ K}/R_{10\ K}$ for the films condensed in quasi-closed volume and in "open" vacuum differs considerably, for the upper it varies from 8 to 30 while for the latter it is about 5 to 8.

The critical current density at a temperature of 0.5 K below the transition temperature of films on polycor substrates with thickness 0.5 to 1.5 μm is about 10^5 A/cm^2 and about 3×10^4 A/cm^2 for the films on sitall substrates. The critical current density decreases with the increase of the film thickness and is about 10^4 A/cm^2 for 2 μm films both on polycor and sitall substrates. For films deposited in "open" vacuum the magnitude of the critical current density is approximately one order less.

Thus the superconducting properties of Nb films deposited by electron-beam evaporation in a quasi-closed volume in a vacuum of about 10^{-4} to 10^{-5} Torr without vacuum annealing are considerably better than those of films condensed in "open" vacuum and are close to characteristics of films deposited in ultra-high vacuum.

References

/1/ L.O. MULLEN and D.B. SULLIVAN, J. appl. Phys. 40, 2115 (1969).

/2/ K.A. OSIPOV, A.F. FROLOV, V.P. DMITRIEV, G.F. IVANOVSKII, and YU.N. LOZINSKII, Fiz. Metallov i Metallovedenie 32, 878 (1971).

/3/ J. SOSNIAK and G.W. HULL, J. appl. Phys. 38, 4390 (1967).

/4/ P. FOWLER, J. appl. Phys. 34, No. 12 (1963).

/5/ I.E. ALEKSEEVSKII, I.V. ANOKHINA, V.A. ERMOLOV, M.M. NIKITIN, and G.N. TARASOV, Fiz. i Khim. Obrabotki Mater. No. 6, 140 (1974).

/6/ E. SEIDOWSKI, V.L. LASKA, V.P. LEDOVSKOI, H.M. LEONTEVA, and M. TIEROK, Kristall und Technik 11, 1237 (1976).

/7/ A.F. MAYADAS, R.B. LAIBOWITS, and J.J. CUOMO, J. appl. Phys. 43, 1287 (1972).

(Received May 30, 1979)

phys. stat. sol. (a) 54, K41 (1979)
Subject classification: 2; 22.1.2

Department of Physics, Heriot-Watt University, Currie, Edinburgh

Interpretation of Structural Anomalies in Hydrogenated Amorphous Silicon

By
D. WEAIRE and F. WOOTEN[1)]

Two recent experimental investigations of the structure of hydrogenated amorphous silicon have yielded results which largely conform to the random network model but show certain anomalous features.

Mosseri et al. /1/ using X-ray diffraction, have reported

(i) an asymmetry of the nearest neighbour peak in the Si-Si radial distribution function (RDF);

(ii) a well defined additional peak at approximately 4.9 Å (also reported by Barna et al. /2/).

In this note we propose a simple explanation of both features. We postulate the inclusion of interstitial SiH_4 molecules in the random network. We have arrived at this explanation after a careful study of the random network itself. In common with previous authors /1/, our initial prejudice lay in that direction. We hoped to find features of the random network of bonded silicon atoms which were induced by the presence of hydrogen and which could account for the observed anomalies, without the introduction of additional ingredients in the structural model. Various difficulties militate against such an approach. Firstly, the distortion of the nearest neighbour peak is rather large, and we would have to invent three-centre bonds or something of that sort to account for it. This is not entirely inconceivable (Fisch and Licciardello /3/), but we prefer to look for something simpler. One would still be left to explain the peak at 4.9 Å, and this is, in terms of network parameters, at a rather arbitrary position, corresponding to a dihedral angle of 35° to 45°/1/. Furthermore, it seems too sharp to occur at this distance in the random network, since angular distortions progressively wash out all structure as distance increases. Admittedly, a much less pronounced feature around 4.9 Å has been seen in unhydrogenated samples, and in some random network models, but it is so small that it could well be dismissed as insignificant and coincidental.

---

1) Permanent address: Department of Applied Science, University of California, Davis, California, USA.

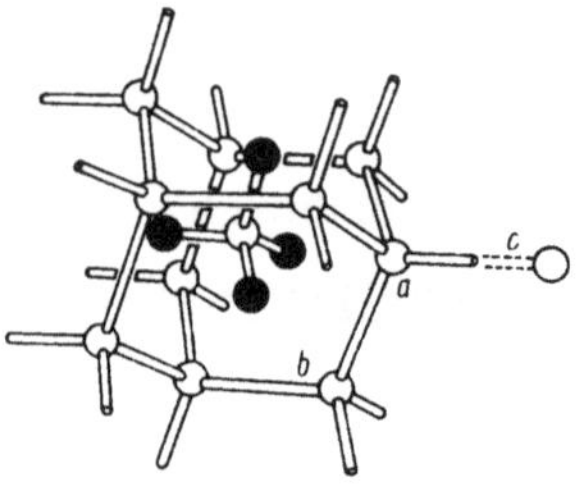

Fig. 1. Proposed interstitial configuration for inclusion of an SiH_4 molecule in a random network. The various near neighbour distances which are discussed in the text relate to atoms a, b, and c

It therefore seems necessary to add some extra ingredient to the random network model.

We have considered a number of configurations, and the most satisfactory of them has proved to be the simplest. In the diamond lattice it is possible to fit an SiH_4 molecule into a tetrahedral interstitial position, as shown in Fig. 1. Such a unit, consisting of an SiH_4 molecule and a cage of ten silicon atoms, is easily imbedded in an otherwise disordered Si network, and it seems energetically acceptable, in terms of the distance between non-bonded atoms.

If we assume for the moment that the Si cage is undistorted, the central Si atom has four non-bonded nearest-neighbour Si atoms, distant from it by precisely the Si-Si bond length (2.4 Å). There are a further six Si atoms at 2.8 Å. An overlapping subsidiary peak in the RDF at this distance, with both peaks thermally broadened, would produce an asymmetry similar to that which has been observed.

Directly in line with each of the four neighbours at 2.4 Å is another at 4.8 Å, if they are tetrahedrally bonded (assuming that the bond in question is not taken up by a hydrogen atom). Because of the straight-line relationship between the central atom and these two neighbours, even a moderate angular distortion, due to incorporation in a random network, has only a small effect on the 4.8 Å distance. The observed sharp peak is at a slightly larger distance than this. This may be attributed to a small expansion of the cage due to the inclusion of the interstitial. This, in turn, has a small effect on the nearest neighbour peak, further enhancing the asymmetry already discussed.

Similar interatomic separations are generated if SiH_4 is placed at the centre of a wurtzite-like cage of twelve silicon atoms, so a mixture of both types could be considered.

We have yet to make a detailed quantitative analysis of the RDF in terms of these ideas, which may entail rather elaborate model-building exercises. However, the following preliminary conclusions may be drawn.

From the size of the observed anomalies, we would estimate that roughly 3% of the Si is attributable to SiH_4 molecules in the sample of Mosseri et al. /1/. Our interpretation also implies that a significant fraction (of the order of 20%) of the hydrogen is incorporated in this form. Infrared absorption studies should be capable of detecting this, although the frequencies of the modes produced will not be those of the isolated molecule.

We would admit that the proposed explanation is speculative, but aspiring model-builders will find it not at all easy to give such a unified interpretation of the features in question. The model here proposed therefore seems worthy of experimental test. In particular it will be interesting to scrutinise existing infrared spectra with this in mind, but a parallel study by infrared and diffraction methods, on the same samples, is what is really needed.

This research is supported by the Science Research Council. We wish to thank R. Mosseri for sending us preprints of his work and M. Tricker and P. John for helpful chemical insights.

References

/1/ R. MOSSERI, C. SELLA, and J. DIXMIER, phys. stat. sol. (a) 52, 475 (1979).

/2/ A. BARNA, P.B. BARNA, G. RADNOCZI, L. TOTH, and P. THOMAS, phys. stat. sol. (a) 41, 81 (1977).

/3/ R. FISCH and D.C. LICCIARDELLO, Phys. Rev. Letters 41, 889 (1978).

(Received June 5, 1979)

phys. stat. sol. (a) 54, K45 (1979)
Subject classification: 10

Laboratory for Metallurgy, Delft University of Technology[1)]

The Probability of Visiting a New Site in a Random Walk on a B.C.C. Lattice

By
C.M. VAN BAAL, R. FASTENAU, and P. PENNING

In theoretical investigations of the kinetics of particles moving on a regular (crystalline) lattice that contains randomly distributed, but immobile trapping sites, an important part is played by the quantity Q_n, the probability that a randomly moving particle at its n-th jump, will reach a site that it has not visited before. This of course, stems from the fact that a site visited before has already been found to be "safe", whereas a new site might be "dangerous". For a few simple cases the behaviour of Q_n for large n has been derived and a summary of these results is given in /1/. Now in a previous paper /2/, treating more complicated cases a.o. by simulation experiments, we mentioned that, in testing our computer program with a random walk on a b.c.c. lattice, we found a discrepancy between the simulation results and the theoretical expression for Q_n, derived by Montroll and Weiss /1/. It is the aim of this short note to examine this case more closely.

Let $P_n(\vec{k})$ be the probability that a random walker on a lattice, starting at the origin, will reach the lattice point $\vec{k}$ in n steps and let $Q_n(\vec{k})$ be the probability that it does so for the first time. We then have

$$Q_n(\vec{k}) = P_n(\vec{k}) - \sum_{n=1}^{n} Q_{n-m}(\vec{k})\, P_m \tag{1}$$

in which $P_m \equiv P_m(0)$. Summing both sides of (1) over all $\vec{k}$ gives

$$Q_m = 1 - \sum_{m=1}^{n} Q_{n-m}\, P_m \tag{2}$$

in which Q_n is the probability that the random walker, after n jumps, is at a lattice point never visited before ($Q_0 = 1$).

Now for the b.c.c. lattice $P_{2n+1} = 0$ and so $Q_{2n+1} = Q_{2n}$. Thus, instead of (2), we have

---

1) Rotterdamseweg 137, 2628 AL Delft, The Netherlands.

$$Q_{2n} = 1 - \sum_{m=1}^{n} Q_{2n-2m} P_{2m} . \tag{3}$$

Also for the b.c.c. lattice we have that P_{2m} is equal to the third power of the analogous quantity for the linear chain

$$P_{2m} = \left\{ \frac{(2m)!}{2^{2m}(m!)^2} \right\}^3 . \tag{4}$$

(We might remark in passing that for the linear chain $Q_{2m} = P_{2m}$.) Developing the factorials with the help of Stirling's formula, we find

$$P_{2m} = \sum_{s=1}^{\infty} p_{2s+1} m^{-s-1/2} = (\pi m)^{-3/2} \Big\{ 1 - \frac{3}{8} m^{-1} + \frac{9}{128} m^{-2} + $$
$$+ 7.2^{-10} m^{-3} + 165.2^{-15} m^{-4} - 957.2^{-18} m^{-5} + 7325.2^{-22} m^{-6} +$$
$$+ O(m^{-7}) \Big\} . \tag{5}$$

We assume now, in an obvious extension of the work of Montroll and Weiss /1/, that Q_{2n} can be developed in a power series in $n^{-1/2}$

$$Q_{2n} = \sum_{r=0}^{\infty} q_r n^{-r/2} . \tag{6}$$

Inserting (5) and (6) into (3) gives us

$$\sum_{r=0}^{\infty} q_r n^{-r/2} = 1 - \sum_{r=0}^{\infty} q_r \left[\sum_{s=1}^{\infty} p_{2s+1} \left\{ \sum_{m=1}^{n-1} (n-m)^{-r/2} m^{-s-1/2} \right\} \right]$$
$$= 1 - \sum_{r=0}^{\infty} q_r \left\{ \sum_{s=1}^{\infty} p_{2s+1} G_{r,2s+1}(n) \right\} \tag{7}$$

with $G_{r,s}(n) = G_{s,r}(n) = \sum_{n=1}^{n-1} (n-m)^{-r/2} m^{-s/2}$.

For $G_{r,s}$ we find

a) for $r = 0$, from /3/:

$$G_{o,s} = \zeta(s/2) - n^{-(1/2)s+1} \left(\frac{2}{s-2} - \frac{1}{2} n^{-1} + \frac{s}{24} n^{-2} - \ldots \right) , \tag{8}$$

where $\zeta(x)$ is the Riemann zeta-function;

b) for odd s and even $r(>0)$, with the help of the Euler-Mc Laurin sum formula /4/ and a table of integrals /5/:

$$G_{r,s} = a_{r,s}\, n^{-(1/2)(r+s-2)} \ln(n) + b_{r,s}\, n^{-t/2} + c_{r,s}\, n^{-t/2+1} + \ldots \qquad (9)$$

in which $t = \min(r, s)$;

c) for r and s both odd and positive using the same technique:

$$G_{r,s} = a_{r,s}\, n^{-t/2} + b_{r,s}\, n^{-t/2+1} + \ldots \qquad (10)$$

with the same definition for t.

Because of the logarithmic term in (9), we drop all terms with even values of r in (6), except $r = 0$. By explicitly working out the first two terms in (8) and (10), we find from (7)

$$q_o = (1 + A)^{-1}, \quad q_1 = 2\pi^{-3/2} q_o^2 \qquad (11)$$

with $A = \sum_{s=1}^{\infty} p_{2s+1}\, \zeta(2s+1)/2$. Using 6 terms in A we find

$$q_o = 0.7174, \qquad q_1 = 0.1849 \; .$$

We have also calculated Q_{2n} directly from (4) and (4) for $n = 1$ to $n = 1000$ and have determined, by a least squares procedure, the best values of the coefficients q_r in (6) to fit Q_{2n} from $n = 15$ to $n = 1000$.

This resulted in:

$$q_o = +0.71777001105\,, \qquad q_1 = +0.1850443397\,, \qquad q_3 = -0.03278987\,,$$
$$q_5 = +0.0044677\,, \qquad q_7 = +0.00131\; .$$

The value for Q_2 found with the help of these coefficients is 0.8758, whereas the exact value is 0.875. The value found for q_o is in excellent agreement with the value $q = 4\pi^3 \{\Gamma(1/4)\}^{-4}$ given by Montroll and Weiss /1/, whereas that of q_1 agrees as well as can be with (11). Both also agree with the values found from our simulation experiments /2/.

The value of q_1, however, is 4 times that given by Montroll and Weiss /1/, but so far we have not been able to trace the source of this discrepancy.

References

/1/ E.W. MONTROLL and G.H. WEISS, J. Math. Phys. 6, 167 (1965).
/2/ R.H.J. FASTENAU, C.M. VAN BAAL, P. PENNING, and A. VAN VEEN, phys. stat. sol. (a) 52, 577 (1979).

/3/ JAHNKE-EMDE-LÖSCH, Tables of Higher Functions, B. G. Teubner, Stuttgart 1960 (p. 38).

/4/ G. A. KORN and TH. M. KORN, Mathematical Handbook for Scientists and Engineers, Mc Graw-Hill, New York 1961 (p. 122).

/5/ J. J. TUMA, Engineering Mathematics Handbook, Mc Graw-Hill, New York 1970 (p. 243).

(Received May 10, 1979)

phys. stat. sol. (a) 54, K49 (1979)
Subject classification: 11 and 14.3.4; 22.1.2

I. V. Kurchatov Institute of Atomic Energy, Moscow

Electrical Properties of Silicon, Irradiated by Neutrons and γ-Quanta, in Strong Electric Fields

By

V.A. ZHUKOV, V.V. SAVYALOV, B.V. KOBA, V.L. LITVINOV, V.P. SADIKOV, and N.A. UKHIN

The purpose of this work was to study the electric field influence on characteristics of irradiated n-Si. The Si crystals were prepared both by the pulling technique and the floating zone method; their initial resistivity was about 10 and 70 Ωcm at room temperature. The samples were cut from Si slabs perpendicular to the [111] direction. They had no special orientation, but in neutron and γ- experiments it was the same. The samples had the form of rectangular bars (with dimensions $\approx$ (0.3 to 1)x4x10 mm^3) or bridges with massive ends which are provided with electrodes on both sides from which carriers are injected. The dimensions of the massive ends are chosen so that their resistance is a small fraction of the total resistance of the samples (less or equal to 1:22). Also the ends are made sufficiently long so that within the duration of the pulse the injected carriers cannot travel this length. Besides, the resistance modulation of the samples in the initial state was small and decreases under irradiation. All the contacts to the samples were made by ion implantation or alloying from Au+Sb Alloy.

The following measurements were performed:

1. Resistivity, Hall effect, and magnetoresistivity at direct current for weak electric fields and a constant $\approx$ 4000 Oe magnetic field without and with white light illumination at RT and 77 K.

2. Current-voltage I-U characterististics in the field strength range 0 to 1.0 kV/cm at RT and 77 K without and with white light illumination and in constant $\approx$ 4000 Oe magnetic field without illumination.

The dc measurements were performed by the usual technique. The illumination was effected by an incandencent lamp at constant current.

The I-U characteristics were measured with square pulses of 0.3 and 1.0 s duration and $\approx$ 40 Hz repetition rate at two current directions. The current pulses were produced by a generator provided with a thyratron and a discharge line. Increase in pulse duration or repetition rate did not influence the results. The current and voltage amplitudes were measured with digital pulse voltmeters. The relative error of measurements did not exceed $\pm 2.5\%$. Constancy of the form of voltage and current pulses was controlled on an oscillograph. The samples were irradiated by reactor neutrons at room temperature and by γ-quanta of Co^{60} at $\approx 60\ ^{o}C$.

Resistivity, carrier concentration, and mobility were calculated using currently accepted formulae. The carrier mobility as a function of the electric field has been calculated assuming that the carrier concentration is field independent. According to /1/ the carrier concentration for non-irradiated semiconductors (Ge and Si) becomes field dependent for $E \geqq 10^4$ V/cm which considerably exceeds the field strengths used in the present investigations. Quantitative estimations of the Poole-Frenkel effect, avalanche multiplication, and tunneling show that the carrier concentration can be changed at field strengths of 10^5 V/cm.

The determination of the carrier concentration as a function of the electric field for irradiated samples is far from being simple. Usually the concentration is measured by the Hall effect. But, firstly, it is not easy to measure the Hall voltage for irradiated semiconductors by the pulse method. Secondly, for the calculation of the carrier concentration it is necessary to know the field dependence of the Hall factor ($r_H = \langle \tau^2 \rangle / \langle \tau \rangle^2$) whose exact form for irradiated Si is not known. In non-irradiated Ge, in accordance with /2/ r_H begins to increase at field strengths $\geqq 500$ V/cm.

Instead, magnetoresistance measurements are simpler, but the resistance change for Si and Ge samples of "prolonged" form is small in magnetic fields of the order of some 1000 Oe. Moreover, we must know the dependence of the magnetoresistivity factor $r_m = (\langle \tau^3 \rangle / \langle \tau \rangle^3)^{1/2}$ on the electric field which is known for irradiated Si.

We estimated the field dependence of the carrier concentration by using I-U characteristics in magnetic field of samples irradiated by neutrons and γ -quanta. The formula for determining the concentration was derived from the results of /3/ assuming that r_m does not depend on the electric field. According to these calculations the carrier concentration of the neutron irradiated samples is constant up to a certain field strength, and then it somewhat increases for higher field strengths. The field independent concentration range decreases with increasing neutron flux; when the resistivity increases by 3 to 5 times it is greater than ≈ 500 V/cm. The change of I-U characteristics of γ-irradiated samples in the same magnetic field was small and remained within the experimental error.

In addition we estimated the field dependence of carrier concentration from the illumination influence on I-U characteristics; just as in the experiment by Ioffe /4/. For small current changes (at fixed voltage) due to illumination, we can write

$$\frac{i^*_{light} - i_{dark}}{i_{dark}} = \frac{\Delta i^*}{i_{dark}} = \frac{\Delta \mu^*}{\mu_{dark}} + \frac{\Delta n^*}{n_{dark}} .$$

For neutron irradiated samples this value decreased with the flux, but was constant or increased slightly with the electric field. It is difficult to assume that mobility change and concentration change differ in sign and their absolute values are equal for the different fluences. Therefore it is reasonable to assume that the concentration was independent of the electric field strength. The change of the I-U characteristics due to illumination in γ-irradiated samples was small and usually remained within the experimental error.

Taking into account all these arguments we assumed that the concentration was independent of the electric field strength and the fluences used in our experiments.

The main results are the following:

The I-U characteristics of unirradiated samples at RT obey Ohm's law and at 77 K they are non-linear. This is also valid for irradiated samples (Fig. 1). The I-U characteristic changes for samples after neutron and γ-irradiation are similar but not identical. All these statements are valid

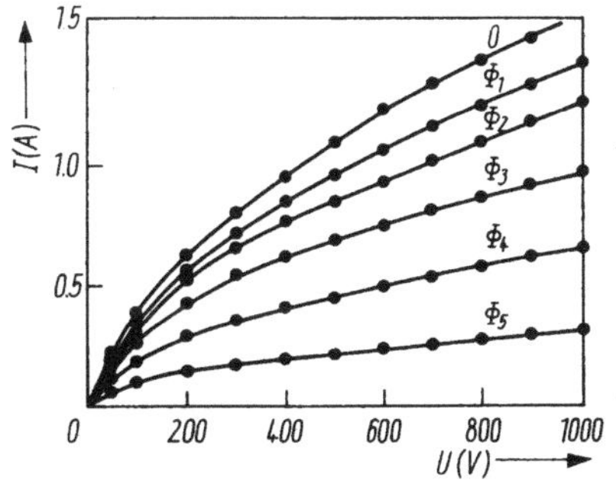

Fig. 1. Change of I-U characteristics of a Si:P sample with ϱ_o = 70 Ωcm irradiated by neutrons of different fluence. Temperature during the measurement was 77 K

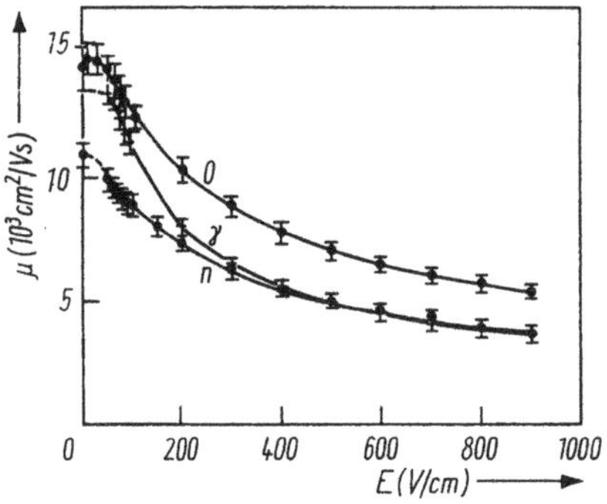

Fig. 2. Change of carrier mobility after neutron and γ-irradiation of Si:P with ϱ_0 = 70 Ωcm for ϱ/ϱ_0 = 3

for both resistivities and sample preparation technologies. The neutron irradiated samples were very sensitive to illumination. The absolute current difference increased with fluence and electric field strength. In γ-irradiated samples the current difference was small.

The carrier mobility change at small neutron and γ-quanta fluxes is qualitatively different. The change of mobility for neutron irradiated samples decreases with increasing field strength, but for γ-irradiated samples it increases (Fig. 2). The mobility change of samples irradiated by greater γ-fluxes becomes similar to the mobility change of neutron irradiated samples.

The interpretation of these results is supposed to be given in a subsequent paper.

References

/1/ V.I. DENIS and I.K. POZHELA, Hot Electrons, Vilnius 1971.

/2/ D. CHATTOPANHYAY and B.R. NAG, Phys. Rev. B 4, 1220 (1971).

/3/ I.F. SOKOLOV and B.G. STEPANOV, Preprint IRE No. 6, Moscow 1973 (p. 136).

/4/ V.I. FISTUL, Introduction to Semiconductor Physics, Moscow 1975.

(Received January 25, 1979)

phys. stat. sol. (a) 54, K53 (1979)

Subject classification: 14.2; 11; 21

Institute of Metal Physics, Academy of Sciences of the USSR, Ural Scientific Centre, Sverdlovsk

Electrical Resistivity of V_3Si and Nb_3Sn under Neutron Radiation

By

A.E. KARKIN, V.E. ARKHIPOV, V.A. MARCHENKO, and B.N. GOSHCHITSKII

The compounds with A-15 structure possessing high transition temperatures into the superconducting state T_c ($\approx$17 K for V_3Si and $\approx$18 K for Nb_3Sn) are of great interest since they display on anomalous temperature behaviour of some physical properties (electrical resistivity as well /1/) in the normal state. The connection of these anomalies with T_c was demonstrated recently in /2, 3/ for V_3Si and in /4, 5/ for Nb_3Sn. In this paper we present results and an interpretation of the temperature dependence of resistivity for polycrystalline V_3Si and Nb_3Sn specimens irradiated by fast neutrons with $4x10^{20}$ cm^{-2} fluence and after subsequent isochronal annealing in the temperature interval 200 to 900 ^oC. The measurement and the irradiation procedures were described earlier /5/.

For the first time the temperature dependence of resistivity of Nb_3Sn was described in /6/ by the empirical law

$$\varrho(T) = \varrho_o + bT + d\exp(-T_o/T) \qquad (1)$$

which rather well depicts also the data for V_3Si /7/, however, the physical meaning of this law is not clear yet.

To explain the anomalous dependences of the susceptibility, resistivity, etc. it has been assumed:

(a) the existence of a narrow peak in the density of states of d-electrons near the Fermi surface (ε_f = 100 K) /8/,

(b) some peculiarities of the Fermi surface /9/,

(c) an unusual temperature dependence of the phonon density of states /10/,

(d) a short electron mean free path /11, 12/.

In a recent paper /13/ the temperature dependence of the electrical resistivity of Nb_3Sn was qualitatively described assuming s-d-hybrid scattering which appears when the narrow d-band comes near (ε_f = 10^3 to 10^4 K) to the

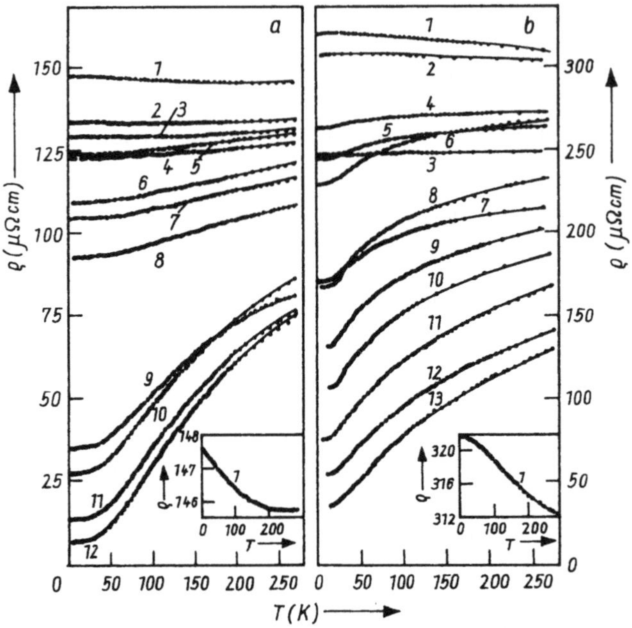

Fig. 1. Temperature dependence of ρ for a) V_3Si and b) Nb_3Sn irradiated by fast neutrons with $4x10^{20}$ cm^{-2} fluence and after isochronal annealing (1 to 13); ● present paper, —fitted by (2). The designations by numbers are from Table 1. Curves (1) in the insets are given in a different scale

Fermi surface. The account of the hybrid scattering leads to a logarithmic deviation from the Bloch-Grüneisen law. In the case of tight hybridization the expression for the resistivity looks as follows /13/:

$$\varrho(T) = \frac{\varrho_0 + a\,\lambda\,(T/\Theta)}{1 + b\,\ln(1 + T^2/\Gamma^2)}\,, \tag{2}$$

where Γ is the damping in the s-d-s-hybrid approach; $\lambda(T/\Theta)$ is the Bloch-Grüneisen function /14/; $\varrho_0 = \varrho(T = 0)$; $\lambda(1) = 1$; $\lambda \sim T^5$ at $T \ll \Theta$; $\lambda \sim T$ at $T > \Theta$; Θ Debye temperature; a and b are expressed through the electronic parameters of the theory /13/. Here τ^{-1} from /13/ is presented as $\varrho_0 + a\,\lambda(T/\Theta)$.

Fig. 1 shows by points the temperature dependence of resistivities of V_3Si and Nb_3Sn in different defective states. The solid curves show the results of fitting according to (2). The best agreement was obtained with Θ = 250 K and 100 K for V_3Si and Nb_3Sn, respectively, and the parameters a, b, and Γ presented in Table 1. The experimental data for V_3Si (curves 6 to 12) can be described also by the expression (1) where the fitting parameter T_c agrees with that presented in /3/.

It is necessary to note that it is impossible to explain in the framework of the model (c) to (d) the observed negative thermal resistivity coefficient for V_3Si and Nb_3Sn in the highly defective state (Fig. 1), while the fitting according (2) gives good agreement (see insets in Fig. 1) either for V_3Si or for Nb_3Sn. In the weakly defective state where the phonon scattering prevails over the impurity one in the low temperature region the expression (2) describes the experimental results worse since the temperature dependence is determined here mainly by $\lambda(T/\Theta) \sim T^5$, while for V_3Si and Nb_3Sn the T^2-law is observed /14/. It should be noted that the resistivity of almost all transition metals which do not possess any anomalies at high temperatures are described by the T^2-law /15/.

The obtained results for the values of the Debye temperatures Θ = 250 K for V_3Si and 100 K for Nb_3Sn turn out to be approximately two times less than those obtained from the heat capacity data /16/, 500 and 310 K, although the acoustic measurements give Θ = 300 K and 130 K, respectively. The structure of the formula (2) is such that the inflexion point ($T_o/2$ in expression (1)) is determined mainly by the magnitude Θ and is equal to 0.4 Θ, while the dependence upon Γ and upon the other parameters is weaker. Thus the characteristic temperature of $\varrho(T)$ anomalies is determined mainly by the phonon subsystem but not by the electron one, as, for example, in /8/. From here the correlation (discovered in /7, 17/) between T_o and the low frequency peaks in the phonon density of states corresponding to acoustic phonons near the zone boundary seems to be clear. Hence the acoustic phonons rather than the optical ones participate in the electron scattering.

Table 1 represents the values of the fitting parameters ϱ_o, a, b, and Γ for V_3Si and Nb_3Sn. It is necessary to note that with increasing ϱ_o a decrease of T_c and a and b, determined mainly by the parameters of the hybrid interaction g /13/ takes place.

Table 1

Values of fitting parameters for V_3Si (upper part) and Nb_3Sn (lower part)

N	T_{ann} (°C)	T_c (K)	ϱ_o (μΩcm)	a (μΩcm)	b	Γ (K)
1	-	1.55 ± 0.02	147.3 ± 3	6.4 ± 1	0.013 ± 0.01	70 ± 10
2	200	1.90	133.5	8.7	0.010	60
3	250	1.90	128.4	13.4	0.013	60
4	300	2.30	123.6	13.2	0.014	60
5	350	2.90	124.3	32.3	0.034	80
6	400	3.82	109.5	30.6	0.017	80
7	430	4.52	104.5 ± 1	32.1 ± 3	0.035 ± 0.02	400 ± 10(
8	470	5.45	92.8	46.8	0.098	375
9	650	13.30	35.3	150	0.617	300
10	700	14.20	27.75	164	0.704	350
11	750	15.15	13.96	185	0.911	385
12	800	15.85	6.94 ± 0.1	215 ± 10	0.961 ± 0.1	355 ± 2(
1	-	2.17 ± 0.02	322 ± 30	0.7 ± 0.2	0.014 ± 0.01	100 ± 1(
2	300	1.71	309.6	3.1	0.017	90
3	350	1.66	247	2.5	0.013	150
4	400	1.40	264	14.5	0.103	200
5	450	1.38	245	18.8	0.118	200
6	500	3.65	230	36.9	0.240	210
7	550	6.95	170	59.1	0.359	150
8	600	9.75	167	65.0	0.397	160
9	650	11.35	132	61.8	0.402	180
10	700	12.82	106 ± 10	67.0 ± 10	0.437 ± 0.05	150 ± 2(
11	750	13.99	76.2	59.5	0.351	170
12	800	15.10	54.0	48.0	0.401	240
13	850	16.79	34.0	48.0	0.401	260

N means the curve number from Fig. 1; T_{ann} is the annealing temperature, T_c the transition temperature into the superconducting state, ϱ_o the residual resistivity, a, b, and Γ are the fitting parameters in (2) for V_3Si and Nb_3Sn after fast neutron irradiation with $4x10^{20}$ cm^{-2} fluence. For T_c, ϱ_o the measurement errors are shown and for a, b, and Γ the uncertainty of the fitting.

Thus it is ascertained that the Debye temperature describing the resistivity turns out to be two times less than the value determined from the heat capacity data and it agrees with the results of acoustic measurements. The anomaly of the resistivity is apparently a consequence of the s-d-hybrid interaction, although the characteristic temperature is determined by the Debye temperature. A successful description of the resistivity of V_3Si and Nb_3Sn with different defective states gives evidence, in our opinion, to that $\varrho(T)$ and T_c are determined by the parameter of the hybrid interaction.

References

/1/ L.R. TESTARDI, M. VEGER, and I. GOLDBERG, Superconducting Compounds with A-15 Structure, Mir, Moscow 1977 (in Russian).

/2/ A. GUHA, M.P. SARASHIK, F.W. SMITH, and L.R. TESTARDI, Phys. Rev. B 18, 9 (1978).
R. VISWANATHAN and R. CATON, Phys. Rev. B 18, 15 (1978).
D.E. COX and J.A. JARVIN, Phys. Rev. B 18, 22 (1978).

/3/ B.S. BROWN, J. nuclear Mater. 72, 76 (1978).

/4/ J. Physique, Colloque C 6, Suppl. 1, 39, (1978).

/5/ A.E. KARKIN, V.E. ARKHIPOV, B.N. GOSHCHITSKII, E.P. ROMANOV, and S.K. SIDOROV, phys. stat. sol. (a) 38, 433 (1976).
A.E. KARKIN, B.N. GOSHCHITSKII, V.E. ARKHIPOV, E.E. VALIEV, and S.K. SIDOROV, phys. stat. sol. (a) 46, K87 (1978).

/6/ D.W. WOODWARD and C.D. CODY, Phys. Rev. 136, A166 (1964).

/7/ M. MILEWITS, S. WILLIAMSON, and H. TAUB, Phys. Rev. B 13, 5199 (1976).

/8/ R.W. COHEN, B.D. CODY, and J.J. HALLORAN, Phys. Rev. Letters 19, 840 (1967).

/9/ L.P. GORKOV, Zh. eksper. teor. Fiz. 65, 1658 (1973).
L.P. GORKOV and O.N. DOROKHOV, J. low-Temp. Phys. 22, 1 (1976).

/10/ P.B. ALLEN, J.C.R. NUI, W.E. PICHETT, C.M. VARMA, and Z. FISK, Solid State Commun. 18, 1157 (1976).

/11/ H. WIESMANN, M. GURWITCH, H. LUTZ, A. GHOSH, B. SCHWARTZ, MYRON STRONGIN, P.B. ALLEN, and J.W. HALLEY, Phys. Rev. Letters 38, 782 (1977).

/12/ G. GRIMVALL, Phys. kondens. Materie 11, 279 (1970).

/13/ A.M. ATOYAN, A.F. BARABANOV, and A.A. MAKSIMOV, Zh. eksper. teor. Fiz. 74, 2220 (1978).

/14/ V.A. MARCHENKO, Fiz. tverd. Tela 15, 1893 (1973).

/15/ F.J. BLATT, Physics of Electronic Conduction in Solids, Mc Graw-Hill Book Co., 1968.

/16/ G.S. KNAPP, S.D. BADER, H.V. CULBERT, F.J. FRADIN, and T.E. KLIPPERT, Phys. Rev. B 11, 4331 (1975).
G.S. KNAPP, S.D. BADER, and Z. FISK, Phys. Rev. B 13, 3783 (1976).

/17/ A.I. GOLOVASHKIN, Zh. eksper. teor. Fiz., Pisma 26, 81 (1977).

(Received May 3, 1979)

phys. stat. sol. (a) 54, K59 (1979)
Subject classification: 10 and 12

Sektion Physik der Pädagogischen Hochschule Dresden

A Simple Model for an Inhomogeneous Dislocation Structure

By
H.-J. BURMEISTER and H. HERMANN[1)]

A simple model is used to explain inhomogeneous dislocation structures which develop during unidirectional or cyclic plastic deformation in metals. We investigated the stability of dislocation multipoles which doubtless give important contributions to the formation of inhomogeneous structures (e.g. /1/ in the case of unidirectional and /2/ in the case of cyclic plastic deformation). Other mechanisms (e.g. cross slip) are not taken into account.

Model; basic equations The crucial condition for dipole formation consists in the existence of dislocations in parallel slip planes. Therefore, as a model we consider a body infinitely extended containing straight edge dislocations in equidistant slip planes (slip planes parallel to the zx plane, slip plane distance h, dislocation lines in z direction). Each slip plane contains dislocations of only one sign. The dislocation signs are different for neighbouring planes. The stress in the l-th slip plane is

$$\sigma_l(x, t) = \sigma^a + \sum_j \int_{-\infty}^{+\infty} dx' \varrho_j(x', t) K(x - x', h|l - j|) \tag{1}$$

(σ^a external shear stress, $\varrho_j(x, t)$ dislocation density in the j-th slip plane), where

$$K(x, y) = \frac{\mu b}{2\pi(1 - \nu)} \frac{d}{dy} \frac{x y}{x^2 + y^2} \tag{1a}$$

(μ shear modulus, ν Poisson's ratio, b absolute value of Burgers vector). The velocity $\vec{v}_l(x, t) = v_l(x, t)\vec{e}_x$ of the dislocations and the stress $\sigma_l(x, t)$ are connected by the empirical relation

$$v_l(x, t) = (-1)^{l-1} A|\sigma_l|^m \frac{\sigma_l}{|\sigma_l|}, \qquad \varrho_l \geqq 0 \quad \text{if } l \text{ odd} \tag{2}$$

(in the following we take m = 1). Finally the local dislocation conservation is

1) Now at Zentralinstitut für Festkörperphysik und Werkstofforschung der Akademie der Wissenschaften der DDR, Dresden.

given by

$$\dot{\varrho}_l(x, t) + \frac{\partial}{\partial x}\left[\varrho_l(x, t) v_l(x, t)\right] = 0 . \tag{3}$$

Equations (1 to 3) are the basic equations of the problem. Eliminating σ_l and v_l we obtain a system of nonlinear singular integro-differential equations:

$$\dot{\varrho}_l(x, t) + A(-1)^{l-1} \frac{\partial}{\partial x}\left[\varrho_l(x, t)\sigma^a + \varrho_l(x, t) \sum_j \int_{-\infty}^{+\infty} dx' \varrho_j(x', t) \times \right.$$
$$\left. \times\ K(x - x', h\,|l - j|)\right] = 0 . \tag{4}$$

If the subsidiary conditions were known $\varrho_l(x, t)$ could be obtained from (4).

<u>Stability</u> In the usual way (e.g. /3/) we analyze the stability of a homogeneous dislocation distribution with respect to periodic spatial perturbations (Λ length of periodicity) within a set of N sequent slip planes. We suppose

$$\varrho_l(x, t) = (-1)^{l-1}\, b\, \varrho(x, t) , \qquad \varrho(x, t) \geqq 0 \tag{5}$$

and

$$\varrho(x, t) = \varrho^0 + \delta\varrho(x, t) , \qquad \delta\varrho(x, t) = \delta\varrho^0 \exp(i(kx - \omega t)) ; \tag{6}$$

ϱ^0 homogeneous distribution, $\delta\varrho$ perturbation, $k = 2\pi/\Lambda$ wave number, ω frequency. Inserting (5) and (6) into (4) and linearizing with respect to the perturbation the dispersion relation

$$\omega = A\,k \left\{(-1)^{l-1}\sigma^a + i\varrho^0 \frac{\mu b^2}{2(1-\nu)} \frac{d}{d(kh)}\left[kh\,\varphi_l(kh; N)\right]\right\} \tag{7}$$

follows with

$$\varphi_l(kh; N) = \frac{1}{1+w}\left[-1 + w + (-w)^l + (-w)^{N-l+1}\right] , \quad w = e^{-kh} . \tag{8}$$

The condition for perturbation amplification is $\mathrm{Im}\,\omega > 0$, i.e.

$$\frac{d}{d(kh)}\left[kh\,\varphi_l(kh; N)\right] > 0 . \tag{9}$$

<u>Results</u> The condition (9) has been analyzed using a computer. If N = 3 and N $\geqq$ 5 (N finite) there are always slip planes showing perturbations increasing in time. In the Fig. 1 a typical situation is shown (N = 33). There are

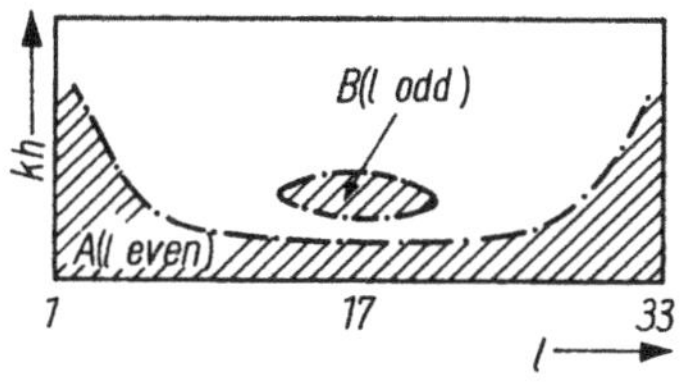

Fig. 1. Result of analyzing condition (9) for N = 33, kh = $2\pi h/\Lambda$, h slip plane distance, Λ length of perturbation, l slip plane number. For points (l, hk) lying within domains A and B, respectively, the homogeneous dislocation distribution is unstable

two domains of instability A and B corresponding to different values Λ_A and Λ_B ($< \Lambda_A$), respectively. In structure B only a few slip planes are affected. But Im ω is much greater for A than for B. Therefore the model shows the tendency of a structure transition A $\rightarrow$ B.

The main features of the model are confirmed by experimental experience concerning cyclic plastic deformation. If h is identified with the greatest distance possible in a stable dipole $\Lambda_A \approx 1\ \mu$m is obtained within our model ($\sigma^a/\mu = 10^{-3}$, N = 33, b = $3\text{x}10^{-10}$ m, $\nu = 0.3$). This is the order of magnitude observed by transmission electron microscopy (TEM) /4/. The B-structure of the model is connected with the early stage of persistent slip band formation observed by TEM. In the early stage they appear as narrow channels within the A-structure ($\Lambda_B \approx \Lambda_A/2$) /4/.

References

/1/ S. IKENO, phys. stat. sol. (a) 36, 317 (1976).
/2/ A. T. WINTER, Phil. Mag. A37, 457 (1978).
/3/ V.I. VLADIMIROV, and B. PEGEL, phys. stat. sol. (b) 56, K105 (1973).
/4/ K. MECKE, phys. stat. sol. (a) 25, K93 (1974); Thesis, Pädagogische Hochschule Dresden, 1972.

(Received May 2, 1979)

phys. stat. sol. (a) 54, K63 (1979)
Subject classification: 14.4; 22

Department of Electrical Engineering,
Indian Institute of Technology, Bombay[1)]

Dielectric Properties of Tantalum Oxynitride Films

By
P.K. REDDY and S.R. JAWALEKAR

Recently tantalum nitride (TaN) has attracted considerable interest for thin film capacitors. TaN capacitors are preferred to β-tantalum capacitors for hybrid integrated circuits as they have low dielectric losses (tan δ), a low temperature coefficient of the capacitance (TCC), and they are more stable /1 to 3/. However, the optimum fabrication conditions of TaN capacitors are still under investigation. The present report contains some studies on TaN capacitors using aluminium as counterelectrode instead of conventional NiCr-Au alloy /4/. The results are explained with existing theories.

TaN films (0.3 μm) are deposited on chemically and ultrasonically cleaned glass and alumina substrates at a rate of 0.012 μm/min. Films are deposited by reactive sputtering in a mixed atmosphere of ultra-high pure argon and nitrogen. TaN films of different N_2 concentration are obtained by varying the nitrogen pressure from 0.2 to 1% of the argon pressure which is maintained at 10^{-2} Torr during sputtering. Sputtered films are baked at 200 °C for 3 h. Capacitor patterns are defined by using the standard photolithographic process and the films are anodized to 150 V in 0.01% citric acid solution at a current density of 1 mA/cm^2. To complete the fabrication of the capacitor aluminium (99.99%) is deposited (0.08 to 0.1 μm thick) as counterelectrode by vacuum evaporation at $2x10^{-5}$ Torr.

For samples containing different nitrogen concentration the leakage current, TCC, tanδ, and the capacitance at 1 kHz are measured before and after heat treatment. The results are presented in Fig. 1 and 2.

Capacitors (≈ 1000 pF) are subjected to leakage current (I_L) tests. For samples containing 0.2 to 0.8% nitrogen the leakage current at 15% of the formation voltage is found to be less than $2x10^{-9}$ A. For undoped samples I_L is about five times higher. This is expected because in case of doped films the presence of nitrogen compensates the defective states which are responsible

1) Powai, Bombay-400076, India.

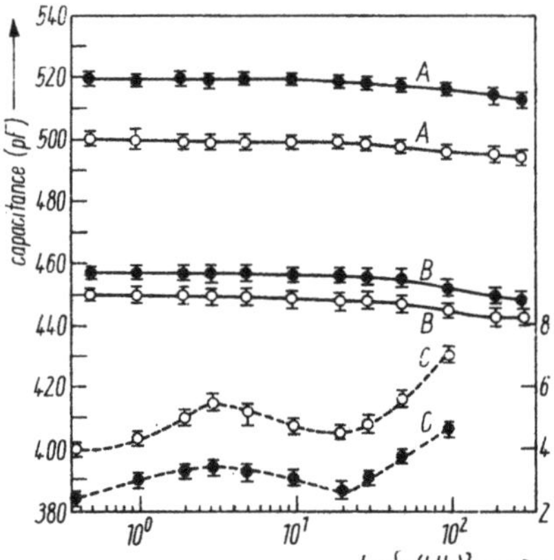

Fig.1

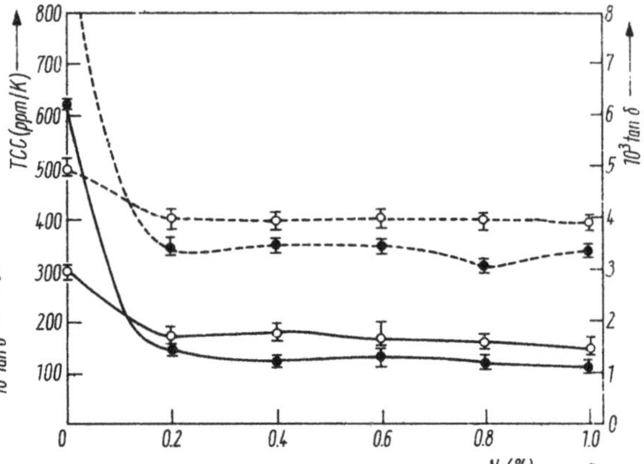

Fig.2

Fig. 1. Dependence of capacitance (——) and tan δ (---) on frequency. (A) Undoped samples, (B), (C) 0.6% nitrogen doped samples; ⌽ before heat treatment, ⏺ after heat treatment

Fig. 2. Variation in TCC (——) and tan δ (---) as a function of the nitrogen doping level (% of argon pressure); ⌽ before heat treatment, ⏺ after heat treatment

for dc conduction in the oxide. These values of I_L are slightly higher than those reported by earlier authors using NiCr-Au as counterlectrode /3/.

The variation in capacitance with frequency is shown in Fig. 1. About 1% fall in capacitance is noticed when the frequency increases from 10 to 100 kHz. The decrease in capacitance with increase in frequency is common for all capacitors of this type. To some extent, the frequency at which the capacitance falls, depends on the porosity of the oxide film and the processing factors. After heat treatment at 300 °C for 2 h, the frequency dependence of the capacitance remains almost the same. However, for samples containing 0.6% nitrogen the capacitance increases by 1.5%, whereas it rises by about 4% in case of undoped samples.

The dependence of the dielectric loss (tan δ) on frequency is also shown in Fig. 1 for 0.6% nitrogen doped samples. The dielectric loss increases with frequency in the range 500 Hz to 3 kHz. However, it falls when the frequency increases from 3 to 20 kHz. This might be attributed to a decrease in the electronic polarizability /5/. According to Westwood et al., the fall in tan δ seems to be a characteristic property of the aluminium counterelectrode /4/. Beyond 20 kHz, the series resistance becomes more predominant and this leads to a rise in tan δ with increase in frequency. After heat treatment tan δ falls by about 35%.

The dependence of TCC (between -80 $^{\circ}$C and +100 $^{\circ}$C) and tan δ on nitrogen concentration is shown in Fig. 2. Our results show that typical values of TCC and tan δ are 300 ppm/K and 0.005, respectively, for capacitors made from β-tantalum. For nitrogen doped samples it is observed that both TCC and tan δ decrease. For example, in case of capacitors with 0.2% N_2 content, TCC is found to be 175 ppm/K and tan δ is 0.004. Further increase in the N_2 content shows a negligible influence on these parameters. When these capacitors are subjected to heat treatment at 300 $^{\circ}$C for 2 h, for capacitors formed from undoped films TCC and tan δ increase from 300 to 625 ppm/K and from 0.005 to 0.012, respectively, whereas for nitrogen doped samples TCC and tan δ fall by 50 ppm/K and 0.001, respectively.

Now the results will be explained in the light of the existing theories and experimental results. In case of capacitors made from β-tantalum films the increase in TCC and tan δ may be attributed to a conductivity gradient formed in the Ta_2O_5 layer due to heat treatment. According to Smyth et al. /6/ oxygen, extracted from the oxide layer diffuses through the Ta_2O_5-Ta interface into the tantalum film. As a result, vacancies are formed in the oxide layer. Therefore, the conductivity of the oxide is very high at the tantalum-oxide interface compared to the conductivity at the oxide-counterelectrode interface. Due to this conductivity gradient the effective thickness of the dielectric decreases with a corresponding increase in TCC. At the same time, the series resistance increases resulting in higher tan δ. In case of N_2-doped Ta_2O_5 films, studied here, during heat treatment oxygen cannot be easily extracted from the oxide layer as most of the interstitial sites in the tantalum lattice are already filled by nitrogen and the dielectric is comparatively homogeneous. Thus, in $Ta_2O_5N_x$ films, no increase in TCC and tan δ should be noticed except for a slight increase in capacitance. However, the small decrease in TCC and tan δ shown in Fig. 2 may be due to a removal of the ions of citric acid and H_2O from the oxide layer during heat treatment. This explains the results shown in Fig. 1 and 2.

In conclusion, it appears that the dielectric properties of tantalum oxynitride films are superior to those of tantalum oxide films. After heat treatment capacitors made with tantalum oxynitride as dielectric exhibit the most favourable values of tan δ and TCC namely, 0.003 and 120 ppm/K, respectively, at nitrogen doping corresponding to 0.8% of the argon pressure. Also, for these capacitors the dc leakage current is less than 2×10^{-9} A and the breakdown voltage lies between 75 and 80% of the formation voltage. These values are obtained by using aluminium as counterelectrode and they may find applications in hybrid integrated circuits.

The authors are thankful to the Director, I.I.T., Bombay, for permission to publish the data. One of the authors (P.K.R.) is also grateful to the Indian Institute of Technology, Bombay, for the award of a Research Scholarship.

References

/1/ R.D. HUTTEMAN, J.M. MORALITO, and D. GESTENBERG, IEEE Trans. Parts, Hybrids, and Packaging 11, 67 (1975).

/2/ J.M. MARTINEZ-DUART, J.L. VELILLA, J.M. ALBELLA, and F. RUEDA, phys. stat. sol. (a) 26, 611 (1974).

/3/ M.H. ROTTERSMAN, M.J. BILL, and D. GESTENBERG, IEEE Trans. Components, Hybrids, and Manufacturing Tech. 1, 137 (1978).

/4/ W.D. WESTWOOD, N. WATERHOUSE, and P.S. WILCOX, Tantalum Thin Films, Academic Press, London 1975 (p. 361).

/5/ J. YAMAZAKI, M. NAKAMURA, and Y. NISHIMURA, Fujitisu Sci. Tech. J. 7, 135 (1971).

/6/ D.M. SMYTH, G.A. SHIRN, and T.B. TRIPP, J. Electrochem. Soc. 110, 1264 (1963).

(Received June 5, 1979)

phys. stat. sol. (a) 54, K67 (1979)
Subject classification: 14.3.3; 22.1.2

Centre for Systems and Devices, Indian Institute of Technology, Madras[1)]

Improvement of Breakdown Characteristics of a Planar Microwave P-I-N Diode by the Bevelling Technique

By
P.R. VAYA and D. KAKATI

Introduction It is well known that the breakdown characteristics of a semiconductor diode depend both on the characteristics of the bulk and those of the surface controlled region of the devices /1/. There exists an electric field greater than the maximum bulk field just below the semiconductor surface /2, 3/. It is this sub-surface field which controls the breakdown properties of p-n junctions commonly used for high voltage structures such as P-I-N diodes. It is desirable to realise a device breakdown characteristic nearer the bulk breakdown by reducing this maximum surface field.

The electric field on the surface of a reverse biased p-n junction can be reduced by as much as an order of magnitude by proper levelling of the surface /1/. For a bevelled P^+-I-N^+ junction with a given applied voltage sufficient to spread the space-charge layer from the P^+ to the N^+ region, a unique level angle exists which will minimise the peak surface field /1, 4, 5/.

Mechanical techniques have been successfully utilised to bevel the large plane junction devices /1, 4 to 6/. These techniques are not suitable for planar junction devices. A new technique of moat etching, as suggested by Temple et al. /7/ has been used for the study of the work reported in this paper.

Theoretical background The sub-surface electric field can be reduced by forcing the depletion region on the heavily doped side to spread-out. This can be achieved by reducing the depletion layer charge stored on either or both sides of the junction in the neighbourhood of the point at which the junction reaches the surface. Using moat etching techniques, as well as other bevelling techniques, some material is removed from near the junction.

In Fig. 1a the removed charge associated with this material is Q_1 and Q_2. The difference $(Q_1 - Q_2)$ must be balanced by additional charges Q^+ and Q^- in

---

1) Madras-600 036, India.

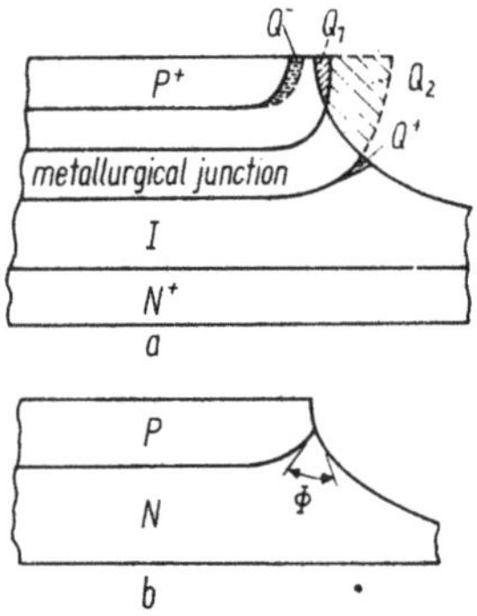

Fig. 1. a) Section of a bevelled planar P-I-N diode, showing the charges Q_1 and Q_2 removed by bevelling. Q^+ and Q^- are the additional charges in the new depletion region shown by dashed lines; b) supplementary bevel angle between the junction and the surface which controls the breakdown characteristics of the device

the new depletion region as shown in Fig. 1. The amount of Q^- and so the extension of the depletion region on the highly doped side depends on the amount of charges removed from the depletion region by etching, which is responsible for reducing the sub-surface electric field. If $Q_1 > Q_2$, then the surface field will tend to reach a peak bulk field value.

Davies and Gentry /1/ have shown that the surface field distribution depends critically on the bevel angle, and that it can be reduced to a minimum by introducing an optimised bevel angle (positive or negative as the case may be), Fig. 1b.

Experimental Planar microwave P-I-N diodes were fabricated on an n-type 400 Ωcm silicon wafer of <111> orientation by selectively diffusing boron on one side to a depth of 10 μm and uniformly diffusing phosphorus onto the other side of the wafer to a depth of 7μm. After the diffusion, aluminium was evap-

Table 1

Experimental results showing the dependence of the breakdown voltage on bevel angle

etching depth (μm)	supplementary bevel angle	breakdown voltage (V)
0	0	200
1	10°	250
2	22°	320
3	30°	400
4	38°	550
5	45°	500
6	53°	450

orated (1.5 μm thick) and patterened using photolithographic techniques. The aluminium served as an etch mask for etching a moat in silicon.

The moat etching was repeated for different etching times resulting in different bevel angles.

Results and discussions The breakdown voltage of a planar P-I-N diode and other bevelled P-I-N diodes were measured. Table 1 shows the dependence of breakdown voltage on bevel angle. The maximum breakdown obtained was for 38° supplementary bevel angle. It, therefore, may be concluded that proper bevelling can reduce the surface electric field of a planar microwave P-I-N diode and thereby improve breakdown characteristics.

References

/1/ R.L. DAVIES and F.E. GENTRY, IEEE Trans. Electron Devices ED-11, 313 (1964).

/2/ R.H. KINGSTON (Ed.), Semiconductor Surface Physics, University of Pennsylvania Press, Philadelphia (Pa.) 1957.

/3/ P.J. HOLMES (Ed.), The Electrochemistry of Semiconductors, Academic Press, New York (N.Y.) 1962.

/4/ J. COMU, IEEE Trans. Electron Devices ED-20, 347 (1973).

/5/ M.S. ADLER and V.A.K. TEMPLE, IEEE Trans. Electron Devices ED-23, 956 (1976).

/6/ J. CORNU, S. SCHWEITZER, and O. KUHU, IEEE Trans. Electron Devices ED-21, 181 (1974).

/7/ V.A.K. TEMPLE, B.J. BALIGA, and M.S. ADLER, IEEE Trans. Electron Devices ED-24, 1304 (1977).

(Received May 16, 1979)

phys. stat. sol. (a) 54, K71 (1979)
Subject classification: 10.2; 22.1.2

Institute of Physical Metallurgy, Czechoslovak Academy of Sciences, Brno[1)] (a) and Institute of Metal Physics, Academy of Sciences of the Ukrainian SSR, Kiev[2] (b)

Investigations of Anomalous Passing of Positrons through Si Crystals

By
A. ČÍŽEK (a), I. YA. DEKHTYAR (b), and A.I. MOSKALEVSKII (b)

Introduction It is well known that in solids the positrons interact with lattice defects in such a way that they are trapped by those having a negative excess charge. This effect observed firstly by Dekhtyar et al. /1/ qualifies the positron annihilation method as useful for the determination of the concentration of imperfections, for the estimation of the formation or migration characteristics of defects as well as for investigating the electronic structure of defects and their configurations in any internal process during annealing or plastic deformation of the crystal investigated.

In all cases, the sensitivity of positrons to the defects is very important. If the defects[3)] are assumed to be uniformly distributed throughout the volume of the crystal, the sensitivity is determined by the mean free path l_o of their motion which depends on the diffusion rate of thermalized positrons D and on the annihilation time τ as $l_o = 2\sqrt{D\tau}$. In metals /2/ $D \approx 10^{-2}$ cm^2/s, $\tau \approx 0.2$ ns and, therefore, $l_o \approx 3x10^{-6}$ cm; in semiconductors /2/, however, $D \approx 0.2$ cm^2/s, $\tau \approx 0.2$ ns, and then $l_o \approx 10^{-5}$ cm. Thus, the minimum concentration of defects to be detected by positrons is about $4x10^{16}$ cm^{-3} and $5x10^{14}$ cm^{-3} in metals and semiconductors, respectively.

Naturally the question arises: What is the reason of such a discrepancy in sensitivities of positrons to lattice defects in metals and semiconductors? We can imagine that the reason lies in the different character of positron migration in both substances /2, 3/.

---

1) Žižkova 22, 616 62 Brno, Czechoslovakia.

2) Prospekt Vernadskogo 36, 252 142 Kiev, USSR.

3) Point defects: vacancies or foreign atoms.

As to semiconductors, it has been shown that an anomalous passing[4)] of positrons through crystals may be observed, e.g. in Si. This transmission is realized by quantum tunneling of positronium atoms (Ps) which are generated in the crystal investigated /4/; at room temperature, the migration may occur by a barrier mechanism. Regardless of the favourable transmission conditions of Ps in the crystal lattice, however, the migration is generally caused by the presence of lattice defects in the crystals /3/.

In this paper, the anomalous passing of positrons through Si is investigated.

Experimental The single crystals, P doped to $\approx 10^{15}$ cm^{-3}, were prepared by the zone refining or Czochralski method. The crystals of $\approx$ 40 mm diameter were cut perpendicular to [111] into platelets which were treated by grinding and etching to the 14th class of roughness according to the Soviet standard (GOST 2789-51). The final thickness was (300 ± 5) μm, the (111) plane was oriented with an accuracy of $\pm 1^{\circ}$.

As the crystals were dislocation-free, the transmission of Ps may be supposed to be caused by foreign centres of concentration $\approx 10^{15}$ cm^{-3}.

The transmission coefficient was estimated by measuring the intensity of the angular distribution of annihilating photons (ADAP) from a single-crystalline Si and a polycrystalline Sn platelet. Both were of the same diameter; the thickness of the Sn platelet was about 0.2 mm and was not transparent for positrons.

The measurements were carried out using the long-slit geometry with a slit-width of 4 mrad at the moving detector and 1 mrad at the other one. The annihilation angle was Θ_0 = 0 mrad, i.e. the picacountrate was measured.[5)] The geometry was identical at all measurements, i.e. the distance between the positron source and the surface of the sample was kept constant. The statistical error of intensity measurements was about 0.5%.

---

4) If the effective absorption coefficient of particles is substantially less than that evaluated e.g. from the relation of Brandt /2/, one speaks about an anomalous passing of particles. In other words, it is such a real passing that cannot be expected from calculation; e.g. in our case $\bar{\mu}_c \approx 99$ cm^{-1} is calculated for Si and positrons from a ^{22}Na source, Table 1 gives a value of $\bar{\mu} \approx 30$ cm^{-1} for the real crystal.

5) Its value was of the order of (25 to 30)x10^3.

Results Let us denote the intensity of ADAP at the geometry used as I_{Si} for Si and I_{Sn} for Sn. The fraction of the transmitted particles is given by the ratio of the number of transmitted particles to that of the incident ones. Of course, the countrate is influenced by a partial absorption of the γ-photons in the sample and, therefore, the knowledge of the fraction of absorbed photons is necessary for the estimation of the above value. Then, the transmission coefficient P_{Si} of the particles may be written as

$$P_{Si} = 1 - \left[\frac{I_{Si}}{I_{Sn}} \frac{k_{Sn}}{k_{Si}} (1 - R_{Sn}) + R_{Si}\right] , \tag{1}$$

where k_{Sn} and k_{Si} are the fractions of γ-photons absorbed in Sn and Si, respectively. These quantities depend on the absorption coefficient μ of γ-photons in the material investigated and on the radius r of the sample,

$$k \approx \frac{1}{\mu R} \int_0^1 \exp(-\mu r \sqrt{1 - z^2}) (1 - \exp(-\mu r \sqrt{1 - z^2})) \, dz . \tag{2}$$

In our case, i.e. for r = 2 cm, $k_{Sn}/k_{Si} = 0.34$, and the reflectivities $R_{Sn} = 0.545$ and $R_{Si} = 0.22$ /5/, the results listed in Table 1 were obtained. Here $\bar{\mu}$ is the effective absorption coefficient of particles estimated from $P = \exp[-\bar{\mu} d]$, where d is the sample thickness. This coefficient appears to be substantially smaller than that of positrons in ^{22}Na /2, 3/.

Discussion The results in Table 1 verify the values given in /3/ and indicate the existence of an anomalous passing of positrons through crystals. It may be useful to discuss the mechanism of the observed effects from a quantum mechanical viewpoint.

It has been shown in /3, 4/ that the behaviour as well as the state of positrons in Si depend considerably on temperature. At low temperatures this state is connected with the formation of Ps the migration of which may be characterized by quantum tunneling /6/. At high temperatures the over-barrier diffusion mechanism is more probable.

At $T \leqq \Theta_D$ (Θ_D Debye temperature of the crystal), a specific migration mechanism of Ps may be supposed; Ps is considered from a quantum mechanical viewpoint as a mass-fluctuation wave /7/. The mechanism is induced by vacancy waves the energy band, Δ_v, of which is much greater than that, Δ, of the mass-fluctuation waves; $\Delta_v \gg \Delta$ /7/. The diffusion coefficient D of Ps is then

Table 1

Characteristics of monocrystalline Si platelets

sample No.	1	2	3	4	5
P_{Si}	0.300 ± 0.005	0.340 ± 0.005	0.420 ± 0.006	0.420 ± 0.006	0.310 ± 0.005
$\bar{\mu}$ (cm^{-1})	40.0 ± 0.5	36.0 ± 0.5	29.1 ± 0.05	29.1 ± 0.05	38.9 ± 0.05

determined by the probability of inelastic scattering, $D \approx a^2 \nu$, where a is the interatomic distance. The frequency of inelastic scattering events is $\nu = N_v v \sigma_{in}$, where N_v is the volume concentration of vacancy waves, v their velocity, and σ_{in} the cross-section of inelastic interaction of vacancy waves with mass-fluctuation waves. At high temperatures $\Delta_v \lessapprox kT$, $N_v \approx a^{-3} \exp\left[-E_v^f/kT\right]$, and $v \approx a\Delta_v/\hbar$, where k Boltzmann constant, $\hbar$ Planck constant, and E_v^f vacancy formation energy. The diffusion coefficient of Ps may be written as

$$D \approx \frac{\Delta_v \sigma_{in}}{\hbar} \exp\left[-E_v^f/kT\right]. \tag{3}$$

If one supposes that the Ps diffusion is realized by Ps migration through the interstices in the Si lattice then $E_v^f = \Theta$ and $\sigma_{in} \approx a^2$. Thus

$$D \approx \frac{\Delta_v \sigma_{in}}{\hbar} \approx \frac{kTa^2}{\hbar} \tag{3a}$$

At room temperature (T = 300 K) and $a_{Si} = 2.35$ Å (3a) yields $D_{Ps} \lessapprox 0.02\ cm^2 s^{-1}$. In /2/ the value $D \approx 0.2\ cm^2\ s^{-1}$ was found for positron diffusion in defective Si; the author has supposed no formation of Ps-like states in Si. It is seen that both estimates are relatively close.

There is evidence that, in presence of trapping centres for migrating particles in Si, the effective diffusion coefficient is changed because the cross-section of inelastic scattering is altered. This provokes a change in the effective absorption coefficient as well as in the transmission one. From this reason, the observed deviations (Table 1) in the effective absorption coefficient may be elucidated by an inhomogeneous distribution of trapping centres in the individual samples.

Conclusions The experimental results yield the following conclusions:

a) The passing of positrons is anomalous and varies in samples cut from the same Si monocrystal.

b) This may be explained by an inhomogeneous distribution of trapping centres with regard to the migrating Ps particles.

c) The mechanism of Ps migration in Si is connected with an inelastic interaction of mass-fluctuation waves with vacancy ones.

d) The diffusion coefficient of Ps differs only about one order from the estimate in /2/ for the positron diffusion in defective Si.

References

/1/ I. YA. DEKHTYAR, D.A. LEVINA, and V.S. MIKHALENKOV, Dokl. Akad. Nauk SSSR 166, 795 (1964).

/2/ W. BRANDT, Appl. Phys. 5, 1 (1974).

/3/ I. YA. DEKHTYAR, E.G. MADATOVA, V.I. SILYANTEV, and V.T. ADONKIN, Zh. eksper. teor. Fiz., Pisma 23, 691 (1976).

/4/ I. YA. DEKHTYAR, V.T. ADONKIN, E.G. MADATOVA, V.I. SILYANTEV, and G. DLUBEK, Zh. eksper. teor. Fiz., Pisma 26, 305 (1977).

/5/ J.K. MC KENZIE, C.W. SHULTE, T. JACKMAN, and J.L. CAMPBELL, Phys. Rev. A 7, 135 (1973).

/6/ A.F. ANDREEV and I.M. LIFSHITS, Zh. eksper. teor. Fiz. 56, 2067 (1969).

/7/ A.F. ANDREEV, Uspekhi fiz. Nauk 118, 251 (1976).

(Received June 5, 1979)

phys. stat. sol. (a) 54, K77 (1979)
Subject classification: 2 and 18.2; 21.1.1

Laboratoire de Magnétisme, C.N.R.S., Meudon-Bellevue[1)]

Magnetic Properties of Amorphous $Fe_{38}Ni_{38}B_{24}$ Alloy

By

R. KRISHNAN, M. TESSIER, and J. SZTERN

Ferromagnetic glassy metal alloys are receiving ever increasing attention in many laboratories as they help us to comprehend better magnetism in the amorphous state besides holding good promise for device applications as exemplified by the recent review by Luborsky /1/. In order to study the influence of Ni on the magnetic properties of Fe we have prepared amorphous ribbons of $(Fe_{1-x}Ni_x)_{76}B_{24}$ alloy by the melt spinning technique /2/ and here we report on the room temperature magnetization and ferromagnetic resonance (FMR) properties of the alloy with x = 0.5 in both as-quenched and annealed states.

The alloys from metals of purity better than 99.9% were first melted in a vacuum of about 10^{-6} Torr with an induction heater. Then bits of the alloy were taken in a 12 mm diameter quartz tube fitted with an orifice of 0.6 mm diameter. The charge was melted with a platinum wound furnace. The molten metal was then ejected on to a spinning copper wheel of 100 mm diameter capable of spinning at speeds up to 12000 r.p.m. using argon pressures in the range 0.5 to 1 bar. The present alloy had a melting point of about 1180 $^{\circ}$C and was ejected with a pressure of 0.8 bar with the wheel spinning at 5000 r.p.m., the angle at which the jet impinged was 10°. The ribbon width was 1.3 mm and the thickness (40 ± 3) μm. Their amorphous state was checked with X-rays.

Several small samples were examined and their properties showed a scatter of about 10%. Magnetization measurements by a vibrating sample magnetometer gave an average σ of 103 e.m.u. g^{-1}, corresponding to a magnetic moment of 1.12 μ_B per transition metal atom. This value compares very well with 1.13 μ_B obtained by Becker et al. /3/ on $Fe_{40}Ni_{40}B_{20}$ at room temperature. More quantitative interpretation should await low temperature measurements.

---

1) 92190 Meudon-Bellevue, France.

T a b l e 1

Typical room temperature properties of the as-quenched and annealed samples

as-quenched						annealed	
magnetometer		FMR				FMR	
σ (e.m.u. g^{-1})	$4\pi M_s$ (G)	$4\pi M_s^{eff}$ (G)	H_a (Oe)	$\Delta H_{\parallel}$ (Oe)	$\Delta H_{\perp}$ (Oe)	$\Delta H_{\parallel}$ (Oe)	$\Delta H_{\perp}$ (Oe)
103	9800	9550	250	400	600	250	500

FMR spectra were taken in the 17 GHz range with the applied field perpendicular ($H_{\perp}$) and parallel ($H_{\parallel}$) to the ribbon plane, which enabled us to calculate the effective magnetization

$$4\pi M_s^{eff} = 4\pi M_s - H_a .$$

Knowing $4\pi M_s$ from magnetization measurements one could then estimate the anisotropy field H_a. The dull rolled surface was stuck against the cavity wall which has been shown by us to give well defined resonance modes /4/. However, the perpendicular spectra in some samples consisted, besides the large mode, of additional small modes forming a shoulder at the low field side. The parallel resonance mode had smaller linewidths. Table 1 shows the properties. A small perpendicular anisotropy was found in all the samples which varied in the range 250 to 300 Oe. This anisotropy could as well be induced by the stresses present in the sample particularly knowing that the magnetostriction in such alloys could /5/ be on the order of $15x10^{-6}$. Such anisotropy has been normally reported by others in as-quenched ribbons. The variation in H_a from sample to sample would indicate a non-uniform distribution of stresses.

The samples were subjected to an annealing in argon at 310°C for 3 h and reinvestigated. Magnetization was unchanged but FMR properties were drastically changed as shown in Table 1. The increase in $4\pi M_s^{eff}$ indicates the suppression of the anisotropy field H_a which is due to stress relief by annealing. What is most spectacular in our opinion is the marked reduction of up to about 50% in the linewidths which led to well defined FMR spectra in both parallel and per-

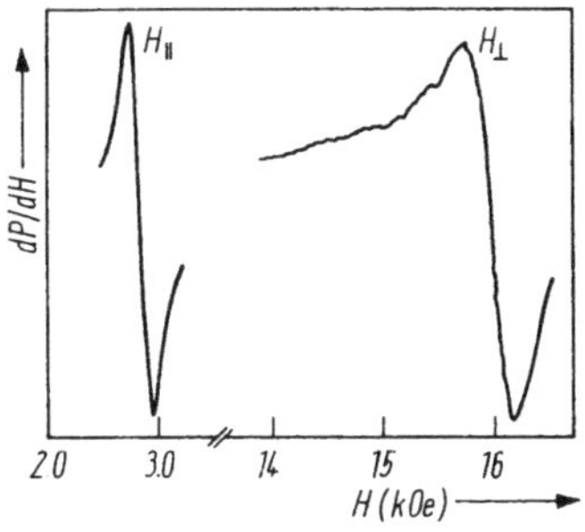

Fig. 1. FMR spectra of Na annealed sample; f = 17.38 GHz

pendicular configurations as shown in Fig. 1. Probably some sort of chemical and hence magnetic homogenization is taking place though the sample was still verified by X-rays to be in the amorphous state, or the suppression of the stresses which were initially distributed in a non-uniform fashion as indicated earlier by the scatter in the anisotropy field.

A more detailed report, particularly of the low temperature properties, will be published in the near future.

References

/1/ F.E. LUBORSKY, J. Magnetism magnetic Mater. 7, 143 (1978).

/2/ H.H. LIEBERMANN and C.D. GRAHAM, IEEE Trans. Magnetics Mag-12, 921 (1976).

/3/ J.J. BECKER, F.E. LUBORSKY, and J.L. WALTER, IEEE Trans. Magnetics MAG-13, 988 (1977).

/4/ R. KRISHNAN, SHIVA PRASAD, and K. BRANSKA, to be presented at 3M-Intermag. Conf., New York, July 1979.

/5/ R.C. O'HANDLEY, Solid State Commun. 21, 1119 (1977).

(Received June 17, 1979)

phys. stat. sol. (a) 54, K81 (1979)
Subject classification: 13.4; 20.3; 22.4.1

School of Physics, Robert Gordon's Institute of Technology, Aberdeen

Shallow Traps in ZnS Single Crystals

By
M.A.S. SWEET and D. URQUHART

Introduction Thermally stimulated luminescence (TSL) has been used extensively to characterise deep lying impurities in both crystalline and powdered ZnS, see for example /1 to 4/. Rather less work has been carried out on shallower levels which appear in the liquid helium to liquid nitrogen temperature range. The purpose of this note is to confirm the presence of previously reported /5/ shallow traps in ZnS and extend the analysis by examining their behaviour with various heating rates.

Experimental The ZnS crystals were in the form of platelets, approximately 5mmx5mmx100 μm, and exhibited an orange fluorescence at room temperature. Their preparation and composition have been described elsewhere /6/.

The cryogenic system used for the present investigations comprised an Oxford Instument CF100 continuous flow cryostat with DTC2 temperature controller. We found that the simplest and most reliable method of achieving linear temperature rise rates in the low temperature range (4 to 50 K) was to use a motor-driven helium flow valve and various settings of heating voltage. The voltage drop across the forward-biased base-emitter junction of a Fairchild C111 transistor was used to monitor the sample temperature. The C111 was first calibrated against a standardised GaAs diode. The platelet was secured in good thermal contact with the transistor can with 'dag' 915 high conductivity paint and the assembly mounted on the cryostat cold finger.

Light emission from the sample was measured with an EMI 9804B photomultiplier, the signal being fed to a Nuclear Enterprises SR5 ratemeter and recorded on a chart strip and data logger along with temperature data. Excitation of the crystal with the Hg 365 nm line for 10 min maximised the glow output. In Fig. 1 the glow curves obtained at four different heating rates are presented; three peaks at (14.7 $\pm$ 0.1), (19.6 $\pm$ 0.1), and (21.0 $\pm$ 0.1)K are resolvable at the slowest rate of warming (0.041 K s^{-1}).

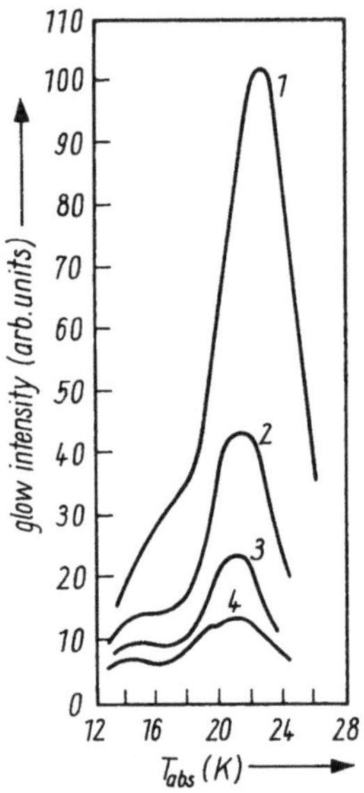

Fig.1

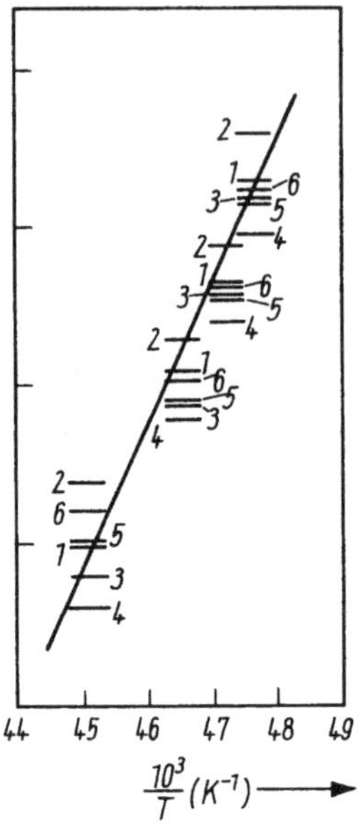

Fig.2

Fig. 1. Glow curves for the ZnS crystal with four different heating rates. (1) 0.451 Ks^{-1}, (2) 0.151 Ks^{-1}, (3) 0.076 Ks^{-1}, (4) 0.041 Ks^{-1}

Fig. 2. A plot of various functions versus reciprocal temperature for the glow curves of Fig. 1. Experimental results together with error bars are shown; points 1: $\ln(T_m^2/w)$, method of Hoogenstraaten; points 2: $\ln(I_m)$, method of Bube et al., points 3: $\ln(w)$, method of Böer et al., points 4: $\ln(I_m/T_m^{3/2})$, method of Boiko et al., points 5: $\ln(I_m/T_m^{7/2})$, method of Boiko et al., points 6: $\ln(I_m T_m^4/w)$, method of Chen and Winer. The line drawn is the least squares fit to the Hoogenstraaten points. The symbols used and the nomenclature of the methods are as defined by Kivits and Hagebeuk /7/

Discussion Most of the methods used to evaluate the activation energy from TSL data have been summarised by Kivits and Hagebeuk /7/. The most reliable of these, particularly when overlapping of peaks distorts the shape of the glow peak, involve the shift of the luminescence maximum as heating rate is changed. The results of applying these methods to the dominant 21 K peak are shown in Fig. 2 and the corresponding least squares fitting results are given in Table 1. As may be anticipated with such small temperature shifts considerable scatter

Table 1

Results of the methods involving shift of glow peak with heating rate. The % column shows the % variation between the result given by the method and that given by the Hoogenstraaten method. This, together with the nomenclature of the methods follows from Kivits and Hagebeuk /7/

method	energy (meV)	%
Hoogenstraaten	78 ± 5	-
Bube, Hearing, and Adams	79 ± 8	+1.0
Boiko, Rashba, and Trofimenko 1	76 ± 8	-1.7
Boiko, Rashba, and Trofimenko 2	75 ± 5	-3.6
Chen and Winer	76 ± 3	-2.0
Böer, Oberländer, and Voigt	82 ± 5	+4.8

is present in the data; and this, together with the overlapping contribution from the shallower trap (19.6 K at 0.041 $K s^{-1}$), throws doubt on the applicability of the method developed by Kivits /8/ of examining the significance of the discrepancies between various methods. Nevertheless the trend of results for the remaining methods is clear - they give large underestimates. The class 1 methods numbered 10, 13, 16 to 18, 26, and 30 by Kivits and Hagebeuk /7/ all indicate energies in the range -83.1% to -77.9% with respect to the Hoogenstraaten figure whilst application of class 2 methods gives Chen 7 (-73.9%), Chen 8 (-80.4%), and Chen 9 (-69.1%) with respect to the same base. When these percentage figures are compared with Table 2 of Kivits and Hagebeuk /7/ a possible conclusion is that the retrapping ratio, δ , is high ($\delta > 3$). This seems likely for such shallow traps however a factor which would reduce the effective δ is the large concentration of holes at activators at the lowest temperatures following activation, due to the presence of deeper filled traps. An alternative explanation that the underestimates of these methods are due to the influence of the shallower trap is currently under investigation.

We wish to thank Dr. B. Ray, Lanchester Polytechnic, for helpful discussions and encouragement.

References

/1/ W. HOOGENSTRAATEN, Philips Res. Rep. 13, 515 (1958).

/2/ F.J. BRYANT and S.A. HAMID, phys. stat. sol. (a) 2, 597 (1970).

/3/ F.J. BRYANT and R.F. KYDD, Solid State Commun. 15, 1929 (1974).

/4/ N. RIEHL, J. Lum. 1/2, 1 (1970).

/5/ M.A.S. SWEET and C.J. WAREING, phys. stat. sol. (a) 3, K297 (1970).

/6/ G.F. ALFREY and D.B. MEADOWCROFT, Brit. J. appl. Phys. 16, 1675 (1965).

/7/ P. KIVITS and H.J.L. HAGEBEUK, J. Lum. 15, 1 (1977).

/8/ P. KIVITS, J. Lum. 16, 119 (1978).

(Received April 9, 1979)

Pre-Printed Titles

Papers to be published
in "physica status solidi (a) - applied research"
Vol. 54, No. 2

Original Papers

M. VON SANDE, J. VAN LANDUYT, and S. AMELINCKX, Antwerpen
Direct Imaging of the Structure and Structural Defects of Rhombohedral γ-Brasses

CHUN CHIANG, Taipei
A Theory of Threshold Switching in Amorphous Thin Films

V.G. CHUDINOV, N.V. MOSEEV, B.N. GOSHCHITSKII, and V.I. PROTASOV, Sverdlovsk
On the Influence of Irradiation Type on the Formation of Radiation Defects in Nb-Based Superconductors with A-15 Structure

P.M. RAO, A. GERARD, and F. GRANDJEAN, Sart Tilman
A Mössbauer Study of the Effects of the Substitution of Fe by Cr in $SrFe_{12}O_{19}$

L.S. KHAZAN, L.A. MATVEEVA, G.N. SEMENOVA, and YU.A. THORIK, Kiev
Account of Dislocation Structure at Stress Measurements in Heteroepitaxial Systems (I)

E. LILLEY and J.E. STRUTT, Brighton
Bulk and Grain Boundary Ionic Conductivity in Polycrystalline β''-Alumina

J. MUCHA and J. RAFAŁOWICZ, Wrocław
Influence of the Lattice Component on the Thermal Conductivity Minimum of Aluminium

X. AYMERICH and F. SERRA-MESTRES, Bellaterra
Zero Bias Anomaly Due to Elastic Tunneling in MIM Structures

V.I. OSINSKII and A.Y. PESHKO, Minsk
Anomalous Temperature Dependence of Recombination Radiation of Gallium Arsenide

V. GOTOVAC and B. LOVREČEK, Split and Zagreb
The Temperature and Compositional Dependence of the Electrical Conductivity of Sintered Aluminium Oxide with Metal Excess

A. ŠLĘZAK, J. LECH, and O. BÓJKO, Czestochowa and Opole
Electron Paramagnetic Resonance of Mn^{2+} Ions in Aragonite Single Crystals

A. AMAMOU, Strasbourg
Mössbauer Effect and Short Range Order in Fe-P-B and Fe-P-C Amorphous Alloys

J.P. FARGES, H. GRASSI, A. BRAU, and P. DUPUIS, Nice and Nancy
Phase Transition in the TCNQ Complex Salt Formed from the Cation (Methyl--1-N-Ethyl Benzimidazolium)

T. SAWADA and H. HASEGAWA, Sapporo
Small-Signal Admittance Study of GaAs Anodic MOS System

M. GUYMONT, D. GRATIAS et A. BISSON, Orsay, Vitry sur Seine et Saclay
Fluctuations prétransitionnelles dans AuCu

J.S. BONHAM, Brisbane
On the Theory of Space-Charged-Limited Current with Diffusion for an Exponential Trap Distribution

V.D. VERNER, D.K. NITCHUGOVSKII, and V.Y. KOVYASIN, Moscow
Investigation of Internal Friction in GaP

O.A. TROITSKII, V.I. SPITSYN, M.M. MOISEENKO, and A.D. TSIGANOV, Moscow
Fractionation of Stable Isotopes and Impurities during Zone Recrystallization of Metals and Preparation of Single Crystals Rich in Individual Isotopes

Y. KAMIURA and F. HASHIMOTO, Okayama
Degree of Copper Contamination in Germanium Heat Treated at High Temperatures

A. IIDA, Tokyo
Applications of X-Ray Triple Crystal Diffractometry to Studies on the Diffusion-Induced Defects in Silicon Crystals

M. YANO, T. TAKASE, and M. KIMATA, Tokyo
Heteroepitaxial InSb Films Grown by Molecular Beam Epitaxy

Papers to be published
in "physica status solidi (b) - basic research"
Vol. 94, No. 2

Original Papers

R.M. NAYAK and B.D. PADALIA, Bombay
Chemical Shifts of the K-Absorption Edge of Mn in Some Mixed Oxidic Spinels

V. MAY, K. HENNEBERGER, and F. HENNEBERGER, Berlin
Renormalization of Polaritons Due to Virtual Formation of Biexcitons at High Densities of Excitation

K.B. TOLPYGO and S.M. ZUBKOVA, Donetsk and Kiev
On the Indirect Exciton Energy Levels in Crystals of Diamond and Sphalerite Type with Spin-Orbit Interaction

A.A. O'CONNEL-BRONIN and V.G. PLEKHANOV, Tartu
Resonant Raman Scattering in Crystals with Self-Trapping Excitons

C. WEIGEL and C.A.J. AMMERLAAN, Würzburg and Amsterdam
Iterative EHT Calculations for the Positive Divacancy in Silicon

H.Y. LAM, R.E. JONES, Jr., W.J. KEELER, and V.V. PARANJAPE, Thunder Bay, Ontario
Pressure Dependent Conductivity in InSb

B. ESSER and D. SPRENGER, Berlin
On the Theory of Absorption and Luminescence Spectra in Amorphous Semiconductors

19. Magnetic Resonance

20. Optical Properties
    - 20.1 Spectra. Optical Constants (X-Ray Spectra See 20)
    - 20.2 Lasers
    - 20.3 Luminescence (X-Ray Spectra See 20)

21. Metals. Alloys
    - 21.1 Transition Metals (Group Sc to Zn) and Their Alloys
        - 21.1.1 Fe and Fe Alloys
    - 21.2 Alkali Metals
    - 21.3 Alkali-Earth Metals
    - 21.4 Rare-Earth Metals
    - 21.5 Actinides
    - 21.6 Noble Metals
    - 21.7 Semi-Metals (Semiconducting Alloys 22.7)

22. Semiconductors and Ionic Crystals
    - 22.1 Elements
        - 22.1.1 Germanium
        - 22.1.2 Silicon
        - 22.1.3 Group VI Elements
    - 22.2 III–V Compounds
        - 22.2.1 Arsenides
        - 22.2.2 Phosphides
        - 22.2.3 Antimonides
        - 22.2.4 Mixed Crystals
    - 22.3 IV–IV, II–V, and III–VI Compounds
        - 22.3.1 Mixed Crystals
    - 22.4 II–VI Compounds
        - 22.4.1 Sulphides
        - 22.4.2 Selenides
        - 22.4.3 Tellurides
        - 22.4.4 Mixed Crystals
    - 22.5 Halides
        - 22.5.1 Silver Halides
        - 22.5.2 Alkali Halides
        - 22.5.3 Alkali-Earth Halides
        - 22.5.4 Mixed Crystals
    - 22.6 Simple Oxides (CdO, ZnO See 22.4)
        - 22.6.1 Mixed Crystals
    - 22.7 Semiconducting Intermetallic Compounds
        - 22.7.1 Mixed Crystals
    - 22.8 Tertiary and Higher Compounds
        - 22.8.1 Salts of Oxy-Acids (Carbonates, Phosphates, Silicates Including Mica etc.)
        - 22.8.2 Spinels, Garnets, Ferrites
    - 22.9 Organic Semiconductors

23. Solidified Gases

Attention

The preceding classification scheme is closely akin to the scheme used by Physics Abstracts to classify the various fields of solid state physics. A paper will either be classified under one or more of the subsections or, if none of these are relevant, it will just appear under the general section heading (i. e. sections and subsections are equivalent in this sense).

The *subject classification* given at the head of each paper begins with the main subject matter of the paper, and this is followed, if necessary, by figures referring to subsidiary subject matter and the type of solids investigated.

The papers published in this issue are listed in the order of their subject classification at the end of the table of contents.

82914

Manuscripts and letters for physica status solidi (b) — basic research and physica status solidi (a) — applied research

should be sent to

DDR-102 Berlin, Neue Schönhauser Str. 20

or

Internationaler Buchversand GmbH, 1 Berlin 19, Eschenallee 21.

The object of "physica status solidi (b)" and "physica status solidi (a)" is to publish rapidly important and topical results concerning fundamental research in the field of solid state physics and aspects of application and experimental techniques of fundamental solid state research, respectively. The issues of "physica status solidi (b)" and "physica status solidi (a)" will be published on the 1st and the 16th day of each month, respectively. Either series will contain the following categories:

1. Review articles written by arrangement with the editors. They will be remunerated.
2. Original papers selected from those which have not been submitted for publication elsewhere, after being reviewed. (The latest date for acceptance of contributions is the 8th and the 23rd day of each month for "physica status solidi (b)" and "physica status solidi (a)", respectively, 50 days before the delivery of the journal.)
3. Short notes published in the off-set part of the journal. They are, if necessary after being reviewed, selected by the editorial staff. (The latest date for acceptance of manuscripts for this part is the 12th and the 27th day of each month for "physica status solidi (b)" and "physica status solidi (a)", respectively, 16 days before the delivery of the journal.)
4. Pre-printed titles of papers to be published in the next issue of both series.

Editorial Information

Manuscripts should be submitted preferentially in English. Besides, papers in German, French, and Russian are admitted. The final decision on the incorporation of the papers in the series (b) or (a) is taken by the editorial staff after consultation of the referee. Authors are asked to observe the following rules:

1. Manuscripts have to be written as concisely as possible. As most, 20 and 4 pages (28 lines per page; abstracts, tables, references, figures, and captions of figures included, where 3 figures are counted as about one page) are admitted for original papers and short notes, respectively. All manuscripts should be type-written on one side only. Two copies are required. Badly arranged manuscripts are returmed.
2. The author's name, initials, and institution should be given.
3. One abstract of about ten lines in English and one in German, French, or Russian are required for original papers.
4. Because of lack of space conclusions (if any) must not have the character of a summary (i.e. substantially a repetition of the contents of the abstract).
5. Formulae should not be written too small and not with pencil. Separate lines for formulae are desirable. Indistinct writing will lead to a return of the manuscript.
6. Drawings should be enclosed on separate sheets of paper, not exceeding the maximum and minimum sizes of 20×30 and 10×15 cm^2, respectively. Photographs should be submitted, on white paper, only for review articles and original papers. Duplicates are required.
7. Captions for the figures should be submitted on separate sheets of paper, and special instructions with regard to the size of illustrations should be given on these.
8. References should be numbered in order of appearance and should be given, on a separate sheet, in the form used in physica status solidi.
9. SI-units should be used.
10. Each paper should contain a proposal for the subject classification according to the classification scheme appearing on the second and third cover of each issue.

The authors will be sent galley proofs of review articles and original papers. Proof reading of short notes and pre-printed titles will be done by the editorial staff.

For each original paper and short note a total of 75 reprints is available free of charge. The number of reprints free of charge per review article is 25. Requests for additional reprints (against payment) should be given together with the submitted manuscript. **The deadline for orders of additional reprints is the 15th and the 30th of each month** for "physica status solidi (b)" and "physica status solidi (a)", respectively, two weeks before the delivery of the journal.

Bei Fragen zur Produktsicherheit wenden Sie sich bitte an:
If you have any questions regarding product safety,
please contact:

Walter de Gruyter GmbH
Genthiner Straße 13
10785 Berlin
productsafety@degruyterbrill.com